MODERN
POWER STATION PRACTICE

Third Edition
(in 12 volumes)

Incorporating Modern Power System Practice

MODERN
POWER STATION PRACTICE

Third Edition

Incorporating Modern Power System Practice

British Electricity International, London

Volume G
Station Operation and Maintenance

PERGAMON PRESS

OXFORD · NEW YORK · SEOUL · TOKYO

U.K.	Pergamon Press plc., Headington Hill Hall, Oxford OX3 OBW, England
U.S.A.	Pergamon Press, Inc., Maxwell House, Fairview Park, Elmsford, New York 10523, U.S.A.
KOREA	Pergamon Press Korea, KPO Box 315, Seoul 110-603, Korea
JAPAN	Pergamon Press, 8th Floor, Matsuoka Central Building, 1-7-1 Nishi-Shinjuku, Shinjuku-ku, Tokyo 160, Japan

First edition 1963

Second edition 1971

Third edition 1991

Library of Congress Cataloging in Publication Data
Modern power station practice: incorporating modern power system practice/British Electricity International.—3rd ed. p. cm.
Includes index.
1. Electric power-plants. I. British Electricity International.
TK1191.M49   1990   62.31'21 — dc20   90-43748

British Library Cataloguing in Publication Data
British Electricity International
Modern power station practice.—3rd. ed.
1. Electric power-plants. Design and construction
I. Title II. Central Electricity Generating Board
621.3121.
ISBN 0-08-040510-X (12 Volume Set)
ISBN 0-08-040517-7 (Volume G)

Printed in the Republic of Singapore by Singapore National Printers Ltd

Contents

Colour Plates

(between pp 34 and 35)

(between pp 434 and 435)

Foreword

G. A. W. Blackman, CBE, FEng
*Chairman, Central Electricity Generating Board
and Chairman, British Electricity International Ltd*

For over thirty years, since its formation in 1958, the Central Electricity Generating Board (CEGB) has been at the forefront of technological advances in the design, construction, operation, and maintenance of power plant and transmission systems. During this time capacity increased almost fivefold, involving the introduction of thermal and nuclear generating units of 500 MW and 660 MW, to supply one of the largest integrated power systems in the world. In fulfilling its statutory responsibility to ensure continuity of a safe and economic supply of electricity, the CEGB built up a powerful engineering and scientific capability, and accumulated a wealth of experience in the operation and maintenance of power plant and systems. With the privatisation of the CEGB this experience and capability is being carried forward by its four successor companies — National Power, PowerGen, Nuclear Electric and National Grid.

At the heart of the CEGB's success has been an awareness of the need to sustain and improve the skills and knowledge of its engineering and technical staff. This was achieved through formal and on-job training, aided by a series of textbooks covering the theory and practice for the whole range of technology to be found on a modern power station. A second edition of the series, known as Modern Power Station Practice, was produced in the early 1970s, and it was sold throughout the world to provide electricity undertakings, engineers and students with an account of the CEGB's practices and hard-won experience. The edition had substantial worldwide sales and achieved recognition as the authoritative reference work on power generation.

A completely revised and enlarged (third) edition has now been produced which updates the relevant information in the earlier edition together with a comprehensive account of the solutions to the many engineering and environmental challenges encountered, and which puts on record the achievements of the CEGB during its lifetime as one of the world's leading public electricity utilities.

In producing this third edition, the opportunity has been taken to restructure the information in the original eight volumes to provide a more logical and detailed exposition of the technical content. The series has also been extended to include three new volumes on 'Station Commissioning', 'EHV Transmission' and 'System Operation'. Each of the eleven subject volumes had an Advisory Editor for the technical validation of the many contributions by individual authors, all of whom are recognised as authorities in their particular field of technology.

All subject volumes carry their own index and a twelfth volume provides a consolidated index for the series overall. Particular attention has been paid to the production of draft material, with text refined through a number of technical and language editorial stages and complemented by a large number of high quality illustrations. The result is a high standard of presentation designed to appeal to a wide international readership.

It is with much pleasure therefore that I introduce this new series, which has been attributed to British Electricity International on behalf of the CEGB and its successor companies. I have been closely associated with its production and have no doubt that it will be invaluable to engineers worldwide who are engaged in the design, construction, commissioning, operation and maintenance of modern power stations and systems.

March 1990

Preface

"It is easy after discovery to say 'How obvious and how simple', but how difficult is any step of advance when shrouded by unknown surroundings." These words are as true today as they were when they were spoken by the greatest of all British power engineers — Charles Algernon Parsons — in 1909. They apply equally to the operation and maintenance of large generating units as they do to the design and manufacture of those units.

The daily operation of the British Electricity Supply System, involving as it does the peak-load operation of 500 MW and 660 MW units integrated with the extended operation of similar units, has only come about after many years of developing highly-specialised plant operating and maintenance techniques to give the required flexibility, reliability and high thermal efficiency. The principal objective of this volume is to convey many of the lessons learned in developing these techniques to a new generation of power engineers throughout the world.

The volume deals, as its title suggests, with the practical aspects of operating and maintaining large power stations, especially those with large generating units. The activities of planning and thermal-efficiency monitoring are described in detail, since they are entirely complementary to the work of the operating and maintenance engineer. Considerable attention has also been paid to describing the theory of large generator design, together with aspects of the operation of the generator and its auxiliary circuits, the importance of which cannot be overstated. Special emphasis has been paid to safety and to describing methods of ensuring safe working practices.

The authors are all practising engineers, engaged daily in the work they describe. The experience they impart has been gained over many years of dealing with the CEGB's largest units.

It must always be remembered that, where detailed descriptions are made of operating or maintenance procedures, these must be taken as typical only. The reader must, when considering his own plant, refer to specific procedures provided by the plant manufacturer. Where these do not exist, great care must be taken to prepare suitable documents, perhaps based on the lessons contained in this volume.

L. C. WHITE
Advisory Editor — Volume G

Contents of All Volumes

Volume A — Station Planning and Design
Power station siting and site layout
Station design and layout
Civil engineering and building works

Volume B — Boilers and Ancillary Plant
Furnace design, gas side characteristics and combustion equipment
Boiler unit — thermal and pressure parts design
Ancillary plant and fittings
Dust extraction, draught systems and flue gas desulphurisation

Volume C — Turbines, Generators and Associated Plant
The steam turbine
Turbine plant systems
Feedwater heating systems
Condensers, pumps and cooling water systems
Hydraulic turbines
The generator

Volume D — Electrical Systems and Equipment
Electrical system design
Electrical system analysis
Transformers
Generator main connections
Switchgear and control gear
Cabling
Motors
Telecommunications
Emergency supply equipment
Mechanical plant electrical services
Protection
Synchronising

Volume E — Chemistry and Metallurgy
Chemistry
Fuel and oil
Corrosion: feed and boiler water
Water treatment plant and cooling water systems
Plant cleaning and inspection
Metallurgy
Introduction to metallurgy
Materials behaviour
Non-ferrous metals and alloys
Non-metallic materials
Materials selection

Volume L — System Operation

Volume M — Index

Evan John Davies

Emeritus Professor of Electrical and Electronic Engineering at Aston University in Birmingham, died on 14 April 1991.

John was an engineer, an intellectual and a respected author in his own right. It was this rare combination of talents that he brought to Modern Power Station Practice as Consulting Editor of seven volumes and, in so doing, bequeathed a legacy from which practising and future engineers will continue to benefit for many years.

Introduction

1 Introduction

This volume deals with activities within a power station. In the main, the comments are appropriate to the largest stations of the Central Electricity Generating Board, which utilise 500 MW and 660 MW units. The subjects covered include power station operation and maintenance, together with the related planning and monitoring activities necessary to ensure compliance with the overall objectives. The safety of both staff and equipment is a prime requisite; such is the importance of safety that a chapter is included on this subject alone.

These subjects, then, form the main content of this volume. For many power engineers, their daily work will very largely consist of applying the lessons contained within. However, few engineers have the luxury of being involved in technical matters in isolation. There are fringe activities, many of which are worthy of a chapter, which have a greater or lesser bearing on the individual. In total, all of these activities form the objectives of the location manager.

1.1 Objectives of the location manager

The overall task of managing a large power station is to utilise resources, plant, staff, materials and finance to provide a safe and secure supply of electricity at minimum cost. In carrying out this work, it is necessary to comply with legal and statutory obligations at all times, and to observe accepted Codes of Practice in all disciplines.

The target availability requirement and expected output from the plant is made known to the manager by higher authority in advance of any given financial year, based upon expected plant behaviour, overhaul requirements, etc. Following on from target output, a target thermal efficiency can be agreed, from which the estimated annual cost of fuel can be determined. However, the fuel budget is not the responsibility of station management, since the output of the station is regulated by others to meet the demands on the system.

The attainment of high availability at each of the stations in an undertaking is an important requirement. Considerable sums of money can be justified in improving the reliability of plant, since this is much cheaper than providing additional generating capacity. In considering this alternative an important factor is the duration of plant outage necessary to carry out the modification. Identifying shortcomings in plant design and engineering solutions to improve reliability is an important task for station engineering staff.

Prior to the financial year, the annual overhaul plan is agreed, to allow the parent organisation to meet its loading commitments at optimum cost. Given the outage programme, the cost of carrying out essential work is determined and added to the revenue costs of operating and maintaining the station throughout the year. Having built up the financial requirements in this manner prior to the start of each year, it is then necessary to ensure that the level and incidence of expenditure is according to plan. In carrying out major overhaul work or modifications, it is necessary to write specifications, obtain tenders and place contracts well in advance of the outage date.

In controlling expenditure, it is important to ensure that the correct sums are spent in carrying out the planned work. At the planning stage, the decision will have been made to invest these sums of money so as to maintain or improve plant performance. Failure to carry out the work, due perhaps to an overspend in other areas, could result in objectives not being attained. Reducing expenditure and not carrying out the work is almost as bad, since once again objectives may not be achieved; in this case, the money will have been made available for the work at considerable cost but not used. If there are good reasons for not carrying out the work, the parent organisation should be informed as soon as possible so that the money can be made available to another location, if necessary. Unexpected plant failure will often lead to the need for additional funds, contingency sums for such failures form part of a centralised budget.

In addition to achieving high availability and high thermal performance, it is necessary to be able to operate the plant with a high degree of flexibility. The operation of large units at reduced load results in much reduced thermal efficiencies, with consequential increased costs of generation. Whilst certain units are specifically designed for two-shift and peak-load operation, many stations operate at base-load until a point is reached in their working lives when their position in the merit order dictates that it is uneconomic for them to operate at periods of reduced demand, such as overnight or at weekends. On these occasions, it is necessary that suitable techniques exist to ensure that two-shift operation can be carried out safely, efficiently and to programme. It is current practice in the CEGB for 500 MW and 660 MW oil-fired plant to be operated with maximum flexibility, with 2000 MW stations varying in output from zero overnight to full-load during the day; units are sometimes synchronised twice in one day, as required within this programme.

At all times, the effect of the power station on the environment must be minimised. Discharges from the chimney, cooling water system and drainage systems have to be maintained well within legal limits. Similarly, noise levels have to be acceptable at all times, but particularly at night, and in urban areas. Special efforts are necessary to ensure that good relationships exist with the public. The avoidance of nuisance is paramount and any complaints received should be acknowledged and investigated. More positive measures can be taken to invite the public to visit the station, or to encourage organised visits from school children, students and the wide range of organisations that exist. Similarly, station staff can visit such organisations to describe the workings of the electricity supply industry.

There is an important need for management to ensure that adequate training facilities exist appropriate to all members of staff. Induction training is necessary for all newcomers and specialist training has to be identified and implemented so as to improve the skills of each employee.

A further pre-requisite of good management is the fostering of good working relationships with the staff. By using the formal meetings of negotiating and consultative machinery, together with less formal approaches, an understanding and a spirit of co-operation can be promoted between all employees so as to maintain morale at the highest possible level. In the same vein, housekeeping standards in all areas of the station site have to be maintained at a high level for internal and external environmental reasons; particularly, since there is a relationship between good housekeeping and safety.

1.2 Station staffing

In comparison with many electricity supply undertakings throughout the world, CEGB power stations may appear at first sight to be more than adequately staffed; the reasons for this are both historical and geographical.

Staffing levels in British power stations have always been based upon the principle of allowing the location to be very largely self-sufficient in terms of providing day to day maintenance and supporting services. Staff numbers for these are calculated to ensure that each person is fully utilised and that there is little or no need to buy-in services that can reasonably be carried out in-house. So British station manning lists include a very wide range of occupations that may not be found when comparing them with other undertakings; such occupations include catering and security staff, gardeners, painters, training staff, and so on.

Use is only made of outside agencies for dealing with excessive workload, such as occurs during major overhauls or for providing infrequent specialised services.

Within the CEGB, less use is made of mobile maintenance resources than in some undertakings elsewhere. In countries where, due to climatic conditions, plant overhauls can take place throughout the year, and where stations are sufficiently close together, a large part of a station's maintenance effort can be planned from a central location and executed by staff who are not station personnel; this results in a corresponding decrease in apparent station staff numbers.

The operation of a Pay and Productivity Scheme between 1971 and 1980 did a great deal to ensure satisfactory productivity rates for industrial staff in the CEGB. This, and the closure of very many old and small power stations during the same period, resulted in the overall number of employees in the CEGB per megawatt of declared net capability reducing from 1.43 in 1971 to a figure of 0.95 in March 1985, the industry by then comprising 81 power stations with a net capability of 51 127 MW.

Current developments (1986) are aimed at achieving a standard organisation structure at the CEGB's 2000 MW power stations which will again ensure correct staff levels, particularly amongst technical staff. The organisational structure at such a power station is shown in Fig 1.1. This reflects the organisation at higher levels within the CEGB, allowing functional working relationships to be maintained between locations and Headquarters, and between one location and another.

Within this organisation:

● *The Production Manager* is responsible to the Station Manager for the day to day operation and maintenance of the plant. The accent is on man management, since the whole of the industrial staff complement, together with supervisory technical staff, is in the Production Department. The organisation of the technical staff in the Production Department of a 2000 MW coal-fired power station is shown in Fig 1.2.

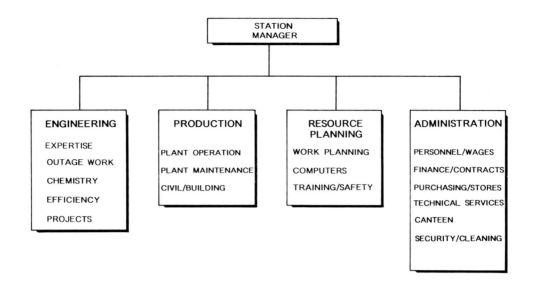

FIG. 1.1 Organisational structure of a 2000 MW power station

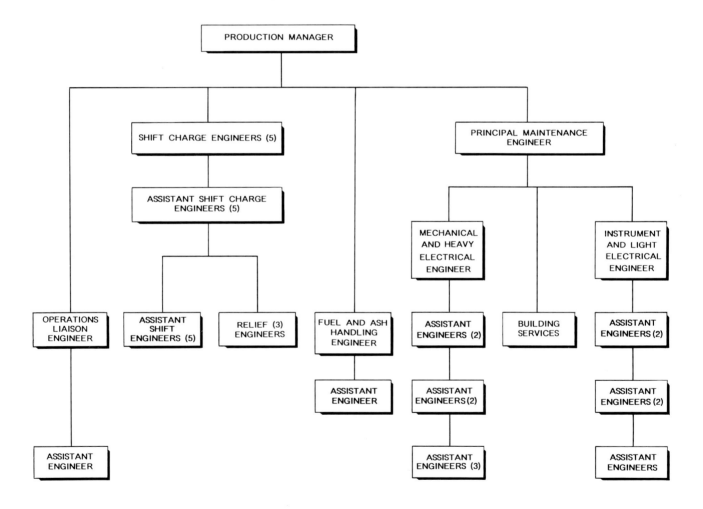

FIG. 1.2 Production department technical staff organisation chart for a 2000 MW coal-fired power station
(typical only)

● *The Engineering Manager* is responsible to the Station Manager for the formulation and enforcement of engineering policies and standards, and for recommending plant operation, maintenance and repair strategies. He can call upon whatever external assistance he considers appropriate from within the CEGB or from manufacturers to ensure plant safety and

the attainment of the plant availability, reliability, output and cost targets. Fig 1.3 shows the organisation of the Engineering Department in a 2000 MW power station.

● *The Resource Planning Manager* is responsible to the Station Manager for all aspects of the commer-

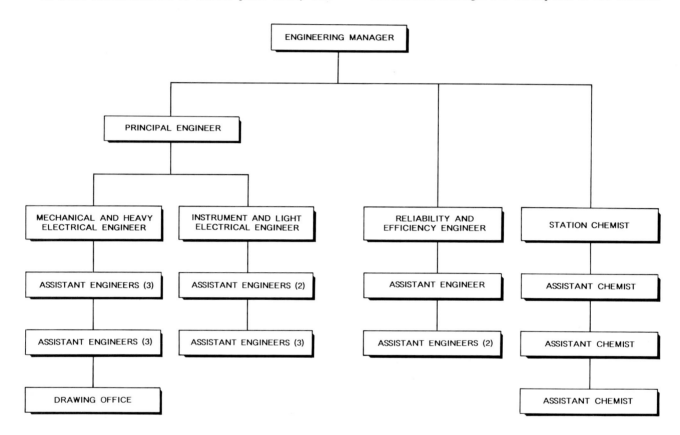

FIG. 1.3 Engineering department technical staff organisation chart for a 2000 MW power station (typical only)

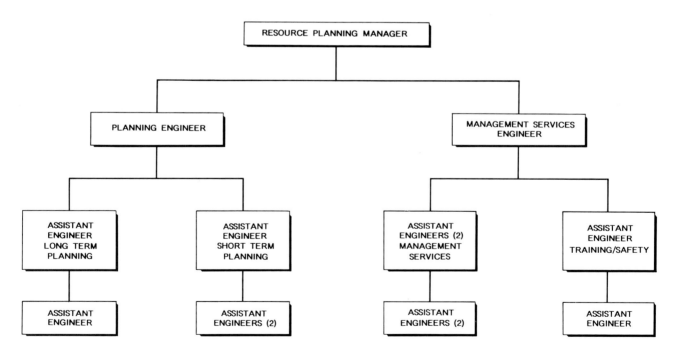

FIG. 1.4 Resource planning department technical staff organisation chart for a 2000 MW power station (typical only)

cial performance of the station. The accent is on ensuring that the resource needs of the entire station are met, determining the priorities between alternative projects and managing the ongoing allocation of resources. The central role is the management of the station's corporate planning process, which involves all departments and includes day-to-day work planning, major overhaul planning and the compilation of the long-term plans for the station. The organisation of the Resource Planning Department for a 2000 MW power station is shown in Fig 1.4.

● *The Station Administration Officer* is responsible for providing a variety of services necessary to ensure the successful functioning of a power station. Figure 1.5 shows the organisation at a 2000 MW power station, which includes stores management, catering services, personnel and wages, monitoring of departmental budgets and contracts, and station security.

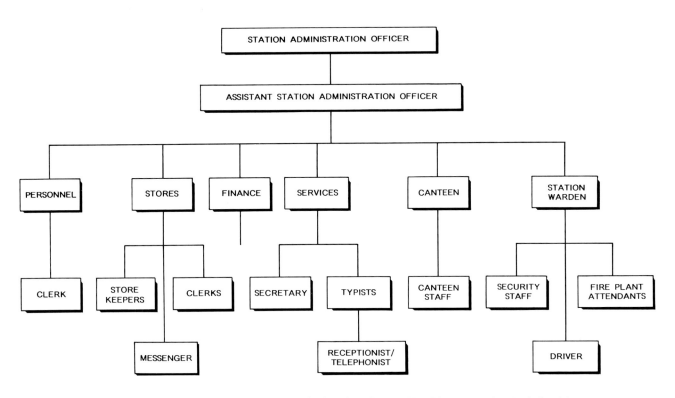

FIG. 1.5 Administration department organisation chart for a 2000 MW power station (typical only)

CHAPTER 2

Power plant operation

1 Boiler plant — normal operation (coal-fired)

1.1 Gas-side considerations

1.1.1 Introduction

Power station coal is a relatively crude, untreated material taken almost directly from the mine and containing a wide range of inorganic material (ash). It results from the degeneration of dead organic vegetable matter over millions of years and in the absence of oxygen. Consisting mainly of hydrogen and carbon as a mix of hydrocarbons and pure carbon, it also has traces of all the other elements essential to vegetable life. The amount of hydrogen varies from 3–4% for anthracites — the older deposits — to 5–6% for the more usual bituminous coals, which were formed more recently. The lower hydrocarbon content of anthracite, which typically contain only 9–13% of volatile matter, results in a coal that is somewhat difficult to ignite and for which more time is needed to complete the process of combustion. In the UK, the major anthracite deposits occur in South Wales, with lesser amounts in Kent. Throughout the rest of the country, the vast majority of coals mined fall into the bituminous range which, with their higher volatile content (typically around 30%), are relatively easy to ignite and quick to burn.

To ensure the complete combustion of anthracite, the pulverised fuel (PF) particles need to remain in the furnace zone for longer than those of the bituminous range. One practical way of achieving this is to employ a downshot-fired furnace (Fig 2.1), in which the down/up flow path of the burning particles not only gives the longer residence time necessary for full combustion but helps to stabilise ignition of the incoming fuel.

Also, with this arrangement, it is only necessary to supply a small proportion of the combustion air at the burners, the remainder being supplied through ports in the boiler walls. This allows the slow-burning flame typical of a low volatile coal to be 'fed' progressively with air and avoids the initial chilling of the flame front that may affect ignition stability.

Bituminous coals are eminently suited for firing in any of the other three types of furnace employed within the CEGB: these frontwall-fired, opposed-fired or corner-fired units are depicted in Figs 2.2 and 2.3. Typical frontwall-fired furnaces have up to 32 burners set in several tiers up the front wall, opposed-fired furnaces have burners set into two opposite walls (again in several tiers), whereas corner-fired furnaces have burners grouped in vertical banks in each corner and positioned to direct the fuel tangentially into a vortex of flame at the centre of the combustion chamber. In such furnaces, the corner-mounted burners can be tilted up or down about a neutral position, thereby

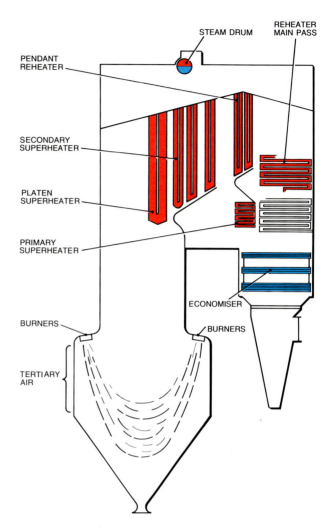

FIG. 2.1 Downshot-fired PF boiler
An unusual, but effective, design for burning low-volatile coals. The U-shaped path taken by the burning particles increases the residence time in the furnace by up to 50% and ensures a complete burn-out before the particles pass through the platen superheater banks.

raising or lowering the position of the fireball in the combustion chamber. This facility is particularly useful in controlling superheated and/or reheated steam temperature, though the burner tilting mechanisms introduce the possibility of mechanical seizure or malfunction which is absent in the static burners of the frontwall-fired or opposed-fired furnaces. A feature of the UK bituminous coal geology is that coal properties vary significantly over short distances. Boiler design and operation needs to take account of this if satisfactory high-load operation is to be assured.

Table 2.1 shows the diversity that can be expected in as-received properties of coals supplied to the CEGB. It will be seen later that the presence or absence of constituents such as chlorine and sulphur greatly affects the ease of burning such coals whilst sustaining high boiler loadings.

On being injected into the furnace, each particle of PF immediately becomes subject to the intense

7

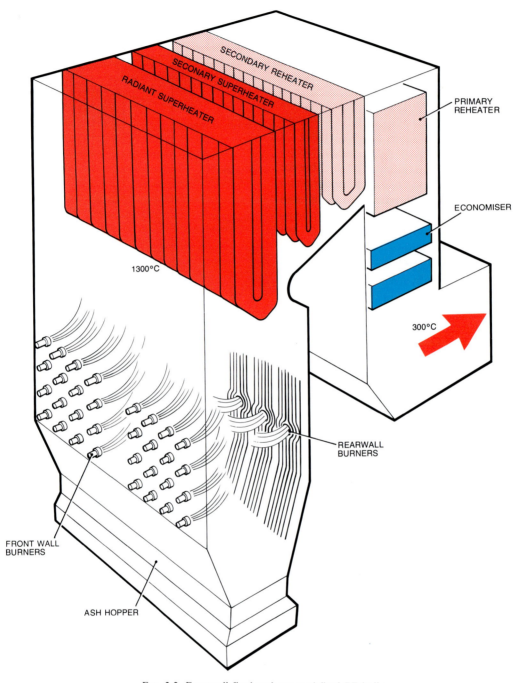

FIG. 2.2 Frontwall-fired and opposed-fired PF boilers

The illustration shows an opposed-fired boiler. Deleting the rearwall burners would convert it to a frontwall-fired design. One disadvantage of opposed firing lies in the need for very long PF piperuns which are vulnerable to PF fallout in horizontal sections, leading to the danger of fires or explosions.

radiation prevailing and quickly swells, giving off its volatile content which ignites and burns. In the local absence of sufficient oxygen (which is normal in the fuel-rich zone around the burner), some of the hydrocarbons pyrolise to form sub-micronic particles of soot. Providing oxygen levels are maintained in the furnace generally, almost all this soot burns off with the yellow, highly-radiant flame characteristic of the combustion of bituminous coal (or oil) and the coal particle develops into a boiling sphere of tar-like sub-stance which continues to lose hydrogen until only the carbon remains. This is then a particle of *coke*. Thereafter the reaction time is that of the carbon-oxygen chemical reaction, which is slow compared with the initial phase and seems to depend on the ability of oxygen atoms to diffuse into the carbon.

As the coal particles only remain in the combustion zone for a few seconds at most, complete burn-out is only possible if they are extremely fine and intimately mixed with the combustion air. The former

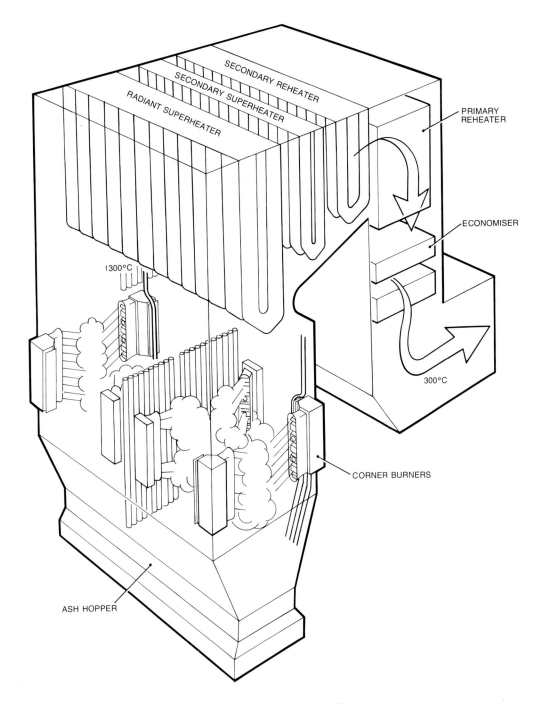

FIG. 2.3 Corner-fired PF boiler
The burners mounted in the middle of front and rear walls are provided only where a central steam or water cooled division wall effectively splits the large combustion chamber into two.

TABLE 2.1
Typical 'as received' properties of coals delivered to the CEGB

	Minimum	Mean	Maximum
Moisture %	6	12	26
Ash %	5	16	34
Volatile matter %	8	27	33
Calorific value kJ/kG	13 000	24 200	32 000
Sulphur %	0.6	1.6	4.2
Chlorine %	0.02	0.25	0.6

requirement is discussed in Section 1.3 of this chapter and the latter in Section 1.1.2.

However good the burner design and operation, unless sufficient air is admitted to the furnace in the correct place and at the right time, combustion will be incomplete. The theoretical quantity of air required to allow complete combustion of the fuel is called the stoichiometric quantity, but because of imperfections in the fuel/air mixing, extra air must be admitted or some fuel will not be completely burnt. With oil, the atomisation of the fuel is more efficient and this excess air should not amount to more than the equivalent of 0.5% oxygen. Pulverised-fuel particles are much coarser than atomised oil droplets and, even in a well designed modern furnace operating at or near full output, some 2% excess oxygen levels are needed. At much lower values, carbon monoxide will be detected in the flue gases, denoting incomplete combustion, while excess oxygen greater than 2% or so will produce inefficient boiler operation due to the increasing loss of heat to the chimney.

One of the aims of the combustion control equipment is to maintain the set fuel/air ratio at all times whatever the demand on the boiler.

During the later stages of combustion, the coke particle splits up into coke and particles of incombustible mineral ash. Most of these ash particles melt and become spherical. Some are hollow, containing trapped combustion gases, and form the 'floaters' seen on ash ponds. Hollow or not, most of the ash particles pass harmlessly from the combustion zone with the other combustion products, cooling and solidifying in the process. Most will be trapped by the ionised electric field of the electrostatic precipitator and deposited into the dust hoppers. The operator can influence to some degree the efficiency of this collection and this will be discussed later. Whilst the ash particles are molten, there is always the possibility that individual particles may adhere either into lightly-bonded structures, termed *sintered deposits*, or worse, into more solid heavily-fused masses of *slag*. These processes are unlikely to develop in the free conditions of the combustion chamber itself but occur readily if the ash particles are impinging onto furnace tubes or superheater elements.

Unless combustion conditions rapidly improve (within hours) and/or flame impingement onto the tubes ceases (and this is the single most likely cause of heavy slag formations), the deposits can continue to grow at a rapid rate, remaining molten and gravitating down the tubes to solidify in the lower (cooler) regions of the combustion chamber. Huge deposits can build out into the furnace space, especially if tube distortion allows convenient 'keying-in' points to exist. Eventually the deposits may fall heavily onto the sloping tubes leading to the ash hopper throat, causing severe mechanical damage (and often tube leaks). Occasionally they may bridge the ash hopper throat, forcing unwanted plant outages. Even if they remain attached to the tubes, unacceptable mechanical loadings may develop on the tube suspension arrangements and, of course, sooner or later they will have to be dislodged. Many boiler outages, officially recorded as being due to tube failure, actually commence with slag damage.

Other than ash, the major products of combustion are nitrogen (N), carbon dioxide (CO_2) and water (H_2O). Minor products include oxides of nitrogen and sulphur, traces of carbon monoxide (CO) and traces of soot. The effect of the sulphur content of the raw coal is especially interesting. A high content results in high sulphur dioxide (SO_2) levels in the flue gases and experience proves that precipitator performance levels improve and therefore total stack emissions fall. Unfortunately, the sulphur trioxide (SO_3) levels also increase, as some 1% of SO_2 is further oxidised to SO_3, which reacts with moisture to form sulphuric acid. Severe corrosion would then occur on any section of the plant permitted to operate with flue-gas temperatures below 120°C: this therefore represents the lower limit to which heat recovery can be allowed to proceed.

Fortunately, in a PF-fired furnace, the alkaline nature of the ash 'mops up' this acid to a large extent. In an oil-fired furnace, the absence of this neutralising ash demands strict control of permitted oxygen levels on which the further oxidation of SO_2 to SO_3 depends.

Typical 'impurities' in a coal supplied to the CEGB are shown in Table 2.2. Other than sulphur, the quantities of chlorine, iron, potassium and calcium are important. The chlorine content is related to fireside corrosion whereas iron, potassium and calcium can result in an increased risk of slagging.

TABLE 2.2
Impurities in coal

'Impurities' in coal (PPM by weight) (Typical power station coal, 15% ash by weight)			
Sulphur	15 000	Zinc	60
Nitrogen	12 000	Phosphor	1 000
Silicon	26 000	Chromium	60
Vanadium	145	Cobalt	47
Iron	13 200	Manganese	85
Nickel	25	Copper	130
Calcium	9 500	Lead	50
Potassium	2 300	Selenium	7
Aluminium	25 500	Cadmium	3
Sodium	1 470	Antimony	2
Chlorine	3 400	Arsenic	10
Magnesium	3 700	Mercury	0.3

Figure 2.4 shows the complete boiler combustion system. The pulverised fuel (PF) is transported to the burners by *primary air* (PA) amounting to about 20–25% of the total air required for complete combustion. The air velocity must be high enough to prevent the possibility either of a 'blow back' from the furnace to the mill or the deposition of any larger PF particles in the long horizontal sections of PF pipework. In addition, the temperature of the primary air needs to be high enough to dry out the raw fuel entering the mill or a wet, sticky PF product results which may adhere to pipe walls and agglomerate into large particles, causing combustion difficulties.

The remaining *secondary air* is ducted to the windbox and admitted round the primary air/PF mixture in a manner dependent on the burner design. Since it is desirable, if not absolutely essential, to match the

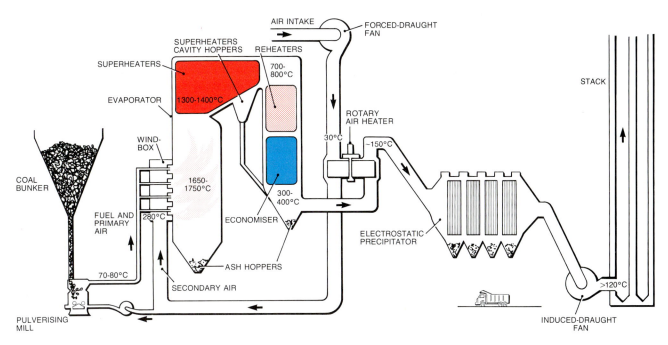

FIG. 2.4 PF-fired boiler — schematic arrangement of the combustion system

secondary air quantity to the burner fuel-through-put, some designs provide for separate windboxes on each burner group, so that the secondary air admitted depends on the air pressure set in each windbox. Obviously this facility is not available in a common arrangement: moreover, unless air is supplied in a balanced fashion (preferably from each end), pressure gradients and flow inequalities are likely, resulting in quite serious mismatches from burner to burner.

Combustion is aided by the temperature of the secondary air, frequently around 200–280°C, as delivered from the air preheater (more commonly referred to as the air heater) though even this elevated temperature may need to be increased if the wettest of fuels delivered to the coal mills are to be dried adequately. One way of doing this is to use separate main airheaters for secondary air and mill air heaters for primary air. The mill air heater is provided with an additional 'hot gas tap' to take hotter flue gases from before the economiser rather than after it, its progressive use allowing an increasing PA temperature. At one power station, a further increase in PA temperature is being contemplated by the provision of propane-fired burners situated in the mill airheater air-outlet ducts — by their use it is hoped to maintain full-load with the very wettest of fuel.

The tubes in the evaporative section of the boiler are kept reasonably cool by the steam/water mix ascending through them, despite the intense furnace radiation and temperatures (1650–1750°C) prevailing. Maximum steam temperatures are limited to 568°C in modern CEGB coal-fired boilers in order to maintain fireside superheater tube metal temperatures below

620°C: above this temperature rapid fireside corrosion can commence. Such temperatures demand furnace-zone exit gas temperatures of to 1250°C, and great care is needed in operation to limit flue-gas temperatures to the absolute minimum necessary or there will be an increasing risk of molten ash particles adhering to superheater tubes to form 'birdsnesting', i.e., large deposits of slag hanging from platens. Equally important here are the volatile potassium and sodium salt contents of the flue gases. These salts condense out onto superheater and reheater tubes to form a surface glaze on previously deposited particles of pulverised fly ash (PFA); this results in stickiness, increased depositions and a real risk of fireside corrosion.

In the remaining sections of the boiler, the flue gases are cooled by water and steam flowing in the tube banks of the various convective heat exchangers before finally entering the airheaters. These are invariably of two designs — either the air hoods rotate in unison above and below the stationary heat-exchange elements (called the *matrix*) or the air hoods are stationary and the matrix rotates. The matrix of the air heaters has the closest clearances in the entire gas system, so it is essential to protect the air heater from any coarse grits or deposits carried over from the superheater region. The continued, efficient operation of air heaters depends in great measure on the efficiency of the sootblowing arrangements and also on the means of sealing adopted between fixed and moving components.

Finally, to complete the combustion system, the PFA burden is removed in banks of electrostatic precipitators, allowing the clean gas to be discharged, via the induced draught (ID) fans, to the stack.

1.1.2 Pulverised-fuel burners

PF burners can be divided into two groups:

- Short-flame turbulent burners having swirl induced by the manner in which the secondary air is applied. These are the burners fitted to frontwall and opposed-fired furnaces.

- Non-swirled burners fitted to corner-fired or downshot-fired furnaces.

Frontwall burners (Fig 2.5)

These burners are inherently self-stable by virtue of the intimate mixing of the PF and air at the burner and of the recirculation of hot furnace gases back into the ignition zone induced by the swirl of the secondary air. In particular, such burners show a considerable ability to maintain stable ignition and flame-shape despite wide variations in the velocity and the strength of the fuel/primary-air mixture (the fuel/air ratio). This is fortunate, for a rapid response to a required load change can be achieved by an initial quick change of the PA flow (and thereby of the fuel input), maintaining the change by adjusting the raw fuel input to the pulverising mills. Altering the fuel input by the latter means alone will negate rapid firing changes due to the stored and recirculating coal in the mills. To prevent the deposition of burning PF onto the furnace wall opposite the burners (for frontwall-fired furnaces), the maximum PA velocity is limited to about 30 m/s. Similarly, a minimum velocity of about 18 m/s is set by the need to prevent PF fallout in the burner tube itself. To give a reasonable fuel turndown ratio, therefore, some variation in the fuel/air ratio may be required, giving in effect a richer mixture at the higher end of the fuel input range. Whilst the velocity and quantity of PA has little effect on flame stability, this is less true of secondary air conditions. Too great an angle of vane opening results in long, narrow non-turbulent flames that lack stability. The point of ignition may 'advance and retreat', causing pulsations in the furnace, and flame impingement onto the rear furnace wall is likely. Too small an opening induces excessive swirl, the flames being short and wide, stabilising well within the burner quarls: overheating and slagging of the area may occur. Impingement of the flames onto the side walls can cause rapid fireside corrosion (see Section 1.1.5 of this chapter). Each installation has an optimum setting for the vane angles which is determined by rig or boiler trials. Some designs of secondary air register have set out to control both the swirl and quantity of secondary air by adjusting the vane angles, but it follows from the above discussion that this is not the ideal approach unless

separate windboxes are fitted to each burner or mill group. With this arrangement, the vane angles can be set to the optimum value and variations in air quantity to suit burner or mill loading can usually be achieved by adjusting the windbox air dampers. With a common windbox, the quantity of air admitted to individual burners cannot be altered in this way and it is necessary to impose a rigid policy with regard to equally-loading burners (and mills). Unfortunately, it is no easy matter to be sure that mills are equally loaded, let alone burners, and fuel mismatches between burners of 10–20% are by no means uncommon. Faced with such problems, one approach has been to set, and then weld, the secondary air vanes at the correct openings, the quantity of air being regulated by sleeve dampers installed around the vanes. A simpler and very cheap alternative solution was adopted at one station facing the same problem. As installed, each vane could be adjusted from closed to fully-open (0–90°), though early trials showed that a setting of 45° was desirable; i.e., the provision of adjustment over the full range was unnecessary. Remote control and indication (0–100%) was also provided, but this proved difficult to maintain and could not be relied upon to give even a reasonably accurate indication of actual opening, whilst local to the actuators the vane position could not be easily nor accurately checked. In addition, the original method of setting up the vanes was done from the fully-closed (or fully-open) condition, thereby ignoring the considerable backlash inherent in the design. Not surprisingly, wide variations in burner fuel/secondary air ratios occurred, causing poor combustion. Operators became convinced of the need for local adjustments by 'eye' making, in effect, the best of a bad job. The problems were completely cured by:

- The provision of a viewing window at each burner, positioned so that the vane operating-linkages could be observed (Fig 2.6) whilst on load.

- The removal of the 'infinite' vane control and substitution of a simple 'open' and 'closed' scheme, with green and red lamps to indicate whether open or closed. Mechanical stops were also fitted at the desired 45° open position, as checked by viewing.

The operators have come to have great confidence in the system and local adjustments have become a thing of the past. Should doubt arise at any time, it is a simple matter to view the vane linkages to ensure a correct positioning.

Because of the common windbox arrangement at the station, however, it is still necessary to aim for equal mill loadings.

Although optimum vane-angle settings will vary from plant to plant, experience suggests that if flame stability is to be assured under conditions of maximum fuel turndown, windbox pressures should be at least 6 mbar at full-load with the normal number of PF

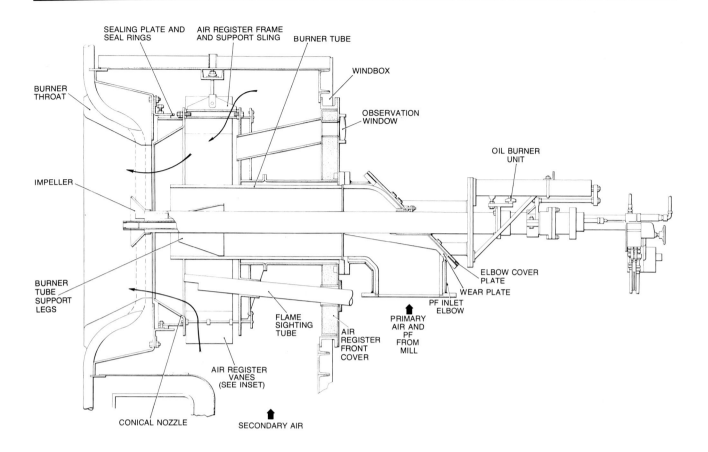

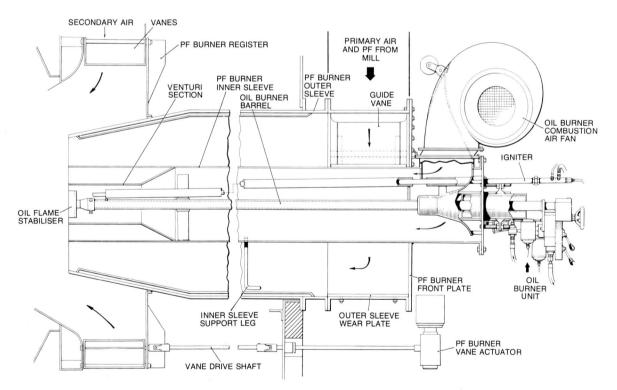

FIG. 2.5 Alternative frontwall-fired PF burners

burners firing. Note that there can be 2–3 mbar difference between the windbox pressure if measured absolutely (to atmosphere) compared with the differential pressure between the windbox and the furnace suction, and installations vary as to the indications provided. Often, with poorly maintained plant, difficulties arise in holding such pressures and, in the past, the recourse has been to restrict secondary air vane-openings further. This improves secondary air pressure but aggravates burner conditions by further

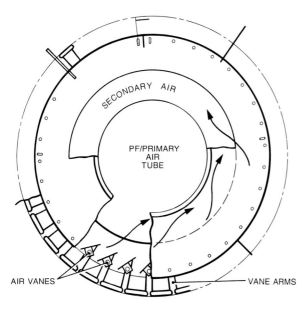

(a) General view

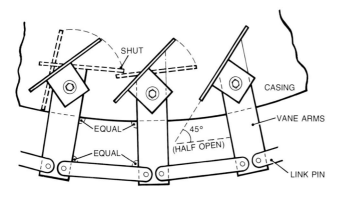

(b) Detail view of air vanes and operating linkage

FIG. 2.6 Typical secondary air vane arrangement for
frontwall-fired boiler

(a) Shows a general view, as visible in part from a window
set in the windbox front wall.

(b) Is an enlargement of the vanes and operating linkages, and attempts to illustrate how easily the human
eye can pick out the symmetry of the figures formed
by the casing, vane arms and link-pins; these figures
standing out clearly in the glare of the fire beyond.
Fortunately, in this case the preferred 45° vane-
opening yields symmetrical figures.

restricting airflow, and shortages of air at firing burners
can lead to reducing flame conditions with the increased risk of fireside corrosion and slagging (see
Sections 1.1.4 and 1.1.5 of this chapter). It is probable
that, if low windbox pressures are a problem, they
arise from an acceptance of furnace air inleakage,
normally referred to as 'tramp air', resulting in a
reduction in the percentage of combustion air supplied
by the windbox, assuming that target oxygen levels
in the flue gas are to be maintained. It is by no means
uncommon for tramp air to account for 20% or more
of the total combustion air requirements and such

values seriously reduce windbox pressures. Additionally, the tramp air reduces boiler efficiency by increasing the flue-gas losses as it contributes only to the
gas-side heat balance of the airheaters.

When searching for possible areas of air ingress
into the furnace zone it should be noted that the
suction levels increase significantly beneath the flame
zone due to the chimney effect, with maximum values
in the ash hopper and boiler bottom deadboxes.

Corner-fired burners (Fig 2.7)

Corner-fired non-swirled burners operate by mixing
the coal and air in the body of the furnace itself, as
a result of the vortex formed by the interacting burner
jets. Little mixing occurs local to the burners and
the recirculation process that ensures the self-stability
of the frontwall burner is missing. The burners consist
of a number of interposed PA/coal nozzles fed from
different mills and secondary air ports. The whole
arrangement is angled to fire tangentially into an
imaginary circle and can also tilt above and below a
neutral position to assist with the control of superheated and/or reheated steam temperature. Because
of the latter requirement there are large gaps around
the secondary air ports which admit leakage air into
the furnace. Probably because of this and the comparatively cool conditions existing in the furnace corners
behind the issuing fuel jets, corner-fired burners show
little tendency to overheating and slagging. Should the
burner alignments be incorrect, however, resulting in a
large diameter 'firing circle', then furnace-wall corrosion
and slagging can be expected.

As slag builds up on the furnace walls and heat
absorption decreases, the burners will progressively
and automatically tilt downwards to contain the increasing furnace-gas outlet temperatures (and superheated or reheated steam temperature). This gives a
useful indication as to when to initiate furnace sootblowing. As the slag deposits are blown clear and
the heat absorption rate recovers, the downward movement of the burner tilt stops and reverses. It may be
necessary to limit the upward tilt above a certain
angle to counter the tendency for the flame to be
carried upwards with the combustion gases into the
superheater zone, with likely loss of ignition.

Downshot-fired burners

These are of limited application within the CEGB,
being designed to burn low volatile coal. All designs
are similar in principle but vary in respect of coal
nozzle shape, and secondary and tertiary air supply
arrangements, Fig 2.8 shows a typical arrangement.
As the volatile matter in the coal contributes significantly to the ease of ignition and flame stability, it
is necessary, when burning low-volatile coal, to assist
flame stability by means of a hot refractory radiation
zone at the arch (see Fig 2.1) and the recirculation

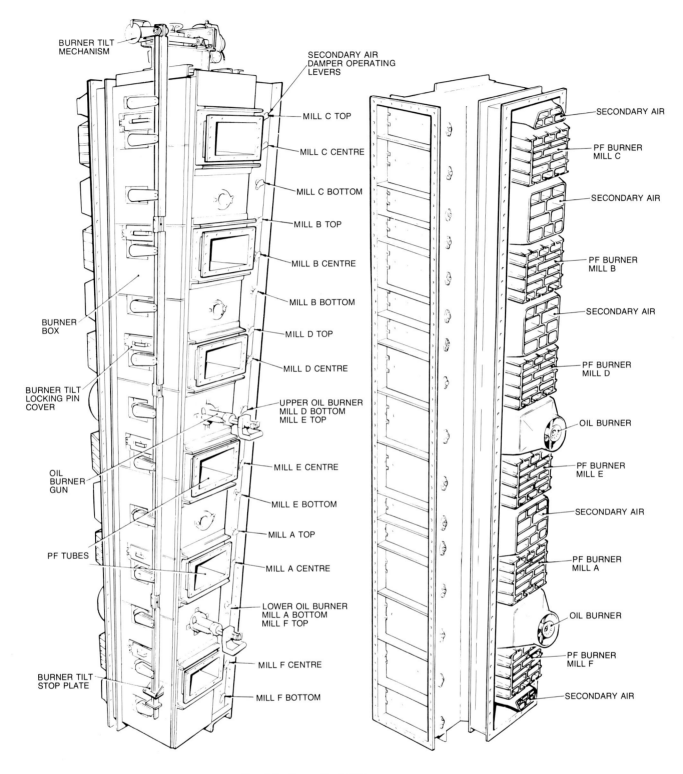

FIG. 2.7 Corner-fired PF burner arrangement

of hot gases from the tail of the U-shaped flame back in to the incoming coal jets. Unlike swirling and corner-fired burners, only a small percentage of combustion air is allowed to enter at the burners, as the need to prevent flame chilling is paramount. The remainder (termed secondary air) is supplied through ports in the furnace walls, typically being some 70% of the total. Of the remainder, about 10% is sup-plied with the fuel as primary air and the balance as tertiary (arch) air. The satisfactory combustion of low-volatile coal demands a greater degree of flexi-bility of air admission (and incidentally, a much finer PF product) than when burning bituminous coal and it is not possible to be so specific about the require-ments of downshot-fired burners as it is with the other types. Each burner system must be set up individually

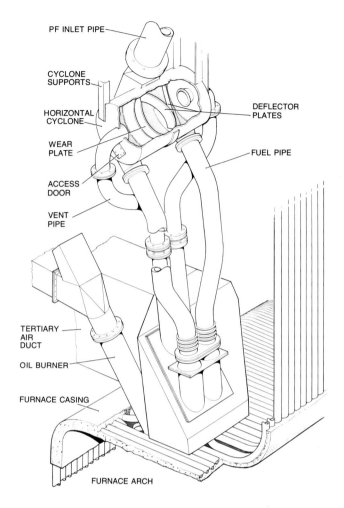

FIG. 2.8 Downshot-fired PF burner

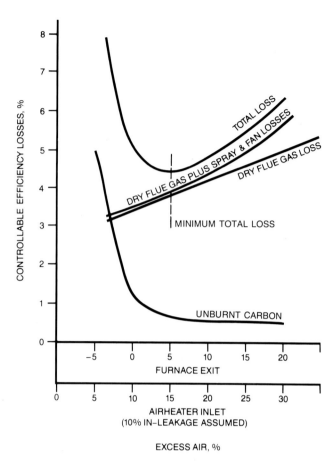

FIG. 2.9 Controllable combustion efficiency losses for a
PF-fired boiler

to obtain the best compromise between stable ignition, a deep U-shaped flame, satisfactory steam/temperature control, and minimum gas and carbon losses.

1.1.3 Combustion control

In general, combustion control schemes have to satisfy two main aims. First, and perhaps obviously, enough heat must be available to satisfy the turbine steam-demand at any instant of time. Secondly, the control scheme must endeavour to maintain minimum flue-gas losses by seeking to balance any tendency to excess airflow on the one hand (leading to high wet/dry flue-gas losses), and a deficiency of air on the other resulting in a rapidly increasing unburnt-carbon loss and an enhanced risk of boiler slagging and fireside corrosion.

The first objective is met by adjusting the input from the pulverising mills, though the manner in which this is done, or is possible, will be seen later to affect the response time of the boiler significantly; the second is met by maintaining some fuel/air ratio to hold the minimum loss point shown in the curves of Fig 2.9. Before the advent, within the last decade (1986), of reliable oxygen and (more recently) carbon monoxide

in flue gas monitoring systems, cost control schemes fell well short of this ideal for a variety of reasons not the least being the need for high load, high availability operation dictated by the then load demands on the CEGB system. Such a need inevitably resulted in stations setting liberal excess-air rates (high O_2 levels) knowing that availability would thereby be best assured. More recently, with the emphasis shifted to the more economical operation of plant, excess air levels have been drastically reduced over a span of several years, typically from 4% O_2 after the economiser to 2%. Several million tonnes of coal have already been saved in consequence.

The relationship between the total flue-gas losses and excess-air values of Fig 2.9 is well known and 'oxygen in flue-gas' monitoring has been used as a parameter in combustion air control schemes for many years since the reliability of the sampling systems improved. Its use, however, has several shortcomings. Firstly, significant uncertainty will always be present when setting target levels, as they will vary with coal quality, load, the number of mills in service, etc. Secondly, severe stratification of flue gases occurs across large ducts so that short of installing a large number of sampling probes (up to 12) large errors occur about the true mean oxygen level. Initially, it would be impossible to set a meaningful target level and plant tests would be necessary to enable the variation in oxygen content across the duct to be plotted. Thereafter, by careful siting of, say, four probes across the width of the boiler, a mean oxygen level would be obtained which would not, at least, be hopelessly unrepresentative of the true mean value, though errors of ±15% can be expected. Of course, effective mixing of the stratified flue-gas flow occurs across the rotary airheaters and, in particular, across the induced draught (ID) fans and it might seem worthwhile to sample after either point. Unfortunately, as many problems are created as are solved, since air inleakage effects increase rapidly across the airheaters (due to leaking seals) and beyond (due to the increasing negative pressures under which these zones operate) and the monitored O_2 levels become difficult to interpret. For example, a 2% concentration at the air heater inlet could be seen as 5% after the ID fans and this increase is dependent on load, draught conditions and time, due to the slowly-deteriorating seal efficiency of the airheaters. Any target set for the ID fan O_2 content is unlikely to correspond to, or remain at, the minimum loss point. Since the introduction of zirconium oxide 'in the duct' O_2 probes, a marked improvement in the reliability of the monitoring systems has been apparent and the present policy is to monitor after the economiser using two to six carefully sited (by field trials) probes. (Incidentally, zirconium oxide cells were developed as part of a research programme into fuel cells for spacecraft and illustrate yet another 'down to earth' spin-off of such development.) They operate by generating a voltage across the two sides of the cell when each side is exposed to atmospheres of differing O_2 concentrations such that the larger the difference in O_2, the greater the voltage generated. Therefore when measuring the relatively low O_2 content in flue-gas, air (20.95% O_2) is used as the reference gas to maximise the voltage, and hence the sensitivity, of the cell. As the voltage generated also varies with the absolute temperature of the gas (and cell), it is necessary to heat the probe (typically to 850°C) and then maintain it at this temperature so as to avoid temperature-induced inaccuracies. 850°C was chosen, by one manufacturer, to maximise the signal strength available without the need for the exotic construction materials that would be required for much higher temperature levels.

On the latest systems, the probe is inserted permanently in the duct, thus avoiding the need to extract and condition the sample gas (by cooling, demoisturing and cleaning); this development has eliminated the time constant of the sampling system and allowed the output to be fed into an automatic combustion-air control scheme rather than be displayed for manual intervention. Notwithstanding all modern developments, however, automatic combustion-air control schemes based on O_2 trimming usually result in the boiler operating well away from the point of minimum loss,

and attention has turned again to carbon monoxide (CO) monitoring.

At the temperatures prevailing in the furnace (> 1000°C), it can be shown that CO can only exist in significant amounts (>100 PPM) if the oxygen content approaches zero, since with a sustained oxidising atmosphere the CO is rapidly burnt off to CO_2. This is not true when gas temperatures fall below 1000°C — under these conditions the reaction rate is low and any CO present will remain. It follows that any CO concentration present at the boiler exit (where temperatures have fallen to 1000°C) above a background level of some 50 PPM, arises as a result of operating the furnace at, or below, stoichiometry. Moreover, the amount of CO present is proportional to the unburnt carbon loss so that the minimum-loss operating point of Fig 2.9 corresponds well with the onset of significant CO production (Fig 2.10).

Whilst spot measurements of CO in flue gas have been carried out for very many years on PF-fired boilers, on the basis that the presence of CO indicated generally 'poor' combustion, reliable continuous

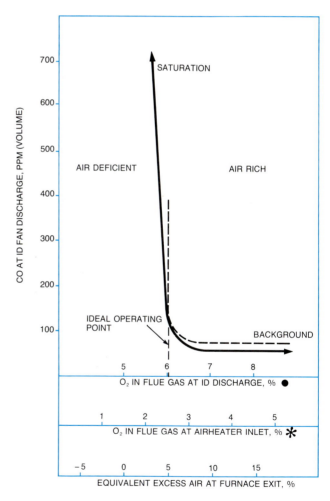

ASSUMES

✱ 10% CASING IN-LEAKAGE

● 15% AIRHEATER + PRECIP. IN-LEAKAGE

Fig. 2.10 Plot of CO in flue gas against O_2 in flue gas

CO monitoring systems have only been introduced relatively recently. The most modern of such systems are non-sampling, commonly referred to as 'in-duct sampling', and indicate the presence of CO in the flue gases without the need to withdraw and condition the gas sample (by cooling and cleaning, for example). As a result, they are fast-acting and maintenance-free, and can provide, together with an O_2 signal, one of the parameters necessary for an automatic combustion-air control scheme. Typically, they operate by detecting the infra-red radiation absorbed by the CO concentration in the flue gas, CO having a well defined spectral absorption in the range 4.5 to 5.0 μm. The equipment consists of an infra-red source operating at 650°C positioned in one duct wall, and a receiver containing a photodetector arranged to view the source through the flue gas. In the receiver are two gas-filled cells, one containing a CO standard as a reference cell and the other (a dummy cell) filled with N_2 which is insensitive to the flue-gas CO concentration. The two cells are alternately switched into the sight path of the detector to provide, by difference, a measure of the CO concentration present (refer to Volume F Chapter 4).

Fortunately, the actual CO levels in the flue gas are not greatly affected by air inleakage effects (unlike O_2 determinations) and there is every incentive to sample after the ID fans where well mixed and relatively clean gas is available, allowing single-probe sampling to be used.

Note If the CO level at the boiler exit is (say) 200 PPM and total air inleakage by volume is 20%, then the CO level indicated at the ID fans is 200 × (100/120) = 167 PPM.

Whilst 'sampled' CO systems remain in use at some stations and continue to give good service, they suffer from the disadvantage of the relatively long time-constants of the sampling system and can only be used in conjunction with recorder or VDU displays as operator aids, rather than as inputs for an automatic control system. Whichever display is chosen, it makes good sense to combine it with the O_2 determinations, so that the operator is presented with the clearest picture of current flue gas conditions at all times; the alternatives are shown in Fig 2.11 (a) and (b). By these means the operator can easily see if the CO breakpoint is tending to occur at elevated O_2 levels as it will be, for example, if milling plant is overloaded, if burner(s) are incorrectly set or as time takes its toll on furnace casing and airheater seals.

An obvious development with CO monitoring, once reliable, stable and fast-acting analysers became generally available, was to incorporate the CO signal into the automatic combustion-air control scheme and this has now been done; several 500 MW units within the CEGB have such control arrangements, whereby the 'set' O_2 levels are automatically varied to suit the CO concentration present in the flue gas. CO

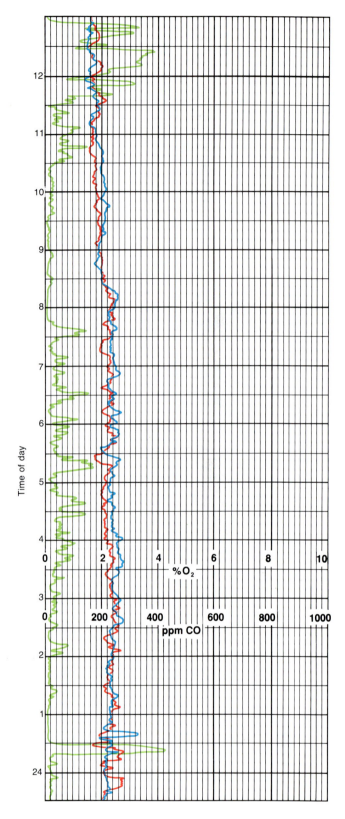

FIG. 2.11 (a) Typical recorder chart from 500 MW boiler showing CO and O_2 levels at full load

The boiler concerned has a divided furnace (by virtue of a central division wall). The chart indicates the conditions in *A* furnace and a similar recorder is provided to monitor *B* furnace.

The O_2 levels (red and blue traces) are derived from Zirconia O_2 probes inserted in two positions in the *A*-side economiser outlet-duct, these positions being chosen by field trials as being reasonably representative of mean O_2 levels in the vicinity. The green trace gives the CO content after the A-side ID fan.

A transient fall in the red O_2 trace at 0030 h (bottom of chart) resulted in a CO level peaking at 420 PPM, showing that combustion excess-air levels were by no means excessive. During the latter part of the night shift, conditions appear to have been maintained at near optimum values but by 0800 h, either as a result of a more equal loading of the milling plant or as a result of a change of coal, it was found possible to reduce excess air by the equivalent of 0.5% O_2 without causing CO production. By midday, however, the operating O_2 levels can be seen to be too low. The chart provides very clear evidence of the benefit of CO-in-flue-gas monitoring in maintaining minimum dry flue-gas losses.

- Furnace zone slagging (including ash hopper bridging and burner 'eyebrows').

- Radiant superheater 'birdsnesting' and convection superheater bonded-deposits.

- Economiser fouling and bonded-deposits.

- Airheater fouling.

Furnace zone slagging

It is in this zone of the boiler that the largest, hardest deposits can quickly form should combustion conditions deteriorate. Under normal conditions, even at

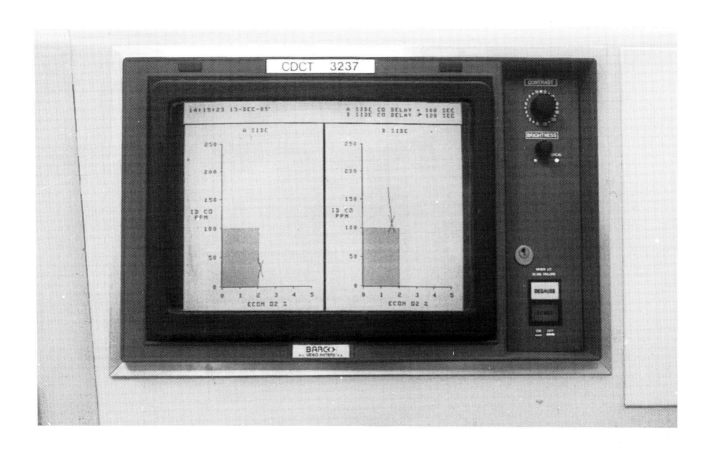

FIG. 2.11 (b) CO-O_2 'window' on a 500 MW unit panel VDU display

(see also colour photograph between pp 34 and 35)

concentrations are chosen that are just above the background count for that particular station (typically 50–100 PPM). The systems are virtually automatic versions of the CO-O_2 displays shown in Fig 2.11 (b), and have built into them the necessary safeguards to ensure stability of combustion. They are likely to be increasingly used.

1.1.4 Fireside slagging and fouling

Figure 2.12 shows those areas of a large PF-fired boiler in which slagging and fouling can occur. They may be conveniently summarised as:

maximum continuous rating (MCR), there is adequate capacity within the furnace zone to absorb the available radiation and many months of high-load operation will be possible especially if the furnace dimensions are generous, thereby making flame impingement onto the furnace waterwalls unlikely. Any deposits formed are loose, lightly sintered and easily removed by regular sootblowing. However, 'sensitive zone' sootblowers must be kept in a condition of high availability and be blown regularly to ensure that deposits do not build up (rather than blown to remove deposits that have already formed). Many stations have found that

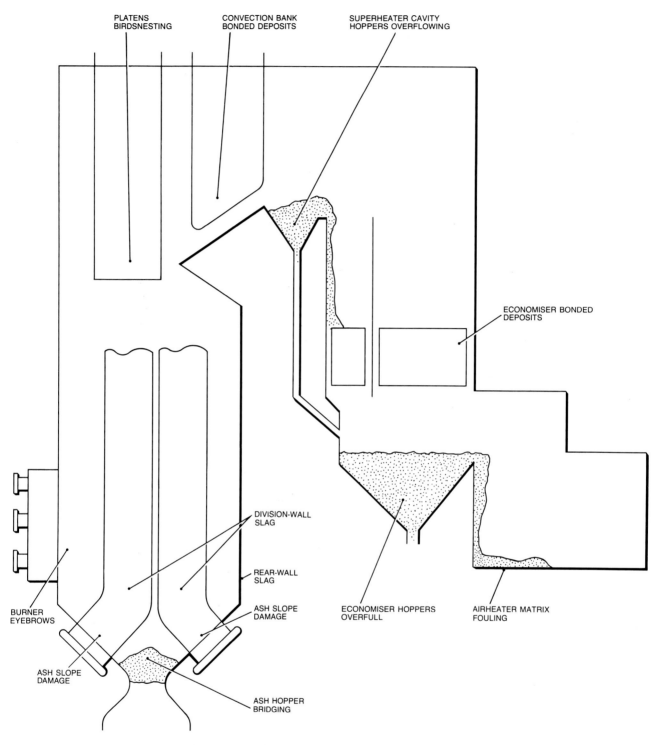

FIG. 2.12 Diagram of typical boiler fouling locations

setting-up a regular sootblower repair party greatly improves sootblower availability; but whatever arrangements are established for sootblower repairs, it should be understood that the longer a particular sootblower is out of commission, the more difficult it is to clear accumulated deposits on its return to service. Indeed, for particularly sensitive sootblowers repairs may be necessary on an emergency basis.

Despite a high sootblower availability, from time to time abnormal deposits may occur which can seriously reduce heat transfer to the furnace waterwalls and increase general furnace-gas temperatures to the point where molten deposits are formed. These deposits will 'slip' slowly down the walls, freezing at the lower, cooler levels. Waterlance sootblowers (see Section 1.1.7 of this chapter) are particularly effective in removing running slag. Another effective way of removing the more heavily-fused deposits is to alternately reduce and increase furnace temperatures (by changing load, or even by shutting down completely, say, overnight): the deposits will weaken and fall under the induced contraction and expansion forces. This method has its cost in lost availability. When considering the subject of slagging, 'prevention' is always better than 'cure'. The art of such prevention takes many forms, not the least being regular furnace inspections and ash hopper checks. The provision of suitable inspection ports is essential. An appreciation of any difficulties arising from the emptying of the boiler ash hoppers is more likely to be known and acted upon if the ashing party is part of, rather than separate from, the main shift operating team. Early warning of deteriorating ash quality is invaluable and must not be ignored. Similarly, the progressive failure of furnace zone sootblowers often indicates running slag conditions. An efficiently run power station depends upon well defined, conscientiously executed routines and furnace operations are no exceptions; if regular furnace inspections detect the onset of slagging, then a costly boiler outage may be prevented.

The tendency of a furnace to slagging depends on many factors, not the least being the boiler design and, in particular, the furnace dimensions. It is important for the boiler operator to appreciate this, so as to be able to concentrate on areas vulnerable to deposition.

Table 2.3 shows the approximate leading dimensions of the present CEGB range of modern PF-fired boilers. Designs A and B might be expected to suffer from rearwall slag deposits in view of the relatively limited depth between the PF burners and rearwall, and this has been so in practice. Design B has limited height between the burner zone and radiant superheaters and, on occasion, suffers from superheater 'birdsnesting'. Design C appears to be limited in furnace width which might encourage sidewall deposits and fireside corrosion effects. Finally, it might be expected that a boiler of design D would be virtually immune from slagging in view of the relatively generous dimensions throughout. Indeed, at the time of writing (February 1986), a CEGB station containing such boilers has recorded a 97% load factor for the 8 months April–November. Even on these boilers, whilst heavy rearwall and superheater slagging

is virtually unknown, the centre partition walls cannot be viewed easily, have no sootblower facilities and are vulnerable to slagging.

The person to whom furnace inspections are entrusted should have the authority to order corrective action to be taken, even if this ultimately requires some load reduction. Whilst there is normally every financial incentive to maintaining full-load with low excess flue-gas oxygen levels, this is counter productive if it causes downtime for off-load deslagging operations.

Occasionally, heavy falls of slag can even bridge across the ash hopper throat gap, and the removal of such lumps is no easy task. Figure 2.13 shows a large lump of slag firmly wedged in the ash hopper of a 500 MW boiler which, by virtue of its generous furnace sizing, might have been expected to be immune from slagging. Whilst load reductions may result in the lumps cracking, it should not be assumed that the weakening will be sufficient to cause collapse. Some stations have applied water jets with success, but there is a real danger of tube failures occurring if the natural circulation process is halted by the copious water quantity running from, or being deflected off, the lumps. Higher in the furnace where radiation is intense, tubes that have suffered circulation failure will overheat, swell and burst. Mechanical prodding has been used, sometimes utilising a specially adapted truck, but there are obvious dangers in trying to dislodge large lumps from beneath, with the boiler on-load. Often the only recourse is to take the boiler off-load, cool it and then scaffold-out to gain access to the lumps in order to break them up. The method employed in the construction of the scaffolding must be carefully considered so as to avoid danger if the lumps should settle or slip. Explosives have occasionally been used to deal with exceptionally large obstructions.

Once in the ash hoppers, the normal extraction-jet water will be quite incapable of removing the larger pieces and internal entry into the hoppers will be necessary in order to remove them manually (Fig 2.14). All in all, this was a very clear case where prevention would have been better than cure.

An appreciation of the characteristics of the coals likely to be burnt is equally important and advances have been made over the last decade in isolating and defining those constituents likely to give rise to slagging. The once widely-held view that low ash-fusion-temperature coals exacerbate slagging is no longer

TABLE 2.3
Leading dimensions of PF-fired boilers

DESIGN	A	B	C	D
Furnace type and rating	All frontwall-fired, 500 MW			
Furnace width, m	29.6	29.6	24.4	25.6
Furnace depth, m	9.2	9.5	11.1	11.3
Height, hopper to roof, m	36.9	29.6	35.7	39.3
Height, basement to drum, m	51.8	50.9	53.7	61.9

FIG. 2.13 Photograph showing a large slag lump firmly wedged in the ash hopper throat

(see also colour photograph between pp 34 and 35)

FIG. 2.14 Ash hopper debris resulting from the eventual fall of the lump shown in Fig 2.13

(see also colour photograph between pp 34 and 35)

believed to be wholly correct, and emphasis has moved to the mineral content of the ash. Many deep mined coals are low in calcium, sodium and iron, and experience shows that they are relatively low in slagging propensity.

High calcium coals, in particular, are now thought to have a high risk of slagging, especially if iron is present. Such ash compositions often occur in opencast coals and these need to be burnt with care. Operationally it may be necessary, though difficult and costly, to mix such fuels and/or to bunker them so as to feed PF mills that fire away from areas likely to slag. If this is impractical, then the only alternative is to arrange for the operators to be informed when they are bunkered, so that increased furnace checks can be instigated. Forewarned is hopefully forearmed. Other than coal composition, the likelihood of slagging is greatly influenced by localised reducing conditions, i.e., a localised deficiency of oxygen in the burner zone. Such conditions are dependent on, amongst other things, PF grading which is adversely affected by the wet, sticky oversized PF product that occurs if the raw coal being supplied to the mill is not adequately dried out and individual particles agglomerate in the PF pipework. Localised reducing conditions are also more likely to occur if fuel/air ratios at individual burners are grossly mismatched and it has occasionally been necessary to fit CEGB CERL-designed riffle boxes into pipe bifurcations to improve the PF distribution from one mill serving, typically, four or more burners. Mill loadings should also be balanced, though the difficulties of doing this will be appreciated when this logic is discussed (see Section 1.3 of this chapter). Similarly, the settings of secondary air vanes is extremely important, as discussed earlier. Providing that these factors are satisfactory (and maintained so by combustion control), then slagging should not be a serious problem if a measure of operator common sense prevails. This aspect should not be overlooked; for instance, a boiler producing enough steam to maintain a turbine-generator output of 500 MW with a winter condenser back-pressure of some 40 mbar would be producing only enough steam to maintain 475 MW if condenser back-pressure rose to 85 mbar (not all that uncommon at an inland cooling tower station in high summer). Loading up to 500 MW under such conditions may make combustion conditions marginal with regard to slag formation.

Accumulations of slag around, and particularly over, PF burners are referred to as 'burner eyebrows'. Such deposits are common in many boilers, being formed by the continuous recirculation of a very small part of the fuel/air mix issuing from the burner. Providing that the eyebrows do not slip and interfere with the fuel/air stream they are not operationally troublesome, but can result in mechanical damage to the quarl refractory.

They have, from time to time, been linked to the following causes:

- Poor PF distribution, leading to reducing conditions.

- The use of oil overburn, especially on burners already firing PF.

- Low ash-fusion-temperature coals.

- Incorrectly-set secondary air vanes.

For obvious reasons, it has been unusual to site sootblowers around PF burners and, in the unusual event of eyebrows actually interfering with the fuel/air stream, it has been necessary to shut down the offending mill group and then re-introduce a cold airflow through the burner, causing contraction and thus weakening the deposits. Alternatively, mechanical prodding may be used. Nonetheless, eyebrows often cause more trouble off-load than on-load as it is necessary to remove them before allowing access into the lower areas of the furnace.

Superheater deposits

Since the introduction of widely spaced (0.7 to 1 m) radiant-zone platen elements and generous clearances between the platens and the flame burn-out point, superheater 'birdsnesting' (as the deposits on pendant radiant superheaters are called) has become less common, providing that sootblowing is effective. The availability of sootblowers operating near the bottom of the platens is critical and a lack of commitment in this area will encourage the formation of large hard deposits. Whilst seldom growing to the point where adjacent loops are bridged, they can, nevertheless, cause considerable damage to ash hopper slope tubes when they fall. The provision of cross-tube or loop-bracing spacers, or operation with badly distorted platens (thereby allowing 'keying-in' points) does much to aggravate any tendency to slag. Operationally then, the need is to encourage early burn-out of the PF by ensuring satisfactory PF fineness, correct secondary air settings and combustion-air quantities, etc., in fact, similar attention to the detail necessary to prevent furnace deposition. Convection superheater bonded-deposits have long been linked to the volatile alkali salt content of the flue-gas PF ash (PFA). In particular, the potassium and sodium salts form low melting point glazes on the PFA particles which then readily adhere to the relatively high metal-temperature tube surfaces.

Unless corrosion occurs under these deposits they rarely affect operation other than by reducing heat transfer rates. Even then it would be extremely unlikely that superheat or reheat outlet steam temperature would suffer because of the excess of superheat and reheat surfaces invariably built into the modern boiler to ensure satisfactory steam temperatures under adverse/lower load conditions. Efficient combustion and adequate and effective sootblowing offer the best solution.

Superheater slope deposits can become quite voluminous, especially on boilers where the designed slope angle is much less than 30° (the angle of repose of PFA is around this figure), and can result in delays in gaining access to combustion chambers should the need arise. At worst, they also encourage high velocity 'gas-laning' in the areas which may induce excessive metal temperatures on any tubes subjected to the increased velocity gases. Incidentally, it is worth encouraging operating staff to inspect boilers internally prior to cleaning, as much can be learnt about the efficiency of combustion and the effectiveness of sootblowers. Finally, the direction of operation of superheater slope sootblowers may be worth considering; if similarly rotating units are installed on either side of the boiler, then one will tend to blow deposits back down the slope and into the furnace whereas the opposite unit will tend to throw them over into the back-passes of the boiler.

Economiser fouling and bonded-deposits

Some boilers have finned-tube economisers and these are particularly prone to the build-up of bonded dust deposits, especially if constructed as a single 'block' of tubes rather than having tiers of tubes separated by access spaces. Experience indicates that high calcium coals (about 7% CaO in ash) give rise to very firmly bonded-deposits which are also insoluble in water and therefore especially difficult to remove by off-load water lancing.

Carry over of deposits from overflowing superheater cavity hoppers is often a cause of severe fouling in the economiser sections beneath: the sootblowers in the affected zones must always be operational and effective. Visual inspections are necessary from time to time to confirm this effectiveness. In a few instances, finned-tubes have been replaced by plain ones, especially if the original design provided for additional banks as required (thereby allowing the lower heat transfer of plain tubes to be offset), and sootblowers have been either modified or resited to improve their effectiveness. Bonus effects have been better steam temperature control and less restriction to flue-gas flow, both improving the combustion process in general. The lower, more uniform gas-velocities flowing through cleaner economiser banks have also greatly reduced outages due to gas-side erosion of the tubes (see Section 1.1.6 of this chapter).

Airheater fouling

Having the smallest clearances in the entire gas-side system the airheaters are particularly susceptible to fouling.

When higher-temperature sections of the boiler are sootblown, much of the fine debris is carried forward with the flue gas; normally this is collected in the hoppers situated in the superheater cavity and at the bottom of the rear pass of the boiler. Some stations have experienced difficulty in removing the debris from these hoppers and material has spilled into the air

heater causing rapid blockage of the gas lanes.

This material has included pieces of fused, sintered and bonded superheater deposits up to 50 mm across; such large pieces could only have reached the rear pass by saltation up the slope under the pendant elements, by successive sootblower actions.

Various plant modifications have been tried to overcome the debris removal problem. At some stations, the top of the economiser hopper has been plated over, so that there is no effective hopper to be emptied; this obviously cured the hopper emptying problem, but is no solution to that of fouling of the top of the airheater matrix. At others, larger hoppers and chutes have been provided and the change made from dry-dust removal to a wet-dust removal system, in which the dust is discharged by gravity into a sluiceway and thence to the ash disposal plant. Incidentally, gas ducts leading from the economiser to the airheater inlet which slope downwards are conducive to coarse debris falling or bouncing onto the airheater matrix. Yet another solution has been to modify airheater sootblowers to bottom-action only, as top-blowing of the airheater can force pieces of fused deposit into the matrix, causing permanent fouling.

Unfortunately, excessive sootblowing of airheaters can damage seals, with the feedback effect of increasing leakage and reducing the availability of combustion air, thereby increasing the risk of combustion chamber slagging.

1.1.5 Fireside corrosion

Several areas of the boilers are susceptible to fireside corrosion. First, and most important, furnace-waterwall or partition-wall tubes positioned in the burner zone can be affected. Tube corrosion is always associated with localised reducing atmospheres and flame impingement, and is exacerbated by the burning of high chlorine-content (0.4–0.6%) coal. If some or all of these factors are absent, then fireside corrosion can be negligible: alternatively, wastage rates can be so high that continued operation from one statutory survey to the next is impossible. Now that survey periods are being increased from 26 to 36 months on newer plant, the problem is more acute.

Secondly, fireside corrosion can occur in superheater or reheater elements as a result of the deposition of condensed alkali salts onto those tubes operating at the highest metal temperatures. The maximum steam temperature of 568°C set in modern CEGB coal-fired plant arises partly from the need to restrict the gas-side metal temperature of superheater elements to 620°C and preferably less, to contain such corrosion. In oil-fired plant, a lower steam temperature limit of 540°C is set to minimise the particularly aggressive attack of deposits containing alkali sulphates and vanadium compounds.

Thirdly, uncooled boiler components, such as superheater and reheater tube spacers or alignment strips, are prime candidates for high temperature corrosion. Such components may promote local tube hotspots, resulting in enhanced local corrosion via the molten-salt mechanism, and several enforced outages have resulted in one 2000 MW station alone from this cause.

Figure 2.15 (a) shows a stable, slow-growing and protective oxide scale arising from oxidising furnace conditions. Very low metal wastage rates would be associated with such a scale, typically of around 0.15 mm/year, which is low enough to give more than the designed tube life (0.15 mm/year is approximately 15 nm/h). If heavy reducing conditions occur, together with high chlorine content coal and poor grinding, then corrosion rates can increase twentyfold up to 300 nm/h, at which value the affected tubes need renewing every three years.

The most damaging reducing conditions occur when burning coal particles impinge on the tubes, usually as a result of poor milling plant or burner operation. Combustion of these coal particles is retarded on the relatively-cool tube surfaces and so they smoulder, releasing large localised quantities of carbon monoxide (CO), sulphur monoxide (SO) and sulphur-bearing carbonaceous material. Most of the chlorine content of the coal is released in the flame as gaseous hydro-chloric acid (HCl). The combined effects result in rapid growth of non-protective scales and corrosion (Fig 2.15 (b)). Often 'panels' of tubes upwards of 30 tubes wide and 10 metres in height can be affected.

Fortunately, so far as operations are concerned, the attention to detail so essential to prevent undue slagging also minimises fireside corrosion. First, just as it was necessary to consider mixing coals of high and low calcium content in order to lessen the risk of slagging, so it is necessary to consider mixing coals of high chlorine content with less incongruous coals, even though it is costly in labour and operationally restrictive. Alternatively, it may be possible to deliver high chlorine coal to bunkers feeding burners firing away from vulnerable walls.

Secondly, correct fuel/air ratios at each burner are crucial and regular checks on primary and secondary riffle boxes may be usefully combined with mill outages to ensure that satisfactory conditions are maintained. At one CEGB 4 × 500 MW unit station, severe fireside corrosion occurred at an early date. CO levels on the affected walls were found to vary up to 10% and alterations to burners simply moved the CO zones around the walls. Serious mismatch between fuel and air at the 'wing' burners was suspected and investigations confirmed this. CEGB CERL-designed riffle boxes were installed and conditions improved dramatically (Fig 2.16), even though general excess air conditions were reduced.

At another station, at the time of writing, daily side-wall CO scans are taken, thereby alerting the operating staff to any reducing conditions that may be present. At the same station, it has long been realised that the finer PF product arising from the regular use of

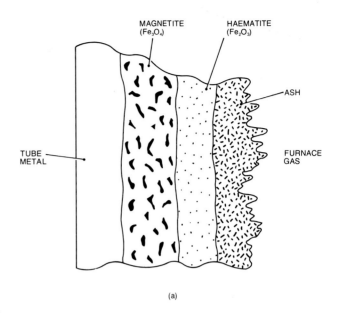

(a)

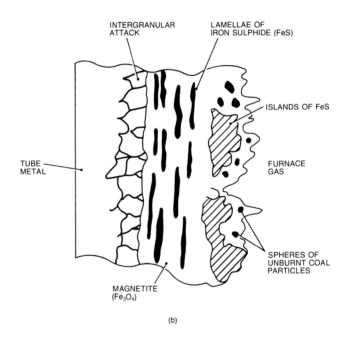

(b)

FIG. 2.15 Oxide scales on furnace tubes

(a) The scale grown under normal oxidising conditions is an inner magnetite layer covered with haematite, and is protective and slow growing.

(b) Under reducing conditions, a rapidly growing magnetite layer forms, containing lamellae of iron sulphide (FeS) and spheres of partially-burnt coal. These smouldering particles release hydrogen chloride gas which promotes intergranular attack on the tube metal.

six mills for full-load (rather than the five mills that are mechanically capable of this load) is so beneficial in avoiding corrosion effects that the small increase in works power and maintenance costs is worthwhile (Fig 2.17).

Regular PF sampling is useful in highlighting a bad grinding mill, even though the sampling cannot be particularly scientific, especially if it is non-isokinetic.

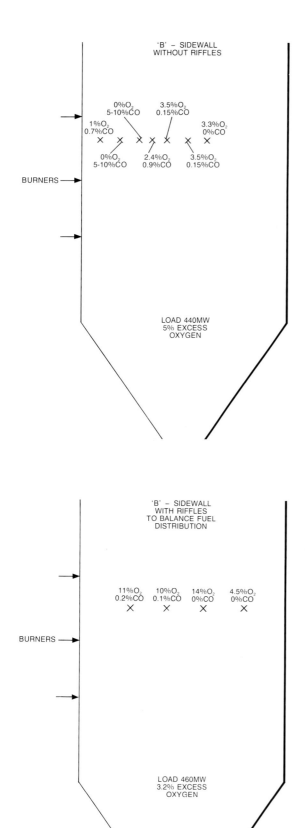

FIG. 2.16 Effect of equalising fuel distribution of CO and O_2 concentrations on a furnace sidewall

If poor PF fineness figures are obtained on a particular mill and repeat checks confirm their validity, it is very worthwhile to establish the cause which may, or may not, be obvious on inspection of the mill. For example, at one station it was found that unexpectedly rapid wear was occurring in the raw coal chute to the grinding elements, where this pipe passed through the outgoing PF product pipe allowing raw coal particles to by-pass the mill completely.

Thirdly, secondary-air control is equally important. What may seem to be a small deviation from normal can seriously affect the flame shape. Once general agreement has been reached at a particular station on the desired settings, these should be rigorously enforced. Figure 2.18 shows the effects of varying the secondary-air vane angles on the sidewall CO levels. It should be noted that the use of oil overburn can produce appreciable CO levels in its own right, let alone when introduced into operational PF burners. Careful attention to the points raised in the preceding discussion, combined with effective furnace and super-heater/reheater sootblowing patterns and a rigorous control of superheater and reheat steam temperatures (to avoid high metal temperatures) should prevent undue corrosion in the convective superheater/reheater sections, although minor attacks on the leading tubes of the hottest-zone elements may occur.

Finally, one important aspect is often misunderstood — the role of excess air. Whilst excess O_2 levels in the flue gases may be quite high and acceptable, severe reducing conditions can be present at the burner zones. The problem is one of a local shortage of air rather than a general one. Similarly, the absence of appreciable quantities of CO in the final flue gases does not mean that values as high as 10% cannot be present around the burners.

If, after having considered all of the above factors, fireside corrosion is still a problem at a particular station, it may be necessary to provide boiler tubing that is more resistant to attack. For furnace tubing, both co-extruded and faceted type tubes (Fig 2.19) have been employed, grouped into panels in the most vulnerable zones (Fig 2.20). Co-extruded tubing has also been developed for application in convection superheater elements.

1.1.6 Erosion

Studies undertaken as early as 1960 indicated that failure of boiler tubes due to erosion was extensive, though some failures in the combustion chambers in particular were found to be due to corrosion rather than erosion, after examination of failed components. However, the vast majority of all failures reported in the rear passes of PF-fired boilers were correctly defined as erosion-induced. Presently, the failure rate of boiler pressure parts due to the combined effects of sootblower and dust erosion exceeds that for the other three principal causes (mechanical design con-

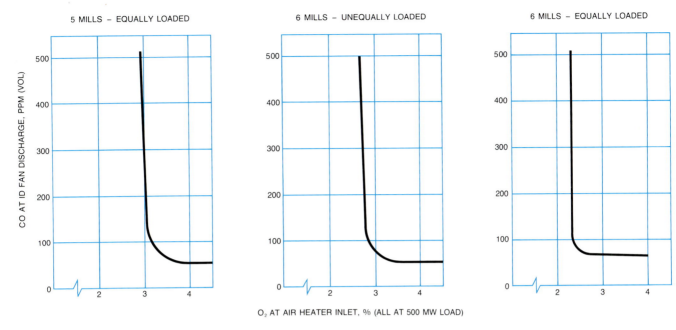

FIG. 2.17 Effect of operating conditions on the CO 'breakpoint' for a 500 MW boiler at full-load

siderations, overheating and defective welds). The subject is naturally receiving much attention, though the options open to the operator in attempting to reduce this failure rate are few and concern mainly the avoidance of sootblower, rather than dust, erosion, since the latter is particularly dependent on design considerations. As was found in the earlier studies the main areas affected remain the furnace, where sootblower erosion only is the problem, and the rear passes, which are vulnerable to both effects.

Tube erosion due to sootblower action is related to the degree of wetness of the steam supply and the impact of any ash particles entrained in the steam or air jets, the former being particularly important. Air-operated sootblowers are generally free from wetness effects and cause many less tube failures than occur with steam systems, especially if the latter are poorly designed and operated. Other relevant factors include the frequency of sootblowing, the blowing pressures selected and the steam/airtightness of the sootblower at rest, all of which are considered in Section 1.1.7 of this chapter.

The boiler tubes at the base of the furnace, forming the sloping section leading to the ash hopper throat, are vulnerable to mechanical damage from heavy falls of slag. The impact can cause the tubes to tear away from their tie-back arrangements, especially at the bend between slope and vertical wall, and they may stand proud into the furnace by 150 mm or more. Such misplaced tubes are obvious candidates to failure by overheating as a result of their somewhat exposed position in the fireball, or by erosion from the action of adjacent sootblowers. However, once the tube (or tubes) have sprung there is a rapid deterioration of the furnace wall skin casing, insulation matting and appearance cleading by burning and this can be ob-

served by operating staff who will be in a position to prevent further erosion damage by reducing or suspending sootblowing in the area until repairs can be effected.

Furnace tube erosion can also occur on ash-hopper slope tubes after an extended period of operation due to the normal continuous falls of ash. Regular sootblowing should minimise any tendency for abnormal deposits to build up and is the only practical means of minimising such erosion effects.

Tube erosion in the rear passes of PF-fired boilers, particularly in the economisers, has been and remains a more serious problem. The higher rates of metal loss cause tube failures within 15 000 h, the velocity of the flue gas and PFA particles being the major factor, as the erosion rate varies with (velocity)$^{3.5}$ (i.e., doubling the velocity will give an elevenfold increase in wear rate). Studies have indicated that erosion is negligible providing flue gas velocities remain below 15 m/s and modern designs attempt to achieve such values. Unfortunately, the progressive blocking of economisers with dust and larger clinkers results in localised high-velocity 'gas-laning' and it is almost always this sort of effect that is responsible for the failures that, all too readily, continue to occur. Even here the role of the operator is limited, for design considerations are again paramount. For example, those economisers mostly at risk employ finned-tubes arranged in deep (often single) banks with tubes running from side to side of the rear pass. The installed sootblowers are often quite incapable of keeping these banks clear and dust banks build up, especially along the front tubes which are vulnerable to deposits carried over from the superheater slopes above. In consequence, gas velocities and erosion rates increase through the rear sections

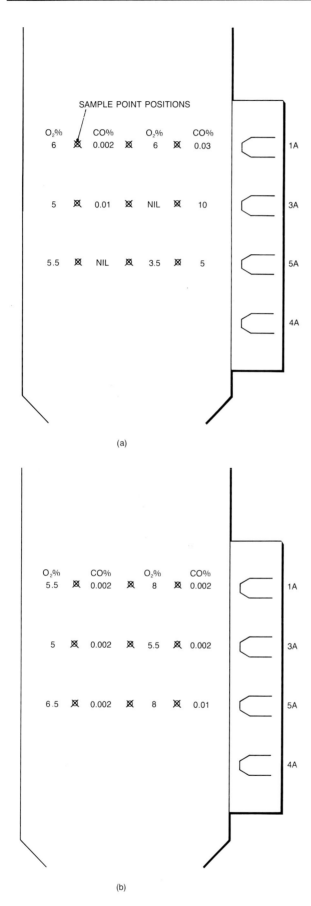

tests), 5A and 4A. In (a), the 5A burner secondary air vanes are restricted to 25% open resulting in a highly-conical, short turbulent flame which is oxygen deficient. Similar low O_2/high CO percentages would probably have been recorded had a sampling point been fitted adjacent to 4A burner. Figure 2.18 (b) shows a return to a normal vane opening of 50%.

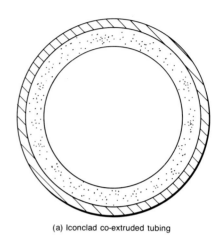

(a) Iconclad co-extruded tubing

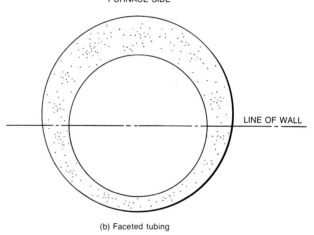

(b) Faceted tubing

FIG. 2.19 Co-extruded and faceted tubing

(a) Shows a cross-section of Iconclad co-extruded tubing. The outer layer is of 50% nickel and 50% chromium for resistance to corrosion. The strength of the tube is mainly in the inner layer, consisting of 32% nickel, 21% chromium and the balance of iron. It is essential to have a good thermal bond between the two tubes.

(b) Shows a faceted tube in cross-section, the extra thickness being sacrificial.

causing replacement tubes to fail, in the worst cases, within a further 15–20 000 h. Erosion shields, Fig 2.21 (a), are often necessary and any gaps between the rearmost tubes and the enclosure walls must be sealed off to prevent similar high-velocity gas flows, Fig 2.21 (b). With such economisers, the propensity to erosion is also aggravated by the line of sootblower travel that is specified. Reference to Fig 2.22 should make the point clear.

FIG. 2.18 Effect of secondary air vane settings on adjacent sidewall CO values. The figures show the six sidewall sampling points (X) on a 500 MW boiler, with the nearest PF burners 1A, 3A (out of service during

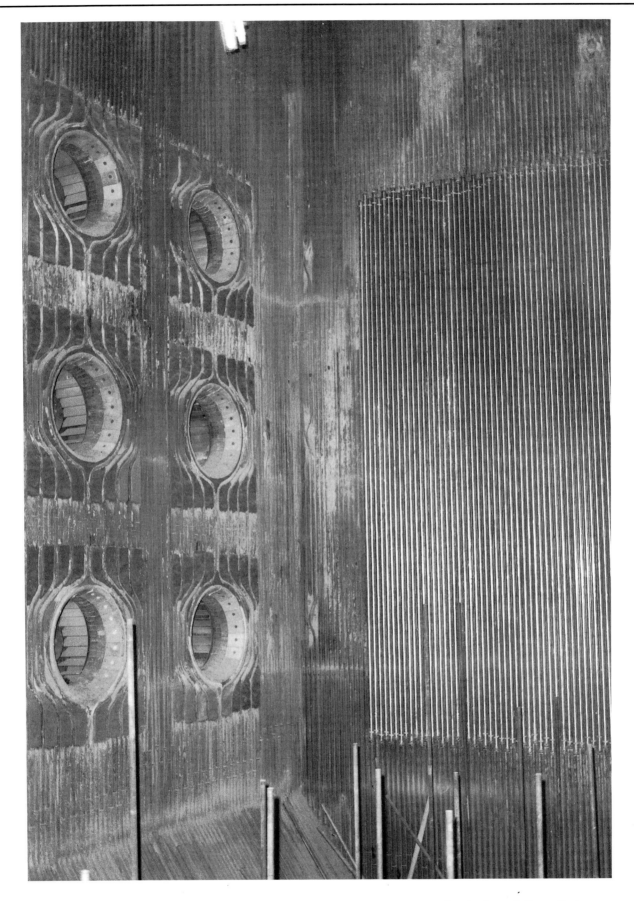

FIG. 2.20 Photograph of a panel of co-extruded tubing installed in a furnace sidewall at Eggborough Power Station
Note the nearness of the end burner to the sidewall, allowing little room for error in setting up the burner conditions.

(see also colour photograph between pp 34 and 35)

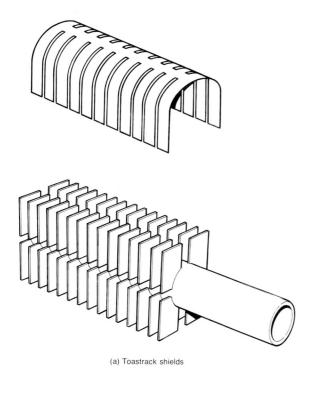

(a) Toastrack shields

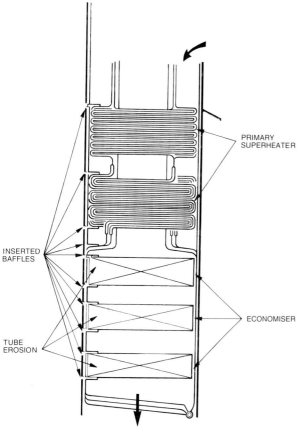

(b) Areas of likely tube erosion

FIG. 2.21 Rear pass erosion

(a) Shows 'toast rack' shields fitted as a preventive measure on finned-tubes located in particularly vulnerable areas.

(b) Shows areas of likely tube erosion and the arrangement of baffles to fill high velocity gaps.

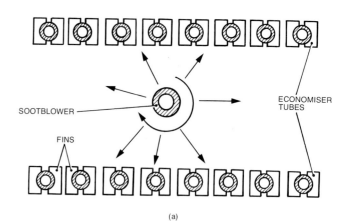

(a)

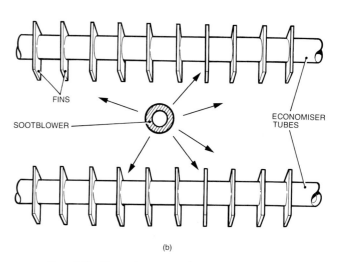

(b)

FIG. 2.22 Illustrating economiser tube erosion due to sootblower action

Arrangement (a) is not preferred, as it leads to increasingly ineffective sootblower action the farther away a tube is situated from the vertical centre line through the sootblower lance. Tube erosion is more likely than fin erosion.

Arrangement (b) is preferred, as the cleaning action continues well past the sootblower lance vertical centre line the fins acting as deflectors channeling the air or steam down through the elements. Also, erosion occurs on the fins rather than the tubes.

Whilst it is necessary for the operator to be aware of deteriorating economiser conditions (by monitoring ID fan suctions, economiser gas-differentials, etc.), there is usually little that can be done, assuming that sootblower operation is as effective as possible. However, it is operationally important to minimise excess air rates to achieve the very lowest economiser gas velocities.

Economisers employing finned-tubes arranged to run from front to back of the rear pass are much less vulnerable to erosion, probably due to the effect shown in Fig 2.22, i.e., the dust particles impinging on the fins rather than the tubes. One problem with such economisers concerns those fitted with separate main and by-pass sections. Whilst no failures of note have ever occurred in the main sections, as these are well away from overflowing superheater slopes or cavity hoppers and seldom block, the same cannot be said for the by-pass sections which are very vulnerable to such effects; indeed at one station deposits at least 4 m high have been noted, extending back into the primary superheater tubes above. On examination of the fouled economiser pass, only about 5–10% of the cross-sectional area remained open to gas flow and erosion was unacceptable. Figure 2.23 shows a failed economiser tube taken from the above pass. Note that the fins have been completely eroded away at one point. De-finning the top section of the three-bank economiser and the installation of more effective soot-blowers between each bank has stabilised the position,

though the elimination of cavity hopper overflowing is the only real solution. Incidentally, an unwanted secondary effect is a rise in reheated steam temperature due to the abnormal gas flows passing through the main pass, resulting in the need to apply emergency 'knock-down' spray.

Finally, plain-tube economisers have proved to be the best operational option, never blocking and seldom failing.

1.1.7 Sootblowing

In introducing the various types of sootblower installed on modern PF-fired boilers, it is convenient to split up the boiler into the following sections, on which completely different types of sootblowers are usually installed:

- The furnace
- Radiant and convection superheaters and reheaters

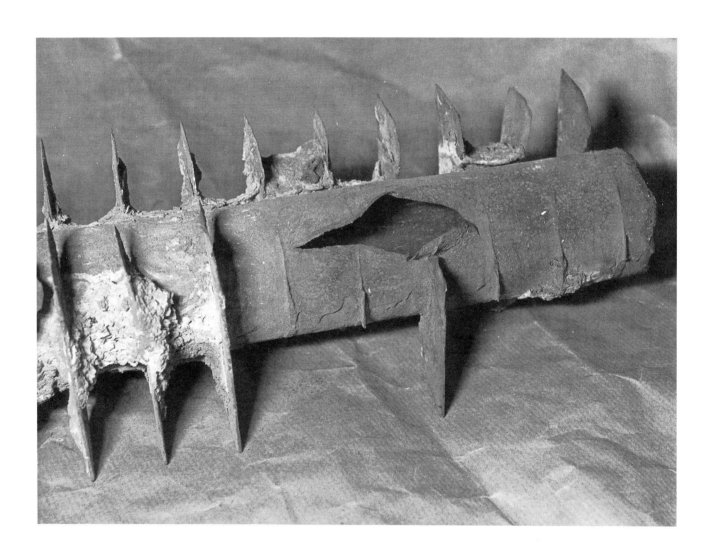

FIG. 2.23 Photograph of a failed economiser tube
High velocity 'gas-laning' had been occurring for some time and has not only completely eroded away the fins at one point, but has eventually caused the tube to rupture due to thinning.
Operating the boiler for even a short time with a leaking economiser tube is unwise. This is because adjacent areas of the economiser and the airheaters beyond are wetted with the resulting spray; PFA deposits can then accumulate and set like concrete, making subsequent removal extremely difficult indeed.

- Economisers

- Airheaters

Furnace sootblowers

Conventionally, all furnace 'gun-type' sootblowers have been similarly designed, typically as shown in Fig 2.24. Dependent upon the blowing requirements and the number of jets in the nozzle, each blower may be arranged to rotate with steam or air applied for several revolutions before retracting to the rest position. Unfortunately, the effective blowing circle is not particularly extensive, having a radius of only about 1.5 m, so up to 60 such blowers may have to be installed to provide reasonable wall coverage. They are most extensively employed on those walls where

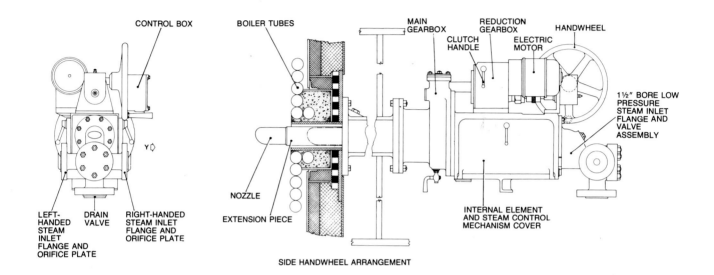

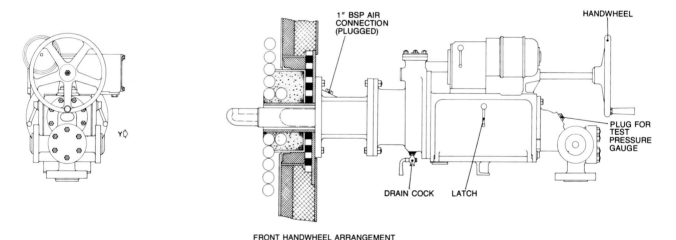

FIG. 2.24 Typical gun-type sootblower

The sootblower has a static inner element which leads into a rotating outer element. Steam is fed from the supply pipework via a mechanically-operated control valve, through the inner element, then into the furnace through a nozzle at the end of the outer element.

The outer element is driven by a stationary motor and gearbox arrangement. A main gearwheel, which is internally threaded to match the spiral on the outer element, provides the traverse and rotating motion via an idler and drive pinion.

Initially, the gearwheel rotates the outer element until the traverse pin contacts the spring-loaded traverse bar. Continued rotation of the gearwheel then causes the outer element to advance into the furnace until the pin reaches a chamfered slot in the traverse bar, which enables the pin to lift the bar against the springs and allows the outer element to rotate again.

Prior to reaching the full travel position, the traverse pin-ring contacts a pivoted clevis which is connected to the valve-operating lever, and opens the steam valve. The valve remains open at the full travel position, whilst the outer element rotates several times. Then the motor reverses and the traverse-pin contacts the opposite side of the traverse bar.

The outer element then withdraws from the furnace and the steam valve closes, the clevis linkage being moved forward by the valve spring compression.

During withdrawal, the traverse pin reaches another chamfered slot in the traverse bar and the outer element rotates again until the motor stops.

slagging propensity is greatest, for example, on the rear walls of frontwall-fired furnaces. Blowing pressures are of the order of 10 to 12 bar and the operating medium is steam or air, the advantages and disadvantages of each being detailed later. Low pressure will result in an ineffective performance, capable of removing only the lightest and most friable of deposits, whereas too high a pressure may exacerbate erosion damage, especially on steam-operated systems.

Furnace-wall sootblowers are operated the most frequently of all types installed, typically between once a day and three times a shift. The latter frequency of use is perhaps surprising but by no means uncommon, and arises from an almost universal need to maximise the absorption of heat by the furnace waterwalls (to prevent excessive superheater and, sometimes, reheater steam temperatures), rather than from any general desire to maintain the walls deposit-free at all costs or because the sootblowers are uniformly ineffective. Availability is generally high, often exceeding 95%, and, providing that the sootblowers are correctly positioned and correctly adjusted, they should be capable of removing all but 'running' slag. The more serious faults are usually confined to the possibility of feed-tube erosion and nozzle cracking, as can occur on poorly-drained steam-supplied systems; minor faults range from passing steam or air shut-off valves to burnt-out nozzles as a result of flame impingement or 'running' slag conditions.

An alternative to the gun blower has been developed within the CEGB and is shown in Fig 2.25. The design combines the operating principles of both gun and lance sootblowers, longitudinal motion being achieved by a pinion driving a rack mounted on the sootblower main beam and rotational motion by means of carriage gearing. Once the blower is fully inserted, steam is applied to the 'three-tier' nozzle, designed to give a more effective clearing action, with less chance of tube erosion, at a radius of up to 3 m. Because the nozzles of conventional furnace-wall sootblowers sometimes overheat and crack whilst the sootblower is retracted (i.e., at rest), the stroke of this design is such that the nozzle is completely removed from the furnace and is protected by a mechanically-operated door.

The prototype sootblower was installed for some considerable time on one large boiler and was found to be both reliable and effective, with the ability to clean a larger radius than the gun blowers installed. Development was discontinued due to the high cost of manufacture and the safety aspects of having the three nozzles exposed outside the furnace when retracted. Whilst the latter could have been overcome the emergence of the water lance sootblower, in particular, precluded further design effort.

Waterlance sootblowers, for furnace application only, have been recently developed and show great promise, being able to deal effectively with running and tenacious slags. They utilise a jet of water as the cleaning medium and remove deposits by a thermal contraction effect rather than by impact as occurs in conventional sootblowers. They have already proved to be reliable in operation and maintenance-free and early fears that they may induce thermal fatigue on boiler tubes appear to be unfounded, though clear operator guidance must

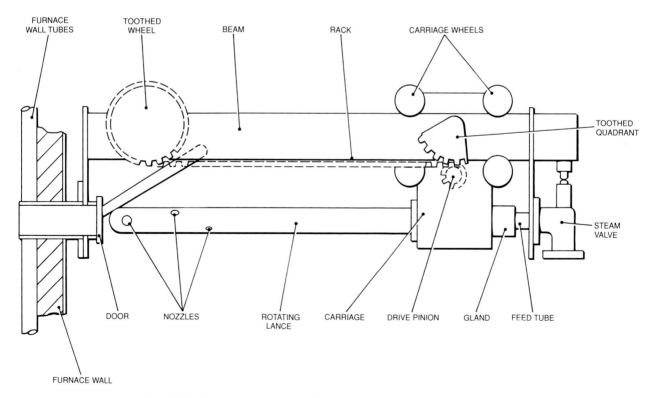

FIG. 2.25 General arrangement of CEGB long-stroke furnace wall sootblower

FIG. 2.13 Photograph showing a large slag lump firmly wedged in the ash hopper throat

FIG. 2.14 Ash hopper debris resulting from the eventual fall of the lump shown in Fig 2.13

FIG. 2.20 Photograph of a panel of co-extruded tubing installed in a furnace sidewall at Eggborough Power Station
Note the nearness of the end burner to the sidewall, allowing little room for error in setting up the burner conditions.

FIG. 2.37 Typical VDU displays of boiler superheater and reheater metal temperatures

PFA particles can cause erosion. One solution has been to provide a by-pass hole in the front of the nozzle which increases lance cooling flow without increasing the blowing forces.

Long-lance sootblower usage varies considerably dependent on the location, but for those units operating in the radiant zone is unlikely to be less than once per shift. The availability of such blowers is always critical, as even short outages can result in strongly-bonded deposits forming which are insensitive to subsequent sootblower action. Low steam/air pressure protection is clearly essential as even partial insertion without effective cooling flows will cause lance failure by bending. The protection may take the form of a pressure switch, armed to operate on a falling steam/air pressure by a protection limit switch, located typically 1 m in from the rest position.

Should the pressure fail to reach the set level by the time the advancing carriage actuates the protection limit switch the sootblower will be retracted. Minimum lance pressures would be 8–10 bar with normal operating pressures of 15 bar, or more.

Economiser sootblower

These can be of several types:

- Multi-nozzle non-retractable sootblowers
- Long-lance rotating sootblowers
- Semi- or part-retractable sootblowers
- Rake sootblowers

The *multi-nozzle non-retractable* sootblower was commonly installed in the relatively small economisers of older plant and consisted of an oscillating lance containing a number of nozzles which remained permanently in the gas system. The *long-lance* type was described in the preceding section. The latter two types are commonly used in modern PF-fired plant as is, to a lesser degree, the long-lance rotating sootblower.

Semi- or part-retractable sootblowers can be most effective for economiser use. Typically, they can be one-half, one-quarter or one-sixth retractable, dependent mainly on the width of the economiser tube banks to be sootblown and the available space outside the economiser. The one-sixth retractable type has six equally-spaced nozzles along the length of the lance so that in operation each nozzle progresses and blows one-sixth of the total travel. To provide more effective cleaning, the sootblower can be designed so that the lance oscillates through an arc of up to 120° instead of rotating fully, thereby concentrating the cleaning action in the appropriate sector.

Rake sootblowers (Fig 2.27) are also extensively used on economisers, especially those employing finned tubes, for which application they were expressly designed. Unless carefully installed they are susceptible both to jamming against the guide rails and to erosion which can be severe unless the rakes have been carefully adjusted so that the jets blow midway between adjacent elements.

Airheater sootblowers

The type of sootblower installed to clean airheater elements depends on the airheater design. The two basic types of airheater are those having rotating air-hoods and those having a rotating matrix; of these the latter has the simpler sootblower, as might be expected.

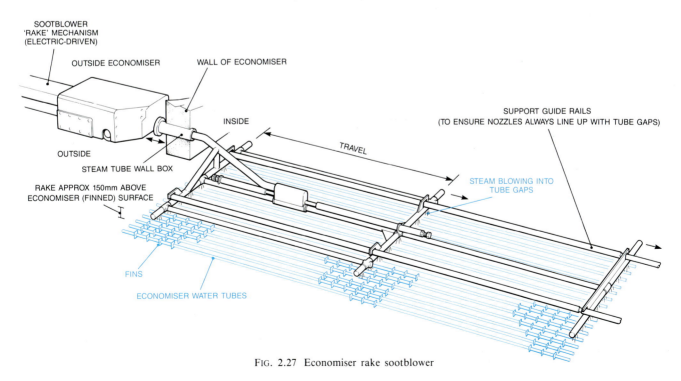

FIG. 2.27 Economiser rake sootblower

The sootblower installed on the rotating air-hood air heater must travel around with the hoods in order that the total area of the matrix is blown. A typical design is shown in Fig 2.28. The sootblower consists of an escapement wheel having 17 pins and a rotating pipe with six nozzles, the assembly being mounted between the hoods. When in operation, an actuator-driven indexing plate is inserted to make contact with a pin on the wheel as it rotates. The wheel is therefore moved around 1/17th of a revolution per revolution of the airheater and the nozzles are moved to blow a different annuli of the matrix. After 17 revolutions of the airheater, the wheel has rotated once and all of the matrix has been covered; typically, this can take up to 20 minutes. The design has proven to be relatively efficient in operation, though stiffness of the unit, due to intermittent rotation, has been overcome at some stations by arranging for the indexing to operate continuously, whether the sootblower is blowing or not.

The sootblowers installed on a rotating matrix air-heater are very simple, consisting of a non-rotating, short-travel lance-type blower. The lance has a number of nozzles and moves across the matrix at a varying speed to ensure effective cleaning, taking into account the varying linear speed of the matrix at varying radii.

Where sootblowers are installed above and below the matrix, it is usual to blow against the gas-flow first (bottom sootblowing) so as to lift and loosen any deposits rather than risk compacting them into the core; some stations that have experienced this problem have even removed the top sootblowers altogether without significantly affecting the cleaning efficiency.

Sonic sootblowers are currently on trial in airheaters at several stations within the CEGB. They consist of a low frequency (20 Hz) sound generator and can be installed in any part of the boiler which may be affected by loose dust deposits. Infrasound is generated at resonant frequency, producing waves that are claimed to clean enclosures up to 15 m^3.

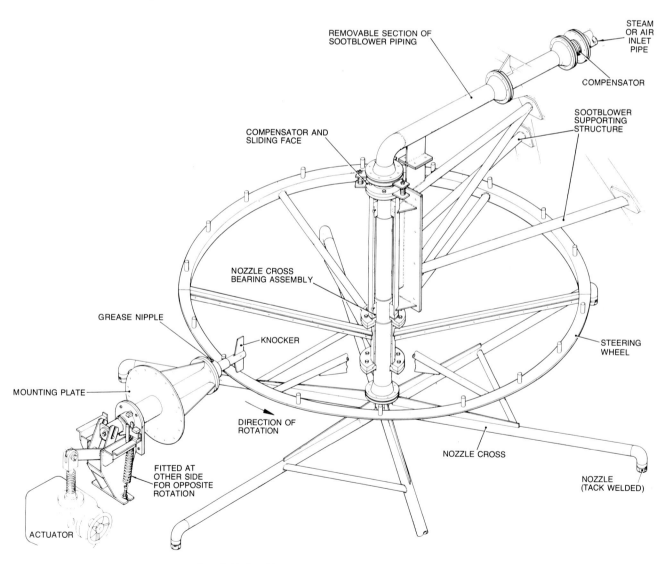

FIG. 2.28 Sootblower for rotating air-hood type airheater

Control systems

The original sootblower control systems installed in modern coal-fired plant were based on electromechanical systems designed to operate, either sequentially or individually (local to or remote from the sootblower selected), the installed sootblowers together with the steam/air supply systems. All the systems, based on uniselectors, were similar and have given satisfactory, if somewhat inflexible, operation to date, though increasing obsolescence has forced some stations to install replacement equipment.

All replacement equipment is based on a microprocessor-based Programmable Logic Controller (PLC). The PLC is ideal for sootblower control: not only is it able to carry out the necessary sequencing and timing operations and necessary safety interlocks, but allows very flexible selection of sootblower sequences. The package normally specified includes:

- A programmable controller, which handles all the logic and sequencing, timing and alarm procedures and can record any malfunctions for later analysis.

- A visual display unit (VDU), presenting system messages visually (alarms, pressures, possibly airflows, etc.).

- A mimic unit showing the boiler in simplified form and the relative positions and states of each sootblower.

- An operator control unit from which the operator can select a particular sootblower sequence.

The PLC system has a major advantage over existing electromechanical systems in that, should the operational need arise for a revised sootblower sequence, this can simply be arranged by a software program change rather than by having to physically rewire, which is sufficiently complex and time consuming to deter all but the most desperate. Dependent on the memory available, a number of alternative sequences can be stored and used to suit the particular needs of the boiler on the day; for example, economiser sootblowing frequencies may be increased if increasing gas-side temperatures indicate that gas-side blockages are building up. Plant monitoring facilities may also be incorporated to provide details of changes in boiler operating conditions following sootblowing, so that the effects of a particular sootblowing sequence can be studied later and, perhaps, optimised.

Miscellaneous aspects

Whilst the present policy of equipping new plant with air sootblowers is likely to continue there seems to be little incentive to convert steam-supplied systems to air, though there are likely to be conversions of furnace sootblowers to the waterlance type. Neither steam nor air systems are cheap to operate and any operational benefits accruing from the use of air are nullified to some degree by the initial capital cost of the sootblower air compressors.

The relative advantages and disadvantages of steam and air systems are detailed below:

(a) Steam advantages/air disadvantages

- The capital cost of the steam supply system, complete with pressure regulating equipment, pipework and drains, is significantly less than the cost of air compressors, motors switchgear and cabling (usually at 3.3 kV at least) and compressor houses. Additionally, the maintenance cost of compressors can be high.

- The incremental capital cost of the boiler and water treatment plant required for steam systems is not high.

(b) Air advantages/steam disadvantages

- With air, effective sootblowing is available right from start-up, this being particularly important with regard to airheater fire risk.

- With steam sootblowing, tube erosion can occur if the steam contains excessive moisture and the provision of an effective condensate drainage system and interlocking scheme to prevent blowing until the steam is of adequate quality, is essential. These requirements are unnecessary with air.

- The maintenance of air sootblowers is considerably reduced due to the absence of condensate in the pipework, thermal shock and erosion on the sootblower mechanism, lance and nozzles. With steam supplied units, the above factors can lead to very heavy maintenance costs, or sootblower non-availability.

The increased risk of tube erosion with steam supplied systems is highlighted by the figures presented in Table 2.4 for eight stations, and is the main reason why air sootblowers are presently specified for new plant since the cost implications of the other advantages/disadvantages set out above result in minimal savings one way or the other.

To maintain effective sootblowing, it is necessary to implement at least some operational routines. For instance, the provision of airflow and pressure indicators/recorders are invaluable for remotely checking sootblower operating conditions and facilitate the diagnosis of faulty poppet valves, burnt-out nozzles and other irregularities which are then checked locally. Such instrumentation is also invaluable in detecting air/steam shut-off valves that are passing or stuck open which, undetected, could easily result in an eroded tube within a short time. This is especially true of part-retractable sootblowers, where some of the nozzles rest permanently within the gas stream.

TABLE 2.4
Table of outage statistics caused by sootblower erosion on 500 and 660 MW boilers from date of commissioning

Sootblower type	Approximate hours steamed per outage	Outages per 100 000 hours	Days outage per annum per boiler
Steam	3 100	32.6	6.5
Steam	11 600	8.6	1.2
Steam	8 400	11.8	2.3
Steam	5 900	17.0	5.1
Steam	7 000	14.2	3.1
Steam	7 700	12.9	4.2
Air	43 600	2.3	0.8
Air	27 400	3.7	1.2

Alternatively, it is relatively easy to diagnose furnace sootblower faults locally by following one or more furnace sootblower sequences. Burnt-out nozzles can usually be correctly diagnosed by the absence of 'rotation noise' reflected back through the furnace wall and such faults as a failure to operate, air/steam valves failing to open or incorrectly-set limits can all be noted as the sequence progresses through. Airheater sootblowers of the rotating, indexing variety, associated with rotating air-hood airheaters, can also be singled out for special attention as it is necessary to ensure that correct indexing of the escapement wheel is taking place. Small inspection doors located adjacent to the indexing gear are, or should be, provided to facilitate such checks.

1.1.8 Precipitators and stack emissions

PF-fired boilers must always employ special collection equipment to deal with the substantial quantities of incombustible ash (PFA) not retained in the furnace zone; these quantities being, typically, 80% of the mineral ash content of the raw fuel. On a 2000 MW (4 × 500 MW) station on base load, and burning 20 000 tonnes of coal per day, this amounts to around 2000 tonnes of PFA. Modern practice is to use plate-type electrostatic precipitators which, by agreement with the Government Pollution Control Inspectorate, are currently designed to retain at least 99.3% of the total weight of solids present at the precipitator inlet. A further current requirement is for the absolute dust burden not to exceed 115 mg/m^3 of gas at normal temperature and pressure (NTP) of 1 bar, 15°C. These two requirements together means a maximum coal ash-content not exceeding around 20–25%, or dust burdens will be exceeded, even at guaranteed precipitator performance.

Plate-type precipitators consist of rows of small diameter discharge wires inserted between collector plates, as shown in Fig 2.29. Such an arrangement is built up to form a treatment zone; in practice, there are up to four such zones in series (forming

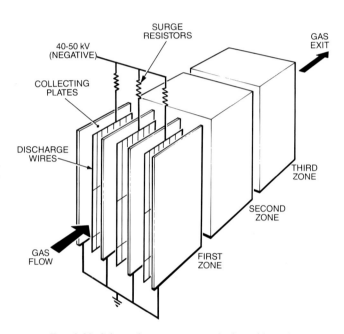

FIG. 2.29 Schematic arrangement of wire grids and collecting plates in an electrostatic precipitator
Solid particles in the turbulent gas flow become electrically charged in the 'high field' regions near the wires and are then attracted to the collecting plates. The amassed layer is periodically dislodged by mechanical rapping (not shown). As the local thickness of residual dust greatly affects the electrical stability of the zone, it is essential that rapping forces are evenly distributed and that wires and plates remain evenly spaced. The latter requirement is more commonly found wanting than may be realised, on account of play and wear in wire-frame interconnecting spacer links.

a stream) and up to three such streams in parallel. This arrangement allows continued operation, albeit at reduced efficiency and/or reduced load, should a zone be out of commission for any reason (usually a broken wire) and, providing isolation facilities are adequate, allows on-load repairs.

The wires are invariably energised at a high negative DC voltage, typically up to 50 kV, thus inducing a powerful electrostatic field and a corona discharge of electrons, which ionises molecules of flue gas passing through the zone. In the turbulent flow, ions may become attached to dust particles which then migrate to the relatively positively-charged collector plates. A negative voltage is applied to induce negative corona as this is more stable than positive corona and allows a higher voltage to be sustained, thus increasing efficiency. Indeed, precipitation efficiency depends so fundamentally on the value of applied voltage (and hence field strength and corona discharge), that the control scheme has to maintain the maximum possible voltage at all times without allowing flashover and power arcs to develop, irrespective of its actual value. This value depends on the mechanical condition of the precipitator (particularly with regard to the maintenance of optimum discharge electrode/collector plate spacing), and on the dust composition and rapping efficiency. *Rapping* is the process of periodically removing the collected dust, usually by mechanically striking the collector plate frames by rapping hammers and, whilst operationally necessary, results in a loss of efficiency. The selection of rapping repetition rate is, therefore, all important. Too frequent a rate introduces unnecessary re-entrainment because, as the dust falls off the plates in a sheet into the hopper below, a small dust storm is created which is carried forward by the gas stream. Momentarily, the high dust-burden affects the electrical stability and flashover may occur. Alternatively, too little rapping encourages re-entrainment of dust into the gas stream from overloaded plates. The optimum rate also depends on the amount of dust collected, which falls off rapidly towards the exit zones as indicated in Table 2.5. Rear-end rapping, in particular, can produce unsightly puffs of dust from the chimney and it is usually preferable to arrange to rap only part of the end-zone plates at a time to minimise this effect. Whilst the subject is extremely complicated, it is perhaps surprising that modern technology has been unable to vary the rapping programme automatically to suit normal day-to-day variations in operating conditions, for example, the outage of a zone due to a broken discharge wire, when dust collection by subsequent zones would greatly increase.

TABLE 2.5

Approximate quantities of dust collected in a precipitator having four treatment zones

Zone	Dust collected tonne/h
Inlet zone A	20
Zone B	4
Zone C	0.8
Outlet zone D	0.1

Other than the requirement to maintain maximum corona discharge, the efficiency of collection is ba-

sically dependent on treatment time (which for a given precipitator is determined by the gas velocity) and, to a lesser extent, on the dust size. In a PF-fired boiler, this size ranges from the submicrometric smoke particles through to agglomerates of 100 μm diameter or more. For particles up to 20 μm, the opposing electrical attraction and viscous drag forces result in a migration velocity that is dependent on the particle size. For larger particles, the collection efficiency is more or less constant at virtually 100%, any small reduction being wholly due to rapping effects. The small percentage of dust that escapes into the chimney tends to be rich in the smaller particles compared with the inlet burden.

Unfortunately, in practice, the collection efficiency is markedly affected by the electrical resistance of the dust particles, which can vary significantly with the sulphur content of the raw coal. During combustion, the sulphur is oxidised to sulphur dioxide (SO_2) and a percentage of this to sulphur trioxide (SO_3). As the temperature of the flue gas falls to the acid dewpoint, a microscopic layer of sulphuric acid is formed as a skin on each particle. It is this layer, or the absence of it, that determines in great measure the electrical resistance of the dust. Low resistance dusts are formed from coals having a high sulphur-content and are easily precipitated. Should their resistance fall too low, however, as can happen if oil is additionally being burnt (resulting in some carry over of unburnt carbon), then the negative charge on the dust particles is rapidly conducted away on reaching the collector plates and they become susceptible to re-entrainment. With high resistance dusts, arising from a low sulphur-content coal, precipitation is again difficult. Being highly insulating, the dust retains its charge on reaching the collector plates and builds up to form an insulating layer that repels incoming particles. The ion flow is reduced and with it the migration of dust. This can be seen, or at least inferred, from the precipitator electrical conditions. For example, on a rectifier set serving one zone of a 500 MW unit, normal full-load precipitator conditions are, typically, 38–40 kV with 500 mA current flow. When handling high resistance dusts, these would change to, say, 40–42 kV at 100 mA current flow (or less) and a worsening of emissions would be noted. Figure 2.30 shows the relationship between summated current flows and emissions at the same station, as noted during joint CEGB/Inspectorate tests. The best precipitation invariably occurs when rectifier voltages and current flows are both at a maximum, always provided that power arcs and flashover are not occurring which would seriously impair corona discharge.

The uneven (sharp-edged) dust build-up on the collector plates can, more or less easily, give rise to a reverse corona which will induce a small percentage of the dust to migrate back to the discharge wires. If left undisturbed, this would build up to several millimetres in depth and would seriously impair the

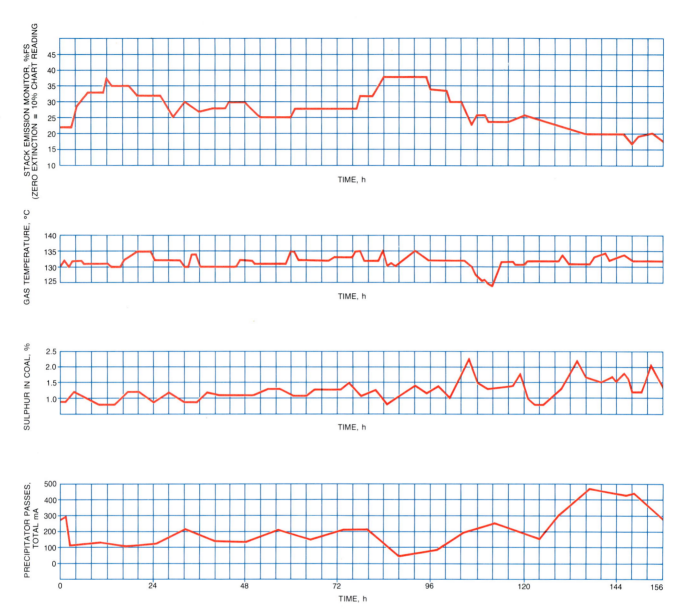

FIG. 2.30 Illustrating the general relationship between stack emissions and total precipitator current flow
Note the general dependence on sulphur.

corona discharge; for this reason, it is essential to periodically rap the wire frames.

Particles emitted from power station chimneys give rise to nuisance in two different ways, dependent on their size. Small particles, in the range submicrometric to about 20 μm, are responsible for the visible plume whereas the larger particles and agglomerates are responsible for local fall out. Both can be objectionable and each modern PF-fired boiler is monitored to ensure that at all times the emissions are within the prescribed limits. Unfortunately, no single instrument presently exists that can monitor both fine and coarse dust, let alone indicate when the emissions have reached the agreed limits in absolute terms (of mg/m^3) and it is necessary to agree alarm levels on the instruments installed by conducting, at least initially, plant emission tests.

Apart from the limit on the actual concentration of solids in the flue gases, there are requirements to be applied regarding the visible plume which is assessed according to the Ringelmann scale of greyness ranging from clear, through several shades of greyness to black. Again, no instruments are available which can measure this 'greyness' directly as it is a subjective quantity depending on the brightness of the sky, on cloud cover and wind speed, to mention but three factors. However, with more or less exactness, it can be related to the optical opacity of the flue gases entering the chimney and several extremely reliable and maintenance-free opacimeters are now available. One such instrument is the SEROP Mk III fine dust and smoke monitor developed by the South Eastern Region of the CEGB.

It basically consists of a withdrawable probe in the form of a slotted tube, carrying integral light-transmitting and sensing elements and purge air facilities to prevent optical fouling (Fig 2.31 (a)). The slots allow the free passage of dust-laden gas thus obscuring the light path and affecting the output of the light-sensitive cell. Another instrument which has performed extremely well on recent trials is the Erwin Sick RM41 smoke/dust monitor. This operates in a similar manner, but has the added advantage of a zero and calibration check available to the operator, which should give enhanced confidence in service.

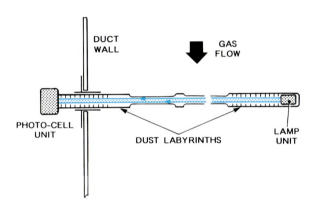

(a) The South Eastern Region's optical probe (SEROP) monitors flue-gas dust by the obscuration of a light beam. The 1.5 m probe is semiportable and can be withdrawn from the duct for zero resetting. Like other instruments depending on light transmission or reflection by the flue gas, its response increases strongly with decreasing particle size and it is therefore particularly sensitive to smoke.

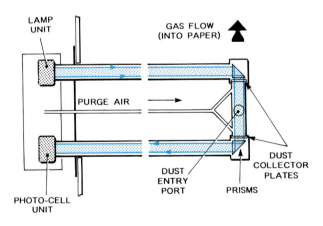

(b) Larger dust particles are preferentially sampled and detected with the instrument devised by CERL. Particles with sufficient inertia can negotiate the entry port and settle on the glass collector plates, whose optical obscuration is integrated over 15-minute intervals before an air purge re-establishes the zero condition.

FIG. 2.31 Dust monitors

Monitoring of coarse dust is possible by using the CERL Mk II dust monitor developed by the Central Electricity Research Laboratories. Dust is collected by deposition on a glass plate (Fig 2.31 (b)) mounted within a cell that is specifically designed to offer easier access to the larger dust particles. A drawback with the instrument is that it requires very careful siting in the duct, for the larger dust particles possess sufficient inertia to make them particularly prone to uneven distribution and readings can vary as a result. Ways of achieving a representative sample have now been developed and the latest version of the instrument, the Mk III, has also shown itself to be remarkably maintenance-free in recent trials.

All of the instruments described above have recently been on trial at selected stations within the CEGB and Fig 2.32 is included to show a typical recording taken off a 500 MW unit, and gives some idea of the sensitivity of current instrumentation available to the operator.

Whilst such factors as the sulphur content of the raw fuel are obviously outside the control of the operator, there are several ways in which he can influence, for better or worse, precipitator efficiency. Firstly, and most importantly, the performance depends on the treatment time, as stated earlier, and hence the gas velocity. The operator can keep this to a minimum by minimising excess air rates and maintaining the lowest possible airheater gas-outlet temperatures (by combustion control and sootblowing). Incidentally, a low flue-gas temperature also helps the formation of the sulphuric acid skin that is so beneficial to easy precipitation.

Effective hopper-emptying routines are essential and some care is necessary to ensure that all hoppers are regularly emptied, as there may be up to 18 hoppers on each 500 MW unit. Overfull hoppers quickly reduce precipitator efficiency because the flue gases scour the dust off the surface and deposit it into the next zone or, in the case of the last zone, into the chimney. If the levels build up into the electrical compartment proper, a short-circuit will result and precipitation will cease altogether. Also, any internal doors which may be fitted to aid maintenance access from one zone to the next will pass significant quantities of dust if left open.

Finally, regular checks on rapping gear are usually possible even if limited to checking whether the gear is running or not, by observing motor current. More sophisticated checks will become possible if microprocessor control of rapping becomes established.

1.2 Steam and waterside considerations

1.2.1 Introduction

Compared with the relatively complex subject of combustion and the associated gas-side effects, both dis-

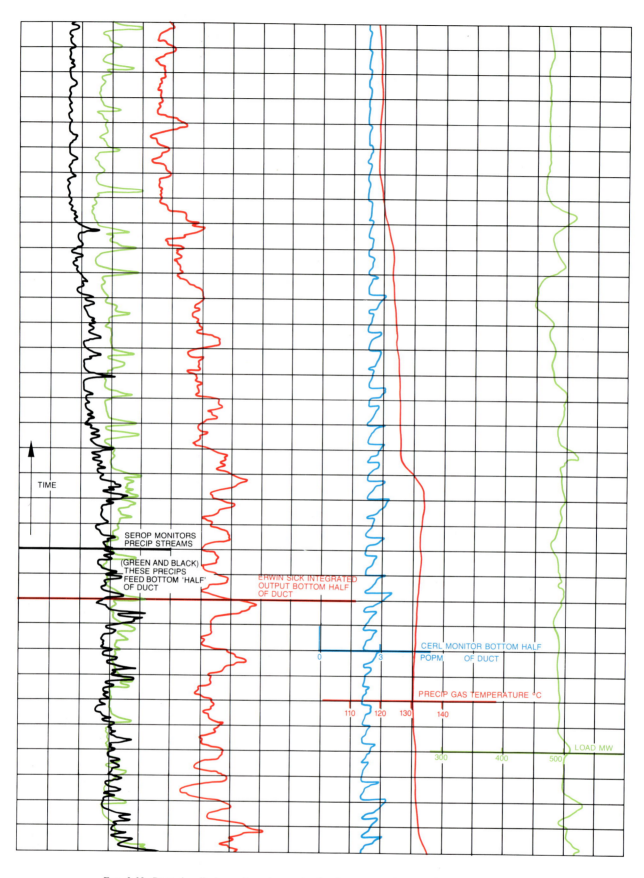

FIG. 2.32 Recorder displays of smoke monitoring instrumentation on trial on a 500 MW unit
The various traces have offset zeros to avoid over-inking. Minor difficulties on the unit resulted in a gradual fall-off in
output to about 470 MW. The reduction in smoke (Erwin Sick/Serop) and coarser dust (CERL) burdens is quite clear;
indeed, this started earlier with reducing precipitator gas temperatures. Whilst both smoke monitors are very sensitive
to rapping spikes, the Erwin Sick was found to be particularly so and the output was integrated to
provide a more meaningful display.

cussed in the last main section, steam and waterside corrosion processes are relatively straightforward and well understood; probably for this reason, the incidence of steam and waterside corrosion failure on modern, high pressure plant is low. Nonetheless, operating staff need a general understanding of the various steam and waterside corrosion processes that can occur if they are to appreciate the need for careful, shift-by-shift, monitoring of boiler chemical conditions.

Iron is soluble in water and reacts with it, the reaction being initially rapid under the conditions prevailing within the boiler. Ironically, but fortunately, the very reason that boiler steel can survive is because of the end product of the reaction process itself, namely iron oxide in the form, normally of magnetite (Fe_3O_4), or unusually, of haematite (Fe_2O_3). Under normally-reducing conditions, with no free oxygen present in the boiler water, magnetite is the natural oxide. It develops into a stable, protective film, typically 10–15 μm thick, its protective ability stemming from its insolubility, adhering property and continuous nature which, once complete, virtually prevents further attack. The formation of magnetite is represented by the equation:

$$3Fe + 4H_2O = Fe_3O_4 + 4H_2$$

The film produced under oxidising conditions is haematite, Fe_2O_3 the reaction being represented by the equation:

$$4Fe + 3O_2 = 2Fe_2O_3$$

Oxygen is consumed in producing haematite and no free hydrogen is evolved; this reaction is associated with oxygenated feedwater conditioning and is discussed later in Section 1.2.6 of this chapter.

Note that in the absence of free oxygen in the boiler water the reaction between steel and water is always accompanied by the production of free hydrogen which, under certain conditions, can induce the corrosion process variously called *hydrogen corrosion* or *embrittlement*, or *hydrogen-induced decarbonisation*.

The maintenance of the magnetite film is so vital to trouble-free operation that it is necessary to be clearly aware of those factors that can lead to its breakdown. They include:

- Mechanical attack: under this heading are included the processes of scratching by abrasive debris circulating with the boiler water; by erosion, as a result of excessive water velocities, especially at bends and in forced circulation, once-through boilers; and by overheating, brought on by excessive local rates of heat transfer and/or by internal deposits. Also surface cracks in the steel, due to any of the fatigue processes, will automatically produce discontinuities in the film.

- Electrochemical attack, due to electrolysis of dissimilar metals. It is for this reason that any transportation of metal oxides (copper, iron or nickel usually) from the feedheaters or condenser is so detrimental to boiler chemical health and much of the process of feedwater control is aimed at minimising such transportation.

- Chemical attack: on the one hand induced by acid-forming salts (e.g., chlorides, which enter the system as a result of condenser leaks) and/or by generally acidic boiler water and, on the other, as a result of gross and often localised, alkalinity. The former process is usually referred to as *acid chloride attack*, whereas the latter is termed *caustic corrosion*. A further process, termed neutral chloride/dissolved oxygen corrosion has free oxygen as an important ingredient. Both of the chloride attacks are often accompanied by hydrogen corrosion, which may become dominant.

Summarising, the best aqueous environment in which boiler steel can form and maintain a stable oxide film occurs when the boiler water is maintained, firstly, in an extremely pure condition so as to contain no abrasive solid debris or acid-forming salts and, secondly, to a controlled degree of alkalinity by periodic dosing with volatile or non-volatile alkalis. For drum-type boilers suitable non-volatile alkalis are sodium hydroxide or sodium phosphate, whilst for drum-type or once-through boilers volatile alkalinity may be produced by hydrazine or ammonium hydroxide dosing, or by other less well-known amines such as morpholine or cyclohexalamine. Within the CEGB, the non-volatile and volatile alkali treatment regimes are usually abreviated to NVT (Non-volatile Alkali Treatment) and AVT (All-volatile Alkali Treatment) respectively.

Sodium hydroxide, NVT dosing provides the best guarantee of avoiding acidic conditions at heat transfer surfaces under normal operating conditions. Additionally, its use affords better protection under start-up conditions when a high level of dissolved oxygen in boiler feedwater is a common, if transient, occurrence. However, it may itself cause corrosion if concentrated locally, when a phenonemon known as 'hide-out' occurs. This is when chemicals which are added to the boiler water immediately begin to reduce, as shown by boiler water analysis. Once the boiler is taken off-load, the chemicals redissolve and show positive chemical reaction. This indicates that chemicals were removed from solution somewhere in the boiler and solid, concentrated salts existed under normal boiler pressure and temperature. Sodium hydroxide can also be carried over into the superheaters.

Sodium phosphate is less strongly alkaline than sodium hydroxide under normal operating conditions and so cannot form such highly aggressive concentrations locally. However, its solubility decreases with temperature, it is, therefore, more prone to 'hide-out' and is more difficult to control chemically.

Volatile alkalis cannot concentrate at metal surfaces, they quickly evaporate with the steam and the possibility of steam-side corrosion, induced by carryover, is eliminated. It is still uncertain as to what extent they can neutralise acidic salts at the steam-generating surfaces and their use demands careful application, especially if the risk of chloride ingress is high. It is clear that both AVT and NVT regimes have advantages and disadvantages.

Finally, the boiler steel, and in particular the austenitic steels of high temperature superheater sections, can suffer from stress corrosion, a form of inter-granular corrosion induced by both stress and sodium hydroxide deposition, the carryover of sodium hydroxide must therefore be avoided.

1.2.2 Hydrogen corrosion

Considering the large numbers of high pressure water-tube boilers in operation, hydrogen corrosion is a relative rarity and is usually associated with gross acid chloride or neutral chloride/dissolved oxygen attack when the rapid reaction with, and growth of, the magnetite film produces significant quantities of hydrogen.

Hydrogen damage of carbon-steel boiler tubing is characterised by brittle, thick-edged fracture. Due to its small size, the hydrogen atom can diffuse through steel. The hydrogen arises from the steel/water reaction discussed earlier, and the diffusion of hydrogen into the steel surface results in a reaction between the hydrogen and the carbon in the steel to form gaseous methane (CH_4). The steel is thus progressively denuded of its carbon content, hence the earlier reference to the decarbonisation of steel. The larger methane molecule is trapped between the grain boundaries and the resulting internal pressure produces discontinuities across the steel grain. In its advanced state, any substantial stress can cause fracture along the partially-separated grain boundaries and a 'window' of the metal can blow out of the affected tube.

The rate of reaction depends on the partial pressure of the hydrogen compared to the working pressure within the boiler. Studies have shown that the hydrogen content in steam in many high pressure boilers rarely exceeds 0.02 PPM (20 μg/kg), which represents a partial pressure of less than 0.1 bar. However, in local areas beneath scale deposits (for example, downstream of weld protrusions or in rapidly growing porous-oxide films), the hydrogen concentration can increase manyfold, resulting in partial pressures of up to one-tenth of the boiler pressure, and it is in such locations that hydrogen corrosion can occur.

From a practical point of view, hydrogen corrosion is somewhat more objectionable than ductile corrosion, for the ruptures are usually more violent and unexpected, and the rapid loss of water presents additional hazards. Partially-embrittled tubes cannot be readily detected in a large boiler and intermittent failures often continue after the corrosion process has been arrested. Such tubing may have sufficient strength to withstand not only normal high temperature operation but also hydraulic testing and only fail as a result of the extra cyclic stresses induced as a result of two-shifting. Improved boiler internal cleanliness is invariably the most effective corrective measure, which may mean off-load chemical cleaning in the worst cases, followed by better management of the water circuit after recommissioning the unit. Low boiler water pH conditions are also particularly conducive to this corrosion because of the increased risk of acid chloride attack.

1.2.3 Caustic corrosion

As mentioned in the introduction, one way of conditioning boiler water is to adopt the non-volatile (NVT) regime and dose with sodium hydroxide (up to a maximum level of 3000 μg/kg on a 160 bar boiler) so as to maintain a surplus of hydroxyl ions over any adventitious acid-forming salts present at all times. The result is boiler water containing up to 3000 μg/kg of sodium hydroxide in solution, a value which is by no means particularly aggressive to boiler steel. During the process of steam generation, however, it has been found that concentrations of sodium hydroxide many times the average value can be raised locally as shown by samples from the boiler drum or from downcomers. The result of this concentration is a very highly corrosive film, which induces rapid corrosion.

Such concentrations are most likely to occur on the heat-absorption side of furnace waterwall tubes, especially if weld intrusions or scale deposits have led to flow irregularities and violent local boiling.

Failures arising from caustic corrosion are normally ductile and result from loss of strength due to metal wastage (unlike hydrogen corrosion), the corroded tube surface having a gouged appearance, with large deep pits.

As heat is transferred to a water/steam mixture a temperature gradient is established and the internal tube-metal temperature rises above the bulk water temperature. Under normal boiling conditions, with internally clean tubes, this gradient would not amount to more than a few kelvin and it is extremely unlikely that high concentrations of sodium hydroxide could be formed. It is only when internal porous deposits exist that the metal temperature is significantly raised above that of the bulk fluid and the potential exists for abnormally rapid local boiling and for sodium hydroxide concentrations to 'dry out' on the metal surface. Since the prime requirement of the boiler is to produce steam, corrosion must be prevented by maintaining internally clean tubes (rather than by reducing the heat flux acting on fouled tubes) and by maintaining a boiler water condition which, if concentrated, will not become highly corrosive. The former is achieved in a drum-type boiler by limiting the total dissolved solids in the boiler water (by blowing down from time to time), by restricting the impurities that reach the boiler in the first place and by off-load acid cleaning as necessary; to achieve the second it may be necessary to reduce the maximum allowable hydroxyl ion concentration and operate on a restricted sodium hydroxide regime or by reverting to volatile alkali treatment, AVT.

In a restricted sodium hydroxide regime the allowable sodium concentration is reduced to <750 μg/kg; this is specified in high pressure oil-fired boilers on account of their very high furnace-wall heat-absorption characteristics. Obviously, with such a regime, the allowable chloride level is similarly reduced to 500 μg/kg and this is likely to result in an inability to tolerate much condenser leakage. If AVT dosing becomes necessary, even minor chloride ingress cannot be tolerated and allowable levels would be further reduced to as little as 200 μg/kg. In practice, the preferred regime on all drum-type high pressure boilers (and mandatory on seawater-cooled stations, where the increased risk of heavy chloride ingress from condenser leaks is very real) is to dose with sodium hydroxide and then to monitor the sodium inventory of the plant carefully to ensure that sodium is not 'hiding-out' in the water circuit. If hide-out is a problem then, on all but seawater-cooled stations where, as stated above, the option is not allowed, dosing would revert to the AVT regime.

The essential difference between once-through and recirculating-drum boilers is that, in the former, all water entering the boiler tubes is evaporated to steam in a number of parallel paths, each tube consisting of a continuous economiser, evaporator and superheater in one. All particulate matter and dissolved salts which enter with the feedwater and do not pass over with the steam flow will accumulate in the boiler tubes and may lead to tube erosion/corrosion, flow instabilities or overheating. In contrast to the drum-type boiler, where any concentration of non-volatile salts remaining in the boiler water can be be reduced by blowdown, no such facility exists in the once-through boiler and it is essential to prevent the ingress of adventitious matter by the use of purification and by conditioning the feedwater with volatile alkalis only.

1.2.4 Carryover

Whilst all solid or dissolved impurities circulating in the evaporator circuit of a drum-type boiler should be retained in that circuit by virtue of the steam separators, scrubbers and driers built into the drum, this is by no means so in practice. *Carryover* is the term used to describe the progression of such impurities into the steam circuit of the boiler and beyond.

As sodium hydroxide is a major impurity, albeit intentionally introduced, its behaviour deserves close study and understanding. Just as its hide-out in the evaporator section can cause caustic corrosion of steam-generating tubes, carryover into the superheaters can give rise to corrosion there. In particular, the austenitic steel used in the final convection superheater banks of modern high temperature boilers is very

susceptible to stress corrosion induced by hydroxide deposition, and strict limits on the sodium content of the saturated steam have been introduced in recent years, typically less than 10 μg/kg. Of the three factors contributing to the carryover of impurities into steam, the volatility of salts, desuperheat spray water quality and mechanical carryover, the last is by far the most important. As little as 0.1% mechanical carryover can contribute as much impurity to the steam as can normal salt volatility and, since spraywater is only a small percentage of total steam flow, it would require spraywater sodium content of 60 μg/kg to produce the same overall concentration of 2.5 μg/kg sodium in steam. The problem of carryover initially is one of design but in this case should the original specification not be met in practice, it would prove expensive and difficult to carry out modifications, even to steam separators and driers, in view of the large amount of general equipment installed within the drum (Fig 2.33).

Fortunately, carryover is very sensitive to drum level, especially at or near maximum evaporation, and operating staff can often play a very useful role in minimising carryover by ascertaining the highest drum level that can be tolerated before significant carryover commences and then by running consistently below this level. In practice, on the more susceptible boilers, this level is often not much, if at all, above drum centre level. Once an agreed operational drum level has been set, it should be possible, with the aid of a sodium in saturated steam monitor mounted on the Unit Control Panel, to maintain adequate purity on all but the worst designed boilers. Incidentally, it follows that drum water level indicators must be reliable and capable of clear interpretation by the operator. Whilst this may seem to be obvious, difficulties have arisen in the past in fully understanding differences in drum level shown by multiple instruments installed on the one drum.

Reference to Fig 2.34 will show that instruments taking their water-leg reference from different parts of the drum will naturally indicate varying levels if the water surface is not level throughout the length of the drum. Indeed, the only time that the water level is constant along the length of the drum occurs when the boiler is not being fired or at low loads. At, or near, maximum evaporation, the steam/water mixture entering the drum from riser tubes can vary in energy and quantity dependent on where in the furnace it originated. For example, more rapid boiling may well occur in tubes serving the burner area and the discharge of this extra volume into one part of the drum can raise the water level at that area (compared with areas relatively unaffected by steam/water discharge) by 50–100 mm and the sheer volume of general equipment housed within the modern drum will tend to maintain such differences.

Drum water level variations can also result from unbalanced firing in a multi-cell furnace, or from the sudden release of large areas of slag, or from soot-blowing, and intermittent carryover may occur unless

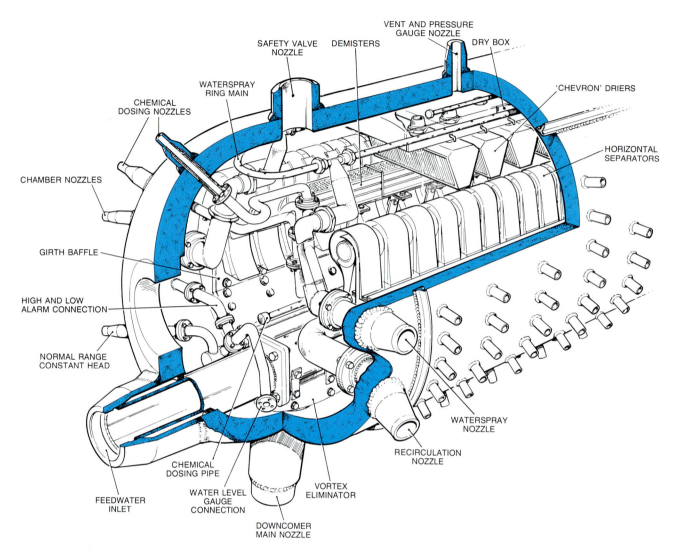

SAFETY VALVE NOZZLE
DEMISTERS
VENT AND PRESSURE GAUGE NOZZLE
DRY BOX
WATERSPRAY RING MAIN
'CHEVRON' DRIERS
CHEMICAL DOSING NOZZLES
HORIZONTAL SEPARATORS
CHAMBER NOZZLES
GIRTH BAFFLE
HIGH AND LOW ALARM CONNECTION
NORMAL RANGE CONSTANT HEAD
WATERSPRAY NOZZLE
RECIRCULATION NOZZLE
FEEDWATER INLET
CHEMICAL DOSING PIPE
WATER LEVEL GAUGE CONNECTION
VORTEX ELIMINATOR
DOWNCOMER MAIN NOZZLE

FIG. 2.33 Steam drum internals

careful control is maintained at a relatively low drum level. Obviously better control of carryover is possible if adequate and reliable drum level instrumentation is provided and maintained. In this context, the relatively recent introduction of the Hydrastep water level gauge has helped. The instrument consists of a column into which are set up to 12 electrodes, which either pass an electrical current or not, dependent on whether or not they are immersed in water. The visual indicator takes the form of a duplicate column of lamps, red for steam and green for water. The main advantage stems from the elimination of the television monitor, invariably used in the past to monitor a conventional gauge glass, with its heavy maintenance and constant 'setting-up' requirements.

Supplementary drum level monitoring equipment can take the form shown in Fig 2.34. As the principle involved depends on being able to compare a relatively small head difference between the two columns, against a background of full boiler pressure, it is immediate-

ly obvious that any small leaks on the high pressure leg have the potential to cause serious instrument error if the constant head column cannot make up the loss. A similar error occurs if a sudden fall in drum pressure causes the water in the constant head column to 'boil-off'; until conditions stabilise and condensation restores any loss of level, the instrument reads high. For such reasons, it is normal to rely on the absolute gauge glasses (e.g., the Hydrastep, or equivalent) at times of serious boiler pressure disturbance.

Sodium is not the only constituent of carryover. Whilst silica plays no significant part in waterside corrosion, its presence in steam can result in turbine blade deposits and efficiency loss. A target value of 20 $\mu g/kg$ is currently set for all plant within the CEGB. This target is based on solubility considerations and provides a safe figure below which turbine deposition will not occur; it can generally be met without difficulty except after overhauls, etc., when it is considered permissible to run at up to 50 $\mu g/kg$ for several

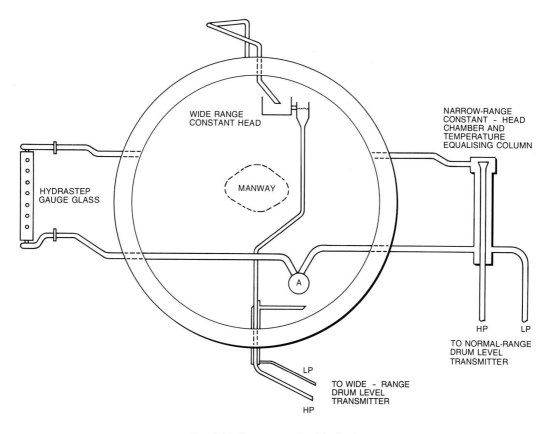

FIG. 2.34 Drum water level indications
The diagram shows a cross-section near to one end. The drum is 36 m long. Pipe A is 12 m long, being open-ended at the
inner end and blanked-off at the other. The hydrastep and drum normal range level indicators take their waterside
references from this pipe, and therefore respond to the water level in the open end of the pipe,
i.e., one-third way into drum.
By comparison, the wide range drum level indicator takes its water and constant head references from a point only 0.5 m
into the drum from the end and can, on occasion, indicate a different level.
This is a typical arrangement only and operating staff should become familiar with the systems employed on
their own plant.

hundred hours of intermittent operation. Such operation often occurs especially after major recommissioning, and can even be useful in providing a means of washing-off any blade deposits formed during normal operation. As the volatility of silica in boiler water is dependent on boiler pressure, a useful control parameter to ensure that the concentration in saturated steam is held below the target value is to reduce the boiler pressure (Fig 2.35), though this cannot be maintained indefinitely. As there is no generally accepted way of removing silica in boiler water chemically, the efforts of operating staff are directed to keeping it out of the feedwater, or costly boiler blowdowns will be necessary.

1.2.5 Reheater corrosion

Steamside corrosion of reheater tubes, both ferritic and austenitic, which has occurred at several stations, has usually been associated with sulphate deposits and has so far been attributed to off-load corrosion. The source of the sulphate can be either from flue gas and ash deposit ingress (as can occur during reheater tube-leak repairs or, if tube leaks exist, during low-load operation when the reheater will be under vacuum) from reheat desuperheat spraywater injection or from boiler drum carryover. It follows that the same operational care needed to contain sodium hydroxide carryover should reduce, or eliminate, sulphate carryover.

A concentrating factor, so far as the reheater is concerned, is the likelihood of superheater and turbine deposits being 'washed' into the reheater on start-up. The prime conditions for this type of corrosion are the presence of moisture, oxygen and salt concentration, the elimination of any one of which arrests the corrosion process. This means that on-load corrosion is extremely unlikely; hence the need, off-load, to maintain the reheaters in a dry condition or if wet, to prevent the presence of oxygen. Whilst the admission of nitrogen to the reheater during the off-load period is usually possible, it is impossible to maintain the nitrogen 'cap' unless the reheater can be isolated, which is unlikely. For example, many designs

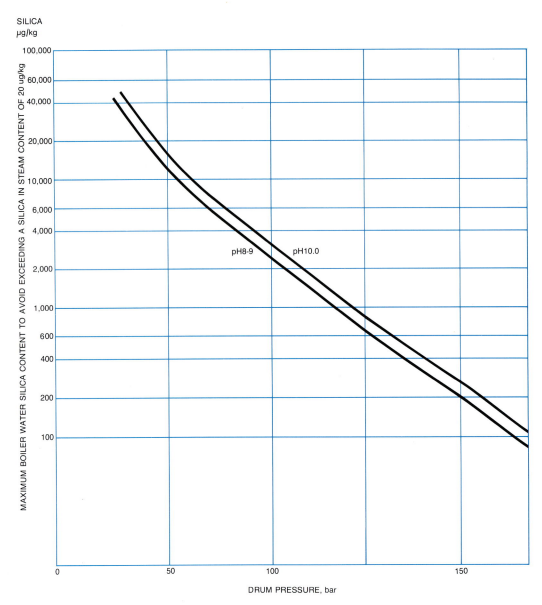

FIG. 2.35 Illustrating the variation of boiler water silica content with the pressure required to avoid exceeding a silica-in-steam level of 20 μg/kg, the current steady state limit

provide for HP turbine/reheater steam dumping in the event of a unit trip, via steam release valves. It is usually not possible to maintain these valves closed once the turbine trip gear has tripped and the exhaust arrangements induce a powerful 'chimney effect' on the reheater circuit (as any operating engineer who has had to destroy this effect, to allow welding repairs to reheater tubes to proceed, is all too aware).

So far as maintaining reheaters dry is concerned, even here there are problems. Horizontal, self-draining reheater sections invariably contain pools of water due to sagging loops and it is at these water/air interfaces that the pitting and corrosion usually occurs. Some stations, badly affected by reheater corrosion, have opted to install dry-air blowing equipment in an attempt to maintain the reheaters moisture-free during off-load periods, though it is too early to reach de-finite conclusions as to its benefits.

1.2.6 Erosion corrosion

It has relatively recently been recognised that the phenomenon of erosion corrosion can have a major impact on the chemical integrity of once-through boilers and whilst this is only really a major consideration for the nuclear side of the industry, the CEGB does operate one coal-fired once-through supercritical unit of 375 MW. Erosion corrosion is the term given to the local dissolution of the protective magnetite film in regions of high turbulence, such as can occur, for example, at bends after orifice plates or downstream of weld protrusions. Whilst the problem is at its most severe in any high velocity areas of feedheating plant, there may be regions in the evaporative section of the boiler where similar high turbulence flows can occur. Resulting corrosion rates can be as high as 1 mm/year and are dependent, amongst other factors, on the solubility of the film.

Oxygenated feedwater conditioning is being introduced more widely in Western Europe as an alternative to the traditional AVT regimes employed on high pressure once-through boilers. The main advantage claimed for such conditioning is that, provided the purity of the feed water is extremely high, it results in lower rates of corrosion of steel in feedheater and boiler components. The reaction between steel and water in an environment containing free oxygen produces haematite as the stable oxide rather than magnetite. Provided that it forms a continuous layer, haematite reduces the erosion of the protective film in vulnerable areas as it is significantly less soluble than magnetite. The feedwater purity must be compatible with that leaving the 100% duty condensate polishing plant; for this reason its use in drum-type boilers is not possible.

1.2.7 Feedwater conditioning

The basic approaches to feedwater conditioning within the CEGB on drum-type boilers are:

- Physical de-aeration, supported by the addition of hydrazine as a residual oxygen scavenger.

- Ammonia dosing to control the feedwater pH.

In the latter, the partial breakdown of hydrazine in the boiler drum into ammonia (and water) results in an increase in the condensate pH leaving the turbine condenser, as the ammonia is volatile and is carried over into the turbine with the steam. The amount of ammonia dosing of the feedwater is, in consequence, somewhat reduced. Incidentally, a fair proportion of this ammonia, together with other incondensable gases is extracted from the condenser steam space by the rotary air pumps and can dissolve and concentrate in the seal water circuit causing severe corrosion to the heat-exchanger tubes, especially those of copper-bearing alloy which suffer from stress corrosion. Similar attack can take place on copper alloy condenser tubes and, in particular, to those in air extraction sections.

Targets for dissolved oxygen are set at both the extraction pump discharge and economiser inlet. Stations having direct-contact feedheaters are specially vulnerable to adventitious oxygen ingress, though a properly designed and operated de-aerator feedheater will be able to reduce the oxygen levels to the required economiser inlet targets. Whilst the greatest oxygen ingress naturally occurs in those low pressure circuits operating at sub-atmospheric pressures, a small percentage can be drawn in through feedpump and heater drains pump glands against the leakage flow, due to venturi or 'corkscrew' effects. To guard against such minor

ingress, the hydrazine target concentration is set at twice the dissolved oxygen level leaving the de-aerator.

1.2.8 Oil

Normal levels of oil ingress, almost always arising at LP turbine bearing-to-gland interfaces, do not increase the risk of boiler tube corrosion, though overheating and subsequent damage has been attributed to slugs of emulsified oil. The concentration of oil in feedwater should be as low as possible and, in any event below 500 μg/kg, although in an emergency levels up to 1000 μg/kg can be permitted for several days.

1.2.9 Chemical control operating limits

The steady state operating limits set at one 500 MW station (160 bar, coal-fired, inland cooling tower) are detailed in Table 2.6, and are offered as a general guide to current operating practice rather than an absolute set of limits to be rigorously applied. In addition, some (secondary) targets not discussed in the text (for example, the conductivity and sodium levels at the condensate extraction pump, and water treatment plant outlet conditions) are included for completeness.

It is stressed that the operating levels refer to steady state operation; more liberal limits may apply to one or two parameters during the unit start-up regime.

1.2.10 Miscellaneous

Boiler tube metal-temperatures

The superheater and reheater tubes used in modern high temperature steam circuits are constructed from a grade of alloy steel suitable for the intended application and especially for the steam temperature, which will vary from near saturation (about 350°C at 150 bar) to 568°C at final superheater or reheater outlet. These alloys are susceptible to the condition known as 'high temperature creep'. When a tube is subjected to an internal pressure at a high temperature, it slowly increases in size (i.e., it creeps) whilst the wall thickness diminishes. Consequently, a steadily increasing stress is applied until the tube finally ruptures. For a given material, the time rate of creep is determined by the stress and operating temperature of the tube. Tubes situated in the gas flow (as opposed to those in the deadspace) are subjected to a varying temperature gradient across the tube wall dependent on their position within the boiler, and it is usual, when determining the rate of creep, to consider the midwall tube temperature. Tubes are usually designed for an operating life of 100 000 h, which is referred to as Life Factor 1 (LF1).

An increase in midwall temperature reduces the life and raises the life factor in accordance with the formula:

Life factor, LF = 100 000/hours to rupture

TABLE 2.6

Interpretation of chemical conditions

CHEMICAL INSTRUMENT PARAMETER	NORMAL LEVEL OR READING	OBSERVATION	POSSIBLE CAUSE OR FAULT	ACTION REQUIRED
Boiler drum water Conductivity before cation column (Normal range 0-50)	10-30 µS/cm	Decreasing	1 Boiler water tube leak 2 Boiler Blowdown in service 3 Boiler Drain or Blowdown Valves passing 4 Possible loss of sample flow or instrument fault	1 Confirm leak HP Dose boiler with Caustic Soda to maintain a minimum Before Column Conductivity of 10 3 Check Valves 4 Check sample flowing through column
		Increasing	1 Boiler HP Dosing in service 2 Condenser leakage 3 Possible loss of sample flow or instrument fault	2 Confirm condenser leaking 3 Check sample flowing through column
Conductivity after cation column (Normal range 0-50) Alarm at 15	5-15 µS/cm	Decreasing	1 Boiler Blowdown in progress 2 Boiler Water Tube Leak 3 Boiler Drain or Blowdown Valve Passing	2 Confirm leak and check Boiler Drumwater before column conductivity 3 Check valves
		Increasing	1 Condenser Leakage 2 Cation Column exhausting or exhausted 3 Contamination of make-up from Water Treatment Plant 4 Possible loss of sample flow or instrument fault	1 Confirm condenser leaking If condenser leaking, Refer to Operating Instructions 2 Contact chemist 3 Check Mixed Bed and Final treated-water-outlet conductivities 4 Check sample flowing through column
Extraction pump discharge Conductivity after cation columns (Normal range 0-1.0) Two traces, after first and second cation columns. Alarm at 0.5	0.1-0.3 µS/cm	Increasing ONE trace ONLY	1 First cation column in sample path exhausting or exhausted	1 Do NOT change sample flow direction through columns. Remove the first and last columns and reverse their order, Replace columns and inform chemist
		Increasing BOTH traces	1 Condenser Leakage 2 Possible Contaminated make-up 3 Both the first and second columns exhausting together (rare occurrence) 4 Loss of sample flow through columns	1 Confirm condenser leaking and then refer to operating instructions 2 Check Water Treatment Plant conductivities 3 Check outlet from third column with laboratory conductivity meter Inform chemist 4 Check flow
Sodium (Normal range 0 to 10)	0.3 to 1.0 µg/kg	Reading at 0.1 or offscale LOW	1 Suspect instrument fault	1a Check meter selected to correct range b Contact chemist
		Reading between 1.0 and 5.0	1 Suspect condenser leakage, (small) 2 Contamination from water treatment plant make-up 3 Possible instrument fault or loss of sample flow	1a Confirm condenser leakage – if due to condenser leakage, refer to operating instructions b Check that condenser inter – tube plate drainage pumps working correctly 2 Check Water Treatment Plant conductivities 3 Contact chemist
		Reading approaching 10 or OFF SCALE HIGH	1 Condenser leakage (large) 2 Gross contamination of treated water outlet 3 Possible instrument fault or loss of sample flow	1a Confirm condenser leakage – If due to condenser leakage refer to operating instructions b Contact chemist 2 Contact chemist if confirmed 3 Contact chemist if instrument fault

TABLE 2.6 (*cont'd*)

Interpretation of chemical conditions

CHEMICAL INSTRUMENT PARAMETER	NORMAL LEVEL OR READING	OBSERVATION	POSSIBLE CAUSE OR FAULT	ACTION REQUIRED
Extraction pump discharge (Continued)...		Reading OFF SCALE HIGH (above 10)	1 Condenser leakage (severe) 2 Severe contamination of treated water outlet 3 Possible instrument fault or loss of sample flow	1a Confirm condenser leakage if due to condenser leakage. Refer to operating instructions b Contact chemist 2 Contact chemist if confirmed 3 Contact chemist if instrument fault confirmed
		Reading above 100	1 Severe condenser leak	1 Reduce or take off superheater spray water but reinstate as soon as sodium falls below 100
Boiler drum saturated steam Sodium (Normal range 1.0 to 100) Alarm at 5.0	0.2 to 0.8 µg/kg	Reading at or above 5.0	1 Automatic calibration sequence on instrument 2 Carry-over of boiler water and chemicals into steam 3 Instrument fault 4 Loss of sample flow	1 Check recorder chart for timing of previous calibration (every 24 hrs) 2 Check boiler drum level and correct condition 3 Contact Instrument Maintenance Department 4 Check flow to instrument
Silica (Normal range 0 to 50) Alarm at 20	1.0 to 20 µg/kg	Reading at or above 20	1 Boiler WATER silica content excessive, causing level in steam to increase above alarm level due to: a) contamination of treated water outlet by exhausting resin bed b) high level of silica in boiler water following a start up 2 Instrument fault 3 Loss of sample flow	1a Control silica level in saturated steam by blowdown and subsequent HP dosing a Check flow
Boiler drum water Sodium chloride (Normal range 50 to 5000)	100 to 1000 µg/kg	Increasing	1 Condenser leakage (rate of chloride increase dependent on size of condenser leak) 2 Contamination from water treatment plant make-up 3 Automatic calibration sequence on instrument 4 Instrument fault or sample flow failure	1a Confirm condenser leakage if confirmed, refer to operating instructions 2 Check Water Treatment Plant conductivities 3 Occurs once every 24 hours 4 Check flow
		Decreasing	1 Boiler blow-down in progress 2 Instrument fault or sample flow failure	2 Check flow
		Reading very low	1 Instrument selected to Reserve Feed Water instead of sample	1 Change sample valves over
Sodium (Normal range 100 to 10000)	1000 to 2500 µg/kg	Increasing	1 Boiler HP dosing in service 2 Condenser leakage or Water Treatment Plant contaminated make-up 3 Automatic calibration sequence	2 Check which
		Decreasing	1 Boiler blowdown in service 2 Boiler water tube leak 3 Boiler drain or blowdown valves passing	2 Confirm boiler tube leak and check boiler drum water conductivity before cation column 3 Check valves

TABLE 2.6 (*cont'd*)

Interpretation of chemical conditions

CHEMICAL INSTRUMENT PARAMETER	NORMAL LEVEL OR READING	OBSERVATION	POSSIBLE CAUSE OR FAULT	ACTION REQUIRED
Boiler drum water (Continued)... Silica (Normal range 0 to 1000)	50 to 150 µg/kg	Reading of 200 or above	1 A high silica level in boiler water may lead to an increase in steam silica 2 Contamination of treated water outlet with subsequent contamination of reserve feed water	1 Refer to silica in saturated steam instrument and if silica in steam is in excess of 0.020, initiate blowdown and dosing programme to control silica in steam
pH (Normal range 6 to 11)	9.5 to 10.2	Increasing	1 Boiler HP dosing in service	
		Decreasing	1 Boiler blowdown in progress 2 Boiler water tube leak 3 Boiler drain valve or blowdown valve passing 4 Contamination from water treatment plant – possibility of resin or regenerant loss	 3 Check valves 4 Contact chemist
Economiser inlet Dissolved oxygen (Normal range 0 to 20) Alarm at 5	0 to 5.0 µg/kg	Reading greater than 5.0	1 Air ingress into feed system above level at which deaeration and LP chemical dosing cannot remove 2 Failure of LP chemical dosing 3 Deaerator operation suspect 4 Possible instrument fault or loss of sample flow	1 Investigate air ingress 2 Contact chemist 3 Check, especially venting arrangements to main condenser 4 Check
pH (Normal range 6 to 11)	8.8 to 9.2	Reading steady below 8.8 but above 7.0	1 Failure of LP chemical dosing 2 Instrument fault or loss of sample flow	1 Contact chemist 2 Check
		Reading falling towards 7.0 or below 7.0	1 Suspected contamination of treated water outlet by resin loss or regenerant loss 2 Possible instrument fault or loss of sample flow	1a Contact chemist b Check boiler water conductivities c If Water Treatment Plant contamination confirmed, refer to operating instructions 2 Check
Conductivity after cation columns (Normal range 0 to 1.0)	0.1 to 0.3 µS/cm	Increasing One trace only	1 Column exhausting	1 Refer to 'Extraction pump discharge conductivity after cation columns' section
		Increasing BOTH traces	1 Possible cause or fault as per 'Extraction pump discharge conductivity after cation columns' section	1 This instrument is only a back-up for extraction pump conductivity
No. 1 DC heater Dissolved oxygen (Normal range 0 to 20) Alarm at 15	5.0 to 10 µg/kg	Reading above 15 or OFF SCALE, HIGH	1 Air ingress at extraction pump 2 No. 1 DC heater emergency make-up valve passing 3 Possible instrument fault or loss of sample	1 Chemist to confirm 2 Check valve and 'jack' shut 3 Check
Hotwell conductivity (Normal range 0-10)	2.0 to 4.0 µS/cm	Reading 2.0 or below	1 LP chemical dosing failure	1 Contact chemist
		Reading 4.0 or above	1 Excessive LP chemical dosing 2 Condenser leakage	
		One reading a lot higher than the other	1 Indication of condenser leakage on one side of condenser	1 Confirm condenser leakage by reference to extraction pump conductivity and sodium
Make-up treated water outlet Conductivity (Normal range 0 to 1.0) Alarm at 0.2 Trip at 0.4	0.02 to 0.1 µS/cm	Reading above 0.1 and rising	1 Stream in service is exhausting 2 With two streams in service one mixed bed exhausting	1 Take exhausting steam out of service before conductivity reaches alarm level
Silica (Normal range 0 to 50)	2.0 to 15 µg/kg	Reading above 15 and increasing	1 Mixed bed resin exhausting 2 With two streams in service one mixed bed exhausting 3 Contamination of a good stream by water from a stream previously regenerated in which the regeneration was not satisfactory	1 Take stream out of service 2 Take exhausting stream out of service 3 Take previously regenerated stream out of service and identify regeneration fault N.B. It is possible for Silica contamination to occur whilst mixed bed or treated water outlet conductivity is unaffected and low.

Therefore, operating at a life factor of 4 reduces the life to 25 000 h and so on, see Table 2.7. In practice, the calculated midwall temperatures corresponding to various life factors have been found to be unduly pessimistic, being based on stress data biased more to the minimum strengths of material property scatter, rather than the mean, and containing allowances for determining the real midwall temperature rather than that calculated from thermocouples set at the tube surface (Fig 2.36 (a)). These allowances have been found to be too low, resulting in unnecessary restrictions being imposed on operating levels, which occasionally reflected back as load restrictions. On one 500 MW boiler, an allowance of 35°C was given to account for the difference between the measured tube-wall temperature and the midwall temperature at a particular location. Tests confirmed the need not only for larger allowance but a varying one dependent on the actual heat flux, on conduction through the thermocouple block or pad, on the slagging present on the block and on the sootblowing undertaken. The allowance needed to vary between 40°C and 70°C dependent on the conditions existing at any instant of time. A great improvement has been the introduction of Chordal-hole thermocouple terminations (Fig 2.36 (b)). Whilst the drilling of the thermocouple hole calls for accuracy, the temperature displayed is the exact midwall temperature and all allowances are eliminated, allowing confident operation up to the stipulated limits. Unfortunately, tube thicknesses much less than 7 mm cannot accommodate such thermocouples. In practice no reheater tubes can therefore be drilled, though the allowances necessary for conventional wall surface-mounted thermocouples are neither particularly large nor variable in the convective zones where these tubes are installed.

Figure 2.37 shows the control room VDU dsplays on a 500 MW boiler and are self-explanatory. Note that the use of such a display should aid steady state operation rather than start-up conditions, when metal temperatures will be increasing more or less rapidly and tube metal rates-of-rise would be a better parameter to monitor, either in the form of 'time/temperature' VDU display or by recorders. The benefit of the display forms shown in Fig 2.37 arises from the ability, provided enough tubes are instrumented, to study the temperature gradient across the particular tube banks chosen from one side of the boiler to the other. By these means, it may be possible to ensure a more even flux distribution by variations in firing patterns. Also, the effect of sootblowing can be studied.

Acoustic monitoring of steam leaks

Continued operation of the boiler with steam leaks in areas such as the top deadspace, with its multitude of feeder tubes serving downcomers and riser tubes carrying steam and water mixes back to the drum, is extremely risky. Even if the leaking tube is small-bore, say terminating in a vent valve on the deadspace roof, secondary damage can occur if the leaking (dust-entrained) steam impinges on adjacent major components. Large diameter feeder tubes have also failed due to gross defects of manufacture or as a result of inadequately designed expansion arrangements.

Fortunately, almost all failures are preceded by a period of low crack-growth prior to total failure. For all the above reasons, it has become the practice to check vulnerable areas of the boiler by means of acoustic monitors. About four microphones are installed

TABLE 2.7
Life factors

Section of superheater	Temperatures for Life factors of 1, 4 and 10 (at pressures corresponding to final outlet of 160 bar)					
	Midwall temperatures °C			Thermocouple temperatures °C		
	1	4	10	1	4	10
Radiant wall	529	551	566	583	605	(612)
Platen superheater Stage 1	520	541	555	552	573	587
Platen superheater Stage 2	530	552	566	562	584	(602)
Platen superheater Stage 3	573	598	614	(598)	(608)	(608)
Secondary superheater	644	673	692	(654)	(666)	(666)

Notes:
- Figures in parentheses denote a corrosion-limiting temperature
- Typical only for one 500 MW boiler
- 'Reduce firing immediately' limited 650°C thermocouple temperatures

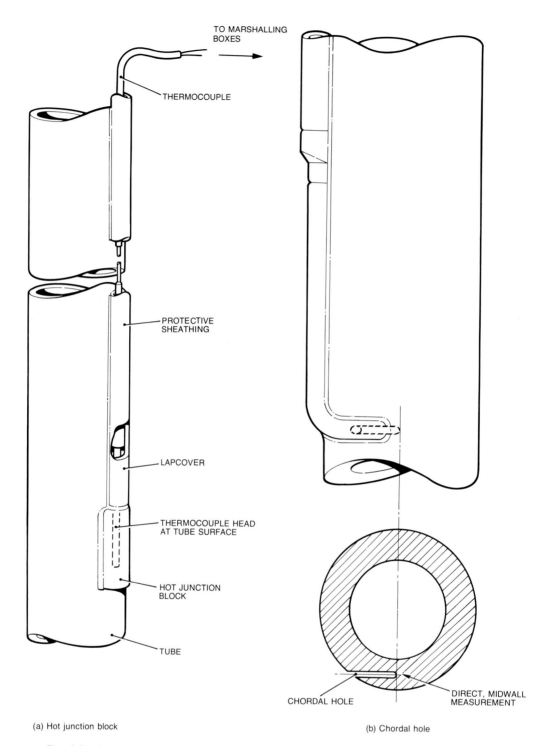

(a) Hot junction block

(b) Chordal hole

FIG. 2.36 Superheater tubes — hot junction block (a) and chordal hole (b) attachment of metal temperature thermocouples

around the deadspace and these are capable of picking up the noise of a steam leak from a 3 mm hole at up to 10 m, above the general background noise. Their output is usually relayed to a multichannel recorder. They have proved invaluable in alerting operating staff to the existence of a leak before much, if any, consequential damage has occurred.

1.3 Milling plant

1.3.1 Introduction — types of mills and systems

An analysis of availability returns from large PF-fired plant within the CEGB shows that performance is greatly dependent on the strength of the installed mill-

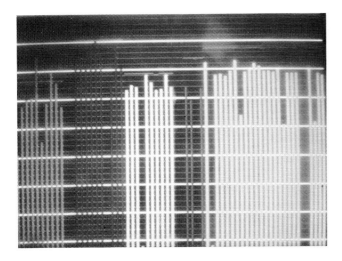

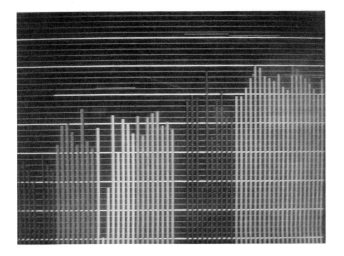

FIG. 2.37 Typical VDU displays of boiler superheater and reheater metal temperatures

(see also colour photograph between pp 34 and 35)

ing plant. If the total complement of mills is such that full unit output can be comfortably carried with one, or even more, mills out of commission, then unit availability will be high, since not only can a rolling programme of major mill (or coal feeder) refurbishment be planned throughout the year but the odd breakdown should not result in much, if any, loss of load. Further, the mills remaining in service will not normally have to be overloaded in order to maintain load. Conversely, if the capacity or reliability (or both) of the milling plant is poor, then unit operation will always be difficult. Not only will all major overhauls have to be completed within the unit survey period but breakdowns, even for such relatively minor and common faults as PF-leaks and coal feeder chain repairs, will probably result in loss of load or will require the remaining mills to be overloaded, with consequential poor grinding and risk of boiler slagging and fireside corrosion.

A number of factors, mainly associated with the properties and condition of the raw coal, affect the performance of the mills, whatever their other attributes. The first consideration is the hardness of the coal expressed, within the UK, as the Hardgrove Index. Before examining the effect of this index, a word of explanation as to its derivation and use may be worthwhile. The Hardgrove machine is basically a miniature vertical-spindle ball mill, and is the simplest and most commonly used of a number of known test procedures for determining the hardness of a coal and therefore the ease with which it can be ground in a mill. To determine the index of a particular coal, a sample of predetermined size and weight is introduced into the machine which is then rotated through a given number of revolutions. The sample is removed and screened: the proportion passing through the sieve is proportional to the grindability index, i.e., soft coals have high Hardgrove indices. The majority of bituminous coals in the UK have an index in the range 45–80. In general, British bituminous coals with a low volatile content are the softer coals. For example, some of the coals from Kent and the South Wales semibituminous coalfields have Hardgrove indices in the region of 80–100, these are the softest coals in the country but form only a very small percentage of the CEGB's supply. The trend is common to the bituminous range of coals but the converse applies to coals outside the bituminous range. Pure anthracites have a very low volatile content and are very hard with an index in the order of 35–40. Obviously, a mill designed to handle a coal having a Hardgrove Index of 50 will have a greater output when handling coal with an Index of 65 and, conversely, a reduced output if harder coals are used. Since the vast bulk of bituminous coals supplied to the CEGB have an index in the range of 50–55, manufacturers usually design their plant for full mill duty with an index of 50.

The moisture content of the fuel, which is made up of 'inherent' (i.e., that within the coal substance) and 'free' or 'surface' moisture, has an important bearing on the mill output. In a well designed mill in which the drying is carried out by the passage of hot air through the mill, typically at between 250–280°C, coal having a total moisture content of about 15% can be handled and all the free moisture and up to half of the inherent moisture removed whilst still maintaining a coal/air (or product) outlet temperature of about 75°C. Coals having a higher moisture content can still be handled, but either the product outlet temperature would fall or the mill throughput would have to be reduced unless the temperature of the primary air (PA) can be increased. A common arrangement employs an airheater specifically for the air requirements of the mills (i.e., a primary air or mill airheater), with facilities to boost the incoming gas-side temperatures at times of need, thereby raising the PA temperature to 300–320°C.

In considering mill output, another important factor is the required fineness of the PF product, a typical specification being shown graphically in Fig 2.38. Whilst overgrinding is wasteful on power, most stations would find it difficult to meet the specification consistently and operation can, and does, continue with relatively oversized PF but the dangers of slagging, etc., as already discussed, will be present. Incidentally, it is not generally realised that the grading of the raw coal affects the power requirements and output of the mill; if a choice of raw fuel exists, the finer the better.

source of ignition, an explosion may occur. A very high coal/air ratio would certainly be well outside the explosive range but the forward velocity of the mixture towards the burners may fall below the recognised minimum of 18 m/s at which PF may fall out of suspension in horizontal piperuns. Within these two extremes, a varying ratio may be desirable to ensure a reasonable turndown of the mill (Fig 2.39 refers).

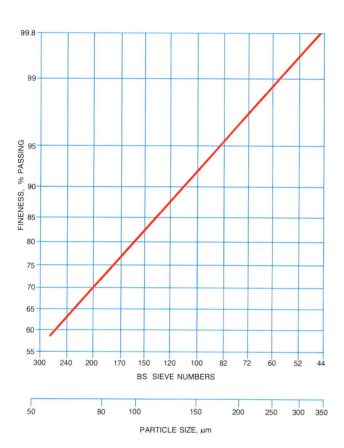

FIG. 2.38 Typical PF fineness specification line for a
vertical spindle mill

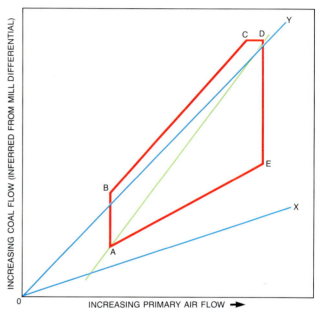

FIG. 2.39 Illustrating mill operating limits

Operation anywhere within the figure ABCDE is permissible. AB represents the PA flow condition below which minimum PF pipe velocities could not be assured; BC the mill choke line, outside of this there would be insufficient airflow to transport the coal away from the mill, and the differential would increase more and more until the mill was 'solid' with coal. The small line CD may represent a maximum feeder throughput condition, and DE a limit in PA capacity (or, for some frontwall-fired furnaces having little 'depth' of furnace, a limit to prevent rear wall slagging). AE is the all important mixture strength line.

It is common to plot a clean air line OX, from which a reasonable mill coal/air ratio can be selected (say, OY) which the automatic controls would seek to maintain. To obtain the absolute maximum turndown potential, it would be necessary to control to a varying coal/air ratio and run along line AD.

As the objective of the PF system is not only to grind the coal to the requisite size but to transport it safely to the PF burners at a velocity and temperature that ensures a good combustion, mills are designed and operated so that there is a definite relationship between the quantities of coal and air passing through the mill. This is the *coal/air ratio* and has an important bearing on the safe operation of the mill. Should the ratio fall below some minimum value, the weakness of the mixture would approach that value at which, given a suitable

Air flow through the mill and PF pipework is maintained at the required level either by a primary air (PA) fan, installed upstream of the mill, or by an exhauster fan positioned after it. Whilst the latter allows the mill to operate under a slight suction, thereby avoiding the necessity for sealing table drive-shafts, reject-disposal doors and instrumentation lines against PF egress, it is no longer the preferred arrangement for several reasons.

First, the exhauster fan requires periodic and, arguably, excessive maintenance as the impellers are handling the coal/air mixture. For the same reason, the impellers need to be substantial and made of wear-resistant material if they are to have a reasonable life. As a result the performance of an exhauster fan is not particularly efficient compared with the PA fan where the impellers are operating in relatively clean air and can be aerodynamically designed to give a very high efficiency. As a matter of interest, whilst the air leaving the PA fan may be clean, the hot air output from the regenerative airheater, will be contaminated with PFA because, for every revolution of the airheater, a complete volume of flue gas is trapped in the matrix by the advancing airhoods and redirected back into the airside. This has repercussions on the hot air system if leaks are present since they result in a considerable dust nuisance around the plant.

Secondly, it is not necessary to provide one PA fan per mill; normally, two large and highly efficient PA fans can supply the total primary air requirements of the mills via a hot air busmain (supplemented by a cold busmain for temperature control). This is not possible if exhausters are specified. Finally, the speed of an exhauster fan needs to be variable to allow a measure of mill output turndown, as damper control of a PF/air mixture is not practicable. AC variable-speed commutator motors have often been employed for this purpose, but they require periodic brushgear and commutator maintenance; a rugged maintenance-free induction motor is usually specified for the PA fan, with its air output controlled relatively efficiently by inlet guide-vane control.

The high speed impact or hammer mill, which was previously in common use in the CEGB, has virtually disappeared because of the rapid increase in individual mill throughput, which is more easily met by the other types of mill; namely the medium-speed vertical-spindle and the low-speed tube ball mill.

The three types of large medium-speed mills currently in use are:

- The Babcock 10E ball mill.

- The NEI/International Combustion roller mill, now superseded by the Loesche mill.

- The PHI Engineering roller mill, a development of the MB and NPS German-designed mills (Fig 2.40), whilst the low-speed tube ball mill is represented by the Foster-Wheeler version.

Illustrations of the first two mills can be found in the boiler maintenance section of Chapter 5 of this volume.

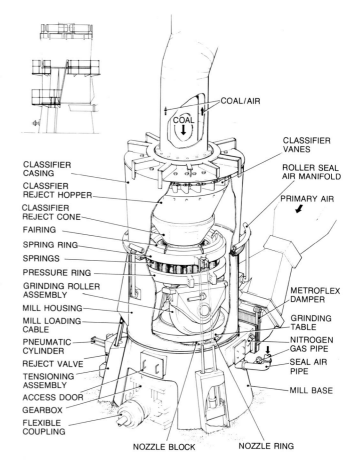

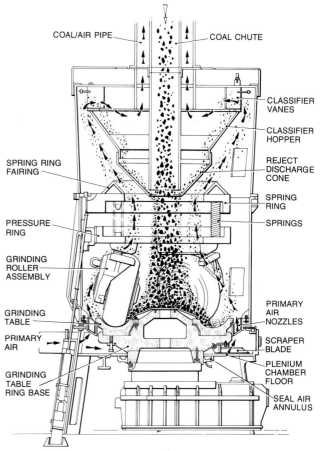

FIG. 2.40 PHI medium-speed, vertical-spindle mill

Whilst the general mechanical operation of a medium-speed roller or ball mill is fairly straightforward in that coal is supplied to the mill grinding-table (by the coal feeder), where it is crushed between the driven table and rotating rollers or balls and then fluidised by the passage of hot primary air, the action of the tube ball mill is more novel. The low-speed tube ball mill is represented by the Foster-Wheeler D9 mill, shown in Fig 2.41. The raw coal enters the mill via the end-mounted static classifiers and is propelled into the drum proper, together with any recirculating classifier rejects, by the ribbon conveyors. Rotation of the drum, which contains a heavy charge of iron alloy balls (not shown), imparts a combined breaking, grinding and crushing action to pulverise the coal, the process being aided by the drying effect of the hot air.

One difference between the two types of mill is the absence of a rejects or pyrites disposal system on the tube ball mill which, in effect, devours anything presented to it. This can be a blessing, as rejects systems consume both operational and maintenance resources unless they are soundly designed and simple in operation.

So far as it is able, the classifier determines the product fineness. Two kinds are in common use: the static type and the rotating, dynamic type, used in the NEI/ICL mill. The static variety imparts a circular swirling motion to the PF/air mixture by either a circle of adjustable guide vanes, as in the PHI mill (Fig 2.40), or a series of circular scroll plates, as in the Foster-Wheeler tube ball mill (Fig 2.41). In both, the separation of the coarser PF particles occurs as a result of the ensuing centrifugal action.

In the dynamic separator, the speed of rotation is variable and the degree of classification can be readily altered causing, as a secondary effect, a change in mill output. With such ease of control, however, lies the temptation to opt for an increased mill output from time to time at the expense of product fineness, especially as the latter cannot be monitored as a matter of course.

A fairly high pressure drop occurs across the classifier shown in Fig 2.40, the pressure within the rejects cone (classifier outlet) falling typically 10 mbar below the inlet conditions. Because of this, the automatic operation of the classifier-cone rejects-flaps is important if an upwards, by-passing flow of PF/air mix through the flaps is to be avoided. Such a flow would, of course, seriously impair the classifier action.

With static, vane-type classifiers, there is merit in adjusting the vanes to give the optimum PF quality and then fixing them in this position.

1.3.2 Bunkers and coal feeders

Coal feeders are nowadays invariably of the drag-link type, shown in Fig 2.42; two such feeders may

be specified for tube ball mills, allowing the mill to continue in operation with one out of service. (For similar reasons, two exhauster fans are often employed on suction-type tube ball mills.) They are relatively reliable in operation and need to be, particularly if associated with a pressurised mill, as safety precautions demand that the mill is shut down, isolated and depressurised before the feeder can be opened up to stimulate coal flow, for example. They are connected to the bunker by transition chutes, usually incorporating bunker shut-off spades. Some stations have experienced difficulties with the reliability of these spades and have resorted to the provision of a series of manually-inserted coal cut-off rods, usually referred to as 'walking sticks' or 'poke rods'. These give a good isolation, provided that they are correctly inserted so that the inner tip of the rod engages fully into the receiving slot or ledge, on the opposite internal wall. In a running mill system, they must also be inserted without undue delay or the coal flow will draw the rod downwards, trapping and bending it. Once this has happened, not only will the isolation be uncertain but internal access is necessary to burn out the bent remains.

Problems can occur with coal 'hanging up' in the bunker or in the bunker-to-feeder transition piece. For the former, the usual remedy is to run the bunker level down as low as possible with the mill in service, trimming the hung-up deposits with air lances. Final emptying via the emergency bunker-emptying conveyor (arguably an essential requirement, certainly an extremely useful one) into lorries is then the preferred final solution, with the mill out-of-service. Bunker-to-feeder transition piece hang-ups can often be cleared by manipulation of the bunker outlet spade or poke rods.

For both problems, the provision of *bunker blasters* has proved invaluable in maintaining coal flow. These are, typically, air-operated systems having a receiver of compressed air mounted adjacent to the point of application and discharged into the bunker when required, the blast of air dislodging any static or caked coal-masses. Whilst care in operation is necessary, especially if bunker levels are low (when coal particles can be blown upwards into the area of the bunker coal conveyors and personnel access ways), they have proved invaluable in maintaining unit load. If a bunker is to stand full for any length of time, however, the air receivers should be isolated and blown down, or there may be an increased danger of bunker fires as a result of passing (leaking) air valves. The blasters (typically three) are positioned on the bunker wall against which coal is most likely to hang up. In this respect, it is not generally realised that the flow of coal occurs preferentially down the bunker face adjacent to the feeder-chain entry point, for it is here that the incoming and empty chain quickly takes on a full charge of fresh coal. Towards the exit face,

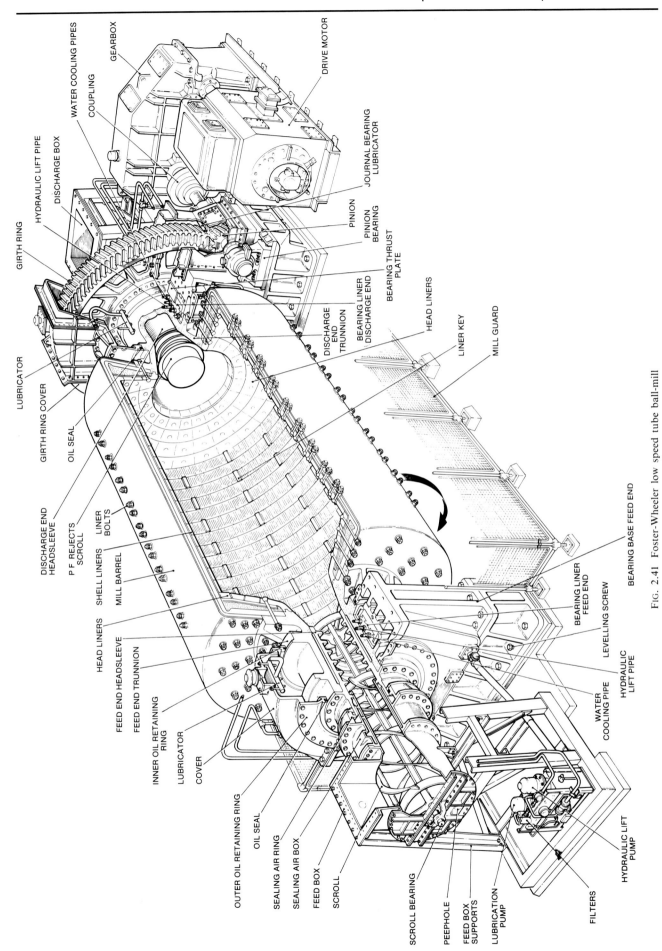

GEARBOX

WATER COOLING PIPES

COUPLING

DRIVE MOTOR

HYDRAULIC LIFT PIPE

DISCHARGE BOX

JOURNAL BEARING LUBRICATOR

GIRTH RING

PINION

PINION BEARING

BEARING THRUST PLATE

BEARING LINER DISCHARGE END

HEAD LINERS

LINER KEY

MILL GUARD

DISCHARGE END TRUNNION

LUBRICATOR

GIRTH RING COVER

OIL SEAL

DISCHARGE END HEADSLEEVE

P F REJECTS SCROLL

LINER BOLTS

SHELL LINERS

MILL BARREL

HEAD LINERS

FEED END HEADSLEEVE

FEED END TRUNNION

INNER OIL RETAINING RING

LUBRICATOR

COVER

OUTER OIL RETAINING RING

OIL SEAL

SEALING AIR RING

SEALING AIR BOX

FEED BOX

SCROLL

SCROLL BEARING

PEEPHOLE

FEED BOX SUPPORTS

LUBRICATION PUMP

BEARING LINER FEED END

LEVELLING SCREW

BEARING BASE FEED END

WATER COOLING PIPE

HYDRAULIC LIFT PIPE

HYDRAULIC LIFT PUMP

FILTERS

FIG. 2.41 Foster-Wheeler low speed tube ball-mill

59

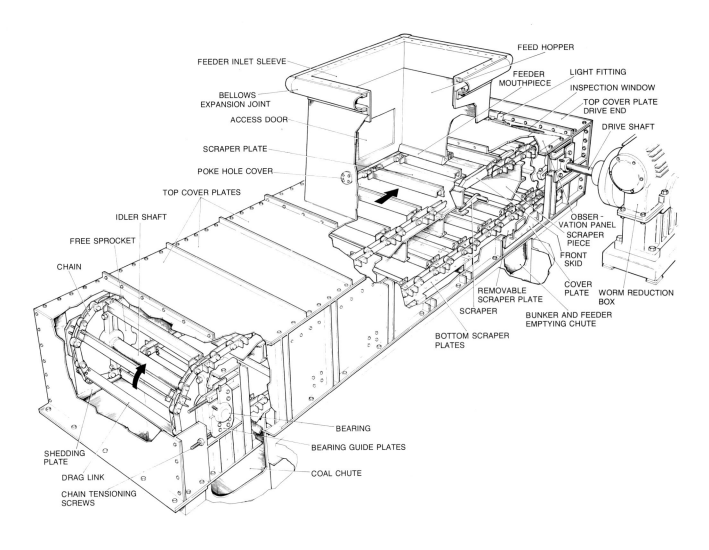

FIG. 2.42 Typical drag-link coal feeder

there will be little coal movement as the chain will be full of coal; because of this, some stations employ a regular and progressive programme of partial poke-rod insertion to stimulate a more even flow.

In theory, a reversible feeder-chain would seem to offer considerable advantages in promoting an even coal flow; so far as is known, none exists.

The tensioning arrangements for coal feeder chains vary between simple screw-tensioners and more sophisticated gas-precharged and grease-filled hydraulic accumulators. Operationally, there is little to choose between them, provided there is some means of ascertaining the tensioner movement left before a chain link, or links, needs removal; to leave this to chance is to risk unwanted load shortfalls, often at inconvenient times.

Speed control of the feeder has traditionally been achieved by the use of a mechanical speed converter in the form of a variable-speed gearbox of one type or another. Recently, thyristor control of the motor voltage has been employed and shows promise.

1.3.3 Mill operation

Explosion risks and fires

Why it has been felt necessary to commence this section with a discussion on explosion risks and fires may be questioned, especially within the overall concept of the normal operation of PF-fired boilers. No apologies are offered as, in the normal day by day operation of milling plant, there is an ever present risk of explosive mixtures and fires developing within the plant, especially when mills are being started or shut down, which is a frequent occurrence in otherwise normal, steady operation. If the general principles necessary to avoid these dangers (which are now clearly established) are understood, then any remaining aspects of mill operation will be almost of secondary importance.

Raw fuel on its own is perfectly safe to handle. However, once it is ground to a fine powder and intimately mixed with the transportation air, it becomes as highly flammable as gas. The operation of PF plant demands understanding and care on the part of all staff if fires and explosions are to be avoided either in the mill or in its associated equipment; in particular, combustion-chamber explosions are to be avoided at all cost. Nowadays, every PF-fired station within the CEGB has a locally produced code of practice for the safe operation of PF equipment tailored to suit the needs of the particular plant installed, although general national guidance will have been incorporated. Operating staff are required to be fully conversant with the fundamental concepts of the code and receive regular refresher training to maintain their awareness.

As little as 30 g of coal dust, uniformly distributed in 1 m^3 of air can cause an explosion if ignited. About 500 g/m^3 is considered to be the concentration above which an explosion is highly unlikely. The first fundamental rule then is to avoid weak PF/air mixtures at all times. Relating the minimum concentration of 30 g/m^3 of air to a typical tube ball mill having a free internal volume of 40 m^3, it can be seen that as little as 1.2 kg of coal dust can be a potential hazard. In normal operation, the coal/air ratio would be many times this value and safe operation is assured. It is only when temporary losses of coal feed occur or when mills are being started or stopped, that the mixture strength falls into the explosive range. If, under these conditions, the forward flow of air is significantly reduced or stopped and a source of ignition is present, then there is a real risk of explosion, although even here the risk varies from plant to plant. For example, Babcock 10E ball mills are apparently immune from damaging explosions. The scientific basis for this safety record is not clearly established but a possible explanation is the so-called 'classifier hypothesis'. A particular feature of the 10E mill classifier system is a high reject ratio, i.e., only a small proportion of the PF entering the classifier passes through, the remainder being recirculated back into the grinding zone. It is postulated that during rundown, then the mixture in the mill body finally enters the explosive range, the mixture in the PF pipework downstream of the classifier is already too weak to explode. Hence a flame igniting in the mill is rapidly extinguished downstream of the classifier and pressure cannot build up in the PF pipework. Thus the basis of the classifier theory is that explosive mixtures do not occur simultaneously in the mill body and in the downstream piping.

A promising and more general approach to the prevention of explosions arises from the observation that explosions are relatively rare, although the incidence of weak PF mixtures and fires within mill or PF pipework occurring together is much more common. It has been suggested that mixtures known to be explosive in static conditions may not explode in a fast-moving stream when exposed to a static ignition source, either because there is insufficient time of exposure to the source for the mixture to reach its ignition temperature or because the heat of combustion is dissipated too rapidly by turbulence for the deflagration to intensify. Where air velocities in mill bodies are high, such as in the 10E mill, ignition sources may not initiate explosions.

Nonetheless, because of the potential dangers that exist around coal feeders, mills and PF pipework, personnel access is often subject to restriction, especially when weak mixtures are known to be likely or when general furnace conditions become unstable. A warning system consisting of klaxons and flashing lights, or similar, is usually installed and initiated by the operator at times of risk — all personnel then withdraw from the areas covered, only returning when the system is cancelled.

Other than loss of coal feed, or during start-up or shutdown operations, weak mixtures can also occur as a result of excessive mill turndown for load regulation purposes or as a result of faulty regulators or unauthorised adjustments to the controls. Additionally, on suction mills, careful attention must be given to the maintenance of ducts, casings and rejects doorseals as weak mixtures can easily build up as a result of dilution by air ingress.

The development of an explosive mixture in the combustion chamber of the boiler is, however, abnormal and is covered in the section on Emergency Operation. Suffice it to note that it follows the failure to ignite an incoming cloud of fuel or loss of ignition during operation and time exists for an explosion to be avoided if prompt operator action is taken to prevent a build-up of unburnt fuel.

Sources of ignition in mill systems vary from mechanically-produced sparks, flashback from the furnace or as a result of mill fires developing. First, it is important to dispel the misconception that PF in suspension can be ignited by the primary air. Any credible air temperature will not ignite the fuel in suspension although the product temperature is normally limited to about 95°C to prevent fuel softening and sticking to PF pipework and for other mechanical reasons. Flashback from the furnace is possible but unlikely if the recommended flow velocity of 18–20 m/s through the PF pipes is maintained; PA flow turndown-limiting devices are in common use to prevent this cause and mills can be arranged to trip if the value falls below an agreed minimum level. It has also been suggested that sparks arising from metal-to-metal, or metal-to-mineral contact may initiate an explosion or promote a fire. It is unlikely that such sparks will have sufficient energy to initiate a PF explosion unless the mill is in internal mechanical distress. If they occur, the mill must be running and the sparks are likely to occur in the coal-rich zones around the rollers or balls.

It is fortunate that the risk from this source is low, as action to eliminate it seems impracticable. However, if sparks are actually seen anywhere in a running mill system, then this must be taken as an indication of the presence of a fire.

Mill fires are a much more likely and common cause of mill explosions and can develop in many ways. If uncooled deposits of coal are allowed to stand in a mill system they are likely to catch fire eventually due to spontaneous combustion, which progresses at a rate determined by the temperature of the fuel and surrounding air. Classifier rejects are particularly vulnerable being already dry and hot. For this reason it is usual, on medium-speed vertical-spindle roller or ball mills, both of which contain relatively small amounts of coal within the mill, to empty the mill during shutdown. Additionally, whilst mills are standing, it is usual to limit temperatures to a maximum of 50°C. Emptying a vertical-spindle mill of coal during shutdown is relatively simple and quick, but tube ball mills contain a much larger coal charge and complete emptying extends the shutdown significantly. For this reason, it is usual to leave the mills shutdown with a normal coal charge in the mill unless maintenance is to be carried out or the shutdown period is likely to be long. As a precaution against spontaneous combustion, some tube ball mills are periodically rotated during the shutdown period, especially during the early stages.

Mill fires can also develop as a result of feeding hot or burning coal into the mill from the coal feeder or bunker. If it is required to empty a bunker containing coal that is on fire, every effort should initially be made to drain the coal directly into lorries for disposal, quenching as necessary. If this is impracticable and the burning coal must be passed through a running mill, the minimum number of personnel, suitably attired, should be employed: they will need to retire from the area immediately if coal feed is lost or the mill is to be shut down for any reason.

Mill fires can also develop as a result of design deficiencies or generally because of mill 'type'. For example, the tube ball mill shown in Fig 2.41 is susceptible to fires developing in the hot-air inlet elbow and special water injection points have been developed. On many vertical-spindle mills, vulnerable areas are in the reject boxes and/or hot-air inlet plenum chambers, as a result of the rejection of larger pieces of coal. Additionally, if the rejects plough is not working efficiently, or has been bent or broken off, then large deposits of coal can build up in the plenum chambers and serious fires develop.

Mill fires can be detected in several ways. First, a high product or mill outlet temperature may indicate a fire if it becomes serious, especially if it is accompanied by fluctuations in mill airflow and draught indications. Secondly, high outside-casing temperatures or visual indications may be present. Finally, but by no means least, fires are detected at a very early stage by CO monitoring equipment. While installations vary to suit the particular vulnerabilities of the protected plant, a typical scheme would monitor for CO at the mill outlet and (for mills having rejects systems) at the rejects box. Careful siting of the probes is essential, especially for those fitted to the rejects system of pressure mills as CO (and other undesirable products of a smouldering fire) can be blown into the mill-bay atmosphere through leaking door seals without triggering the monitor if the probe is positioned too high in the box and the seat of the fire is against the door itself. Occasionally mill inlet conditions are monitored for CO in order to detect the background level usually present in the incoming hot air. This is derived from boiler combustion deficiencies (or planned operating levels) and brought about by mixing in the airheater. Monitoring at this point is not essential as the background count can be inferred from the general printout of all the mills, always assuming all do not contain fires!

Alarm levels, initially set at 100 PPM, have been reduced to as low as 50 PPM (or less) as a result of incidents and experiences.

Mills should never be put into service with a fire indicated because the sudden rush of hot air into the mill can easily disturb enough PF to induce an explosion; even if the airflow becomes safely established, the start-up of the rollers or balls will disturb any fire smouldering on the grinding table, with similar results.

If a mill is on-load and a fire develops, the first actions, provided that there appears to be no mechanical danger, are usually to increase the coal feed to the maximum practicable (to ensure a rich mixture) and to apply maximum tempering air to cool the mill. The maximum practical coal feed is that beyond which the mill will choke or start to reject raw fuel. The latter must be avoided as the spillage would feed any fire present in the rejects box or in the plenum chamber. After about 15 minutes, the situation is reassessed and hopefully the fire will show signs of burning out or being smothered. If the intensity of the fire appears to be increasing, or the mill is in mechanical danger or coal feed is lost the mill will have to be shut down. However this is done, there is a danger of explosion and instructions are necessary based on the type of plant installed. Where PF pipe-purging facilities are provided, it is usually best to isolate the mill quickly from the system, whilst at the same time applying purge air to each PF pipe. An explosive mixture may occur for a short time during settlement of the suspended dust but the danger zone will be traversed quickly. In addition, the cessation of airflow will tend to cause a fire to smother itself. The mill is then isolated tightly to starve the fire of air and is left for 30 minutes, or so, before reassessing the situation. If an extinguisher can be injected into the isolated mill body from a remote point, this should be done immediately. It is often possible to tee into the bulk nitrogen supply system which is frequently installed to protect turbine, feedheater and boiler drum steam spaces against oxygen attack during shutdown periods. This is a great help, as the mill can be charged with nitrogen and sealed by the use of a 'trickle charge' for as long as it takes for the fire to subside. Carbon dioxide (CO_2) has been used in the past but should be avoided, as the oxygen in the CO_2 can actually feed the fire at elevated temperatures through dissociation. Sometimes it is possible to inject a fine spray of water by providing suitable connections at vulnerable areas, although there are obvious potential dangers in spraying cold water onto hot metal parts. Each problem must be considered on its merits.

If no purge facilities exist, then no single method of shutting down can be devised which will protect both mill and PF pipework and the absence of remotely-operated burner-isolating dampers will increase the risk.

Instrumentation and control

Bearing in mind the need to maintain both the PF velocity and strength as discussed in the previous section, the primary instrumentation required on any mill system will be primary airflow and PF quantity; the latter is usually inferred from either the product outlet temperature and/or the coal supplied to, or contained within, the mill.

PA flow is usually derived from the pressure drop occurring across a venturi section placed in the hot-air inlet duct to the mill. The instrument can be scaled either to read the pressure drop directly, in mbar, or be marked to indicate the mass flow, kg/s. For the latter reading to be strictly correct, a means of correcting for density change with hot-air inlet temperature must be provided.

PF mixture strength is more difficult to measure and has to be inferred from one or more parameters. First, the product outlet temperature, which varies significantly and fairly linearly with flow, is simply obtained and is free from process 'noise'. The sensors must be of the quick-response type, or serious (and possibly dangerous) discrepancies will occur between the actual peak product-temperature reached as a result of a temporary loss of coal flow and that indicated on the instruments. Long time constants will also exist, aggravated by any heavy shielding applied around the probe in an attempt to prevent erosion by PF. Quick-response probes were developed when investigations into mill explosions revealed the degree of error that could occur in the measurement of PF product temperatures.

It is convenient to stress here the need to maintain the availability of tempering (cold) air at all times that the mill is in use. Whilst indiscriminate use of tempering air is costly in efficiency terms, as it by-passes the airheaters, it must always be available. One possible solution, if separate primary air or mill airheaters are installed, is to maintain the desired product outlet temperature by controlling the mill airheater duty whilst leaving the tempering air damper closed. If the setpoint of the tempering air damper controller is adjusted to, say, 5°C above the normal product outlet temperature, cold air is automatically introduced on a temporary loss of coal flow to prevent excessive temperature rise.

Secondly, PF strength can be inferred from the coal flow supplied to, or contained in, the mill. The former can be quite accurately assessed from the coal feeder speed, provided that coal flow irregularities are not being experienced, whereas the latter is obtained by measuring the mill differential. This is a widely used parameter, making use of the fact that for a given PA flow, the pressure drop across the mill is dependent on the quantity of coal being processed.

Compared with a medium-speed vertical-spindle mill, the tube ball mill contains much more coal at any instant. Moreover, this quantity needs to be carefully regulated; too little gives rise to overheating due to the mechanical friction between the cascading balls, whereas too much causes the grinding action to cease altogether. It is necessary, therefore, to be aware at all times of the coal level within the mill drum. A typical scheme is shown in Fig 2.43.

Of the other mill instrumentation, some is provided to ensure that seal-air pressures are adequate before mills are commissioned. Other instrumentation is useful, rather than essential. For example, variations in mill input power, which are usually presented as mill motor current, can be very misleading to inexperienced personnel. In a tube ball mill, motor current varies little with load, whereas in vertical-spindle mills it can rise steeply by an amount depending on the mill type, the tensioning applied to the rollers or balls, or the mechanical condition of the mill. In the case of the PHI mills installed in one station, 'mill running hours since last overhaul' can be predicted with some certainty from observed motor currents.

A change in mill output is usually initiated by a change in PA flow, either by damper or vane control or exhauster speed variation, over a limited range, as discussed in the last section. It is important to permit as much PA variation as possible so that a good response to random system frequency movement or to instructed load changes is effected. This is because the response time of a large PF mill of any type is poor, if the fuel output is varied solely by fuel input change. Some 5 minutes or more would elapse, in practice, between a step-change in feeder speed and a boiler output response. However, if the PF product velocity is varied in each PF pipe by 25%, then a correspondingly rapid change in the fuel input to the boiler would occur and be reflected quickly into an output change. Obviously, to maintain the change, the coal feeder speed needs to be adjusted or the mills either choke or run empty of coal. Despite the above, because of minimum PF pipe-velocity considerations on the one hand, limited PA resources on the other, and flame stability concern on both, coupled with natural operator caution in setting

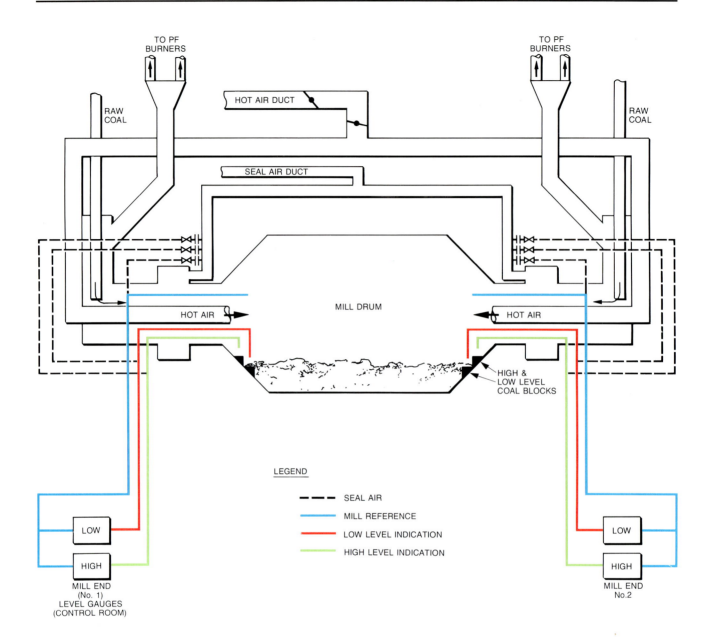

FIG. 2.43 Tube ball-mill level control scheme

The quantity of coal in each end of the mill is indicated by a high level and low level contents gauge. Each responds to a differential obtained between a stable drum reference pressure, taken from a probe remote from the coal bed and supplied with clean seal air, and the same pressure fed to the probe adjacent to a high and low coal level block. The seal-air connections are necessary to prevent PF from entering and blocking the instrument lines, but since the same pressure is fed to each system, it does not alter the changes in differential that are occurring between them. When the mill is empty, the pressure at the gauge connections on both low and high level indicators is the same and therefore the differential, and mill level reading, is zero. As the coal level rises, the air escaping from the probes adjacent to the level blocks will become progressively restricted, first on the low level probe and subsequently on the high, the differentials will increase as will the instrument readings until, when the airflow is virtually sealed off by the increasing depth of coal, the instruments will rise to full scale reading.

combustion change rates, the response time of most large boilers is slow.

As stated earlier, the control of coal feeder speed is widely achieved by linking it to changes in mill differential pressure. Initially, it is necessary to plot a 'clean air line' as shown in Fig 2.39, which shows the relationship between mill differential and airflow alone. Once coal is introduced into the mill, the dif-

ferential, for a given PA flow, increases above the clean air figure and a value can be selected for control purposes that ensures both stable mill conditions and satisfactory coal/air ratios. Unfortunately, in some mills, the bulk of the differential occurs on clean air alone and the limited change arising with increasing coal feed can result in poor and coarse control. For this reason, a better means of control might be achieved by regulating fuel input against PF product (outlet) temperature, which is very sensitive to coal flow. It would be necessary to install outlet-temperature bias control to allow the operator to deal with varying raw-coal mixtures and mill inlet-temperature levels. The main advantages of this control, which is, in effect, a heat balance across the mill, lies in the relative ease with which equal mill loadings can be achieved, provided that fuels of similar moisture content are simultaneously delivered to each mill. As this may not always be true, other mill parameters such as feeder speed and mill differential, need to be retained in order that the fullest possible picture is presented to the operator so that he can choose the correct operating bias.

1.3.4 PF pipework and dampers

The ideal mill PF pipework layout is one in which an even distribution of coal is obtained to all the burners at all loads and with any combination of mills. This condition is extremely difficult to fulfil, mainly on account of the unavoidable differences in length between pipe runs and routing possibilities, so compromises are necessary. Even though pipe runs can be purposely lengthened in some areas to try to ensure similar pressure drops between a particular mill and all its associated burners, distribution can be very uneven and the use of riffle boxes, distributors or deflectors in fuel lines is essential, if any sort of an even split is to occur at bifurcations.

Figures 2.44 (a) and (b) show the arrangements specified for 500 MW corner-fired and frontwall-fired PF boilers, respectively, and give some idea of the complexity of the PF pipework. Note that particular care has been taken to ensure that even firing occurs into each half of the split-furnace designs, whichever mills are in service.

Deflectors and distributors cannot normally be relied upon to give good distribution at all mill loads: if adjustment facilities are provided, they should normally be locked in the best position as determined by field trial.

Properly designed riffle boxes, such as the CEGB CERL design shown in Fig 2.45, offer the best solution provided that the relatively large pressure drop across them can be accommodated, although even here the fuel distribution is not likely to be perfectly balanced at all mill loads on account of the varying fuel/air ratio usually present across the load range.

Velocities have to be maintained in PF pipework at values that not only prevent PF deposition in the most vulnerable sections of a particular system (usually the top bends) but also prevent any flash-back of flame from the furnace. This can take place if the velocity of flame propagation in the PF pipe is greater than the forward velocity of the fuel/air mix; past experience has shown that a minimum value of 18 m/s should prevent either possibility. In the past, some

stations have maintained minimum pipe velocities under mill turndown conditions by shutting off one or more PF pipe dampers. Whilst this achieves the desired effect, there are dangers inherent in the practice. First, proper purging of the PF pipework should always be done as soon as the selected damper is shut and this can be difficult whilst the mill remains in service. Secondly, the overall firing pattern may become unbalanced and lead to drum level control problems unless a corresponding leg is shut off. Thirdly, the closed PF pipe damper may cause enhanced local pressures to develop during an explosion, as a result of the full reflection of the shockwave from the closed damper. Finally, there is a danger of incipient seepage of PF across the closed dampers unless they are sealing particularly well, which could lead to a PF pipework fire. All in all, it is usually best to accommodate unit load reductions by taking complete mill groups out of service, having taken up any available mill turndown.

PF pipe dampers have progressed in recent years from simple, automatically-operated non-return flaps, remaining more or less in the PF stream and vulnerable to severe erosion, to quite sophisticated designs where the flap is power-operated and is moved completely out of the PF stream, when open. Other designs include spade dampers which, whilst giving a perfect isolation, have been prone to sticking (usually closed) on a mill trip. Incidentally, the possibility of wear by erosion affecting the isolation properties of any PF damper being used as isolation for work purposes must always be considered most carefully before allowing work to proceed and, if necessary, blanks may need to be fitted. Damper control systems vary, but include hydraulically-powered systems backed up by hydraulic accumulators. Such accumulators are trouble-free and need little operational attention other than to ensure that the pre-charge gas (usually nitrogen) pressure is maintained. Ideally, this requires the accumulator to be taken out of service and drained down on the fluid side, but an indication of a loss of nitrogen pressure can be inferred from an increasing cut-in frequency of the hydraulic pump(s). It is important to maintain the correct gas pressure if the full potential of the accumulator is to be available should a firing trip require the simultaneous closing of all mill safety dampers. The safety dampers installed in one particular layout are shown in Fig 2.46; these are automatically triggered on a mill trip, not only to ensure the mill is isolated from the hot air supplies and furnace but also that the PF pipework is adequately purged, although on a firing trip this may have to be inhibited to prevent dangerous accumulations of PF from entering the furnace.

From Fig 2.46, it can be seen that when PF pipe purging is complete, manual closing of the burner-isolating dampers is recommended to prevent the pipework from overheating as a result of recirculating furnace gases — in the extreme, this could raise

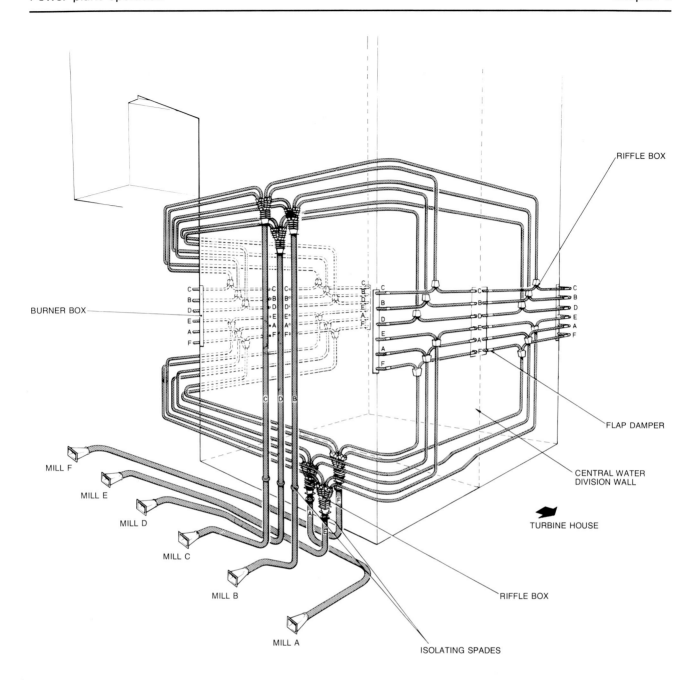

RIFFLE BOX

BURNER BOX

FLAP DAMPER

CENTRAL WATER
DIVISION WALL

TURBINE HOUSE

MILL F

MILL E

MILL D

MILL C

MILL B

MILL A

RIFFLE BOX

ISOLATING SPADES

FIG. 2.44 (a) PF distribution arrangements for a split corner-fired furnace

the PF pipe temperature to several hundred degrees and lead to the possibility of weakening pipework components (e.g., couplings) and fires.

In the late 1960s, a new kind of mill explosion was identified in which the pressures generated far exceeded the hitherto specified containment value of 3.45 bar and in which damage was incurred by the PF pipework rather than by the mill and its associated plant as had happened previously. Following much investigation by working parties and enquiries both within and outside the industry and from observations on plant, it emerged that pressures of up to 28 bar, static equivalent, might occur in the more vulnerable areas of the plant and that, given the

conditions needed to induce shockwave reflection, values up to 150 bar were theoretically possible. Containment up to this value is not practical, so primary emphasis must remain on sound operational procedures, systematic training of plant operators and instrumentation to prevent, so far as possible, the potentially dangerous conditions from occurring. Nonetheless, PF pipework and plant is now categorised as 'shock' or 'non-shock', based on the PF pipework geometry, type of fuel, mill type and on the operating experience with the plant in question. The new standards for PF pipework installed in 'shock' plant specify the full use of ductile materials (to prevent the fragmentation that has occurred with cast iron

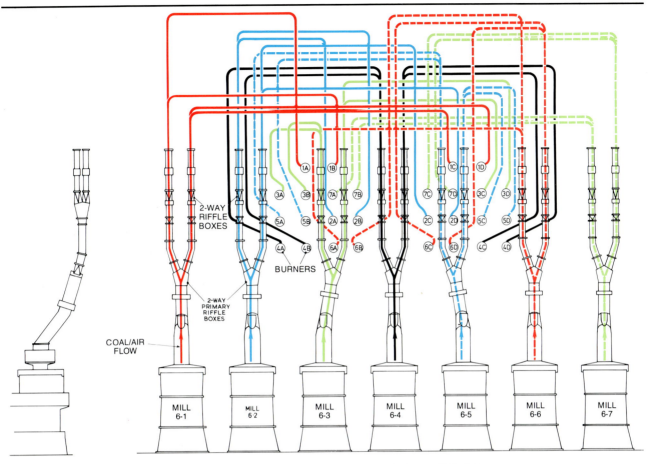

Fig. 2.44 (b) PF distribution arrangements for a split frontwall-fired furnace

components) designed to withstand 13.8 bar, static equivalent. Other requirements include:

- The minimum number of joints.

- An operating temperature of up to 110°C.

- Sufficient flexible couplings, constrained to prevent axial dislocation in the event of an explosion, to ensure that no excessive loads are transmitted through the system.

- The piping to take account of mill vibrations, which can be severe under fault conditions.

Whilst discussing the standards for PF pipework, it is important to ensure that all pipework (at intervals), dampers and fittings are adequately identified, as it is extremely easy to mistake one line for another at locations having a high density of pipe runs.

The use of inert flue gas as the transportation medium for PF can, in principle, provide complete protection from PF explosions but for economic and engineering reasons is not an attractive option.

1.3.5 Oil burners

Traditionally the role of fuel oil in PF-fired plant has been for lighting-up and for PF flame stabilisation at low loads or during periods of flame in-

stability. The plant provided for this was (and is) simple in concept and inexpensive to operate. Oil consumption was low, storage tanks were small and the use of 950 seconds (Redwood No.1) fuel oil, commonly referred to as boiler lighting-up fuel oil (BLUFO), enabled satisfactory transfer to be made from storage tanks to pumping units and oil heaters without the need for elaborate trace-heating systems. Within the UK, fuel oils are still commonly referred to by their viscosity at 37°C (100°F) as expressed on the Redwood No. 1 scale; BLUFO has a viscosity characteristic roughly equal to curve 6 on Fig 2.47.

Whilst oils of such viscosity were freely available some years here ago, refinery processes have now advanced to the point that residual fuel oils contain less of the lighter fractions and strongly resemble tars at normal temperatures, having viscosities of 3500 seconds Redwood No. 1 (curve 3) at 37°C. Such residual fuel oils are commonly referred to as RFO and require heating at all stages from off-loading to firing before they can be satisfactorily pumped.

During the early 1970s, a surplus of RFO became freely available to the CEGB at an attractive price. At that time, some of the large PF-fired plants were experiencing major difficulties with milling plant and many MWh of generation were being lost. Consideration was given in some stations to the wholesale replacement of the mills with more reliable designs and in others to the provision of additional mills,

ARRANGEMENT OF PRIMARY
DEFLECTORS SHOWING
CONTRA PF FLOW

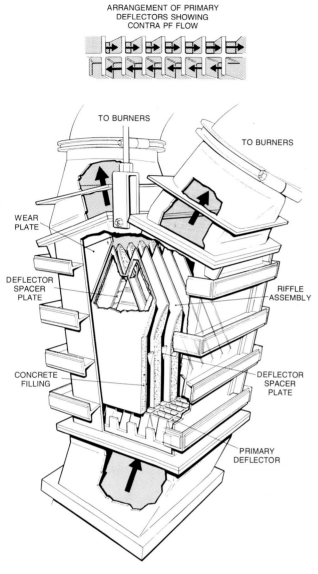

FIG. 2.45 PF riffle box

but neither option was particularly attractive for many reasons.

The restoration of these load shortfalls, using cheap RFO in the form of oil overburn, was an attractive alternative and schemes designed to provide a 100 MW oil-overburn capacity on selected 500 MW units were proposed and, in many cases, installed. Ironically, as these schemes were coming to fruition, the 1973 oil crisis was unfolding; the subsequent redistribution of energy prices in favour of coal and the gradual elimination of most milling plant problems has meant that overburn installations have seen little use in recent years in the role for which they were designed.

In comparison to the simple lighting-up installations, oil overburn plant is complex. First, several large (up to 8000 tonne) storage tanks are needed, together with rail and road unloading facilities which, in the case of rail tankers, need to include off-

loading pumps. Trace heating, usually steam, is also required from tanker discharge to storage tank inlet whilst the tanks themselves need to be maintained at around 55–60°C to keep the contents in a pumpable condition. Secondly, pumps are needed to transfer the oil to the heating and high pressure pumping units if, for site layout reasons, the heating and pumping complex is situated at some distance from the storage area. Finally, as the oil consumption of, say, 2 boilers each on 100 MW is around 15 kg/s, substantial heating is required to raise the temperature of the oil to 120°C, which is a typical design temperature for a pressure-atomised burner (see Fig 2.47). Such heating is most economically accomplished by the use of steam, either generated by purpose built auxiliary boilers (often fired on gas-oil) or drawn from the main plant. For the latter, facilities need to be provided in case all main units are off-load. Some stations, initially provided with auxiliary boilers, have found it economical to interconnect with the main plant in any case, or to redesign the auxiliary boilers to be fired on cheaper fuel oils. In the former, steam can be taken from the reheater where suitable temperature and pressure conditions for the oil heaters exist, or from the boiler primary steam circuit. One station has conveniently tapped into the 'live steam to de-aerator heating' system, gaining de-aerator off-load heating interconnection facilities at the same time.

Further complexities on RFO overburn plant include condensate and heat recovery systems, control-air compressors, driers and associated equipment, and suitable fire fighting facilities. Some condensate recovery systems are difficult to justify, especially under the present low utilisation regimes; additionally, there is the danger of severe oil contamination of feed or condensate tanks and de-aerators should heater tube-leaks develop, and reliable oil-in-condensate monitors are essential to intiate an automatic dumping of the condensate should they operate. One station has found it more economical to allow a proportion of the hot oil circulating around the boilers to be returned back to the storage tanks at times of little or no demand. This operational procedure has allowed all steam trace heating to be shut down, except during periods of inclement weather or when fuel oil deliveries are being received; a method of operation that has already saved a considerable amount of money. Whilst such cost-saving methods are acceptable in general, no effort should be spared in maintaining adequate RFO temperatures after the main oil heaters. Long pipe-runs are usually needed to transport the hot oil from the pumping and heating units to the boiler house and it is important to maintain the trace heating on these sections in working order if oil burner combustion is to be acceptable.

Two types of oil burner in common use today are:

- Pressure- or mechanically-atomised burners

- Steam-atomised burners

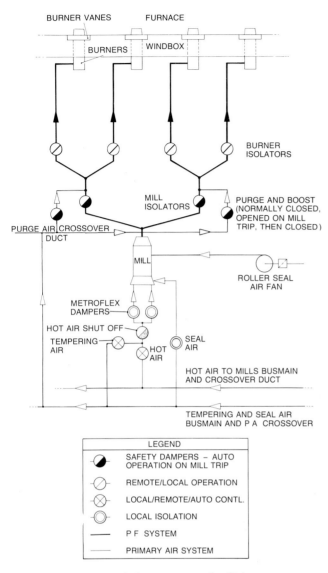

FIG. 2.46 Typical arrangement of mill dampers

Pressure-atomised burners

In operation, oil under pressure is supplied to the burner gun and enters the atomiser through slots cut tangentially to the periphery of the atomiser (Fig 2.48). The tangential slots have an area relationship to the tip orifice and are sized to cause a relatively high pressure-drop as the oil passes through the slots; the pressure drop, measured at maximum flow rate is known as the 'natural differential pressure' of the atomiser. This pressure drop is converted into velocity, causing the oil to enter the atomiser tangentially at very high speed thus establishing a rapid rotation of the oil as it passes through the tip orifice and hence a uniform, hollow, conical spray. The fineness of the spray is dependent upon the speed of oil rotation in the atomiser, which is a function of the

pressure drop across the tangential slots and the viscosity of the oil. On burners designed for a turndown, typically 2/1 between high and low fire modes, a removable restriction in the spill return line effectively alters the pressure drop across the slots. Figure 2.48 shows the means by which this is done and also illustrates the principle of tip recirculation which is employed to ensure that oil at the correct temperature is available right up to the tip, whether the burner is operating or not. This materially aids rapid and correct ignition of the incoming oil. A secondary benefit of tip recirculation is the ability to fix the oil burner into the firing position at all times, the recirculating flow rate being chosen to keep the tip cool. By this means, the insertion/retract mechanism of the conventional burner is dispensed with. Note that the tip shut-off valve is held closed by the differential pressure acting on it, rather than by the spring, which is provided to keep the pintle valve 'loosely' on its seat when the oil is shut off and the barrel is being inserted or removed. On the particular installation shown, the relief holes in the self-seal couplings were provided after several barrel over-pressurisation incidents had occurred after barrels had been only partly removed from the furnace — the radiation continuing to act on the isolated barrel being responsible for the pressure build-up within. This incident illustrates clearly one aspect of tip-recirculation burner operation; barrels should either be in service with oil circulating, or withdrawn. Should oil pressure fail or fall to the point where no differential exists, no cooling of the tip occurs. This illustrates the second operating concern on such burners. Providing that the associated mill group is in service, no harm will come to the tip as the incoming clouds of raw PF will keep it cool. Tests have shown that the tip metal temperature stays at a temperature similar to that of the PF, viz. 75–80°C, despite the fact that the PF ignites soon after. Should the mill group be out of service, the full furnace radiation is applied to the uncooled tip and temperatures of up to 600°C have been recorded within a minute or so. When inserting oil burners whose PF burners are out of service, no time should ever be lost in applying oil; similarly, such burners need retracting fully immediately the oil is isolated.

For all pressure-atomised burners, oil viscosities of 80–120 second Redwood No. 1 are required, demanding oil temperatures of 120°C for oils having viscosities similar to curve 3 of Fig 2.47.

Steam-atomised burners

The difficulties in turndown range which apply generally to pressure-atomised burners are largely overcome in the steam-atomised burner, where turndown ratios of 10/1 are possible. In this burner (Fig 2.49)

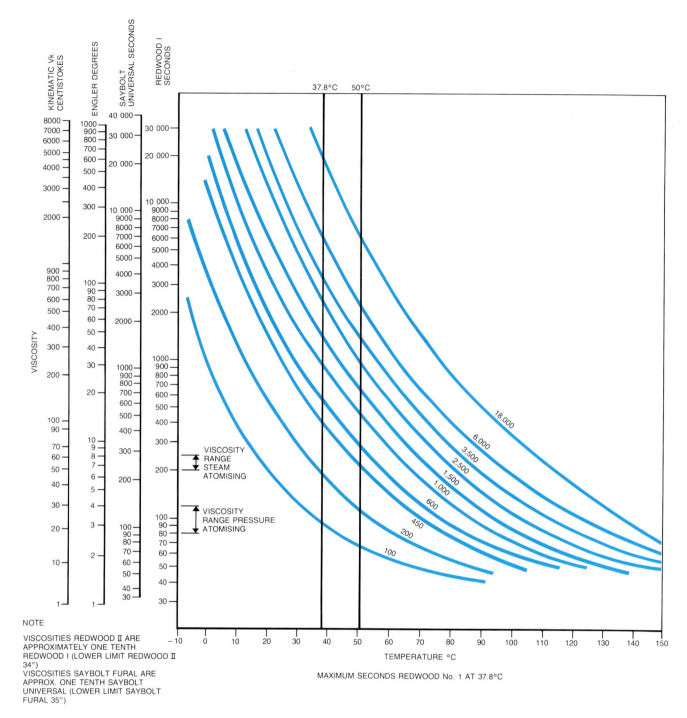

FIG. 2.47 Viscosity/temperature curves for fuel oils

atomising is by pressurised steam which is fed through the central tube and issues from drillings to meet the oil, which has passed along an annular space between the central steam tube and the concentric outer tube. In this type of burner, the oil temperature need not be so high as with the pressure-atomised version, as additional heat is imparted to the oil in its passage down the central tube. The steam-atomised burner produces a particularly fine spray which is advantageous when used to light up

a cold furnace; steam consumption, however, is far from negligible and has to be provided from adjacent boilers and/or by other means.

Whichever type of burner is installed, the conditions which give the best results when firing oil burners in test rigs, for example, prior to manufacturers freezing their designs, do not necessarily do the same when the oil burner is installed in the boiler proper and many man-hours of effort have been necessary throughout the CEGB in conducting

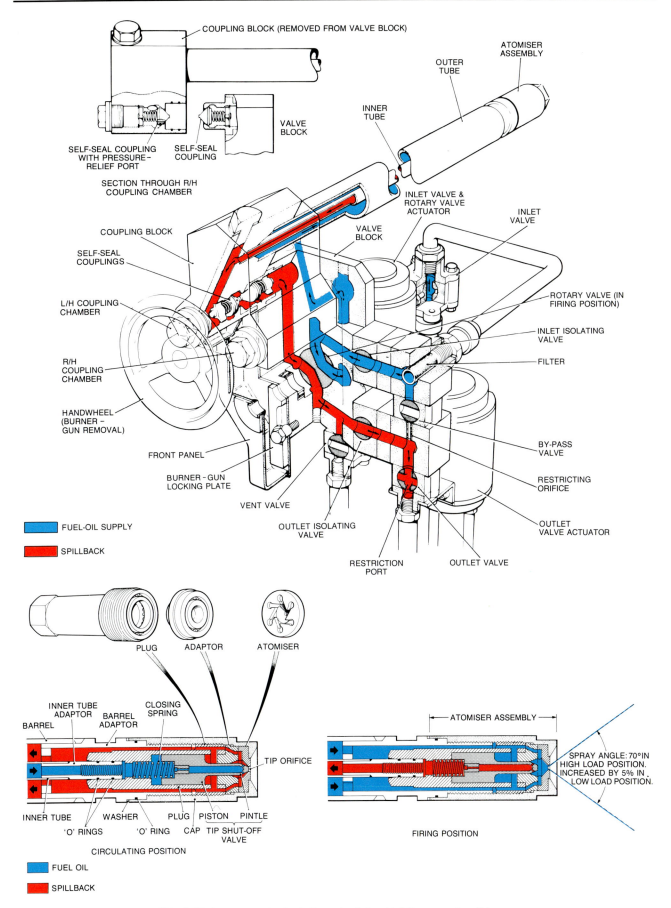

FIG. 2.48 A pressure-atomised, tip-recirculating high/low capacity oil burner

A turndown of about 2:1 is achieved by restricting the flow of oil recirculating whilst firing. This is done by operating the rotary outlet valve to bring the restriction port on-line. During tip recirculation, the outlet valve should be set to give full recirculation flow to keep the tip as cool as possible.

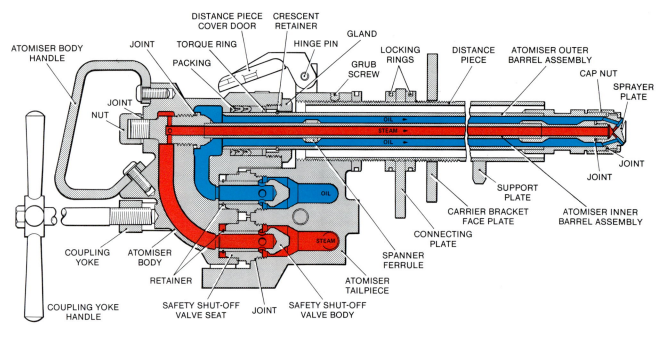

FIG. 2.49 Steam atomised burner

trials and 'tuning' systems to arrive at the best possible solutions. Factors that have been shown to have a strong bearing on the oil flame shape and its performance in respect of combustion efficiency range from the igniter system chosen, the absence (or otherwise) of PF burner 'eyebrows', windbox and secondary air vane settings and, as importantly, whether or not the mill is in service. Even more important, perhaps, are such factors as the condition of the oil-burner swirler or stabiliser and whether the correct amount of combustion air is available, preferably supplied by the burner's own PA fan. The control of PA is particularly difficult on oil burners that have large turndown ratios. Indeed, some stations, tired of attempting to achieve good combustion under all conditions, have settled for a fixed oil burner output and tuned the rest of the support services to suit this one condition.

Finally it is worth pointing out that oil burners should never be operated without the protection of flame failure devices. Whilst early claims for such devices were generally optimistic, in particular those that suggested flame failure devices could adequately differentiate between oil and PF flames, equipment does now exist that is reliable and reasonably maintenance-free and its use has undoubtedly prevented many potential furnace explosions.

2 Boiler plant — normal operation (oil-fired)

The principal problems associated with oil-fired boilers arise from the presence of the complex organic compounds of vanadium, sodium and sulphur in the fuel. The normal range of residual fuel-oil ultimate analyses is shown below.

Carbon	83.5	— 86.5%
Hydrogen	10.0	— 13.0%
Sulphur	trace	— 5.0%
Chlorine	0.004	— 0.017%
Vanadium	0.002	— 0.06%
Sodium	0.003	— 0.01%
Phosphorus	trace	— 0.005%
Ash	0.03	— 0.10%

The problems manifest themselves in four areas — creep, high temperature corrosion and slagging, cold-end corrosion and fouling, and finally air pollution.

The use of austenitic steels, with their improved creep strength, is not possible where the presence of vanadium will catalyse complex mechanisms of high temperature corrosion with the nickel present in austenitic steel. Accordingly, chrome steels are used in the construction of oil-fired boiler pressure parts. Creep strain is a strong function of temperature (i.e., creep $= k\theta^3$, for example) and, consequently, the inability to use austenitics requires a lower limit on the maximum ratings of component metal temperatures, typically leading to the reduction of the maximum final superheater outlet temperature from 565°C to 540°C.

The sodium and vanadium constituents in the fuel produce complex compounds of low melting-point ash within the furnace. This ash will tend to deposit upon the cooler surfaces within the furnace. The amount of ash production is a function of the levels of sodium and vanadium; the greater the quantities present in the fuel, the greater the deposition. Being of relatively low melting-point, the deposit will fuse to a liquid slag, which will adhere to furnace water-tubes, radiant superheater and final superheater tubes. Ash and other impurities present in the boiler gases will adhere to the vanadium/sodium-rich layers to produce a deposit which will impair heat transfer and reduce the areas of gas passage, leading to higher impingement forces and even greater deposition. Such deposits, when cooled, are extremely hard and difficult to remove by manual cleaning techniques. Some success has been achieved with water-lancing techniques, particularly with cold water whilst deposits are hot, but extreme caution must be exercised during such operations to avoid thermal quench-cracking of boiler tubes. The steady state volume of slagging is governed by a reaction which maintains an equilibrium between the vapour, liquid and solid phases and will change as a direct function of the proportions of sodium and vanadium in the fuel. In the molten state, complex vanadium oxide compounds catalyse the corrosion of chrome-steel alloys, the rates of wastage being a function of temperature. Sulphates also leach out iron, causing pitting and wastage.

Sulphur compounds in the fuel are largely oxidised to sulphur dioxide, very little oxidation to sulphur trioxide occurring in the furnace. The production of SO_3 is greatly increased by the catalytic action of vanadium oxides in the temperature range of $600-750°C$, typically the primary superheater and reheater section gas temperatures. Condensation of SO_3 to form sulphuric acid will occur on the cooler back-end of the gas-pass, leading to severe airheater, grit arrestor, ducting and ID fan corrosion risk. The dewpoint temperature is an inverse function of the concentration of SO_3 present.

When flue-gas temperatures approach the acid dewpoint and condensation occurs, ash particles of an acidic nature tend to agglomerate and will adhere in regions of lower gas velocity. There is considerable risk that such deposits will be loosened and carried out of the stack with the gas-velocity changes inherent in start-up and loading. There is no simple way of preventing the formation of acidic deposits, given that it is generally prohibitively expensive to burn very low sulphur (i.e., $<0.3\%$) fuels. On the contrary, the great escalation of fuel oil costs since 1973 has increased the requirements to cope with higher sulphur, heavier and higher asphaltenes, fuel oil. The last characteristic tends to increase flue-gas dust burdens and provide nuclei for the formation of acidic smuts. Palliatives include the undesirably expensive measures of deliberate elevation of final flue-gas temperature, particularly low excess-air operation, the injection of neutralising additives, such as magnesium oxide, heating of the top section of stacks and the regular cleaning of stack flues.

Modern oil-fired boilers are designed to operate with low excess-air (typically equivalent to $0.2-0.4\%$ excess oxygen) in order to achieve close to stoichiometric combustion and to reduce flue-gas losses. The production of SO_3 is also a direct function of excess oxygen, further increasing the pressure to operate ever closer to stoichiometric conditions. In order to achieve the full benefits of both these objectives, it is necessary not only to achieve accurate metering of fuel and air and equality of fuel/air ratio to each of many oil burners (up to 32 in a large modern unit), but it is also essential to reduce the ingress of air to those parts of the gas pass operating at sub-atmospheric pressure. This requires the maintenance of leak-free furnace casings and ductings, pipework and structural support penetrations, low leakage at rotary airheater seals, grit-arrestor fittings and ID fan shaft-seals, to name but a few of the multiplicity of points on a large unit where such ingress can occur.

Low excess-air operation requires high quality combustion control and monitoring systems. The problem of obtaining accurate and truly representative quantitative analyses of boiler flue-gas still represents a formidable challenge. Even though considerable improvements have been achieved over the last few years in the design and serviceability of oxygen and carbon monoxide analysers and dust and grit monitors, a heavy commitment to the maintenance of such equipment is necessary to achieve adequate assurance on the quality of display of information, mainly because of the inevitably hostile environment surrounding, and the corrosive sample gas contained within, such equipment. It is a simple matter to extract a representative sample from a gas duct which may have an area of many square metres. In spite of the best attempts to implement a grid network of sampling points within such large ducts on current units of up to 660 MW rating, considerable imperfections in measurement accuracy continue because of the tendency of gas flow to stratify and for sampler aspiration points to become blocked. Even with a heavy programme of routine maintenance, it is necessary to equip modern oil-fired units with a multiplicity of gas analysis equipment in order that redundancy will overcome the susceptibility to individual system errors. It is necessary to equip such units with oxygen analysers at varying points along the flue-gas path, since there is inevitably tramp-air ingress. The minimum requirement for a large modern unit is for analysis as soon as is technically possible after the furnace exit, typically at economiser location, and at the ID fan discharge. The former point principally indicates the combustion ratio condition, whilst the latter, in addition and by comparison with the first, gives objective information on tramp-air ingress.

In order to achieve near-stoichiometric combustion, it is necessary that good equality of distribution of fuel and air to the very many burners exists. Even with the best practicable means currently possible, it is still inevitable that unburnt carbon will still exist at significant levels, even at excess oxygen values of 0.5%. The unburnt carbon is best diagnosed by modern carbon monoxide analysers, which provide the best and most sensitive means of assessing combustion. A typical current 500 MW oil-fired unit will produce CO at around 250 PPM, currently with excess oxygen at 0.3%. Very small variations in this value of excess oxygen will produce large variations in CO level. Modern CO analyser systems are becoming sufficiently reliable to be incorporated within automatic boiler combustion-control equipment, this being achieved on a CEGB 500 MW unit oil-fired station in 1980.

Given reliable and accurate analysis and display of combustion conditions, boiler firing control systems capable of quick and accurate automatic response to varying conditions are essential. Current techniques of distributed digital-computer control are well suited for rapid, accurate integration of the complex processes involved in the operation of large boilers. Following boiler construction, burner air registers, quarls and dampers must be accurately calibrated under pre-commissioning procedures to ensure that the best possible equality of distribution of air to individual burners is achieved; this is not easy in large units with their necessarily-large windboxes. The provision of modulating dampers, which can be preset, is of considerable advantage towards the better achievement of this objective, although such techniques generally require special tests at pre-commissioning, or after return from major overhauls. It is not yet generally feasible to obtain an analysis of individual burner combustion conditions on load, although recent specialised photometry techniques hold out the possibility of achieving this. Above all, the characteristic required of a burner air door is that it should give good shut-off capability.

The main requirement when firing residual fuel oil is the achievement of good atomisation of the fuel at the point of combustion. Complete control of fuel oil temperature and pressure and combustion air volume and temperature are necessary to achieve this objective. Given that these measures are maintained, it is essential that no less diligence is exercised in the control of burner oil-sprayer equipment, particularly the sprayer tips. The latter are generally machined to very close tolerances, and experience has demonstrated that the slightest damage or fouling at tip orifices will lead to significantly degraded atomisation and throughput rate. Regular cleaning and

testing of sprayers and tips must be maintained to a high standard of performance. The effectiveness of individual burner steam purging must also be maintained at a uniformly high standard. Such measures will also maintain equality of oil distribution between individual burners. It is not possible to overstress the importance of such requirements; any shortfall in the oil throughput of one burner, compared with the others in service with it, will produce apparently incompatible and confusing indications on combustion monitoring equipment, with concurrent high excess O_2 and CO readings as shown in Fig 2.50.

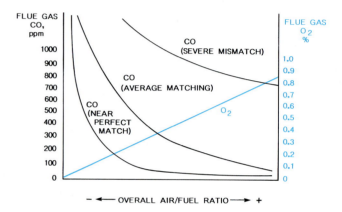

FIG. 2.50 Graph showing the effects of matching air/oil distribution between individual burners on a 32-burner 500 MW unit

Apart from grossly poor burner performance, ultraviolet radiation monitoring systems do not commonly give much information on the quality of an individual flame. Modern infrared flicker-radiation monitoring techniques offer more reliable indications in this direction, as well as providing improved performance in terms of the sensitivity versus discrimination conflict, which is a typical drawback associated with UV monitoring. The basic reason for this is that a dense oil mist (present at the base of the flame cone) is much more opaque at higher radiation frequencies.

The achievement and maintenance of the best possible combustion conditions will minimise, but not totally eliminate, the generation of dust, be it acidic or neutral, and ash. The economic pressures to burn increasingly undesirable residual fuel oils will continue to counter improvements in combustion control. Consequently, the requirements for on-load cleaning and dust extraction are likely to be of even greater importance in the future.

The provisions for effective sootblowing of super-heaters, reheaters, economisers and airheaters should be maintained in an effective manner. Even if early operational experience should indicate that certain sections do not require extensive or regular use, changing fuel oil specifications are very likely to alter that position. Accordingly, all sootblowing equipment provided should be regularly used. Such equipment is placed in an arduous environment, and experience has shown that simple neglect can bring about considerably more loss of serviceability than will be caused by a programme of reasonably regular exercising. This stricture applies to almost any equipment associated with the acidic and dirty conditions of oil-fired boiler gas-passes — given long periods of inaction on-load, dampers will seize, sampling points will become blocked and dust extraction equipment will become ineffective.

Airheater sootblowers are of particular significance on oil-fired plant. Even with the normal provisions of steam preheating equipment before the gas air-heaters, temperature and dust burden considerations will lead to increasing airheater fouling, producing poorer heat-exchange performance and imposing differential pressure losses that will increase fan power and seal-losses on rotary types. These factors can increase dramatically as the onset of fouling will lead to ever-increasing fouling rates. When such conditions prevail, frequent airheater washing may become necessary.

The many necessary requirements associated with the good operation of oil-fired boilers are summed up in the following list of CEGB recommendations:

1 Boilers should be capable of steaming at 100% MCR, with an 0.4%, oxygen in the flue gas measured at the furnace exit. At loads between 70% and 100% MCR, the flue-gas oxygen should not exceed 0.5%.

2 When operating in the above conditions, the following should be met:

- The average SO_3, measured at the economiser outlet, should not exceed 5 PPM.

- The solid dust burden in the flue gas, measured at the airheater outlet, should not exceed 0.115 g/m^3 at NTP, dry. The shell smoke-number, measured at the economiser outlet, should not exceed 5.

3 The flue-gas oxygen should not increase by more than the following values between combustion chamber and ID fan outlet:

- 0.5%, for boilers with plate or tubular airheaters.

- 1.5%, for boilers with rotary airheaters.

4 Ignition equipment should ignite oil within 2 seconds of commencement of oil flow, with a failure rate not more than once in 20 operations. Each burner must have its own igniter, and each should automatically shutdown and steam purge on failure to ignite within the stated time limit.

5 No air should be admitted to the combustion chamber through burners which are not firing.

6 The air/fuel ratio should be the same at each burner and there should be provision on each burner for adjustment to obtain this.

7 The complete burner register should be removable, without any necessity to enter the combustion chamber.

8 Isolating dampers after the ID fans should be gas-tight.

9 It is essential that adequate and good-quality instruments are installed on boiler plant to measure the quality of combustion and the emission from chimneys. These instruments must be maintained to the highest standard to give continuous and accurate information to operators.

10 Instruments required for emission control in the UK consist of smoke and coarse-dust monitors. They are required not only for operation, but also for compliance with statutory requirements in the UK.

11 For close combustion control, it is necessary to have installed oxygen and CO meters monitoring the flue gas. These meters should sample the flue gas at a point as near to the combustion chamber as practical. This is normally at some point on the economiser. The sampling and analytical arrangements should be such that the average O_2 and CO level at the plane of sampling is displayed to the operator.

12 Portable standby oxygen and CO instruments should be available to cover any instrument outage of more than a short duration.

13   TV flame-monitoring is advisable for combustion control as additional instrumentation (this facility being of special importance during light-up operations).

14   The stack plume should be continuously monitored by CCTV, with display in the main control room.

15   Operating instructions should be in writing, and should be readily available to all staff whose duties concern the operation of oil-fired boilers.

16   The operating instructions should include the figures of oxygen and CO in flue gas which would be attainable at a number of steam loads on each boiler. They should also give guidance on parameters such as windbox pressures, atomising-oil pressures, atomising-oil temperatures and furnace pressure. These values should be determined by technical staff and should take into account any relevant data obtained on recent tests. The information should be kept under constant review and should incorporate the effects of the current state of deterioration of the boiler and boiler components.

17   The combustion conditions on an operating boiler should be such that the likelihood of acid-soot emission is negligible. The oxygen in flue gas level at the furnace exit should be as low as possible, commensurate with an acceptable level of solids burden and CO, without the formation of smoke.

18   Care should be taken to see that, as far as possible, surfaces within the gas-pass are maintained above the acid dewpoint.

19   The importance of maintaining oil pressures and temperatures at levels high enough to ensure good atomisation is fundamental. This is particularly relevant when oil supply viscosities change. Limits of oil pressure and temperature should be included in the operating instructions.

20   During load changes, extreme care should be exercised in keeping combustion conditions correct during the complete operation. Sudden changes in combustion conditions should be avoided, as these may lead to increased carbon, increased SO_3 or re-entrainment of dust by flue gas.

21   The combustion chamber should be kept under slight pressure to ensure that no air inleakage occurs. If this is not practical, the suction should be as low as possible.

22   If it is necessary to reduce boiler load to low values, the oxygen in flue gas should not be allowed to rise above full-load values, unless other problems exist which make this necessary.

23   Cleaning of surfaces on-load should be carried out with extreme care and only when off-load methods are not practicable.

24   Operation procedures for lighting-up and shutting-down should be in writing and readily available to all staff whose duties involve these boiler operations.

25   The prime objective in establishing lighting-up and shutting-down procedures must be to control conditions in the combustion chamber and gas-passes, so that emission of acid-soots, either during the operation or later, is avoided.

26   It is important that the lighting-up and shutting-down processes should be completed in the minimum time.

27   When a boiler is brought on-load, it is necessary to bring the boiler into full oxygen control as quickly as possible after reaching a steady steam load. It is suggested that oxygen control should be established within five minutes of reaching a target load.

28   Operation procedures should ensure that the surface temperatures within gas-passes are kept as high as possible, preferably above the acid dewpoint.

29   During shutdown periods, terminal isolating dampers should be shut, so that no undue cooling of gas-passes takes place.

30   A senior member of the technical staff should take steps to confirm at intervals that the written operating procedures are used in practice.

31   Instructions for the cleaning, servicing and maintenance of all components of oil-fired boilers

should be in writing and should be readily available to those staff who are involved in those activities.

32 Cleaning, servicing and maintenance of boiler components should be carried out in such a manner, and with such a frequency, that the boiler is able to operate at all times at its optimum combustion conditions.

33 It is particularly important to ensure that burner sprayer assemblies are manufactured correctly to the appropriate specification and maintained in good condition. Adequate checks are required to eliminate worn tips or carbon build-up.

34 Sprayer assemblies should be taken out of service for cleaning and inspection at intervals determined by experience. Cleaning must be carried out by methods approved by the station management. Cleaning methods must be such that no damage occurs to the tip during the process.

35 Inspection should be carried out of the sprayer assemblies at intervals determined by experience. The inspections should be carried out by competent staff, who should approve the condition of the sprayers for further service and make records of inspection results. Inspection of sprayers should be carried out by methods approved by station management. Visual inspection with the naked eye is not sufficient and the use of a test rig capable of demonstrating flow rate, spray angle and pattern is recommended.

36 The complete burner assembly, including register and quarl, should be maintained according to specification, design or any other approved documents within the station. Care should be taken to see that:

- All parts of the burner register and quarl are concentric with the sprayer tube.

- Each burner on the boiler has equal dimensions to ensure that all burners are matched.

- Critical dimensions which affect air flow, oil flow or flame stability are set up according to drawings, or any other station instruction.

- The shape of the refractory quarl is accurately formed.

- Air shut-off dampers, where provided, operate correctly.

37 It is of prime importance that air leakage into the gas-passes is reduced to a minimum. Air is likely to leak into the gas-passes at many points, but experience suggests that the following are the most common source of major leakage, although they are not necessarily listed in order of importance:

- The area around registers

- Unlit burners

- Combustion chamber corner-seals

- Seals adjacent to headers and drums

- Expansion joints in ducts

- Pipe penetrations into gas-passes

- Explosion doors

- Access and inspection doors

- Hopper valves

- Airheater seals

All points where air leakage is possible should be given close attention during overhauls and, where possible, should be given maintenance at other times.

38 Before any major overhaul on the boiler, a test should be carried out with the boiler hot to determine the position of air leakage. This can be done with the boiler shut down by pressurising the gas-passes and noting the points where air escapes. It may be necessary to introduce smoke into the boiler to assist with identification. A similar test should be carried out after maintenance to satisfy operating engineers that the boiler is suitable for operation.

39 During the period between major overhauls, every effort should be made to keep gas-passes tight. Small leaks on rigid surfaces may be sealed with materials such as Vermiculite or Calcofoam and Calcoseal. Pressure or smoke tests may be carried out to assist with this work.

40 It is essential that every opportunity is taken both during overhauls and between overhauls to maintain the cleanliness of gas-passes. Any accumulated dust is liable to be picked up by flue gas while the boiler is steaming and particularly when gas velocities are changed. Consideration

should be given to the modification of any ductwork, where dust is known to collect.

41 Off-load cleaning may be required from time to time on superheaters, economisers and airheaters. Records should be maintained of the cleaning and of the draught losses, so that optimum cleaning programmes can be established and reviewed.

42 When off-load cleaning is carried out on gas-passes, superheaters, economisers or airheaters, care should be taken that no deposits remain on the metal surfaces. Such deposits will form a key for further build-up and inadequately cleaned surfaces are of little long-term value. Operating engineers should inspect the result of any cleaning operation and should satisfy themselves that it has been correctly carried out.

43 All hoppers where dust can collect should be completely cleared regularly. Operating engineers should satisfy themselves by periodic inspection that the cleaning operations are carried out effectively.

44 Chimneys should be inspected internally at intervals determined by station management to look for defects and deposits. Any places where air inleakage occurs should be sealed. With respect to deposits, it is thought that chimney cleaning has a beneficial effect on acid soot emission, but methods and techniques have not been fully developed. Chimneys must be capped during cleaning to prevent the escape of deposits.

45 The lagging of all gas-passes should be kept in good condition to prevent condensation of acid deposits on cool surfaces. Any lagging which becomes detached should be replaced immediately.

46 The use of additives should be considered, where scientifically-based plant tests have demonstrated that they are beneficial.

47 A means of assessing emissions, both by physical area sampling and maintenance of records of any complaints, should be established and maintained.

3 Boiler plant — emergency operation

3.1 Introduction

It is under abnormal and emergency conditions of plant operation that the skill of the operator in par-

ticular and operating staff in general is most severely tested. Therefore it is imperative that the actions required are anticipated and training given, so that the risk of injury to personnel and damage to plant is minimised. The role of plant simulators, now widely in use within the CEGB, has done much to give staff the confidence they need to act quickly and decisively under such conditions. Today all staff engaged in the operation of the larger units receive regular refresher training on an appropriate simulator.

On PF-fired boilers, the major emergency operations that need to be considered are:

- Complete loss of ignition
- Loss of feedwater/drum level
- Excessive overfiring
- Overspraying
- Airheater fires

These are now considered in detail.

3.2 Complete loss of ignition

Initially at least, this condition is treated by allowing the turbine-generator to remain connected to the Grid and allowing the unit transformer to continue to supply the boiler (and other) auxiliaries. This leaves the induced draught (ID) and forced draught (FD) fans supplied from the unit board and running, to effect a complete purging of the boiler gas-passes. Fires must be completely extinguished by operating the ignition trip button which should be arranged to trip both mills and oil burners. Advantage is taken of the stored energy within the boiler to carry out a short-term investigation; provided that the turbine-generator load can be held to a very low level, this period can extend to, typically, 15 minutes on some plants, though adverse casing-to-rotor differentials on high pressure (HP) or intermediate pressure (IP) turbines may curtail these operations prematurely. During this period, it is absolutely essential to confirm that all fires are extinguished and that no fuel oil is entering the furnace, as, for example, a result of faulty oil burner shut-off valves. Each case must be judged on its merits but a useful guide can be obtained from boiler flue-gas or mill CO probe indications, and locally by shining a powerful torch across the furnace. When the mills trip, the associated PF burner secondary air dampers may run closed automatically and will require reopening fully to re-establish the purging flow, as will dampers in any shut-off section of the gas-passes. For example, reheater by-pass dampers may require reopening. Thus if the fault is found quickly, it may be possible to re-establish ignition and restore boiler load.

The skill of the operator is first tested by his ability to confirm that ignition has been lost. Here plant experience is important. Frontwall-fired boilers are extremely unlikely to lose complete ignition, being very stable and tolerant of quite major plant disturbances; corner-fired furnaces are less so. A complete loss of ignition would be confirmed by flame failure indicators, initial high-negative furnace suction (though this would be quickly corrected by the automatic controls), a falling drum water level and falling steam pressure and temperature. In this context, the provision of audible alarms for abnormal furnace pressure conditions are judged to be essential if the operator's attention is to be rapidly concentrated on the abnormality. If the above symptoms are present, then the ignition trip button must be operated without delay and the unit load reduced to as low a level as practicable. Any desuperheat spray flow or reheater desuperheat spray flow in use must be stopped: either manually or, if automatic facilities exist to do this (which is not by any means always the case), they must be checked to have operated. Failure to do this will result in quantities of free water collecting in the boiler superheater banks and/or in the cold reheat lines to the obvious danger of the turbine. The boiler drum water level may initially drop out of sight in the level indicators as the reduction in water volume on cessation of steam-bubble formation is considerable. This is of no concern provided that the fires are out and the level can be gradually restored prior to refiring the boiler.

Note that if facilities for manual purging of PF pipework exist, then this must only be attempted once the furnace is purged of combustible products and oil burners re-established. Several automatic mill PF-pipework purging schemes have had to be modified to prevent the activation of this feature if all the mills are tripped, otherwise large concentrations of PF injected into the furnace after the fire has been lost could ignite off hot refractory or slag lumps, with devastating effects.

Finally, as with most aspects of power plant operation, prevention is better than cure and the sensible use of the appropriate number of oil burners at times of furnace disturbance may prevent a loss of ignition incident altogether.

3.3 Loss of feedwater/drum level

Such an incident can be treated very nearly as a complete loss of ignition. Whilst some boilers are fitted with an emergency low-level trip, the practice is not universal and clear guidance needs to be laid down as to what the minimum level is. Complications can set in all too easily if the loss of drum level is caused by a boiler tube leak, for then unbalanced drum levels can easily occur. To add to the confusion, the constant-head chambers of differential-

type drum level indicators may 'flash-off' if the boiler pressure is falling fast, giving transient, incorrect levels. The standard gauge glass, usually in the form of the hydrastep gauge, should always be taken as the primary indication of drum level. If a furnace tube leak develops, the operator must assess very quickly whether or not the severity of the leak has affected his fire; in the extreme, a ruptured tube may cause a loss of ignition incident as well.

If the situation is judged to be irrecoverable, then the turbine will be off-loaded and tripped. However, should the fault resulting in the loss of feed flow and/or drum level be ascertained and corrected, then the actions outlined in the last section can be followed prior to re-establishing firing and load, though drum level may have to be restored quite gently in order to keep drum metal differentials within the specified limits. Note again that the loss of drum level need cause no concern provided that the fire is extinguished.

3.4 Excessive overfiring

Overfiring occurs whenever the fuel input is greater than that required to meet the instantaneous steam demand. Such conditions occur transiently throughout the life of the boiler, arising from normal governor action and automatic combustion control, but are of short duration and of limited magnitude. Overfiring also occurs during initial pressure raising and unit loading, but is again of limited magnitude. Boiler metal-temperatures are most likely to reach a level which will give rise to significant damage to superheaters and reheaters when the steam flow is reduced by a significant amount compared with the firing rate. In practice, the most likely cause of excessive overfiring is a complete, or almost complete, loss of turbine load brought about, for example, by the operation of the vacuum unloading system, inadvertent operation of the governor speeder gear, the opening of the high voltage circuit-breaker or a grid system fault, etc. If the boiler is operating at, or near, full-load and at normal design pressure, then such a loss of turbine steam flow will result in a rapid lifting of the superheater (and drum) safety valves. The resulting steam flow has traditionally been assumed to be sufficient to keep vulnerable parts of the superheater cool but early experiences with large boiler units cast some doubt on this assumption, especially for boilers having extensive radiant superheater components located in the furnace zone proper, for instance, a furnace steam division wall. A boiler may be operated at reduced pressure, either for chemical reasons or because part-load is required for a considerable period. In either case, although the load on the boiler is not likely to be above 70% MCR, a cessation of turbine steam flow without a corresponding reduction in firing rate will produce a rapid increase in boiler pressure, and in superheater and reheater metal temperatures, because there will be no steam flow provided by safety valve operation. In this respect, reduced pressure operation on a 160 bar boiler is often carried out at a value as low as 110 bar and, under these conditions, catastrophic failure is almost certain to occur in radiantly-heated steam-cooled surfaces before the pressure reaches the safety valve lift-point.

In any of the above some measure of protection is afforded if the boiler firing is on automatic control, in that the mills will be turned down as the boiler pressure increases and, in the event of a turbine trip, adequate protection exists because the boiler would be intertripped with the turbine and a complete cessation of firing would result. Nonetheless, all large boilers have now been categorised dependent on potential risk. On those having little radiant superheater sections, the maximum rate of metal-temperature rise is considered to be sufficiently low to enable reliance to be placed on the operator in protecting his plant. For boilers having extensive radiant sections, automatic overfiring protection schemes have been installed, in the worst cases being operative at any boiler pressure. They operate by sensing a predetermined low steam flow condition inferred either from the turbine governor-valve sensitive (or pilot) oil or fluid pressure, or from the turbine inlet-belt steam pressure, as this is proportional to steam flow, or from both. During normal shutdown or start-up operations, facilities are provided to enable the operator to inhibit the firing trip, the inhibit usually being automatically removed on rising turbine load. Alternatively, an alarm may be arranged to operate whilst the protection is defeated.

It should be noted that the protection does not normally detect a partial load rejection down to the value determined by the pressure switch settings. Such partial load rejections are not common but can occur as a result of a transmission system disturbance leading to a split system with a transient high frequency. Under these conditions, provided that the boiler response time is slow enough to allow the operator to stabilise the firing conditions manually, it is important that the firing is not tripped so that, once the transient high frequency has been dealt with, normal loading can resume.

3.5 Overspraying

Severe overheating and resultant failure of platen superheaters has occurred at several stations due to the excessive use of desuperheater sprays (resulting in waterlogging of the superheater pendants) during conditions of low load and unbalanced firing. It is essential that steam temperatures should not be allowed to fall as low as the saturation temperature corresponding to the operating pressure. In the super-

heater, this is most likely to occur immediately after the desuperheater section and indication of the steam temperatures leaving the final desuperheater section must be provided, together with an indication of saturated steam temperature usually taken from a convenient saturated steam header. Saturation temperatures at 160 bar are approximately 350°C, so if desuperheater outlet temperatures at full pressure operation are maintained above this value by a margin of, say, 20°C, no waterlogging should ever occur, not that it is likely to in any event at full-load. The use of sprays should be avoided on low loads or when firing conditions may become unbalanced, such as the loss of a fan group or if the use of a mill group is being considered on less than its full complement of burners, when the unbalanced gas temperatures resulting across the boiler may automatically initiate an excessive spray flow on one side or the other to balance steam temperatures. Unfortunately, if an over-spraying condition occurs and is confirmed, then the only safe way to ensure that the affected tubes are protected is to trip the firing, followed by a controlled 'boil-off' exercise prior to reloading.

Emergency reheater 'knock down' sprays should also be used with care as similar conditions apply, though the major danger in this case can be of water ingress into the turbine if the spray injection is into the cold reheat mains from the HP turbine to the boiler. At a full-load reheater pressure of about 45 bar, the corresponding saturated steam temperature is 260°C but this varies significantly with load (total pressure change being from 45 bar to vacuum conditions between full and no-load) and it is necessary to have a 'sliding' limit which can most conveniently be provided by a mini-computer programme and VDU display. Many stations now have automatic isolation of master spraywater isolators in the event of a loss-of-load or unit trip condition; for those that do not have this facility it is essential that the operator takes early action to isolate the systems.

3.6 Airheater fires

Over the years, several airheaters have been destroyed by fire. The possibility is extremely remote on boilers that operate predominantly on PF at continuous high loads but is more likely during or after a boiler light-up on fuel oil, particularly from cold or during periods of lengthy low-load operation during which the PF flames are stabilised by fuel oil. Having the smallest clearances in the entire boiler gas-side system, the airheater matrix is vulnerable to the collection of unburnt carbon and soot deposits which, despite the best of efforts, many oil burners are still prone to produce.

A boiler should never be fired on oil unless the airheater sootblowers are available for use, nor should firing continue if sootblowing is suspended for any reason. For stationary matrix airheaters, not only should air or steam supplies be available but the indexing gear must be in working order or sootblowing will be patchy and ineffective. All airheaters should be sootblown including any that are isolated and boxed-in, for dampers seldom seal effectively and a surprising quantity of combustible material can settle out in the matrix. During such sootblowing, the gas outlet damper must be cracked open to allow the sootblown deposits to escape.

If a fire is suspected in an airheater, it is difficult to assess the degree of severity unless the fire is extremely serious and the casing is glowing red hot. In this event no time should be lost in boxing-in the affected airheater as tightly as possible (to starve the fire of oxygen) and in initiating the fire fighting sprays. Any consequential secondary problems, such as turning any dry dust present in the air heater hoppers into a hard, concrete-like mass, will be minor compared to the possibility of the fire spreading throughout and destroying the matrix.

Smaller fires, probably affecting only a sector of the matrix, are extremely difficult to confirm and can even burn themselves out without spreading; indeed when the air heater is inspected during a fortuitous outage, it is not unknown to find that the matrix has suffered a small fire at some time past. In a stationary matrix airheater, however, it is quite possible and beneficial to have a system of thermocouples embedded in the core, their outputs being displayed on the unit control panel. Should several indicate a fire, action can be taken at an early stage to isolate and cool the airheater.

It is not uncommon to find that air heaters cannot be individually boxed-in; for this reason, if a fire is confirmed on such an airheater, it may be necessary to trip the fires and box-in the boiler completely to effect an airheater isolation.

4 Boiler plant — routine testing

On stations that are operating efficiently and well, there is a place for carefully chosen routines, be they of an operational or maintenance nature. For example, as far as maintenance is concerned, most stations having spare milling plant capacity would elect to undertake routine preventive maintenance on the mills and associated plant rather than operate on a breakdown basis; under breakdown conditions, no forward planning is possible and, because of the random nature of things, more than one mill might break down at the same time. Similarly, in the field of operation, there are sound reasons for initiating a regular programme of routine testing which, in its widest form, extends beyond the simple concept of a few mandatory alarm tests (for example, boiler drum water level alarms). Stations naturally vary in their approach to the subject and no firm guidance can, or should, be given. The approach taken by one station has been to divide plant and alarm testing into two groups; those that can be conveniently and easily tested by operating personnel generally and those that cannot. The latter routines have been redistributed to the specialist functions, e.g., high voltage circuit-breaker protection an trip testing to the Electrical Branch and PF sampling to the station laboratory staff. Some of these tests will be discussed later.

For the first group, the approach has been to subdivide it into alarm and plant testing and to consider each carefully prior to initiating a programme of routines. For example, with boiler alarms, the full schedule of every boiler alarm fitted has been studied and categorised as to whether or not each alarm should be tested and, if so, by whom. Those that can be tested conveniently by operations personnel, either from the unit control panel or local to the sensor, have been grouped into like or similar packages of work. Specifications have been written detailing the method of testing, alarm levels and tests results expected and have been timed for work study purposes.

Test intervals have then been allocated and the whole package transferred to computer file from which the work instructions (together with the results sheets) are issued periodically. The test intervals have been standardised at one, four and thirteen weekly intervals for plant tests and at four and thirteen weekly intervals for alarm tests. With alarms, the frequency of testing ought to reflect both the potential danger to the plant should the alarm not operate and its reliability of operation. If any alarm needs to be tested more than four times a year (unless it is extremely important) to ensure that it is still working satisfactorily, then there is something wrong with that circuit and improvements are needed. When work has been completed, any records defined as necessary are taken from the results sheet and committed to computer file, from which it is possible to ascertain the incidence with which the routines are being completed and any persistent failures to operate. The whole system is under the overall general control of one engineer who is responsible for ensuring that the specifications, tests and frequencies remain up to date and meaningful.

Tests are presently further subdivided into on-load or off-load routines. The latter are issued four times a year, en masse (to reflect the present likelihood of units being off-load), so as to always be available in the event of a breakdown.

Plant tests are considered in a similar manner and often combined with associated alarms. A typical format for a mill test is shown in Fig 2.51.

Plant tests needing either specialist services or which, for other reasons, are rather beyond the capabilities of operating staff, are undertaken by the specialists concerned. Typical of these would be:

WORK CONTROL CARD

R2532 (rev. 7/85)

DEPARTMENT					TASK AREA	COST CODE	W.C.C. No.	R029122 A
OPS	OPS	OPS	OPS	OPS	M	0911370000		

SITE	UNIT	PLANT CODE	ITEM	TYPE CODE	UNIQUE PLANT IDENTIFIER	ORIGINATOR/ROUTINE	DATE
SS	06	37000	1	1810/6/02	X ORAT611	ROUTINE	DUE 03MAR86

PLANT DESCRIPTION & LOCATION   6.1 MILL COAL FIRING SYSTEM

JOB TITLE/DEFECT   CARRY OUT MILL SHUTDOWN ROUTINE TEST NO. 611

COMMENTS

PLANNING INFORMATION

HEADER No.	O 3700011A	08
W.I. No.	AT37000030	09
PERIOD	13 S	SUP
LAST DONE		SCH A
GROUP No.		P.C. ?
SPECIAL		S.D. N
OUTAGE		O.P. A
PERT		
SCHEDULE		S S M T W T F
	ON 8611	
PLG PRIORITY	A	

CREDITS	DURATION	MANNING AND TIME
CWA		1OP CWA

TECHNICAL REFERENCES
   0110200031

ISOL REQUIREMENTS & COMMENTS	SAFETY DOCS	REQ.	PFW	ROMP	LWC	SFT	S.D. No.
NO SFTY.DOC.							

SAFETY REQUIREMENTS

```
     CARD      6.1 MILL COAL FIRING SYSTEM
     DETAIL    CARRY OUT MILL SHUTDOWN ROUTINE TEST NO. 611
```

CHECK APPLICABLE SAFETY DOCUMENT BEFORE STARTING WORK

```
                      *** WORK INSTRUCTION ***
CONTACT INSIDE ASSISTANT SHIFT CHARGE ENG. FOR PERMISSION TO CARRY OUT TESTS
INFORM UNIT OPERATOR THAT THE TEST IS TO BE CARRIED OUT.
===========================================================================
            JOB                  !NO!  !  TEST   !PASS=OK!  ! TESTED BY AND DATE
                                 !  !  ! RESULT  !FAIL=X !  !    REMARKS
---------------------------------!--!--!---------!--------!--!---------------------
***ON SHUTDOWN CHECK:-           !  !  !         !        !  !
MILL OUTPUT CONTROL DAMPER       !1 !  !         !        !1 !
TEMPERING AIR DAMPER             !2 !  !         !        !2 !
'COAL/AIR TEMP LOW' ALARM        !3 !  !         !        !3 !
'COAL/AIR TEMP HIGH' ALARM       !4 !  !         !        !4 !
'PRIMARY AIR FLOW LOW' ALARM     !5 !  !         !        !5 !
AND/OR                           !  !  !         !        !  !
'ROLLER SEAL AIR DIFF PRESS      !  !  !         !        !  !
LOW' ALARM                       !6 !  !         !        !6 !
PRIMARY AIR FLOW TRIP            !7 !H !         !        !7 !
AND/OR ESB TRIP                  !8 !H !         !        !8 !
AND/OR ROLLER SEAL AIR           !  !  !         !        !  !
PRESSURE LOW TRIP                !9 !H !         !        !9 !
                                 !  !  !         !        !  !
***ON TRIP CHECK OPERATION       !  !  !         !        !  !
OF THE FOLLOWING DAMPERS:-       !  !  !         !        !  !
HOT AIR SHUT OFF DAMPER          !  !  !         !        !  !
(HASOD)                          !10!  !         !        !10!
A MILL ISOLATING DAMPER          !11!  !         !        !11!
B MILL ISOLATING DAMPER          !12!  !         !        !12!
A PURGE & BOOST DAMPER           !13!  !         !        !13!
B PURGE & BOOST DAMPER           !14!  !         !        !14!
                                 !  !  !         !        !  !
***ON BOXING IN MILL CHECK       !  !  !         !        !  !
OPERATION OF:-                   !  !  !         !        !  !
A,B,C & D BURNER SECONDARY       !  !  !         !        !  !
AIR VANES                        !15!  !         !        !15!
A,B,C & D BURNER ISOLATING       !  !  !         !        !  !
DAMPERS                          !16!  !         !        !16!
'SEAL AIR DIFF PRESSURE LOW'     !  !  !         !        !  !
ALARM                            !17!  !         !        !17!
                                 !  !  !         !        !  !
                      *** TOOLS ***
HISTORY RECORDS-- ITEM NOS "H" SUFFIX  / TEST RESULTS  / FAIL COND. & REMARKS
```

FIG. 2.51 Format of a mill routine test

- Boiler sidewall CO measurements daily

- Oxygen and CO levels in flue
  gas before and after airheaters
  and after ID fans weekly

- Boiler gas-differential trends weekly

- PF sampling from each PF pipe weekly

- Sootblower test runs, etc. to suit
  application

Sootblower test runs are interesting in that, whilst some stations find the need to organise joint weekly test parties from all relevant disciplines, others do little. Much depends on the maintenance organisation set-up and on the importance of the sootblowers installed compared to the unit loading regime. It should be noted that the health of furnace sootblowers is easily ascertained on many plants by locally following around a normal sequence sootblow, during which the trained observer is easily able to note such common faults as poppet valves failing to open, incorrectly-set limits or nozzles burnt off. Further, the installation of an airflow or steamflow recorder opens up the possibility of remote diagnostic checks being made as and when convenient.

With regard to PF sampling, the test results can be very illustrative of deteriorating condtions for whilst the sampling is often not particularly scientific nor isokinetically based (this aimed at ensuring that a representative sample of coarse and fine particles is collected) if the procedure is simply to insert a collecting probe into the selected PF pipe for, say, three minutes, the results collected week by week will certainly show up any consistently coarse grinding and mills may be stopped for examination as a result.

5 Turbine plant — normal operation

5.1 Introduction

Unlike the PF-fired boiler, which requires continual on-load adjustments and operator care if optimum combustion conditions are to be maintained, the normal operation of the turbine is relatively straightforward in that, providing it is correctly designed and reasonably maintained, it will operate for lengthy periods with a minimum of attention. However, during start-up, when the machine is being run to speed, brought up to working temperature and loaded, or during emergency conditions, operations need to be carefully controlled if short or long term damage is to be avoided. Moreover, because of the heavy rotating masses involved (typically 240 tonne for a 500 MW turbine-generator rotor train) and the fine clearances

(often of the order of 1 mm) between rotating and stationary parts necessary to achieve good thermal efficiency, there is always the potential for serious problems to develop whenever the turbine is at speed. The fact that operations continue safely for many years, in the vast majority of cases, reflects much credit on designers, maintenance personnel and operators alike. The dynamic nature of turbine operation inevitably means that most of the aspects to be covered in this section are linked to the rotation of shafts within turbine cylinders. Whilst it is not proposed to discuss turbine theory, it may first be helpful to the reader to review the current scene and the factors that have led to the development of the present range of large machines (see Volume C and Chapter 7 of this volume).

It has long been known that for the greatest efficiency in converting the heat energy available in fuel into useful power, the thermodynamic heat cycle should employ high initial temperature and to develop power from this high-temperature steam, high initial pressure is also necessary. Further thermodynamic advantage can be gained if the steam, after partial expansion in the turbine, is returned to the boiler for reheating back to the initial temperature before it continues to expand through the turbine down to condenser pressure. Additional stages of reheating will yield further, albeit smaller, gains in thermal efficiency but may not lead to an overall reduction in costs, due to the increasing complexity and cost of the additional piping and reheater surfaces in the boiler. Most of the more recently installed fossil-fired plant in CEGB power stations has a power output ranging from 200–660 MW and inlet steam conditions of 158 bar. Superheated and reheated steam temperatures are 566°C for coal-fired plant and 538°C for oil-fired plant, on account of the increased risk of superheater and reheater metal corrosion that can occur with oil firing.

Advanced conditions like these can, in theory, be used on any sized unit, but the need for larger running clearances in high temperature turbines and the greater leakage losses present at high steam pressures mean that sets of high output are more efficient.

With the very largest units, the steam flow through the high pressure (HP) turbine in particular is so great that the relative importance of leakage effects and other losses, arising with the necessarily short and robust blades, is reduced to an acceptable level. Thus there has been a natural progression, in the interests of overall economy, to the 660 MW units now in use. The designed steam pressure of 158 bar is chosen for boiler reasons as, above this figure, the relative densities of steam and water are too close to ensure satisfactory natural circulation and recourse would be necessary to the added complexity and cost of either the assisted circulation boiler (for marginal pressure increase) or the supercritical, once-through boiler for more worthwhile pressure gain. Steam temperatures are limited for reasons of both boiler and turbine design; above 566°C, a rapid fall-off in the creep strength of ferritic steels occurs, together with increasing oxidisation and scaling, and a move towards austenitic steels is necessary. Unfortunately, austenitic steels have many disadvantages; they are difficult to produce and machine in large pieces and are therefore costly. Moreover, the coefficient of expansion of austenitic steels is some 50% greater than that of other steels and the thermal conductivity is appreciably lower so that, in conjunction with the high expansion coefficient, this gives rise to high thermal stresses. There is every inducement therefore to use austenitic steels as little as possible and this has been the guiding principle in the design of high pressure, high temperature turbines and boilers, and the reason for limiting steam temperatures to those currently in use.

Increase in output affects design chiefly in terms of the larger volume of steam and the consequent need for a larger flow area. The impact of this requirement is felt mainly at the low pressure end of the turbine, where specific volumes under vacuum conditions are high. In general, the higher the rotational speed the smaller, and therefore the cheaper, the turbine so there is a strong incentive to go for high speed machines. As the speed of rotation is fixed by the electrical frequency, the highest speed possible with a 50 Hz electrical system is 50 r/s (3000 r/min). If, with this speed, the largest blade-row annulus area that is practicable from a mechanical point of view leads to an excessively-high leaving loss, then two alternatives are possible:

- The speed can be kept the same and the number of blade rows exhausting to the condenser can be increased. This leads to the standard 3000 r/min tandem compound machine with multiple exhausts; for 200 MW units, three exhausts have been used, for 350–375 MW units, four exhausts have been used and for 500–660 MW units six exhausts are used, as shown in Fig 2.52 (a), (b) and (c).

- Drop the rotational speed to a sub-multiple of the frequency, the highest such speed being 25 r/s (for 50 Hz). This means that longer blades on a greater mean diameter could be used, thereby giving the same, or even a greater, total last-row annular area with a smaller number of exhausts and a smaller number of low pressure (LP) turbines and lower initial cost. The whole set could run at this speed but, as a higher speed is perfectly acceptable at the high pressure end, a convenient arrangement would be to run the HP and intermediate pressure (IP) turbines at the higher speed and the remainder at 1500 r/min. This arrangement forms the cross-compound set, as shown in Fig 2.52 (d).

Within the UK, the overall scenario has favoured the tandem-compound set, a typical 500 MW unit being shown in Fig 2.53. Whilst this machine has six exhausts, the development of longer last-row blades and a reassessment of the comparative costs between the initial capital outlay on LP turbines and the increase in running costs incurred in accepting an increased leaving loss from the LP blades (but allowing the use of four rather than six exhausts) has meant that a 500 MW unit having only four exhausts is now feasible. Referring to Fig 2.53, all rotors are solidly coupled to each other and to the generator rotor, the whole assembly being supported in spherical journal bearings designed to locate the rotors correctly within the bores of the cylinders, whilst allowing free axial differential movement between rotors and casings. The control of this differential expansion, especially during run-up, and the measurement and interpretation of bearing vibration and any eccentric rotation of the shafts form an important aspect of turbine operation. One thrust bearing locates the shafts axially and takes the combined thrust of all rotors, though this is not excessive as the IP and LP turbines are arranged for double, balanced flow. In this design, the single-flow HP turbine thrust is balanced by a dummy piston, indeed it is slightly overbalanced to ensure a positive thrust at all times towards the generator, thereby holding the machine steady on this thrust face.

In a turbine with such a range of steam conditions, the design of the rotating blading varies enormously from one end of the machine to the other. High efficiency and safety in operation are the prime requirements applicable to all stages. For this reason, every stage is arranged to have full steam admission at all times, other than when one main governor valve is closed at a time during a routine check on the freedom of movement of all valves, when a load reduction is enforced to limit blading stresses. Other than during valve testing, full admission is achieved by throttle, rather than nozzle, governing. Avoidance of partial admission, particularly in stages where the steam density is high, considerably reduces the danger of failure from overstressed blades.

The HP and IP rotors are of solid construction, Fig 2.54 (a) and (b). This has been recognised as good practice for many years and has completely superseded both the practice of mounting blades on discs shrunk on to a shaft and the use of hollow forgings, as previously supplied for turbines of different manufacture (Fig 2.55 (a) and (b)). Rotors machined from such hollow forgings were susceptible to distortion with increased running hours, a process which was aggravated by the manner in which the forgings were heat treated. There was also a tendency for a non-uniform heat flow into the stub-end forging across the bolting face. Operationally, problems were also caused by an accumulation of a small amount of

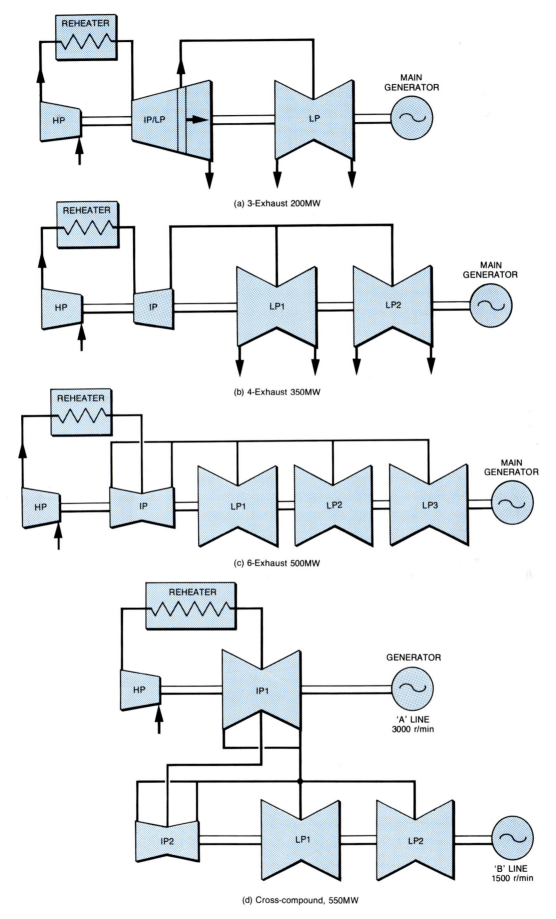

(a) 3-Exhaust 200MW

(b) 4-Exhaust 350MW

(c) 6-Exhaust 500MW

(d) Cross-compound, 550MW

FIG. 2.52 Steam-turbine cylinder arrangements (a), (b) and (c) show tandem-compound machines of increasing output whereas (d) shows a 550 MW cross-compound machine.

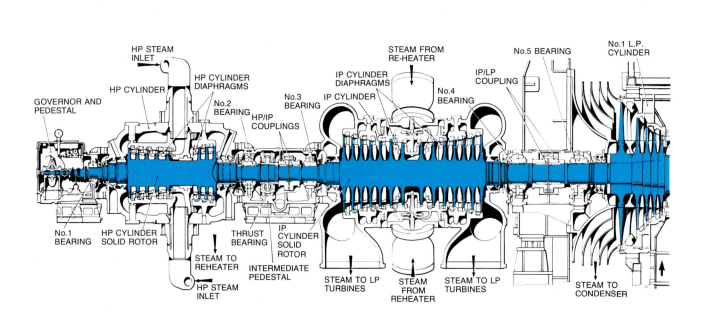

FIG. 2.53 Typical arrangement of 500 MW tandem-compound turbine

moisture in the hollow bore of the HP rotor when cool, as a result of leakage across the bolting face, though this problem was solved by drilling into the shaft at the region of the gland low pressure leak-off port and utilising the vacuum present at this point to keep the main bore evacuated (Fig 2.55 (a)). On another similarly designed machine, unexpected vibrations and eccentricity on cold and warm starts were found to be due to the accumulation of small quantities of oil in the hollow bore. As the rotor temperature increased, the oil evaporated, leaving the machine to run smoothly.

The operational price to pay for the general introduction of the solid rotor with its increased mass, however, has been a reduction in allowable loading rates, typically down from 25 to 10 MW/min, for a hot machine.

In low temperature LP turbines, where shaft diameters are large and blades long, the high centrifugal stresses developed in the discs made it difficult to obtain solid forgings of sufficient strength. Disc forgings in high quality material were easier to obtain and have resulted in the employment of large numbers of LP rotors of built-up construction in which the large discs are mounted (shrunk) onto a shaft forging. Careful design and machining is necessary to avoid 'stress raisers'. These are points of sudden change in section or areas of poor machining with bad tool scratches, at which high stress concentrations are centred, leading to crack propagation. Since the introduction of the present generation of

large machines, many failures of LP turbine shafts due to cracking have occurred throughout the world, especially those of built-up construction though, so far as is known, none have failed in a catastrophic way. Operationally, it is of vital importance to be able to detect the onset and the development of a crack by vibration monitoring: and this is discussed later. As a result of these difficulties, there has been a trend away from shrunk-on discs and back to solid forgings, despite the extra cost. With regard to HP turbine casings, simplification is usually achieved by arranging for the point at which steam is extracted for reheating to occur at the HP turbine exhaust. This avoids the need to return the reheated steam to a different point in the same turbine cylinder from which it left, eliminating sharp temperature gradients within the one casing. As casings are invariably split along the horizontal centre line and bolted together, problems can arise with the fastening arrangements due to low allowable bolt-stresses on account of the high prevailing temperatures. The problem is greatly eased by the introduction of the double-casing design of Fig 2.54 (a), the primary object being to break down the full steam pressure into two steps, thus enabling thinner casings and simpler bolting arrangements to be used. This also eases temperature gradients at start-up, allowing a faster run to speed and subsequent loading. Despite the double-casing construction, stresses on outer cylinder flange-bolts are still high and facilities for steam heating of the bolts and flanges are usually provided; during start-up, the

LP2 TURBINE   LP3 TURBINE

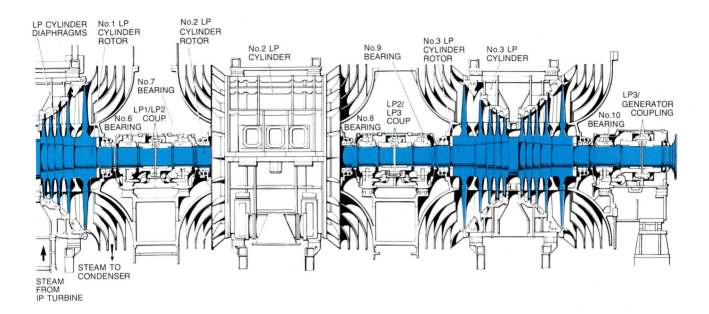

FIG. 2.53 Typical arrangement of 500 MW tandem-compound turbine (*cont'd*)

use of this system equalises the heating rates between the relatively thin-walled cylinders and heavy flanges, thereby minimising cylinder distortion and flange bolt stresses. So successful was the double-casing concept that it has been developed, in the later 660 MW machines, into the triple-casing design as illustrated in Fig 2.56. Naturally these machines also incorporate the very best in current practice, e.g., full steam admission at all times even when valve testing (by virtue of a full 360° steam admission belt and interconnected piping, see Fig 2.57) and the HP, IP and LP rotors are all of solid construction.

Once pressures fall to an acceptable level, a return to a single casing design is possible; in the design illustrated in Fig 2.53, this occurs part-way down the IP turbine (Fig 2.54 (b)).

In the days of low steam pressure and temperature, it was quite common to house steam governing valves in chests built as an integral part of the turbine casings, thus giving a very compact design with small volumes of steam between the control valves and the first stage nozzles. Whilst this eased the problem of speed rise on loss of electrical load, it considerably increased the complexity and possibility of distortion of the turbine casing and it is now usual to provide separate, remote steam chests, so that turbine cylinders can be kept as simple and symmetrical as possible. Figure 2.57 shows a typical arrangement. Note that the steam chests contain not only the governing valves, used to control the load on the machine, but also an arrangement of emer-

gency stop valves which close automatically when the turbine protective system operates. Whilst the steam chests illustrated control the steam admission to the HP turbine, valves are also provided at the IP turbine inlet on account of the large volume of steam available in the reheater circuit to precipitate an overspeed condition. All valves are invariably controlled by a hydraulic system and closed by the action of springs. For many years the source of hydraulic pressure was turbine lubricating oil (being readily available) at a pressure, typically, of 8 bar. Pressure was restricted to this low level because of the ever-present fire risk and this resulted in the need for massive valve-relays in order to open the valves against their springs. For these reasons, the newer 500/660 MW units have valve-relay systems operated by a fire-resistant fluid (FRF), usually of the phosphate-ester type. Because of the significant reduction in fire risk, hydraulic pressures can be raised to 100 bar, resulting in small, compact valve-relays. Operationally, the FRF must be kept extremely clean if relay skirts are not to be damaged by pitting, either from dirt or by electrokinetic streaming effects. Corrosion by phosphoric acid, produced if free water is present in the fluid, is also a possibility.

To minimise the effects of uneven heating of the LP casings due to the churning effects of the long blades in the relatively stagnant flow arising at low loads, a spray cooling system is installed in each exhaust duct. Uneven heating of the casings could result in shaft misalignment.

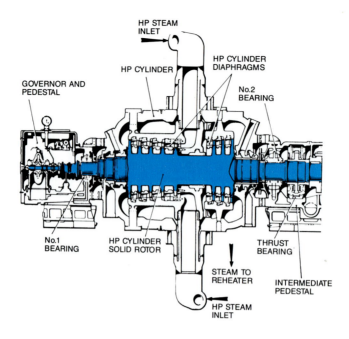

(a) HP Turbine

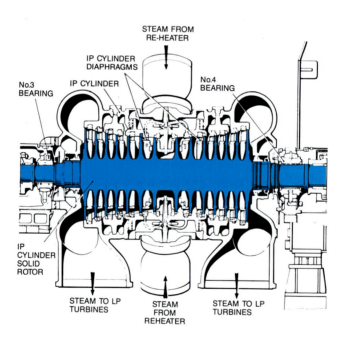

(b) IP Turbine

FIG. 2.54 Enlarged views of the (a) HP and (b) IP turbines of the machine illustrated in Fig 2.53
The rotors are of solid construction, i.e., they are machined from one solid forging. This method of manufacture represents
current best practice.

Where turbine shafts penetrate the casings, steam-sealed glands are provided to prevent air ingress into the LP turbines at all loads and into HP and IP turbines at low loads. These glands also limit the egress of steam from HP and IP turbine shaft penetrations at higher loads. In operation, gland-steam systems often require careful setting-up to prevent an excessive moisture-in-lubricating-oil content on the one hand or oil-in-condensate content on the other.

Finally, condensers and air extraction arrangements are so important as to deserve a section to themselves (see Section 5.5 of this chapter).

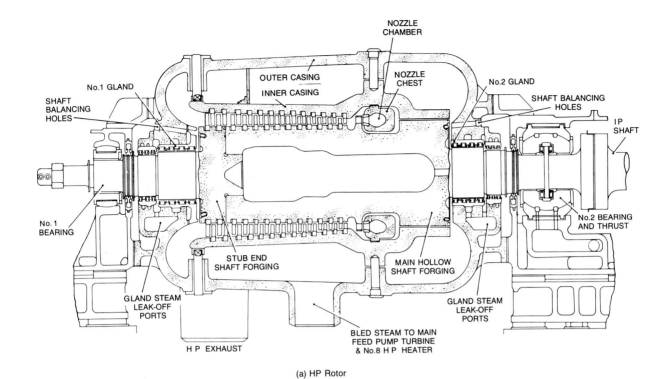

(a) HP Rotor

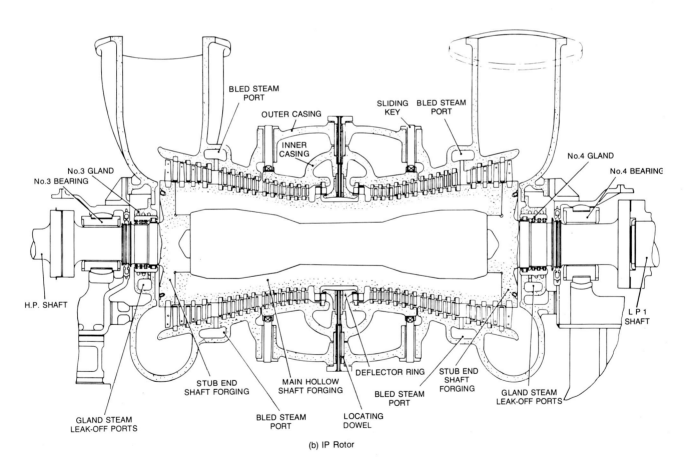

(b) IP Rotor

FIG. 2.55 'Hollow' HP and IP rotors

These have gradually been replaced by the solid type. Such hollow, built-up rotors are no longer favoured, being susceptible to increasing vibration levels in operation, brought on by a slow but irreversible distortion of the main hollow forging as a result of thermal cycling.

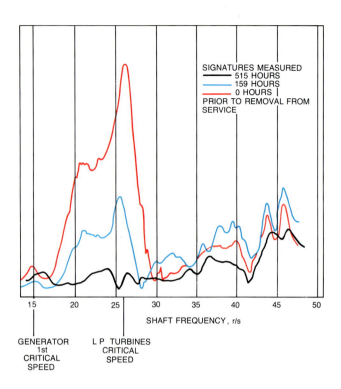

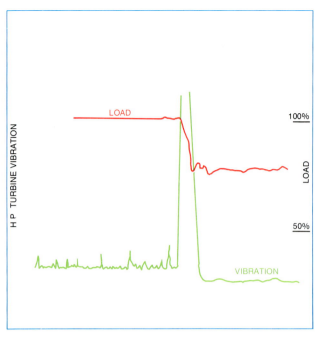

FIG. 2.59 Illustrating steam whirl

FIG. 2.58 Illustrating the effect of a developing crack in a LP turbine rotor on the LP turbine vibration recording

to reduce load and/or speed in order to avoid a high speed, sustained rub. Such a rub would cause intense heating of the shaft which may be so severe as to cause plastic deformation on cooling, resulting in a permanently bent shaft. Once the machine is loaded and heat-soaked, vibrations should settle down to a repeatable pattern between overhauls, though regular monitoring is essential in order to observe any worsening trends.

When operating a new or overhauled machine, however, it is necessary to be aware of the possibility of vibration caused by *steam- or oil-whirl*. Load restrictions due to steam-excited rotor vibrations have become a worldwide problem on some types of highly-rated machine. The vibration is usually confined to the HP turbine and has a characteristic sub-synchronous frequency of about half the fundamental. Above some critical output level the amplitude can suddenly increase, necessitating a rapid load reduction to safeguard the machine. Figure 2.59 illustrates the effect. There is little that operating staff can do as the problem is fundamentally one of design.

A more common sub-synchronous vibration is that of oil-whirl set up in one or more turbine bearings. The problem is usually associated with oversized, lightly-loaded or misaligned bearings and whilst off-load attention may be necessary in the worst cases, when vibration levels can reach unacceptable values, operating staff can often play a useful role in reducing vibration levels by operational adjustments to ensure continued

running. For example, Fig 2.60 (a) shows the onset of oil-whirl on a 500 MW turbine, centred in this case on bearing No 3 (IP turbine, HP end); whilst vibration levels were not severe enough to force an outage, they were undesirable for the medium-term integrity of the affected bearings. It was found that the vibration could be very much reduced by increasing the bearing oil pressure from 1 to 1.3 bar and prevented altogether by maintaining a reasonably-high reheat steam temperature. Subsequent scientific investigation showed bearing No 3 to be low compared with adjacent bearing No 2 and hence lightly loaded. Furthermore, a fall in reheat steam temperature by 15°C contracted the IP turbine casing and tilted the common pedestal bearings Nos 2 and 3 slightly, such as to unload bearing No 3 still further and initiate oil-whirl. As the oil-whirl causes precession of the shaft axis about its proper centre, in the basic form of an ellipse, the adjacent shaft eccentricity detectors also responded to the shaft movement (Fig 2.60 (b)).

To complete this section on vibration, any sudden increase in LP turbine vibration coinciding with an increase in condensate conductivity, confirming a condenser tube-leak, might indicate the loss of part of an LP blade or blade tip erosion shield.

(a) Shows the changes in vibration levels brought about by the onset of oil whirl. All bearings on the five cylinder turbine-generator are continuously monitored for vibration by velocity transducers which measure the peak-to-peak movement of the bearing keeps in μm. Prior to the start of oil-whirl, the vibration levels of the machine are low and stable. The onset of oil-whirl, centred on bearing No 3, affects not only that bearing badly but also associated bearing No 2. Other bearings are hardly affected.

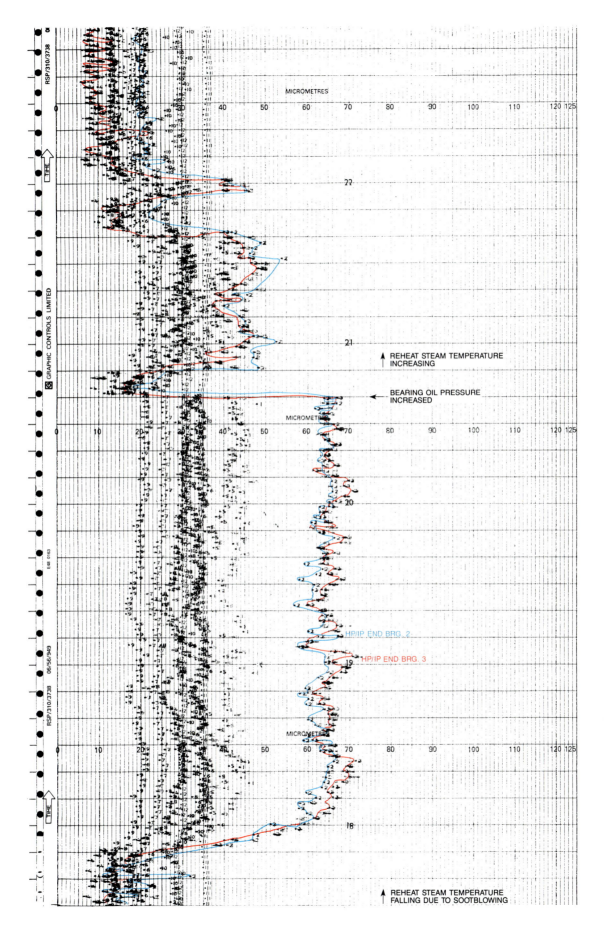

FIG. 2.60 (a) Oil-whirl on a 500 MW machine

(b) Shows the HP and IP turbine shaft-eccentricity recordings for the same whirling event. It is important to realise that the increasing levels coincidental with the start of the oil-whirl do not arise from a bent shaft, but arise from a movement of the shaft within the confines of the bearings.

5.2.2 Bearing temperatures

The oil wedge created in each journal or thrust bearing pad by the combined action of the oil supplied to the bearing and the rotation of the shaft, ensures that the shaft runs freely and safely for long periods without any wear taking place on the white-metal bearing surface itself. Whilst relatively little oil is needed for lubrication purposes, i.e., the creation of the continuous oil wedge, heating effects both from frictional forces within the bearing and as a result of the conduction of heat along turbine shafts demand that additional oil is supplied for cooling. It is the designer's function to ensure that sufficient oil is provided for both purposes.

Temperature limits vary between manufacturers, so it is impossible to be specific, but an oil inlet temperature much below 35°C is undesirable as the increased viscosity of the oil may make the formation of a continuous oil wedge difficult at speed. On the other hand, 45°C represents a reasonable upper limit to avoid an excessively-high bearing metal temperature. The monitoring of the actual bearing metal temperature is interesting; one manufacturer measures this directly by thermocouples embedded in the white metal of the bearing, whereas another measures the oil temperature leaving the bearing and also provides 'inlet to outlet oil temperature-differential high' and 'low flow' alarms. Studies between 500 MW units of different manufacture show surprisingly different limits, for example, at one plant employing direct white-metal temperature monitoring a maximum of 75°C is specified, whereas on another plant operation with an oil outlet temperature of up to 85°C is allowed. Whatever the instrumentation provided and the limits set, it is unusual for readings to vary much between survey overhauls, though the spread of temperature rises across the bearings of a particular machine can be higher than desirable if bearing alignments are incorrect. Indeed it is not unusual to find that one particular bearing runs significantly hotter than the rest and often near the maximum limit. At one station on such a bearing, the oil outlet temperature was reduced by 3°C by applying air hoses to the bearing keep. Whilst it was clear that this was a genuine reduction (as the monitoring was by a sensor measuring the temperature of oil in a bath underneath the main bearing), the mechanism producing it, which was in addition to previous improvements made by increasing the bearing oil pressure, is unclear. Nevertheless, it allowed the machine to continue in operation until its survey overhaul some months later.

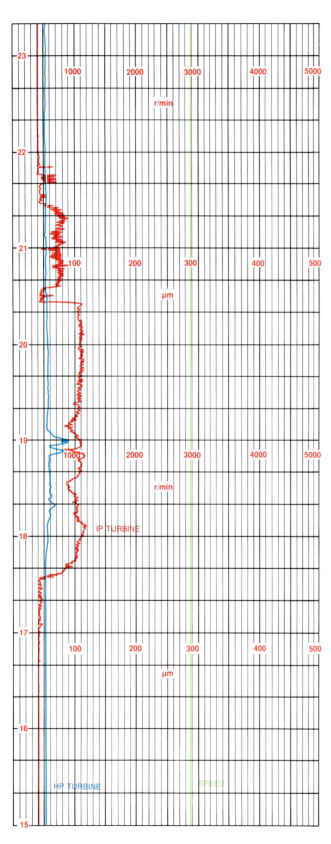

FIG. 2.60 (b) Oil-whirl on a 500 MW machine

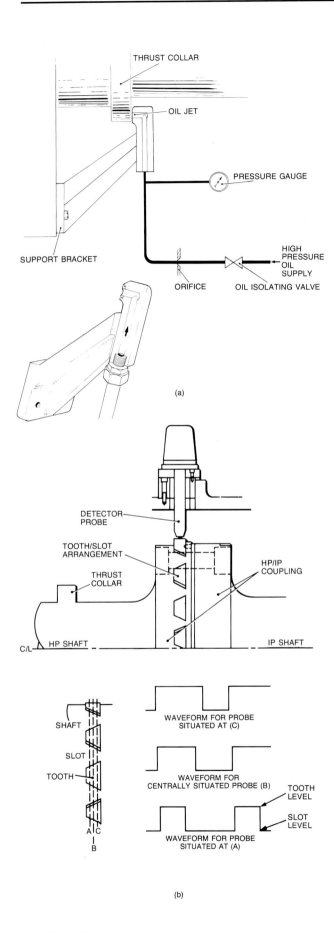

FIG. 2.61 Two methods of measuring thrust-bearing wear and/or shaft position

(a) Shows a system based on changes in hydraulic pressure. The ease with which the oil can escape from the oil jet depends on the thrust collar position. In turn, this affects the oil pressure. The orifice is sized to give a useful variation in oil pressure for the expected thrust collar movement.

(b) Shows an interesting development in which the electrical probe scans a tooth/slot arrangement to give a varying square waveform dependent on the proportion of tooth-to-slot (see inset).

Wear on thrust-bearing pads results in a change in position of not only the shaft thrust-collar but also of all of the turbine rotors in relation to their casings and in the extreme would result in a complete loss of axial clearance at some point or points within the turbines. Thrust-collar position and thrust-pad wear have been measured accurately for many years using 'proximity probes'; two such ways of doing this are illustrated in Fig 2.61 (a) and (b). While instrumentation of this type provides a reliable indication of the amount of thrust wear which has actually taken place, it gives no short-term warning of changes in operating conditions and it is usual to supplement it with thermocouples embedded in the white metal of selected thrust-pads, often on both 'running' and 'surge' pads. This arrangement accurately measures white-metal temperature to provide an early indication of abnormal operation.

5.2.3 Shaft eccentricity

In turbine installations, shaft eccentricity is defined as the out-of-centre excursion of the axis of rotation of the shaft and is measured as the diameter of a locus traced by the centre of the shaft at the point of measurement. Unfortunately, this point is, for practical reasons, external to the turbine casing and usually near to the bearing itself, so that the actual eccentricity along the shaft concerned is unknown and has to be inferred from the movement, or wobble, of the shaft at or near the bearing. On large turbines currently in use within the CEGB, measurement of HP and IP shaft eccentricity is universal and, on some machines, LP turbine shaft eccentricity is also measured. Whilst a detector mounted on the horizontal centre line of the shaft will give the highest reading (as the eccentricities present at all other planes are affected and reduced by the effects of gravity), it is now common practice to monitor for eccentricity in the vertical plane as well so that a vector summation of both traces can be displayed on a VDU to show the locus of the shaft movement. Irregularities, such as oil-whirl, can be readily seen by these means. Tolerable levels of eccentricity vary between manufacturers, as shown in Table 2.8. The differences may reflect the actual position of the eccentricity detector in relation to the bearings and the length and design of the turbine rotor. The figures shown in the table give the reader some idea

TABLE 2.8
Typical large HP/IP turbine eccentricity limits

Manufacturer	A	B	Remarks
Hold speed	76	127	All eccentricities measured horizontally in μm of movement (peak-to-peak)
Run-back	127	203	
Trip machine	254	381	

of current maximum values and must not be used for any particular machine. Incidentally normal running values can be as low as 30 μm. If high eccentricity levels are measured in practice, but result in no associated bearing vibration, it is possible that they are not genuine eccentricities at all. For example, Fig 2.62 shows the eccentricity traces taken from a 500 MW turbine during governor valve testing, a

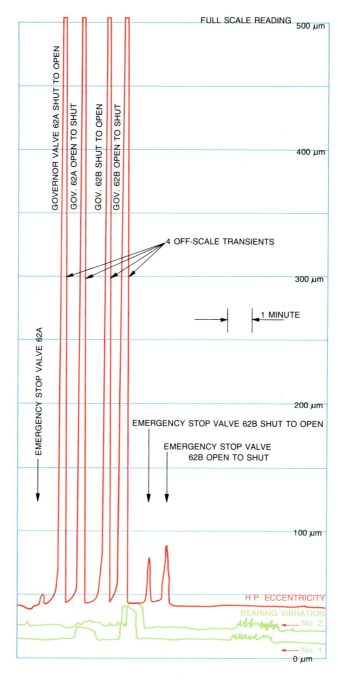

FIG. 2.62 Apparent HP turbine rotor eccentricity excursions during valve testing
Investigations showed that these were not true eccentricity levels at all. The eccentricity measuring equipment, as originally installed, was found to be too sensitive to low frequency shaft movement, and was interpreting a change in shaft position within the adjacent bearing as an eccentricity. It has since been modified and such large, transient peaks are a thing of the past.

weekly exercise entailing shutting one valve at a time and designed to prove, on base-load machines, that the valves are free to close as required. Investigations revealed that the rapid transitory eccentricity peaks to greater than 500 μm were spurious and due to a movement of the shaft within the bearing, rather than the result of eccentricity. The movement was initiated by a change of thrust onto the turbine blades as a result, in this design of machine, of shutting off the steam flow to a sector of the admission annulus.

Air gauges, arranged to monitor shaft (and gland) position, can also be effective in detecting changes in shaft running position, but are usually extremely difficult to maintain should a probe fault occur on account of their close proximity to the gland and rotating shaft. Several stations have found it impossible to maintain them in reliable working order and have dismantled the equipment.

5.2.4 Expansions

The wide variations in temperature that occur throughout the turbine during start-up, shutdown and as a result of load changes give rise to differential expansions and contractions (both transverse and axial) between turbine components, whether stationary or rotating. Special arrangements are needed to allow these components to move freely relative to one another, whilst maintaining correct alignment and preventing excessive stress from arising in any parts of the machine. Although the methods by which the designer maintains correct axial alignment between bearings and accommodates transverse expansions are of limited concern to operating staff, the same is not true of axial expansions, both between rotors and casings and between casings and pedestals, and a clear understanding of the mechanisms at work is essential.

The complete turbine is anchored to the foundations at one point only: in Fig 2.63, this is at the IP/LP1 turbine pedestal, but can be elsewhere. All other contacts between turbine casings and foundations are through bearing pedestals which are arranged to slide longitudinally with the casings. Thus, in the example shown, the total turbine casing expansions are cumulative in both directions from the anchor point, reaching their maxima at No 1 bearing pedestal and the generator end of LP3 turbine, though the latter is not large and is not normally measured. Expansions from cold to hot at the No 1 bearing pedestal are, however, as much as 30–50 mm and it is usual to measure these on both sides of the axis of the machine in order to detect any 'crabbing' movement of the machine in its expansion and contraction, as well as the smoothness (or otherwise) of the

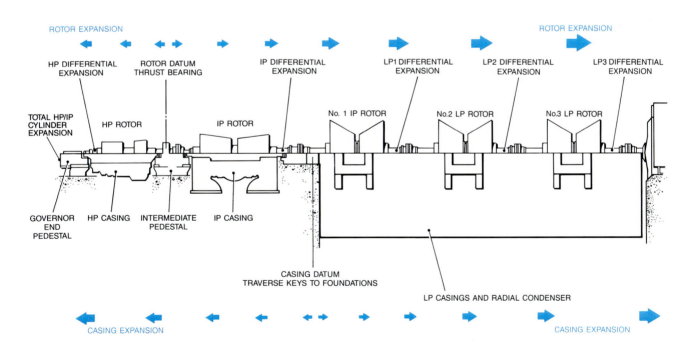

FIG. 2.63 500 MW turbine expansion arrangements and position of measuring equipment
The differential expansion detectors are mounted at the ends of the cylinders remote from the thrust bearing, which fixes the axial position of the rotor train within the cylinders. Any differential expansion occurring between the HP and IP turbine rotors and their cylinders is indicated directly on the associated instrument. The differential expansion between any LP turbine rotor and its cylinder is complicated by the movement in the preceding turbines, the worst case being the differential expansion of LP3 turbine: this is the summation of those of the IP, LP1 and LP2 turbines, in addition to the actual differential expansion occurring within the LP3 turbine itself.
Nonetheless, the total movement measured on LP3 detector is real enough and explains the need for increased clearances between the fixed and moving parts of LP3 turbine to values that will accommodate this total potential movement.

movement. Crabbing results in misalignment and rough running, as well as in unseen stresses, and usually results from a lack of grease on the sliding feet. Though lubrication routines may be completed regularly, the injection of grease into remotely-located nipples is one thing; whether or not the grease reaches the intended destination is quite another. In recognising such practical difficulties, manufacturers now offer sliding feet in a material that requires no lubrication and these can be fitted retrospectively to existing machines should the need arise.

In order to reduce expansion differentials between rotors and casings to a minimum, the thrust bearing, which locates the position of the rotor assembly, is usually placed between the HP and IP turbines, rotor expansions taking place in both directions from this point and contractions occurring towards it. As the clearances between fixed and moving parts of the machine are limited by design, in order to ensure efficient operation, it is necessary to be aware at all times of the relative positions between rotors and casings at points along the turbine length (Fig 2.63), so that limits of movement can be set for safe operation. Figures 2.64 (a), (b) and (c) illustrate three methods by which these differential expansions are measured. Method (b) must be used with caution: although it increases the range over which the relative movements can be measured compared with method (a), the double-cone arrangement can be heated significantly, if used at the hot end of the machine, as a result of the conduction of heat along the shaft from within the casing. Further heating can occur due to windage effects of the adjacent coupling. Both effects expand the cones radially, reducing the detector air gaps. Whilst this is of no consequence if the gaps are all equal, i.e., if the double cone is centralised about the detectors, this is not true when the system is unbalanced and the change in air gaps is seen as a change in differential (longitudinal) movement which, of course, it is not. The other two methods of measurement appear to be free from this effect.

Whichever system of measurement is specified, it is normal to adjust the detector outputs to read zero differential movement when both casings and rotors are completely cold and when the rotors are jacked onto the running pads of the thrust bearing. During a cold start, HP and IP rotors expand faster than their casings (being lighter and completely surrounded by the heating steam) and the differential expansion indication will increase towards a maximum 'rotor longer than casing' value. Whether this is termed a negative or positive differential depends on manufacturer's convention. As the heavier casings reach their equilibrium temperature, this increasing differential trend will stabilise and reverse until a normal operating level is reached. On the other hand, during a hot start, when the metal temperature can initially be higher than the steam inlet temperature due to throttling effects or following a sudden drop in boiler

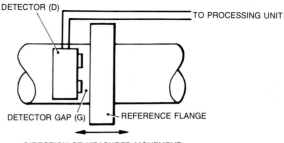

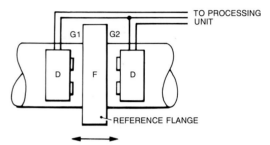

(a) Using 1 or 2 detectors (D)

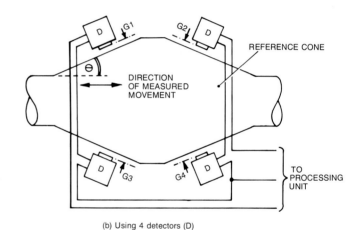

(b) Using 4 detectors (D)

(c)

FIG. 2.64 Three methods of measuring differential expansion between fixed and moving components
The single detector arrangement illustrated in (a) is the electrical equivalent of the oil jet of Fig 2.61 (a). The output from the double detector arrangement of (a) can be fed to an electrical bridge circuit, which is adjusted to be in balance when the reference flange is positioned centrally between the detectors. Movement of the flange one way or the other will unbalance the bridge circuit

to a greater or lesser extent, the resulting unbalanced current flow representing the extent of shaft movement. Method (b) increases the range of the method over (a). Method (c) is identical to that illustrated in Fig 2.61 (b) and is used to infer thrust bearing wear.

steam temperature, the steam will have a cooling effect on the rotors and a reversed differential will occur. This is usually the more dangerous condition of the two as the permissible contracting differential is limited, at least in the HP turbine, on account of the tight clearances between diaphragms or stationary blades and the inlet edges of the moving blades necessary for good efficiency.

In the LP turbines, the cylinders are light and usually of fabricated construction and expand or contract quicker than the rotors, i.e., the opposite effects are at work to those in the HP and IP turbines. This has operating implications at start-up, when it may be necessary to apply gland-steam and raise vacuum earlier than is strictly necessary, in order to induce some rotor expansion prior to run-up. This will negate to some degree the subsequent contracting ('rotor smaller than casing') differentials caused by the rapidly-expanding casings relative to the rotors. As all of the rotors are solidly coupled, the total movement at the outboard end of LP3 turbine is the cumulative effect of the differential movements in the IP, LP1 and LP2 turbines, in addition to that arising within LP3 turbine itself, and can amount to 8 mm in normal operation from cold to hot. The differential movement within the HP turbine is much less, unless abnormal heating or cooling effects are present.

One final but extremely important phenomenon that affects the differential expansions, on large tandem-compound machines in particular, during speed changes is the *Poisson effect*. As the speed of a large diameter shaft increases so do the centrifugal forces, causing the shaft to grow slightly in diameter at the expense of its length. So, during a run to speed, each LP turbine rotor contracts. When these contractions are summed and added to the small one occurring in the IP rotor, a total contraction occurs at the outboard gland of LP3 turbine of the order of 5 mm. A similar total expansion occurs on a turbine trip or run-down. Operating staff need to ensure, therefore, that the running differentials are at all times limited so that the expansions that will occur on run-down can be accommodated without exceeding the absolute (expanded) limits. Similarly, during start-up and before speed can be raised to full value, differential expansions must have progressed to the point that the reversals that will take place due to the Poisson effect can be accommodated without exceeding the absolute (contracted) limits. The movements are well illustrated in the chart recordings taken from a 500 MW turbine and shown in Fig 2.65.

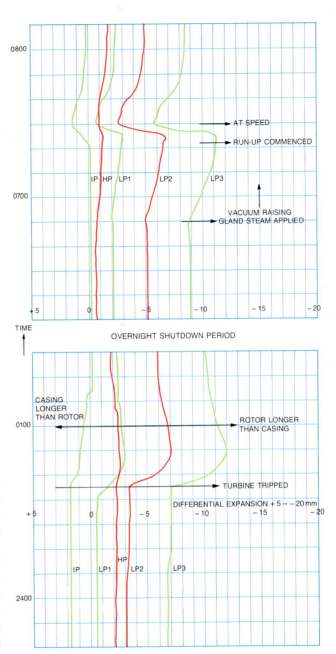

FIG. 2.65 The Poisson effect
The recorder charts of differential expansion are taken from a 500 MW turbine during an overnight shutdown and show the rotor expansions on run-down and contractions on run-up as a result, predominantly, of the Poisson effect. Note that the cumulative movements in the preceding turbines affect LP3 badly, to the order of 5 mm.

5.3 Oil systems

5.3.1 Integrity

The integrity of the turbine lubricating oil system is of vital importance to the safety not only of the main turbine and (usually) the generator gas seals but, in the more recent designs, also of the main boiler feed pump turbine unit; all of these share a common oil tank at least. Operating staff need to be fully familiar with the systems installed on their own plant and of any potential weaknesses present. For example, not all large turbines are protected by an automatic trip on loss of lubricating oil pressure, presumably on account of the substantial provisions installed to prevent such an occurrence. Nonetheless, reasons for a disastrous failure in bearing oil pressure can be envisaged, not the least being an inadvertent low oil level in the main oil tank. The following aspects, drawn from experience, illustrate the manner in which oil systems should be critically reviewed by operating staff from time to time, to ensure that the designed integrity is still present.

First, the operating point of the low oil level alarm should be such that, except for a serious loss of oil, time exists to correct the situation before the possibility of pump cavitation occurs. It is sound logic to employ a second float operating at 50 mm below the low level alarm float and arranged to initiate an alarm 'lube oil level extra low' or similar; on receipt of both alarms in rapid succession, no operator could then ignore the gravity of the situation. At least one level indicator should be of a simple locally-mounted type, for example, a float with an indicator rod engraved with high, low and normal working levels. Whilst remote electrically-operated level indicators may appear to be more modern, a faulty instrument can cause confusion and result in a delay in ascertaining the true oil level in the tank, unless an additional local gauge is provided. Given a choice, the simple local gauge is by far the most practical though its range should be checked to ensure that it continues to function for some distance below the low level alarm point in order that any continuing fall in oil level can be confirmed.

Secondly, all oil system drains that are not routed back into the main reservoir or into an alarmed drains tank are points from which potential serious oil loss can occur. On one design, presumably for aesthetic reasons, oil-filter drains were routed into the turbine house sub-basement and as a result were out of sight and out of mind. At least one generator hydrogen-seal explosion and fire owes its origins to a serious loss of oil from the main oil tank (which continued unnoticed) as a result of such drains being inadvertently left open following a filter element-changing operation. Fortunately, the machine was shut down at the time.

Finally any means provided whereby oil is drawn off the main oil tank and transported to a common, or interconnected, conditioning unit for purification should be critically examined to ensure that, so far as is practicable, inadvertent loss of oil from one system to another cannot occur.

5.3.2 Water in oil

This can be a persistent operational problem, especial-

ly if turbine gland systems are poorly designed or operated, and has resulted in some machines being fitted with bearing air-blowers. Low pressure air is fed to the most vulnerable bearings, usually those on the HP and IP turbines, and seals the outer bearing annulus to prevent the ingress of steam from the adjacent gland.

This can be an effective and low cost way of preventing moisture in oil, though it treats the condition rather than eliminating the cause. Indeed, much money has been allocated in the past to the supply of larger purifiers or oil conditioners to treat contaminated oil instead of preventing contamination in the first place. Unfortunately, this recurring theme is not confined to this aspect of power station operation.

Where centrifugal oil-purifiers are installed, these can be very effective. The oil is heated as necessary by self-contained heaters to ensure a correct viscosity for purification, which can continue, whether the turbine is on-load (and the oil is hot), or not. Because of the moving parts and the difficulty experienced by some stations in maintaining prime on the oil pumps (especially if the purifier is installed above the oil level in the main tank to avoid any inadvertent loss of oil due to flooding) static, coalescer-type oil conditioners have found increasing use in recent years. Not all are provided with heaters and their ability to remove water from cool lubricating oil, as occurs once the turbine is off-load, is limited.

Furthermore, the free water/oil interfaces often present in these units form ideal breeding grounds for algae and bacteria which, once established, are difficult to eliminate and can spread to and multiply in other parts of the oil system where conditions are suitable, such as in hydrogen seals, where the resulting sludges and slimes can seriously affect the lubrication of the seal faces.

Possibly the most undesirable feature of excessive water in lubricating oil is that this water can be released in the generator shaft seals to produce unacceptable 'moisture in hydrogen' levels within the generator. In turn, the moisture condenses out on those parts of the generator which fall to below dew-point temperature, particularly when the unit is off-load, leading to fault conditions when the machine is re-energised. This is discussed further in Chapter 3.

Despite the care taken by designers and operators to prevent water ingress into lubricating oil (and seal oil) systems, some moisture will still be entrained. Water can produce rust and contaminate the oil to promote consequential damage throughout the turbine-generator. Any oil-operated control relays are especially vulnerable. Although the vast majority of entrained water is removed at the main oil reservoir, with its associated purification and vapour extraction systems, a small proportion circulates with the oil. This moisture tends to precipitate in regions of low velocity and gravitates to the lowest possible point. Should it be allowed to collect, then rusting will occur so it is important that all possible collection points are identified. One such region is found in oil-filter compartments mounted on the side of the oil coolers. These compartments are invariably provided with a drainage facility which should be operated regularly to drain off any collected water.

5.3.3 Oil coolers

Three 50% or four 33% oil coolers are commonly provided allowing, in theory at least, one cooler at a time to be taken out of service for cleaning. On return to the standby mode, it is desirable that the oil side is pressurised before the water side, so preventing any possibility of leakage of water into the oil due to leaking tubes. Once the cooler is on-line, this condition is automatically fulfilled as the oil pressure is higher than that of the circulating water (CW). Incidentally, it is bad practice to leave any cooler isolated but charged on the CW side for extended periods, as the stagnant water conditions increase the likelihood of waterside tube corrosion.

In practice, during the summer months on inland tower-cooled stations, it is fairly common to find that all the oil coolers are needed in service more or less continuously and the ability to remove a cooler from service for a thorough CW-side clean is limited. Under these conditions, unless the deterioration in heat transfer is predominantly due to hard scaling, a quick and effective tube clean can be achieved by *bubble cleaning* without unboxing the CW-side covers. Bubble cleaning is achieved by introducing a compressed air supply to the inlet (bottom) waterbox via a suitable connection, with the cooler out of service, the CW inlet and outlet valves closed, and the outlet (top) waterbox vent (which must be at least 50 mm in diameter) opened. The CW side is left full of water. When air is applied, the upward rush of air bubbles through the tubes quickly and effectively removes loose sludges and scales, enabling the cooler to be returned quickly to service without having to remove and replace CW-side waterbox covers. Gross scaling is a different proposition and is best dealt with by preventing scaling conditions from occurring in the first place, see Section 5.5 of this chapter.

Should the heat transfer capability of an oil cooler deteriorate to a marked extent, it is usually because of contamination on the oil side of the tubes rather than on the water side. Realising this, a novel method of servicing a fouled cooler was introduced at one station. Neat 'Polyclense', a proprietary paintbrush cleaner, was circulated through the assembled cooler for several hours using a pneumatically-operated diaphragm pump, followed by several hours of intermittent flushing with water (penetration to the centre of the nest was facilitated by replacing a central tube with a sparge pipe). On return to service, a substantial improvement in the temperature drop across the cooler was apparent. While cleaning by this method may not restore the cooler to an 'as new' condition and care has to be taken to flush the oil side thoroughly before return to service, it certainly saves time. Since the chemicals in paintbrush cleaners do not attack rubber or any of the materials in the cooler, the method may well be of more general interest.

Procedures like this and bubble cleaning must be very clearly thought out beforehand, to avoid danger to personnel or harm to plant. The operations must be carried out strictly in accordance with well documented work instructions.

5.4 Gland-steam systems

Glands are provided on HP and IP turbines to reduce the escape of steam to a minimum when the turbines are on high load and to prevent the ingress of air into the turbines during start-up or low-load operation, when the pressure in the cylinders is less than atmospheric. LP turbine glands need only seal against air ingress as the cylinders run under vacuum conditions at all loads but, whatever the application, steam-packed, labyrinth-seal-type glands are invariably employed (Fig 2.66). The stationary part is divided into two or more sectors, each of which is spring-loaded in an annular groove. If a rub occurs between stationary and moving parts of the gland during operation, the contact pressure will be relieved and little heating will occur. This effectively prevents high speed sustained rubs from causing heating of the shaft such that permanent bending results. Figure 2.66 shows that the gland provides a series of very fine annular clearances in the gap between the knife-edged fins and the shaft, through which the steam is expanded (throttled) and its pressure reduced step by step. In expanding through each clearance, the steam therefore develops kinetic energy at the expense of its enthalpy, but in the space beyond the gap the steam is brought to rest again and the kinetic energy is re-absorbed, restoring the enthalpy to its original value. The result is a loss of pressure at constant enthalpy and almost constant temperature, and a rise in entropy. Due to the high pressures present in HP and IP turbines and the need to design the glands for the dual roles of preventing both steam egress and air ingress, it is usual to employ relatively long glands divided into sections, separated by ports or pockets (Fig 2.66). By this means, if the gland system has been sensibly designed and is in sound mechanical condition, it is possible for the outer section of the gland to perform the function of preventing air ingress at all loads relatively independently of the inner section, which can be designed exclusively to prevent excessive steam egress. Such an arrangement allows a more gradual temperature gradient along the shaft and eases the design of the gland-steam sealing system as conditions at the sealing pocket will vary less from no-load to full-load. It is also usual to put the higher pressure leak-off flows to good use, either to enhance the bled-steam flow to an appro-

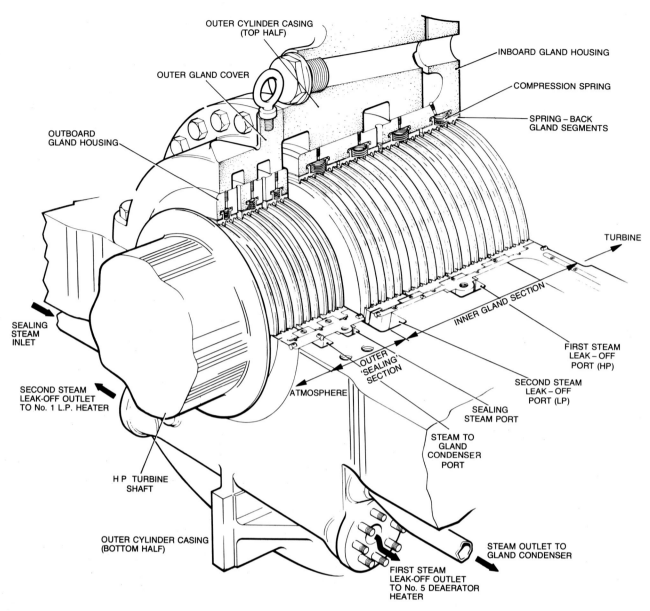

FIG. 2.66 Typical HP turbine gland

priate feedheater or to use one, at higher turbine loads, to seal the LP turbine glands, rather than continually use the live steam supply; this results in a small saving in thermal efficiency. It is also usual to split the gland-sealing system into two parts, serving the HP/IP and LP turbines, respectively. This arrangement allows the different temperature conditions required by the HP/IP turbine shafts, compared with those of the LP turbines, to be met. Operationally, it is extremely important to maintain the gland-steam sealing temperatures specified, as excessively high or fluctuating levels lead to an increased risk of cracks developing in shafts. Typical levels are 320°C for HP and IP glands and 150°C for LP glands, though these vary between manufacturers.

Figure 2.67 shows a gland-steam system for a 500 MW turbine. The trimming valves provided in each gland-steam supply line are important as one particular gland may deteriorate in use more than another and it is necessary to adjust the supply pressures individually from time-to-time to maintain a balance. If, however, a particular gland demands a really excessive pressure at its sealing port, say 500–1000 mbar against a normal 50–100 mbar, then something more fundamental may be at fault. For example, on one such LP turbine gland it was found that the leak-off line to the gland-steam condenser was holed where it ran through the LP turbine exhaust duct; since this incident, similar holes have been found in supply lines and pipe-erosion shields have been strengthened. Incidentally, it is also possible for LP exhaust spray-water to impinge upon LP turbine gland-steam supply pipes, if they pass through the exhaust ducts, and cause unwanted depressions in sealing-steam tempera-

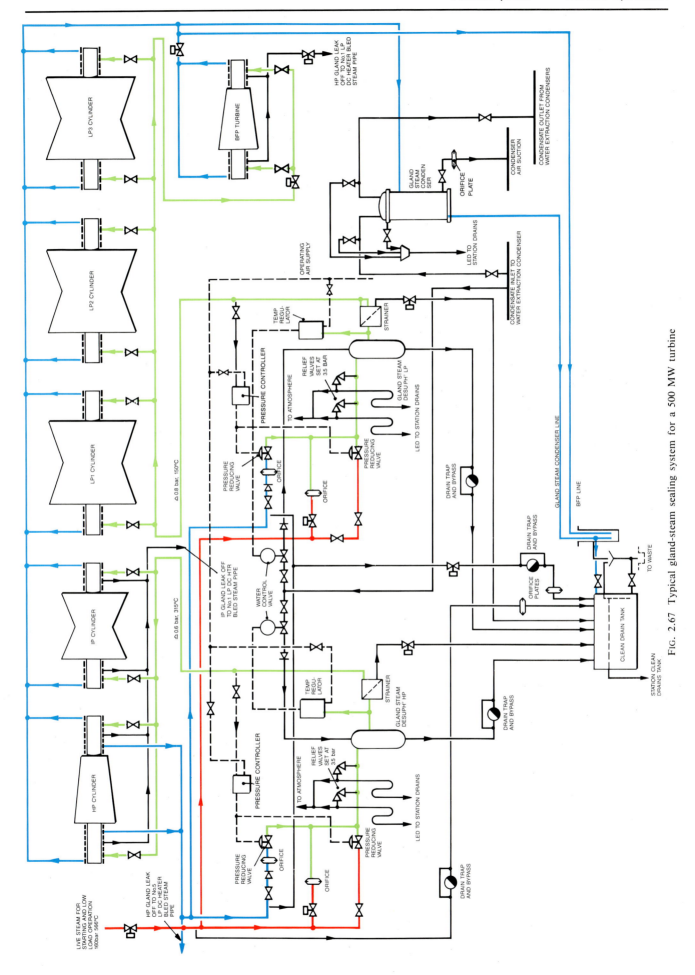

Fig. 2.67 Typical gland-steam sealing system for a 500 MW turbine

ture. Again, referring to Fig 2.67, note that all the outermost pockets are interconnected and led to a gland-steam condenser from which the condensate is recovered. This condenser is usually maintained at a small negative pressure, ideally by the provision of duplicated exhauster fans, but sometimes by the use of an orificed vent line to the main condenser. By these means, a depression is induced in the gland pockets, preventing both steam escape into the atmosphere and moisture ingress into the adjacent bearing. From an operating point of view, exhauster fans are considered to be ideal because they can be specifically designed to give relatively high flow at low suction conditions, thereby ensuring that any reasonable amount of entrained air is removed from the gland-steam condenser, allowing it to operate as designed. With the orificed-vent arrangement of Fig 2.67, one poorly-sealed gland can cause enough air to be drawn into the gland-steam condenser to swamp the venting arrangements and its operating pressure would rise to slightly above atmospheric, resulting in a general 'steaming' condition at all glands. Increasing the orifice size may cure this problem but is likely to result in a small, but unacceptable, deterioration in main condenser vacuum. Finally, whatever system is installed, excessive suction conditions at the glands can encourage the entrainment of oil vapour from the adjacent bearing into the gland-steam condenser and result in a general oil contamination of the turbine clean drains tank and condensate systems.

5.5 Condensers

5.5.1 Introduction

The condensing system is simply a heat-exchange process in which the steam exhausted from the LP turbines is condensed back into water (condensate) before being returned to the boiler. Unfortunately, because of the very large amounts of low-grade heat involved, it has the potential to impair the unit efficiency seriously if poorly designed or operated and is the one item in the turbine house requiring constant, shift-by-shift, monitoring. Almost invariably condensers are of the surface type, i.e., tubed, with circulating water (which is drawn from the sea, river or from the cooling tower ponds) pumped through the tube banks whilst the exhaust steam surrounds the banks. Water has a much smaller comparative volume than steam and a vacuum is created as the steam condenses, allowing the steam to expand down to a very low absolute value and enabling more energy to be extracted, thereby improving the cycle efficiency. Incorporated in the system is the air extraction plant, which must be capable of removing all incondensible gases (mainly air) carried into the condenser with the exhaust steam or via drain or vent lines.

Failure to do this efficiently will seriously affect the condenser performance. Condensers of the radial, pannier or underslung type are all in common use. As the type of condenser installed affects operation only marginally it is not proposed to discuss these designs in any detail, except to illustrate the general arrangements (Fig 2.68 (a), (b) and (c)). As important operationally as the type of condenser is the arrangement of tubebanks, waterboxes and flow paths, as these details can markedly affect the ease with which condenser leaks can be repaired on-load, if indeed this is possible at all. The tubes of under slung condensers can be positioned axially or transversely, whereas pannier and radial condensers have a natural axial alignment. If the total CW flow in such tubes is arranged in one direction from one end of the condenser to the other each LP turbine will have a different operating vacuum, depending on whether it exhausts to the cooler inlet or hotter, outlet end

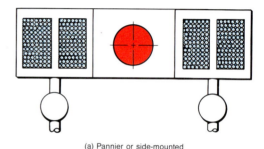

(a) Pannier or side-mounted

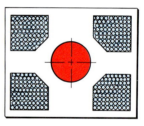

(b) Radial or integral

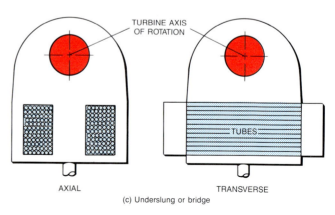

(c) Underslung or bridge

FIG. 2.68 Types of condenser currently in use

of the condenser. Typically, the range can be as much as 20 mbar. A small thermal gain is claimed, due to the improved mean effective heat-transfer coefficient, but in practice this can easily be more than nullified by inefficient air extraction.

In view of the relatively large quantity of circulating water required, water treatment, other than by screening and chlorination, has traditionally been thought to be unjustifiable. More recently, this view has been questioned, as even a small gain in vacuum can be worthwhile. At least one station has embarked upon continuous CW pH control by sulphuric acid dosing as a means of preventing scaling of condenser tubes.

Air extraction systems vary between the provision of steam-operated air ejectors, rotary air pumps (combined with air-operated ejectors) and hydraulic air pumps, or combinations thereof.

More like the PF boiler than the turbine, the condenser is the one item of turbine house plant whose performance can be maintained and occasionally improved by good housekeeping and operation, provided that the design factors necessary for efficient operation have been incorporated, or modifications have been completed towards this end. The prevention of air ingress (or the efficiency with which it can be removed), the management of the CW system and the tube cleanliness are the three factors that concern the operation of surface condensers. Management of the CW system embraces such aspects as chlorination and, as already mentioned, pH control.

Tube cleanliness is particularly important; both internal cleaning and, at some stations, external cleaning can improve performance substantially. Removal of non-condensibles is also important. A recent survey into condenser performance within the CEGB reveals that many plants appear to have worthwhile gains still to be made. For example, air ingress rates vary widely between 50 and 200 kg/h, with an average value nearer 100 kg/h. Air leakage rates quoted for some overseas plants are much lower, though straight comparisons can be misleading. For example, the extensive use of sophisticated direct-contact feedheating plants within the CEGB, with their long lengths of bled-steam pipework, often operating at sub-atmospheric pressures, can significantly increase the potential for air ingress.

5.5.2 Condenser operation

CW considerations

The best possible vacuum depends fundamentally on the CW inlet temperature, this being more or less outside the control of the operator. On an inland, tower-cooled station, however, there are two possible areas of control. First, on interconnected tower systems, the appropriate number of cooling towers must be in service. Often this means all, even though one or more units may be off-load, because a lower pond

temperature will result. However, should the total CW flow fall to the point where the spray pattern across the tower packs becomes broken or patchy, tower performance may deteriorate to the point that pond temperature is not as low as if fewer towers had remained in use. Tests are necessary to determine the optimum arrangements. Secondly, the unnecessary use of anti-icing spraywater can increase pond temperatures by 0.5°C. One station has removed remote operating facilities from the tower CW inlet-isolating valves as these were extremely infrequently closed, and has used the redundant control room switches and indications to operate, remotely, the anti-icing valves. By this means, anti-icing is only applied when really necessary. The cost of the conversion was saved in the first week of operation.

If CW pumping costs were negligible and tube-erosion unknown, it might be thought beneficial to increase the CW flow to the absolute maximum possible; certainly by this means the average CW temperature within the condenser would fall nearer to the inlet level and a gain in vacuum would result. Overall, this approach is neither practical nor even theoretically correct. First, there would exist a danger of sub-cooling the condensate (i.e., of removing some sensible heat as well as all latent heat), with modern regenerative condensers, however, this is normally unlikely, at least on inland stations. Secondly, whilst there may be an improvement in vacuum and hence in condenser performance, this will not be reflected back into an overall improvement in turbine efficiency because, in any turbine, the kinetic energy in the steam leaving the last rows of blades for the condenser is lost, and this increases rapidly as the vacuum improves. Moreover, the lower condensate temperature requires additional bled-steam for the LP feedheaters, causing a reduction in steam flow through the latter stages of the turbine, with a resultant loss of generation.

Consider a lowering of exhaust pressure towards the optimum with constant steam flow. On the one hand, the available energy per kg of steam increases, as does the leaving loss, but there is an increase in output. On the other hand, the lower condensate temperature causes a reduction in turbine steam flow and output because of the increased steam demand of the LP feedheaters. The optimum exhaust pressure occurs when the two effects are equal. It must be appreciated, of course, that the above applies only to optimum conditions for the turbine. For overall economy, the net reduction in heat rate caused by lowering the back pressure must be weighed against the cost of increase of CW pumping power. Applying the above principles, each station should produce an operating nomogram (Fig 2.69), which can be used in the form shown to highlight any deficiencies in vacuum or be simply converted to a computer program and displayed on demand on a VDU screen. By these means, operating staff can always be fully aware of the current target.

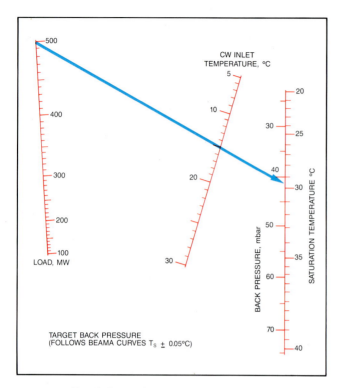

FIG. 2.69 Condenser operating nomogram
A straight line drawn between the load and CW inlet
temperature lines will intersect the target back-pressure line
at the target value, e.g., load 500 MW, CW inlet 15°C
target back-pressure 41.5 mbar.

Having fixed an optimum CW flow, it is essential that it is maintained. Again this is relatively simple to do as it can be related to differential CW pressure across the condenser. With such instrumentation provided in the control room, flow can be readily adjusted to maintain optimum conditions. As an alternative, on base-load stations, it may be acceptable to work to agreed condenser CW outlet valve positions and the number of CW pumps in service varied, dependent on the number of main units off-load.

Provided that CW flows are correct, deviations from target would result from:

● Dirty tubes, to be discussed later.

● Blocked tubes or tubeplates, as a result of the transportation of pieces of tower packing, polythene, scale, etc., into the condenser inlet waterbox. This condition may be assumed if the CW temperature rise across the condenser remains relatively unaffected even by abnormally large opening of the CW valves, assuming that there is no general shortage of water. The problem varies between stations, but the solution is usually to take the offending path out of service (which may mean a shutdown) to pick-off the debris.

● Air ingress, discussed below.

Air ingress

Air ingress is one of the most troublesome aspects of condenser operation, however it must be realised that it is entirely preventable and that considerable effort can be justified in reducing the ingress of air to an absolute minimum. The procedure for determining the quantity of air leaking into the condenser is discussed at length in Chapter 7 of this volume, it is now necessary to detect where the air is entering the system. Clearly, for air to enter it is necessary for the point of leakage to be at sub-atmospheric pressure. Thus, if inleakage is pronounced at low loads but disappears at high loads, then that part of the plant which operates at sub-atmospheric pressure at low loads and above atmospheric pressure at high loads is obviously suspect.

A halogen gas sensor is a very convenient means of detecting air ingress, it is fitted in the air suction pipe and connected to a suitable detector. Typical equipment is shown in Fig 2.70. The sensor reacts to gases of the halogen family, such as *Isceon 12* or *Arklone P*. The fluid is sprayed over the suspect area and, should there be a leak, some enters the system under suction and eventually passes out of the condenser to the air suction pipe, where the sensor indicates that leakage is present. The use of suitable radio equipment allows the search to be conducted by one person. An intimate knowledge of the plant concerned will be invaluable in directing the search to known vulnerable areas.

Typical trouble spots for air ingress include:

● Turbine to condenser joints.

● Turbine explosion diaphragms.

● Turbine atmospheric valve.

● LP glands and gland housings.

● LP bled-steam lines and feedheaters.

● Condenser fittings (gauge glasses, etc.).

● Any drain line passing back to the steam spaces of the condenser, erosion at pipe bends can be a particular problem.

A complementary approach to leakage detection can be made by ensuring that the air extraction plant is working to its maximum capability. The capacity of the plant can be increased by ensuring that air is removed only from the very coldest parts of the condenser and that as little steam as possible is entrained with it. At one station, lengths of butyl rubber were squeezed into the gap between the tubes and neighbouring baffles to prevent steam from entering the air-cooling section. A useful increase in air extraction plant capacity can also be achieved by ensuring that the temperature of the water supply to the coolers, ejectors and pumps is as low as possible.

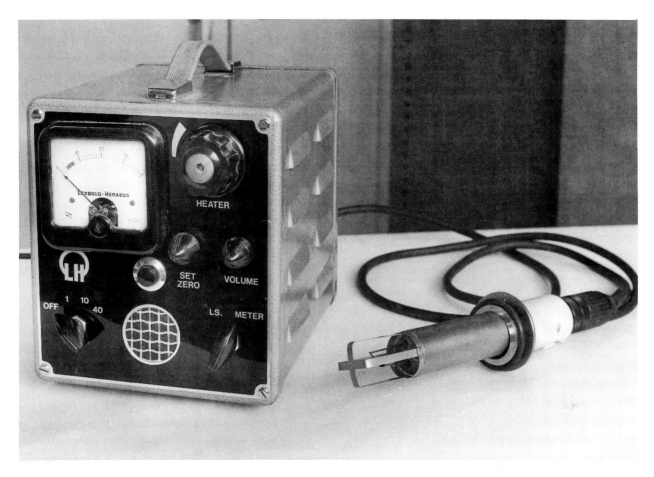

FIG. 2.70 Halogen gas detector

Approximately 5 mbar improvement in pump suction pressure was achieved at one station by adding cold towns water to the pump seal and pre-cooler water supply.

On a number of the CEGB's 500 MW units, the performance of the air extraction equipment has been considerably improved by providing a single-stage air ejector in the common air suction line between the condenser and the rotary air pumps.

Dirty tubes

For optimum vacuum, tube cleanliness is also essential. Traditionally, chlorination has been employed to remove organic growths. The storage and transportation of bulk liquid chlorine, however, is becoming increasingly unacceptable and sodium hypochlorite (which is safer to handle) is likely to find popular use. Another alternative being employed is the production of sodium hypochlorite by the electrolysis of brine; high safety standards are necessary for the plant since hydrogen is also produced.

Considerable savings can be made by ensuring that only the minimum chlorination necessary to provide protection is carried out. Depending upon CW temperature, it may be found that chlorination can be dispensed with completely during the winter and carried out intermittently at other times.

However, chlorination costs are likely to increase and a worthwhile reduction in dosing rate has been achieved by the introduction of the Taprogge tube-cleaning systems (Fig 2.71). This process also removes the phosphate/carbonate scaling of condenser tubes which has become a serious problem on many inland cooled stations.

A different approach to scaling has been adopted at one station by introducing CW pH control by sulphuric acid dosing. Serious scaling only occurs at CW temperatures above 30° (and can be prevented by maintaining a water pH level of around 7.5). Initially, the existing scaling, which had resulted in unacceptable vacuum shortfalls, was removed by sustained pH reduction, using sulphuric acid dosing of the entire CW system. Thereafter, controlled dosing has maintained the pH at about 7.5 during scaling conditions. One dramatic side-effect of the initial treatment was a noticeable improvement in the performance of auxiliary coolers (for example, turbine lubricating and seal-oil coolers) and the cooler running of station compressors, etc. Indeed, one of the benefits of the treatment lies in its ability to maintain scale-free conditions on

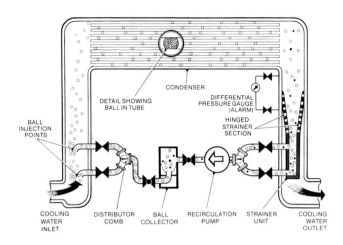

FIG. 2.71 Taprogge condenser tube-cleaning installation
The means by which the abrasive balls are circulated
around the condenser are clearly illustrated. A selection
of balls is available ranging from soft to quite highly
abrasive. A small percentage of the ball inventory
is invariably lost and needs to be made up from
time to time.

all plant connected to the main and auxiliary CW systems. Off-load bulleting, still necessary on the auxiliary coolers of stations employing Taprogge cleaning of the main condensers, has ceased.

5.5.3 Condenser tube leaks

On large modern plant, even very small amounts of cooling water leakage into the steam spaces of a condenser soon prove to be unacceptable. Strict limits on boiler water quality necessitates blowing-down to control impurities and where this proves inadequate there is no alternative to reducing load on the unit and finally shutting it down. It is then that the rapid identification of the cause of leakage assumes great importance.

Cooling water contamination of condensate can arise from a number of causes, the most likely of which is failure of a condenser tube wall, however, time is well spent in detailed studies of the design of a condenser to identify other possible leakage paths. In one design of 500 MW condenser, cooling water passed along the peripheral bolts locating the double tubeplate assembly, and the only cure was to seal-weld around the bolt heads to prevent further leakage.

Returning to condenser tubes, these can fail due to a wide variety of causes and it is important that when a failure takes place, the cause is identified so that measures can be adopted to prevent, as far as possible, a recurrence. Tubes can fail due to steamside corrosion, brought on by ammonia attack, especially in the air-cooling sections; they can also fail at support plates due to vibration or at mid-span due to tubes clashing, again as the result of vibration. Tube fouling by debris, such as concrete, corrosion products or seawater organisms soon leads to tube erosion as

a result of localised increases in water velocity and eddying. An effective solution for this particular problem is to fit plastic inserts at the tube inlets, these serve the dual purpose of not allowing objects of sufficient size to become lodged in the tube to enter them and of protecting the initial length of tube from erosion damage caused by changes in direction of water flow. Internal erosion may also occur if the velocity of the cooling water is allowed to exceed the design limit, for aluminium-brass tubes this is of the order of 2 m/s.

Before considering the repair of condenser tube leaks in detail it is as well to note that hazards accompany the insertion of rubber plugs in failed condenser tubes. It is not unknown for plugs to be drawn into a tube by the vacuum resulting, and ejected against the far end waterbox with such force that the tightening bolt is bent U-shaped.

The gradual adoption of sodium-in-condensate monitoring has highlighted leaking tubes, resulting in the detection of leaks that would have passed unnoticed with conductivity monitoring. It is necessary, therefore, to give clear guidance to operating staff as to when to attempt repairs, if indeed this is possible on-load. If repairs are not possible, a simple and effective way of temporarily sealing a condenser leak is to inject sawdust into the affected pass or simply to scatter one or two bags onto the surface of the water in the CW pump suction dock. It is surprising how effective this process can be.

Whether or not leaking tubes are repaired on-load depends not only on the design of the condenser but on the attitude prevailing at the station. At a station employing pannier condensers, each side having two separate CW flows, load was initially reduced to 100 MW on a 500 MW unit if a pass containing the air-cooled section was to be isolated; the same path on the opposite side was also isolated and drained to maintain even turbine conditions. Naturally, a bad vacuum resulted. Experience has changed this procedure so that only the defective pass is isolated and drained: full-load is maintained, with no obvious ill-effect on the turbine.

At first glance, the task of determining which tube is leaking in a condenser which contains, say, 20 000 tubes seems formidable, but often is not as difficult as it seems. Detection methods fall into two broad categories — those which require a vacuum in the condenser and those which do not.

Detection methods requiring a vacuum include:

- Foam

- Halogen gas

- Bubbler leak

- Indiplugs

- Acoustic meters.

- Conductivity measurement

Those that do not require a vacuum include:

- Fluorescein

- Bubbler leak

- Furmanite guns

- Corrodograph survey

Consider each in turn.

Foam

There must be access to the affected condenser water boxes and there must be a vacuum on the steam side. A stable foam which will adhere to the tube plates for some time is required. The foam is sprayed onto the tube plates at either end of the condenser until all the tube ends are covered. If there is a leaking tube, a foam 'plug' will enter it, so identifying it precisely. Sometimes it is suggested that foam need only be used at one tube plate and the other one covered in plastic sheeting or paper. In practice, this is not very satisfactory.

Water residing in the tubes tends to break the foam plugs and give false indication, so suspect tubes should be checked with a bubbler or Furmanite gun.

Halogen gas

This method will only indicate the level of the affected tube, which must then be located by other means. The condenser is isolated on the CW side and the water level lowered in a controlled manner. Meanwhile, halogen gas is injected into the water boxes above the CW level and, as the tube ends are uncovered, the gas will eventually enter the leaking tube, pass into the steam space and along the air suction pipe. The presence of the gas will be indicated at the halogen detector, thus enabling the level of the faulty tube to be determined (Fig 2.72).

Bubbler leak

This is simple and positive, but laborious. Access is necessary to the tube plates, one of which is sealed. The bubbler is then connected to each tube in turn at the other tube plate, as shown in Fig 2.73. A leaking tube causes a stream of air bubbles to appear in the container.

Indiplugs

These are tapered, rubber plugs which fit into the tube ends. There is a hole along the length of the plug, and the wide end of the plug is covered with a thin diaphragm. The Indiplugs are inserted in the tubes at one tube plate whilst the tube ends at the other are covered or plugged. The faulty tube is located by the distortion of the diaphragm. This is a quick and positive method.

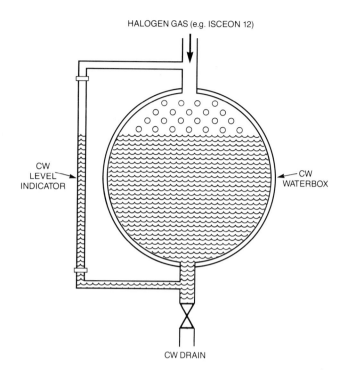

FIG. 2.72 Condenser tube leak detection using halogen gas
One method of determining the level of a leaking tube is to lower the cooling water level slowly and inject Halogen gas above the cooling water. If a leaking tube is uncovered, the gas will pass through the leak into the steam space and be detected by the gas monitor.

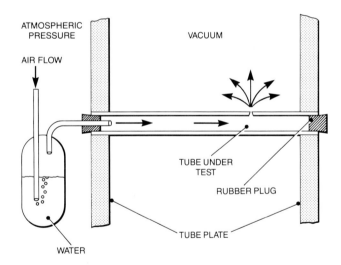

FIG. 2.73 Conventional bubbler-leak apparatus
When the apparatus is connected to a leaking tube, a stream of bubbles will appear in the container.

Acoustic meters

These devices are handheld and respond to the acoustic frequencies associated with air being drawn into faulty tubes in a condenser. A detector is passed over the exposed tube ends in the (drained) water box, the other ends of the tubes being plugged. In practice, there is often interference from extraneous noises so, although the method is very simple, it is not as reliable as some of the others.

Conductivity measurement

This is an on-load method in which the CW level of the condenser being tested is lowered. The 'acid' conductivity of the condensate (i.e., the conductivity after the sample has been passed through a cation exchange column) at the extraction pump discharge is recorded. If there is a condenser leak, the acid conductivity will be higher than normal. The CW level in the water-box is lowered until the acid conductivity returns to normal: this indicates the level of the leaking tube, thus narrowing the search area for the leak by other means. As in the halogen method, a temporary CW level indicator must be fitted to the water box before this method can be used, but it need only consist of a length of plastic tubing suitably connected.

Fluorescein

This is a traditional detection method which still has some use. The steam space is filled with condensate to which fluorescein has been added. Any leakage into a faulty tube will be discharged at the tube ends and, therefore, if an ultraviolet light is shone onto the tube plate, the leakage will fluoresce. In practice, the method has some limitations. For example, if the leakage is small in the first place it may be that the hole will seal because of the cold conditions. Also, if there are several leaks it may be difficult to segregate sound tubes from faulty ones as there is so much fluorescein on the tube plate. On the other hand, a complete tube plate can be scanned quickly, so it is a very easy and convenient method on occasion. This method can only be used on underslung condensers and jacking arrangements must be in position before filling the steam space.

Bubbler leak

This is a modification of the previous bubbler leak apparatus, and is shown in Fig 2.74. With the vent valve shut, the apparatus is brought under vacuum by the vacuum pump. The pump isolating valve is then shut and, shortly after, the balance valve is also shut. The remote end of the tube under test is already plugged, so any leakage will cause a stream of bubbles in the bubbler jar.

Furmanite guns

One end of the tube is sealed. The gun is applied to the other end of the tube, which it seals and pressurises with air. The rate of decay of the air pressure is a measure of the leakage. These guns can also be

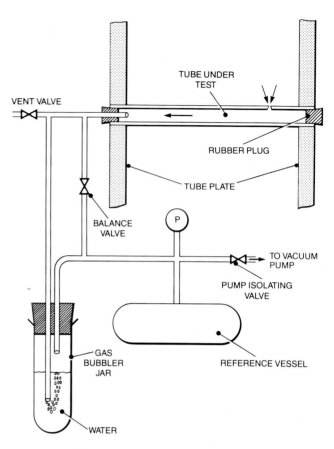

FIG. 2.74 MEL bubble-leak apparatus
This is a modification of the apparatus in Fig 2.73 which permits the detection of leaks when the steam side is not under vacuum.

used for detecting leaks in other heat exchangers, such as feedheater nests.

Corrodograph survey

This will locate tube perforations, but not tube-end leakage. The corrodograph is applicable to any copper, copper-alloy and other non-magnetic tubes used in condensers and it will detect a variety of tube problems such as:

- Pitting, on both the outside and inside of tubes

- Wall thinning

- Inlet-end corrosion or erosion

- Incondensible gas attack

- Steam erosion

- Vibration damage

- Debris-induced metal loss

- Manufacturing defects

- Fatigue cracks

- Stress-corrosion cracking

A specially designed probe, electrically connected to the Corrodograph instrument, is drawn through the tube at a predetermined speed. Variations in eddy currents induced by the probe in the surrounding tube wall are detected by a coil within the probe, from which a trace is produced on a moving strip of sensitised paper. Figure 2.75 shows examples of different traces for various types of attack. Useful work can be carried out during major overhauls by surveying all the condenser tubes on a unit. A history of tube damage can be compiled and tubes having significant defects can be plugged before failure takes place.

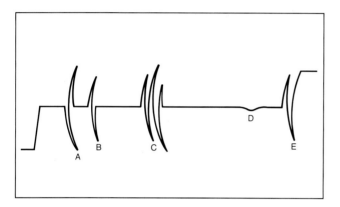

FIG. 2.75 Schematic example of a Corrodograph trace from a tube, showing different types of corrosion

A — normal support plate deflection

B — isolated (pitting) attack

C — attack on outside diameter of tube at support plate position (incondensible gas attack)

D — rubbing of adjacent tubes

E — inlet and impingement attack

6 Turbine plant — emergency operation

6.1 Introduction

A considerable responsibility is thrown on to operating personnel when there is clearly some abnormality but the cause is not apparent. It may be necessary to shut down the turbine at once to avoid serious damage, for the energy in the rotating masses of a large machine is considerable indeed, if put to destructive use. In other circumstances, an immediate shutdown will give no opportunity to locate the source of the trouble and it may be necessary to start up again for this purpose later. To ensure that action is not delayed and to ease the responsibility placed on operating staff, automatic protection is specified to trip the turbine under certain conditions, such as overspeed or low vacuum. The categories of protection trip are now virtually standardised throughout the CEGB; even so, it is essential for operators to

be fully aware of the schemes fitted on their own plant and their limits, so that there is no delay in assuming control of a deteriorating situation.

6.2 Automatic protection

The turbine protection is unavoidably interlinked with that of the generator as energy can be fed into the turbine under fault conditions from either the steam or the electrical end. The protection schemes operate in two distinct ways. Electrical faults on the generator, generator transformer, unit transformer and associated cables and connections can cause serious damage to plant unless the fault is quickly cleared. For this reason, operation of any electrical protection causes both the turbine steam valves and the generator electrical circuit-breakers to trip simultaneously. When such a trip occurs on a unit carrying a high load, however, there is a risk of the turbine reaching a dangerously-high overspeed condition if the closure of the steam valves is not immediately effective; in order to minimise this possibility, only serious electrical faults are arranged to initiate immediate electrical disconnection. For less onerous fault conditions, the turbine steam valves alone initially trip shut and only when the generator output is sensed to have fallen to approximately zero is the electrical disconnection made, thereby safeguarding the turbine against possible overspeed without materially increasing the risk to the machine as a result of the fault. Most turbine faults fall into this category.

6.2.1 Overspeed

Every turbine has elaborate systems to limit overspeed which is potentially the most serious condition of all. Initially, a rise in speed is sensed by the speed-governing system be this mechanical (via flyball governors) or electrical (derived from speed signals obtained electronically) and this results in steam valves being automatically adjusted to limit the rise in speed. Should the speed continue to rise, some form of auxiliary governing is usually provided to act on the governor or, on some plants, on emergency stop valves causing them to shut rapidly, thus limiting the speed rise to that below which the emergency trip gear operates. This may be supplemented by some form of anticipatory action; for instance, if the turbine is operating at a high load, the opening of the electrical circuit-breaker can be arranged to close, transiently at least, the steam valves in anticipation of a speed rise. Should the turbine speed increase by more than 10% of normal, duplicated overspeed bolts are arranged to move outwards from their normal positions under increasing centrifugal force and operate the trip gear, directly closing all steam valves. It is absolutely essential that these bolts are free to operate when required and they are tested regularly, together with all turbine steam valves; see Section 7 of this chapter.

6.2.2 High condenser or hotwell level

This condition also has potentially serious consequences, for if the water level rises to the point that it makes contact with moving LP turbine blades serious shock to, and failure of, the blades is unavoidable. Modern practice is to fit duplicated trip floats to the condenser steam-space or hotwell and to arrange for these to be tested, one at a time, on-load.

6.2.3 Low vacuum

This condition can cause overheating of LP turbine blading and also give rise to a rapid expansion of the turbine casings relative to the rotors leading to fully-contracted differential expansions. Low vacuum un-loading systems are usually installed to ensure that automatic turbine unloading takes places as the va-cuum falls, prior to tripping at a level which may be around 700 mbar of vacuum. It can sometimes be supplemented by LP turbine exhaust casing high-temperature protection.

6.2.4 Miscellaneous

As described earlier, low oil pressure and main bear-ing temperature trip schemes may, or may not, be installed. Whatever is provided, it is essential that operating staff are aware of the maximum laid down limits. In the case of journal bearings, discontinuities in the oil film around the bearing circumference are likely if oil pressure falls below 0.5 bar.

On operation of the automatic protection it is es-sential that the major parameters listed below are checked, so as to be sure that the situation is under control, viz:

● Load has fallen to zero.

● Turbine emergency stop valves have closed.

● Turbine governor valves have closed; this sometimes has to be inferred from relay oil or hydraulic fluid system pressures.

● Electrical circuit-breakers, main and unit transfor-mer, have opened.

● Turbine speed is falling.

● Auxiliary oil pumps have started, maintaining bearing- and seal-oil pressures.

Providing that operators have been trained, ideally by simulator, to operate calmly it is surprising how quickly such an action list can be completed, before proceeding to attend to other less important aspects arising from the trip.

6.3 Manual actions

6.3.1 Vibration/shaft eccentricity

This aspect of operation has been fully covered in Section 5.2 of this chapter. Permanent shaft distortion can occur if actions are delayed beyond laid-down levels of vibration and/or eccentricity.

6.3.2 Loss of load

This can cause unexpected confusion, to say the least, but should not result in an overspeed trip if the govern-ing systems are operating to specification. By following a check list similar to the one shown in Section 6.2.4 of this chapter, however, it should be possible to diagnose the reason quickly. For example, inadvertent closure of the steam valves alone results in a genera-tor motoring condition at 3000 r/min, whereas loss of electrical output is manifested in a running speed above 3000 r/min, at a value determined either by the gov-ernor droop characteristic or by the auxiliary governor setting: the steam valves would be unaffected, except that governor valves would be operating at a very low-load position. Unless automatic provisions are incor-porated it will be necessary, in the first example, to reduce the governor setting immediately to avoid any possibility of unexpected restoration of load and, in the latter, to stabilise speed at 3000 r/min before deal-ing with the causes. Rapid inadvertent restoration of load can seriously overload thrust bearings and blades and should be avoided at all times.

6.3.3 Water ingress

This condition will be dealt with fully in the section on auxiliary plant.

6.3.4 Turbine fires

Many fires have occurred on turbine-generators, some of which have resulted in major damage to the units concerned, leading to extended outages. Historically, the use of mineral lubricating oil for the control and op-eration of steam valve relays has resulted in such a risk of fire that a change to fire-resistant fluids for steam valve use has become widespread. Nonetheless, much older plant remains in service and good house-keeping is important in preventing or containing leak-age. The present policy of the CEGB is that the long-term availability of plant is of paramount im-portance when compared with any short-term gain that may be realised in attempting to continue in operation. For this reason, should there be any doubt about the ability to contain and extinguish a fire out-break, the unit should be unloaded and shut down in a manner appropriate to the circumstances and the fixed-jet fire fighting installation operated; a decision, incidentally, that has often been taken some-what reluctantly in the past by operating personnel, no doubt because of the possibility of quenching and cracking hot components. No operating procedure is ever likely to satisfactorily cover all aspects and even-tualities but may reasonably be based on the categories detailed below:

- *Minor fires* are localised, with little or no flame and would most probably arise from oil-soaked lagging near bearings, bled-steam pipes or drain vessels. Portable appliances are sufficient to con-tain the outbreaks and the machine would remain on load.

- *Substantial fires* are associated with large amounts of smoke and localised flame and lead to rapid-ly deteriorating visibility throughout the turbine house. They would probably arise from a relatively small, but continuous, oil leak and the potential would be present for a major fire to develop. Since they are unlikely to respond solely to portable fire fighting appliances, a decision to unload and trip the turbine is necessary, but before this is taken, it may be reasonable to initiate, in several short bursts, the fixed fire fighting installation in the area of the fire. Dependent on the results, the fire will either revert to the minor category or develop, forcing a decision to shut down and take action as below.

- *Major fires* are obvious by the amount of smoke and flame present. They are likely to be associated with plant malfunction and irregularities and, be-cause of the limited visibility, it is difficult to establish the seat of the fire. The machine must be quickly off-loaded and tripped and steam mains depressurised, allowing the unrestricted use of the fixed fire fighting installation. Vacuum is broken early to reduce the turbine shaft run-down time to a minimum, and oil pumps and bearing oil pressures are reduced to a minimum, consistent with continued bearing lubrication during run-down.

7 Turbine plant — routine testing

The routine testing of turbine plant may follow the logic outlined for the boiler, tests being split up into on-load and off-load groups. It is only proposed to mention the most important of each which, in the case of on-load tests, are undoubtedly those concerning the operation of turbine steam valves and overspeed bolts.

All large machines are provided with facilities (vary-ing from turbine to turbine) that allow valves and overspeed bolts to be checked for freedom of opera-tion, thereby ensuring that they will operate correctly in anger. Once a week, normally, when the turbine

is in a condition to allow testing, steam valves are isolated from their normal control loop and closed in a suitable sequence. The closing speed is often pur-posely restricted by the adoption of an orificed test drain because, if the valve closes smoothly and fully under these conditions, it is not likely to fail to do so under emergency conditions, when the relay oil or fluid is unconditionally dumped to drain. On more recent designs, facilities exist to time the closure rate, giving an additional check against any tendency to stick open.

Overspeed bolts form the last line of defence against overspeed and cannot be allowed to fail. One at a time, each bolt is checked to satisfactorily operate the turbine trip gear by injecting a pressure of oil into the bolt housing, whilst the machine remains on-load. This increases the out-of-balance force acting on the bolt and causes it to operate against its restrain-ing spring at normal turbine speed. The pressure re-quired to operate the bolt is recorded and compared with previous results to ensure that it is not increas-ing. The trip gear is duplicated and provision is made to interlock the complete system so that oil cannot be injected under a bolt until its associated trip gear is by-passed. Failure to isolate the part of the trip gear that will operate as the oil is injected under the over-speed bolt will naturally result in the turbine steam valves being tripped closed.

One other important test is that involving the con-denser or hotwell high-level trip. On later machines, provision is made to employ duplicate float columns with facilities for water injection. By these means, one column at a time can be isolated from the pro-tection system and flooded to ensure that each float physically operates on a rising water level. Such a test should be undertaken monthly.

Off-load testing can be extensive, though much can reasonably be undertaken by operating personnel. Examples would include:

- The automatic cut-in of pumps, both lubricating oil pumps and, in the case of fire-resisting fluid systems, control-fluid pumps.

- The operation of turbine anticipatory devices.

- The operation of turbine exhaust spray system, and so on.

8 Auxiliary plant — normal operation

8.1 Introduction

For the purposes of the present chapter auxiliary plant is deemed to include all plant other than boilers, tur-bines, generators and gas-turbine units. The operation of boilers and turbines has been fully covered in the preceding sections of this chapter and gas-turbine operation is to follow. The operation of generators forms part of the full review of generators and motors covered in Chapter 3 of this volume.

In consequence, the operation of condensate and feedheating plant and feed pumps forms a major part of this section, such plant being of vital importance to the continued, safe operation of the main boiler/ turbine unit. The operating principles of condensate systems, surface-type feedheaters and feed pumps are well established, enabling a general review of such systems to be made with some confidence, despite the wide variations in the details of installed systems. The same is now true of direct-contact (DC) feedheating systems, though this was not so in the early operating days, when a series of design shortcomings and poor operational understanding resulted in a mediocre performance from many of the systems that had been installed in some numbers on a large part of the 500 MW unit and early 660 MW unit plant programmes. Whilst all DC feedheating systems are now operating satisfactorily, there has been a trend back to surface type (tubed) LP feedheating systems in the very latest plant, no doubt partly as a result of the operating experience with DC systems. Certainly the increased risk of water ingress into the turbine, should the protection fail to isolate a flooding heater, and the need for several pumping stages between condenser and de-aerator (dependent on the number of cascades installed) have both weighed heavily in this decision.

So far as the remaining unit or station plant is concerned, no attempt will be made to describe operation in detail; indeed, such an approach would be doomed to failure on account of the wide variations in the plant installed. This would particularly apply to ash and dust collection/disposal systems which can operate in basically different ways. A better approach is to discuss, on a broad front, those aspects of operation that appear to be of general interest and which, based on a long operating experience, seem relevant to the continued satisfactory operation of the plant concerned.

8.2 Condensate, LP feedheating plant and feed pump operation

8.2.1 Condensate extraction pumps

Whilst operation is straightforward, such pumps have a relatively high duty, drawing from condenser vacuum and discharging, in the case of surface-type LP feedheaters, to the de-aerator. As a result, a twin-stage pump is sometimes specified, usually having both stages contained within a single casing, with internal seals and often a water-lubricated internal bearing. This bearing is vulnerable to damage by dirt, despite

being provided with a filtered supply of lubricating water taken from the pump discharge branch. Consideration should be given to the provision of duplicated strainers, to allow on-load cleaning and, in addition, it is sound operational policy to flush out the condenser hotwells after any long outage. The same filtered water supply is often used to seal the extraction pump external glands, this being particularly important for the standby unit which is vulnerable to air ingress, therefore it is doubly important to keep the strainers clean.

Since the condensate extraction pump operates under extremely low suction conditions, a small change in the water level at its suction branch changes the suction head on the pump appreciably and hence the pump output which, in effect, becomes reasonably self-regulating with unit load. Unfortunately, the danger of cavitation or voidage occurring in both coolers and feedheaters may force the installation of a pressure-sustaining valve or hotwell level control scheme despite the increased artificial head imposed on the pump as a result.

8.2.2 Condensate and surface-type LP feedheating systems

It may not be generally realised how many coolers can be installed in the condensate system prior to the start of the LP feedheating section proper and Fig 2.76 may be of interest in this respect. When the prime purpose of the system is heat recovery it may seem strange, at first sight, to employ condensate coolers. These coolers are necessary to ensure that, under all operational conditions, the condensate presented to the following generator hydrogen coolers is cool enough to ensure a satisfactory gas temperature.

During normal operation they are not needed, but during abnormally hot weather at an inland cooling-tower station the CW temperature at condenser inlet may rise beyond 25°C and the resulting high back-pressure will yield a condensate temperature in the order of 35°C. Clearly it is impossible, using such condensate, to maintain hydrogen gas temperature below 35°C (as is normally required) without the added help of the condensate coolers.

Similar conditions can occur on any station if the condenser vacuum deteriorates for any other reason, for example, as might occur during the locating and repairing of condenser leaks when a section of the condenser would be out of service.

Drainage of any part of the condensate or surface-type LP feedheating systems, or continued heat input on cessation of pumping, can lead to the formation of vapour cavities, the collapse of which may result in destructive hydraulic shock on pump restart. Indeed, in the early days of operation of 500 MW plant, several generator hydrogen-cooler condensate-water-boxes were badly damaged by such forces. Potential points of small drainage include, for example, pump

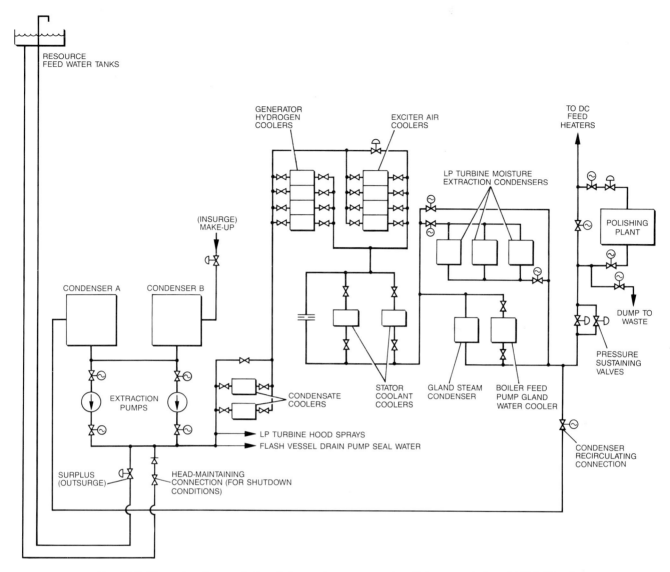

RESOURCE
FEED WATER TANKS

GENERATOR
HYDROGEN
COOLERS

EXCITER AIR
COOLERS

TO DC
FEED
HEATERS

(INSURGE)
MAKE-UP

LP TURBINE MOISTURE
EXTRACTION CONDENSERS

POLISHING
PLANT

CONDENSER A

CONDENSER B

DUMP TO
WASTE

PRESSURE
SUSTAINING
VALVES

EXTRACTION
PUMPS

CONDENSATE
COOLERS

STATOR
COOLANT
COOLERS

GLAND STEAM
CONDENSER

BOILER FEED
PUMP GLAND
WATER COOLER

LP TURBINE HOOD SPRAYS
FLASH VESSEL DRAIN PUMP SEAL WATER

CONDENSER
RECIRCULATING
CONNECTION

SURPLUS
(OUTSURGE)

HEAD-MAINTAINING
CONNECTION (FOR SHUTDOWN
CONDITIONS)

FIG. 2.76 Illustrating the complexity of the condensate system installed on one design of 500 MW unit

gland-sealing connections or LP turbine hood spray-tappings, whilst a point of rapid and unacceptable drainage would be via the condenser recirculating connection, installed to ensure a positive flow of condensate though auxiliary coolers at low unit loads. Drainage via this connection is particularly unacceptable, as it places the system in direct communication with the condenser vacuum which would greatly promote flashing and cavity formation. Whilst each particular system has provisions designed into it to prevent cavity formation on cessation of pumping when, for example, the recirculating connection can be arranged to close and reserve feedwater tank head can be automatically applied, the manual priming and venting of the entire system prior to initial start-up is, naturally, the responsibility of the operating staff. If a reserve feedwater tank head-connection is provided for priming purposes, it must have a non-return valve fitted to prevent the outflow of condensate during normal operation.

An unlikely hazard associated with condenser recirculating connections arises in the event of leakage at a hydrogen cooler when the hydrogen, being at a higher pressure, passes into the condensate to be liberated in the condenser. Cases of hydrogen-in-condenser explosions are on record and the DC feed-heating train of a large overseas unit was severely damaged by a hydrogen explosion.

Whilst tubed LP feedwater heaters impose much less risk of water ingress into turbines than their direct-contact counterparts, the possibility cannot be ignored and a heater by-pass arrangement would usually be provided to allow continued unit operation in the event of a heater tube leak, albeit at reduced thermal efficiency. The designer will have ensured that adequate provision has been made to accommodate the increased drainage occurring as a result of one, or even two, completely severed tubes without causing a dangerously-high condensate level in the heater steam space, though if the provisions are too generous a tube failure may go unnoticed and the slow consequential erosion of healthy tubes, by the leakage water, may eventually result in a multiple tube failure. It is important, therefore, that regular operational checks are conducted to highlight any leaking tubes, should opportunities present themselves. If cooler or feedheater condensate drains lines are provided with test drain facilities (if not, they should be installed), then these can be opened as soon as the main unit is offloaded and whilst the condensate extraction pump remains in service, when a continuing flow from any drain would indicate a tube leak.

One final point deserves mention, namely, that not all heat exchangers may be fitted with condensate-side relief valves. Should any such cooler be isolated on the condensate side there is the possibility of dangerous overpressurisation developing, especially if a degree of heat input continues. In consequence, as soon as such a cooler is isolated, the condensate side vent or drain should be opened.

8.2.3 Direct-contact (DC) feedheating systems

DC feedheaters were installed on many large units because of a desire to obtain an increase in cycle efficiency over that possible with equivalent, tubed heaters: this increase, on a high load-factor station, can approach 0.2%. Such a gain in cycle efficiency arises because the DC feedheater is equivalent to a surface-type heater with a zero temperature terminal difference and additionally because of the absence of condensate drains which, in surface heaters, reject heat to a lower-temperature heater. Some of this potential thermal gain is lost as a result of the drop in bled-steam pressure that occurs in the long bled-steam lines that are a characteristic of DC feed systems unless these lines are made uneconomically large in diameter. DC heaters occupy thermodynamic positions similar to those of the surface heaters which they supplant but, in practice, they have to be stacked vertically, such that the hydrostatic head between each successive heater is more than equivalent to the difference in bled-steam pressures to the heaters. These requirements may conveniently be met by positioning one heater above the next in a continuous train or cascade; in this manner, only one additional pumping stage is required, though the bled-steam pipe lengths to the higher-level heaters are such that the drops in steam pressure along them begin seriously to negate the advantages of the DC heaters themselves. An improvement in this respect is gained by adopting a twin cascade, though additional pumping stages become necessary.

Even with a twin cascade, the vertical separation needed between cascading heaters requires more height than is available in the turbine house; therefore, the whole system is often installed in the de-aerator annexe.

A DC feedheater is essentially a pressure vessel in which the bled-steam and condensate are intimately mixed, so that the steam gives up its latent heat in providing sensible heat to the condensate. Vertically- or horizontally-mounted heaters are in use and, in order assist the completion of the mixing process in the shortest time and in as small a volume as possible, the incoming condensate is atomised by spray nozzles and the resulting droplets fall through the steam atmosphere and may cascade over trays.

Apart from thermodynamic considerations, the three benefits usually quoted when considering DC feedheaters over tubed heaters are:

- The elimination of non-ferrous materials in the feed system obviating copper 'pick-up' and resulting in less metal contamination of the boiler feedwater.

- The continuous de-aeration of the condensate prior to the de-aerator/storage vessel proper, contributing to an overall reduction in system oxygen levels.

- The simplicity of the heater, in terms of cost, often permitting more stages of feedheating to be economically employed.

Figure 2.77 shows various configurations of DC feedheating cascades installed on 500 MW units. In order to understand the operation of a typical cascade better, under both normal and transient conditions, and therefore to more fully appreciate the need for the somewhat elaborate protective systems now installed, it may be helpful to consider the four-heater, twin cascade system (Fig 2.77 (d)) in more detail. The condensate (after passing through generator and turbine auxiliary coolers) is pumped into No. 1 heater by the condensate extraction pump. This must be provided with reliable output control if surges of condensate into No. 1 heater (and possible flooding) are to be avoided, since it can be argued that as many DC feedheating system trips have occurred in the past as a result of input surges as for any reason associated with the systems themselves. Unnecessary starting of the standby pump may also result in a transient excessive flow to the heaters, as may any increase in output of the running pump caused by condenser vacuum decay. Because the extraction pump operates under very low inlet pressure conditions, vacuum decay is manifested as an overpressure at the pump suction and the pump responds by discharging at an excessive rate.

Of the two valves provided in the condensate line of Fig 2.77 (d), one is a quick-acting isolator, which is arranged to trip closed if any heater floods (as a secondary back-up to the bled-steam isolating valves) and the second is a slow-acting control valve necessary to introduce, or reintroduce, condensate in a gradual manner after a system trip. Without this valve, it would be impossible to control the admission into

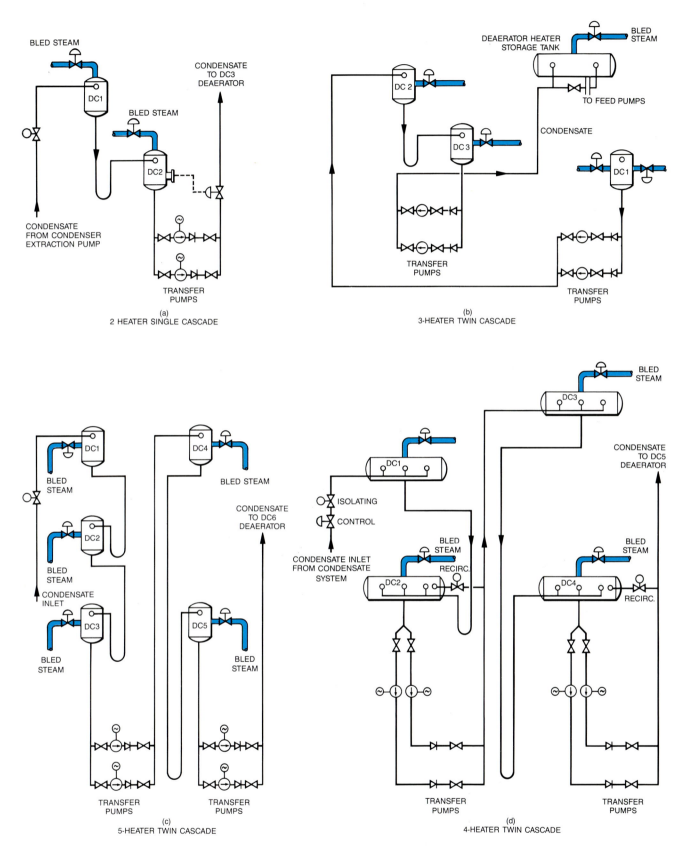

FIG. 2.77 Four direct-contact feedheating possibilities

(a) Shows a simple cascade system, with pump discharge control installed to maintain flooded conditions at the pump suctions at all loads. Such conditions greatly reduce impeller wear induced by the cavitation arising from free suction-head operation.

(b) Illustrates a three-heater twin-cascade system and highlights, in basic form, how the vertical separation between heaters is dependent on the differential steam pressure across the heaters concerned, with allowances for friction and safety.

(c) Shows a more complicated system which was initially designed to operate on free suction-head, i.e., the provision of the pump discharge control scheme of (a) or the recirculating connections of (d) are absent.

(d) Shows an alternative four-heater twin-cascade system, employing horizontally-mounted heaters.

the system of the large volume of condensate built up in the condenser hotwells following a DC feed system trip. After heating in No. 1 heater, the condensate gravitates to No. 2 heater, set far enough below to allow the establishment of a balancing condensate column equivalent in height to the pressure differential between the heaters. This column will reach a maximum steady level at full-load, but the heater separation will exceed this to take into account any variations in bled-steam pressure compared with the design value, to accommodate frictional effects and safety margins, and to withstand any malfunctioning or the inadvertent operation of an individual bled-steam valve. The last aspect is extremely important, as the likely consequences of such operation can be severe, especially at full-load. For example, if No. 1 heater bled-steam valve alone is closed, the increased pressure differential between No. 1 and 2 heaters may require such an increased condensate column as to encroach into the No. 1 heater body itself, causing flooding. Conversely, if No. 2 heater bled-steam valve is closed at high load, the heater pressure will decay rapidly (ultimately to that of No. 1 heater) and the leg of condensate held between heaters 1 and 2 will collapse. Unless sufficient storage space is provided, No. 2 heater may be flooded. On DC feed-heating systems, therefore, it is now common practice to cause all bled-steam valves to trip closed on receipt of a high water level condition in any particular heater.

Care in the operation of bled-steam valves (which, whilst fast-closing, are invariably fitted with devices to limit the speed of opening) is therefore of crucial importance. Operating instructions will most likely require that bled-steam valves be opened in ascending order and may bar operation at loads in excess of half full-load, necessitating a load drop to that level prior to the re-establishment of bled-steam after a tripping incident, though problems with condensate inventory shift (into the hotwells and away from the de-aerator storage tank) may force a more severe load reduction.

In Fig 2.77 (d), note also the U-loop seal in the condensate leg; this is as important as the provision of vertical heater separation. At low loads, the condensate leg is small in height and unless the U-seal is deep enough, a rapid increase in load, or the sudden re-opening of No. 2 heater bled-steam valve, may be sufficient to cause a violent reflux of condensate back into the No. 1 heater. Once the seal is broken, it may not re-establish itself and the continuing upward flow of bled steam will prevent condensate drainage from No. 1 heater altogether.

Condensate is pumped from No. 2 heater via the first pumping stage and the cascade is thereafter repeated, though the increasing bled-steam pressure differentials between Nos. 3 and 4 heaters over Nos. 1 and 2 effectively magnify all of the previously described effects.

Due to the obvious flooding dangers, the primary protection against condensate ingress into the turbine via bled-steam lines is by the provision of bled-steam isolating valves, installed as near to the heaters as possible. Additional back-up is provided by non-return valves and these also protect the turbine against the possibility of overspeed after a turbine trip (as a result of the energy of the steam stored in the bled-steam lines), if they are positioned at the turbine end of the lines. With regard to the pumps, protection schemes differ, some allowing continued selective pump operation following a heater trip, whereas others require that all pumps are tripped. If the latter, the condensate that gravitates to the low level heaters is transferred to the de-aerator by operator action once conditions allow, though both the bled-steam and the condensate inlet isolating valves are inhibited closed until all the heater 'high water trip level' alarms have reset. It is when this manual reinstatement is being done that it is so important that operators are fully aware of the consequences of their actions.

Mindful of the above discussion then, which is fairly typical whatever the actual cascading system installed, the normal operation of any DC feedheating system revolves around the need to avoid sudden changes in load or pump output, or inadvertent bled-steam valve operation. These, together with a full understanding of the processes at work, assure satisfactory long-term operation.

The cascade transfer pumps may be conventional glanded units or be of the fully-sealed non-glanded type, in which the motor forms an integral part of the sealed unit, i.e., the motor runs in and the bearings are lubricated by the pumped condensate, albeit cooled. Several aspects of the latter deserve mention. First, the motor is usually positioned under the pump that it serves so as to maintain prime at all times. Further, it is separated from the pump by a restricted water-cooled neck so as to prevent the downward conduction of heat. Experience with such an arrangement requires that the motor section of the unit be completely leak-free, or hot water may be drawn through the neck in sufficient quantity to negate the cooling and the motor winding insulation will overheat and fail. Should a leak develop on the service pump, the hot make-up water will be diluted and cooled to some degree in the motor-cooling circuit and damage may be prevented; it is better therefore to run a leaking unit rather than leave it on standby (and better still to isolate it and effect immediate repairs). A second point on the same topic arises on those pumps that run continually under vacuum conditions, when the water in the motor-cooling circuit will gradually evaporate through the neck and, in the extreme, uncover the top motor bearing with disastrous results. Here, a solution has been to apply a positive head of cool water (usually from reserve feedwater tanks) to the pump motor-cooling circuit so as to ensure that a positive prime is maintained on the motor at all times, though there will be a small upward flow of water through the neck from motor to pump. Non-return valves must be installed in any such priming line that serves a motor operating under both vacuum and pressure conditions, dependent on unit load, or the increasing pressure on the pump as load increases may cause a flow reversal in the priming line. This will be the equivalent of a severe motor-cooling circuit leak.

Finally, whilst many such pumping units were designed to operate on free suction-head, excessive impeller wear, due to cavitation, has forced the adoption of a level control scheme in most installations.

8.2.4 De-aerators

The basic requirement of the de-aerator is to supply feedwater with an almost-zero oxygen content, typically $5\mu g/kg$ (or 5 parts per thousand million), in order to safeguard boiler pressure parts from internal corrosion. Although this is the figure required of physical de-aeration, it is usual to reduce this further, often to less than $1\mu g/kg$, by subsequent chemical dosing. Another system requirement that may conve-

niently be combined with the de-aeration process is the need for a store of deoxygenated feedwater above boiler feed pump suctions; the result is the typical modern de-aerator/storage tank installation illustrated in Fig 2.78. The reservoir must have a sufficient capacity to meet any sudden demand for feedwater by the boiler, as could occur following a rapid reduction in firing or firing trip, when the reduction or cessation of steam-bubble formation will cause a general shrinkage of boiler water contents. The reservoir may also be used to hold any short-term excess of feedwater flowing in the system which will occur during an increase in the boiler evaporation rate, when the boiler water contents will 'swell' and cause an immediate reduction in the demand for feedwater. The buffering capacity of the reservoir is purposely limited to a relatively small change in de-aerator tank level due to the danger of water ingress into bled-steam lines (high level), or an encroachment into what is essentially an emergency store of water (low level). Typically, this emergency store is equivalent to about 10 minutes operation at full-load and is sufficient to allow the unit to be safely unloaded and shut down

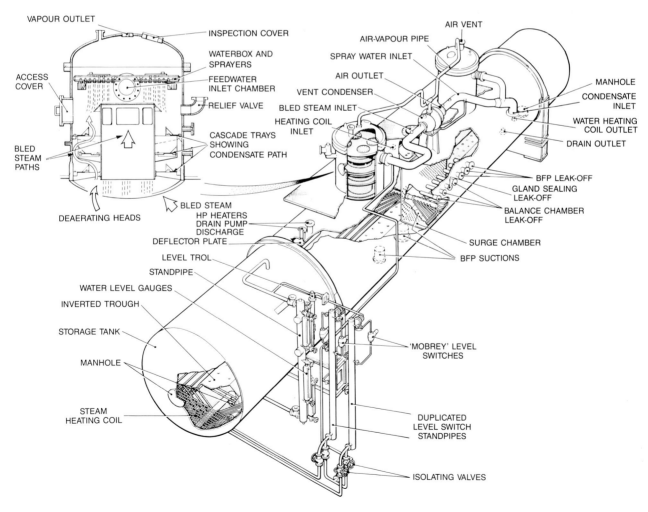

FIG. 2.78 Typical modern de-aerator/storage tank arrangement

under fault conditions. Water level variations greater than the limited buffering range set by the de-aerator storage tank level controller are accommodated by the automatic and progressive opening of a condensate system 'surplus' or 'outsurge' valve (high water level) or of a 'make-up' or 'insurge' valve for low water levels. Outsurge water is pumped to the reserve feed-water tanks and insurge water is drawn from them, usually via the polishing section of the water treat-ment plant.

To maintain the quality of the deoxygenated water for starting purposes, the de-aerator storage tank is kept under pressure during shutdown periods by clos-ing the vents to the condenser and by the use, as necessary, of electric immersion-heating or of steam-heating coils supplied from the boiler. To prevent oxygen contamination of this store, water from the condenser hotwells should not be admitted to the tank during start-up until steam is available from the turbine for de-aeration unless the hotwell contents are stored under nitrogen pressure, during shutdown, when the previous de-aeration effected in the con-denser may result in an acceptable water quality. In no circumstances should make-up water be admitted to the condensers once vacuum has been broken or serious oxygen contamination will occur unless ade-quate atmospheric boiling of the de-aerator storage tank can be accommodated. Incidentally, the atmos-pheric vents fitted to de-aerators are only provided to allow such boiling (and for cold-filling and drain-ing operations) and must never be opened if the de-aerator is under pressure, or the rapid decay and violent ebullition of the storage tank contents may cause feed pump cavitation. In practice, steam-heating coils have been found to be invaluable, not only in maintaining the quality of the stored water during shutdown or in achieving an acceptable quality from cold conditions, but (if the steam-coil facilities have been provided for interconnection with another unit) in reducing the pressure-raising time of a cold boiler or being able more easily to refill a hot boiler drum following, say, tube leak repairs.

The elevation of the de-aerator/storage tank unit should be such that a satisfactory suction head on boiler feed pumps is assured under the worst possible conditions, which will be seen shortly to be on a hot restart, following the failure of condensate extraction pumps or a DC feed system trip. If the unit rejects some, or all, of its load and condensate continues to be supplied to the de-aerating heads then, as the condensate gradually cools (ultimately to condenser conditions) there is an increased attraction for bled-steam which the turbine will be unable to supply. The continuing fall in pressure in the heads promotes a flow of steam from the storage tank steam space and the flashing of the contents, which are stored at the boiling point corresponding to the pressure at the time of load rejection. The rate at which the water in feed pump suction lines gains static head, as it

descends, must be greater than the rate of pressure decay in the de-aerator if flashing in the pump suc-tions is to be avoided. Figure 2.79 illustrates the varia-tion in pressure in a de-aerator and at a feed pump positioned beneath it, following a load rejection. The

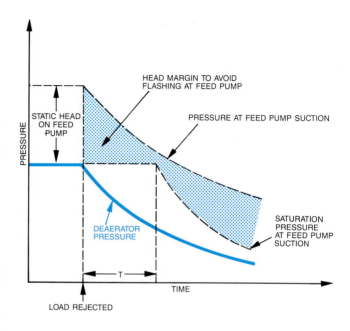

FIG. 2.79 Illustrating the inroads made into boiler feed pump head margins as a result of a rapid decay in de-aerator pressure

curve of pump suction pressure is displaced from the curve of de-aerator pressure by a constant amount equal to the static head of the de-aerator on the pump. As de-aerator pressure decays following a load rejec-tion, the outlet water temperature slowly falls. How-ever, a particle of water leaving at the instant of load rejection will be at the full operating temperature and arrives at the inlet to the pump after travelling for a certain time in the pipe from the de-aerator. The water particle will flash to steam if the total pressure at the pump suction (static head, plus de-aerator pressure, less friction, at that time) is less than the saturation pressure corresponding to the tem-perature of the particle. In theory, flashing will not occur as the dimensions of suction pipes (lengths and diameters) will have been calculated to ensure that a suitable margin of head is available to pre-vent steam-locking of the pump(s) at this critical time. Cooler water arriving subsequently from the de-aera-tor has a greater head margin available. In practice, feed pumps have occasionally been damaged because of several operational factors which had not been adequately considered, as detailed below:

● The emergency start-up of electrically-driven boiler feed pumps on load rejection, due to the fall in output of the steam-driven pump, causing unusually high short-term suction flows from the de-aerator, and hence high pipework friction losses.

• Blocked feed pump suction strainers, or attempting to run at too high a load if one of the duplicated strainers is isolated.

• The additional pressure decay in the de-aerator arising on electric feed pump start if the leak-off lines contain cold water and this is sprayed into the de-aerator steam space to act as a jet condenser. This can be a problem if suction conditions are already near to flashing for other reasons. A remedy here is to arrange for the leak-off lines to end below normal working level; if this is done, however, arrangements have to be made to prevent water syphoning down the leak-off lines when the unit is off-load.

The worst possible case occurs if the unit is off-loaded with an exhausted de-aerator water supply following the total failure of condensate extraction pumps, or as a result of a DC feed system trip. On restart, unless extreme care is taken, the transfer into the de-aerator of the large store of cool condensate that will have accumulated in the condenser hotwells will cause a rapid quenching of the remaining contents and result in an almost total loss of pressure. Should a feed pump be running on low load (possibly on leak-off flow only), the hot water in its suction pipe will be only very slowly replaced by cold de-aerator water and flashing is highly likely. Operationally, it is essential to limit the rate of refilling of the de-aerator to a value that will avoid excessive quenching of the contents and a flow-limiting device, in the form of a de-aerator condensate inlet by-pass line with orifice plate, may be fitted to prevent operator indiscretion. Steam heating coils may be useful in maintaining de-aerator pressure at this time.

Following the unit restart and, incidentally, at any time when the LP feedheating system is operating below designed conditions so far as condensate temperature is concerned, it may be necessary to limit the degree of opening of the de-aerator bled-steam valve, as the cold condensate will attract enough bled steam to damage internal fittings and de-aerating trays, whilst inducing surging of the storage tank water contents. Alarming physical movement of de-aerators is not unknown caused by such effects.

8.2.5 Boiler feed pump considerations

Most 500 MW and 660 MW units employ one 100%-duty steam-turbine-driven feed pump, with one or two 50%-duty electrically-driven pumps provided to facilitate starting and to back up the main pump whilst on load. The electrically-driven pump (or pumps) are usually arranged for automatic starting on loss

of main unit load, when the deteriorating steam conditions available to the feed pump turbine will cause an increasing fall-off in pump output below 60% main unit load, or on loss of the main pump discharge pressure. Whilst an automatic start for the former reason is usually conditional on the establishment of lubricating oil pressure, for example, it is usual to by-pass as many starting checks as possible in the case of the latter in anticipation of a rapidly-deteriorating boiler drum water level. Only if essential pump conditions fail to become established subsequent to the start would the pump be tripped.

Of all the various pumps installed in the power station, feed pumps operate under arduous conditions and are the least accommodating of operator error or neglect. Consequently, they are often provided with relatively elaborate starting interlocks. Unless there is a genuine and continuing will to maintain such interlocking in working order, however, either pumps will not start on demand or will start untoward, without satisfactory pre-start conditions having been met, and reliance on physical checking will become necessary. Such checks must include not only the provision of satisfactory suction conditions, i.e., that all valves between the de-aerator outlet and feed pump suction are correctly positioned and that any installed strainers are on-line and backflushed ready for use, but also important services such as pump leak-offs, balance chamber and gland unloading lines and gland-seal supplies need to be fully available for use. Leak-off arrangements are often duplicated for safety reasons and need to be maintained in full working order. Whilst the automatic opening at some predetermined minimum pump throughput is, of course, essential to preserve pump integrity, even the closing function is important as it is surprising by how much full-load pump duty is reduced if one leak-off valve fails to close.

Pump suction valves must always be opened before, and closed after, the associated operations on the discharge valves, or a damaging over-pressurisation of the pump casing will sooner or later occur due to leaking non-return valves. Whilst this is not a common occurrence, it is not unknown and many stations exercise the option to close the pump discharge valves, on shutdown, before tripping the prime mover.

The operation of variable-speed electric feed pumps at very low boiler pressures may be difficult. On the one hand, the pump needs to be run at minimum speed to keep discharge pressures low and motor cooling may suffer if this is derived from a motor-shaft-driven cooling fan. On the other hand, feed regulating valve control may become coarse as the difference between feed pump discharge and boiler pressure increases. Some stations have found it necessary to apply an artificial head (by only partially-opening feed pump discharge or other isolating valves) and feed system vibration has occurred during feed pump operation at very low boiler pressures if this has not been done.

With feed pumps it is usually beneficial to take periodic readings of important conditions, such as balance chamber or gland unloading pressures, to highlight deteriorating trends.

One final aspect of pump operation that needs to be considered is the avoidance of excessive casing metal-temperature differentials, usually top to bottom, but dependent on pump design. During operation the pump casing remains at an even temperature but, on shutdown, a reversed flow of hot water can occur in leak-off and balance chamber lines, for example, and cause localised heating and distortion of the casing. In the worst cases, the distortion may be such as to take up all running clearances and the rotor will seize. With a turbine-driven pump, the off-load turning of the turbine/pump unit (which may be done for turbine reasons) may have a beneficial effect on the pump, if the speed of rotation is high enough to mix the pump contents adequately, though this is by no means always so. Such potentially damaging leakage flows are usually absent if all or no feed pumps are operating. Conversely, the worst case occurs if one pump is on standby and the others are running, when the hot de-aerator water can be induced to flow down leak-off or balance chamber lines in a reversed manner and into the pump casing by the suction 'pull' of the running pumps. Whilst a variety of methods can be postulated to prevent such reversals, each has its own particular disadvantages and stations must arrive at their own solutions dependent on the severity of the problem. Note, however, that if flow reversals are prevented completely, the initial discharge of the cold water contained in leak-off and balance lines into the de-aerator on pump start may result in a transient, but unwelcome, de-aerator pressure decay.

The steam turbine used to drive the main feed pump may be quite complex, with one or more stages of regenerative feedheating and may be fitted with supervisory equipment to measure differential expansion and rotor eccentricity. Operation is usually easier than with the main turbine on account of the generally low steam conditions available to it, even at full unit load. The worst case arises if the main machine is on a high load and the turbine feed pump needs to be started from cold when it will be necessary, once the glands are sealed and drains set, etc., to permit a lengthy warming period during which the turbine metal-temperatures can rise to approach those of the steam. Sometimes this heating process needs to be accomplished at a somewhat higher speed than might be thought appropriate on account of the need to run the feed pump at a speed sufficient to establish satisfactory hydraulic balance. With the hot (and relatively matched) metal-temperature conditions present following an overnight shutdown, it is often permissible, and is certainly as easy, to run-up the

turbine with the main machine or soon after, when the pump will be available at minimum governor setting to take load as the main machine output increases.

One final point; moisture in lubricating oil can be as serious a problem as on the main machine and is often much worse on account of the extraneous and additional sources of leakage water from feed pump gland housings.

8.3 HP feedheaters

The risk of feedwater ingress into turbines, either as a result of leaking tubes or from a reversed flow of water in cascading drain lines, is comparable to that arising from the use of DC feedheaters, an account of the very high feedwater pressures involved and the energy available in the bled-steam and drainage water. The designer will have ensured that the best practical means have been employed to minimise the risk by providing, for example, high water level alarm and trip floats, duplicated to allow on-load testing; or by the provision of inverted loops in bled-steam lines prior to their entry into the heater shells, in order to increase to a maximum the steam-space volume available for the storage of flood water following its entry into the lines. Quick-acting bled-steam isolating valves backed up by non-return valves are also commonly specified but, despite all these provisions, it is wise to conduct operational checks from time to time to establish whether or not any tubes are leaking, the frequency depending on the susceptibility of the particular installation to tube leakage and on the operating regime of the plant. On base-load plant, the opportunity to make meaningful checks is likely to be limited, it being difficult to detect any increase in drainage flow resulting from a leaking tube (or tubes) until such leakage is severe enough to cause high water level alarm-initiation, frequent automatic starting of the standby heater drains pump, or actual heater-bank trips. This is unfortunate, as a leaking HP feedheater tube has the potential to cause rapid consequential damage by erosion of other tubes in its vicinity. An approach may be made by noting the operating positions of cascaded drain control valves, if fitted, or of drains pump discharge-controllers under steady full-load conditions. Based on experience, it may then be possible to detect unusually high drainage flows, though the presence of variable factors outside the control of the operator may increase the drainage enough to mask any small tube leak present. Occasionally it is claimed that leaking tubes can be detected by an increase in heater noise level and this is, perhaps, a more promising method that deserves development. On base-load machines, it is usually necessary to take advantage of any fortuitous shutdown to conduct a leak check. This is done by maintaining the feedwater side under pressure after the turbine has been shut down, when a continuing flow of water from the steamside drains would indicate a tube leak.

A second aspect concerns the operation of inverted U-tube feedheaters, which are commonly installed. With such heaters, difficulty can be encountered in priming and venting the water side, especially if it is necessary to return a bank of heaters to service from cold, with the main unit on load. Obviously air cannot be displaced naturally from the inverted tube banks and it will be necessary to drive it from heater to heater by the establishment of a significant water flow before final venting at the heater outlet. The situation is similar to the general condition known to promote water hammer (that can occur in any water system as a result of inadequate venting) and damage to heater waterbox baffle plates can result. As these plates separate the water flow entering the tubes from that leaving, any damage will promote a direct flow of water across the waterbox and the heater will cease to function effectively. Some stations have found it necessary to strengthen the plates, damage to which will only become apparent by a lack of feedwater temperature rise across the water on its return to service. Such venting difficulties also cause problems in establishing an initial small warming flow of feedwater in the thick-walled sections prior to recommissioning the heaters, and other means of warming the metal have to be considered. One method is to heat-soak the waterbox and tube plate metal by establishing a small pressure of steam within the heater shell. This should result in a more controlled rise in metal temperature, something that is desirable even in heaters of all-welded construction and is usually essential where bolted joints are employed, if leakage is to be avoided.

A small percentage of the steam entering the heater is vented to the condenser and orifice plates are fitted in the vent pipework to control this flow. Any condensible gases are swept from the heat transfer surfaces of the heater by the carrier steam, finally to be removed by the condenser air extraction plant. Each heater should be vented separately and continuously and any attempt to save heat by the cascading of the vents from heater to heater is misguided, since the risk of air blanketing is greatly increased, particularly in the last heater in the cascade. For the same reason, vent valves should always remain open whenever a heater is in service.

The control of drainage flow from heater to heater may be by the installation of fixed orifice plates in the interconnecting lines or, with heaters having integral drains cooling sections, by the provision of level-controlled drain valves. The automatc drain control valve required for this duty acts as the pressure breakdown device and is fitted at the inlet to the flashbox receiving the drain. Operationally, difficulties occasionally arise with these level controllers and, on base-load machines, it is useful to have an auto/manual selector so that drain valves can be manually

positioned. This may also facilitate maintenance work on the level controllers.

8.4 Miscellaneous plant

8.4.1 Chlorination plant

Despite the widespread introduction of mechanical means of on-load condenser tube cleaning in recent years (Taprogge recirculating-ball systems — see Section 5.5.2 of this chapter), intermittent chlorination is still normally necessary to prevent a slow build-up of organic slimes and sludges. At coastal and estuarial power stations, chlorination is necessary at certain times of the year for mussel control. Lack of attention here can soon lead to gross contamination of the system as it is not unknown for layers of mussels 0.5 m thick to form on culvert walls. On auxiliary CW systems, on-load mechanical cleaning of most, if not all, of the many coolers installed is not cost effective at present and their thermal performance is maintained primarily by chlorination, though it is still necessary from time to time to remove loose sludges and inorganic silts either by 'bubble cleaning' or by utilising the scouring effect produced by increasing the CW velocity above normal. The latter can often only be done with the main unit off-load, when the CW velocity through a particular cooler can usually be increased by isolating all other coolers on the appropriate section of auxiliary main and running more CW pumps than are strictly necessary for the duration of the flushing exercise, so that a higher than normal pressure head is available at the cooler. Unfortunately, none of these methods are effective against hard inorganic scale which must be removed either by acid cleaning, or by isolating and opening up the affected cooler for manual attention with brushes or bullets.

Chlorination of both main and auxiliary systems is automatically effected at intervals decided in the light of early operating experience. The interval is usually between four and twelve hours, the chlorine solution being switched between CW circuits in a timed sequence until all application points have been dosed. The strength of the chlorine solution is also determined by trial, being adjusted to a value that will yield a small residual chlorine content, typically 500 μg/kg, at the condenser or cooler outlet whilst dosing is in progress on the circuit. It is worthwhile establishing a regular monitoring procedure for such residuals for, should a particular point of application valve be passing or fail to close for any reason after a chlorination sequence, then subsequent circuits will receive less chlorine (or may not receive any) dependent on how the total solution flow divides between the parallel paths. This state of affairs can continue for some time if not noticed. The monitoring procedure is not demanding, depending only on observing the colour change produced by a tablet dropped into a sample of the CW collected from the condenser or cooler outlet branch during the chlorination cycle. As such, it can be done by operating staff, who may be better placed to investigate the reasons for any absence of residual chlorine on a particular circuit than are members of the station chemist's staff.

Whilst the storage of bulk chlorine liquid within the CEGB is gradually being phased-out because of the significant risks involved in its transportation, handling and storage, several years will elapse before the process is completed and many bulk storage plants will remain in service for this period. Operations on such plant need to be carefully considered at all times to avoid an accidental spillage of liquid or escape of gas. This implies that a full set of operating instructions or job specifications should always be available and steps be taken to ensure that they represent the best current practice. It is extremely easy for such instructions to become gradually out-dated unless someone is made responsible for updating them. This person could well become the operational 'expert' on the plant. Chlorination plants operate satisfactorily for very lengthy periods with a minimum of attention and it is usually only when the human element becomes involved that the potential for spillage becomes more likely; therefore they should be operated (and maintained) by the minimum number of trained personnel in order to maintain high standards. The biggest risk is incurred when unloading bulk chlorine liquid from a road tanker.

Present policy is to carry breathing apparatus, ready for immediate use, and to work in pairs. It is usual to transfer the liquor using compressed air, and operators need to satisfy themselves that the air is adequately dry, usually not less than that which will give a dewpoint (at 10 bar) of $-30°C$. Note the potential for a serious escape of gas via the compressed air line should pressure fall as a result, for example, if the compressor air dryer's automatic changeover sequence should fail with a vent valve stuck open. Whilst non-return valves are provided and usually operate correctly to prevent a flow reversal down the line, they may not do so positively if the pressure decay is slow; gas has escaped by such a mechanism several times in the author's experience and vigilance is needed. Chlorination plants are provided with means to vent-off any excess of gas pressure that may build up in storage tanks or relief vessels or to facilitate the removal of liquid chlorine from filling lines. Bulk chlorine delivery should never be accepted unless this 'vent gas' system is available for use or it will be impossible to control the pressure in the tank being filled to below a safe level or to clear any slugs of liquor that may remain in the filling line after transfer. These slugs may become trapped between two closed valves, when dangerously high pressures can build up as a result of the high co-efficient of expansion of chlorine liquid.

Finally, the following points are valid:

- Heat must *never* be applied to any vessel containing chlorine.

- The minimum of liquor should be stored, relative to chlorination requirements, ideally split between tanks.

- A receiving 'sink' must always be available, i.e., a CW pump and CW path into which the chlorine solution can be pumped, whilst bulk chlorine is being stored.

8.4.2 Water treatment and hydrogen production plant

These are grouped together, as operations in both plants revolve around the respective safety requirements. In water treatment plants, large quantities of sulphuric acid and caustic soda are stored to be used in the regeneration processes of the ion-exchange resin beds and, whilst direct contact with these chemicals is unnecessary, the potential for spillage clearly exists. Operators need to be fully conversant with the potential dangers and in the action necessary should physical contact with the acid or alkali occur. Panic showers and baths are provided complete with alarms, and must be kept in working order. Specially designed pipe-joint shrouds or covers are available and can be fitted to acid or caustic soda line joints. The covers not only dissipate the energy present in a jet of liquid issuing from a failed joint but have the advantage of being manufactured in a material that changes colour if a small leak is present, thereby giving advance warning of joint failure. They are well worth fitting. So far as the operation of the water treatment plant itself is concerned, this is straightforward though the following points should be noted. First, the quality of the water leaving the plant is of crucial importance to the main generating units, and it is usual to install an automatic trip that isolates any exhausted treatment units should the level of conductivity or silica in the treated water outlet-main rise above agreed levels, typically 40 $\mu S/m$ and 30 $\mu g/kg$, respectively. Secondly, should all the final mixed-resin bed (polishing) units become exhausted and trip, or be manually isolated for other reasons, the continuing demand of the main generating units for make-up water (which is normally supplied to the condenser for de-aeration purposes) may result in a vacuum forming in the condenser make-up line and mixed beds. If the resin traps are partially blocked then the increasing pressure drop across them may result in a resin breakthrough. For this reason, it is usual to provide a by-pass valve across the polishing units, arranged to open automatically on falling pressure in the make-up line. Water from the reserve feedwater tanks would then pass directly into the condenser to satisfy the make-up demand. It is worth ensuring that any such valve is functioning correctly.

Finally, the effluent from the regeneration processes must be carefully neutralised before it is discharged.

The safety requirements necessary in the operation of catalytic hydrogen production plants are understandably onerous on account of the flammable nature of not only the end product, hydrogen, but of the raw fuel, methanol. Liquid methanol is not only flammable but is volatile, the vapour forming an explosive mixture with air. Additionally methanol has narcotic properties, the main toxic effect being on the central nervous system and in particular on the optic nerve. Should actual ingestion of the liquid occur, serious poisoning effects follow, including blindness, mental disorder and, ultimately, death. Operating and maintenance personnel, therefore, need to be fully instructed in the safety requirements necessary to avoid personal danger and to prevent fires. As with so many other similar situations, however, regular refresher training is necessary if high standards are not to fall with time. For personal protection, water supplies and panic showers are provided and must be kept fully available, as must eye-wash bottles for emergency use for splashes in the eyes. Small methanol or hydrogen fires are best dealt with by the use of portable dry powder or CO_2 appliances, whilst larger fires can be tackled with water, preferably in the form of a spray, if other types of extinguisher are not immediately available. Large methanol fires require the use of large quantities of water or the application of an alcohol-resistant foam. Note that, as methanol is completely miscible in water, it is possible to dilute a spillage to a concentration that is no longer flammable, therefore it is normal to find multiple hydrant points installed in the vicinity of the plant in addition to the automatic fixed-jet waterspray installations covering the methanol tanks and hydrogen receivers.

The process of hydrogen production itself is relatively simple. The first stage involves the diluting of the neat methanol with demineralised water to give a mixture strength of 62–64%, by weight, methanol in water. The ratio is quite critical, in particular, methanol-rich mixtures lead to excessive carbonisation of the reaction chambers of the hydrogen generators, as well as to a general waste of methanol. The mixing is effected by combining the output of two, small variable-stroke pumps, one handling the methanol and the other the water. Adjustment of the stroke is by micrometer scale to allow fine tuning, such tuning being necessary once or twice a year dependent on changes in the seasons. It is usual to leave the stroke of the methanol pump near to maximum and regulate on the water pump; if for no other reason, this maximises the combined output and minimises the running time necessary to keep the mixed-fuel tank full. Whilst the effects of small adjustments can be monitored on the microprocessed specific gravity meter/recorder, larger adjustments may be occasionally necessary following pump maintenance, or the processor may be unavailable. To cover these eventualities, means are provided for manually sampling the mixture when, once its specific gravity and temperature have been taken, its strength can be determined from tables. Battery-powered portable density meters are also available giving a direct, digital readout of the sample density and temperature.

The mixed fuel is delivered to the reaction chamber of the generator by another adjustable-stroke pump via a heat-exchanger and pre-heater, in which the fuel is vapourised. By passing this vapour over a heated catalyst, a two-stage chemical breakdown occurs. The methanol first breaks down into carbon monoxide and hydrogen and then the carbon monoxide reacts with the water to release hydrogen and produce carbon dioxide. Pure hydrogen is separated by diffusion in up to twelve silver/palladium diffusers contained in each generator and is then fed to a LP hydrogen storage vessel from which it is periodically compressed for transfer to the main HP storage banks. As not all the reactions may be completed, the hydrogen gas that is produced will be mixed with unwanted water vapour, carbon monoxide, carbon dioxide and traces of methanol. These constituents build up in the reaction chamber, being unable to pass through the diffuser membranes with the pure hydrogen, and are discharged to waste via a pressure regulating valve which, in effect, controls the pressure in the chamber and influences the rate at which pure hydrogen is forced through the diffuser membranes. The other factor influencing the hydrogen flow is, of course, the back-pressure of the LP hydrogen receiver; once this rises to approach that of the reaction chamber, no flow can occur at all. For this reason, the pressure in the receiver is usually restricted to less than 10 percent of that in the reaction chamber.

Operationally, it is necessary to be alert for deteriorating hydrogen purity. Normal purity is greater than 99.99% with a dewpoint, at normal temperature and pressure, of better than −70°C. Should the dewpoint in particular fall, this should be taken as an indication of a diffuser defect, allowing waste water vapour to pass through with the pure hydrogen. Whilst an automatic trip is provided which will eventually isolate the faulty generator, it is usual to intervene manually before that point is reached and to isolate the diffusers, either in groups, initially, or singly when a return to normal dewpoint indicates that the faulty diffuser(s) had been located. Means are provided for venting the generated hydrogen to waste as acceptable dewpoints or purities are being achieved; whilst the unit is being run in this mode, the low dewpoint or purity tripping protection is inhibited. Incidentally, much confusion can be caused when comparing LP and HP hydrogen dewpoints with that existing in the casing of the main electrical generator, as the actual dewpoints depend on the pressure. It is suggested that all dewpoints be converted into the equivalent values that would exist at generator casing pressure as this is constant and the generator is, in any event, the recipient of the hydrogen, sooner or later. By this means, confusion is lessened and straight comparisons are made possible, though a microprocessor is needed for easy conversion.

8.4.3 Compressed air services

Some years ago such services were restricted to two main systems; those supplying the air necessary for unit control and those supplying general station service air, for power tool operation, coal-bunker air stimulation, etc. Additional, and usually separate systems were sometimes installed to serve the air requirements of automatic fire fighting systems whilst the air demands of the ash and dust plant, for example, would have depended greatly on the type of plant installed. Interconnecting facilities were or were not specified, depending to a large extent on whether the quality of the general services air (in terms of its moisture and/or oil content) was even remotely acceptable for use in the delicate internals of unit control system regulators and valves.

In recent years, new uses for compressed air have been specified, and compressed air services are now both varied and extensive. A full list may include:

- Unit control.

- General station services, including (additionally) precipitator hopper dust agitation, cooler 'bubble' cleaning, strainer cleaning, generator casing purge air.

- Sootblowing.

- Turbine forced-air cooling.

- Coal plant, for general use and for the power operation of coal wagon door-release mechanisms.

- Fire fighting services, often extensive.

- Ash and dust plant control and, for dry dust transportation, power air.

- Oil burner control, atomising and/or purge air.

- Residual fuel oil (ORF) plant control air.

Many of these uses are so varied, in terms of pressures and flows, that specific compressor plant is often installed. On a coal-fired station, the result may be a total compressor complement of around 40 machines of various types, shapes and sizes. Whilst there is an increasing preference for machines delivering clean, oil-free air, because of the reduced

contamination to equipment supplied, by no means all of the compressors presently in use operate on that principle, and attention to the lubrication needs of the total complement of compressors forms an important part of their operation, especially with those requiring oil injection. Modes of operation also vary between simple reciprocating designs (which are still extensively used) to the sliding-vane types and high duty, high speed centrifugal machines installed for sootblowing use. As a result, it is impossible to be specific regarding compressor operation, though certain themes recur. First, regular suction-filter maintenance is required, as specified by the manufacturer, if undue compressor wear is to be avoided. Early mistakes in the siting of compressor plant, which was sometimes installed in wet, steamy turbine house sub-basements or adjacent to dust plants, will hopefully not be repeated so that long compressor life is now more than ever dependent on good operating care and attention. Secondly, problems can occur with air driers. In heatless twin-tower designs, the moisture in the incoming air is removed by the special desiccant contained in the towers. Periodically the flow is switched automatically from one tower to the other in order to allow the desiccant to be dried out. This is done by passing through it (to waste) a small percentage of the dry air output of the on-line tower. The automatic changeover sequence and related valve movements are prone to failure and wet air can pass into the air mains. Driers employing heated towers suffer from the additional disadvantage of occasional heater element failure. One station, tired of such failures, has opted to monitor the changeover sequences electrically and has provided, local to each drier unit, a simple flow diagram with neon lamp indicators to show the state of all valves. Timers are incorporated to give a remote alarm should the changeover sequence halt, for whatever reason. Finally, compressors often suffer from cooling water problems, usually associated with scaled cylinder jackets, intercoolers or aftercoolers. The result is an overheated machine, unduly wet air, or both. One solution has been to acid-clean the individual compressor cooling water circuits periodically. A better method would be to control the pH of the cooling water so as to avoid scaling conditions altogether, though this may require the whole of the cooling system being so treated. A final solution that will find favour on those plants fortunate enough to possess a separate closed-cooling water circuit for the more important compressors, is to convert it, usually after acid cleaning, to operate on condensate. This has been done at one station and, whilst it is necessary to exercise a closer control over make-up requirements if an undue loss of high quality water is not to occur, it has resulted in the virtual elimination of compressor problems.

With regard to the compressed air systems themselves, possibilities clearly exist for interconnecting systems of like pressure, providing the air qualities are compatible. Interconnecting facilities are of value in times of emergency, though the temptation to use them as a matter of course to cover up compressor deficiencies will always be present. As a result, overall system integrity can deteriorate rather than improve. A sound policy might involve identifying the interconnecting valves by colour as well as by title and locking them closed. The keys would then be deposited at a central point, affording the opportunity for periodic audit. The use of any interconnecting facilities provided between the control air systems of individual generating units should be undertaken only as a last resort because, in the unlikely event of a major failure on one system, other units may be affected.

To complete this section, a few words on the economics of operation are appropriate. Unlike most other services, compressed air tends to be regarded as a free resource. In actual fact it is expensive, and unnecessary running of the larger compressors in particular should be avoided. A typical sootblowing air compressor, for example, might consume 1600 kW, making any care devoted to its sensible use well worthwhile. At the other end of the scale, general service air system valves can leak heavily with prolonged use and automatic drain traps fitted to receivers (or at intervals around the air mains) are also prone to failure.

No apologies are offered for suggesting, yet again, that one person or a particular shift team be made responsible for conducting regular audits of the system to ensure that defects are reported for attention.

8.4.4 LP water systems

Such systems are not only extensively employed but can be taken very much for granted. Ranging from the main CW system on the one hand, with its associated auxiliary CW circuits and tower pond make-up systems (if applicable), to ash and dust plant collecting and conveying water systems and general swilling-down mains on the other, all systems have some common operating aspects. Any suction strainers that may be fitted require periodic cleaning. This may pose difficulties unless specific operatives are always allocated to the plant: if so, such cleaning will be done almost by instinct, at the appropriate time. With 'common user' plant, reliance must be placed on laid-down routines (which can result in over-cleaning as easily as under-cleaning), or on the observation of differential pressure alarms or gauges. Such gauges do not have a reputation for reliability and direct-reading pressure gauges, positioned before and after the strainers, may be preferred.

Some of the systems are inevitably treated as free commodities; common abuses include:

- The continuous operation of ash and dust slurry pumps as a result of either poor supervisory control or defective automatic pump start/stop controls.

- Oversized ash-sluicing nozzles as a result of wear.

- The unnecessary use of water-operated ejectors.

- Overflowing of CW tower ponds, over and above any requirement to purge.

- Cooling water supplies habitually left open on shutdown (as opposed to standby) plant.

- General leakage.

As a result, a considerable amount of money can be wasted over a period of time, especially in the operation of ash and dust plant water systems, where pump motors may operate at up to 3.3 kV on account of their size. Whilst many reasons will be offered to support the practice of running, for example, ash and dust slurry pumps more than is strictly necessary, they can no longer include the general unreliability of automatic start/stop controls. Pump starting can be ensured by the adoption of the simple pear float, so-named as a result of its general external appearance. This robust device is suspended at the desired operating level by its own electrical cable. It contains an internal mercury switch which operates when the float tilts as a result of a rising water level. Reasonable amounts of floating debris and water heavily contaminated with particulate matter can both be accommodated without problem. Pump stopping is often best achieved by the use of reliable low pump-motor-current detection equipment and whilst the pump will momentarily lose suction before tripping, this is usually of little consequence.

9 Auxiliary plant — emergency operation

9.1 Loss of condensate extraction pumps

This condition is met by rapidly reducing unit load to as low a level as possible, bearing in mind the probable need to prevent a motoring condition of the turbine-generator. Unless steam flow is quickly reduced, a rapid rise in condenser hotwell level will result in a condenser high water level trip, whilst the balancing fall in de-aerator storage tank level necessitates a feed pump shutdown and boiler firing trip. Furthermore, the cessation of condensate flow through the generator auxiliary coolers will require a rapid unloading of the machine to avoid overheating of the generator. Note also that the surplus water held in the hotwells cannot be drained away until an extraction pump is recommissioned or turbine vacuum is broken.

Provided that the fault is quickly located and bearing in mind the need to both prevent water hammer in feed lines on extraction pump restart (if water voidage has occurred) and to avoid quenching the remaining hot de-aerator contents with cold condensate (with pressure loss and feed pump cavitation), it is usually possible to prevent a complete shutdown and load can be gradually restored.

9.2 Water ingress into turbines from feedheating plant

This is a serious eventuality and it is essential that automatic protective devices operate to prevent it from happening. There is very little chance of an operator being able to assess a deteriorating situation quickly to the extent that a judgement could be made as to whether or not water, water droplets or flash steam were present in bled-steam lines. Indeed, an over-hasty manually-initiated turbine trip may initially make matters worse, as the decay in turbine stage-pressures to vacuum conditions would encourage such reversed flows. It is here that the integrity of non-return valves is so vitally important. The initial actions, whether or not a high water level has resulted in the isolation of one or more tubed heaters or the complete DC feedheating system, are to ensure that any actions automatically triggered by the operation of the protection have, in fact, occurred. Obviously, unless operating personnel are completely aware of the detailed provisions of the protection scheme, they cannot do this; regular refresher training in this field can be commended. Bled-steam isolating and drain valve positions should be confirmed and if additional drains are provided in the affected zone, then these should be opened if high water level alarms persist. Continued ingress of condensate or feedwater is equally unacceptable and, if any doubt exists, it would be prudent to use any additional safeguards that may be available. Once the high water level alarms have reset, there is no immediate danger of water ingress via bled-steam lines though these must be fully drained before bled-steam valves are re-opened.

Finally, whilst strictly not associated with water ingress as a result of malfunctioning feedheaters, flooding incidents occur from time to time by other routes. The reheat/desuperheat spray system is a case in point, where the process of injection of cooling water into reheater steam lines has obvious concomitant dangers. Other routes are more obscure. On one large machine, an unexpected route into the IP turbine was via the blowdown provisions on pressure transmitter blowdown pipework, which fed into a boiler blowdown manifold leading to the atmospheric blowdown vessel. Isolation valves had been inadvertently left open after maintenance and, during boiler blowdown, whilst the hot turbine was on turning gear, heavy drainage from main steam pipes caused blowdown

water to back-up these opened instrument drains and enter the turbine, where severe chilling occurred. Drainage systems on power plant are often infamously complex and minor infringements of maintenance or operating procedures can have unexpected major results.

9.3 General station emergencies

These include, for example, chlorine escape, fires and gross spillage of chemicals or oil into station drains. Detailed operating procedures vary from plant to plant but the following general observations can be offered:

- *Chlorine spillage* Immediate steps must be taken to assess the situation, classifying it as *minor* (containable using station resources only) or *major* (requiring outside help). Inadvertent entry into the affected area by personnel unconnected with the escape must be prevented and, using breathing apparatus, the leak is isolated (if possible) and the spillage contained, using sand, ash or even coal. Drains are sealed off for subsequent neutralisation and a foam blanket applied to reduce the evaporation rate. Gas clouds are best dispersed with a fire water-curtain spray. Whilst a start of these actions would be made by station staff, it is likely that the public emergency services would take over general responsibilities for the incident on their arrival at the scene, though station personnel can always offer valuable detailed advice applicable to their particular location.

- *Fires* Unless the fire is very minor, it is usual to enlist the help of fire authorities, if for no other reason that the number of station personnel available may not be sufficient both to fight the fire and to control running plant. Nonetheless, extensive fixed and portable fire-fighting equipment is installed and many quite large fires are extinguished before the arrival of the public emergency services as a result of the prompt actions of station staff. With serious fires, it may be necessary to evacuate non-essential personnel from the main plant buildings; procedures are always available to cover this possibility.

  In severe cases, it may be necessary to take a staff roll call. This usually proves to be much more difficult than might at first appear. Procedures must exist to deal with this eventuality and practised regularly to prove the arrangement and train supervisory staff.

- *Oil or chemical spillages* Gross spillages from bulk storage vessels are rare on account of the bunding arrangements now provided around each vessel. Less serious spillages can still occur and station drains can become contaminated. Means are usually provided to isolate site drains-chambers; whether they are available or not, prime consideration is always given to containing the spillage in as small an area as possible. In this context, it is highly desirable that staff know and understand the site drainage arrangements at their particular location. If there be any danger of contamination of waterways the water authorities may need to be informed: it is sound policy to do so from the start if any doubt exists, as generally a more helpful attitude should then be forthcoming. Serious oil spillages within the confines of the main buildings can usually be more easily dealt with, as the spillage can be contained in the station basement drains pump collecting-chamber, if the pumps are quickly isolated.

10 Auxiliary plant — routine testing

This follows the same general logic that has been described in the sections on main boiler and turbine plant operations and it is unnecessary to repeat it here.

Of all the routine testing associated with auxiliary plant, those tests undertaken to prove the integrity of feedheater high water level alarm and trip floats are as important as any. On direct contact LP and HP feedheaters, duplicate float columns are now specially arranged such as to allow on-load testing of the equipment without effecting an actual trip. With LP float columns, these can be isolated from the system and vented to atmosphere before water is added manually to initiate float operation. A better way is to provide a test water supply, usually taken from the incoming condensate to the heater, so that only manipulation of the appropriate valves is required. HP heater float columns may best be tested by isolating the column steamside connection and then subjecting the top of the column to a pressure reduction achieved by venting off to the condenser. The subsequent flashing-off of the feedwater contained in the waterside connection is usually more than sufficient to operate the floats.

Other auxiliary plant tests may include the following, as appropriate:

- Main boiler feed pump turbine overspeed and on-load overspeed bolt oil-injection testing. On some units it has proved necessary to disconnect the turbine from the feed pump when carrying out overspeed tests, to avoid subjecting the pump to excessive speeds.

- Standby boiler feed pump automatic and/or emergency start tests.

- Fixed-jet fire fighting deluge tests, usually after major outage, the object being to ensure that all scale and rust is removed and that water issues in the correct direction from all nozzles.

- Fire fighting alarm tests.

- Chlorine-leakage emergency klaxon tests and valve exercising.

- Ash and dust pump automatic start/stop tests.

- Compressor tests.

The greatest hazard associated with routine testing is that the protective equipment may be left in the test position and therefore not available for its designed purpose. Testing should be carried out by fully trained staff working to carefully laid down procedures, particularly designed to prevent such an occurrence.

11 Unit operation — run-up and loading

11.1 Pre-start checking

In recent years, not only has the complexity of plant increased but the cost penalties incurred by a late return to service have also risen steeply. As a result, there has been a general tendency towards the adoption of a formal system of pre-start checking. Factors which make the introduction of such a system especially worthwhile include:

- A return to service after an extensive shutdown during which a large amount of maintenance work had been undertaken, necessitating the isolation and draining of multiple systems.

- A high turnover of staff leading to inexperience; this includes those stations having stable staffing but which work to a system whereby plant personnel are subject to job rotation around all the systems, both unit and station, e.g., ashing, dusting and water treatment plant duties, as well as main plant activity. Such arrangements, whilst of great benefit in minimising cover arrangements for sickness and holidays, etc., as any staff fully trained in all plant areas can be reallocated to cover a casual vacancy, are not conducive to the emergence of individual operators who are highly competent in any one particular area.

- Those stations in which the plant runs on baseload. Within the last year on one such station, the total number of starts (both hot and cold and for whatever reason) was only 23, an average of less than 5 per staff shift. Not surprisingly, the operating staff can never be fully confident in start-up activity.

- Those stations having unusual or highly complex units, often of differing manufacture.

Part of a typical start-up network is shown in Fig 2.80, from which it will be seen that the combining of pre-start checking activity and individual actions, such as 'start ID fan', in a logical and systematic way, results in nothing being left to chance in the safe return to service of the unit. In practice, the

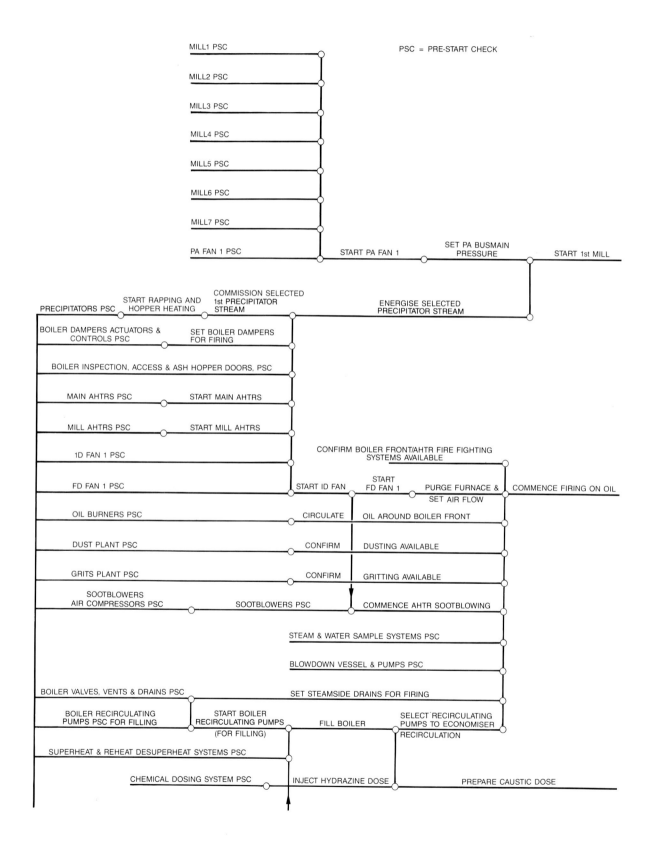

FIG. 2.80 Part of a 500 MW unit start-up network

total display measures some 1 m by 0.8 m and is mounted on a purpose made mobile wooden frame on the unit being returned to service, so that the operating staff in general, and the unit operator in particular, can chart the progress of the return to service. A particularly useful aspect of such a formal

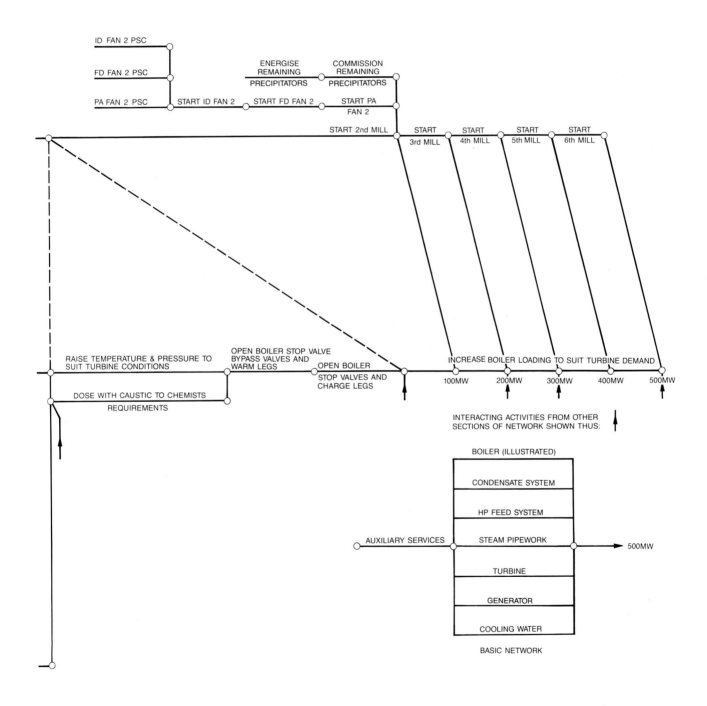

FIG. 2.80 Part of a 500 MW unit start-up network (*cont'd*)

arrangement is that the smooth handover of the unit across any shift-change boundary is assured, pre-start checks either being signed off as complete by the off-going shift for all to see, or awaiting the attention of the incoming shift. There can be no errors caused by communication shortcomings and provided that the operator never starts an item of plant until all activities displayed to the left of that particular action are signed off as complete and the circles inked in and initialled, then nothing can be inadvertently overlooked.

Whilst such a system prevents any forgetfulness, it will not necessarily result in a fast return to service and it is here that possible conflict will emerge. Unless each activity is processed by the work study department and a time allocated, it is impossible in theory to ascertain which is the critical path. Initially some times may be found wanting and need adjustment, but fortunately it does not take long in practice to establish which activities are difficult and take longer than originally thought. For example, on one make of generator currently in use in large numbers within the CEGB, the pre-start checking and priming (including the satisfactory venting-off of all air) of the stator coolant system takes much longer than might be imagined and until this is done the machine cannot be run up; many avoidable delays occurred until 'experience tempered theory'. Similarly, there is little point in theory in pre-start checking that boiler gas-side doors are boxed-up (so as to permit firing) until the boiler is filled with water to normal working level. In practice, this is not necessarily so. If extensive entry has been made during the outage and many doors need to be boxed-up progressively (and possibly have hinges that need freeing), then the boiler can be full and ready for firing well before any draught plant can be started. Nor should it be assumed that every plant attendant can accomplish his allotted task within the stated time; the more experienced may do the job in half the time, the newer attendant may take twice as long, for he is unlikely to sign off the check sheet until he is fully satisfied that all is well and may double-check some items. Provided that the network is correct and the pre-start activities well thought out and regularly revised, they form the basis of a sensible and acceptable approach, and no more. A certain amount of confusion and much frustration occurs if a particularly lengthy pre-start check list is not logical in its approach; priming a section of pipework cannot be done in a 100% method study or labour-saving way as the drains, for example, need shutting first, whether or not two trips to that area are necessary. It is perhaps surprising how often in the past pre-start checks have lacked such basic common sense. Further, the major activities which must be done correctly, or plant damage will automatically result, need to be highlighted in bold type, printed in capital letters or otherwise made to stand out as important or the attendant may become utterly con-

fused between important and less-important activities. Such a style can be used to advantage for units returning to service after short-term outages, where little work has been done. Here, the network can still be of use but a decision may be made to complete only the major items appearing on the check sheets, thereby gaining a rapid return to service with little or no increase in risk. Another approach adopted by one station has been to accept a more-senior operating person's signature directly onto the network display as satisfactory evidence that he has 'looked over' the particular item of plant without recourse to the formal procedure. This can save much time with little extra risk and such an approach is worth considering in any event if some plant items have continued to run during the outage, for example, auxiliary CW systems or an ID fan, when it would be necessary for the operator to acknowledge that fact and by-pass part of the network.

Finally, if such a formal system is to remain effective, a person or persons must be charged with the responsibility of revising the procedures periodically, for it is surprising how often modifications are made (sometimes due to obsolescence) that can affect a pre-start check even on the more successful stations.

None of the above would be either necessary or reasonable in returning a unit to service after a short, say overnight, shutdown during which little or no maintenance had been undertaken. Whilst there may still be a need for a summarised action sheet, or similar, to prompt the operator's memory, there is certainly neither the time nor the need to invoke the formal pre-start checking procedure.

11.2 Boiler considerations

11.2.1 Filling and hydraulic testing

Hydraulic tests on boilers are done for two main reasons:

- To satisfy the boiler insurer's representative at agreed intervals following major refurbishment or overhaul that the boiler is safe for continued operation. The pressure to which the boiler is raised may exceed the normal operating pressure (referred to as the *maximum permissible working pressure*, MPWP) and is called an *overpressure* test.

- As a means of testing for leaks, either before a major overhaul or after minor tube repairs have been effected. Here, the hydraulic test pressure never exceeds the MPWP and is often only a fraction of it, as the aim is to check that no leaks exist rather than to subject any minor repairs to a pressure that will prove their ability to withstand normal operation, a requirement that is fulfilled just as adequately by some form of non-destructive testing, e.g., ultrasonic testing.

The basic operating requirements for hydraulic testing are quite straightforward and are set out below:

- The quality of the water used for the test should be as high as practicable. This is particularly important where austenitic steels are present in the boiler pressure parts, as any appreciable amounts of chlorides present in the water can induce the stress-corrosion cracking of this steel. The requirement is satisfied by the specification for modern water treatment plants, i.e., a conductivity not exceeding 10 μS/m at the plant outlet. Prior to filling, if there is any doubt that the water contained in reserve feedwater tanks may not reach this standard, then it is a sound policy to draw directly from the water treatment plant outlet. In practice this can usually be achieved simply by ensuring that the outlet flow is maintained at such a rate that a small surplus of water flows up to the reserve feedwater tanks, over and above the filling flow, rather than in the reverse direction.

- The temperature of the water must be greater than 7°C at all times in order to eliminate any possibility of freezing.

- Drum differentials must never be exceeded.

- The best practical means of removing air from the pressure parts must be employed. This requirement is paramount for reasons of personnel safety. If an appreciable amount of air is trapped in the boiler pressure parts then it will be compressed as the pressure is raised, and be available as stored energy to eject water forcibly should a failure occur under test. The whole concept of hydraulic testing is designed to ensure that potentially dangerous energy levels cannot build up during pressure raising; should any air (or gas in the form of nitrogen, if the boiler has been stored wet, under nitrogen pressure) be present, except in minimal quantities, then the concept is clearly negated. All boiler vents must, therefore, be operated adequately as the boiler is filled but the process of filling itself must be done with care. If a boiler feed pump is used for this operation it is likely that the leak-off arrangements will be in constant use, causing a certain amount of aeration of the feedwater in the de-aerator and subsequently in the boiler. Similarly, a large amount of air can be entrained in the downcomers during the filling process if the feed flow to the boiler is high. For both these reasons, it is necessary, once the boiler is full, to allow a period of standing during which the bubbles can rise, collect and finally be vented off. If the means exist to fill the boiler using the recirculating pumps, which may have filling connections teed into their suctions, enabling water to be drawn directly from the water treatment plant

outlet, then this should be done. Such connections would be subsequently blanked off prior to returning the boiler to service. It is further suggested that the normal venting arrangements need augmenting by additional precautions, varying from plant to plant and dependent on the layout of auxiliary piping. Boiler steam and water drain-lines and desuperheater spray systems may contain long lengths of horizontal pipework which may not vent off automatically during the filling process. Such eventualities need considering so that a programme of blowing down can be instigated to clear the systems of air during a convenient part of the filling operation. Sample systems are usually quite extensive and would benefit from similar treatment though the total volume of air contained in them is not excessive.

A good indication of the degree of success in removing all air is the ease with which pressure can be raised once the boiler is completely full and vented. If no air remains, a small opening of the feed pump discharge by-pass valve should see the pressure rising steadily towards the desired value. If difficulty is apparent at this stage and use of the main discharge valve becomes necessary, then either there are significant leaks (which may be in the form of passing valves) or much air remains which is being steadily compressed.

- During the pressure raising process, boiler pressure parts should be protected from severe impact: operationally this is fulfilled by ensuring a steady and uniform pressure raising sequence. Experience over many years has shown that it is certainly possible to inflict damage on a sound drum during an overpressure test, in particular, by bad practice.

Finally, if all the above considerations have been acted upon and no major boiler leaks of any description are present, it should be possible to maintain the test pressure for a period with the boiler feed pump isolated from the boiler, or at least using a percentage of the discharge by-pass valve flow only, under which conditions the potential energy stored in the boiler pressure parts is the minimum practicable and the personal safety of the inspection party is assured so far as is possible.

The normal filling of the boiler to working level can be accomplished very rapidly provided that drum differentials allow. Examples are on record where a 500 MW boiler was ready for firing within the hour, the main limitations being the replenishment of the de-aerator and the rapid removal of the large quantities of air displaced from the boiler. During filling it is usual to introduce a charge of hydrazine via the HP dosing pumps to 'mop up' any oxygen present in the feedwater: though solid additives, in the form of caustic soda, must never be injected until firing has commenced and a natural circulation established to thoroughly distribute the additive uniformly throughout the boiler. The drum level must be confirmed by observing the absolute gauge glasses and not from any differential-type instruments fitted which rely on a reference level maintained in a constant-head chamber, because such a reference may not be established until steam production and condensation is established.

Some boilers are designed to operate completely on natural circulation and have relatively large water-wall tube diameters. Others employ much smaller diameter tubes and rely on a measure of assisted circulation provided by boiler recirculating pumps, which must be running prior to firing.

11.2.2 Pressure raising

Drum differentials

Drum differentials are most likely to be exceeded when filling a hot drum with cool water or during pressure raising or forced-cooling operations. Modern drums extend to 36 m or more in length and have wall thicknesses of, typically, 100 mm. Such massive structures can be seriously damaged by repeated metal yielding (leading to fatigue cracking) brought about by thermal stresses set up by differential temperature changes occurring either across drum walls (inner-to-outer surfaces), across drum top-to-bottom surfaces or along the length of the drum. The last is somewhat less usual unless excessive or localised firing patterns result in relatively hot/steam mixtures being injected into the drum at certain positions only, along its length.

To minimise thermal stressing during start-up or shutdown, it is necessary to limit and control the temperature gradients occurring both across the walls and between drum top and bottom surfaces, these being related such that if one is high the other must be kept low, and vice versa. The whole scenario is also dependent on the internal operating pressure. Figure 2.81 shows a typical set of operating curves. Under start-up or during forced-cooling operations, a close approximation to the through-wall differential temperature is given by comparing the steam off-take pipe temperature with the drum-top external temperature or by a comparison of feeder tube temperature with drum-bottom external temperature. Drum-inside metal-temperatures can therefore be inferred with a close accuracy by thermocouples embedded in steam off-take pipes or feeder tubes, a much simpler proposition than metering drum internal temperatures directly. Figure 2.82 shows one practical application. At any given drum pressure, the average through-wall differentials can be compared with the top-to-bottom differentials to ascertain from the graph whether operation is within limits or not.

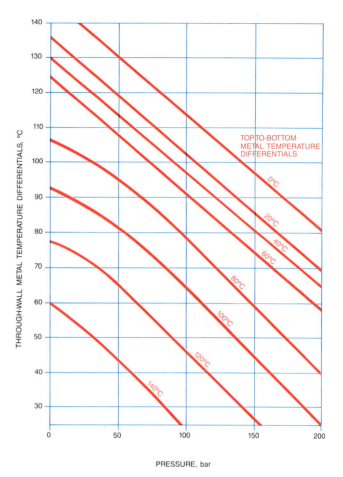

FIG. 2.81 Allowable drum metal-temperature differentials
Typical only, for a boiler having a normal working pressure of 160 bar.

Top-to-bottom differentials can be effectively controlled in some cases by the use of drum spray, especially during forced-cooling activities when no steam is being drawn through the superheaters and so no danger of water carryover can be present; conversely, the use of drum spray during light-up conditions is often restricted whilst boiler steamside drains are open.

In deciding if a hot and empty drum can be filled with cool water, a different approach needs to be taken as internal drum metal-temperatures can no longer be inferred from empty feeder tubes or from off-take tubes having no steam flow. Here, providing that the drum has been empty for some hours, the readings of external metal temperature will approximate more and more to the internal wall temperature because no unnatural heating or cooling is taking place. It is permissible, before filling commences, to compare the drum metal-temperature with the incoming water temperature and read this as a through-wall differential on the graph, thereby fixing the allowable top-to-bottom differential.

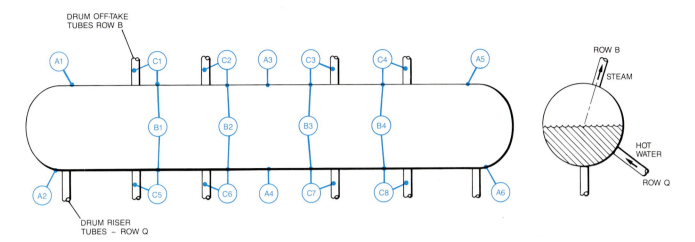

RECORDER CHANNEL A 1-6 DRUM OUTSIDE WALL METAL TEMPERATURES °C
RECORDER CHANNEL B 1-4 DRUM TOP TO BOTTOM DIFFERENTIAL TEMPERATURES °C
RECORDER CHANNEL C 1-4 DRUM TOP INNER TO OUTER DIFFERENTIAL TEMPERATURES °C
(THE INNER METAL TEMPERATURE INFERRED FROM THE STEAM OFF-TAKE TUBE METAL TEMPERATURES)
RECORDER CHANNEL C 5-8 DRUM BOTTOM INNER TO OUTER DIFFERENTIAL TEMPERATURES °C
(THE INNER METAL TEMPERATURE INFERRED FROM THE WATER RISER TUBE METAL TEMPERATURES)

FIG. 2.82 Typical recorder display showing drum metal-temperatures and differentials

Firing and drainage

A cold furnace is preferably fired by the use of oil burners because the heat input can be better controlled than when using PF and because the short, localised radiant oil flame is much better suited to initial pressure raising than the slower-burning PF flame, which can result in the overheating of superheater elements until a good steam flow is established through them. Every effort must be made to obtain good combustion, especially whilst the furnace is cold, and oil burners should be used in combinations that give the best possible mutual support. Faulty or smoky burners should not be used. Adequate control over initial combustion from the control room is very difficult and it is always beneficial to provide a person locally to monitor conditions. Certain oil burner selections may also be necessary in order to keep the heat away from any vulnerable radiant superheater surfaces located in the furnace, unless they are equipped with sufficient and reliable metal-temperature thermocouples to monitor their condition effectively.

Many stations have experienced great difficulty in maintaining such gas-side thermocouples and have installed 'deadspace' thermocouples monitoring the superheater (or reheater) outlet pipes instead. Whilst quite acceptable for on-load monitoring, being free from all those factors which make the determination of accurate mid-wall metal temperatures so difficult in the gas stream, they are of no value at all during start-up conditions until a steam flow has been established.

Pressure raising is usually limited to a rate that gives a 50–80°C/h change in drum saturated-steam temperature. This allows the full boiler pressure to be achieved from cold in 5 h on a 160 bar boiler, always provided that drum differential and superheater/reheater metal-temperature limits are not exceeded. The saturated-steam temperature rise is fixed not only to ensure the integrity of thick-walled components, such as the drum or headers, but also to limit the expansion rate of the boiler pressure parts as a whole, so that differentials between those parts and headers, skin casings or support arrangements cannot result in torn welds or similar damage. During pressure raising, the reheater is particularly vulnerable, as it sees no steam flow at all until the turbine is running; for this reason, it is usual to bypass as much flue gas as possible away from the tube banks.

Superheater circuit drainage varies to suit the needs of the particular installation but conforms to the following principles. Intermediate and boiler final-drains usually perform different functions. The former are installed to allow the removal of condensate which can collect in large quantities in the inlet headers to intermediate superheater stages and, if not removed, can prevent steam flow through some, or all, of the tubes of the following stages, causing overheating. The danger is particularly acute in the case of a steam furnace-division wall. Intermediate drains should only normally be operated sufficiently to remove the condensate; this may entail intermittent operation and/or the use of condensate-in-drain-line detection equipment, such as the Hydratect system.

Final superheat drains perform a different, or at least, an additional function; they promote sufficient

135

steam flow through the superheater as a whole to
keep the various sections 'cool'. In consequence, the
drainage applied, whilst often varying with the type
of start-up, is invariably continuous until steam flow
is transferred to the turbine steam-lead drains or to
the turbine itself.

During the later stages of a cold boiler start, when
the first PF mill may be commissioned, or throughout a hot start-up necessitating the use of a PF mill
from the beginning, the setting on the superheater
final drain may be usefully varied, within reason, to
maintain a steady rise in boiler pressure. This is a
better method than fixing the drainage and attempting
to trim the firing rate, always a difficult process with
PF. It may not be generally realised that one of the
main objectives in any boiler pressure raising operation is to maintain a steady rise to the desired value.
If pressure swings start to develop, drum levels,
superheater metal-temperatures and other parameters will
cycle in sympathy. Whilst the pressure and temperature
conditions required of the boiler for a cold turbine
start-up will be low, typically no more than 60 bar,
280–300°C for a 160 bar unit, the same is not true
for hot turbine starts, for which boiler superheater
metal-temperatures will have to be held quite near to
alarm levels if adequate final temperatures are to be
achieved. For hot starts, the control of steam pressure
is quite demanding. Figure 2.83 shows a record from
a 500 MW boiler hot start-up which illustrates the effect
of not keeping to a steady pressure rise.

One last item may be of interest on this subject.
It is by no means uncommon to find that otherwise
similar, actuator-operated final drains start to pass
steam flow at significantly-different indicated settings.
In general, it is always a wise precaution to note the
setting at which the valve 'cracks' by noting the sudden
change in previously static drain-line or steam-lead
temperatures, adjusting subsequent openings to suit.

Boiler loading procedures are not particularly demanding provided that the introduction of second
and subsequent mills is correctly timed to take advantage of increased steam demand and is carefully
executed, but note the restriction on the use of desuperheater spray at low boiler loads. This can be
quite a problem on some boilers, as the radiant
pendant superheater sections tend to overheat at the
lower loads, being directly above the fire and having
relatively little steam flow. Once sprays can be commissioned, there is seldom any further concern.

11.3 Turbine considerations

11.3.1 Preparation

Pre-start activities can reasonably follow along the
lines discussed in Section 11.1 of this chapter, though
any attempt here to detail activities in any depth
would fail due to the differences between machines

and ancillary systems. Suffice it to say that the
following general principles apply:

- Commission turbine lubricating oil system, paying
  particular attention to the need to raise the oil
  temperature to normal prior to run up; start the
  jacking oil pumps and commission the turning
  gear. Barring speeds vary between machines and
  the jacking oil pumps may have to run continuously thereafter until the turbine run-up is in progress.
  This activity is usually done in conjunction with
  the generator preparation.

- Set turbine drains systems.

- Charge condensers with circulating water and establish a starting flow.

- Commission the pumps of any fire-resistant fluid
  system installed for turbine valve control, raising
  the fluid temperature to normal.

- Set and prove the turbine trip gear, exercising steam
  valves according to instructions. Limitations in valve
  lift may apply whilst valve seats are cold.

- Commission or set auxiliary systems, e.g., condenser
  extraction pumps, gland-sealing and flange heating
  systems and vacuum-raising plant.

- Commission turbine supervisory equipment.

- Ensure that the fire protection systems are correctly
  set. Following a major outage, it is usual to initiate
  the deluge systems to ensure that all nozzles are in
  place and are correctly positioned and operating.

11.3.2 Thermal constraints

Basic principles

A steam turbine is primarily designed to carry its
load at rated speed and under specified steam inlet
conditions. It may be safely assumed, by and large,
that it will do this once stable conditions have been
achieved. Such conditions include the establishment
of normal temperature gradients from inlet to outlet
and the free development of all thermal expansions. If
it were possible to put the machine very quickly
into this condition, then the most rapid of starts would
be possible and desirable. Unfortunately, that situation is only approached when restarting a very hot
machine after the briefest of shutdowns, such as
might occur, for example, after an inadvertent trip
from a high load with an immediate restart. All other
turbine starts fall short of this ideal to a greater
or lesser extent.

To limit stresses in thick metal sections, the starting process resolves fundamentally into a controlled

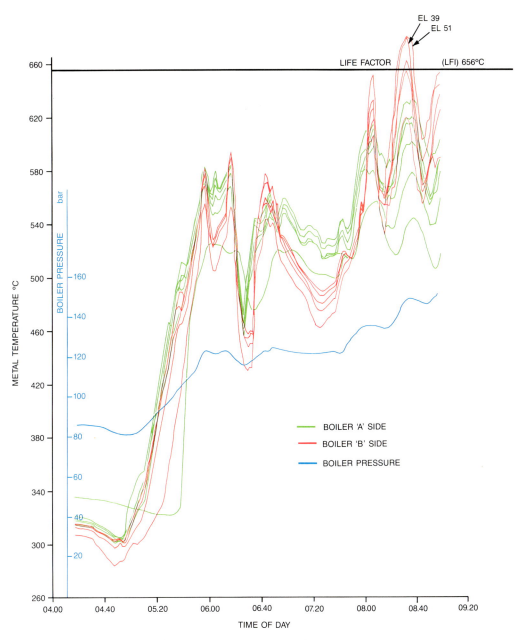

FIG. 2.83 500 MW superheater (secondary inlet loop) metal-temperatures showing the result of poor boiler pressure-raising control

heating exercise of main steam leads, steam chests and turbine components and, as the heating steam also drives the turbine, it follows that the rate-of-rise of turbine speed and load is also controlled (restricted) by the heating process to a degree dependent on the temperature of the metal parts. The object then, in starting, is to control the admission of steam passing into steam heads, chests and turbine casings to effect a run-up in the minimum time consistent with keeping metal stresses within acceptable limits. Experience and research have shown that with the thickest of metal sections currently in use, thermal stresses will be acceptable if the rate-of-change

of surface metal temperatures are below 150–200°C/h, assuming that the sections are heated from one side only. For the more complex, non-symmetrical sections the acceptable rate falls and a general working limit of 120°C/h (2°C/min) is often specified and can be regarded as typical. Not so long ago some turbine manufacturers, in setting down operating instructions for their plant, specified only acceptable steam inlet conditions, run-up times and subsequent loading rates for a range of pre-start HP and IP turbine metal-temperatures, arguing that providing these were adhered to heating rates would be automatically acceptable. There was some merit in this

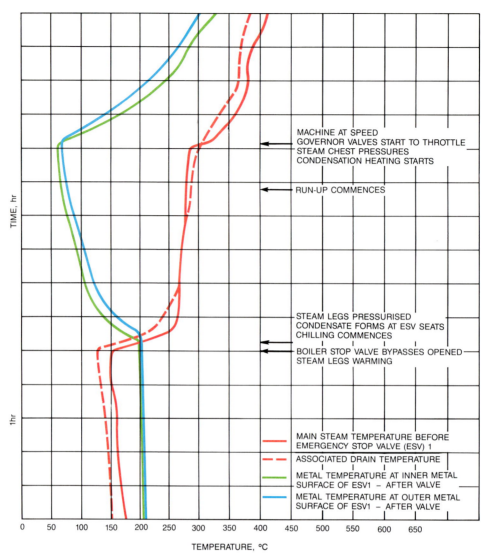

FIG. 2.85 Steam chest chilling

The recorder traces show a bad chilling event on a 500 MW turbine steam chest. The fall in emergency stop valve (ESV) outlet metal-temperature is brought about by a passing valve, and is aggravated by a too-rapid pressurisation of the steam leads, causing condensation to form. The subsequent rapid rise in temperature is as a result of condensation heating of the metal as the chest is pressurised. Both effects are clearly seen on recorder traces 3 and 4. Operating staff need to exercise the greatest care in their start-up operations to avoid such potentially-damaging temperature cycles.

soon as the glands can be sealed and a small vacuum raised. The pressure is allowed to rise slowly to about 7 bar in two hours and is then held at that value for a further two hours. The machine may roll off barring and run at a few hundred revolutions/minute. A small flow of heating steam passes into the IP turbine via the shaft cooling line (this being provided to seal and cool the stagnant central section of the double-flow IP cylinder and rotor with relatively cold HP exhaust (cold reheat) steam at all times). At the end of the process, HP and IP turbine metal-temperatures have risen to about 240/120°C, allowing the machine to be run up to speed and loaded to the warm-start curves, resulting in a small reduction in the overall time taken to reach full-load.

Open-valve starting techniques are finding increasing use and generally result in lower rates-of-rise of metal temperature (and lower through-wall differentials) in all components. The boiler is prepared for firing as normal but the BSVs are opened wide. Turbine checks are completed and the machine is put onto turning-gear whilst the condensers are charged with circulating water and a small flow established. Main steam-lead drains are opened and steam valves exercised and, if permitted, trip tested. Thereafter, both emergency stop and governor valves are opened soon after firing commences and a small vacuum is raised, in case steam passes into the turbine earlier than expected. Steam production proper can be seen by the readings on boiler final-drain and steam-lead temperature indicators rising towards 100°C; at this point, the vacuum-raising plant is commissioned fully and within a few minutes the turbine will roll off barring under steam. The subsequent speed rise may have to be held down to comply with any restrictions and, in any event, must match the initially-poor vacuum conditions prevailing. If facilities are provided to enable the speed to be adjusted by throttling across the governor valves so much the better, as by careful manipulation of these valves and the emergency stop valves, the steam chest pressure can be kept at a value low enough to avoid saturation heating. Indeed, if these adjustments have been done with care, it may be possible to have the steam chest temperature within 20°C of the steam saturation temperature by the time the machine speed is transferred to the automatic speed loop. If manual throttle-valve control is not provided for run-up purposes, the machine should be held at just below governing speed until the steam chest temperature is within 20°C of saturation temperature and this may well extend the total run-up time. The turbine is typically up to full speed in something over one hour from initial rolling, but even when transfer to governor control is permitted, it is still usual to hold it at 3000 r/min until the IP turbine metal temperatures have risen to, say, 200°C. Thereafter, the machine is synchronised to the grid system and loaded to the cold-start curves.

Warm starts

Warm starts are defined as when the HP turbine valve chest and/or turbine metal-temperatures are between 75 and 300°C. In practice, these starts can be more troublesome than cold or hot starts as a result of complications arising with turbine differential expansions. Run-up times may be quoted to suit turbine conditions but in practice may be dictated by the requirement to warm the steam chests to within 20°C of saturated steam temperature before they are pressurised. Steam conditions at turbine stop valves should be adjusted during the vacuum-raising process such that, after throttling to about 4 bar (the initial inlet-stage pressure), they match the HP turbine inner-casing temperature. Once on-load, the machine is loaded to the warm-start curves.

Hot starts

Starts are defined as 'hot' when the HP turbine valve chest and/or inlet metal-temperatures are above 300°C. With such starts, providing that the boiler pressure is held down to a sensible level, saturation heating is impossible. Steam conditions should again be such that, after throttling to 4 bar, they match the HP inner casing temperature. Run-up times range from less than 5 minutes for the hottest conditions to 10 minutes, or so, at 300°C. If the turbine inlet metal-

temperature is much in excess of 400°C, it is unlikely that the inner casing temperature will be matched until the machine is on-load and an initial block load is applied. Such a load would, typically, be 10–20% of rated full-load. Subsequent load increases are to the hot-start loading curves.

11.3.3 Mechanical constraints

Mechanical constraints arise as a result of the heating processes discussed in the last section or they may be independent of heating; from either cause, they deserve close attention if turbine damage is not to result. Fortunately, with cold starts, both in general and after major overhauls in particular, time exists to monitor carefully and react to any abnormalities prior to the run-up starting: for example, in vibration, shaft eccentricity or expansion, though obviously it is vitally important for operating staff to be fully aware of the relevant limits.

So far as differential expansions are concerned, it is permissible and necessary to commence a cold run-up with differentials that would prevent higher-speed operation on account of the contraction of the shaft train with the Poisson effect, knowing that the heating process will expand the shafts sufficiently, in relation to casings, to accommodate this effect during the course of the start. In the past, some stations have adjusted gland-steam temperature controllers to increase the heating effect at the gland areas knowing that this would increase the (expanding) differential expansion, thereby allowing an earlier run-up to speed. The practice should be avoided, however, due to the possibility of cracking gland breathing grooves. With hot starts, by comparison, the usual five minute run-to-speed programme allows no room for complacency nor time to manoeuvre, and differential expansion (and other) conditions must be satisfactory for 3000 r/min operation before the run-up is permitted.

With any type of start, the natural vibrational characteristics of the machine will be known and the actual peak values anticipated, especially at critical speeds. Table 2.9 shows the critical speeds for one particular machine and illustrates the speed ranges, outside those critical bands, in which speed can be held. Whilst there is no necessity for any rapid increase in speed when passing through critical ranges, it is important that speed is not allowed to dwell therein. On the more modern machines, automatic run-up sequences should prevent operation at a speed too high for prevailing conditions but often, once speed is held, control reverts to the operator and he must be in a position to take over.

Should eccentricity levels increase beyond normal, it is likely that heating transients are responsible and the rate of run-up and gland-steam, main steam and metal-temperature conditions should be checked. Ultimately, it is essential that speed be reduced to a

TABLE 2.9
Critical speeds for one particular machine

Component	Critical speed 0–3000, r/min range
HP turbine	Nil
IP turbine	2600
LP turbine	1500
Generator	1000 (1st), 1800 (2nd)

Speed 'hold' ranges, based on above;
> 900   r/min
2000–2300   r/min
> 2700   r/min

level that will prevent permanent bending and such a speed must lie outside the critical range.

Bearing temperatures, and differentials across bearings if fitted, need to be carefully monitored on recently overhauled machines until a repeatable trend has developed.

Finally, if overspeed testing is due, it will be completed and any mechanical auxiliary governor reset before sychronising and loading.

12 Unit operation — de-loading and shutdown

Not unexpectedly, the reason for the unit shutdown will influence the de-loading procedure to be adopted. For example, if the shutdown is for two-shifting purposes, it is important not only to maintain the highest possible turbine metal-temperature prior to shutdown, but also to agree on a target boiler pressure at which to come off load. High turbine metal-temperatures allow rapid subsequent reloading of the turbine, as the temperature loss due to natural off-load cooling will be at a minimum. Additionally, the overall heat requirements of the two-shift will be minimised. The very highest turbine metal temperatures can occur only if the unit is virtually tripped from·full-load and, whilst such a procedure would normally neither be acceptable to the System Operations department nor form the basis of a realistic operating procedure, the nearer that it can be approached in practice the better. A compromise solution may be to reduce load initially to about half-load at a rate to suit System Operations requirements, taking out of service as much of the milling plant as possible. The unit load is then held at this level long enough to regain any loss in steam and turbine metal-temperatures that may have occurred during the initial unloading phase, usually as a result of a planned reduction in boiler pressure, which action too should also be completed at this stage. With some designs of boiler, it may not be possible to maintain full steam temperature at half-load and it may be necessary to pause at a somewhat higher load.

The target boiler pressure chosen at which to come off load is such as to allow the desired start-up steam temperature to be achieved without risking an excessively high boiler pressure during light-up. For example, for the shortest of shutdown periods boiler pressure may have to be reduced, on a 150 bar boiler, to a value as low as 80 bar, as there will be little natural pressure decay during the short off-load period and the heavy firing necessary on starting, to match the turbine metal-temperatures, may otherwise result in an embarrassingly high boiler pressure. For the more normal 6 to 8 h shutdown period a target of 100 bar, again on a 150 bar boiler, would be reasonable. Notwithstanding this, boiler pressure needs to be reduced somewhat in any event, prior to shutdown, or the heat retained in furnace slag deposits may cause the safety valves to lift once the turbine is tripped and the boiler has been 'boxed in' by closing the gas-side dampers. Incidentally, whatever the target boiler pressure chosen, it is important that is is achieved prior to the cessation of firing or it will be necessary to delay the final unloading of the turbine to effect the required response, which will automatically result in significant cooling of the turbine.

During this period opportunity is taken to adjust other unit conditions, such activities may include:

● The completion of any previously-initiated airheater sootblowing sequence.

● The cooling of the remaining in-service mills to avoid any possibility of internal coal deposits from firing during the shutdown period. It is quite common to limit the shutdown internal mill air temperature to 50°C for this reason; without some artificially applied cooling prior to shutdown, the heat retained in grinding-zone components may be sufficient to lift the temperature above this level.

● The effective isolation, whatever this implies, of the superheater and reheater desuperheat spray systems.

● The transfer of the boiler feedwater requirements from the turbine-driven feed pump to the standby, motor-driven unit.

● The transfer of the unit electrical supplies from the unit transformer to the station switchboard.

Once the necessary unit conditions have been achieved and permission to shut down has been granted, oil burners are commissioned (if not already in use) and shutdown procedures are simultaneously implemented on the remaining mills. As soon as the coal flames are extinguished, the oil burners are tripped (usually via the ignition trip button to check its operation) and the unit load is reduced quickly to zero. Generator motoring conditions are proven and the main circuit-breaker is opened as soon as the active and reactive loadings on the generator have been readjusted to zero. Motoring conditions can be observed on the generator electrical metering and give confidence that the turbine speed will not rise after the circuit-breaker has been opened, due to passing (leaking) steam valves. The steam valves must not be tripped-closed to shut down the machine until the generator excitation control has been placed on 'manual' and reduced, or the automatic voltage regulator will attempt to maintain normal machine voltage despite the falling speed and an overfluxing condition will occur on the generator transformer core, resulting in overheating and possible interlamination insulation breakdown. This possibility is real enough for most machines to be fitted with overfluxing protection which operates by comparing the generator voltage with the machine frequency and tripping the generator field switch should the result be unfavourable.

Once the turbine steam valves have been tripped-closed, all that remains to be done is to check that the standby lubricating oil pump starts automatically (as the output from the shaft-driven oil pump falls with speed) and that bearing oil pressure is maintained; also that the turbine speed is falling at the expected rate. Vibration levels increase as the machine speed falls into the critical speed ranges, though it would be unusual for operating staff to notice any worsening of trends due to the transient nature of the resonant vibration. Since increasing vibration at resonant speed is a clear indication of crack propagation in a shaft, it is becoming usual to initiate a full vibration recording during the rundown period automatically, the data being stored on magnetic tape for later analysis and comparison with previous vibration surveys.

Meanwhile, on the boiler, once ignition is tripped and whilst the stipulated furnace purge is proceeding, it is essential to conduct checks (locally, if necessary) to ensure that all oil burner flames are out and that oil is not leaking into the furnace. It may even be good policy to isolate the oil mains, as cases are on record where oil continued to pass unnoticed into the furnace after shutdown and in some of these the resulting vapours ignited, causing structural damage. Boiler fans can now be shut down and gas-side dampers tightly closed to minimise the loss of convected heat and pressure decay.

On the turbine, vacuum may be broken as the speed falls, noting that the longer LP turbine blades may overheat if this is done at too high a speed. Unless maintenance work is planned in turbine, condenser or feedheating plant steam-spaces, vacuum may beneficially be broken with nitrogen, the resulting inert atmosphere limiting off-load corrosion. Care must be taken that the bulk nitrogen and gland-steam supplies are isolated, however, before vacuum is completely lost if inadvertent lifting of the LP turbine-exhaust overpressure diaphragms is not to occur, though a trickle-charge of nitrogen may usefully be maintained throughout the shutdown period to preserve the nitrogen blanket. Thereafter, boiler stop valves can be closed and steam-leads depressurised, condenser CW flows curtailed and CW pumps shutdown.

Should the turbine fail to go on to turning-gear for any reason and if the fault cannot be rectified quickly, it is probable that the shaft train will seize as the HP and IP rotors bend (hog) and make contact with stationary metal sections. The hogging results from the rapid establishment of natural temperature gradients within the steam spaces of the cylinders brought about by rising hot-air currents and was the prime reason for the general introduction, years ago, of the turning-gear.

Whilst the failure may cause concern and may result in the non-availability of the unit for 24 h or more, there is nothing to be gained in trying to force the shafts around, and much to be lost. On no account should steam be reapplied in an effort to induce movement or serious damage will result and whilst attempts should be made, at intervals, to operate the manual barring attachment normally provided on the motor output coupling, no undue force must be used. As the cylinder temperatures slowly fall, the bends will ease and eventually the shafts will be free enough for hand barring. When reasonable freedom of movement is apparent at all angular positions of the shaft, the electric turning-gear can be recommissioned. Initially, the motor current may be higher than normal and if the shaft eccentricity detectors are capable of operating at barring speed, they will respond to the bent shaft(s) and a wide trace will be displayed on the recorder chart proportional to the degree of bend. As the motor current falls and the eccentricity recorder trace reduces to more normal levels, the machine may be tentatively run-up under steam. Great care must be taken to reduce speed should eccentricity rise unduly and in the worst cases a further spell on the turning gear may be necessary.

Other shutdown actions may include:

- The isolation of all condensate make-up supplies to the condenser, as soon as possible and in any event before vacuum is broken to reduce oxygen levels in the system to a minimum.

- The maintenance of lubricating and seal-oil temperatures and hydrogen gas temperatures as appropriate.

- The topping-up of the boiler drum to a level that obviates the need to restart the feed pump during the shutdown period. The level is arrived at by experience, but is seldom less than the maximum reading displayed on the drum wide-range level indicator.

Unloading of the unit for an extended maintenance outage requires a different approach, as the boiler may need to be off-pressure and cool enough for internal entry as soon as possible, whilst the turbine may be required to be cold and off-barring. During the un-loading stage, it is normally possible to reduce steam pressure and temperature to quite low values and in doing so gain valuable initial cooling of both boiler and turbine. Any subsequent forced-cooling programme undertaken on either the boiler or turbine will be completed more quickly in consequence.

Boiler forced-cooling consists basically of maintaining a cooling airflow through the furnace and gas passes by the continued operation of the fans. Airflows need to be set to limit the rate-of-fall of drum saturated-steam temperature as the contraction of the pressure parts is linked to this parameter, whereas the contraction of the boiler skin casing and buckstays, for example, are influenced by other factors. Undue stresses between such components can be set up if cooling proceeds too quickly, giving rise to torn tubes or damaged skin. Falls of up to 100°C may be permissible but can easily be exceeded in the early stages of the forced-cooling exercise (when boiler pressure is falling rapidly) unless care is taken, though later the airflow may have to be increased to near full-load values if the cooling is to continue at the fastest permissible rate. Any airheater gas or air-side by-pass dampers that are fitted should be opened to reduce the negative contribution made by the airheaters in continuing to rotate and function as heat exchangers, and they should be shut down as soon as the airheater gas-inlet temperature falls to an acceptable level. This results in a noticeable quickening of the cooling process. Once the boiler has been ashed-out, the ash doors may beneficially be left open to induce a flow of cold air into the lower regions of the furnace to supplement the main flow entering higher up at the secondary air vanes. On boilers provided with circulating pumps, these should remain in operation to promote a more even and better cooling of the water inventory. On natural circulation boilers, once the firing has ceased the circulation is virtually non-existent and is not promoted to any degree by any amount of cold airflow applied to the furnace. Similarly, any facilities provided for drum spray should be used and will ensure a satisfactory and even cooling of all parts of the drum. As the boiler pressure falls to around 10–20 bar, the water may either be blown down to waste and the pressure parts stored dry or, as the pressure decays to zero, the nitrogen-blanketting system should be put on-line and the boiler either stored full or drained to waste under nitrogen, to suit the station policy or maintenance requirements.

Most large turbines are now provided with forced-air cooling facilities to reduce the barring time required, typically from 96 to 24 h. A compressed air supply is applied to the HP and IP turbines, the hot air being exhausted to atmosphere in a manner applicable to the particular machine. In use, it is necessary to increase the air admission gradually to the specified values whilst being alert for possible adverse through-wall differential temperatures across thick-walled metal

flanges, for example, or contracted differential expansions of shafts in cylinders, the general process being a reversal of the heating-up process when the machine is returned to service.

13 Unit operation — shutdown conditions

Once the unit is safely shut down and is not required for immediate load (rather than out of service for maintenance, when cooling, draining and isolating activities would be necessary), operational activity revolves basically around the routine monitoring of certain unit conditions, often selected from experience, and an informal or formal process of pre-start checking may be considered in anticipation of a return to service, usually depending on how long the unit has been shutdown and whether any work has been undertaken during the shutdown. Both activities can conveniently be summarised as follows:

- The monitoring of the turning-gear motor current so that any tight spots in the rotor train can be noted. Should these occur, a local inspection is advisable so that the causes can be revealed by audible squeaks at a gland or by observing, for example, that the jacking oil pressure to a particular bearing has failed. Whilst the barring speed on many machines is high enough to dispense with continuous use of the jacking oil pumps, this is by no means always so.

- Checking that all turbine and steam-lead drains are correctly closed to avoid local chilling effects and unduly rapid cooling. In this context, the occasional monitoring of selected recorder charts is worthwhile to confirm that cooling rates generally are proceeding as expected. Should a particular metal section be found to be cooling at an abnormal rate, it is worth trying to establish the cause. At best, this may be the result of defective lagging, for example, but occasionally could be the result of water or wet-steam ingress into the affected section, possibly as a result of valves passing or spray water supplies not being correctly isolated and the situation can be corrected before any damage is done.

- The inspection of the boiler for tube leaks, using both instrumentation and local observation.

- The maintenance of a supply of de-oxygenated feed-water in the de-aerator storage vessel, using live steam heating and atmospheric boiling, if required.

- The observation of mill and airheater casings and internal temperature instrumentation for signs of possible fires.

- The cleaning of oil burner tips or barrels.

- General off-load routines, such as oil changes, greasing or plant cleaning.

- Informal pre-start checking would include establishing that turbine steam valve pilot-motors are operating correctly from the control room ready for start-up, the governor speeder gear is set to minimum, the generator excitation system setting is at minimum (normally) and that boiler-fan and mill sequence checks are satisfactory, so far as can be ascertained at this stage, so that the plant will start when required. More detailed guidance on pre-start checking cannot be given with any confidence, though an aide-mémoire, in the form of a single-page operating instruction is worth considering for those stations that experience little two-shift operation, and a system of formal pre-start checks can be invaluable for the longer-term shutdowns if sensibly developed and applied.

14 Unit operation — two-shift and peak-load operation

In order to cover diurnal variations in demand, it is necessary to vary system total generation continuously. This can be achieved by part-loading generating units, or by shutting down some plant as load declines and restarting as demand increases. The latter offers considerable heat and cost savings, since the former implies that the cheaper-fuelled and more efficient units are not fully utilised. In addition, the no-load heat consumption and fuel costs of the more expensive units of generation can be at least partly saved if they can be shutdown periodically, rather than run continuously.

Figure 2.86 shows the typical form of the diurnal variation on the CEGB system on a winter weekday. Ideally, all plant is ranked in cost per MW (or *merit*) order and is placed on-load in the preferred order. The cumulative capacity up to the base-load (the cheapest) will be run continuously at full-load. The theoretical optimum to meet the remainder of the varying load profile is to start generating units in marginal cost (or merit) order and to load them to full capacity until the peak is attained and to shut down in reverse order on the same basis. In practice, this ideal may be constrained by factors, such as transmission system limitations, the implications of very brief load runs in the vicinity of the peak, the converse at the trough and the practical difficulties in achieving system rate-of-change of load with very few machines individually loaded.

The principle of merit order selection of units also ignores the start-up and shutdown costs associated with operating (or 'two-shifting') generating units intermittently. For relatively long-duration load runs, these are negligible when compared to the costs incurred during the generation run, but for shorter (or 'peak-lopping') runs, current practice assigns those costs in a procedure which ascertains the total cost of a block of generation amongst a group of generating units that compete to provide that utility.

In general, the profile of costs associated with the operation of a generating unit takes the form shown in Fig 2.87 and are similar in principle to any continuous process cost characteristics for start-up, run and subsequent shutdown.

For the purposes of considering the relative merits of selection from alternative generating units, the fixed costs are not considered, since such expenditure will be unaffected by the decision. It is only necessary to consider those costs which will be varied by the decision whether or not to run a unit. Those costs may be divided into *semi-variable* and *variable*.

Semi-variable costs include the expenditure on resources needed to maintain the generating unit at a state such that generation may take place. Examples in this category would be the costs associated with the operation of unit auxiliaries, such as boiler fans, feed pumps and CW pumps, fuel burnt to raise and maintain boiler temperatures, to maintain the turbine-generator at synchronous speed and even, possibly, extra labour costs that may have to be incurred to maintain that capability.

The *variable* cost is principally the fuel cost associated with MW generation, varying as a direct function of generating unit load and thermal efficiency. Increased repair and maintenance costs as the direct result of the potentially damaging cyclical stresses of two-shifting could well be included in this category, but such costs are extremely difficult to ascertain rapidly and accurately. Since there is strong evidence that such costs are small, compared with the other operating costs on two-shifting plant, they are ignored.

The central objective in the operation of a generating system is minimum cost. The form of costs for an individual generating run therefore make clear the basic objectives of a two-shift operation. Generating units must be started up with the minimum possible energy expenditure (which usually means minimum time from initial boiler light-up), must be loaded to maximum as soon as possible, and boiler firing and auxiliary plant operation stopped as rapidly as possible after shutdown. In addition to these local objectives, Systems Operations require a specified degree of assurance and repeatability of the timekeeping of the two-shift evolution in order to be able to meet predicted total system demand variations.

The rates at which generating plant can be started and loaded will be a function of the design and construction of the major components. It is almost invariably the thermal stresses on major, thick-walled items, such as boiler drums and headers, and turbine steam chests, casings and rotors that set the limits to the rates at which temperatures can be increased from the shutdown values to the maximum design values at full-load. The increase of general tempera-

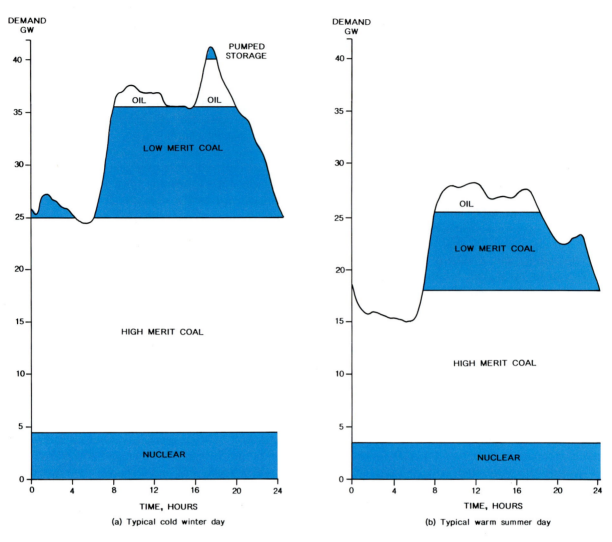

FIG. 2.86 Typical daily load curves

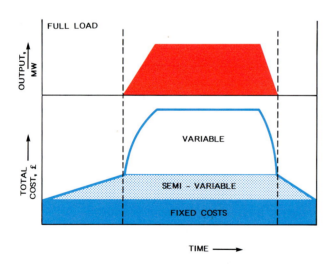

FIG. 2.87 General form of production costs

of plant and these differentials can induce very high mechanical stresses.

An essential pre-requisite to the development of two-shift operation is to perform monitored thermal stress trials on an example of that design of plant in order to assess the rates-of-change of temperature that can be accommodated without exceeding the maximum permissible stresses on these major components.

Where boiler plant is concerned, the drum is likely to be the limiting constraint item and operational maxima for overall rate-of-change of temperature (and hence boiler pressure) and through-wall temperature differential must be specified, measured, displayed and applied. Where turbine steam chests and casings are concerned, similar limits must be ascertained and applied. In general, such limits can be translated after synchronisation into rates-of-change of unit load, since temperature variations throughout the turbine will be a strong, direct function of unit load variation, (e.g., $\theta \propto P^n$, $n > 2$). Such considerations will set the practical limits to the basic location objectives indicated in the general form of two-shift operation

tures during the start-up procedure implies an increase in differential temperature across such items

146

costing shown above. The development of two-shifting techniques at an individual location is likely to be a difficult process, given that the levels of monitoring and control of possibly hundreds of operating parameters required for the rapid changes of state inherent in the fast-changing conditions of start-up and shutdown of large, modern and complicated generating units will require to be of a better capability than for steady state, base-load operation.

The key to achieving such a capability to control and monitor at high levels of speed and assurance lies in the quality of remote and automatic control gear and in the extent and quality of information gathering and display equipment. It has been a requirement in all CEGB 500 MW unit locations, where two-shift operations have been instituted, that there has first to have been considerable upgrading in the quality of instrumentation systems and information displays. In addition, at the first 500 MW unit stations to undertake such duties in 1975, considerable modifications to boiler automatic control gear were required. The later 660 MW units benefited from the recognition of such shortcomings in 500 MW unit plant.

Once the initial stages of design validation have set the practical limits to operational practice, trials must be carried out in order to test the capabilities of auxiliary plant, control gear and operational procedures to withstand the repeated high levels of stress and activity inherent in the rapidly varying states of two-shift operation. Such operational trials are usually carried out on a specified lead unit at a multi-unit location in order that any unforeseen damage to plant as the particular result of those stresses will be limited as far as possible. In addition to proving the capacity of plant and equipment to perform the required duty accurately and repeatedly, this period provides an opportunity for all operating staff to accumulate the high levels of knowledge and experience which is needed for them to cope with the exceedingly complex and heavy workload associated with the safe, but rapid, start-up of an entire multi-unit station.

Initially, unit starts will be carried out at a comparatively leisurely and deliberate pace in order to be absolutely certain of avoiding operational mishaps. The accumulation of experience during the working-up period of two-shifting should lead to an acceleration in the start-up process, whilst at least maintaining the required levels of vigilance and supervision and most certainly allowing no decline in those standards. The duties of all levels of operating staff must be examined and modified, where necessary, in order to ensure that no time constraint inferior to the basic optimum given above is imposed. The only 'risk' that can be tolerated is that concerning the timing of the unit start — indeed, if a large number of trials were to result in a timekeeping record of 100%, this would be evidence to believe that a considerable margin for error is built into the procedure. In other words, an unnecessarily early commencement to the start-up activities is probably being made. Such a practice is not acceptable, because it is not achieving the central objective — minimum cost.

The operational development period is also used to test for the capability to perform multi-unit starts; initially a second unit, but working up to the complete station complement as soon as possible. Supervisory staff have a responsibility for all generating units at a location. If the levels of vigilance and supervision are to be maintained at those set as policy by the management of the enterprise, then the sequence of starting-up units can be compressed but cannot become simultaneous.

The start-up process is undoubtedly the most significant and stressful part of two-shift operations. Many hundreds of items of auxiliary plant must start up and operate reliably. An even greater number of control and interlock sequences must operate satisfactorily. Even with very high levels of reliability, the great multiplicity of such items inevitably leads to a number of individual failures.

Methods to cope with such minor failures must be highly developed. Rapid location, diagnosis and repair or by-pass of such faults is essential in order to avoid delay in the operational sequence. This demands the highest levels of knowledge and judgement from operating and supervising staff. On items of plant with exceedingly complex sets of interlock sequences, such as boiler burner central-interlock units, where there may well be very long series-chains of limit switches on numerous dampers or valves, the provision of fault (or hold-point) diagnostic and indication equipment is necessary.

Where multi-unit start-up is concerned, great judgement is needed in coping with faults in order to prevent delays on one unit having a 'knock-on' delaying effect on succeeding units in a station start-up sequence. The decision to postpone a faulty unit start and switch it to the end of the sequence of incoming units will almost always have to be made within a very few minutes of the onset of the problem.

The fundamental conditions required for the two-shifting start-up of a generating unit will be similar to those previously defined for hot start on that type of plant at the location. Broadly, these will consist of a precise definition of the turbine ESV steam temperature and the pressure required before run-up to minimise subsequent thermal stresses. Considerations of turbine differential expansion, eccentricity and vibration levels will affect the run-up and loading rates: these will, of course, be peculiar to the particular design of machine. The emphasis in two-shift practice should be to analyse such characteristics and, where practically and economically viable, reduce times as far as possible.

The total activity network for a unit start-up should be analysed and modified, wherever possible, to ensure that the critical path is formed solely by the unalterable constraints intrinsic in the design of major plant items, rather than by way of auxiliary plant operations or other procedural activities. The ideal critical path should consist of the following sequence; boiler light-up, boiler pressure and temperature raising, turbine steam leg warming, turbine run-up and subsequent unit loading to maximum. These activities are necessarily sequential by their nature. Each has a maximum rate which is fundamental to the design of that equipment and improvement is likely to be economically impractical. Peripheral activities, such as turbine vacuum-raising, CW system start-up, condensate system commissioning and many others, should not be allowed to intrude into the critical path. If, for example, a unit design is such that turbine gland-steam supply is inadequate for vacuum-raising until boiler pressure-raising is almost complete, thereby introducing this activity into the critical path, then modification of the gland-steam supply system to provide its capability at lower boiler pressures would be required and almost certainly would be economically viable.

The effect of considerations on unit start-up are fairly obvious. Other considerations can be far more subtle. For example, several agencies are involved in the planning and execution of a two-shift start. First, the schedule for synchronising will be recommended by the appropriate Systems Operations load despatcher, given his requirement for the time at which the unit is to be at full-load. Secondly, the Shift Charge Engineer will set a timetable for boiler light-up, steam leg charging and turbine run-up. Lastly, the unit operator will execute the operational sequence. It is a natural tendency for each to add his own time margins to the minimum possible in order to ensure that his own schedules are met, since the perception of success or failure inevitably becomes 'timetable'-orientated to the 'milestone' events of unit start. Such tendencies are to be vigorously resisted, since they clearly represent sub-optimisation and fail to meet the central objective of minimum cost.

Cost is an operating parameter rarely visible in the plethora of information displayed on a modern unit control desk. Current computer developments allow for a great deal of rapid manipulation of operational data and the translation and calculation of cost from the traditionally supplied parameters, such as fuel flow and auxiliary plant power consumption, for a unit start-up is entirely feasible. Such a facility allows for the immediate assessment of the quality of a two-shift process, because the following set of questions is answered:

● How much did the start-up cost?

● Were vital operating parameters constrained within specified operating limits?

● Did synchronisation and the subsequent load profile match optimum and meet the schedule?

The answers to these questions constitute the acid test of the quality of two-shift operation. The more easily and rapidly the answers can be fed back to the operating agencies, the more effective will be the control and improvement of the economic return for the activity. These returns can be very high indeed — at the then-current fuel price levels, two 2000 MW stations in the CEGB saved approximately £5 million in heat costs during 1980 by means of some 1600 individual 500 MW unit two-shift cycles. The much greater rates-of-change of loading that the later generation of 660 MW units has displayed have led to even greater cost savings.

15 Gas turbine-generators

15.1 Introduction

Among the financial considerations to be examined, when planning the installation of generating plant, are cost per kW installed, cost per unit generated and the maintenance costs associated with the degree of reliability and flexibility required.

Hydro and nuclear installations are capital intensive, but produce electricity cheaply. To recover the investment, these plants must operate at high load factors over many years.

Fossil-fuel plant is generally less expensive to build, but the running costs are higher. Financial returns are planned over a shorter period, as this plant must to be more flexible than hydro or nuclear, so producing higher rates of plant depreciation.

Size will offer economies on all types of plant, but the constant variation in demand throughout the day and year means that there will be a large capacity of plant at the margins of operation which will work at very low load factors. It is here that gas turbine-generators are worth considering. Every power station needs electrical supplies for starting its generating plant and these supplies will be drawn from the interconnected transmission system. Once the generator is on-load, it is normal to supply auxiliary power from its own unit transformer, making it independent of the Grid System. However, if alternative electrical supplies are not available at a time of shutdown or when a generator trips from load, it becomes unusable until supplies can be restored. It is also possible for hot plant to be damaged if items, such as lubricating oil and cooling supplies, fail. Auxiliary generators are thus an insurance against loss of power supplies.

Starting power for large generators is in excess of 10 MW and to install auxiliary generators for this purpose only would be expensive. However, combining this duty with the need to provide generators to meet peak demands provides an acceptable financial compromise. The main requirements of gas turbine plant are high availability, high flexibility, output in excess of 10 MW and minimum maintenance cost. The output of a gas turbine installed as an auxiliary to a main generating unit is determined by the auxiliary power requirements of the main unit. Generally speaking, the cost of the power generated is of secondary consideration. The gas turbine-generator operating on the open cycle without the efficiency refinement of interstage cooling and waste heat recovery fits all the above requirements.

Two aero-engines manufactured by Rolls Royce and modified for ground operation are in common use in the CEGB. They are the *Avon*, with an output capacity of 10 MW, and the larger *Olympus* with its twin-spool compressor capable of up to 25 MW. The efflux from these engines is discharged into a turbine, known as the 'free power turbine', which is directly coupled to a generator and exciter.

It is possible to exhaust two engines into the same free power turbine and early units used two Avon engines in this mode. The output of both engines must be matched throughout the operating range to prevent stalling or compressor surging. Balancing both engines at start-up is difficult to achieve and tripping devices are installed to prevent engine damage from reversed gas flow. Where possible, this arrangement should be avoided, as the start reliability is less than for single-engine units. The Olympus engine has been widely adopted and, for larger outputs, additional engines and free power turbines are directly coupled to the generator (see Fig 2.88).

Current designs for a gas-turbine power station are a single 70 MW generator directly coupled to two free power turbines at each end, with each turbine driven by one Olympus engine (see Fig 2.89). Operation at part-load is possible by controlling the number of engines in service.

An advantage of this system is the ease of providing a generator for synchronous compensation.

15.2 Working cycle

The working cycle of the gas turbine is similar to that of the four-stroke piston engine, but in the gas turbine engine combustion occurs at a constant pressure, whilst in the piston engine it occurs at constant volume. Both engines cycle through induction, compression, combustion and exhaust to produce power (Fig 2.90).

In the piston engine, the cycle is intermittent, with one power stroke in four. The turbine engine, in contrast, has a continuous cycle and contains separate compressor, combustion, turbine and exhaust systems. The continuous cycle and absence of reciprocating parts gives a smoother-running engine and enables

more energy to be released for a given engine size. As previously mentioned, in the gas turbine engine combustion occurs at constant pressure and produces an increase in gas volume (or velocity), therefore the peak pressures which occur in a piston engine are avoided. This allows the use of lightweight, fabricated combustion chambers and low octane fuels. The higher flame-temperatures produced need special materials and cooling arrangements to prolong the life of combustion chambers and other hot zone components. Since the gas turbine is a heat engine, the higher the temperature of combustion the greater is the expansion of gases, and hence the greater the thermal efficiency.

However, this temperature must be controlled at the maximum value consistent with the materials used in turbine and exhaust components. The use of bled air to cool blades and other components allows the engine to operate with gas temperatures up to 800°C.

15.3 Starting

The description which follows is applicable to one installation, using 17.5 MW machines, and must be regarded as typical only. Manufacturer's specific instructions must be complied with when operating gas turbine plant.

To set the gas turbine-generator in motion, a starter motor rotates the HP compressor and gas turbine assembly. The HP compressor induces air into the LP compressor, which also rotates. At an HP compressor speed of 800 r/min, the igniters are energised and two seconds later the fuel boost pump is started and the fuel servo-valve opens.

Before stable combustion can be achieved, it is necessary for the engine compressor to develop sufficient air pressure to provide air for combustion and for airflow to cool combustion chambers and other components at light-up. The compressor must reach a minimum speed of 1000 r/min before these conditions are met and fuel admitted.

Engines are equipped with duplicate ignition systems, using 'high energy' igniters. In a multicombustion chamber system, burners will ignite from adjacent chambers.

In the starting sequence, the compressor speed will be at 1000–1200 r/min within 15 to 20 s and fuel is admitted with ignition 'on'; within a further 10 to 20 s, all combustion chambers should be lit and the engine exhaust gas temperature rising rapidly. To protect the plant from the hazards of unburnt fuel, a maximum of 45 s from the beginning of the start sequence to the detection of exhaust gas temperatures in excess of 200°C is allowed. Failure to achieve this temperature operates the engine protection and shutdown will occur.

A successful 'light-up' will increase the energy in the gases leaving the combustion area and passing into

the turbine blades. The turbine blades will now extract energy from these gases to drive the free power turbine-generator up to a speed of 1200 to 1500 r/min. Increase in engine power is now controlled by the 'throttle' and, as the throttle is advanced, greater volumes of gas are discharged into the free power turbine, together with some increase in temperature. The turbine-generator speed will increase until its speed governor becomes active. A second control now takes over, the governor datum. The throttle may now be advanced to 100% without further increase in engine power.

The governor datum will have been set at minimum which corresponds to a turbine-generator speed of approximately 2300 r/min. 'Minimum governor' state has now been achieved. The turbine-generator can now be raised to synchronous speed (3000 r/min) by increasing the governor datum. At synchronous speed, the governor datum will be approximately 80% and the engine exhaust gas temperature about 300°C. On synchronising, the machine can be loaded by increasing governor datum at a rate determined by the free power turbine design.

Engine exhaust temperature increases rapidly during loading, reaching 600 to 610°C at full-load. It is this rapid rise in temperature which constrains loading rates; times to full-load vary from 2 to 5 minutes.

This simple process is suitable for sequence control by relay or microprocessor systems. Included in the process will be control of auxiliary lubricating pumps, cooling fans, fuel boost pumps and fuel supply pumps to top-up head tanks.

To maintain optimum efficiency and acceptable environmental conditions, it is essential that fuel/air ratios are accurately controlled over the full range of operation. The microprocessor-based engine management system is ideally suited to this purpose, as it can be programmed to match the characteristics of fuel delivery systems and variations in compressor performance. Modern control systems will also provide plant condition monitoring, alarm scanning, data processing and fault diagnosis.

15.4 Shutdown

The shutdown process is the reverse of starting, except that some designs of free power turbine require a controlled cooling period to reduce thermal shock when the engine is tripped. The natural draught caused by the hot ducting and stack will draw cold air through the engine and power turbine for some time after shutdown, producing high thermal stresses. Operating at 2 to 3 MW for 5 minutes will reduce gas temperatures entering the power turbine to 300–325°C, thereby greatly reducing thermal shock on shutdown. An alternative arrangement is to close the engine air-intake by operating a damper on shutdown. Because of the interlocks associated with this damper, start reliability is impaired.

In common with current practice on large steam-driven turbine-generators, a period of shaft barring is incorporated into the shutdown sequence.

15.5 Protection

The equivalent of the steam turbine emergency stop valve on an aero-engine is the *high speed shut-off cock* (HSSOC). This spring-operated device is latched open at the start of a run and admits high pressure fuel to the engine burners. It is tripped by initiating a shutdown under controlled or emergency conditions.

A second fuel shut-off valve is positioned before the engine fuel pumps and regulator valve and is known as the *low pressure cock* (LPC). This valve operates with the HSSOC on all occasions, although manual operation of the LPC is provided for priming the engine fuel system following maintenance.

Engines are housed in acoustic enclosures which are strengthened to protect personnel from possible engine explosions. These enclosures are fitted with CO_2 fire protection operated from fire detection systems. Operation of the fire detectors will trip the machine and close fire valves positioned in the fuel supply line.

The precautions taken during starting to guard against failure to ignite fuel must continue throughout normal operation. Flame-detection devices are generally unreliable and engine exhaust temperature has proved to be the most reliable method of detecting 'flame-out'. Minimum fuel settings would normally produce exit-gas temperatures of about 250°C, so if this temperature falls below, say, 200°C, it can be assumed the engine has 'flamed-out' and protection will trip the machine.

As previously mentioned, the gas temperature entering the free power turbine at full-load will be 600–610°C, depending on the ambient air temperature and compressor performance. To protect turbine components from excessive gas temperature, engine fuel supply will be progressively reduced by a temperature control unit operating in the range of 615–625°C. Failure to control this temperature will lead to an engine trip at 635°C. Precise settings are determined by turbine design.

The engine has a self-contained lubricating oil system, from which a low oil pressure trip will be operated. All engine bearings are sealed with air bled from the compressor and faults in the sealing system will also trip the engine.

Low fuel pressure at the engine will also initiate a trip.

Conventional electrical and mechanical protection is provided for the turbine-generator.

15.6 Black start

The term *Black start* describes a condition where normal AC electrical supplies have been lost and the gas turbine is required to start, aided only by battery-driven auxiliaries, and to achieve a stable condition from where it is capable of carrying full-load for sufficient time to re-establish normal AC electrical supplies.

The minimum requirements to start a gas turbine are:

• Battery supplies to drive the engine up to firing speed, operate fuel igniters, power instruments and control equipment.

• Fuel supply available to the engine — usually stored in a header tank.

• Battery supplies to drive turbine-generator lubricating oil pumps and engine fuel boost pumps when fitted.

A 110 V battery is normally used for engine starting, control and instrumentation; a 28 V tapping is provided for ignition.

240 V station batteries supply the larger power requirements of fuel pumps and lubricating oil pumps.

Starting is automatic on the loss of busbar volts, or if the frequency falls below a predetermined value. On reaching synchronous speed, the main circuit-breaker would be unable to close if the busbar volts are lost, so special measures are necessary to override synchronising interlocks if it is deemed essential to close automatically, as may be necessary with nuclear plant.

In fossil-fuel stations, manual control is taken over when the gas turbine reaches synchronous speed. Power stations have local operating instructions prepared to indicate the plan for restoring electrical supplies, consistent with the capacity of gas turbines and the anticipated requirements of main plant, but high in priority will be restoring battery-charger supplies and the replenishment of gas turbine fuel stocks.

Other urgent requirements will be associated with main turbine-generator lubricating oil and seal-oil systems and the establishment of turbine barring. Regarding boilers, the first step is to establish cooling supplies to vulnerable plant, such as boiler circulating pumps. Finally, boiler draught plant must be run to purge the furnace and gas passes of explosive gases, before relighting.

When starting large induction-motor-driven auxiliaries, it is advisable to raise the gas turbine speed to, say, 51 Hz in preparation for the initial drop on starting.

15.7 Routine tests

The most important aspect of the aero engine is its reliability and the maintenance necessary to achieve this. Modern engines are designed to operate at least 2000 service hours between major overhauls. The operating regime in the electricity supply industry differs from aviation by operating continuously at ground level and therefore subjecting components to maximum stress when full power is applied. The number of starts will also be greater, although generally the period of each operation is shorter. Ground operation stresses are reduced by derating the engines, and the Avon capable of developing 12.5 MW is rated at less than 10 MW in most installations.

Plant condition monitoring has been widely adopted in the ESI for the management of plant maintenance and it is expected that periods between major overhauls will be extended beyond 2000 h.

The two main areas which reflect the engine condition are bearing vibration levels and the debris to be found in the bearing lubricating oil scavenge filters.

At 500 h intervals, inspections of air-intake areas, compressor blade conditions and the fuel drain system should be carried out to ensure the continuance of safe operation. Routine testing of protection systems would be at intervals ranging from one to six months.

Environmental conditions determine the frequency of compressor-blade cleaning. This should be performed before conditions seriously affect engine performance. The injection of an approved solvent into the compressor intake while the engine is running at cranking speed is then sufficient to loosen dirt and grease. Following a soak period, a spray of demineralised water, again at cranking speed, will flush the compressor clean. A short run at idle speed is then recommended to leave the engine in a dry condition.

As previously stated, start reliability is important and routine tests are necessary to check starting equipment. Facilities are usually provided to perform a start in which the burner ignition remains switched off. This is called a 'Wet run'.

A 'Dry run' operates with ignition on but prevents the fuel valve from opening. These facilities avoid full starting stresses being applied to the engine when carrying out routine tests. Dry runs should be performed following a Wet run or a false start to purge the engine of unburnt fuel.

This routine should be carried out weekly, provided that the engine has not operated in this period.

The free power turbine and generator are of conventional design and routine maintenance schedules will follow practices described elsewhere in this volume.

CHAPTER 3

Performance and operation of generators

1 Introduction

With the increase in size of generating units to 500 MW and above, maximum use was made of the material properties in order to minimise both plant size and weight. These units were also directly connected through transformers to the 400 kV main transmission system, with the net result that operators are now required to operate these units nearer to their limits of design than ever before. This means that the operating staff needs to be more knowledgeable in the behaviour of large generators, both under healthy and abnormal system conditions, since from knowledge stems confidence, and hence sound judgement in making correct engineering decisions when things go wrong.

It has been found, from experience, that lack of understanding of the physical behaviour of the turbine-generator under normal and abnormal conditions has led to the wrong use of certain machine constants and of misunderstanding concerning performance, stability, etc. In addition, the number of ways in which machine theory is explained in various text books tends to add to confusion, rather than understanding, because of the very large choice of alternative ways of looking at the same problem.

This chapter is essentially practical in approach and should be regarded as supplementary to other published works on this subject.

2 The synchronous machine

2.1 Introduction

The synchronous machine, in its most basic form, consists of a two-pole rotor excited by a winding around the poles fed from a DC supply rotating in a stator containing a winding (or armature winding) which can act as a generator when driven by a prime mover, or can drive a mechanical device, when receiving power from the supply system. A single synchronous generator supplying power to an impedance load acts as a voltage source, whose frequency is determined by the prime-mover speed: frequency $f = pn$ Hz, where p is the number of pole pairs and n is the prime

mover speed in revolutions per second. The type of rotor defines the machine performance and characteristics; there are two main types of synchronous machine; the round or cylindrical rotor and the salient-pole type.

2.2 Synchronous machine rotors

In a salient-pole machine, the poles are constructed as projections on the rotor body which carry the field windings, as shown symbolically in Fig 3.1. The windings are also provided on the pole yokes as a constructional necessity if more than four poles are required. Modern large turbine-generators are high speed machines having cylindrical rotors, departure from roundness (in the magnetic sense) arises from the need to accommodate the rotor winding in slots in the rotor body. Reference to Fig 3.1 shows lines drawn radially through the middle of the pole centres; such lines define the *direct axes*. Similar lines drawn midway between the direct axes are called the *quadrature axes*.

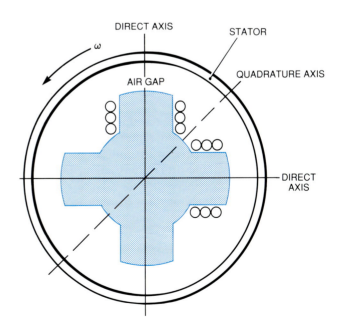

FIG. 3.1 Salient-pole generator rotor projections from shaft provide poles to wind turns but disturbs symmetry of rotor

3 Armature reaction

3.1 Introduction

When the armature winding of a three-phase alternator carries current, it is acted upon by two distinct sources of magnetomotive force (MMF); that due to the field system, which can be regarded as constant, and that due to the armature itself which can also be regarded as constant for a given armature current. The armature MMF reacts on the field MMF, hence the term 'armature-reaction MMF', and distorts the main-field flux, so producing a resultant flux differing from that due to the main-field alone. The useful flux on-load, therefore, differs both in magnitude and position from that existing on no-load.

Consider a single conductor in a magnetic field normal to the field direction and carrying a current as shown in Fig 3.2 (a). Standard conventions for field direction and current flow are used. The interaction of the magnetic fields results in a force in the direction shown, tending to drive the conductor sideways — motor action. This force can be visualised as being due to the distortion of the magnetic field, i.e., bunching and stretching of the field lines to one side of the conductor — armature reaction. If these field lines are regarded as rubber bands, the stretching action causes them to press sideways against the conductor, setting up a driving force in an effort to resume their former relaxed condition.

In the generator, Fig 3.2 (b), it is an experimental fact that if a conductor with its ends joined together to form a closed loop is pushed through a magnetic field, it will mechanically oppose the driving force as if cutting through butter. This occurs by a current being set up to flow in a direction such that its own self-created magnetic field interacts with the main field to build up the flux ahead and weaken it behind the moving conductor. The direction of current flow in the conductor is as shown, which is the same as for the motoring action, since the motion is in the opposite direction.

In both cases, armature reaction may be considered as mechanical reaction rather than the usually implied notions associated with volt drop. Thus it can be seen that without armature reaction, there can be no opposing forces necessary to balance the energy equation; and a motor or generator could not perform its purpose of interchanging mechanical and electrical power.

Figure 3.3 (a) is a developed sketch of the armature and field windings for a cylindrical rotor generator. The space fundamental MMF produced by the field winding is the sinusoid F; this wave may also represent the corresponding component flux-density wave on open-circuit. The MMF wave created by the stator current, i.e., the armature reaction MMF, is shown as A, also a sinusoid. The resultant MMF is

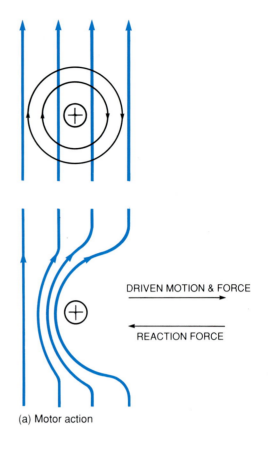

(a) Motor action

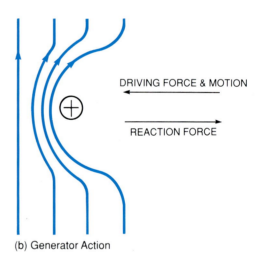

(b) Generator Action

FIG. 3.2 Interaction between conductor and magnetic field — armature reaction opposes mechanism of change

the sum of the two components produced by the field current and armature reaction. The resultant MMF wave R is obtained by adding these components (i.e., A and F), R will also represent the resultant flux density in the generator. Because sinusoids can be conveniently added by phasor methods, the same addition can be performed by means of phasor diagrams.

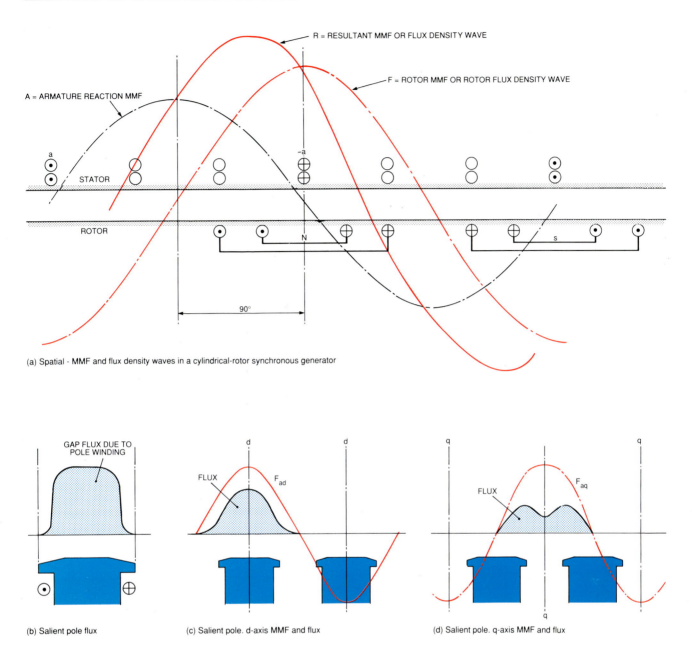

(a) Spatial - MMF and flux density waves in a cylindrical-rotor synchronous generator

(b) Salient pole flux

(c) Salient pole. d-axis MMF and flux

(d) Salient pole. q-axis MMF and flux

Fig. 3.3 Spatial MMF and flux density waves in a synchronous generator

For salient-pole machines, Fig 3.3 (b), the method of equating MMFs is not valid because the gap is not uniform over the pole pitch. Equivalence has to be based on the fundamental fluxes that the three-phase and salient-pole windings would separately produce in the air gap. The component of three-phase winding MMF wave that is directed along the inter-polar axis has no equivalent in the field winding. The MMF component in this the quadrature-axis is applied to the interpolar gap which represents a wide-ly varying permeance, so that the quadrature-axis flux distribution has a large spatial third harmonic. The fundamental of the q-axis flux does, however, generate an electromotive force (EMF) of fundamen-tal frequency in the three-phase winding, giving the

q-axis reaction the nature of a leakage reactance. Evaluation of the q-axis flux fundamental from the profile of the air gap is based on flux plotting or the application of gap-permeance coefficients.

3.2 Machine on open-circuit

The armature conductors carry no current on open-circuit, hence there can be no armature reaction and therefore no driving torque will be necessary to gen-erate an EMF, if windage and friction are ignored.

Considering Fig 3.4 (a), this shows the rotor direct-axis on the vertical centre line of the stator. It is

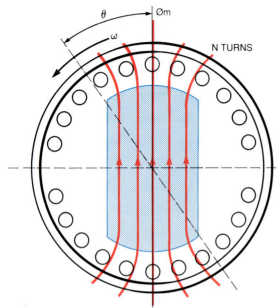

(a) Position of rotor at time t = O, $\theta = \omega t$

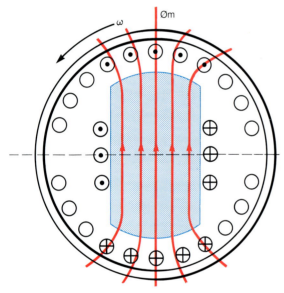

(b) Polarity of EMFs in stator conductor

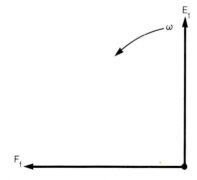

(c) Relationship between field MMF and generator voltage

FIG. 3.4 Rotor position and field MMF for generator
on no-load

assumed that the rotor is being driven at a constant speed of n r/s, i.e., an angular velocity of $\omega = 2\pi n$ radians/second. The rotor direct-axis is on the stator winding centre line and maximum flux density will occur at the conductors on this winding axis. In an idealised machine, the distribution of flux round the rotor is sinusoidal. (In practice, it is trapezoidal.) Thus on the axis at right angles to the direct-axis, it is zero, and at any angle θ to left or right of the direct-axis is given by:

$$\phi = \phi_m \sin[(\pi/2) - \theta] \text{ Wb/m}^2$$

$$= \phi_m \cos \theta \text{ Wb/m}^2$$

Since the rotor is rotating at a uniform speed of ω radians/second, after a time t, the direct-axis will have advanced to $\omega t = \theta$. Hence, if ϕ is the total flux in one pole pitch, the value of flux cutting the centre line coil will be given by:

$$\phi_m \cos \omega t \qquad (3.1)$$

However, as it is the rate-of-change of flux linked by this stator coil which is responsible for the generation of the EMF, it will be that value of flux lying on the cross-axis whose rate-of-change is required and this is given by:

$$\phi = \phi_m \cos \left(\omega t + \frac{\pi}{2}\right) \qquad (3.2)$$

and if N is the number of effective stator turns on this coil, then the instantaneous value of EMF generated will be given by:

$$e = -N(d\phi/dt) \text{ V}$$

$$= -N(d\phi/dt)[\phi_m \cos (\omega t + \pi/2)]$$

$$= N\omega\phi_m \cos \omega t \qquad (3.3)$$

Equations (3.1) and (3.3) show that the maximum EMF occurs in the conductor which is being cut by the direct-axis of the rotor flux. This enables a polarity to be assigned to conductors as shown in Fig 3.4 (b). There is an important concept that should be grasped at this stage and it is one which bothers most practical engineers, i.e., the relationship between space and time phasors. The above reasoning shows the spatial relationship between maximum flux position and position of maximum EMF axis, whilst Equations (3.2) and (3.3) show that the *flux-linkage time phasor is 90° ahead of the generated voltage time phasor*, a well known property of electromagnetic induction.

The equations also show that, at no-load, the rotor direct-axis is coincident with the conductors carrying the maximum EMF and that *this can be considered to represent a zero rotor-angle as a convenient reference for no-load.*

Equation (3.2) shows that, at t = 0, the flux linking the stator conductors is a MINIMUM, but from Equation (3.3) that the *rate-of-change of the linkages is a MAXIMUM*. This gives a relationship as depicted in Fig 3.4 (c), where it will be seen that the generated voltage is shown as lagging the effective airgap ampere turns (flux linkages) by 90°.

3.3 Machine at unity power factor

With a unity power factor load connected, the stator current must be in phase with the generated EMF and the action of this current is to produce a cross-axis MMF F_{ar} relative to the effective position of the rotor direct-axis MMF F_f, as shown in Fig 3.5 (a).

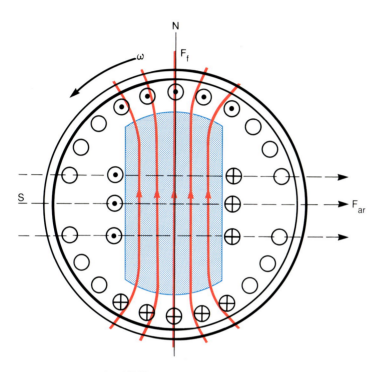

(a) Armature reaction MMF at unity power factor load

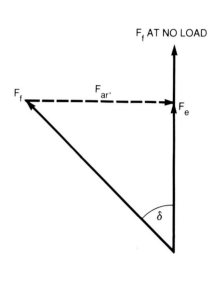

(c) Effect or regulation drop in terminal voltage

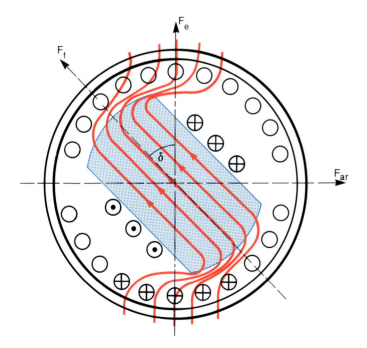

(b) Position of rotor to deliver unity power factor load

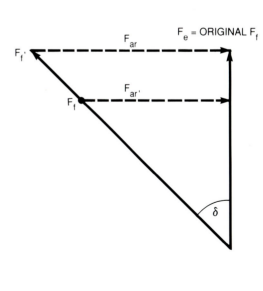

(d) Need to increase F_f to $F_{f'}$ to support required F_{ar}

FIG. 3.5 Armature reaction MMF influences the position of the rotor to deliver unity power factor load and produces regulation drop in terminal voltage, giving a need to increase the rotor current

This cross-MMF tends to reduce the resultant value of flux and causes a reduction of voltage. It is this quality of causing a voltage drop, which is more popularly associated with armature reaction, that is represented by the cross-MMF. As the load builds up, the conductor current increases and force must be exerted against the flux, as already explained.

This can only be achieved by the rotor being pushed ahead in its direction of rotation, so that it occupies a position, relative to the axis of the armature reaction MMF, to derive a component of its own MMF to just balance this. Figure 3.5 (b) shows the rotor advanced to the required position and making an angle δ with the stator maximum EMF axis (i.e., the zero-load position of the rotor direct-axis). The interaction of the fluxes produces the configuration shown, which is suggestive of the stretched rubber bands analogy. Figure 3.5 (c) is a diagram showing the position of the various MMFs, from which it will be apparent that if the magnitude of the field MMF is not increased, the eventual effective airgap flux will be less than its original value. This will cause a reduction in load proportional to F_{ar} at rotor angle δ.

In order to deliver the required power, proportional to F_{ar}, the rotor excitation will have to be increased so that F_e is restored to its former no-load value of F_f; Fig 3.5 (d) shows this.

Thus, armature reaction sets up the opposing torques necessary for the required conversion of shaft torque into electrical power and also causes a regulation drop in the output voltage. It can, therefore, be regarded as behaving like a reactance.

Conversion of the MMF axes into the equivalent voltage phasors as in Fig 3.4 (c), shows that when referring to Fig 3.6 (a), the armature reaction equivalent voltage V_{ar} can be regarded as a voltage drop due to the passage of the load current I through an equivalent reactance, since the armature reaction drop is in quadrature with the current causing it.

It is this quality of causing a voltage drop which is popularly associated with the term armature reaction, and the reactance X_{ad} is called ARMATURE-REACTION REACTANCE. The voltage drop IX_{ad}, when combined with the voltage due to flux E_f, gives terminal voltage E_t, as in Fig 3.6 (b).

From the previous reasoning, the following conclusions can be drawn:

- Power can only be delivered by an advance of the rotor angle in the direction of rotation.

- The rotor angle δ can be seen to be a measure of delivered power since, at no-load, δ is zero.

- Armature reaction behaves as a reactance and is denoted by X_{ad}.

The complete voltage diagram for the machine takes into account a further reactance and additional re-

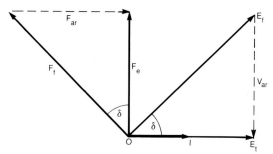

(a) MMF axes and voltage phasor diagram

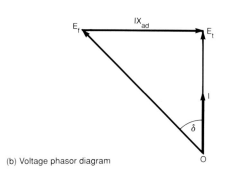

(b) Voltage phasor diagram

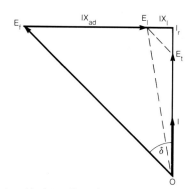

(c) Total machine internal impedances including resistance

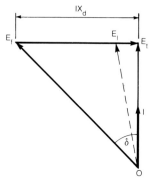

(d) Synchronous reactance (ignoring stator resistance)

Fɪɢ. 3.6 Voltage phasor diagrams for unity power factor load, indicating the voltage drop effect of armature reaction

sistive drop. The former is a true leakage-reactance and is due to the stator fluxes which do not link with the rotor fluxes, such as the stator overhang and the leakages across the stator teeth, and the leakage into the air gap. This total leakage-reactance, which is denoted by X_l, is usually small for a typical 500 MW machine, being slightly lower than the subtransient reactance X_d and about a sixteenth of the synchronous reactance.

The resistive drop in the stator resistance is usually so small, even when compared with the leakage reactance, as to be ignored. A typical value is one sixteen hundredth of the synchronous reactance. It is, therefore, ignored in most calculations where rotor angle or voltage problems arise, but has sometimes to be included in stability problems and has always to be taken into account where time constants are concerned. The leakage-reactance X_l has the same phase-angle as the armature-reaction reactance X_{ad} and is, therefore, usually added to it and the combined reactance denoted by X_d. It is called the SYNCHRONOUS REACTANCE, see Fig 3.6 (c) and (d).

3.4 Machine at zero power factor lagging

From the conclusions in Section 3.3 of this chapter, it would be expected that the rotor angle δ would be zero for a zero power factor load, since no driving torque would be required to deliver the generated current: reference to Fig 3.7 shows this to be so for a wholly-lagging load.

Here, the armature-reaction MMF is seen to be entirely demagnetising in action and from the line-up of the fluxes due to the field and the stator conductors, there exists no torque between them. To maintain the original value of terminal voltage, the field MMF F_f will have to be increased to just cancel the armature-reaction MMF F_{ar}. For a typical modern large generator, the effectiveness of the stator current MMF is roughly 2.5 times the main field MMF in terms of the machine rated-load, i.e., less than 0.5 p.u. current being circulated at zero power factor would produce an air gap MMF to equal the main field MMF. If excitation for no-load were put onto the machine with its terminals short-circuited, less than half full-load current would flow after the machine had settled down to the steady state, since this current would just cancel out the field MMF.

The modern machine is often referred to as a 'copper' type due to the predominant effect of the armature ampere-turns and the relatively weak effect of the field system. This became necessary with the large machines because of the mechanical difficulties of having to keep down the rotor sizes and weights, which resulted in relatively smaller air gaps than would exist if bigger rotor windings could be permitted. Older and smaller machines with synchronous reactance va-

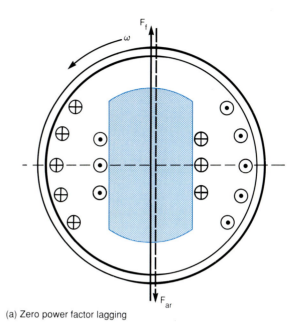

(a) Zero power factor lagging

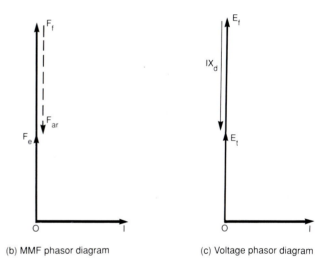

(b) MMF phasor diagram   (c) Voltage phasor diagram

FIG. 3.7 Generator on zero lagging power factor load

lues of 150%, or even less in some cases, have relatively bigger and more effective fields and are hence known as 'iron' machines.

Large X_d machines have an inherent large voltage regulation and, in consequence, it has become necessary to regard the automatic voltage regulator as an essential part of the machine rather than a desirable adjunct.

3.5 Machine at zero power factor leading

Reference to Fig 3.8 shows again that, as there is no power transfer, the rotor angle δ is zero, but that the effect of the load current is to increase the effective airgap flux. Thus, to maintain the original terminal voltage, a weaker field is required when the

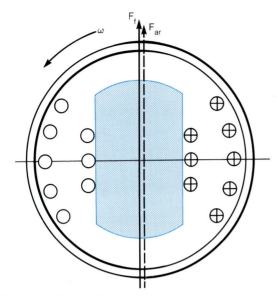

(a) Zero power factor leading

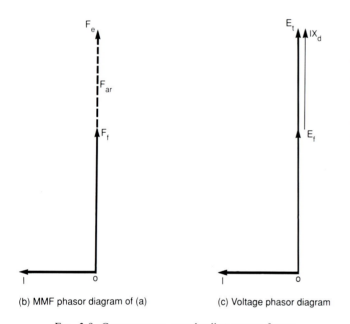

(b) MMF phasor diagram of (a)

(c) Voltage phasor diagram

FIG. 3.8 Generator on zero leading power factor
load — the armature-reaction MMF is
wholly demagnetising

machine is on-load, for if the field excitation were maintained, the terminal voltage would rise on application of the load and drop after its removal.

In zero power factor lagging, the reverse occurs and there is a rise of terminal voltage on removal of the load.

So, for leading power factor, the capacitative load on the generator causes the armature reaction to assist the excitation. It is therefore receiving excitation from the system, whereas for lagging load this demands further excitation from the machine, i.e., it is delivering excitation to the load through the power system.

It is the fact of receiving excitation from the sys-

tem which can produce operational difficulties in an interconnected network, particularly where there is a large cable content. During light-load periods at night, the predominantly capacitive nature of the system and low inductive load demand a weak excitation, and can still give rise to high system voltage beyond the control of the connected generators, so other means, such as shunt reactors, are needed to absorb the exciting MVArs generated by the network.

From a consideration of the operation of the machine at zero power factor, the following conclusions can be drawn:

(a) A machine operating on a lagging load requiring more excitation than its no-load value, will deliver its excitation to the system in the form of lagging MVAr; it has negative voltage regulation.

(b) A machine operating on a leading load, requiring a weak excitation, will receive excitation from the system in the form of leading MVAr; it has positive voltage regulation.

3.6 Machine at steady load, basic phasor diagram and equivalent circuit

Extending the work of Sections 3.3 to 3.5 of this chapter, and including for the stator resistance, the voltage diagram of Fig 3.9 (a) represents the steady state conditions on the machine in which:

E_f = equivalent excitation required

E_t = machine terminal voltage

I_R = stator resistance drop

IX_d = synchronous reactance drop

IZ_s = synchronous impedance drop

θ = synchronous impedance angle

δ = load angle (rotor angle)

ϕ = power factor of load current

I = load current

Z_s = synchronous impedance

R = stator resistance

Referring to Figure 3.9 (a), the voltage diagram for a machine on-load, it can be seen that this is also the voltage diagram for the equivalent circuit of Figure 3.9 (b), i.e., two AC voltages E_f and E_t connected by an impedance Z_s through which the current I flows. The power delivered through the impedance to E_t is:

$$P_t = E_t I \cos \phi \qquad (3.4)$$

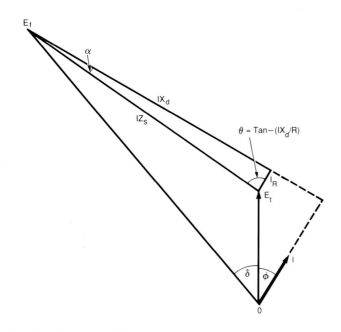

(a) Voltage diagram of machine on load

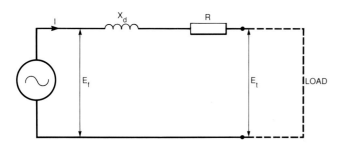

(b) Equivalent circuit of steady state generator

FIG. 3.9 Phasor diagram and equivalent circuit of generator including stator resistance

The phasor current $I = (E_f - E_t)/Z_s$ (3.5)

If the phasor voltages and currents are expressed in polar form,

$$I = \frac{E_f \angle \delta - E_t \angle 0°}{Z_s \angle \theta} = \frac{E_f \angle \delta - \theta}{Z} - \frac{E_t \angle 0°}{Z} \quad (3.6)$$

The real part of phasor Equation (3.6) is the component of I in phase with E_t.

$$I \cos \phi = (E_f/Z_s) \cos(\delta - \theta) - (E_t/Z_s) \cos(-\theta) \quad (3.7)$$

Using the fact that $\cos(-\theta) = \cos(\theta) = R/Z_s$ and substituting Equation (3.4),

$$P_t = (E_t E_f/Z_s) \cos(\delta - \theta) - (E_t^2 R/Z_s)$$

$$= (E_t E_f/Z_s) \sin(\delta + \alpha) - (E_t^2 R/Z_s) \quad (3.8)$$

Where $\alpha = \pi/2 - \theta = \tan^{-1} R/X$ and is usually a small angle. Similarly, the power at the source end E_f of the impedance is

$$P_t = (E_f E_t/Z_s) \sin(\delta - \alpha) + (E_f^2 R/Z_s)$$

Now since the resistance and therefore α are negligibly small

$$P_f = P_t = (E_f E_t/X_d) \sin \alpha \quad (3.9)$$

This equation is of considerable importance, since it describes a great deal about the performance of the machine:

- Maximum power output occurs at $\delta = 90°$, when $P_f = E_f E_t/X_d$.

- Power output is directly proportional to excitation.

- Power output is directly proportional to machine terminal voltage (busbar voltage).

- Power output is inversely proportional to the impedance of the machine.

4 Machine steady state performance

Consideration of Equation (3.9) suggests the basis of a steady state electrical performance diagram and points the way in which such diagrams can be constructed.

It also forms the basis for an explanation of steady state stability, a study of the criteria which govern this mode of operation and of the action of the automatic voltage regulator.

It should be noted, however, that this only gives a simplified understanding of these phenomena, since the development of the theory ignores such factors as saturation and saliency, both of which are important when considering certain abnormal modes of operation, such as zero excitation, reluctance torque or high rotor-angle running. The effect of these are discussed in later sections.

4.1 Steady state performance loci

Figure 3.10 reproduces the simplified voltage phasor diagram of Fig 3.9 (a) for a generator operating on

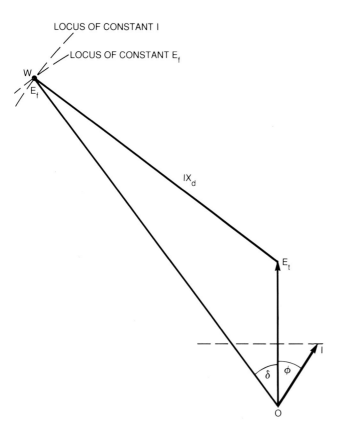

FIG. 3.10 Voltage diagram and working point —
Note: locus of constant I centre E_t, and locus of
constant E_f, centre O

a steady state load I at a power factor cos ϕ relative to the terminal voltage E_t; stator resistance is ignored, so the triangle of Fig 3.9 (a) reduces to that of Fig 3.10, which satisfies Equation (3.9).

The apex W of the voltage triangle OE_tE_f is often referred to as the working point of the machine, since excursions of W describe the loading conditions on the machine. This is reasonable when it is remembered that the terminal voltage is usually assumed to be constant, as also is the synchronous reactance X_d. The operating point W will vary with load current, power factor, power, reactive power and excitation.

If any one other quantity is fixed, assuming that the terminal voltage is fixed also (machine operating onto an infinite busbar), then the point W will describe a locus for the behaviour of the other quantity.

For example, if the magnitude of the load current is held constant, then the point W will describe a circular locus about the point E_t having a radius of $E_fE_t = IX_d$.

Similarly, if the excitation is held constant, W will describe another circle about the point O of radius equal to $OE_f = E_f$, since E_f is proportional to the excitation current applied to the field.

The first example implies that the steam input and excitation must both be adjusted if a constant current input is required; in the second, the steam input is

the only other adjustment and will cause variation of delivered power and power factor when changed.

At any particular instant of time, the intersection of these loci describes the working point of the machine and this can be used when trying to envisage what happens when changes to steam input and excitation take place.

As an example, imagine that the turbine governor setting is allowed to deliver more output from the machine and that the exciter output is controlled by hand. Reference to Equation (3.9) will show that the rotor angle must increase, since E_t and E_f are both fixed and X_d is a constant of the machine design (V = E_t, in this case). Referring to Fig 3.10, the point W must move along the constant excitation locus in a downwards direction to cause δ to increase and will stop at the point which satisfies the required power output. This will then make the load current the dependent variable; it will obviously have to increase slightly and improve in power factor to meet the new demand.

Thus, increasing steam input without adjusting the excitation causes the machine to deliver a higher power-factor load than before, i.e., a drift towards unity power factor. It will be apparent from this discussion that the most useful loci are the constant-power and constant-excitation loci, the former being developed from a study of Fig 3.11.

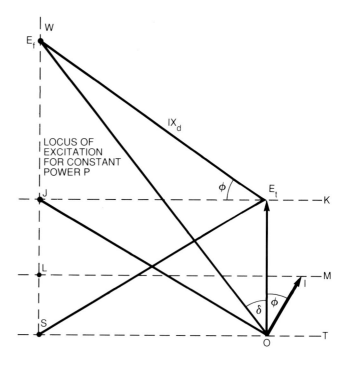

FIG. 3.11 Constant power locus, i.e., variation of
working point with constant power and varying
excitation

The line LM in Fig 3.11 is the locus of constant power output from the machine, since it describes the various positions which can be occupied by the current phasor OI for the same component in-phase with the terminal voltage phasor OE_t. By geometry, since E_fE_t is normal to OI, it follows that, if the line JK is made at an angle ϕ to E_fE_t that JK is parallel to LM, and if the line ST through O is also made parallel to LM, all are at right angles to OE_t.

Thus, the line joining E_f with its projection onto ST at S must be the locus of the excitation for constant power and is usually knowns as the *CONSTANT POWER LINE*. It is in fact the locus that the excitation must follow in order to maintain constant power output at a constant terminal voltage.

A study of Fig 3.11 will show that, at unity power factor loading, the excitation required will be OJ and that, if the steam input is not altered but the excitation is, then weakening the excitation makes the power factor less lagging, increasing it makes it more lagging. This is indicated by the movement of the point W along the constant power-line locus.

The point S and the line OS is the excitation required to produce a rotor angle of $\delta = 90°$ (angle KOS $= 90°$). From a reference to Equation (3.9), this also represents the maximum power output which the machine can deliver at this excitation, since $\sin 90° = 1$, and V (i.e., E_t) and power input are both held constant. Inspection of Fig 3.11 also shows that if the excitation is made less than OS, the point S cannot fall on the constant power output locus unless a new locus line is drawn nearer to the OE_t phasor, i.e., at the expense of reduced electrical power output. It therefore follows that, should a reduction in excitation be attempted below the minimum required to support the load originally set, the electrical output must fall below the steam input, there is a net surplus of power and hence the rotor must advance and begin to accelerate. The machine will, in fact, fall out of step and enter the instability region.

The line ST is usually known as the pull-out line and this is the locus of all operating points at the maximum theoretical stability limit. The line ST is also sometimes called the *THEORETICAL STABILITY LIMIT LINE* or *PULL-OUT LINE*.

It will be obvious that a number of lines parallel to SE_f must represent different power loadings and can be labelled as such, and that the circles drawn about O as centre with radii such as OE_f must represent different excitation values. The intersection of any two loci must, therefore, define the working point of the machine in terms of excitation, power factor, power output and rotor angle. Figure 3.12 combines these two sets of loci to give the basis of the usual performance diagram.

4.2 Steady state performance diagram

Use can be made of the above discussion to construct a practical performance diagram of considerable value to power station operating staff and this will be

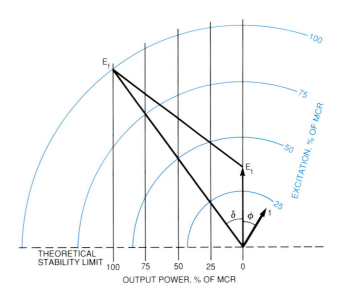

Fig. 3.12 Generator on steady load performance loci

developed below. Figure 3.12 forms the basis of the method.

It should be remembered that Fig 3.12 is a voltage diagram but, if the various voltage quantities are multiplied by the two constants, terminal voltage (because of infinite busbar) and the inverse of the synchronous reactance (a design constant), then a Power-MVA diagram will result. Figures 3.13 (a) and (b) show the development of this reasoning. Thus:

Multiplying IX_d by V/X_d gives VI (α MVA)

Multiplying OE_t by V/X_d gives V^2/X_d

Multiplying OE_f by V/X_d (if E_f replaces OE_f) gives VE_f/X_d

All have the dimensions of MVA, if voltages are expressed in kV (Fig 3.13 (b)).

From previous work, the lines JK and ST are normal to OV, so that VI must be at angle ϕ to JK, if the load current I is at a power factor $\cos \phi$.

Hence, the projection of E_f on the axis JK represents the power, its projection on OH represents the reactive power lagging, and the other axes VO and VK can therefore be scaled similarly as reactive power leading and negative power (motoring), respectively. This construction leads to the performance diagram: this can either be on a per unit basis, or directly labelled in MW, MVAr and excitation (rotor current), which is then of immediate practical use to the operator, Fig 3.14.

It should be noted that the diagram scaling is only correct for rated machine terminal voltage, and that all values must be appropriately adjusted for different values of terminal voltage, i.e., they must be multi-

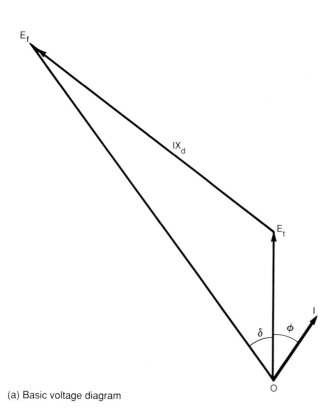

(a) Basic voltage diagram

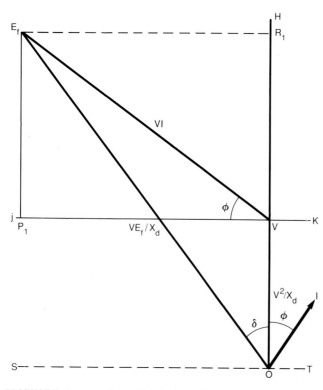

(b) MW/MVA diagram after multiplying the voltage phasor by V/X_d

FIG. 3.13 Conversion of phasor diagram to power

plied by V^2, so that if the terminal voltage were, say, 0.9 p.u. then all MW and MVAr scalings would have to be multiplied by $(0.9)^2 = 0.81$, although the excitation scaling would remain the same.

A quick reference to this last argument will show that as the pull-out line ST is at OA MVAr for 1.0 p.u. terminal voltage, these MVAr will be only 81% as great for 0.9 p.u. terminal voltage. Hence, a machine low terminal voltage produces a less stiff or less stable machine, since the pull-out line is closer to the MW axis.

It should also be noted that the excitation loci are circles, since saturation is ignored; if saturation is taken into account, the circles become flattened at the top and tend towards ellipsoids.

4.3 Practical performance diagrams

Various restrictions will apply to the operating regime of a generator other than that of stability, as already stated. Stator core-end heating can provide a limit to leading power factor operation. At lagging power factors, the heating of the stator and rotor conductors can limit the operating regime.

The generator can be connected to the system directly, or through a reactor, or through a transformer, which may itself have on- or off-load tappings. A unit transformer may be connected to the generator terminals to provide auxiliary supplies. However, performance charts are plotted depending only on whether the generator terminal voltage stays constant or varies throughout its operating range.

Variation of the load or excitation, with a transformer-coupled generator, will cause a change in generator terminal voltage which can be corrected by altering the transformer tapping. Should the generator transformer have an on-load tapchanger, it will be possible to operate at constant terminal voltage anywhere within a range of MW and MVAr loading for various values of HV voltage but, if the taps can only be changed off-load, a similar operating range will only be achieved by allowing the terminal voltage to vary.

The chart used for directly-connected generators assumes that the terminal/system voltage can be regarded as constant. While this may not be strictly true, it will be found that the chart will show high, normal, and low system voltage limits, so that it is possible to interpolate for intermediate voltages, if the voltage varies.

The generator reactor arrangement is treated as if the generator had a higher reactance.

Figure 3.15 shows the various limitations depicted on a generator performance chart; these are often also marked on the panel vector-meter. Obviously not all of the limits indicated will apply to one machine, but are shown here for completeness. These limits are discussed below:

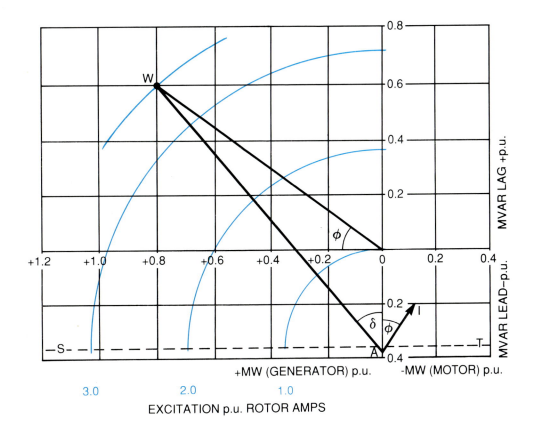

FIG. 3.14 Basic performance diagram

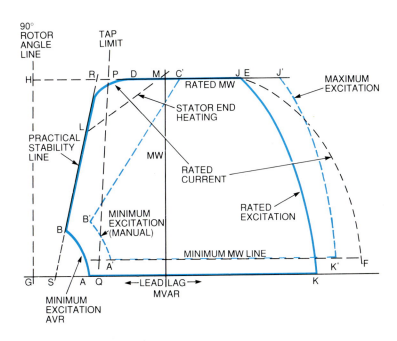

FIG. 3.15 Performance diagram showing the various limiting factors that can be applied in practice

Minimum excitation The portion AB is the minimum excitation locus, which depends upon the minimum generator rotor current that can be achieved, with the main field switch closed; generally, this will be the same for both AVR and manual excitation control, but if this differs, then a curve such as A'B' will also be drawn for the minimum manual excitation. Care should be taken to ensure that this minimum value is not limited by incorrect setting of the main-exciter negative field, where fitted, and that the brushes are of suitable grade for these low values of rotor current.

Practical stability limit Lines BC and B′C′ are the arbitrary stability limits for a machine under AVR and manual control, respectively. The stability margins used depend on the excitation control system and, for the British Grid System, are as defined in CEGB Generation Design Circular No 14, as follows:

Category 1

- Continuously acting automatic regulators (having no deadband) with VAr limiters.

- MW margins 4% at full-load; 10% at no-load.

Category 2

- Automatic regulators having some deadband and with VAr limiters.

- MW margins 10% at full-load; 10% at no-load.

Category 3

- Any type of regulator which is not fitted with a VAr limiter, or has a limiter NOT in operation.

- Any excitation control system which is not automatic (e.g., the 'hand' or 'manual' control position on an excitation system).

- MW margins 20% at full-load; 20% or 10% at no-load.

These lines are established by taking the excitation on the theoretical stability limit line for full-load GH, plus the percentage MW margin, as the radius of a curve and drawing the locus of this back to cut the full-load line HE at R. Next, the excitation for 10% of full-load excitation along the theoretical stability limit (i.e., 10% of GH) and drawing the locus to cut the zero MW axis at S, the practical stability limit is then obtained by joining the two points R and S, as indicated in Fig 3.15.

Whatever the category of AVR, the arbitrary practical stability limit line can only be applied once it has been shown to be achievable in practice, i.e., the setting of the VAr limiter has to be proven to control the excitation to this line, or else its actual setting line must be shown.

It is possible that the AVR control ranges provided may make it impossible to follow the arbitrary line. Similarly, it may not be possible to make the excitation limiter match the arbitrary line throughout its range, or there may be stator temperature limits that

prohibit operating the generator to these leading power factors: a line such as LM would be typical of this type of limitation. The actual practical achievable would normally be indicated.

Tests have shown that the use of high speed, continuously-acting AVRs can allow stable steady state operation at rotor angles in the region of 130–140°. However, transient stability considerations, i.e., faults on the system, change of network impedance by switching, the effect and disposition of other generators on the system, etc., make operation at these high angles unacceptable and the arbitrary limits stated above give margins to cater for transient stability. It should be noted that, for generators connected at the ends of long feeders, it is possible that even these margins may not be adequate and special stability studies are needed to determine a suitable limit line.

Rated current Lines CD and EF are portions of the locus of the rated current for the generator stator or transformer and they are circles about the origin, whichever is the smaller. Generally the point E will coincide with the rated excitation current point J. It should be noted that for connections where the generator terminal voltage can vary, the diagram will no longer be true, since it is based on constant voltage.

Rated MW output The portion DE is the rated MW output limit of the main generator, either the nameplate rating or any re-rating that may have been applied.

90° rotor angle The line GH represents the 90° load angle of the generator to its own terminal voltage phasor and is used in the construction of the practical stability limit line. This line does not represent the theoretical stability limit line of the generator at constant excitation or under AVR control, because the generator terminals are not connected to infinite busbars but to a finite power system with varying generating demand and system conditions. The line will not be vertical for generator arrangements where the terminal voltage varies.

Maximum excitation limit The line EK represents the maximum rotor current and is the excitation current for rated load and power factor, including the effect of saturation. It represents the thermal limit of the rotor and needs to be checked on commissioning the generator. Should the possibility exist of increasing the hydrogen pressure and hence of increasing the rotor cooling, then a line such as J′K′ may be drawn with the hydrogen pressure relating to that, stated.

Minimum MW output The portion AF is a minimum MW line, below which sustained operation of the generator is not possible. Some of the factors which prevent operation at lower loads are as follows:

- Low turbine vacuum, resulting in high temperatures of turbine LP stages.

- Expansion and clearance of prime mover parts.

- Rate-of-reloading.

- Boiler minimum steaming conditions; varies with range or unit construction.

- De-aerator problems associated with permissible oxygen content of feedwater.

Tap limit A condition may arise where a generator operating at constant terminal voltage, with an on-load tap transformer, cannot achieve full MVAr absorption at a given HV system voltage, because the transformer has reached its extreme negative tap (minimum turns ratio). This restriction is represented by line PQ.

Operation at a more leading power factor would reduce the terminal voltage, in turn causing the practical stability limit line BC to move to the right: eventually an optimum operating-point will be reached.

Generator transformer overfluxing limit Operation at maximum generator MVAr absorption and constant terminal voltage entails tap changing to reduce the turns ratio of the transformer (i.e., removal of turns from the HV winding). Dependent on the HV voltage and the nominal working flux-density of the transformer, a tap position may be reached at which the transformer iron becomes progressively more saturated, causing high iron losses, high iron/oil temperature gradients, or excessive core-bolt temperature.

This can be calculated from a knowledge of the transformer design and working flux-density and the system operating voltage; if this restricts the leading power factor capability, it will be shown on the operation diagram.

5 The effects of saturation

In the development of the simplified theory for the machine on steady load, no account has been taken so far of saturation in the iron circuit of the rotor and stator. In general, efforts are made in the design of any electrical machine to avoid the saturation of its iron parts, but because of economics and the sheer technical difficulties with the very large machines, it is neither economic nor metallurgically possible with present materials to design for low levels of flux density under rated operating conditions.

It becomes necessary to work fairly close to the knee of the saturation curve of the rotor iron and,

in consequence, the effects of this have to be allowed for.

5.1 Open- and short-circuit curves

The open-circuit curve for a typical machine is shown in Fig 3.16. This is the open-circuit voltage/excitation current curve and therefore reflects the B/H characteristic. The percentage saturation is shown as the ratio of the extra excitation current required to excite the machine to rated open-circuit voltage (1 p.u.V to the excitation required for this voltage on the airgap line), i.e., the linear portion of the open-circuit curve, where the iron is unsaturated, extended. Here, the figure is about 10%.

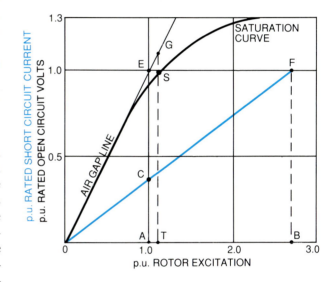

FIG. 3.16 Open- and short-circuit curves

The short-circuit current/excitation current curve is plotted on the same graph as the open-circuit curve, Fig 3.16. This has been obtained by plotting the current flowing in a short-circuit applied to the terminals of the generator against excitation current up to the rated current. The result is a straight line since only a small airgap flux is required to generate the small voltage needed to pass full-load current through the reactance of the stator and saturation does not occur. Most of the MMF of the rotor is used to overcome the demagnetising effects of the stator.

Short-circuit and open-circuit tests are performed on site during initial commissioning and use is made of the two tests to determine X_d.

5.2 Unsaturated synchronous reactance

Since, with the machine short-circuited, the terminal voltage is zero, it follows that the generated voltage is such as to overcome the armature reaction and leakage reactance drop. If the stator resistance is neglected, the ratio of the normal open-circuit voltage on the airgap line AE to the short-circuit current for the same excitation AC will give an impedance and, if the curves are plotted in p.u. values, will give the p.u. value of the *UNSATURATED SYNCHRONOUS REACTANCE* X_d. This is the correct figure for the reactance at low levels of airgap flux, but is incorrect for heavily-lagging power factor currents of the order of rated current at normal rated terminal voltage.

Referring to Fig 3.16:

$$\text{Unsaturated } X_d = \text{AE/AC} = 2.7 \text{ p.u.}$$

In practice, and as given in BS4296 'Methods of test for determining synchronous machine quantities', X_d is obtained by taking the ratio between the excitation current to circulate rated short-circuit current to that required to generate normal open-circuit voltage as measured on the airgap line, then unsaturated X_d = OB/OA = 2.7 p.u.

5.3 Calculation of excitation and rotor angle

When making accurate field-current calculations and for constructing accurate performance diagrams, several methods are available, offering varying degrees of accuracy according to what extent the effects of saturation are taken into account.

Two methods are really worthy of note and give reliable results. The first, which is known as the *AIEE method*, gives the most accurate results but, being geometric, is seldom used today.

The second, which from experience has produced good results when calculated and measured figures are compared, is known as the *POTIER VOLTAGE* or *ZERO POWER FACTOR* method. It lends itself to computer application and needs only the open-circuit curve and the manufacturer's stated Potier reactance.

5.3.1 Potier voltage and Potier reactance

Figure 3.17 shows the open-circuit curve of Fig 3.16 and, in addition, the zero power factor (ZPF) rated current curve as obtained from works tests, the machine being loaded at zero lagging power factor for the purpose.

The voltage phasor diagram of Fig 3.18 shows the additional internal voltage denoted by s (in terms of additional excitation) required to produce the effective terminal voltage E_t.

The voltage drop is made up of that due to the linear quantity X_{ad} (armature-reaction reactance) and the further drop IX_p, where X_p is known as the *POTIER REACTANCE* and IX_p the *POTIER DROP*. From previous work in Section 3.3 of this chapter, the true leakage reactance X_l was regarded as an additional reactance on top of X_{ad} to make up the total effective reactance of X_d.

In fact, it is found by test that the value of X_p is significantly greater than X_l, this being the ap-

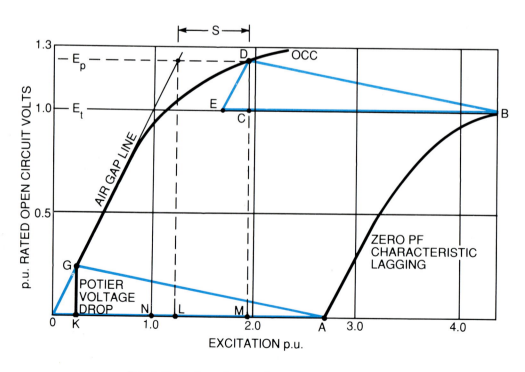

Fig. 3.17 Potier voltage and zero power factor curves

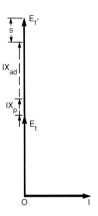

FIG. 3.18 Voltage diagram showing the effects of
saturation, armature-reaction drop and Potier voltage

parent leakage reactance which presents itself rather
than X_l, in producing the correct airgap voltage to
support a constant terminal voltage for *slow and relatively small changes* of machine loading.

Referring to Fig 3.17, the excitation OA is required
to produce rated short-circuit current and is that needed
to overcome the armature reaction and the Potier reactance drop, which is initially unknown.

The ZPF curve obtained by works test displays
a constant displacement from the OCC by the sloping
line DB. This fact, when originally observed experi-
mentally, led to the discovery that the reactance X_p
operated rather than X_l for small loading changes
and hence is used for deriving the value of X_p.

DC is then the Potier voltage drop due to rated
current, i.e., IX_p, and CB is the armature reaction
(flux demagnetising) voltage.

Thus, DM is the airgap voltage (flux) required to
produce the terminal voltage E_t, DC being the Potier
drop and the rest, armature-reaction drop.

Since OL is the excitation required to produce E_p
with no saturation and OM is the excitation required
allowing for it, it follows that (OM-OL) is the saturation in terms of excitation current (i.e., additional
rotor MMF) needed at this loading.

The triangle OGA is made by drawing OK equal
to EC, GK = DC and joining GA. Then KA represents the armature-reaction reactance, X_{ad}, and OK
the Potier reactance, X_p, the total OA = OK + KA
being proportional to the synchronous reactance X_d.

On the p.u. basis $X_p = OK/ON$

$$X_{ad} = KA/ON$$

$$X_d = OA/ON$$

5.3.2 Calculation of excitation from Potier voltage

The stages in the method are first to calculate the

Potier voltage E_p and, from this, to determine the
additional excitation required to supply the difference
due to saturation.

Figure 3.19 is the initial voltage diagram and reference
to it will show that, for a particular loading current,
power factor and terminal voltage, IX_p can be
calculated. Thus, E_p can be calculated, this being the
required airgap voltage, from which the additional
excitation due to saturation can be derived.

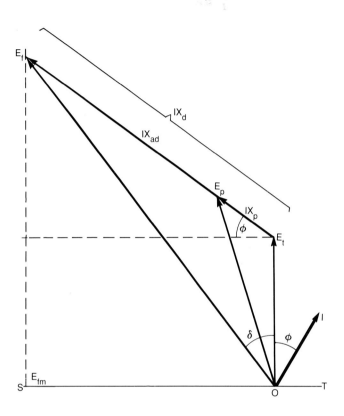

FIG. 3.19 Voltage diagram for generator on-load,
excluding saturation

Note that E_f and δ are the excitation and rotor
angle respectively which would be required if saturation were ignored.

Figure 3.20 is identical to Fig 3.19 for OE_t, I, E_f
and E_p. If, from this point, it is regarded as an excitation (MMF) triangle, then the point A and the
excitation OA is that required to generate E_p taken
from the saturation curve, i.e., OM in Fig 3.17. That
is, the additional MMF needed to deal with the saturation of the iron is S or AE_p in Fig 3.20 and is
LM in Fig 3.17. AB is then the field current needed
to counter armature reaction and so AB must be parallel to $E_t E_f$ and of the same magnitude as $E_p E_f$,
i.e., IX_{ad} or F_{ad}.

This gives the point B and the phasor OB is then
the required excitation to support this load.

Note that the rotor angle is now less and the excitation greater than if the unsaturated synchronous

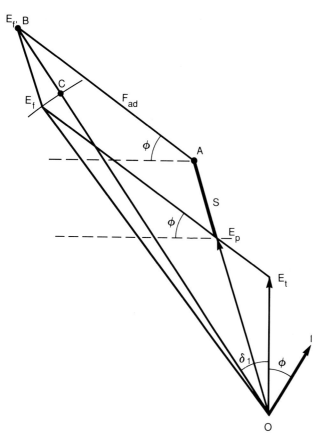

FIG. 3.20 Voltage and excitation diagram for generator
on-load allowing for saturation

reactance were used. The machine is made stiffer
and more stable because of saturation.

5.4   Short-circuit ratio

A more general figure used for dealing with loading
in the lagging region is the *SHORT-CIRCUIT RATIO*
(SCR), which is almost the inverse of the synchronous reactance, but modified by the factor S, depicted
in Fig 3.18. Thus:

$$SCR = OT/OB = 1.1/2.7 = 0.41 \text{ p.u.}$$

The short-circuit ratio is often used as a rough and
ready figure for calculating excitation for various
loadings: this is described in the next section.

6   Steady state stability

The term *STABILITY*, as used for the behaviour
of a generator, defines its ability to remain in synchronism with the power system to which it is connected. The two regimes of stability are 'transient'
and 'steady state'.

The *STEADY STATE STABILITY* of a generator
defines its ability to remain in synchronism under
conditions of steady state operation. In practice, this
also includes load drift, voltage drift and small fairly-
slow load changes (of lesser intensity than those due
to a rolling mill, for example).

TRANSIENT STABILITY defines the ability of
the generator to return to its former state of stable
operation, after having been disturbed from this state
due to sudden changes of loading or voltage conditions. The type of changes included here are those
due to power system faults, load swings, sudden tripping of sections of the power system network, etc.

Although a fairly general statement to make, it is
true to say that if a machine is operating with a fair
margin of safety within its steady state stability limit,
it will rarely become unstable under transient conditions, unless the disturbance is very violent and
sustained, such as a close-up three-phase fault, which
is also slow to clear.

Conversely, a machine operating close to its limits
of steady state stability is less likely to cope with a
moderate system disturbance.

6.1   The power angle equation

If reference is again made to Equation (3.9), it is
seen that the power output for a generator connected
to an infinite busbar (volts held constant), is given
by the simple relationship:

$$P = (VE_f/X_d) \sin \delta \qquad (3.10)$$

This equation holds for all values of V, E_f and X_d,
provided saturation is neglected (saturation makes E_f
non-linear) and resistance is also neglected. Since it
is true for all values of X_d, it is also true for any
reactance value over and above the design value for
the machine and, therefore, for additional reactances
between the machine terminals and the infinite busbar
(Fig 3.21), such as that due to the generator transformer and any interconnecting system to a point
considered as the infinite busbar.

If this result is plotted as a relationship between
the power P and the load angle δ, then a family of
sine curves represents the power angle characteristics
of the machine for different values of excitation,
busbar voltage or total (machine + system) reactance, i.e., $(X_d + X_s)$, where X_s is the total system
reactance.

6.2   Dynamic instability

The power angle diagram of Fig 3.22 shows the excitation set at a value to deliver full output at the

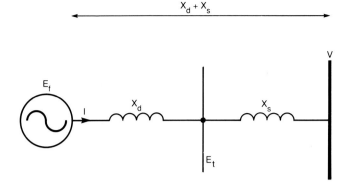

FIG. 3.21 Single line diagram of generator on load, with system reactance to infinite busbar included

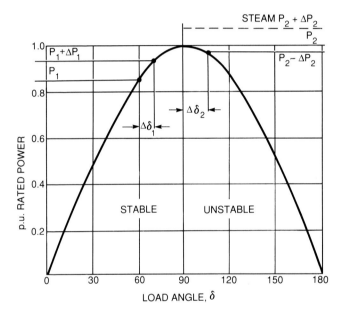

FIG. 3.22 Power angle diagram and effect of increasing the input power to the generator

theoretical stability limit — not a practical working condition, but it serves to illustrate several points:

- If the machine is operating at a power level P_1 and rotor angle of 60°, an increase in steam to $P_1 + \Delta P_1$ will cause the rotor angle to advance to $60° + \Delta\delta_1$. This will also cause the electrical power output to increase — an obviously stable state of affairs.

- If the machine is now operating at the limit of stability at the point P_2 and a rotor angle of 90°, and the steam input is increased to $P_2 + \Delta P_2$, the rotor angle must also be increased to $90° + \Delta\delta_2°$. This will cause the machine to operate at the point $[P_2 - \Delta P_2]$ which can be seen to be at a lower level of electrical output than both the

original steam input and the increased output. The rotor angle will, therefore, further increase since there is a surplus of input power available to accelerate the rotor. The resultant effect will be for the rotor to continue advancing at an accelerating rate and the machine will pull out of step and continue pole-slipping, unless remedial action is taken.

- From similar considerations, all operating points to the right of the maximum power limit are in the unstable region and this is sometimes called the *DYNAMIC REGION OF STABILITY*.

It will be apparent that several remedial measures are available to arrest the pole-slipping condition even after this has commenced, although several complete pole-slips will have occurred before this action can be taken.

There are also several measures that can be adopted to forestall the condition, as suggested by the power angle equation, and these will be briefly examined:

(a) If the busbar volts are increased by, for example, tapping-up an incoming transformer circuit, the maximum power output will be increased in direct proportion, i.e., point A to B on Fig 3.23. If operating at a particular load angle δ_1, this has the effect of reducing it to, say, δ_2.

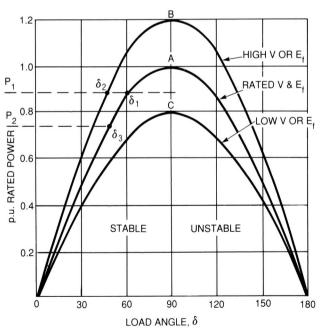

FIG. 3.23 Power angle diagrams for varying VE_f indicating change of angle for increasing VE_f

(b) Increasing the excitation will have the same effect as (a). This suggests that an automatic excitation regulator would be an ideal stabilising device.

(c) Reducing the total reactance. This can only be effected so far as the reactance external to the machine is concerned. Switching in another feeder between the power system and generator would achieve this, and the effect would be similar to (a) or (b), above.

(d) Reducing the steam input. This is the most effective measure that could be adopted, but it would have to be known that a condition of instability was imminent. Thus, if operating on curve A, a reduction of power input from P_1 to P_2 would allow the rotor angle δ_1 to drop back to δ_3, hence increasing the margin of stability. This is unquestionably the most direct and effective way of restoring stability after pole-slipping has commenced, especially when the reason for the condition may not be immediately apparent.

It should be noted that the effect of downward frequency drift is one of the greatest dangers likely to lead to instability, since this condition is nearly always attended by a corresponding voltage drift, also downwards. The sloping characteristic of the governor, which is normally set at 4% between no-load and full-load, results in an increase in the loading on the machine as the frequency sinks without the steam input being manually adjusted. (It is not always CEGB practice to allow all machines to act in this way, see Volume L.)

The combined effects of loading increase and voltage reductions are divergent as far as the stability of the machine is concerned and, if not arrested in time, could become cumulative and result in the collapse of the total generation at a power station, or even of a complete section of the system.

It should also be noted that an upwards frequency drift has an opposite and, therefore, beneficial stabilising effect.

7 The effects of saliency

Consideration has so far been given to the steady behaviour of synchronous generators, on the basis of the rotor being truly cylindrical, i.e., that not only is the airgap assumed to be uniform, but also the reluctance of the magnetic paths.

7.1 The round-rotor machine — reluctance torque

Referring back to Section 3 of this chapter, it will be remembered that the reluctance of the magnetic path at right angles to the direct axis of the main flux is greater than that in the direct axis. This is because of the presence of the winding slots and

windings themselves. Furthermore, the slots are not usually closed by ferrous wedges for obvious reasons, so the rotor reluctance paths are far from uniform. A flux is established in the direct axis by passing DC through the stator coils on the quadrature or cross-axis, the rotor will be cut by this flux passing through its direct axis. It is also assumed, for the following argument, that there is no current in the rotor coils and therefore no rotor excitation, and that the rotor is at a standstill.

If now the rotor were to be pushed to say the right, an opposing force would immediately be felt: this would be greatest when in the direct axis and least when turned through 90°.

Releasing the pressure, the rotor would spring back towards the former position on the direct axis, after performing several damped oscillations. This effect is known as *RELUCTANCE TORQUE*: it is caused by the magnetic field taking the lowest reluctance path, or using the flux-axis concept, for this to try to occupy the shortest route.

If now, in place of the fixed DC current in one set of stator coils, a rotating field is set up by the machine being connected to the three-phase system on no-load, but with the rotor field circuit open, then, provided the machine is still locked to the system, the rotor would also be rotating at synchronous speed, locked with its direct axis to the axis of the rotating flux.

If now steam were admitted to the turbine, there would be a torque applied to the rotor body. This would go on increasing until it was in excess of the reluctance torque previously described. Thus, electrical power could be delivered up to the limit of this reluctance torque, after which the machine would pull out of synchronism and run as an asynchronous generator.

The presence of reluctance torque gives what appears to be something for nothing in that limited power can be generated without any excitation at all. The amount that can be so supported depends upon the ratio of the direct/quadrature axis reluctance, which in most large modern machines varies between 3% and 6% of rated MVA. This phenomenon can be of considerable value since it provides an inbuilt margin of safety for steady state stability. It increases the safety factors in operating generators on weak field as synchronous inductors, i.e., MVAr absorbers, a valuable feature for dealing with the light-load periods at night, when the capacitance of the 400 kV system causes high system voltages.

7.2 Direct and quadrature axis synchronous reactance

7.2.1 Direct-axis synchronous reactance

Consider the conditions shown in Fig 3.7 with the machine on zero power, but rated MVA at ZPF lagging. The armature reaction F_{ar} lies on the direct axis and is wholly demagnetising: therefore, the machine presents a reactance drop directly proportional to F_{ar} and equal to $I(X_{ad} + X_l)$, when converted to an actual voltage drop.

Since this value is measured by short-circuiting the machine terminals when operating at synchronous speed, the measured value of synchronous reactance X_d is called the *DIRECT-AXIS SYNCHRONOUS REACTANCE* and denoted by X_d, the significance of the subscript 'd' now being obvious.

It should be noted that with the machine on zero leading power factor, the armature-reaction MMF is again on the direct axis, although its effect is now magnetising. Thus, in this position, the direct-axis synchronous reactance is presented. If the machine were operated at synchronous speed on no-load, locked to the system but with the field switch open, it would be operating in this manner and the equivalent circuit diagram would be similar to Fig 3.9 except that $E_f = 0$ and E_t would be shown connected to a source of EMF, i.e., the other generating plant on the system. This is shown in Fig 3.24 (a), and Fig 3.24 (b) is

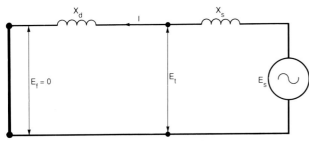

(a) Equivalent circuit for zero excitation

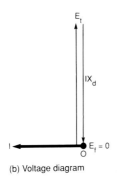

(b) Voltage diagram

FIG. 3.24 Generator on zero load, zero excitation

the associated phasor diagram. It will be seen that, when operating in this mode, the generator would *draw* a 90° lagging current, or deliver a 90° leading current, the value being determined by E_t/X_d. In fact, the machine would be maintaining its airgap flux at a

value sufficient to generate E_t by drawing its magnetising current I from the system. It is operating as a *synchronous inductor*. The value of lagging MVAr *import* to the machine (i.e., generation of leading MVAr export) is given by (rated MVA/X_d) which, from Fig 3.14, is the dimension OA, i.e., 0.37 p.u. MVAr lead.

7.2.2 Quadrature-axis synchronous reactance

From the above discussion and referring to Fig 3.7, it will be seen that when the machine is generating at zero lagging power factor, it is the reluctance of the direct-axis magnetic path which will determine the magnitude of the demagnetising flux set up by the stator current MMF.

If it is now imagined that a small input power is admitted to the turbine when operating as above, there will be a small output of electrical power and the rotor angle will advance by a degree or so. At this stage, it is convenient to consider the influence of the small in-phase stator current separately. This will set up a small MMF in the direction of the quadrature axis, see Fig 3.6 (a), and almost at 90° to the rotor direct-axis (actually $90 - \delta$).

The reluctance of the magnetic path in the quadrature axis is higher than in the direct axis, so it will be apparent that the flux which the stator current MMF sets up must be less, *pro rata*, than in the direct axis. This means that the armature-reaction reactance in this axis must be less, and hence also the effective voltage drop. If it could be measured, it would be less than the value IX_{ad} of Figs 3.6 (b) and (c) formerly used in the round-rotor theory. The value is designated *QUADRATURE-AXIS REACTION REACTANCE* and is denoted by X_{aq}. In Fig 3.25 (a), the machine is shown on general load; it is obviously convenient to resolve the current along the two axes in order to calculate the equivalent reaction in the two axes taken separately.

In the practical example of Fig 3.5 (b), it will be apparent that a reluctance somewhere between these due to the direct- and quadrature-axes will be presented to the armature reaction MMF F_{ar}, and that this will be less than the equivalent voltage drop IX_{ad} of Fig 3.6.

The method of calculating this equivalent voltage drop is not straightforward, since the effective reactance will vary between X_{ad} and X_{aq} according to the rotor position and loading.

It should be noted that a method for measuring X_{aq} is to operate the machine when connected to the system at slight positive slip, say 1 revolution every 5 minutes, the field switch being open. The stator current and stator voltage are then measured as the rotor slowly slips through the airgap flux supplied from the system. This is a very safe test to perform provided the machine will govern easily down to values of load of, say, 3–6%. The fact that the machine rotor is slipping will be obvious from the pulsation of the stator current which will vary between $\pm 3\%$ to $\pm 6\%$ of its average value. The value of X_d and X_q will then be given by (E_t/I). Remembering that $X_d = X_{ad} + X_l$, it follows that $X_q = X_{aq} + X_l$ and that, when the above test is carried out, it is the total value of the reactances which are being measured, i.e., X_d and X_q.

7.3 The two-reaction theory

The previous Section 7.2 showed how the non-roundness of the rotor and the variation of the reluctance of the rotor iron circuit affected the synchronous reactance value to be used in calculating machine performance. It was also shown that the value to be used depended upon its particular loading.

The obvious difficulties in analysis presented here — where the solution is required in order to choose a suitable value of reactance to find the solution — are resolved by the development of the two-reaction theory. This can be recognised in part as being the basis of the theory used in the development of Park's equations and the 'Generalised Theory of Electrical Machines'.

It is essential to understand the theory thoroughly, if interest lies in the asynchronous operation of generators or their behaviour under small, but sudden, system transients. The general principles used are identical to those used in the simpler round-rotor theory.

7.3.1 Basis of the two-reaction theory

The device is used, of replacing the actual windings with two equivalent sets, respectively on the d and q axes and resolving the current I into two fictitious components, I_d giving its effect on the direct axis of the rotor and I_q giving its effect on the quadrature axis. The resultant MMF of the resolved currents and coils giving the actual stator MMFs.

The reason for this will be apparent from a study of Fig 3.7 (a), where it is the current occupying the stator coil with its axis along the direct axis whose MMF will be influenced by the direct-axis reluctance and, from Fig 3.25, the current occupying the stator coil with its axis in quadrature with the direct axis whose MMF will be influenced by the quadrature-axis reluctance.

These are currents which are, respectively, in-quadrature with and in-phase with the driving MMF and, therefore, the rotor direct axis.

Figure 3.25 (a) is an illustration of this procedure, but it will be apparent that difficulty will be experienced, since the rotor position has to be known before the current can be resolved along and across it, and these values are required to calculate the rotor angle. One method is to use successive approximation, but a better method is to use geometrical construction

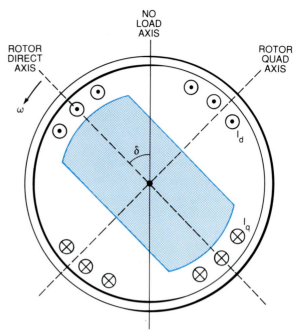

(a) Rotor position at load angle δ

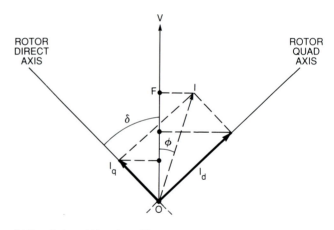

(b) Resolution of I into I_q and I_d

FIG. 3.25 Resolution of load current along the direct and quadrature axes of the rotor

which gives a rigorous solution. This will be described in the next section. Referring first to Fig 3.25 (b), the current I is resolved into the direct and quadrature axes, it will then be necessary to put I_q and I_d in terms of I, and ϕ.

Whence $I \cos \phi = OF$

$$= I_q \cos \delta + I_d \sin \delta \qquad (3.11)$$

7.3.2 The voltage triangle, using two-reaction theory and its construction

Figure 3.26 is the voltage triangle for the machine at a particular loading. It assumes that the correct rotor angle position has already been obtained.

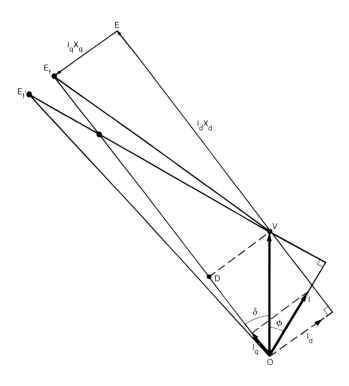

FIG. 3.26 Voltage phasor diagram allowing for saliency

The total voltage drops due to the two reactances X_d and X_q will be given from the reasoning used in the previous section, by the two phasors I_dX_d at right angles to I_d and I_qX_q at right angles to I_q.

The point E_f is the working point of the machine, OE_f is the excitation required and E_fV an equivalent voltage drop due to the combined effect of the imaginary volt-drops I_dX_d and I_qX_q component currents. It is important to note that E_fV is not at right angles to I, unless X_d is equal to X_q. If $X_d = X_q$, then E'_fV will be at right angles to OI and will be equal in magnitude to IX_d — this being the limiting case of the round-motor machine.

It will be seen, from Fig 3.26, by comparing the voltage triangle OVE_f, allowing for saliency, with the voltage triangle OVE_f', when $X_d = X_q$ (round-rotor), that the effects of saliency are beneficial, because the machine operates at a lower rotor angle and requires less excitation to support the same loading conditions.

The equation for the generator on-load, allowing for saliency, but neglecting resistance and saturation, is derived from Fig 3.26, as follows:

$$P = VI \cos \phi$$
$$= V (I_q \cos \delta + I_d \sin \delta) \qquad (3.12)$$

by substituting for $I \cos \phi$ from Equation (3.11).

Since the equation is required in terms of V, E_f, δ, X_d and X_q, it is necessary to eliminate I. For this purpose a normal, VD, is dropped onto OE_f to cut OE_f at D.

Now since I_qX_q is normal to OE_f and so is DV, and since EV is parallel to OE_f, then $DV = I_qX_q$, but $DV = V \sin \delta$, therefore $V \sin \delta = I_qX_q$,

so
$$I_q = \frac{V \sin \delta}{X_q} \qquad (3.13)$$

and since $E_fD = E_fO - OD$ and $E_fD = I_dX_d$, then substituting for E_fO, E_fD and OD, $E_f - V \cos \delta = I_dX_d$

so
$$I_d = (E_f - V \cos \delta)/X_d \qquad (3.14)$$

Substituting these values into Equation (3.12)

$$P = V \left[\frac{V \sin \delta \cos \delta}{X_q} + \frac{E_f - V \cos \delta \sin \delta}{X_d} \right]$$

Recognising that the middle term is the expression for the power of a round-rotor machine and collecting together similar terms:

$$P = \left[\frac{V E_f \sin \delta}{X_d} \right] + V^2 \left[\frac{\sin \delta \cos \delta}{X_q} - \frac{\sin \delta \cos \delta}{X_d} \right]$$

Since $2 \sin \delta \cos \delta = \sin 2\delta$, then $\sin \delta \cos \delta = (\sin 2\delta)/2$ and substituting in the above equation:

$$P = (VE_f \sin \delta/X_d) + V^2/2 \, (1/X_q - 1/X_d) \sin 2\delta \quad (3.15)$$

This is the classical equation for the salient-pole generator derived from the two-reaction theory and several of its interesting features will be discussed in the next section.

In the meantime, it is necessary at this stage to describe the device used for overcoming the difficulties previously mentioned for deriving the two-axis voltage drops. The method, which is geometric, is suggested by a consideration of Fig 3.27 and the previous discussion:

● From V, draw $VE_{f'}$ at right angles to OI and equal in magnitude to IX_d.

● Mark off M on $VE_{f'}$ such that $MV = IX_q$ in magnitude.

● Join OM and extend to E_f, such that $E_{f'} E_f$ is perpendicular to E_fO.

● Then, OE_f is the excitation voltage and $\angle MOV$ the rotor angle δ.

This can easily be proved since $\angle DVM = (\delta + \phi)$

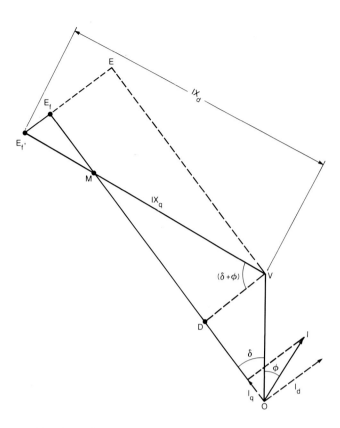

FIG. 3.27 Geometrical construction of voltage triangle

Then $E_f D$ $= (E_{f'} V) \sin (\delta + \phi)$

$\qquad\qquad = I \sin (\delta + \phi) X_d$

$\qquad\qquad = I_d X_d$ because $EV = E_f D$

Similarly $DV = I X_q \cos (\delta + \phi)$

$\qquad\qquad = I \cos (\delta + \phi) X_q$

$\qquad\qquad = I_q X_q$

7.4   Effects on operation due to saliency

Equation (3.15) fully describes the effects which various constants and variables have on the performance of the machine, when saliency is taken into account.

It should be noted again that resistance has been neglected on account of its very small value and the fact that the accuracy of calculations is not seriously impaired by omitting it, but the clarity of understanding the theory is greatly enhanced. Saturation effects can be taken into account just as easily and in precisely the same manner as previously discussed in Section 5 of this chapter.

Equation (3.15) is restated below:

$$P = (V E_f \sin \delta / X_d) + V^2/2 \,(1/X_q - 1/X_d) \sin 2\delta \quad (3.15)$$

It will be seen that if the excitation E_f is reduced to zero, the expression $V E_f \sin \delta / X_d$ vanishes, but the expression $V^2/2 \,(1/X_q - 1/X_d) \sin 2\delta$ remains. The first expression corresponds to the theoretical round-rotor machine which, it will be remembered, will not support any load at zero excitation. The second expression shows mathematically why the practical round-rotor machine, which does possess 'saliency', will support load and gives the magnitude of the power in terms of the machine terminal volts (busbar volts) and the two constants X_d and X_q. This expression $V^2/2(1/X_q - 1/X_d) \sin 2\delta$ is known as the *RELUCTANCE TORQUE* or *RELUCTANCE POWER* equation.

It will be seen also that if $X_d = X_q$, the generator will be unable to support any load at all at zero excitation, since both the first and second expressions will both vanish. Conversely, the greater the saliency, i.e., the more different X_q is from X_d, the greater the power that can be generated whilst still in synchronism without any rotor excitation.

The second expression reaches a maximum when $2\delta = 90°$, that is when the actual rotor angle is at 45° for a 2-pole machine, 22.5° for a 4-pole machine. When the rotor angle is at 90°, this is a maximum for the 'round-rotor' power but the saliency power will now be zero, since $2\delta = 180°$. For rotor angles greater than 90°, the reluctance power will be negative.

Thus, reluctance power due to saliency has the effect of superimposing a 'second harmonic' power component on the true round-rotor power angle curve and this is illustrated in Fig 3.28.

The effect of saliency is to shift the maximum rotor angle axis to a value less than 90° but also to increase the maximum power at the stability limit slightly. The curve shows the effect for 10% saliency power (i.e., a saliency factor of 1.2), but for a true salient-pole machine, such as the water-wheel generator, a more marked increase in the maximum power level is obtained. For a 20-pole diesel alternator of 1.2 MW rating (1.5 MVA) with $X_d = 0.68$ p.u. and $X_q = 0.38$ p.u., the reluctance power limit is 1/2 $(1/0.38 - 1/0.68) \times 1.5$ MW, i.e., 0.87 MW for a 1.2 MW machine, which is very considerable.

The effects of saliency, therefore, are beneficial in increasing the inherent stability of a machine, but this is not very marked with typical 2-pole designs. The principal benefit lies in being able to float or to generate light load on reluctance torque at zero excitation, thereby enabling some small generators, with turbines capable of operating at no-load, to behave as an invaluable synchronous compensator by absorbing MVArs at night time, when the grid volts tend to run high. It is obvious that 'motoring' the generator would be even more economic but the difficulties with the LP rotor heating would require a high vacuum in the LP casing, and possibly special measures such as rotor blade-tip cooling by means of water sprays.

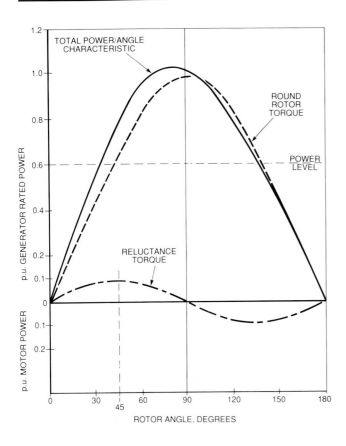

FIG. 3.28 Power angle diagram for generator with saliency operating at 0.6 p.u. power

7.5 Effects of saliency on synchronising power

Using the same method as in Section 2.8 of this chapter, the value of the synchronising power at any rotor angle δ, allowing for the effects of saliency, is given by differentiating Equation (3.15).

Thus, $$\frac{dP}{d\delta} = \frac{V E_f}{X_d} \cos \delta + V^2 \left[\frac{1}{X_q} - \frac{1}{X_d} \right] \cos 2\delta \quad (3.16)$$

At the limit of stability, there will be no increase of electrical power output with increase of rotor angle and the synchronising power available to restore equilibrium after a disturbance will be zero. Hence, at the stability limit; $(dP/d\delta) = \delta$ and

$$(VE_f/X_d) \cos \delta + V^2 (1/X_q - 1/X_d) \cos 2\delta = 0$$

The equation is used later for the construction of the theoretical stability limit, allowing for the effect of saliency.

7.6 Construction of performance diagram, allowing for saliency

It is not possible to construct a single equivalent-circuit

diagram for a salient-pole machine, as it was for the theoretical round-rotor machine, with the result that there is no true circle diagram. An equivalent voltage diagram, which is a modified form of the cylindrical-rotor circle diagram, can be constructed, however, and this adequately represents the action of the salient-pole machine.

From this diagram, the performance diagram allowing for the effect of saliency can be constructed, as for the round-rotor machine. The method adopted is geometrical.

Referring back to the voltage phasor diagram of Fig 3.27, using the two-reaction method for the generator operating at a particular loading $VI \cos \phi$ and rotor angle δ.

If this is reproduced as in Fig 3.29, but modified (as shown) by drawing WC equal and parallel to E_fO such that E_f and $E_{f'}$ coincide, then WC will be equal in length to OE_f. This gives the points V, W and C.

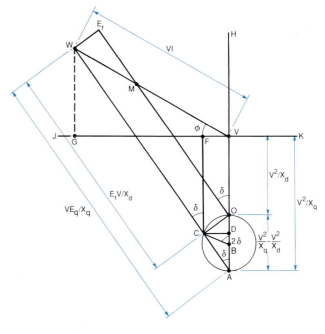

FIG. 3.29 Performance diagram for a generator with saliency derived from the voltage diagram Fig 3.27

G is the projection of W on the horizontal axis JK through V and F is the projection of C on this axis. The axis HVO is at right angles to JVK. If now each of these phasors is multiplied by the constants V and $(1/X_d)$, then $WV = VI$, $WC = (E_fV)/X_d$ and $VO = (V^2/X_d)$. This can be recognised as identical to the round-rotor solution, except that there the triangle formed by the points V, C and W was closed by C coinciding with O.

The disparity is due to the reluctance-power term in Equation (3.15), which is restated below, and should now be re-examined.

$$P = (VE_f \sin \delta / X_d) + V^2/2 \, (1/X_q - 1/X_d) \sin 2\delta \quad (3.15)$$

The dimension $GF = VE_f \sin \delta / X_d$, from Fig 3.29. Therefore, the dimension FV must be the second part of Equation (3.15), viz:

$$V^2/2(1/X_q - 1/X_d) \sin 2\delta.$$

Recognising that $VO = (V^2/X_d)$, then if a further point A on the VO axis $= (V^2/X_q)$ is constructed, the centre point B of $(V^2/X_d) - (V^2/X_d) = OA$ must be such that $BO = AO = V^2/2 \, (1/X_q - 1/X_d)$. Since the second part of the equation requires that the additional power due to reluctance torque be equal to $V^2/2 \, (1/X_q - 1/X_d) \sin 2\delta$, the suggested construction is that FV must equal the dimension OB $\sin 2\delta$ since $OB = V^2/2 \, (1/X_q - 1/X_d)$.

This suggests that the line $CB = OB$ be drawn at an angle 2δ to BO to satisfy this requirement, since $CD = CB \sin 2\delta = FV$.

Thus, A, C and O must all lie on a circle of radius equal to $V^2/2 \, (1/X_q - 1/X_d)$, since it is a property of circles that the angle $\angle OAC = \frac{1}{2} \angle CBO = 2\delta/2 = \delta$. Since CA is at angle δ to AO, it follows that AC and CW are one straight line.

It should be noted that, as for the round-rotor, the two axes JK and HO are the MW and MVAr axes, respectively, in the converted performance diagram.

The working point of the machine W for various excitation and power values no longer lies on simple circle loci. The total output power loci will still be straight lines passing through W and parallel to the HO axis, but the equal-excitation loci will be complicated by the fact that their centres lie on the circumference of the circle OCA.

Note that, if $X_d = X_q$, the points A and O will coincide, i.e., A will approach O and the saliency circle collapses for progressively greater parity between these two quantities.

Figure 3.29 is the basis of the performance diagram for the Salient-pole generator.

7.7 Construction of stability limits

A method of constructing the theoretical stability limit, allowing for saliency, will now be described.

Referring to Fig 3.30 (a), the method is as follows:

- From A, draw AL to cut the circle at L. AL continued will be the excitation locus leading to the working point.

- Draw LM parallel to AO to cut the circle in M.

- NM is drawn perpendicular to OA, through M, to cut AL continued at S_1. S_1 is then a point on the theoretical stability limit line.

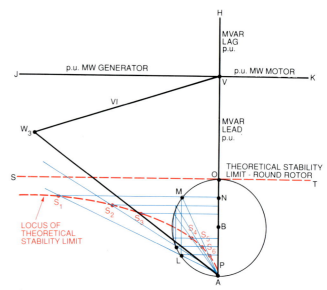

(a) Construction of theoretical stability limit

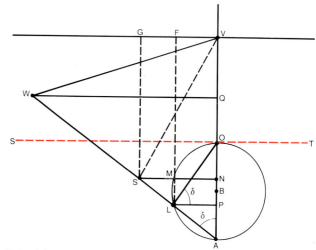

(b) Proof of construction of theoretical stability limit

FIG. 3.30 Theoretical stability for generator with saliency

    (a) Construction of theoretical stability limit

    (b) Proof of construction of theoretical stability limit

- The point S_6 is fixed by dropping LP perpendicular to OA; where this cuts MA defines the point.

- S_5, S_4 and all points inside the circle are constructed in this manner. It should be noted that points inside the circle define the working point of the machine with *negative excitation* applied.

From Equation (3.16) it can be shown that, by appropriate rearrangements and substitutions in terms of different variables and constants, $(dP/d\delta)$ is also given by the equation:

$$\frac{dP}{d\delta} = \frac{VE_q}{X_q}\cos\delta - V^2\left[\frac{1}{X_q} - \frac{1}{X_d}\right]\sin 2\delta \quad (3.17)$$

At the stability limit $\dfrac{dP}{d\delta} = 0$

Reference to Fig 3.30 (b) shows that $WA = VE_q/X_q$. Thus, the first term in Equation (3.17), i.e., $(VE_q \cos \delta)/X_q = WA \cos \delta = AQ$. Q is the point marking the reactive loading on the machine VQ, which in this case is heavily leading, the machine being underexcited and nearing the stability limit.

Taking the second part of Equation (3.17), i.e., $V^2/2(1/X_q - 1/X_d) \sin 2\delta = (AO \sin \delta) \sin \delta$. But $AO \sin \delta = LO$ and $PO = LO \sin \delta$, therefore $V^2/2$ $(1/X_q - 1/X_d) \sin 2\delta = PO$

Rewriting Equation (3.17) in geometrical form:

$$\frac{dP}{d\delta} = AQ - PO$$
$$= AQ - AN, \text{ because } PO = AN \text{ by symmetry}$$

At the stability limit, Equation (3.17) becomes zero, so $dP/d\delta = 0$, i.e., $AQ - AN = 0$, or $AQ = AN$. Thus, the stability limit must be such that the projection of W on AV, i.e., Q, coincides with N, that is, W coincides with S. Then, S must be at the limit of stability and is a point on the theoretical stability limit locus.

GV will be the total output power, FV the reluctance power and LS the minimum excitation, to supply this total power.

For zero excitation, the points M and S coincide at L, giving a total output power equal to the maximum value of $V^2(1/X_q - 1/X_d) \cos 2\delta$. Since $\cos 2\delta = 1.0$, this will occur at a rotor angle of 45° for a 2-pole machine ($22\frac{1}{2}$° for a 4-pole machine).

It will be apparent from the position and shape of the theoretical stability limit line of Fig 3.30 (a), that the presence of saliency improves the 'stiffness' or inherent stability of a generator, particularly at the lower steaming levels, but only gives small improvement at high outputs.

This feature is helpful, however, if zero-excitation running is contemplated, since it ensures that restoration of synchronism will be assisted after a system disturbance.

8 The automatic voltage regulator

This section does not attempt to deal with the mechanical and electrical details of various types of voltage regulator, since it is considered to be of more importance for the power station operating engineer to grasp the fundamental principles of voltage-regulator action and its indispensability in the safe and successful operation of large generating units.

The practical details can more easily be appreciated if the basic principles are understood.

It is more usual on static systems, such as bulk supply transformers, to regard the automatic voltage regulator (AVR) as a device for maintaining a constant busbar voltage, and this is indeed its only function.

When applied to a dynamic interconnected system, such as the terminals of generators, this is only an incidental part of its function.

If the limiting case is taken of a generator connected to an isolated system with no other synchronous plant present, then the function of the AVR is solely to maintain a constant busbar voltage, as in the static system.

When applied to a multi-generator system, then voltage regulation forms a part of its function, but its main effect is to maintain machine rotor angle and, therefore, to assist in maintaining steady state stability.

In fact, the AVR forms an integral part of the design and operation of modern large generators since they have a large synchronous reactance and, therefore, are weak on inherent stability. The AVR restores the machine 'stiffness', which is so necessary on a commercial power system. It is no exaggeration to state that the machine and its regulator are inseparable, and large generators should only be operated up to maximum rotor angles determined by the Category 3 practical stability limits (see Section 4.3 of this chapter) if the voltage regulator is out of action.

8.1 Machine directly connected to an infinite busbar

The term 'infinite busbar' implies that it is a 'bottomless pit' for the receipt or delivery of load under whatever conditions of power factor, and that it maintains a rigid and invariable voltage at all times.

In practice, the 'infinite bus' can adequately be regarded as the rest of the power system beyond the first busbars out from the generator busbar: it is usual to represent the generator as connected to this infinite busbar by an equivalent impedance. (This would be given by calculating the fault level of this bar to the rest of the system, leaving out the local generation, and then converting the fault MVA to an equivalent impedance.)

A generator connected directly to an infinite busbar is considered first, although this is hypothetical. Figure 3.31 is a schematic arrangement of the system.

Considering Fig 3.32, if the generator is operating at the loading level shown, with an excitation to give E_{fl} and the demand on the machine from the infinite bus is increased to a higher level constant stator current, but at about the same MVA, the rotor angle will be advanced from δ_1 to δ_2, and the phasor

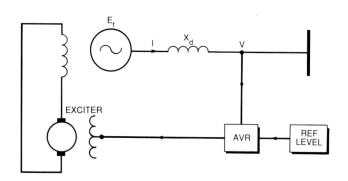

FIG. 3.31 Schematic of AVR circuit and generator
connected to infinite busbar

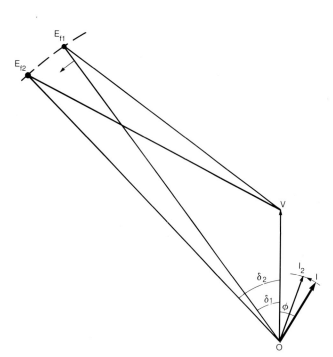

FIG. 3.32 Phasor diagram showing effect of power
increase

no change to the busbar voltage.

In order to enable the regulator to perform a useful function, it is necessary to make the control element of the AVR sensitive to power factor. This is achieved by injecting into the voltage signal an additive voltage proportional to the load current, but in quadrature with it, i.e., the quantity shown as jI in Fig 3.33. One method would be to inject a sample of the current from the yellow phase into a resistance R connected in series with the R-B voltage and the AVR control coil (see Fig 3.34).

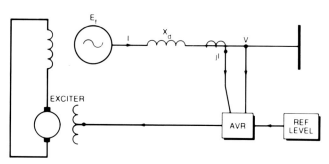

FIG. 3.33 Schematic of AVR with power factor
compensation with generator connected to
infinite busbar

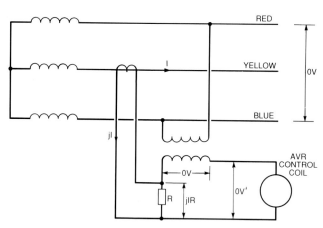

FIG. 3.34 Quadrature power factor compensation
circuit, injecting a compounding signal into the AVR
measuring circuit

E_{f1} will move along the locus of the circle having OE_{f1} as radius. The rotor angle will settle at a value to meet the new power level and, since there can be no signal from the AVR (because it is connected to the infinite busbar and, therefore, 'sees' the constant signal proportional to OV), the new excitation position E_{f2} will lie on the constant excitation locus at a new rotor angle, but be of the same magnitude as formerly, i.e., E_{f1}, and will, therefore, advance in phase to δ_2. The net effect will be an improvement in power factor and movement of the machine working point towards the stability limit.

Such an arrangement is unsatisfactory and the regulator in fact performs no useful function. This is because the system forms an 'open-ended' control loop since a condition change at the machine causes

The effect of this is seen in Fig 3.35, whereupon an equivalent voltage of OV′ is then applied to the control coil instead of OV. The magnitude of this injection jIR will obviously depend upon the secondary current jI and the resistance R, but a typical value for jIR is about 10% of OV.

Referring to Fig 3.36, the AVR will now see an apparent voltage of OV′ and, if the power level is

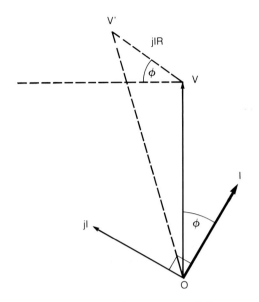

FIG. 3.35 Effect of quadrature current injection on control coil voltage

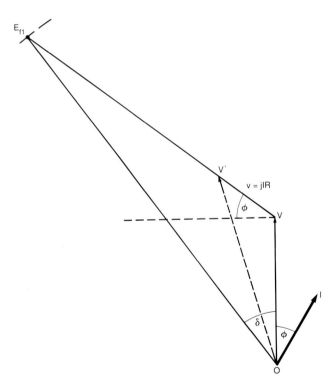

FIG. 3.36 The effect of quadrature current injection on the equivalent voltage diagram of the generator

increased by reducing the power factor angle ϕ, the angle of $v = jIR$ will reduce and OV' will reduce. This will call for an increase in excitation and the tendency will be to restore OV' to its original magnitude and the power factor to its former value.

This now forms the basis of a stable system and its action tends to maintain the machine at a constant power factor. It is, therefore, known as *power factor compensation*. The effect is to produce an overall droop of the machine terminal voltage with increase of load, which makes for stable sharing of MVAr loadings between all the machines connected to one common busbar.

8.2 Machine connected to an infinite busbar through a generator transformer

This is the more practical condition and, in any event, is representative of a machine operating on an undertaking busbar, which is far from behaving as an infinite bus because of the impedance between it and the rest of the system. The operation of the AVR in this case will, therefore, be studied in more detail.

Referring to Fig 3.37, this represents a generator connected to the infinite busbar through to a system of reactance X_s, which can be the reactance of the generator transformer or of the interconnecting system. Figure 3.38 is the voltage triangle for the machine operating initially at a power factor ϕ_1, a load current I_1, an excitation E_{f1} and load angle of δ_1. The busbar voltage OV is constant, this being the infinite bus.

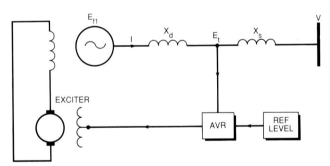

FIG. 3.37 Schematic diagram of AVR and connections for a generator connected through a transformer, or some reactance, to the infinite busbar

If there is now an increase in power output by admitting steam to the generator, I_1 will advance toward OV and ϕ_1 will reduce. This will advance E_{f1} towards E_{f2} on a circle of radius equal to OE_{f1}, also δ_1 advances toward δ_2. This will cause E_{t1} to fall towards E_{t2} and, since this is the voltage signal controlling the AVR, an increase in excitation will be called for. Thus, E_{f2} will now tend to move vertically upwards on the new P_2 power line towards E_{f3} and E_{t2} towards E_{t3}, until E_{t3} is restored to the same magnitude as E_{t1} originally.

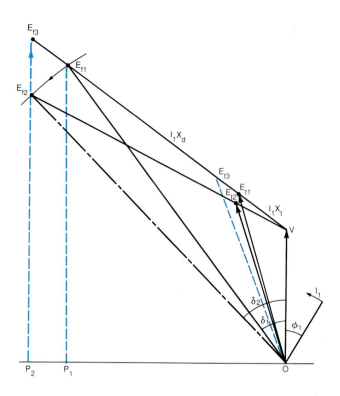

FIG. 3.38 Simplified explanation of AVR action

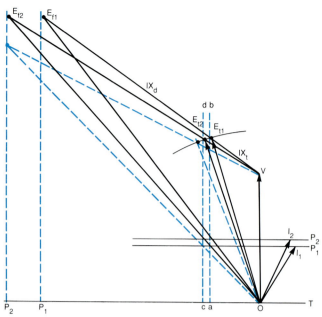

FIG. 3.39 Exact construction of excitation and rotor
angle loci under AVR action

The combined effect will be for the machine to operate at a higher power level, slightly higher power factor and at roughly the same rotor angle. In fact, E_{f3} will not settle exactly at the position shown in Fig 3.38, nor will E_{t3}, since an inspection of Fig 3.38 will show that, for the conditions depicted, E_{t3} is bigger than E_{t1}. A more exact analysis of these effects is discussed in Section 8.3 of this chapter.

The AVR acts as a maintainer of rotor angle as well as of busbar voltage and the former is its more important function. It is this function which is suggested under Section 6.2 of this chapter that makes the AVR an inseparable companion to the generator on load for stability.

8.3   Exact construction for action of an AVR

Referring to Fig 3.39, the original voltage triangle is first constructed and the two load lines P_1 and P_2 added. The locus of E_t is next constructed by drawing the arc about O with radius equal to OE_{t1}, the setting on the AVR. Since the ratio of IX_t to IX_d must be maintained for all working points of the machine (as X_t/X_d is a constant of the system), it follows that the new value and position for E_{t1} must lie on the arc of radius OE_{t1}, for E_{t1} must lie on the arc of radius OE_{t1}, and the locus cd which is parallel to OV and displaced by the amount $(O_a/O_c) = (P_1/P_2)$. Thus the point E_{t2} defines the

new operating point for the machine terminal voltage and E_{f2} must be the new machine working point in which, from geometry:

$$\frac{E_{t2}V}{E_{f2}E_{t2}} = \frac{E_{t1}V}{E_{f1}E_{t1}} = \frac{I_1X_t}{I_1X_d}$$

which, as has previously been stated is a constant. (X_t/X_d is a constant of the system)

E_{f2} is fixed by producing VE_{t2} to cut the P_2 excitation load line; the intersection defines E_{f2}

8.4   The AVR in the control of steady state stability

It should be noted that the presence of the reactance between the machine terminals and the infinite busbar is almost identical in its effect to the artificially produced reactance drop discussed under Section 8.1 of this chapter. Thus, the presence of a generator transformer, whose impedance can vary between 12.5% and 16% for the very large machines, is to introduce automatically a device to enable the AVR to function as a stable control system. The effect is identical to that achieved by power factor compensation and, when the latter is provided with an AVR installation, particular care should be exercised before including it in the scheme, to avoid excessive compensation.

For large impedance transformers, it may even be necessary to connect the power factor compensation negatively.

The presence of an AVR not only assists in maintaining steady state stability, it also aids post-fault stability and makes possible the stable operation of the machine at large rotor angles in excess of the theoretical stability limit, i.e., in the unstable or 'dynamic zone'. Tests performed at Stella and Cliff Quay Power Stations have shown that stable running up to rotor angles of 120° are possible, the limit being set by the response speed of the exciter-rotor circuit, rather than the speed of the AVR. Taking into account these factors, the maximum achievable rotor angle is about 135°.

The manner in which 'forward rotor angle' running is achieved is seen by referring to Figs 3.40 and 3.41.

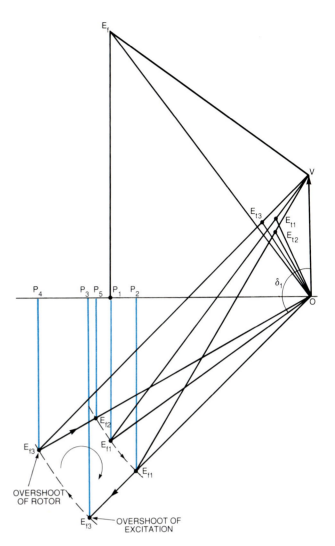

FIG. 3.41 Action of AVR at large rotor angles illustrating rotor oscillations near stability limit

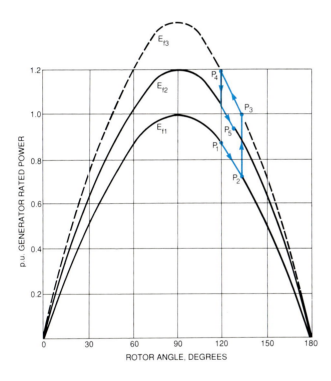

FIG. 3.40 Power angle diagram — illustrating dynamic operation under the control of an AVR in the unstable region of the power angle diagram.

Assume that the machine is operating in the unstable region at power level P_1 on the power angle curve for excitation E_{f1}. If there is an increase of steam input to P_5, the rotor will advance to the point P_2 of electrical power output on the E_{f1} power angle curve. This will reduce the voltage applied to the AVR from E_{t1} to E_{t2} and the regulator will then increase the excitation. If the regulator is inadequately damped, there will be an overshoot. Suppose that the excitation therefore increases to E_{f3}; then from Fig 3.40, the output electrical power will increase to P_3 and

be in excess of the steam input. This will arrest the advance of the rotor, which will then begin to decelerate along the new E_{f3} power angle curve to the point P_4, where the electrical output will increase still further, causing further deceleration. This will increase the voltage applied to the AVR to E_{t3}, which will then reduce the excitation to, say, E_{f2} and the rotor deceleration will be arrested and come to rest after several oscillations about the point P_5 at an excitation E_{f2}. If the regulator/machine system is well damped, the overshooting and undershooting described above will be very small, but when the machine is operating at the limit set by the speed of response of the exciter-rotor circuit in relation to the mechanical inertia of the complete turbine-generator rotor system, a continuous oscillation can be observed. In fact this is a feature of operation in the unstable region and at high rotor angles.

A locus can be calculated for the theoretical stability limit as determined by the exciter rotor response.

In any circuit having no EMF, Ri + L (di/dt) = 0 describes the instantaneous condition, where L (di/dt) is the flux linkage term and Ri represents the volt drop in the resistance of the circuit, which gives rise to the losses in the circuit.

If the circuit has no resistance, then the expression becomes

$$L \frac{di}{dt} = 0 \quad \text{or, since } L = \frac{N\phi}{i}, \text{ then } N \frac{d\phi}{dt} = 0$$

This equation describes the fact that the flux linkages must remain constant and also defines that if an attempt were made to change the flux through the coil (inductance L) by introducing some other source of magnetisation, then a current would have to flow in the coil to maintain the original status quo, i.e., *in a direction to oppose the change*. In a resistance-free coil, the current would flow indefinitely and the flux remain constant indefinitely, in spite of the external influence, because there would be no losses to dissipate the energy. Now that resistance-free conductors are no longer a textbook fantasy, the truth of this theorem has already been proven in connection with superconductivity experiments, where it has been demonstrated that a magnetic flux can be trapped in a coil.

The action of the current flowing everlastingly in the resistance-free coil sets up a 'barrier wall' of flux to prevent interaction with any other external fluxes. It produces a kind of magnetic insulation.

In practice, this 'ideal' situation could not arise since the presence of the resistance would result in an i^2R loss, so that the 'cancelling' current would slowly die and thereby permit the flux to change gradually to its new value. The delay would be a simple exponential of the form $\exp(-Rt/L)$, where the time constant is given by the value of time t which makes the index of the exponential $(R/L)t = 1$, i.e., $t = L/R$. The ratio L/R is known as the *time constant* and is usually denoted by τ.

10.2 The application of a sudden short-circuit

It is perhaps preferable to examine the actual test results of suddenly short-circuiting a generator first, and to analyse them before building up a supporting theory.

It is assumed throughout all the subsequent discussions that the machine is on open-circuit excitation (no-load) at normal synchronous speed and that stator resistance is ignored, except when considering time constants.

Figure 3.44 (a) is an oscillogram of the stator current of one phase of a generator subjected to a three-phase short-circuit and the phase chosen is that which

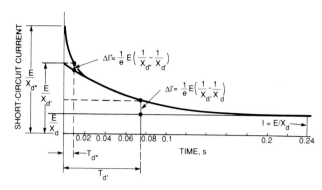

(a) Short-circuit current envelope machine on no load excitation

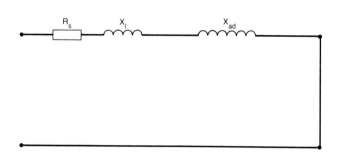

(b) Equivalent circuit in synchronous state

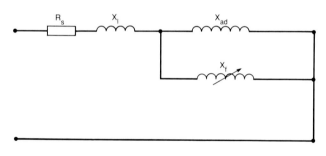

(c) Equivalent circuit in transient state

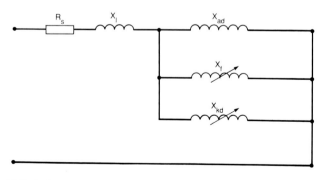

(d) Equivalent circuit in subtransient state

FIG. 3.44 Three-phase short-circuit, equivalent circuit and reactances

has no DC asymmetry in it (i.e., the short-circuit occurs at maximum stator voltage in this phase; the other two phases would have equal and opposite amounts of DC component in them).

One immediate difference can be recognised between the short-circuit current/time curve of the generator and that of static plant, such as a transformer or line. In the former, there is a very significant current decrement. In the latter, the current remains substantially constant.

An examination of Fig 3.44 (a) shows that the current decays in two stages, the first being a very rapid decrement, usually about 0.05 s; the second stage takes about 1.0 s, or less.

The decrement curve is in fact a double exponential, the first decrement being defined by one exponential, the second by another. The final steady state current will be defined by the excitation on the machine at the time of the short-circuit and the synchronous reactance will determine its magnitude, as already discussed under Section 3.2 of this chapter, i.e., $I_d = E_t/X_d$.

Analysis of the short-circuit envelope is quite easily carried out for the 'symmetrical' current case and a detailed description of the procedure is set out in the AIEE Test Code for Synchronous Machines, Pub. No. 503.

Figure 3.44 (a) shows that there is intially a very steep decrement: this is known as the *SUBTRANSIENT* period and the current is known as the *SUBTRANSIENT SHORT-CIRCUIT CURRENT*. There follows a more prolonged period known as the *TRANSIENT* period and the current flowing is the *TRANSIENT SHORT-CIRCUIT CURRENT*.

It is, therefore, as if there were a variable leakage reactance between the machine terminals and the short circuit, this varying as a function of time according to this double exponential.

10.2.1 Synchronous reactance — steady state stage

Probably the best way to envisage what happens when the short-circuit occurs is to work backwards from the final steady state to the initial stage at the actual instant of short-circuit.

As previously suggested, the reasoning will consider the flux linkages involved since these, or more accurately the flux (the stator coil turns and rotor coil turns being fixed), are the independent variables in the problem.

When the current has settled down to the steady state, the airgap flux is just sufficient to overcome the stator leakage-reactance drop I_dX_l, the remaining initial flux being opposed by the armature-reaction ampere-turns F_{ar} caused by IX_{ad}, which is wholly demagnetising, since the fault current is wholly lagging. The flux conditions in the machine are then as shown in Fig 3.7, with airgap flux denoted by

ϕ_L, i.e., that needed to generate enough voltage to overcome the leakage reactance X_l.

The value of this current will be given not by the actual no-load voltage E_t, but by the value of open circuit voltage which the no-load excitation would give on the airgap line. In Fig 3.16, this is the voltage GT given by extending TS to cut the airgap line at G. The reason for this is that there is no saturation involved, since the actual airgap flux ϕ_L to support the short-circuit conditions is only that to generate I_dX_l, which is about 10% of E_t.

Thus, the steady state short-circuit current I_d is given by $I_d = E_t/X_d$ taken from airgap line. The value of X_d is determined almost entirely by the armature-reaction reactance X_{ad} and a suitable equivalent circuit to describe this stage is shown in Fig 3.44 (b). From this figure:

$$X_d = X_{ad} + X_l \qquad (3.19)$$

where X_{ad} = armature-reaction reactance

10.2.2 Transient reactance — transient stage

Before application of the short-circuit, the rotor would be establishing a total flux of, say, ϕ_T. Part of this would not link the stator because of leakage paths and if ϕ_L denoted the leakage flux, ϕ would be the effective flux linking the stator winding to generate E_t on open circuit, i.e., $\phi = \phi_T - \phi_L$.

Remembering the theorem of constant flux linkages discussed under Section 10.1 of this chapter, it will be recognised that between the instant of short-circuit and the final steady state, there has taken place a severe reduction in airgap flux. As the rotor itself carries a highly inductive winding, it will be obvious that the flux change cannot take place instantly and that currents must, therefore, flow in the rotor winding to uphold the initial total flux linkage associated with the rotor winding of ϕ_T.

With the increased rotor currents, giving a considerably increased rotor MMF, more leakage flux must be diverted into the leakage paths and, since ϕ_T remains constant, it follows that the effective flux crossing the airgap must be reduced to, say, ϕ'.

This will, therefore give a reduced voltage at the generator terminals and the short-circuit current will be lower than if this phenomenon had not occurred, i.e., it will be less than E_t/X_l. This effect is as if reactance additional to the leakage reactance had been introduced at the instant of short-circuit, limiting the current to the initial transient current. This reactance is called the *TRANSIENT REACTANCE* and is denoted by X_d', and is in fact a combination of rotor winding leakage-reactance and the synchronous reactance. Thus, transient current $I_d' = E_t/X_d'$.

It is greater than X_l and a suitable equivalent circuit for this stage is shown in Fig 3.44 (c). From this figure,

$$X_d' = (X_{ad} X_f/X_{ad} + X_f) + X_e \quad (3.20)$$

$$= X_f' + X_l$$

where

X_f = leakage reactance of the rotor winding

X_f' = effective leakage reactance of rotor winding

The slightly-reduced effective airgap flux ϕ' cannot be maintained indefinitely because of the presence of resistance (r) in the rotor winding, so that the additional rotor current will slowly fall, as the energy is absorbed by the I^2R losses in the rotor resistance. This compensating current will fall at approximately the natural time-constant of the rotor winding (given by L_f/r_f) and so, therefore, must the stator fault current. This must decay exponentially and the transient reactance can conveniently be regarded as a 'function of time' reactance which obeys the following law:

$$X_d'(t) = X_d' \exp(t/T_d')$$

where $T_d' = 1/\omega r_f [X_f + (X_{ad} X_l/X_{ad} + X_l)]$

The *TRANSIENT SHORT-CIRCUIT TIME-CONSTANT (T_d')* is the time which it takes for the current to decay to within (1/e) of the difference between the initial transient short-circuit current and the final steady state curve, see Fig 3.44 (a).

10.2.3 Subtransient reactance — subtransient stage

It will be remembered that reference has been made in a number of places in this chapter to the rotor damper winding. In the practical 2-pole machine, this takes the form of solid copper bars which run along the rotor winding slots just under the slot wedges. Similar bars also run under slots in the pole faces in some designs, the latter slots being fitted with steel wedges to maintain an unbroken magnetic path at the pole face. The ends of the bars are effectively short-circuited when the rotor is at speed, due to centrifugal force pressing the bar-ends against the rotor end-bells, so forming an equivalent squirrel-cage.

Thus, the damper bars occupying the interpole area, i.e., with their axis on the direct axis, will behave as an equivalent subsidiary short-circuited rotor winding which links with the total rotor flux ϕ_T. At the instant of short-circuit, the initial fault current would be determined solely by the stator leakage reactance X_l, if the compensating current flowing in the damper windings were sufficient to maintain the effective airgap flux ϕ at its initial value. As previously, this does not occur, since the compensating currents which now flow in the damper windings increase the effec-

tive rotor MMF to maintain the total flux ϕ_T, a greater proportion is diverted into leakage paths than that giving rise to X_l. The resultant effective airgap flux ϕ'' will be slightly smaller than ϕ but larger than ϕ'.

This will give a slightly reduced voltage at the generator terminals but higher than in the transient condition, and will, therefore, give rise to a lower current than that determined by the stator leakage reactance, but higher than that due to the transient reactance.

As previously, this slight voltage drop due to leakage flux is as if an additional reactance to the stator leakage reactance had been introduced at the instant of fault and since this occurs before the transient stage, it is called the *SUBTRANSIENT REACTANCE* and is denoted by X_d''. Thus, $I_d'' = E_t/X_d''$.

This reactance is greater than X_l but smaller than X_d' and a suitable equivalent circuit to describe this stage is shown in Fig 3.44 (d). From this:

$$X_d'' = \frac{X_{ad} X_f X_{kd}}{X_{ad} X_f + X_{kd} X_f + X_{ad} X_{kd}} + X_l$$

$$= X_{kd} + X_l \quad (3.21)$$

where

X_{kd} = leakage reactance of damper winding

X_{kd}' = effective leakage reactance of damper winding

As in the transient stage, the damper bars possess resistance, so that the damper currents must eventually decay. As the resistance of the damper circuit is higher compared with the inductance than that of the field winding, the damper currents will decay much more rapidly than the transient currents. In most practical machines, the time constant is of the order of 0.02 s and will be given approximately by the ratio L_{kd}/r_{kd}. The stator current must, therefore, decay exponentially at approximately this same rate and the subtransient reactance can again be conveniently written as a function of time:

$$X_d''(t) = X_d'' \exp(t/T_d'')$$

where

$$T_d'' = \frac{1}{\omega r_{kd}} \left[X_{kd} + \frac{X_{ad} X_l X_f}{X_{ad} X_l + X_{ad} X_f + X_l X_f} \right]$$

The *SUBTRANSIENT SHORT-CIRCUIT TIME-CONSTANT (T_d'')* is that time which it takes the current to decay to within (1/e) of the difference between the initial subtransient short-circuit current and the transient short-circuit current, see Fig 3.44 (a).

10.2.4 DC asymmetry — point-on-wave of fault

Still reasoning in reverse, there remains one more 'initial' condition to be considered. This is the condition of the short-circuit itself. It will be remembered from well known circuit theory, that the point-on-wave of voltage at which a short-circuit occurs is of prime significance. If the small additional reactance to the leakage-reactance at the instant of fault is disregarded for the moment, since this only comes into play immediately after fault current begins to flow, then it will be the stator leakage-reactance X_l and the stator resistance R_s which will define the initial conditions for the fault current. Since the value X_l is never attained, however, it will be the subtransient reactance X_d'' and the stator resistance R_s which will define the initial conditions for the fault current.

The equivalent circuit for a short-circuit at any instant of time t, when the applied voltage is given by $E_m \sin(\omega t + \psi)$ is given by Fig 3.45 (a) where R_s is the stator resistance and X_l is assumed to be equal to X_d''.

The descriptive equation for this circuit is given by:

$$E_m \sin(\omega t + \psi) = R_s i + L \frac{di}{dt}$$

or, in Laplace form:

$$E_m \sin(\omega t + \psi) = (R_s + Lp)i$$

from which the solution is:

$$i(t) = E_m/Z \sin(\omega t + \psi - \phi) + E_m/Z \sin(\phi - \psi) \exp - (R_s/L)t$$

Since $\phi = \tan^{-1} L/R_s$ where $\omega L = X_d''$ and, as R_s can be ignored, except in the exponential term, then the equation simplifies to:

$$i(t) = (E_m/X_d'') [\cos \psi \exp - (R_s/L)t - \cos(\omega t + \psi)]$$

where ψ is the point-on-wave of voltage at which the fault occurs. For values of $\psi = 0, \pi, 2\pi$ the transient term (which is unidirectional, since $\cos \psi$ is a constant and not a continuously varying function of time) will have its maximum value given by E_m/X_d''. This will be the DC component, when the fault occurs at zero voltage. If there were no decrement of the first term in the bracket, the current would reach a maximum value of twice the normal symmetrical value at the value of t which makes $\cos(\omega t + \psi)$ negative, i.e., at $\omega t + \psi = \pi$ after the application of the fault and it would remain at this value.

The second term $(E_m/X_d'') \cos(\omega t + \psi)$ gives the instantaneous value of the AC component of fault

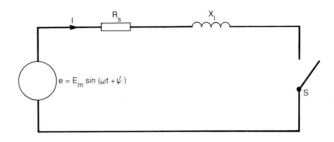

(a) Equivalent circuit for switching onto inductive circuit

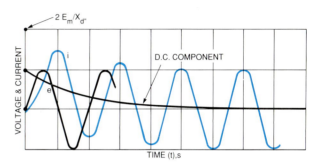

(b) Short-circuit for circuit (a) maximum offset point-on-wave for e = 0

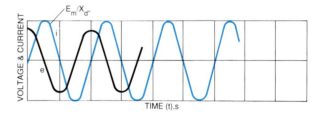

(c) Short-circuit for current (a) no offset point-on-wave for $e = E_m$

FIG. 3.45 Effect of point-on-wave of fault voltage

current, which if sustained would have the maximum value of E_m/X_d'' or $\sqrt{(2E_t/X_d'')}$, where E_t/X_d'' is the RMS value of the initial symmetrical fault current I_d''.

The value of the DC component of the fault current is given by the first term:

$$(E_m/X_d'') \cos \psi \exp - (R_s/L)t$$

i.e., $(\sqrt{2}E_t/X_d'') \cos \psi \exp - (R_s/L)t$

where the value L is given approximately by X_d''/ω and R_s is the stator resistance.

Thus, $I_{DC} = (\sqrt{2}E_t/X_d'') \exp - (R_s/L)t \cos \psi \quad (3.22)$

$= (\sqrt{2}E_t/X_d'') \exp - T_a \cos \psi$

where T_a is the time constant of the DC component and is given approximately by $X_d''/\omega R_s$.

This is the well known DC offset effect and the DC component has to be added to the AC component of fault current given from the previous sections, see Fig 3.45 (b).

If the point of wave occurs at $\psi = \pi$, 3π, etc., i.e., at maximum voltage, the first term in the bracket is zero and there is no exponential component, as shown in Fig 3.45 (c).

10.2.5 Complete equivalent circuit for the generator

From a consideration of the previous four sections, an overall equation can be written which defines the short-circuit current as a function of time in terms of the terminal voltage E_t at the instant of the application of the short-circuit (E_t is the value taken from the airgap line, if rated no-load excitation is the base value chosen), and the initial values of the three reactances X_d, X_d' and X_d'' treated as constants, obtained from the oscillogram of Fig 3.44 (a), or by calculation from Equations (3.19, 3.20 and 3.21).

Thus, adding Equations (3.19, 3.20, 3.21 and 3.22):

$$I = E_t\,[1/X_d + (1/X_d' - 1/X_d)\exp - (t/T_d') + (1/X_d'' - 1/X_d')\exp - (t/T_d'') + (\sqrt{2}/X_d'')\exp - (t/T_a)\cos\psi] \qquad (3.23)$$

where T_d' is the transient SC time constant

$\quad T_d''$ is the subtransient SC time constant

$\quad T_a$ is the armature time constant

This equation shows that the reactance so-called 'constants' are, in fact, functions of time and this necessitates computer solution of problems, where R_s is taken into account, such as when deriving the DC component. In this case, the quantities E and I are also functions of time, requiring the solution of non-linear differential equations.

The equivalent circuit of Fig 3.46 represents the machine reactances as two adjustable reactances being variable as exponential functions of time and their time constants in terms of the more usually quoted machine constants of X_d'', X_d' and X_d, rather than those used to develop Equations (3.19, 3.20 and 3.21), i.e., X_{ad}, X_f and X_{kd}.

This alternative representation has been included to help visualise the behaviour of the machine under fault conditions and to give a diagrammatic interpretation of the current decrement curve of Fig 3.44 (a) and of Equation (3.23) in practical terms.

The circuit of Fig 3.46 could be used to analyse the behaviour of the machine on short-circuit, provided that the ohmic impedances were replaced by

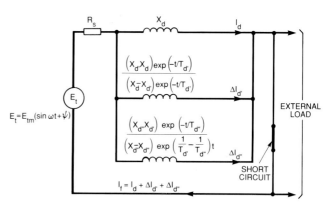

FIG. 3.46 Equivalent circuit for symmetrical three-phase fault using reactances X_d'' X_d' and X_d

the equivalent operational impedances.

Figure 3.47 is the equivalent direct-axis circuit for a generator in which operational impedances are used and where the applied voltage is given as a rate-of-change of a flux linkage, i.e., $P\Psi_d$, where P is equivalent to d/dt and Ψ_d is the flux linkage on the stator direct axis.

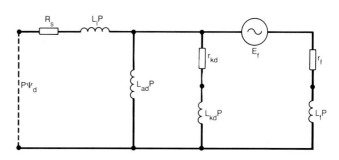

FIG. 3.47 Equivalent circuit for direct axis using operational impedances: d = direct axis flux linkages

It should be noted that Equation (3.23) is based upon an 'ideal' machine, i.e., one in which the influences of body currents flowing in the solid forging of the rotor have been greatly simplified by assuming that they all flow in the damper windings. In the practical machine, currents flow not only in the damper winding but also in the rotor body during the subtransient stage and this offsets the magnitude of X_d'' and of the subtransient decay time.

The more accurate equation for the AC component, taking into account these factors can be found in [4].

It will be seen that it is only the subtransient component of current which is affected by the rotor body currents and that the value is determined by the stator leakage reactance X_l rather than X_d''. The exponential gives a more rapid decrement than in Equation (3.23) and this lines up with test results taken of machines subjected to sudden short-circuits.

10.2.6 Summary of fundamental machine constants

From previous work, it will be recalled that the fundamental time constants of the various machine reactances are as follows:

$$T_d' = [1/\omega r_f][(X_f + X_{ad}X_l)/(X_{ad} + X_l)]$$
$$= X_f/\omega r_f, \text{ approximately}$$

$$T_d'' = [1/\omega r_{kd}][X_{kd} + (X_{ad}X_f X_l/X_{ad}X_f + X_f X_l + X_l X_{ad})]$$
$$= X_{kd}/\omega r_{kd}, \text{ approximately}$$

Other time constants can be derived, such as the transient open-circuit time constant T_{do}', the stator short-circuit time constant T_d, and the subtransient open-circuit time constant T_{do}''. They are as follows:

$$T_{do} = 1/\omega r_f (X_f + X_{ad})$$
$$= X_d/\omega r_f, \text{ approximately}$$

$$T_{do}'' = 1/\omega r_{kd} [X_{kd} + (X_{ad}X_f/X_{ad} + X_f)]$$
$$T_d'' = 2X_d'' X_q/\omega R_s$$
$$= X_d''/\omega R_s, \text{ approximately}$$

From these equations, the three machine reactances can be explained in terms of each other and their time constants. These can easily be proved by rearranging the time constant equations and comparing them with the previously derived values of Equations (3.19, 3.20 and 3.21).

Thus:

$$X_d = X_{ad} + X_l \text{ already given as Equation (3.19)}$$

$$X_d'' = \frac{X_d T_d'}{T_{do}'} = X_l + \frac{X_{ad} X_f X_l}{X_{ad} + X_f} \quad (3.24)$$

$$X_d'' = \frac{X_d T_d' T_d''}{T_{do}'' T_{do}''}$$
$$= X_l + \frac{X_{ad} X_f X_l}{X_{ad} X_f + X_f X_l + X_l X_{ad}} \quad (3.25)$$

Some typical values of machine reactances are given in Table 3.1.

10.3 Significance of reactances

10.3.1 Synchronous reactance X_d

The significance of this constant has been fully covered in previous Sections 3.1 to 3.6 of this chapter. It has been shown that it indicates the steady state stability of the machine and governs its behaviour for slow changes, such as drift of frequency, voltage and load.

10.3.2 Transient reactance X_d'

Since it is the rate-of-change of stator current as an RMS quantity which influences the total machine flux, comparatively rapid changes will involve the rotor winding, where these are of the order of the time constant of the rotor field winding, i.e., changes which occur in the range 0.1 to, say, 3 seconds.

Such changes occur during asynchronous operation, load oscillation of the rotor, and post-fault conditions; so X_d' is used for the conduct of transient stability studies.

10.3.3 Subtransient reactance X_d''

This determines the initial current values following a system disturbance, such as a fault or sudden load switching.

It is used in assessing the rupturing capacity of circuit-breakers, especially their ability to withstand closure onto a fault, when fully asymmetrical conditions can also exist.

10.4 Negative sequence reactance X_2

When a generator is subjected to an unbalanced three-phase load, negative sequence currents flowing in the stator conductors set up a corresponding counter-rotating airgap MMF. Unlike the positive-sequence currents, where the MMF rotates in synchronism with the rotor MMF, this counter-rotating MMF reacts with the rotor at twice synchronous speed, thus producing 100 Hz currents in the rotor body and in the conductive components of the rotor. There tends to be almost complete reflection in the rotor surface of the stator current and the rotor currents flow in the surface of the rotor and close circumferentially. The flux due to the stator current is forced into paths of low permeance, which do not link any rotor circuits. These paths are the same as for the subtransient reactances. Since the MMF wave moves at twice synchronous speed with respect to the rotor, it alternately meets the permeances of the two rotor axes, corresponding to subtransient reactances $X_{d''}$ and $X_{q''}$. Hence the value of X_2 lies between $X_{d''}$ and $X_{q''}$. The choice is normally taken between $(X_{d''} + X_{q''})/2$ or $\sqrt{(X_{d''} X_{q''})}$: the CEGB uses the arithmetic mean [5].

10.5 Zero sequence reactance X_o

When fourth-wire currents flow between the machine neutral and the phase terminals, they have no time phase-difference, being zero sequence. They are, there-

TABLE 3.1
Typical generator constants

MW	MVA	kV	Stator resistance Ω/phase	Rotor resistance Ω	X_d'' p.u.	X_d' p.u.	X_d p.u.	T_d'' s	T_d' s	T_{do}' s
30	37.5	11.8	.0047	0.685	0.105	0.165	1.58	0.04	0.82	8.0
60	75	11.8	.00499	0.729	0.11	0.16	1.87	0.025	0.58	8.0
100	125	13.8	.0016	0.218	0.18	0.25	1.87	0.022	0.65	7.0
120	150	13.8	.0015	0.176	0.14	0.22	1.95	0.02	0.7	6.2
200	222	16.5	.00185	0.138	0.18	0.23	1.36	0.015	0.77	4.5
350	412	18.0	.00156	0.179	0.18	0.26	2.14	0.022	0.59	4.9
500	588	22.0	.00165	0.0705	0.21	0.28	2.56	0.015	0.38	3.5
660	776	23.0	.0013	0.0903	0.29	0.34	2.43	0.015	0.6	7.5
313	330*	18.0			0.143	0.216	1.15	0.037	2.6	12.3

* 12-pole pumped storage machine with $X_q = 0.633$ p.u.

fore, in phase with each other and since the stator windings carrying them have a space/phase difference of 120°, it follows that the net airgap flux due to these currents must be zero.

The only reactance which can influence these currents must, therefore, be that due to the stator overhang and slot leakages and it is more or less equal to the stator leakage reactance X_l.

10.6   Quadrature-axis reactances X_q' and X_q''

In Sections 2.2 and 9.2 of this chapter, it has been shown how the presence of saliency, i.e., the difference in the reluctances of the quadrature and direct-axis rotor iron circuits, affects the value of the synchronous reactance on the two axes. X_d can be different from X_q by as much as 80% for a salient-pole machine and between 6% and 12% for a 2-pole cylindrical rotor machine.

In view of the transient reactance X_d' being a function of the rotor winding leakage-inductance, it will be apparent that there can be no equivalent quadrature-axis transient reactance, since there is no rotor winding on this axis.

Since there are damper bars on the pole shoes as well as in the interpole spaces (in fact, completely round the rotor for a 2-pole machine), there will obviously be a quadrature-axis subtransient reactance X_q'' but this will in general be lower than the more familiar direct-axis subtransient reactance because of the greater reluctance afforded by whatever rotor body currents are flowing. X_q'' is the symbol which denotes this reactance.

11   Generator transient stability

It has already been explained under Section 10.3.2 of

this chapter that, for fairly rapid changes of conditions, the transient reactance determines the circuit behaviour.

Therefore, following a short-circuit condition after the subtransient period has decayed within the first 2 cycles, and provided all the changes take place within the time period 0.1 to, say, 1 second, the transient reactance is operative and, therefore, is normally used in any computations involving transient stability.

Remembering that the power equation for a machine on steady-load is given by $P = (E_f V/X_d) \sin \delta$, it follows that the equivalent equation under transient conditions will be given by:

$$P' = (E_d' V/X_d') \sin \delta' \qquad (3.26)$$

The voltage E_d' is the transient voltage and is equal to E_t, if the machine is on no-load when the disturbance takes place, but if the machine is on-load, $E_{d'}$ is a potential voltage waiting to come into operation once the change takes place. It is a measure of the transient airgap flux ϕ' as previously discussed under Section 10.2.2 of this chapter, and is higher than E_t according to the loading conditions on the machine.

It is obtained from a consideration of the voltage triangle of Fig 3.48 and is calculated, provided that the power factor of the load current is known, from the phasor equation:

$$E_d' = E_t + IX_d' \qquad (3.27)$$

A justifiable approximation for practical lagging-load power factors is given by $E_d' = (E_t + IX_d') \sin \phi$, where $\cos \phi$ is the load power factor.

11.1   Transient power angle diagram

Reference to Equation (3.26) shows that the transient electrical power output of a generator under transient

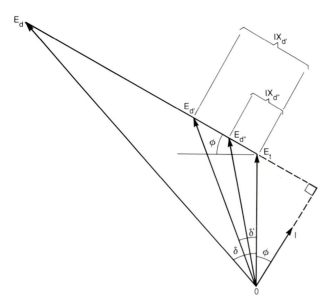

FIG. 3.48 Voltage phasor triangle for a generator on-load

conditions is considerably greater for a particular rotor angle than in the steady state. Figures 3.49 and 3.50 show the two power angle diagrams, plotted on the same axes.

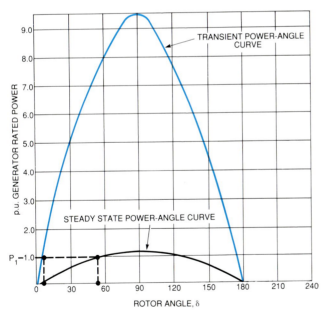

FIG. 3.49 Comparison between transient and steady state power angle diagrams

The maximum power at a transient rotor angle of 90° will be given by the ratio X_d/X_d', which is usually of the order of $2.6/0.3 \simeq 9$ times as great as for the steady state maximum power limit.

A particular difficulty is often met when considering the problem of the behaviour of the machine immediately before and after a sudden disturbance takes place. For example, up to the point of the fault, the machine will be operating on the steady state power angle diagram at a rotor angle of 50°, say. After the instant of fault, it will be operating on the transient state power angle diagram where the rotor angle will still be at the pre-fault value. The difficulty lies in deciding upon how to correlate these two conditions, since from the transient power angle diagram, the pre-fault rotor angle should give an equivalent electrical output many times greater than before the fault.

On the other hand, the equivalent rotor angle when the machine is in the transient state, to give an unchanged pre-fault electrical power output, will be considerably less than the actual pre-fault rotor angle of 50° (from Fig 3.49, about 8–10°).

Since this cannot and does not happen, the problem is to resolve these two incompatible facts: the difficulty is resolved by viewing the facts in the correct light. Remembering that the machine cannot respond in accordance with the steady state power angle diagram when in the transient state, it will be apparent that any rotor excursion after this state is established is quite independent of the steady state. Provided the machine can be shown to pull back into step during the transient state, it will in fact restore to its previous rotor angle in the steady state, provided there have been no external circuit changes. If there have been, the rotor will assume the new steady state angle it would have attained if there had been no transient condition in between.

The solution lies in superimposing the two power angle diagrams (as shown in Fig 3.50) by displacing

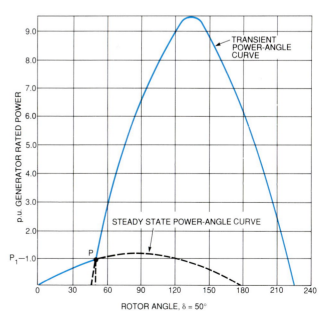

FIG. 3.50 Superimposition of transient and steady state power angle curves, for transient stability study

the transient-state diagram horizontally to intersect with the steady state diagram at the common point P, which satisfies the correct rotor angles and power level for both diagrams.

11.2 Transient stability — the equal-area criterion

It can be shown that the area under the power curve over a particular duration of rotor angle swing is proportional to the total energy involved in this rotor excursion.

Referring to Fig 3.51, this is a transient power angle diagram for a machine. At the instant of fault, the rotor angle was at, say, 40°; it is assumed that a dead three-phase short-circuit occurs. Since the electrical power output must be zero during the fault ($V = 0$, so that the power equation $P' = (VE_{d''}/X_{d'}) \sin \delta$ becomes zero), it will be obvious that the rotor must accelerate, since the steam input P_1 is being maintained constant.

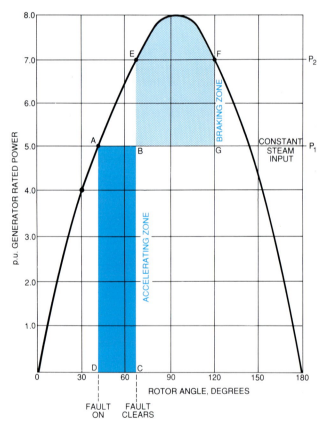

FIG. 3.51 Rotor swings for three-phase fault from 40° to 120°, $E_t = 0$ during fault

If now the fault clears when the rotor has advanced to say 70°, the total accelerating energy imparted will be proportional to the area under the constant power input line, i.e., ABCD.

After fault clearance, since the rotor angle has advanced, the generator will now be delivering a higher electrical output at P_2. Rotor advance will still continue however, and the electrical output must follow the power angle diagram. The advance will continue until the area under the curve BEFG is equal to the area ABCD during the former accelerating period. This area must be a braking zone, since the electrical output is always in excess of the steam input. Once the total braking energy equals the accelerating energy, the rotor will stop advancing and will begin to decelerate towards its former position of equilibrium before the fault. Some overshooting will occur and the rotor will eventually settle back to position A at an angle of 40° by a series of damped oscillations.

The damper bars will help the damping of this oscillation and so the values of $X_{d''}$ and $X_{q''}$ will influence the effectiveness of this.

This technique is known classically as the *EQUAL-AREA CRITERION* and gives a fairly good approximation of transient stability problems.

Transient stability studies using this classical technique tend to be overpessimistic, but recent work by Shackshaft, and others [6, 7] has made possible more accurate analytical solutions, which agree, very closely with actual test results. These newer techniques allow for the effects of damping, which have perhaps the greatest influence.

Not all faults result in a complete collapse of the machine terminal voltage. Here, the power angle diagram is drawn for the value of power output given by putting the reduced value of V into the transient power angle equation, i.e., $P' = (V'E_{d'}/X_{d'}) \sin \delta$. Figure 3.52 illustrates the construction for this purpose.

It is not intended to develop this section beyond this point since machine-system stability analysis forms a very extensive subject in its own right, which would require detailed coverage well beyond the scope of this chapter [7, 8 and 9].

An introduction to this subject has been included, however, to give operating staff an appreciation of the problems involved in operating a system safely under normal and abnormal conditions, such as faults.

Large machines, because of their inherent design and comparatively smaller frame size/MVA, have larger values of synchronous and transient reactances, in addition to which they have progressively smaller moments of inertia. The combined effect of these divergent qualities is more 'lively' (i.e., less stiff) machines and, in consequence, they are inherently less stable.

The importance of the automatic voltage regulator as an aid to, at least, steady state stability will therefore be apparent.

12 The induction motor

At this stage, it will be useful to set aside for the moment, the subject of the operation of turbine-generators

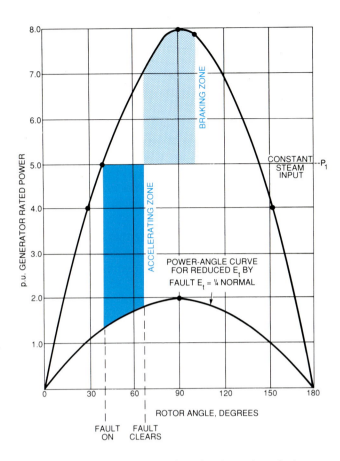

FIG. 3.52 Rotor angle swings for three-phase fault from 40° to 100°, $E_t = \frac{1}{4}$ normal during fault

and to review some of the better known facts about the commonest of the rotating electrical machines, the squirrel-cage motor. This will be of considerable help in considering asynchronous operation in normal turbine-generator plant, since there are no complications of non-linearity of the rotor constants (such as inductance and resistance), as there are with the solid forged rotor of the turbine-generator. This is because the conventionally-designed induction motor uses a laminated rotor. The squirrel-cage induction motor is the familiar workhorse to be found almost everywhere in the power station and in industry, wherever a robust rotary drive is required. Fortunately, it also provides a good basis upon which to explain the behaviour of the asynchronously-operating turbine-generator.

12.1 Rotational torques

The basic induction motor comprises a three-phase wound stator to which the supply is connected and a laminated cylindrical rotor, which is slotted to receive heavy-section copper bars welded or brazed at the rotor ends to produce shorting rings, this arrangement forming the well known squirrel cage.

The operation of the induction motor results from

the three-phase stator winding setting up a constant magnitude flux ϕ_{ag}, which rotates at a speed of 3000 r/min for a 2-pole machine. The action of the airgap flux cutting the cage bars, generates an EMF which causes currents to flow in them in the spatial position shown in Fig 3.53, which should be regarded as a frozen picture of the airgap flux position at the time when it is vertical. As in the synchronous generator, and using the same conventions for voltage and current direction, maximum EMF must occur in the rotor bars coincident with the axis of the maximum airgap flux distribution, see Section 3.1 of this chapter and Fig 3.4. As in the generator, the flux distribution is sinusoidal round the airgap and so, therefore, is the rotor EMF, zero volts being induced in the cross-axis conductors.

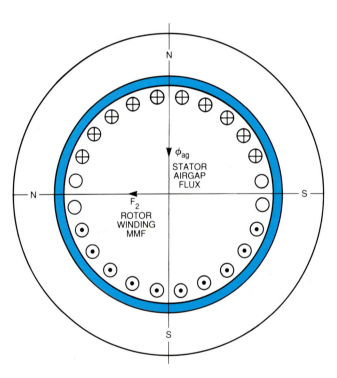

FIG. 3.53 Generation of rotor EMF and establishment of fluxes (rotor currents at unity, induction motor at standstill)

For the current distribution to be identical with that of the generated EMF in the rotor winding (as shown in Fig 3.53) the current must be in phase with the EMF, i.e., at unity power factor, a state of affairs which cannot be achieved in the practical induction motor, but one which assists the reasoning. This produces the ampere-turns F_2 as shown, which is similar to armature reaction as described in Section 3 of this chapter and Fig 3.5 (a), except that here it is the rotor which produces it and not the stator.

At a later instant of time, the stator airgap flux will have advanced in the direction shown in Fig 3.54,

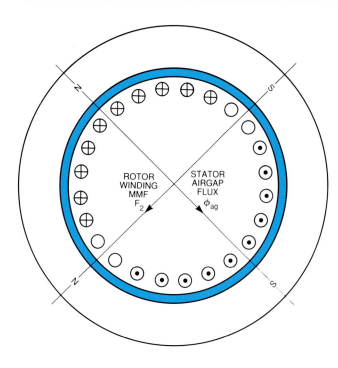

FIG. 3.54 Position of stator airgap flux and rotor an instant of time later than that shown in Fig 3.53

and with it also the rotor currents and ampere-turns F_2. In other words, the induced rotor currents set up a rotating MMF in the rotor, which will be 90° ahead in space of the stator airgap flux and keeping exactly in step with and ahead of it, if the current is at unity power factor in relation to the rotor EMF. The interaction of these rotor ampere-turns and the airgap flux sets up a mechanical torque exactly as in the generator (shown in Fig 3.2 (a)) except, of course, that motor action results instead. Since it has been assumed, in the first instance, that the rotor is at standstill, the torque will cause the rotor itself to begin to turn in the direction of the stator airgap flux, in effect dragging the rotor with it. This is because the torque is experienced by the rotor conductors and since they are embedded rigidly in the rotor laminations, must transmit their thrust to the body of the rotor itself.

As the rotor speed rises, the rate at which the cage winding cuts the airgap flux will fall by an amount equal to the speed difference between that of the rotor itself and the airgap flux, and with it the rotor EMF and current will also fall. This will reduce the torque and the process will continue until it just balances the opposing torque of the load the motor is driving, when the speed of the rotor will settle at a value to satisfy a state of equilibrium.

If the torque, due to the mechanical load, is increased, the rotor speed must reduce in order that the speed difference is increased sufficiently to produce a rate of cutting the airgap flux by the cage winding to generate the required rotor EMF, causing the rotor current to flow to produce the new torque.

The reverse must occur as the mechanical load is reduced, until at no-load the rotor is not required to develop any torque at all apart from windage and friction loss, which is being ignored at present. To satisfy this, the rotor speed must rise until there is no speed difference between the cage winding and the airgap flux, resulting in no EMF being generated and hence no rotor current or torque. The rotor would be at synchronous speed. This brings out the important fact that an induction motor cannot deliver and torque at synchronous speed, so that, for the practical unloaded machine, the small torque developed by windage and friction must require the machine to run at a slight slip.

It is important to note throughout the above argument that, although there is a speed difference between the rotor and the airgap flux, the rotor MMF axis is *ALWAYS EXACTLY AT SYNCHRONOUS SPEED AND IN STEP WITH THE AIRGAP FLUX* at an angle fixed by the phase difference between the rotor current and generated EMF, whatever the rotor speed or speed difference.

This is because the speed of rotation of the rotor MMF, relative to the rotor, is equal to the speed difference between the rotor and that of the airgap flux and since the rotor is itself rotating, it carries the rotating MMF with it, making the combined speed, relative to the stator, identical with that of the airgap flux. As a result all the normal quantities in the rotor will be at slip frequency.

This speed difference of the rotor is known as *SLIP* and is usually denoted by s and is expressed as a fraction of synchronous speed. Thus, a slip of 1 is a speed difference of 3000 r/min, i.e., the rotor is at standstill, a slip of 0 is equal to synchronous speed of 3000 r/min, in 2-pole machines. Thus: Rotor speed = (1 − s) times Synchronous speed. This emphasises an important and fundamental difference between the behaviour of the asynchronous machine and that of the synchronous machine described in earlier sections.

In the former, an increase in mechanical loading causes *no change at all in rotor angle*, i.e., between the axis of the rotor MMF and airgap flux *but results in a speed reduction*, whereas in the latter, *it causes an increase in rotor angle* between rotor MMF and airgap flux *but no change in speed*.

This reasoning has assumed that the cage winding was entirely resistive, hardly a practical proposition, since it will be obvious that the presence of the rotor slots, into which the cage bars are embedded, must result in it being substantially inductive. This will result in the rotor current always lagging the rotor EMF and in consequence the rotor MMF axis can never be ideally at 90° to the airgap flux. Figure 3.55 shows the flux and MMF relationship of a typical induction motor, as it would be at standstill, where the phase angle between the flux and rotor MMF axis is shown at 75°. This means that the standstill

$$X_s = \omega L_2 = 2\phi \text{ (sf)} L_2$$

where s = slip

f = supply frequency, Hz

X_s = reactance at slip s, Ω

L_2 = rotor inductance, H

As the resistance of the cage remains constant, the total impedance falls and the phase angle increases. This results in an increase of the rotor current and an improvement in the power factor, both effects producing a net increase in torque, as seen from previous reasoning. Figure 3.56 shows the flux and rotor ampere-turns relationship at speed and when the motor is loaded, where the phase angle of the current has improved to 30°.

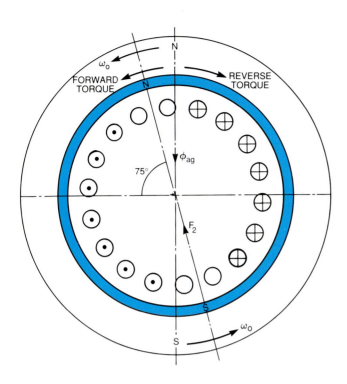

FIG. 3.55 Stator and rotor fluxes at standstill (rotor currents lagging 75°)

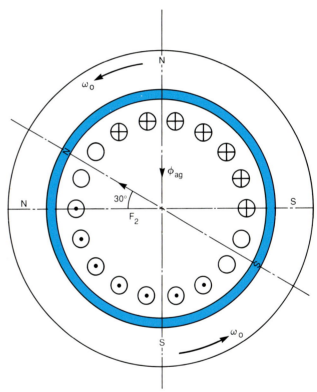

FIG. 3.56 Stator and rotor fluxes at speed (rotor currents lagging 30°)

or 50 Hz impedance presented by the rotor winding to the generated EMF will have a lagging phase angle of 75° or, put another way, the ratio of the resistance to the reactance will be cos 75° or $R_2 = 0.26 X_2$; where X_2 is the rotor reactance at 50 Hz and R_2 is the rotor resistance.

The poor power factor of the current has the effect of reducing the starting torque, since the torque-producing component, i.e., the component F_2 at right angles to the airgap flux, is now quite small, i.e., equal to F_2 cos 75° = 0.26 F_2. Another way of looking at this, by referring to Fig 3.55, is to consider that the rotor MMF F_2 will produce poles as shown on the rotor surface. These poles to the right of the airgap flux axis will produce reverse torques, those to the left, forward torques. The effective starting torque is then the difference between the forward- and reverse-acting torques, with the former predominating. It should be noted, whilst viewing things in this way, that if the rotor current were at unity power factor, since the rotor MMF axis is at 90° to the airgap flux axis, all the rotor poles will be to the left of the airgap flux axis and will all be torque-producing, so there will be no reverse-acting torques.

Assuming that the rotor begins to run, as the rotor speed rises, so the slip speed and frequency of the currents in the cage winding will fall since the frequency in the rotor is always sf. This will bring about a reduction in the reactance of the cage winding which is directly proportional to frequency since

Summarising, as the rotor speeds up from standstill, the torque increases and, therefore, the rotor accelerates since the torque produced is greater than the load, giving a surplus of torque to accelerate the rotor. This obviously cannot continue indefinitely since, as already stated, as the rotor speed approaches synchronous speed, so the slip approaches zero and so do the rotor currents and torque. As would be expected, the torque developed as the motor speeds up, reaches a maximum at a certain slip, after which it successively reduces until at synchronous speed it reaches zero. The operating point will be where the load curve crosses the speed torque curve such as point L on Fig 3.57, which gives typical torque/slip curves of an indiction motor for different values of rotor resistance/inductance ratios. The curve for a rotor having a resistance of 0.26 of the 50 Hz reactance is shown, so also are the values of slip at maximum torque.

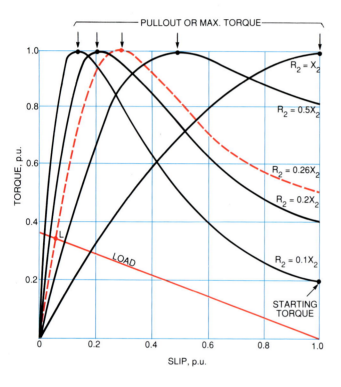

FIG. 3.57 Typical induction motor torque/slip characteristics

Several important conclusions about the behaviour of the induction motor can be drawn from inspection of Fig 3.57 as follows:

● The torque/slip curve (for all cases of rotor resistance less than reactance) from standstill to maximum torque, represents unstable behaviour, since the speed of the motor increases with torque.

● All values of slip less than that at maximum torque show a decrease of torque with increase of rotor speed, a stable behaviour since, as the mechanical torque is increased, the speed will reduce and the motor settle to a value of slip suitable to meet the new torque. Thus, the 'front-end' of the torque/slip curve is the stable running part of the curve.

● The rotor must follow the torque curve shown from standstill until equilibrium torque is reached. The

excess of electrical torque over mechanical load torque accelerates the rotor, a healthy state of affairs for rapid getaway.

● Increasing the rotor resistance improves the starting torque; when it is equal to the reactance at standstill, it gives maximum possible starting torque.

● Increasing the rotor resistance, on the other hand, gives a greater speed regulation with change in mechanical loading, i.e., the front end of the torque/slip curve is less steep.

● If the mechanical load is continuously increased from a certain steady-running value, the slip will follow the torque/slip characteristic. When the mechanical load is exactly equal to the maximum torque capability of the motor, the motor will be on the verge of stalling, for if there is now a small further increase in the mechanical loading, the slip will increase further, but the torque output will reduce. The motor is now operating on the unstable part of the characteristic and will cumulatively increase the slip, further reduce the output torque until it finally shuts down to a standstill. This emphasises that the 'back end' of the torque/slip characteristic is unstable.

12.2 Transformer action

So far, only the rotor performance has been considered and, recognising that the mechanical power to deliver the mechanical torque to the load must ultimately come from the supply system in the form of electrical power, it will be evident that the currents which flow in the rotor to set up the necessary torque must ultimately be supplied from the stator, as is also the other torque producing component, the airgap flux.

Transformer action is responsible for this current transference taking place between the stator winding on one side of the airgap and the rotor winding on the other: the stator can be considered as the primary and the rotor as the secondary winding of a transformer. The airgap flux and stator/rotor iron circuit are then considered with the common linking magnetic core and its magnetising field, as in the transformer.

12.3 Equivalent circuit of the induction motor

The equivalent circuit of the static transformer can be used to represent the induction motor, provided certain adjustments are made to allow for the effect of the rotational voltages generated in the rotor windings at the slip speed.

12.3.1 Rotor circuit

With the rotor at standstill and the stator winding energised by a supply voltage of E_1, the induced rotor EMF will be E_1. Assuming that the rotor runs up to a slip s (i.e., the actual rotor speed will be $(1 - s)$ synchronous speed), when the rotor-generated EMF $= sE_2$.

At a slip frequency of sf, the rotor winding reactance will be sX_2, where X_2 is the standstill reactance.

Then the rotor current is given by:

$$I_2 = sE_2/\sqrt{[R^2_2 + (sX_2)^2]} \qquad (3.28)$$

where R_2 = rotor resistance, Ω

   I_2 = rotor current, A

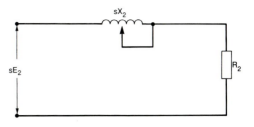

(a) Equivalent rotor circuit using rotational voltage concept

The equivalent circuit is shown in Fig 3.58 (a).

This is inconvenient in this form for equivalent circuit representation, since it contains a 'rotational' term sE_2, which is obviously not compatible with static transformer concepts.

Rearranging Equation (3.28), by dividing the top and bottom by s, the rotational voltage sE_2 becomes a constant voltage E_1, that of the supply:

$$I_2 = E_1/\sqrt{[(R_2/s)^2 + (X_2)^2]} \qquad (3.29)$$

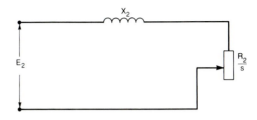

(b) Equivalent rotor circuit using constant voltage & rotor resistance as F(s) concept

This has converted the equation into a form such that the rotor reactance is now a constant, represented by the standstill value, and the rotor resistance now artificially contains the rotational component. Since this is the only resistive component of the current in Equation (3.29), it must obviously represent, at least in part, the power component of the rotor. This power component is seen to vary inversely with the slips.

This is also the equation for a series circuit, supplied by a constant voltage E_1, having a fixed reactance X_2 and a variable resistance R_2/s, whose value varies inversely as the slip frequency s. The performance of such a circuit can be described by the well known current locus diagram, where the locus of the current formed by varying the values of R_2/s, describes a semicircle, having as diameter the maximum current magnitude E_1X_2 when $R_2/s = 0$, and passing through the point $E_1/\sqrt{[(R_2/s)^2 + X_2^2]} = 0$, when $s = 0$, i.e., at synchronous speed, when $R_2/s = \infty$.

Figure 3.58 (b) shows the rotor portion of the equivalent circuit and Fig 3.58 (d) the current-locus diagram, where AE represents the voltage phasor at 90° to the semicircle diameter AB and points, such as AP', represent the current values and phase angles for different values of (R_2/s). The 'rotational' value of rotor resistance at standstill (s = 1) is given by $R_2/s = R_2$ and, since no useful mechanical power

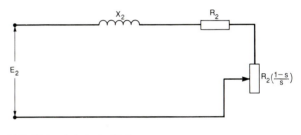

(c) Modified equivalent rotor circuit

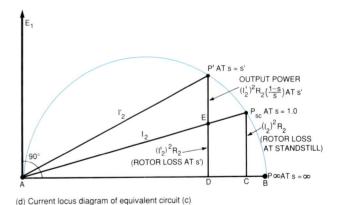

(d) Current locus diagram of equivalent circuit (c)

FIG. 3.58 Induction motor rotor equivalent circuits and current locus diagram

is being produced because the machine is at standstill, $(I_2)^2 R_2$ will be entirely rotor loss.

Figure 3.58 (d) gives the rotor loss at standstill as CP_{sc}, since this is the in-phase component of I_2 relative to the voltage AE_1. Thus the equivalent rotor resistive component to represent the power output on load can be replaced by the quantity $R_2 (1 - s)/s$ and the total effective rotor resistance by $R_2/s = R_2 + R_2 (1 - s)/s$ represented in Fig 3.58 (c).

201

12.3.2 Stator circuit

The equivalent circuit has so far only been considered from the rotor winding side which, since it delivers the eventual load, can be represented as the secondary loading of an equivalent transformer circuit. This is now possible because the rotational voltage sE_2 has been replaced by the constant supply voltage E_1 and the rotational term incorporated very conveniently into the rotor resistance component.

This equivalent rotor circuit can be connected across the secondary winding of an equivalent transformer so that the primary winding can now be added as shown in Fig 3.59 (a). Also, the magnetisation of the transformer core must be equivalent to the magnetisation of the airgap in the induction motor, since it is this magnetic field in the airgap ϕ_{ag}, which is the mutual coupling between stator and rotor in the actual induction motor. Since the transformer and also the induction motor must take a magnetising current and there are magnetising iron losses, for the case of rated voltage only, these can be represented as a shunt reactance and resistance connected across the primary winding as shown in Fig 3.59 (b).

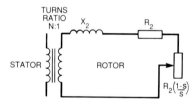

(a) Equivalent circuit of rotor by including stator/rotor transformer winding

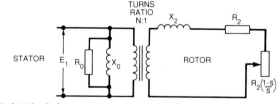

(b) Equivalent circuit of stator and rotor circuit by including magnetising admittance

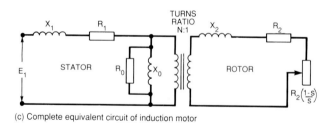

(c) Complete equivalent circuit of induction motor

Fig. 3.59 Development of equivalent circuit of induction motor

The equivalent circuit is now almost complete and if the stator winding had no leakage reactance or resistance, the diagram would be completed by merely adding the supply voltage E_1 to the transformer pri-

mary terminals. Since the stator does have leakage reactance and resistance (the latter being larger than in a transformer because the main flux has to cross an airgap in the induction motor, thereby encouraging greater leakage), these impedances must be connected between the supply voltage terminals and the terminals of the equivalent transformer primary winding. If these are designated X_1 and R_1, and the magnetising reactance and iron loss resistance as X_o and R_o, respectively, the complete equivalent circuit is as represented by Fig 3.59 (c).

12.3.3 The complete equivalent circuit

It is now possible for the transformer to be deleted by including the effect of its turns ratio, either by converting the rotor impedances from the actual values to the values 'seen' on the stator side, in which case all rotor impedances will be multiplied by N^2 and the applied supply voltage will be conveniently E_1, or, by referring to the rotor side (which is very rarely done), when the stator impedances would be multiplied by $(1/N^2)$ and the supply voltage would then be (E_1/N).

Figure 3.60 (a) shows the former conversion: moving the magnetising arm results in the approximate equivalent circuit of Fig 3.60 (b), shown on basis of stator side, i.e., $X_{2'} = X_2N^2$, $R_{2'} = R_2N^2$ and $R_{2'}(1 - s/s) = R_2N^2(1 - s/s)$.

12.4 Interpretation and use of the circle diagram

Figure 3.60 (c) is the approximate circle diagram based on the equivalent circuit of Fig 3.60 (b), but has added to it other locus lines which are used in making calculations of various performance factors; these include:

- *Construction of torque line* The point P_{sc} defines the starting current, both in magnitude and phase, and is obtained from the 'short-circuit' or 'locked rotor' test. The point A defines the magnetising current and is obtained from the 'open-circuit' or running light test. The point C on the normal drawn from P_{sc} onto AB at D divides $P_{sc}D$ in the ratio $(P_{sc}C)/(CD) = R_2/R_1$. Then, AC produced to the locus circle is the *TORQUE LINE* or power input into the rotor. It also yields the shaft output torque.

- *Construction of output line* The line AP_{sc} is the *OUTPUT POWER LINE*, or useful power at the motor shaft, i.e., the product of torque and speed.

- *Construction of slip line* The line $P_{sc}L$ is drawn parallel to the torque line AC to cut the perpen-

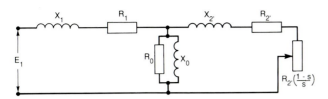

(a) Equivalent circuit of induction motor omitting coupling transformer

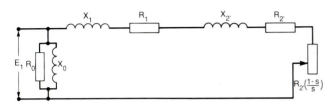

(b) Approximate equivalent circuit of induction motor

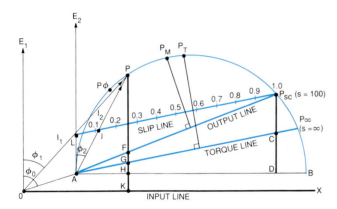

(c) Approximate circle diagram performance locus lines

FIG. 3.60 Equivalent circuits and circle diagram: induction motor

This diagram yields almost all the significant *STEADY STATE* performance data of the induction motor and the diagram is interpreted as follows.

From the working point P, drop the perpendicular PK onto the input line and mark the intercepts F on the output line, G on the torque line and H on the circle diameter AB. Join A and P and mark the intercept J on the slip line, join O and P. Let E_1 be the supply voltage, kV (phase/neutral). This gives:

- *INPUT POWER* $= E_1 \times PK$ kW per phase, where PK is converted to amperes.

- *OUTPUT POWER* $= E_1 \times PF$ kW per phase, where *(SHAFT POWER)* is PF converted to amperes.

- *OUTPUT TORQUE* $= E_1 \times PG$ kW per phase, where PG is converted to amperes.

- *ROTOR LOSSES* $= E_1 \times FG$ kW per phase, where FG is converted to amperes.

- *STATOR LOSSES* $= E_1 \times GH$ kW per phase, where GH is converted to amperes.

- *SLIP* $= LJ = 0.12$, for this example.

- *INPUT POWER FACTOR* $= \cos \phi_1$.

- *LIGHT-LOAD POWER FACTOR* $= \cos \phi_0$.

- *ROTOR POWER FACTOR* $= \cos \phi_2$.

- *STATOR CURRENT* $= OP = I$, A (supplied by system).

- *ROTOR CURRENT* (in cage bars) $= AP \times N = I_2 \times N$ in amperes where N is the stator/rotor turns ratio.

- *WORKING POINT AT* $=$ Point $P\phi$, which makes *MAXIMUM POWER*.

- *MAXIMUM POWER FACTOR* $= OP\phi$ a tangent to the circle.

- *MAXIMUM OUTPUT POWER* $=$ Point P_M which makes the perpendicular from P_M onto the output line a maximum.

dicular AE_2 in L. $P_{sc}L$ is then sub-divided into, say, 10 or 100 equal parts. Each sub-division, commencing at L, which is at zero slip, is then the appropriate fraction of the slip up to the maximum at P_{sc} of $s = 1$. In Fig 3.60 (c), the line LP_{sc} is divided into ten equal parts, such that each division is 0.1 p.u. slip. The line LP_{sc} is the *SLIP LINE*.

- *Construction of input line* The line OX, is the *INPUT POWER LINE*, i.e., it yields the electrical power supplied from the system.

- *Construction of motor working point* Consider the point P on the circle locus: then, AP is the equivalent rotor current, OP the stator current, $\angle E_1OP$ is the power supply input power-factor angle, $\angle E_1OA$ the light-load or magnetising-current power-factor angle.

● *MAXIMUM TORQUE* = Point P_T which makes the perpendicular from P_T onto the torque line a maximum.

It will be obvious that, in constructing the diagram, a scaling will have to be determined which equates the dimensions quoted above with actual values of current. Since these will be phase values, and the voltage is also the phase value in kV, the product must be multiplied by 3 to obtain the total three-phase value. Since this is an 'electrical' diagram, it will also be necessary to convert to mechanical quantities, taking the speed into account, to obtain practical performance figures.

13 The laminated-rotor induction generator

In considering the induction generator, the fact of reversibility between the generating and motoring condition, previously mentioned in Section 3 of this chapter, will form a useful basis from which to construct a mental picture of the operation of the induction generator from that of the induction motor.

Consider a squirrel-cage induction motor, such as that described in the previous section, shaft-coupled to a DC drive motor and to a band brake to provide mechanical loading, the induction motor being supplied from the supply system; not a very profitable engineering set-up!

If now the motor is run up to speed and then loaded mechanically by the brake, it will settle down to a steady slip, having followed the electrical torque/slip curve of Fig 3.57, during run-up. If the mechanical load is reduced to zero, the motor will run at a small slip to support the windage and friction losses of the assembly. If the input is further increased it will be found that, as the induction motor is driven above synchronous speed, it will start to *deliver* load to the AC system and the DC drive motor will then be supplying this output electrical power in addition to the total windage and friction losses.

The induction motor is now operating as an INDUCTION GENERATOR. If values of torque, slip and power output are now measured, it will be found that a torque/slip characteristic similar to that of the induction motor will result but this will, of course, be the inverse of the motoring characteristic. Slips will be negative when operating in this regime, if motoring slips are regarded as positive. Figure 3.61 is a typical torque/slip curve for the motor over the range of $s = -1$ (generating) to $s = +1$ (motoring). The actual speed of the induction generator at a slip of $s = -1$ is obtained from:

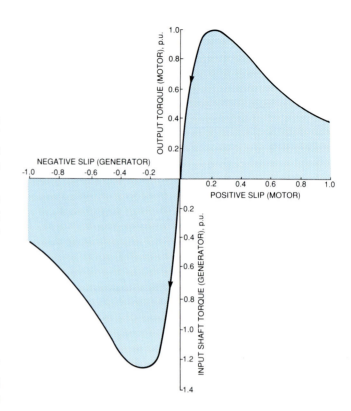

FIG. 3.61 Torque/slip characteristics: induction motor-generator

Rotor speed = $(1 - s)$ (synchronous speed)

= $(1 + 1)$ (synchronous speed)

= 2 (synchronous speed)

In the approximate equivalent circuit for the induction motor, as shown in Fig 3.60 (b), or the accurate equivalent circuit of Fig 3.60 (a), it is simply necessary to replace the rotational variable s by $-s$, so the equivalent load component $R_2(1 - s)/s$ becomes $-R_2(1 + s)/s$, i.e., R_2 becomes a negative resistance and, therefore, the 'load' becomes an output. This is the significance of the negative resistance, although it is often regarded as a hypothetical quantity.

13.1 The multicage induction machine

The simplicity of the ordinary squirrel-cage induction motor has established its place firmly throughout industry. It is robust, cheap and uncomplicated, but it can suffer from two main snags if special design measures are not taken:

(a) High starting currents

(b) Low starting torques

The evidence of (a) can often be found in power stations in motor drives to milling plant, crusher plant, etc., where the driven load is subject to mechanical jamming and stalling, and where the motor is also started direct-on-line. This evidence takes the form of distorted cage bars, or sometimes fractured bars which can cause consequential damage to the stator windings.

Further evidence of (a) is found in stator winding movement, loose end-winding packing blocks which can eventually result in overhang-to-stator core faults. Both are the result of too frequent restarts after stalling, and are the direct consequence of heavy starting currents which principally damage the rotor bars, by causing overheating. The high starting currents also cause vibration of the stator windings, eventually causing loosening of the windings in the slots and ultimate insulation damage at the point of emergence of the winding from the slots.

The stator-winding insulation fault is less prevalent than that of the damaged rotor bar, so it is the heavy rotor starting currents problem which has received the most attention from design engineers. The resolution of this problem has also improved the low starting torque, and so brought all round benefits. The methods developed range from using rotor bars of special shapes to control the current flow from start-up to running speed, or the use of several separate cages at various depths in the rotor.

13.2 Double-cage windings

The performance of the multicage-winding machine also helps in the understanding of the behaviour of the solid rotor turbine-generator under induction generator action, so the broad outline of this type of machine will now be considered.

The most common example is the double-cage winding which comprises two separate cages, concentric with each other: an inner cage of low-resistance copper bars and an outer cage of higher-resistance bars. Under standstill conditions, the outer bars take the greater share of current because their reactance is lower than that of the inner bars on account of the latter being more deeply embedded in the rotor iron, so having greater self-leakage flux. Another way of looking at this is to consider that at the high slip frequency at standstill, the flux does not penetrate deeply into the rotor because of the screening effect of the eddy currents flowing in the outer cage. Since the resistance of the outer cage is higher than for a normal single-cage winding, a greater starting torque results, in addition to a lower starting current. As the rotor speeds up and the slip frequency falls, the inner cage takes a greater share of the rotor current and torque. Since at small values of slip under normal load conditions, the reactance of both cages is negligible, so the rotor current shares between the cages are proportional to their resistances and the inner

cage makes the greatest contribution to the output torque finally developed by the motor.

13.3 Equivalent circuit of the multicage winding

Since there are two cages, the equivalent circuit of the rotor would be expected to comprise two separate shunt-connected impedances, one for the outer cage and one for the inner cage. Figure 3.62 (a) shows the complete simplified equivalent circuit constructed on this basis where R_2, X_2 are the resistance and reactance of the outer cage, and R_3, X_3 of the inner cage.

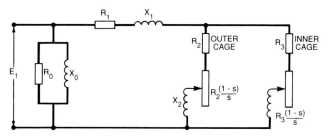

(a) Simplified equivalent circuit of double cage induction motor/generator, s is positive for motor and negative for generator

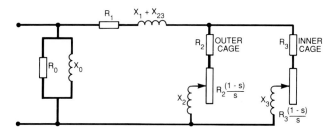

(b) Approximate equivalent circuit of double cage induction motor/generator

FIG. 3.62 Double-cage induction motor generator equivalent circuits

This equivalent circuit is accurate only if the main airgap flux links both cages completely, but since this does not occur, due to mutual coupling between the two cage windings, the mutual reactance should be included for absolute accuracy. Figure 3.62 (b) is the approximate equivalent circuit which yields results sufficiently accurate for practical purposes and this includes the mutual reactance X_{23} between cages 2 and 3.

13.3.1 The skin-effect cage winding

Another type of cage arrangement, which gives a similar characteristic to that of the double-cage, is the 'skin-effect' conductor design. This comprises a wedge-shaped conductor or tee-bar of considerably greater radial depth in the rotor body than the conventional cage bar. The widest part of the wedge is nearer to the centre. This gives effectively a graded resistance and reactance conductor where, as in the double-cage, the resistance at the peripheral end of the bar is higher than at the inner end. The reactance of this bar will also be graded between being lower at the peripheral end and higher at the inner end because of the gradually increasing proportion of leakage flux towards the centre of the rotor body; this is the 'skin-effect', which is greatest at standstill, where the current tends to be driven towards the outer radius of the conductor, i.e., to occupy the periphery of the cage.

There is no simple equivalent circuit for this type of cage winding, since there is an infinite gradation of *inductance and resistance*, which also depends upon the slip speed, quantities which are constants in the case of the discrete multicage winding, but which are now variables.

14 Asynchronous operation of the generator

14.1 General considerations

It is appropriate to include a section on the asynchronous operation of turbine-generator plant, not because this is a normally accepted mode of operation — indeed no generator is designed in the first instance to deal with asynchronous running — but because some plant has operated routinely in this manner at night and at weekends during the light-load periods to help deal with a national problem of high grid voltage, due to the surplus of reactive power generated by the 400 kV supergrid system at these times. Although reactive compensation has now dealt with this problem, a situation may yet arise where it is considered that asynchronous operation would alleviate the situation, or where it would occur as a fault condition, therefore an appreciation of the effects of this mode is given.

Before developing the theory behind asynchronous operation it would be appropriate to clear up what is meant by the word 'asynchronous' and then to deal with certain basic facts concerning this. Used in this context, asynchronous means other than synchronous and refers to operation of the machine when not locked to the system and running at synchronous speed.

Thus, asynchronous as applied to any rotating AC electrical machine, such as a simple squirrel-cage induction motor, refers to the fact of its non-synchronous operation, at a small speed difference or slip

characteristic of the induction motor. An alternator which is normally operating synchronously, i.e., locked to the system, can be described as asynchronous when it is pole-slipping after having become unstable. Asynchronous operation as applied to a turbine-generator is, however, confined to describing its *operation out-of-synchronism but with the rotor winding unexcited*. The term 'pole-slipping' is reserved in Britain to imply *running out-of-synchronism with the rotor excited*.

The subtle difference in the use of the two words lies in the very decided dissimilarity which the two modes of operation have on the supply system to which the machine is connected. With pole-slipping, the presence of excitation on the rotor causes severe voltage, power and reactive MVA surges to take place and the effect, as viewed from the control room, can be described, at the least, as distressing. Asynchronous operation, with the field unexcited, is usually an innocuous occurrence and results in only small system fluctuations, perhaps ±10% current and even smaller voltage swings.

An ordinary induction motor, although running asynchronously, causes no fluctuations at all on the system after the initial voltage dip during run-up to speed, since the rotor possesses no saliency to cause it.

It is now appropriate to examine some of the basic facts associated with asynchronous operation.

14.2 Basic considerations

To deliver full-load output from a 500 MW generator while running at 3000 r/min, a shaft torque of about 1.6×10^6 Nm has to be developed by the turbine. Obviously, an asynchronously operating machine, if it could successfully deliver its full output (which it cannot, because of certain serious practical limitations to be discussed later), would also require a torque of the same magnitude to be developed. Figure 3.11 and Section 4.1 of this chapter, dealing with the steady state performance, shows that under conditions of minimum excitation, the smallest rotor current to support full-load MW output, i.e., OS, is approximately 2.25 p.u. rotor current, when referred to Fig 3.14. Under these conditions the whole of the rotor flux (produced by the rotor ampere-turns) is used to produce MW, and it can be seen that there is no component of this current to produce MVArs, so that MVArs are supplied from the system under minimum excitation conditions in order to magnetise the air gap by drawing MWArs proportional to OA, i.e., 0.4 p.u. of the rated machine MVA.

The machine power output must be supported by the torque set up between the whole of the rotor MMF and its interaction with stator airgap MMF.

It can be concluded from this that it is always the component of rotor MMF at right angles to the airgap MMF which produces the torque, the other component along the airgap MMF axis being non-torque-producing and is responsible only for the generation of voltage or reactive VA. For example, in Fig 3.14, the rotor current is proportional to WA, and has two components, one along the MVAr axis equivalent to 0.6 p.u. MVAr lagging, which is non-torque-producing and delivering only MVArs, and the other along the MW axis equivalent to 0.8 p.u. MW, which is wholly torque-producing and delivering MW.

In the case of minimum excitation being considered, where the rotor angle δ is 90°, reference to Fig 3.14 shows that the value of the torque-producing flux is 2.25 p.u. and is the result of the rotor ampere-turns acting against the reluctance of the rotor-stator iron circuit which, if saturation is ignored, is proportional to it.

If, therefore, the rotor current were reduced to zero and the field switch opened under the steaming conditions of full load MW (a practice not to be encouraged with the machine on commercial load), it follows that an MMF must be self-generated in the rotor to interact with the stator MMF, to produce the necessary torque between turbine and generator to support the full-load MW. Furthermore, this will have to be as great as when the rotor was excited in the normal way in order to maintain the necessary torque. Thus, for the case of Fig 3.14, if it were assumed the machine had a rating of 500 MW, it follows that the rotor body must have induced in it a current at least equal to the ampere-turns needed to support full-load with the rotor normally excited, i.e., in the case being considered, the rotor is wound with 224 turns, a typical practical figure for a 500 MW machine, and 1 p.u. excitation is 1120 amps. Therefore, the MW component of rotor = 112 × 2.25 × 1120 ampere-turns at full-load = 282 240 ampere-turns.

Examining this figure of approximately 300 000 ampere-turns for a moment and recognising that the rotor comprises a solid steel forging, it will be seen that there can only be one effective turn so the current must flow in the surface of the rotor body some of which will embrace the axial length of the rotor, closing across the end bells, Fig 3.63. The value of the current must be of the order of 300 000 A to effect the necessary ampere-turn equality.

Further thought will show that this cannot be a direct current as in the normally-excited case, since an internally-generated EMF must be established axially around the rotor in order to circulate the required current in the rotor body and this can only occur as a result of the rotor slipping through the airgap flux, i.e., the rotor must be above synchronous speed. The current must, therefore, be alternating, the frequency being the slip frequency or difference between the rotor speed and synchronous speed of 3000 r/min. It will also be seen that, since the rotor-body currents must be circulating through the whole length

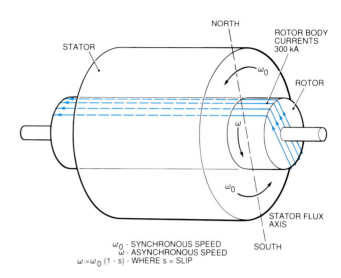

FIG. 3.63 Generation of rotor body currents

of the rotor against the electrical resistance and inductance of the steel, the voltage needed must depend upon the rate at which the rotor cuts the stator airgap flux and the slip speed must, therefore, rise until it reaches a value sufficient to generate this voltage.

In this mode, the generator is operating as a *SUPER-SYNCHRONOUS GENERATOR* or, as an INDUCTION GENERATOR. Summarising at this stage, the following facts can be listed:

● If a normal generator is delivering load with the rotor winding open-circuited, body currents of a high order must flow in the rotor to establish the necessary torque with the stator airgap flux. The value of the current for a typical 500 MW machine, if it were operating asynchronously under open-circuit field conditions at full-load, would be about 300 000 A.

● The stator airgap flux is supplied by the MVArs delivered to the generator from the system, i.e., the machine is absorbing MVArs from the system. For the 500 MW machine considered, having a synchronous reactance of 2.56 p.u., this is about 588 × 0.39 = 230 MVAr.

● The generator must run at a sufficiently high speed above synchronism to generate an adequate axial voltage in the rotor to circulate the required minimum of 300 000 A along the length of the rotor body.

● The machine operating as above would be behaving as an induction generator or as an asynchronous generator.

14.3   Practical limitations

From the previous section, it will be seen that very heavy body currents must flow in the rotor to support a substantial generator power output when the machine is operated asynchronously and unexcited. A single-turn figure of 300 000 A has just been calculated for a typical 500 MW, 588 MVA, 22 kV machine, operating at full-load. This is a very substantial current and the following considerations will establish the obvious limitations which are imposed by the design of the machine rotor upon such modes of operation at high loading. It is, in fact, the high loading limitations which have established the reputation that all asynchronous operation is damaging and should not be permitted under any circumstances. It will be shown that asynchronous operation at low loadings for a 60 MW machine is quite safe, provided certain safeguards are observed.

14.3.1   High loading asynchronous operation

The principal limitations are those imposed by the rotor-body currents, although there are others such as stator core-end heating which are discussed later. Not only is the current high under conditions of high loading, but the slip speed must also be relatively high and, therefore, also the slip frequency. The effects of a relatively high slip frequency current is to cause this to be driven outwards towards the periphery of the rotor by skin-effect and to flow along the teeth, damper bars and slot wedges, instead of using the full cross-section of the main forging.

For a 60 MW machine, the rotor induced current would be of the order of 94 000 A and a typical rotor is some 0.97 m in diameter, but slip frequency currents of the order 2–3% slip, i.e., 1–1.5 Hz, are most probably concentrated in the outer 50–80 mm, or so, of the rotor periphery.

This not only produces rotor losses in the form of heat, which are higher than those due to normal excitation, but the forcing of the current paths into slot wedges and damper bars causes local hot-spots to develop at points of poor electrical contact. Furthermore, since the currents must return via the ends of the rotor in order to complete the current loop, they must take whatever paths the end bells themselves offer, in addition to the damper-bar shorting arrangements under the end bells.

Figure 3.64 shows a typical part cross-section of a rotor and Fig 3.65, a side view. The method of interconnecting the damper bars at the ends of the rotor to form an effective end-ring varies with different machine designs; some are welded into a continuous band, others rely on centrifugal forces pressing the damper bar ends against each other and the underside of the end bell. Here again, poor contact can result in the development of local hot-spots under the end bell or even arcing in some cases, in addition

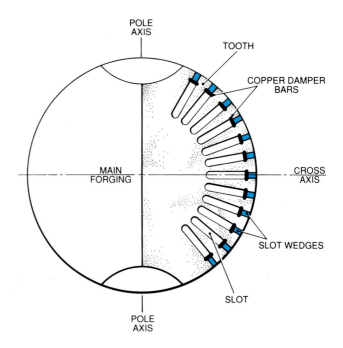

FIG. 3.64 Cross-section of a typical rotor

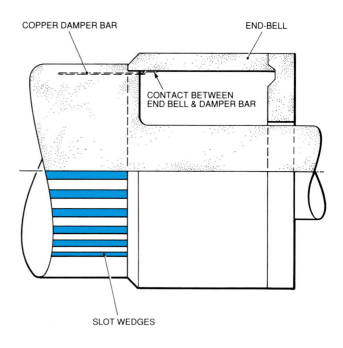

FIG. 3.65 Side elevation of rotor end-bell arrangements

to a general local heating due to the smaller cross-sectional area of the end bell.

The results of the asynchronous running tests at Stella Power Station reported that slight blueing of the rotor end-bells was found after the machine, a 60 MW generator, had been operating asynchronously for only about 30 seconds at a high loading and a slip of about 2.85%. The machine was operating in the unstable portion of the torque-slip characteristic, being held there by the action of the governor.

14.3.2 Low loading asynchronous operation

Still discussing a 60 MW machine, if the power loading were reduced to 10% of full load, i.e., 6 MW, the torque to be developed would be correspondingly reduced therefore, the rotor-body current required to establish the necessary torque with the airgap flux would be reduced to 10% of the previous value, i.e., 9 400 A. In addition, if the effective resistance and inductance of the rotor body were constant at this loading and slip speed, the voltage generated within the rotor body to drive this current around it, would now only have to be 10% of the full-load value. On these assumptions, the slip frequency would also be reduced to 10% of the previous full-load value, say 10% of 2.85% or 0.2 – 0.3% of 3000 r/min. This in turn would make the rotor-current slip frequency 10% of the full-load value: hence, the skin-effect would be reduced, resulting in a deeper penetration into the main body of the forging, which would reduce the effective resistance and the inductance, tending to cause a greater current to flow, or in practice, a reduction in the self-generated rotor voltage needed to establish the necessary rotor current to support the new light-load torque.

The net result of these effects is that the slip frequency is much lower than the *pro rata* fraction just given. In fact, measured values of slip speed and frequency on the 60 MW machine tests at Stella Power Station and other stations show that at 10% power loading, the slip speed is of the order of one revolution in 2 minutes or 0.0167%, which is considerably less than 10% of the measured full-load slip speed of 2.85%.

Figure 3.66 shows the power-slip curves of the 60 MW generator at Stella Power Station which was operated asynchronously from full-load by opening the field switch, but leaving the rotor connected across its field-discharge resistor. This is a typical curve and can be seen to be very similar to that of an induction motor. The additional shaded portion is the torque supported by currents flowing in the rotor winding, which reach a peak at about 1% slip and then rapidly decay to zero at about 1.5% slip. This is because, at the higher slip speeds, the flux cannot penetrate deeply enough to cut the rotor winding and only those currents which flow on the outer skin of the rotor body make any torque contribution.

Summarising, asynchronous operation at small power loadings is practical on small generators up to 60 MW rating because:

- The deep penetration of the body currents relieves the slot wedges and damper bars from high current-density heating and therefore the creation of localised hot-spots is eliminated.

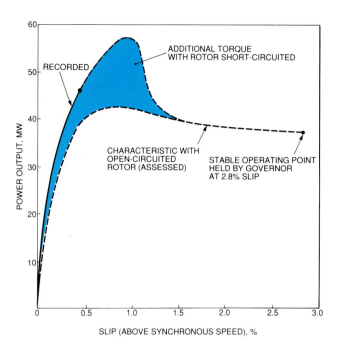

FIG. 3.66 Torque (MW)/slip characteristic of 60 MW turbine-generator operating asynchronously

- The general rotor heating, due to rotor losses, will be more nearly equal to the normal synchronous-operation rotor losses at that loading.

- The localised heating of the end-bells will be considerably reduced, because the deeper penetration of the rotor current into the main forging will cause the return end-currents to flow more in the rotor end itself, rather than attempting to take the path of the end-bells, which presents a much higher impedance.

This shows low power asynchronous operation to be a reasonably practical proposition: in fact, as already mentioned in Section 12.1 of this chapter, such operation has been performed routinely at certain power stations in the CEGB for extended periods during the night. This mode of operation required confirmation that there were no other limitations, due to peculiarities of rotor construction, than those already mentioned, or due to stator core-end heating. In any case, before it was decided to operate regularly on commercial load in this way, careful thought was given, in consultation with the manufacturers, to design features and possible weaknesses.

Before concluding this section, it is important to recall that whatever the slip frequency or the power loading of the asynchronous-operating machine, the rotor-body currents must always be constant for this particular loading, i.e., 94 000 A for 60 MW, 9 400 A for 6 MW and *pro rata* values of rotor current at different power loadings, irrespective of the slip frequency. This fact is a useful reminder of the 'can't get something for nothing' axiom which governs the physical world. At the low slip speeds, rotor losses will be higher than, but more nearly equal to, the normally-excited machine losses for a particular loading, and at high slip speeds these losses must of necessity be higher, because of skin-effect.

15 Generator systems

Associated with the generator are the excitation, cooling and lubricating systems necessary for the operation of the generator unit and these are now briefly described.

15.1 Excitation

The action of the excitation system in controlling the voltage and MVAr loading has been dealt with previously; however, the methods of achieving this excitation have not been discussed. Shaft-driven DC generators were used in the past to provide the excitation, but with the higher current loadings required by the larger generators, AC machines are used, with diode bridges converting the AC to the required DC levels. The diode bridges can be static, or rotating on a stub shaft, so supplying the required DC without the use of sliprings. Self-excitation can also be used from a transformer connected to the terminals of the machine, supplying either a diode or thyristor bridge.

Operationally all the above systems are highly reliable, provided that normal maintenance is carried out, e.g., inspection of brushgear and cleanliness of sliprings, etc. The static excitation elements have built-in redundancy; they can tolerate the loss of bridge arms and still maintain output.

15.2 Hydrogen cooling

To achieve the high outputs of modern large generators economically and with minimum weight of components, they are intensively cooled. The stator core and rotor conductors are cooled with hydrogen at a pressure of 4 to 5 bar. This hydrogen may be produced on-site, or obtained from commercial sources (see volume C). It is important that, whichever way the hydrogen is obtained, its purity is maintained to the recommended levels. Associated with the hydrogen system, there will also be purging-gas systems of carbon dioxide or nitrogen. It should be noted that whilst carbon dioxide is a good gas for purging the generator casing it is not inert and is *NOT* recommended for storing a generator for long periods.

Variation of hydrogen pressure directly affects the rotor cooling and hence the rotor-winding temperature and stator-core temperature. If operation at lower than rated hydrogen pressure is contemplated, due to some operational problem, then a rotor current limit will need to be considered and the machine performance diagram amended. The hottest core thermocouple should be continuously monitored and a recommended temperature not exceeded. This condition can therefore introduce limitations at lagging power factors (rotor current limit) and at leading power factors (core temperature limit). Since the hydrogen pressure is kept above the stator coolant pressure for the above condition, the stator coolant pressure needs to be reduced, so giving rise to less coolant flow and a rise in stator copper temperatures. This could limit the allowable stator current and needs to be considered.

15.3 Stator conductor cooling

Modern high-output generators have water cooled conductors in order to achieve a maximum rating with the minimum weight and frame size. The water used has to be kept within stringent limits to preserve the integrity of the generators. The conductivity of the water has to be lower than 200 μS/m (2 μS/cm)[‡] and should be kept as free of oxygen as possible to prevent oxides forming and being carried round the circuit.

The coolant circuit should have been supplied free of ferrous materials to prevent the formation of magnetite. Filters should be kept in the stator circuit and the condition of these maintained (see Volume C).

Since the purpose of the water circulation is to cool the stator conductor bars, it is useful if the operator has a knowledge of the effects on cooling of factors such as:

- Total or partial loss of pumping capacity.

- Total or partial blockage of water passages due to debris.

- Reduction of coolant flow due to air/hydrogen locks.

- Leakage of stator hydrogen into the water system.

- Deliberate blocking of some subconductors to prevent hydrogen-to-coolant leaks.

---

‡ *Footnote*: In the past, the CEGB conductivity monitors were calibrated in *Dionic Units*. These represented μS/cm. On changing to SI, it was convenient to re-label the monitors in μS/cm, keeping the same scales, so the CEGB still tends to quote its conductivities in the SI-related unit of μS/cm. The monitors normally sound the alarm at 500 μS/m (5 μS/cm).

Limits for any of these factors are supplied by the manufacturer, or are calculated by computer and applied. However, the object here is not to present these limits, since they vary between manufacturers and machine sizes, but to indicate how they affect operation.

The flux which exists in the slot induces eddy current losses in each subconductor. These induced losses tend to be the greatest at the top of a slot and least at the bottom since the flux crossing the slot is greatest at the top, diminishing towards the bottom (see Volume C). A top-of-slot bar has more induced heating when the other bar in the slot is in the same phase, as then the cross-slot flux is greatest. These effects mean that thermally, there are three distinct types of bar:

(a) Bottom-of-slot bars.

(b) Top-of-slot bars, where the other bar in the slot is in a different phase.

(c) Top-of-slot bars, where the other bar in the slot is in the same phase.

Since the flow affects the temperature and the resistivity is temperature dependent, changes of flow to a bar will affect the losses in the bar. A reduction in flow causes a rise in resistivity and a rise in temperature. This will have a negligible effect on the stator current and so the transport losses will increase with a reduction in flow. However, the eddy current losses are inversely proportional to resistivity and so will decrease with a reduction in flow. The net effect of small variations in flow on losses in fact turns out to be very small, typically a 10% reduction in flow gives an increase in total losses of less than 0.2% for a bottom-of-slot bar type (a) and less than 0.07% for a type (c) bar.

Reduction of flow can occur for various reasons, giving increased conductor-bar temperatures. If this occurs to such an extent that the water reaches boiling point, then severe pressure fluctuations can be expected, with the possibility of the total flow to at least one bar being reduced even further. Because the water in the stator coolant system is fed from a raised header tank, the water pressure in the conductor bars is above atmospheric pressure. Also, the pressure at the inlet end will be higher than at the outlet (or hot) end of the bar. For this reason, the water will not actually boil until its temperature (at the hot end) reaches approximately 120°C.

Computations show that, for a particular machine, this will occur for a type (c) bar (at rated machine conditions) when the water flow to the bar falls below 4.5 litres/min or 26% of rated flow. For type (a) and (b) bars, there are slightly more margins (17% and 23%, respectively).

Where the total flow to a bar is suddenly lost — due to, for example, debris, pump failure or an air lock — the temperatures in the bar would start to rise, and boiling would begin in approximately one minute for a type (c) bar (for type (a) and (b) bars, it would take twice as long).

If the machine were not tripped, temperatures would continue to rise and the water in the bar would be expelled and replaced by steam. With no water flow, all the thermal loss is by heat conduction through the main wall insulation of the bar. Assuming that the insulation remains in place and its thermal conductivity is constant, then typically 10 minutes after the loss of cooling, the bar could reach 600°C. After this time, the temperature would rise more rapidly, until within 20 minutes of the fault, the melting point of copper can be reached. The increased rate-of-rise is caused by the increasing resistivity of the copper, which gives higher electrical losses.

Circumstances can arise where certain subconductors are starved of water due to, for example, a hydrogen leak, the presence of debris, or the deliberate blocking-off to stop a hydrogen leak. The blocked-off subconductors are then no longer directly water cooled, but rely on heat conduction to their neighbouring subconductors to dissipate their electrical losses. Consequently, their temperatures tend to rise above their neighbours. This means that, although the average temperature may be quite acceptable, local insulation temperatures close to blocked subconductors may be high and temperature differences between subconductors can give rise to stresses caused by different thermal expansion. During normal machine operations, different stator currents will give rise to varying bar temperatures, so a bar will tend to expand and contract within the slot. Whether it does, or not, depends on the friction between the bar and the slot. If all the subconductors have similar temperatures then they will expand and contract together and no individual subconductor will be subject to significant thermal stress. However, if one subconductor is starved of water, then it is likely to be stressed compressively, as it is hotter than the others. It has been estimated by one research worker that a temperature differential of 30°C or more is sufficient to give a compressive stress in the hot subconductor which exceeds the compressive field stress of copper.

Computations of subconductor temperatures show that the maximum differential depends on exactly which subconductor is blocked. The worst case is where the subconductor is blocked at the top of the slot at the beginning and end of the slot. Its temperature at the hot end of the slot, with the machine at rated conditions, has then been calculated to be some 40°C above the others, so exceeding the 30°C quoted above. In this example, to avoid compressive yielding, the stator current would have to be reduced to 83% of rated current.

Although 40°C causes excessive thermal stress, there is little risk of inflicting insulation damage at that level of temperature.

If two subconductors were blocked, temperatures approaching 170°C could be expected, even though the maximum water temperature in any subconductor would be less than 80°C. Both the Class B and Class F insulation temperatures are exceeded (130°C and 155°C); under these conditions, the insulation in the vicinity of the hot subconductor will be well above its recommended operating temperature. Consequently, it can be considered that blocking off two subconductors will accelerate the ageing and degradation of parts of the bar insulation.

Should the flow to two subconductors be restricted, but not stopped altogether (debris, subconductor collapse or a hydrogen leak), then the temperature of the water in both of them can reach boiling point and so steam can form. Calculations indicate that this could occur if the flow to these two were reduced to 3% of normal; however, this will vary widely between different designs of generator.

One possible cause of reduced water flow to a conductor bar is inleakage of hydrogen. If a crack develops, or a brazed joint or a seal develops a leak, the pressure difference between the stator hydrogen and the conductor water will tend to force hydrogen through the crack into the water system. Apart from the possibility of gas locks building up somewhere in the system, the presence of gas in the subconductors will tend to impede the flow of water. Hydrogen will also become dissolved in the water and, since the solubility of hydrogen in water is lower at higher temperatures, it is likely to come out of solution and form bubbles in the hotter regions of the bar, thus resembling boiling.

For small hydrogen leak rates, as the hydrogen percolates through the insulation and subconductor crack, its pressure will fall to the local water pressure and a flow of water will be maintained. For the hydrogen to stop the water, its flow rate must be high enough to raise the pressure in the subconductor. If the leak is near the outlet end, the flow resistance experienced by the hydrogen is low and a very large hydrogen flow rate would be necessary to stop the flow. A leak near the inlet has a higher flow resistance and a lower flow rate of hydrogen is needed: if it is right at the inlet end, the hydrogen flow rate needed to stop the flow is at its minimum. It is this last value that is calculated and used as a basis for the maximum allowable hydrogen-to-coolant leak rate for an operational generator.

15.4 Hydrogen seals

Individual manufacturers use different designs of hydrogen seal, so requiring different guidelines for maintenance and operation. However, there are some basic engineering standards that can be of benefit and similar operating options in the event of problems. The hydrodynamically-generated oil film that separates the seal white-metal from the mating rotor-collar varies between 0.013 mm to 0.038 mm when operating at normal speed. This film approaches a measurable thickness when rotational speed exceeds 400 r/min (on 3000 r/min machines). Thus for barring and sub-400 r/min operation, the white metal and collar are working under boundary lubrication conditions. There is, therefore, a need for the mating faces to be of good surface finish, flat and in the same plane. With attention paid to these points, there is one further criterion — a good oil supply.

With the very small running clearances in an operational hydrogen seal, any debris present should ideally be smaller than the clearance. Under barring conditions, the situation is worsened as the situation changes to one of boundary lubrication, consequently any clearances existing are extremely small. To cover this, filters would need to remove particles down to at least 1 μm, since the seal oil system is sometimes an appendage on the end of the main lubricating oil system that may only be filtered to 38 μm. The seal oil filter needs, therefore, to be a compromise between the smallest reasonable particle removal and rate of filter blockage. Extra filters need to be fitted on returning the machine after overhaul until a good quality of oil is obtainable (it should be noted that a 10 μm filter can remove 98% of 3 μm particles and 100% of 10 μm and would be the recommended filter). Good filtration will certainly prolong the life of the seals.

The seal body has to be capable of moving axially to follow the expansion and contraction of the shaft system. In order to provide oil and gas sealing to the whole seal body as it moves axially, secondary elastomer seals are used. These are either a lip-seal or O-ring format, sometimes a combination of the two. The fit of the elastomers and the surface finish of the housing they slide against affects the frictional forces needed to initiate whole body movement. Lip seals can be over-compressed and the compression should be controlled by spacers or shouldered bolts. Only the correct size of O-ring should be used and the surface against which the elastomers operate should be of a very high finish.

The seal support system should allow complete axial freedom for the seal body to follow shaft movements. The movement that the seal undergoes is a bulk movement which causes the elastomers to slide over the contact faces and is caused by large axial expansions or contractions, as in run-up or shutdown, or major steam temperature excursions on-load. Also, small movements need to be accommodated solely by elastomer deflection caused by small thermally-induced movements when loaded. Supporting keys and slides must be of the correct material combination; simple changes carried out due to expediency can lead to unsatisfactory material combinations, which result in larger forces having to be applied to the seal body to initiate movement. Wear of the seals is generally signalled by high white-metal temperatures and excessive leakage of oil or hydrogen, although some seal types only show fluctuating and progressively increasing metal temperatures rather than oil or gas leaks.

An approximate metal-temperature guide is that the seal material should not exceed 120°C on *its collar contact face*. With most seals, there is some 4.5 mm to 6 mm of metal between the contact face and the thermocouple: there is a temperature drop across this material of approximately 20°C to 25°C, which gives a convenient top operating temperature of 100°C *thermocouple reading*. With wear, the distance between the front face and thermocouple position will reduce, and hence the temperature will drop, such that if there is good reason to believe wear has taken place, an allowance can be made. Even with good filtering any mild-steel pipework between the filter and seal can give rise to debris in the presence of water in the oil.

The operational problems that tend to recur most frequently on hydrogen seals can generally be considered as:

● Hydrogen leaking past the seal

● Oil leaking into the generator

● Excessive white-metal temperatures

Before operational procedures can be used to overcome these problems, an understanding of the force balance acting on a seal is required.

In simple terms, the 'holding-on' force generated by oil pressure, spring and gas pressure is balanced by the hydrostatic and hydrodynamic thrust generated across the seal face. Thus reductions in 'holding-on' forces result in an increasing rotor-to-seal gap, and vice versa. Further fine tuning can be carried out on some systems where the hydraulic 'holding-on' forces are generated in two chambers behind the main seal body, the pressure in each being individually adjustable. Thus the 'holding-on' force can be varied without directly affecting the pressure of the lubricating oil supply to the seal face.

With gas leakage, the tendency should be to reduce the collar/seal gap, whilst with high metal-temperatures it is advisable to increase it, or to provide better cooling. If the problem is oil leakage, the oil differential should be reduced, but this could lead to gas leakage. General advice is difficult since it need not be correct for specific problems; for instance, any of the previously mentioned symptoms can be generated by a temporarily-sticking seal, where moving the seal away from its sticking point by expansion or contraction of the generator/turbine shafting would probably be a better solution. This can be achieved by

small variations in the degree of hydrogen cooling or the temperature of the LP steam.

Such operational procedures can be used for isolated problems but, if the symptoms regularly return, it must be considered that seal state is changing and that wear may be taking place. Extreme care should be exercised during palliative actions to avoid exceeding the normal operating limits of the parameters being varied.

From the foregoing, the seal metal-temperature is clearly very important. It can indicate changes in seal gap (increased gap allows higher oil flow across the seal and drops the temperature, and vice-versa), warn of the impending failure of a seal (slow but sure temperature rise, which continues despite corrective measures) or clearly indicate a sticking seal (cycling of first one seal temperature, then the other). Thus, before any operational changes are made, a full knowledge of the seal metal-temperature records is important.

16 Condition monitoring

In order that safe operation and fault conditions can be recognised early and action taken to prevent a catastrophic failure, various condition monitors are fitted to generators as well as protection systems.

16.1 Temperature monitoring — thermocouples

Thermocouples are fitted to the generator, covering inlet/outlet hydrogen gas to coolers, inlet/outlet stator coolant to coolers, core temperatures and (as already mentioned) seal-face temperatures.

Thermocouples can only be placed at discrete positions in the core and are therefore only capable of monitoring the average core temperature. Thermocouples placed between stator conductor bars, together with the inlet and outlet water coolant temperatures, give indication of the state of the conductor coolant circuit. A rise in temperature of one inter-conductor thermocouple, without corresponding rises on other thermocouples, gives warning of a blockage in one of the conductor bars in a slot.

The thermocouple can only respond to a fault condition which is producing heat after the heat has been conducted through the materials between the hot spot and the thermocouple; this introduces a time delay, typically of the order of 30 minutes.

Temperature limitations are based on the thermal capability of the insulation systems and although the conductor-bar insulation may be of Class F materials, the core varnish could be Class B. Furthermore, in interpreting thermocouple readings, it should be remembered that water cooled conductor bars are cooler inside than outside whereas, with solid conductor bars, the heat has to be transferred across the insulation to the coolant gas on the outside of the bar.

It is not possible to monitor the rotor conductor-bar temperatures directly, but the average rotor-bar temperature can be deduced from an ohmmeter connected to the current through the rotor and the voltage across the sliprings, calibrated in temperature for the resistance of that rotor.

16.2 Hydrogen gas analysis

When any organic insulating material is heated to a temperature in excess of about 200°C, both an aerosol, composed of sub-micrometer particles of high boiling point liquid, and gaseous products are produced at its surface. In a generator, these products become entrained in the hydrogen coolant, where their presence can be used as an indication of an insulation fault. Two types of instrument are used as on-line station instrumentation. These instruments analyse either the gases involved or the particulates generated, when the insulation is heated (see also Volumes F and C).

A core fault is the name given to a fault originating in the stator core, typified by the melting of a relatively large volume of core laminations and leading to a generator outage. There is no built-in protection on generators against core melting and faults are only detected electrically when the core damage spreads to the stator winding and a stator earth fault occurs. An *incipient core fault* is defined as damage that shorts laminations together (e.g., a scratch, or metallic debris). It may be discovered because it is near a thermocouple, or located from flux tests. The damage is not large enough to cause an outage, but can potentially grow under operating conditions to produce a major core fault. The condition most likely to cause the growth of a core fault is that of pole slipping, either from instability or during the decay of the rotor flux under loss of excitation. Electromagnetic studies have shown that, under such conditions, transient high interlamination voltages are produced that can break down interlamination insulation, causing growth of the damaged area. Several such disturbances may be needed to produce a catastrophic fault. Foreign material, if present in the airgap, can weld itself to the laminations and cause a failure, thus highlighting the need for extreme cleanliness and control of entry to any generator opened up for maintenance or other work, in addition to that required in the original build, both before and after fitting conductor bars.

Core fault detection techniques need to be very sensitive in order that a fault is detected at an early stage. Operating conditions can then be controlled to arrest the further development: for example, the core temperature is kept to a minimum by operating at lagging megavars and reduced terminal voltage.

Stator winding faults, involving thermal degradation of insulation by such mechanisms as failure of subconductor insulation, loss of cooling water, or restriction of coolant flow, are only detected electrically when a large volume of insulation has failed and an interwinding or an earth fault occurs. Since there is much more insulation on a conductor bar than on the core plate, the method of detection does not have to be so sensitive.

The ion-chamber particle-detector, also known as the 'core monitor', fulfils the first requirement, as it is very sensitive.

Hydrogen from the generator enters a chamber which is lined with either thorium oxide or americium pellets. This lining emits alpha particles which ionise some of the hydrogen molecules. The ionised hydrogen then passes between electrodes, across which a voltage gradient is maintained. The gradient is sufficient to drive all free ions to the collector electrode, where they produce a small electrical current in an electrical circuit. When an aerosol of sub-micrometre particles enters the chamber, the ion current is reduced. Insulation degradation is therefore indicated by a reduction in electrical output. Oil mist, produced either thermally or mechanically inside a generator, can also give a spurious reading and alarm. Any filter capable of removing this mist will also remove particles produced by overheating insulation. A heated ion-chamber or a preheat to 100–130°C will vaporise the oil mist, but not the particles derived from the solid insulation, but there is a loss of sensitivity. Figure 3.67

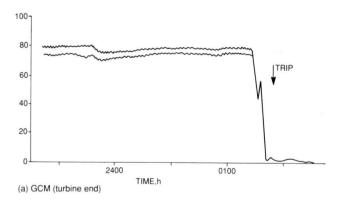

(a) GCM (turbine end)

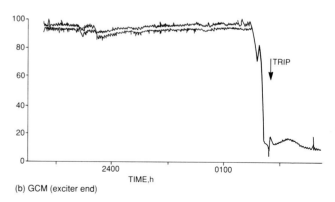

(b) GCM (exciter end)

Fɪɢ. 3.67 Generator core monitor response to generator core fault

shows a generator core monitor (GCM) response to a generator core fault and Fig 3.68 the procedure flow diagram on receipt of an alarm.

On-line gas analysis detecting organic thermal degradation products is not as sensitive as the particle detector, but is useful in monitoring for winding faults and can also be designed to detect specific substances that may be in the varnish, for example, sulphur or a special 'tagging' substance introduced deliberately into material used in the construction of the generator to indicate a fault and the position of such a fault.

High merit plant can be fitted with a complete system, comprising particulate monitor (GCM), gas analysers for organic compounds and sulphur, the flow diagram indicating an interpretation of alarms from the monitors is given in Fig 3.69.

16.3 Hydrogen dewpoint monitoring and control

The control of moisture in hydrogen in generators is important to prevent electrical flashovers, particularly on endwindings where there are exposed metal parts, and for the prevention of aqueous stress-corrosion cracking of austenitic end rings. Maintaining the correct dewpoint is particularly important for generators having direct-cooled stator windings. The prime objective of generator moisture control is to ensure that, under all plant conditions, condensation cannot occur within the generator. Table 3.2 summarises the requirements for the various operating conditions of a generator: during normal operation, the hydrogen driers will be in service and capable of maintaining the required dewpoint unless the moisture-in-oil concentration exceeds that stated. It is also important to recognise that dewpoint increases with increasing pressure, for example, a dewpoint of $-10°C$ at atmospheric pressure corresponds to a dewpoint of $+7.9°C$ at 3.1 bar (gauge), as shown in Fig 3.70. When operating a generator, the dewpoint at frame pressure should always be used, to avoid confusion.

The effect which a moist environment has upon the insulation of a large machine is dependent on both the condition and type of insulation system. Where this is well bonded, penetration is restricted to the surface layers. Epoxy-bonded systems, for example, are particularly resistant to bulk penetration. If, on the other hand, there is delamination of mica, as might be found on an older machine, there could well be penetration deep into the bulk of the insulation wall.

During off-load periods, where water has penetrated into the insulation, the bulk electrical properties such as dielectric loss and volume resistivity will be affected. Where there is delamination, the electric strength could be significantly reduced. Under these conditions, a dry-out would be necessary, prior to returning the stator to service, to prevent a possible electrical puncture of the main wall insulation.

The aim of any dryout technique is to remove the moisture from all surfaces subjected to electrical potential. Heating the conductor does not remove this moisture (it only drives any moisture to, but not from, the surface); 'windage dryout' merely circulates gas already in the casing (whether it be damp or dry). Blowing heated station-air from a fan heater directed at the windings is likely to cause condensation of this damp hot air on to the colder parts of the winding, thereby making matters worse. Conductor heating and windage dryouts are only successful where they are used with a change of stator gas — and it is the change which is important. The only effective method of removing moisture from the surface is to replace the existing damp gas in the casing with dry gas and to continue this until equilibrium is reached between the dry gas and the insulation surface.

16.4 Vibration monitoring

All large turbine-generators are provided with extensive turbine supervisory equipment to monitor vibration and other parameters, and are continuously supervised. The vibrations measured by the transducers are normally the velocities of bearing pedestals. The overall levels of the corresponding displacement are displayed and recorded on charts. This type of monitoring can give warning of an impending problem and a maximum amplitude can be stated, beyond which damage to connecting pipework and bearings could be expected. However, to make diagnostic use of vibration data additional test equipment is fitted to the bearing pedestals and recordings made, so that both amplitude and phase information can be obtained.

Continuous recording of run-up and shutdown vibrations gives a signature for the shaft-system response over a wide range of exciting frequencies. This data is analysed for both once and twice per revolution components, for both phase and amplitude. This type of analysis can indicate if the vibration is due to mechanical imbalance, out of alignment, oil-whirl or even a cracked shaft. Thermal imbalances can be detected and off-load balancing adjustments made to offset the thermal vector change, such that the operating regime is maximised (Section 19.7 of this chapter).

17 Operational limitations

17.1 Temperatures

As stated previously, the temperature of the component parts of the generator, together with inlet and

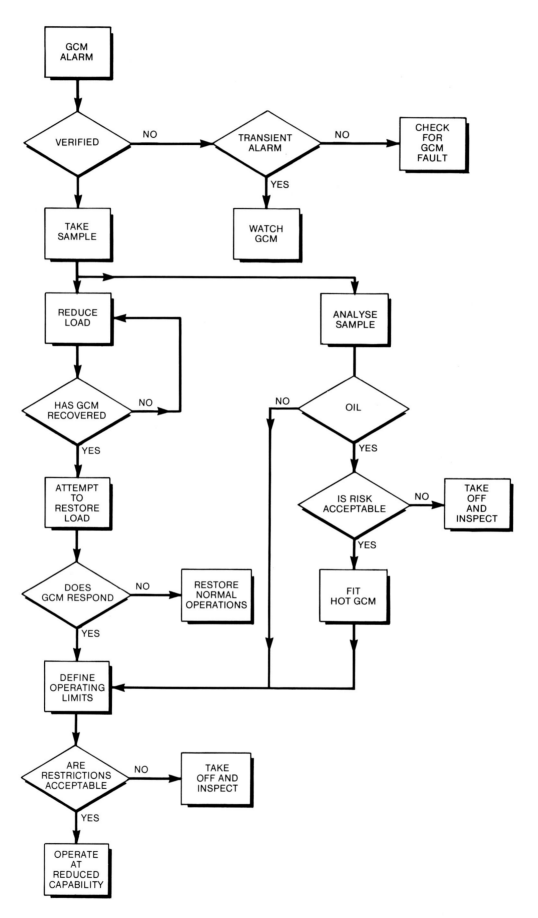

FIG. 3.68 Procedure following a generator core monitor alarm

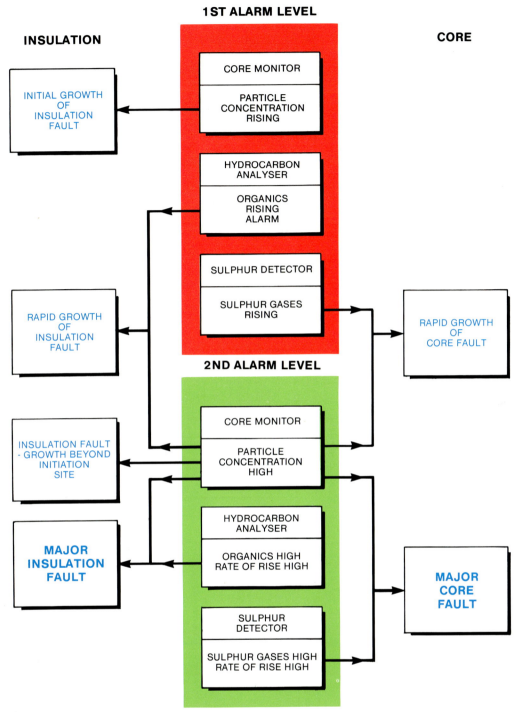

FIG. 3.69 Overall alarm system display

outlet temperatures of the cooling media, are monitored and recorded. An indication of the temperature limitations for stator-coolant water has already been discussed in Section 15.3 of this chapter, where it was stated that the limitation on this temperature was the avoidance of boiling in the subconductors.

In the core, the limitation is that of damage to insulation, both of the core insulation and of the conductor-bar insulation in the core slot. The highest core temperatures are at the ends of the core, either just at the base of the stator teeth or in the teeth. These temperatures are produced by currents flowing within the plane of the laminations induced by an axial flux at the ends of the core. To control these currents, slits are cut in the teeth at the ends of the core for about 0.5 m into the core. Outside the core body, aluminium rings are mounted as flux screens to control the axial leakage flux. Although all these methods help to control the leakage flux and the temperatures are reduced, the highest temperatures are

TABLE 3.2

Dewpoint requirements for hydrogen-cooled generators operating at 4 bar hydrogen pressure

Generator Operation \ Parameter to be monitored	(a) Stator Coolant Temperature and Control	(b) Hydrogen Dew point and Shut down Criteria	(c) Liquid in Stator Casing and Hydrogen Cooler Drains	(d) Drier Operation and Control	(e) Drier Blower Operation	(f) Moisture in Oil Concentration and Control
(a) Starting up and After prolonged Shut down	Temperature maintained at or above 30°C and at least 5°C above the cold hydrogen temperature measured at the hydrogen cooler outlets.	The dew point must be better than –18°C at 1 atmosphere, (equivalent to better than 0°C at frame pressure, 4 bar) with the set spinning (> 2200 r/min) and immediately prior to excitation.	Remove, identify and log the volume of any liquid from stator and hydrogen cooler drains. Investigate cause. Providing shut down dew point was satisfactory rely on dew point immediately prior to excitation for decision on whether or not to load machine. Note: Rotor Earth faults should be regarded as an indication of the presence of water in the frame until proved otherwise.	In service and operating satisfactorily. Check efficiency during dry out after prolonged shut down.	In service continuously.	The target moisture in oil concentration should be <0.05% w/w. The oil centrifuge should be in service and operating satisfactorily.
(b) Running	Temperature maintained at or above 30°C and at least 5°C above the cold hydrogen temperature measured at the hydrogen cooler outlets.	The target dew point should be better than –18°C at 1 atmosphere (0°C at frame pressure, 4 bar). Remedial/investigative action should be taken to reduce moisture ingress if the dew point is worse than the above value. Immediate shut down if the dew point reaches –5°C at 1 atmosphere (e.g. to +18°C in frame).	Remove identify and log the value of all liquid from drains once a shift. Investigate cause of any moisture ingress – ensure dew point measurement is correct since liquid accumulation would not be expected on load with dew point (at frame pressure) better than +18°C.	In service. Daily check for efficiency. Dew point measurements in and out. Increase frequency if dew point deteriorates from normal.	In service continuously.	In addition to the above weekly water in oil check providing H2 dew point is normal. Increase test frequency if H2 dew point deteriorates. Investigate the cause if H2O in oil >0.05% since this will effect the H2 dew point. Speed of action is determined by extent of deterioration of H2 dew point.
(c) Running Down	Temperature maintained at or above 30°C and at least 5°C above the cold hydrogen temperature measured at the hydrogen cooler outlets.	Maintain dew point and Remedial/Investigative action as in "running".	No action.	In service.	In service continuously.	Maintain centrifuge in service and operating satisfactorily.
(d) Shut down for one week or less	Temperature maintained at or above 30°C and at least 5°C above the cold hydrogen temperature measured at the hydrogen cooler outlets.	As above to avoid prolonged dry out on start up.	Drain, identify and log the water volume removed from the drains once a shift. Investigate the cause of any water ingress.	In service. Check daily for efficiency.	In service continuously.	Maintain centrifuge in service and operating satisfactorily to keep seal oil dry.
(e) Shut down for longer than 1 week	Isolated and drained. Seal oil system in service.	Degassed	Drain, identify and log the water volume removed from the drains once a shift. Investigate the cause of any water ingress.	Degassed and out of service. Dew point measurement in and out. Increase frequency if dew point deteriorates.	Degassed and out of service.	Maintain centrifuge in service and operating satisfactorily to keep seal oil dry.
(f) Generator Inspections	Isolated and drained. Seal oil system also shut down.	Degassed and kept warm with hot air blowers.	Drain, identify and log the water volume removed from the drains once a shift. Investigate the cause of any water ingress.	Degassed and out of service.	Degassed and out of service.	Maintain centrifuge in service and operating satisfactorily to keep seal oil dry.

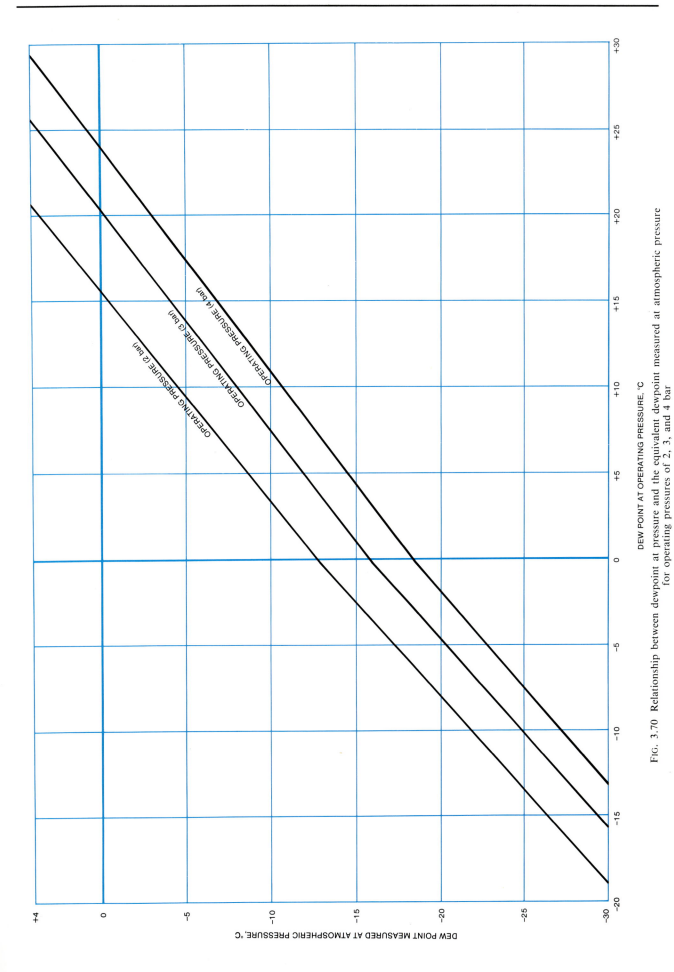

DEW POINT AT OPERATING PRESSURE, °C

FIG. 3.70 Relationship between dewpoint at pressure and the equivalent dewpoint measured at atmospheric pressure for operating pressures of 2, 3, and 4 bar

still found at the ends of the core. This axial flux increases with rotor angle and the maximum temperatures occur at leading power factor; to lower the core-end temperatures, the generator needs to run at a more-lagging power factor. Since the temperatures in the core are due to fluxes, then reducing them will also reduce the core temperatures, i.e., reducing the terminal voltage. Obviously there is a limit to which this can be done because of the tapping range of both the unit and generator transformers, but where a core defect and a localised hotspot are known to exist, the leading power factor capability can be restricted and the terminal voltage reduced from, say, 22 kV to 21 kV in order to keep the generators in service. Such a defect is still prone to propagate, if the generator is subjected to a loss of stability pole-slipping event or a loss of excitation, both of which increase the axial flux, as the rotor angle advances, and hence the eddy currents and core-end temperatures increase.

17.2   Stability

Generally speaking, the design of the Grid System, and the protection and circuit-breaker operation times are such that there is no likelihood of instability occurring. However, during light load conditions and with circuit outages for maintenance, generators may have to be restricted from operating to the normal stability limits shown on the operation diagrams and vector meters. Treating the generator as one end of a transmission network and considering the combined sending-end and receiving-end diagram, a stability curve can be drawn relating the generator impedance to the system impedance (or fault level).

It is important for a control engineer to be familiar with the system interconnections to a power station and a knowledge of the fault levels on the station bars will give him a feel for the likelihood of a stability problem.

In Section 3.6 of this chapter, the power equation for a machine was derived (Equation (3.9)), which was itself derived from the voltage diagram. Then in Section 4.2 of this chapter, the performance diagram was developed and Figure 3.14 (b) obtained. Referring now to Fig 3.71, consider the power circle-diagram for the transmission of power through an impedance Z; this differs from Fig 3.14 (b) in that resistance has been included.

Received power, P_R =

$$\frac{E_S E_R}{Z} \cos(\delta - \theta) - \frac{E_R^2}{Z} \cos \delta \text{ and}$$

Sending power, P_S =

$$-\frac{E_S E_R}{Z} \cos(\delta - \theta) - \frac{E_S^2}{Z} \cos \delta$$

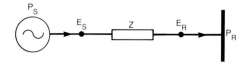

FIG. 3.71 Circuit diagram for transmission of power through an impedance Z

Where Z is the impedance of the circuit and θ is the angle of displacement between the terminal voltages E_S and E_R. Circle diagrams can be drawn from P_S and P_R and Fig 3.72 (a) illustrates the power locus for the receiving-end power where, for an impedance Z, the centre of the locus is at the co-ordinates $-E_R^2 X/Z^2$, $-E_R^2 R/Z^2$. Compare this with the sending-end diagram (Fig 3.72 (b)), where the centre of the locus is now $E_S^2 R/Z^2$, $E_S^2 X/Z^2$. Assuming that R ≪ X, which is so in practice, then $E_R^2 R/Z^2$ and $E_S^2 R/Z^2$ both become negligible and the centres of the circles are offset along the Q axis only: Z will then equal X, which was the case considered for the generator performance diagram (Fig 3.14).

Considering a machine connected through a reactance to the infinite busbar as depicted in Fig 3.73, it is possible to combine both diagrams for the position X in the figure. The point X can have a receiving-end diagram to represent the power received from the generator and a sending-end diagram for the power being sent to the infinite busbar and, since they are for the same point, they can be amalgamated onto one diagram, Fig 3.74.

For the point X, $P = (E_t^2/X_d) + (E_t E_g/X_d)$

With the generator terminal voltage fixed, the families of receiving-end circles and the sending-end circles are concentric about fixed points on the power axis. The receiving-end circles then have radii $E_t E_g/X_d$ with this centre on the Q-axis, offset from O by $-E_t^2/X_d$: the sending-end circles have radii $E_t V_i/X_S$, with centre on the Q-axis offset from O by E_t^2/X_S.

The generator and system reactance power angles δ_g and δ_s can be read directly from the composite diagram. Each component has a power limit of 90° but, coupled together, the power limit of the composite system is a system power angle of 90°, i.e., $\delta = \delta_g + \delta_s = 90°$

The steady state operating point is confined to a different sending-end circle for each value of busbar voltage; on each of those circles, there is an operating point which defines the limit of stable generator operation. The curve which is the locus of all such operating points is called the *static stability limit curve* and, since the included angle $\delta_g + \delta_s = 90°$, is a semicircle based on the Q-axis between the points

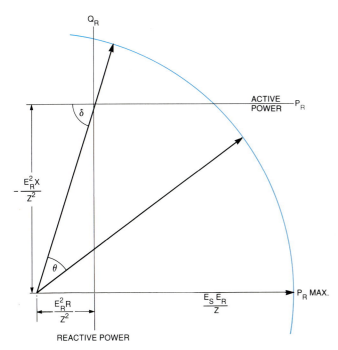

(a) Receiving end power circle diagram

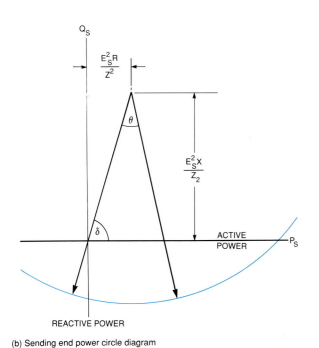

(b) Sending end power circle diagram

FIG. 3.72 Power circle diagrams for transmission of power through an impedance Z

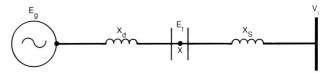

FIG. 3.73 Generator with reactance X_d connected to infinite busbar through reactance

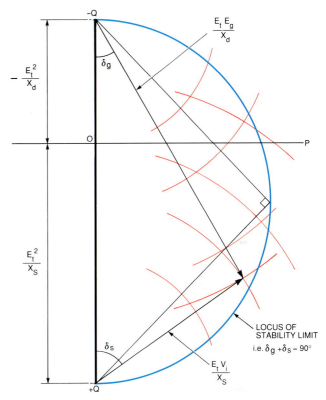

FIG. 3.74 Combined receiving and sending diagram

E_t^2/X_S and $-E_t^2/X_d$, as shown by the boundary line. Changes in the value of X_d markedly affect the size of the stable operating point in the leading

reactive-power region. In general, the size of the generator reactance is larger than the tie-line reactance, and in such circumstances the size of the set of stable operating points in the leading power factor region is significantly smaller than the lagging. By letting $(E_t^2/X_d) = 1$, it is possible to draw a generalised diagram (Fig 3.75) giving the locus of the stability limit for various ratios of X_S/X_d. Notice the curve drawn for a typical generator transformer reactance X_t and the VAr limiter setting line. Methods of reducing the reactance of machines to improve stable operation at leading power factor can only be achieved by making machines larger, resulting in larger short-circuit forces. Figure 3.76 shows a curve produced from computer studies, giving the dynamic stability limits for 500 MW generators connected through generator transformers to the 400 kV supergrid system for differing post-

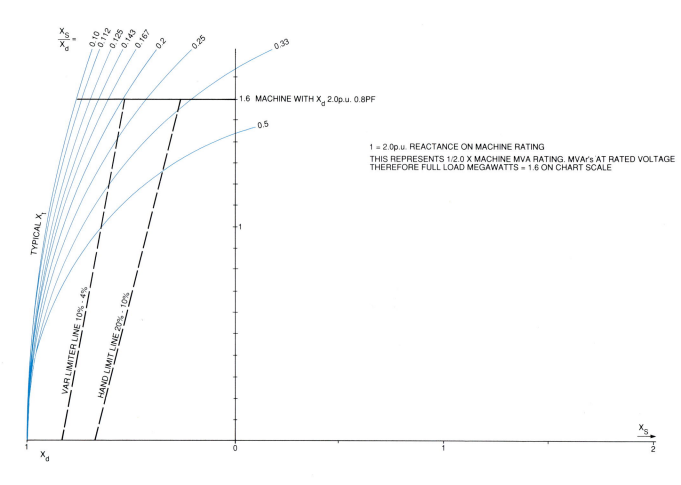

$$\frac{X_S}{X_d} =$$

0.10
0.112
0.125
0.143
0.167
0.2
0.25
0.33

1.6 MACHINE WITH X_d 2.0p.u. 0.8PF

0.5

1 = 2.0p.u. REACTANCE ON MACHINE RATING
THIS REPRESENTS 1/2.0 X MACHINE MVA RATING. MVAr's AT RATED VOLTAGE
THEREFORE FULL LOAD MEGAWATTS = 1.6 ON CHART SCALE

TYPICAL X_t

VAR LIMITER LINE 10% - 4%

HAND LIMIT LINE 20% - 10%

$\frac{X_S}{}$

1

X_d

0

1

2

FIG. 3.75 Normalised diagram, showing stability limit for ratios of X_s/X_d

fault infeed levels on the local busbars per generator and, although the circles have been flattened, it does bear some relationship to the static stability curves just derived. These curves are based on 500 MW generators with an X_d' of 0.305 p.u. and a fault clearance time of 0.14 s. It was also assumed that the ratio of pre-fault to post-fault infeed was 1.6.

17.3 Voltage control

For satisfactory control and system operation, voltage criteria have been established giving maximum voltages, step changes, etc. In order that this can be established, static compensation plant is fitted at appropriate nodes in the network, but generating plant needs to provide continuous control. The assessment of the capabilities of the generating plant connected at a node again needs a knowledge of the fault MVA, or system impedance at that point.

Taking the equivalent circuit of Fig 3.77, for a generator supplying a load of P + JQ, through an impedance R + jX, which is similar to that considered for the generator previously, and the phasor diagram of Fig 3.78. It can be seen that $I_P R$ and

$I_Q R$ disappear if the resistance is neglected, and $I_P X$ and $I_Q X$ determine the voltage positions and magnitudes. From this, approximate formulae can be derived:

P = VE/X sin δ (as previously derived and used to consider $E^2 = (V + \Delta V)^2 + \delta V^2 = (V + IR\cos\phi + IX\sin\phi)^2 + (IX\cos\phi - IR\sin\phi)^2$ steady state stability) and therefore $E^2 = [V + (RP/V + XQ/V)]^2 + (XP/V - RQ/V)^2$. Hence $\Delta V = (RP + XQ)/V$ and $\delta V = (XP - RQ)/V$. If then, $\delta V \ll V + \Delta V$; $E - V = (RP - XQ)/V = \Delta V$ and, if R is negligible,

$$E - V = XQ/V \qquad (3.30)$$

i.e., the active power transmitted is proportional to the angular difference between the sending and receiving end voltages, and the scalar voltage difference determines the reactive power flow.

If the voltage needs to be corrected at a node, then either the injection or absorption of MVArs is required at that point, the value depending on the magnitude of voltage change needed. Differentiation of

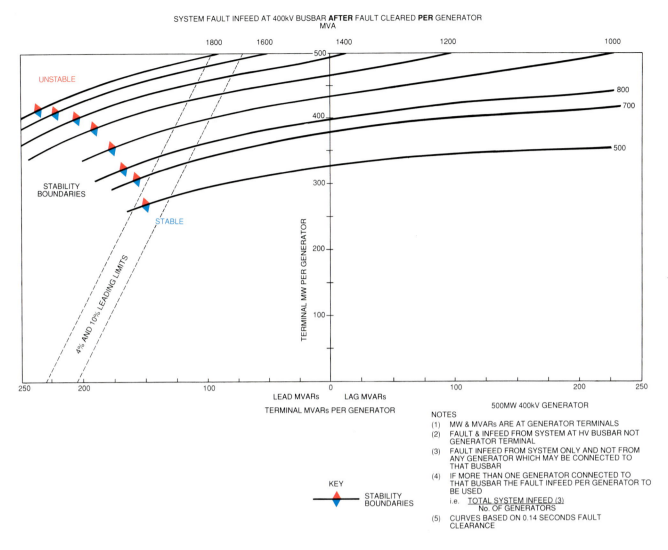

SYSTEM FAULT INFEED AT 400kV BUSBAR **AFTER** FAULT CLEARED **PER** GENERATOR
MVA

NOTES
(1) MW & MVARs ARE AT GENERATOR TERMINALS
(2) FAULT & INFEED FROM SYSTEM AT HV BUSBAR NOT GENERATOR TERMINAL
(3) FAULT INFEED FROM SYSTEM ONLY AND NOT FROM ANY GENERATOR WHICH MAY BE CONNECTED TO THAT BUSBAR
(4) IF MORE THAN ONE GENERATOR CONNECTED TO THAT BUSBAR THE FAULT INFEED PER GENERATOR TO BE USED
  i.e. $\dfrac{\text{TOTAL SYSTEM INFEED (3)}}{\text{No. OF GENERATORS}}$
(5) CURVES BASED ON 0.14 SECONDS FAULT CLEARANCE

FIG. 3.76 Stability limits for 400 kV 500 MW generator

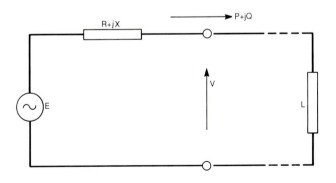

FIG. 3.77 Generator supplying a load P + jQ through impedance R + jx

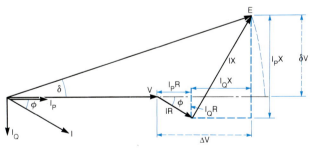

FIG. 3.78 Phasor diagram for the voltages at source and load

Equation (3.30) produces a relationship between the change of MVAr and the change of voltage, i.e., (dQ/dV) = (E − 2V)/X. For small differences, as in the voltage control of the grid system:

$$\frac{dQ}{dV} = -\frac{V}{X} \qquad (3.31)$$

where V/X is the short-circuit current at the node considered. The short-circuit level is normally given

223

in MVA and a series of formulae for each grid voltage level aids calculation, viz:

- For 132 kV, $\text{MVAr/kV} = \dfrac{\text{MVA Fault Level}}{1000} \times 7.6$

- For 275 kV, $\text{MVAr/kV} = \dfrac{\text{MWA Fault Level}}{1000} \times 3.64$

- For 400 kV, $\text{MVAr/kV} = \dfrac{\text{MVA Fault Level}}{1000} \times 2.5$

17.4 Hydrogen leakage

The leakage of the hydrogen into the conductor water-coolant system has already been discussed in Section 15.3 of this chapter, where it was seen that the limits are fixed by the blanketing effect of the hydrogen on the coolant system.

Hydrogen leakage from the frame presents its own problems with respect to explosion and fire. Leakage into the slipring enclosure through radial seals on the upshaft leads is not tolerable and hydrogen leakage detectors should alarm for this condition.

Under emergency conditions, it may be decided that the hydrogen pressure should be lowered to reduce hydrogen leakage from the frame. The first consideration on water-cooled winding generators is the differential pressure between the hydrogen and the conductor coolant, as the hydrogen must be maintained at a higher pressure than the water coolant. Therefore the water coolant flow will need to be decreased, thus affecting the current rating of the generator stator.

The rotor conductors are directly cooled by the hydrogen so a reduction of hydrogen pressure will cause the rotor current to be limited. Obviously not all designs of generator are the same and the extent to which a reduction of hydrogen pressure affects particular designs is extremely variable.

Derating of generators with hydrogen-only cooling for a decrease in hydrogen pressure is generally possible to the extent that the generator can be run in air (at 1.03 bar) at about 60% of the kVA rating in hydrogen. Figure 3.79 shows the derating of a generator with reduction in hydrogen pressure for one design of generator.

Manufacturers will generally quote a rating for a lower operating pressure (say 30% reduction of pressure), but lower pressures need to have ratings calculated and proven by test.

The stator core temperatures are not the best guide to the thermal rating, but the limits imposed on core temperatures should not be exceeded. Conventional generators of 60 MW capacity are generally liberally rated and full-load can be met at reduced hydrogen pressure, but the MVAr capability at lagging power factor is restricted by a rotor current limitation. Where a rotor-temperature indicator is fitted, this needs to be

monitored: it should be borne in mind that they are based on measuring the rotor resistance and are therefore monitoring the average temperature. Generators above 60 MW are normally designed with direct cooling of rotor conductors and are more severely restricted.

18 Abnormal operating conditions

Faults can occur on operational plant requiring either less flexible operation of the plant, due to replacement plant of the correct rating not being available, or a modification to allow the plant to run until a suitable outage or a spare becomes available. It is not possible to consider all the possibilities that can occur. A few examples are given of the type of problem that has occurred and how it was solved.

18.1 Fixed tap/different ratio transformer

Following the failure of a generator transformer, it is possible that a transformer having the same ratio may not be available.

Generally speaking, generator transformers have tapchangers to enable a generator to cover its full MVAr range and system operating voltages with a reasonable size of exciter and fixed terminal voltage. Consider a generator of 22 000 V terminal voltage operating with a 432 kV/23.5 kV generator transformer and a $+2\%$ to -16% tapchanger, which when connected to a generator with a terminal voltage of 22 kV effectively becomes 405 kV/22 kV $+2\%$ to -16% and means operating the transformer on the $+2\%$ tap, i.e., tap 1. Allowing the generator terminal voltage to vary from 0.95 to 1.05 per unit and operating the generator transformer on a fixed tap, i.e., tap 1, then Fig 3.80 shows the operating range of the generator unit. The diagram was derived as discussed in Section 4.3. of this chapter.

18.2 Excitation from separate DC supply

The excitation chain of a generator normally consists of a pilot exciter feeding the main exciter, which, on large generators, feeds a diode bank that can be either static or rotating. Failure of the pilot exciter could lead to a generator being out of service. In an emergency, the pilot exciter can be replaced by a DC supply from batteries with float charger and a means of varying the DC voltage, or some other variable DC supply, such as a battery charger fed through a continuously-variable-ratio transformer. Under these circumstances the AVR would be out of service and care would need to be taken in the operation of the unit. Under these circumstances, the excitation should be increased before increasing load on the unit, to ensure that the generation remains in synchronism after

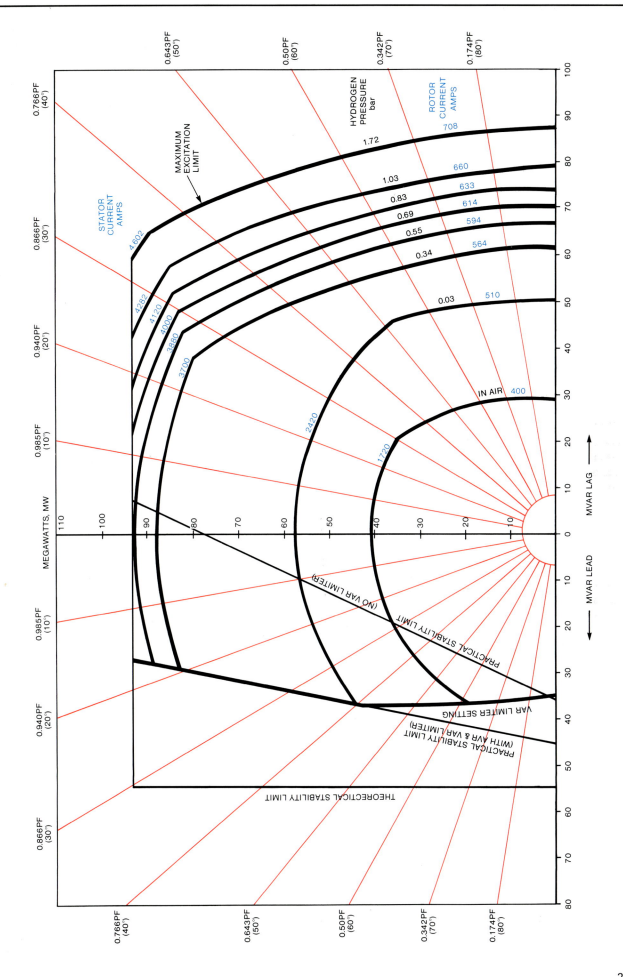

FIG. 3.79 Derating of 94 MW generator with reduced hydrogen

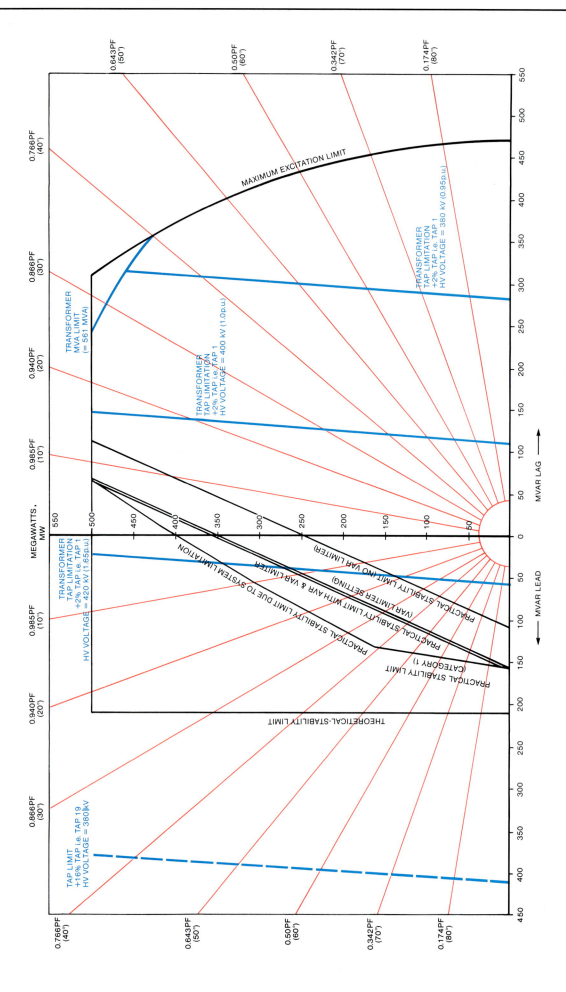

FIG. 3.80 Generator operating a generator transformer of incorrect ratio, i.e., 432/23.5 kV instead of 432/22 kV

the load pick-up due to the increase of rotor angle (Fig 3.81, increasing rotor current from (2) to (1)). However, it should also be noted that, under free-governor action with a falling frequency, the governor will attempt to increase the power on the unit and can cause the generator to go unstable.

Consider a unit with the main exciter fed from a battery charger, with a variable-ratio transformer in the primary supply. Control of the transformer ratio was from the field regulator control on the Unit Control Desk with the generator transformer overfluxing-protection wiring modified to include the run-to-minimum feature should this protection operate. The unit operated successfully at loads up to 450 MW without any difficulty over a period of time until an incident during one night, when the unit was carrying 385 MW and 100 MVAr lag (point (2) on Fig 3.81). A load pick-up of 50 MW was carried out without an increase in excita-

tion and resulted in the generator operating at 435 MW and UPF (point 2A). A further increase in load then occurred due to free-governor action, with the system frequency falling. The terminal voltage dropped, due to the change in MVAr, so causing a drop in voltage on the supply to the battery charger connected to the unit board as indicated by points 4, 9, 10 in Fig 3.81. A cumulative effect then occurred, resulting in the generator eventually going unstable. Figure 3.82 depicts the terminal voltage and resulting rotor currents at various times, corresponding to the points on Fig 3.81.

The instability was caused by the combination of additional load pick up and reduction of excitation current on a reduction of unit voltage caused by the change in rotor angle.

On at least one occasion, a diesel-electric locomotive has been used to supply the excitation of a generator, illustrating that there are many ways of supplying the

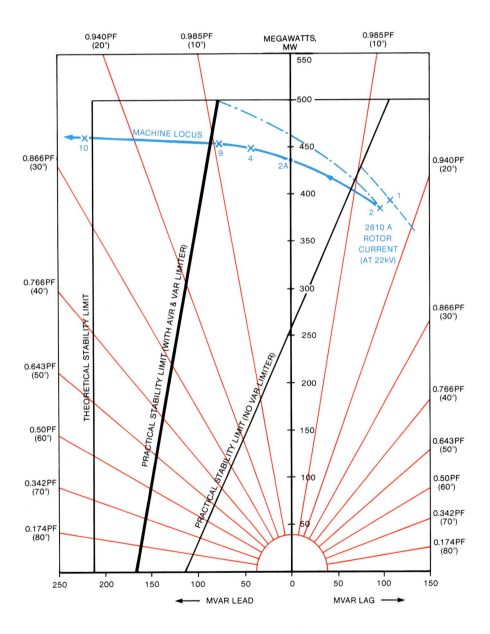

FIG. 3.81 Operating chart of generator at time of incident

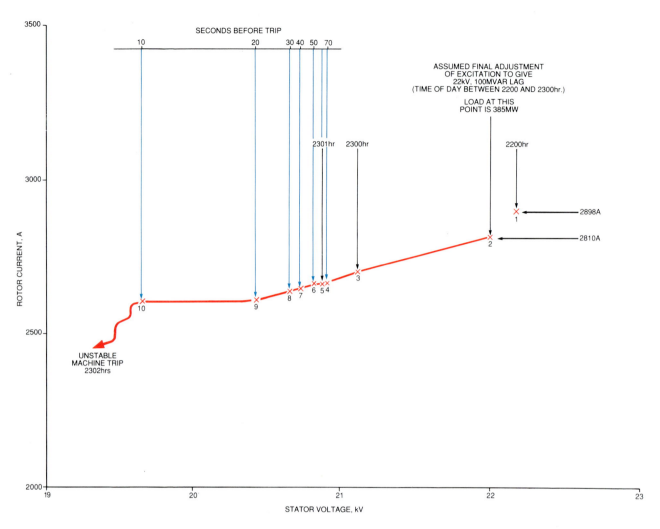

FIG. 3.82 Stator voltage/rotor current curve plotted from computer printout

required excitation in an emergency. The point to note is that the full characteristics of any supply must be investigated to prevent an event, such as that described, occurring.

18.3   Modified stator with reduced conductors

As an example of an emergency situation that can occur, consider a generator with an earth fault in an inaccessible position on a bottom conductor bar. If the fault is in an accessible position, a temporary repair can be considered, removing the damaged insulation and scarfing in new insulation. To replace the faulty bar, removing some 20 or more top conductor bars would be necessary to gain access. An alternative approach that was successfully carried out on a generator with cracked subconductors adjacent to the scarf joint at the slot exit of a bottom conductor bar involved reconnecting the winding to take this bar out of service.

The generator was rated at 588 MVA, 500 MW, 22 kV, 0.85 p.f., with its three phases each made

up of two parallel windings. The initial fault was a hydrogen-to-coolant leak. Eventually the leak was discovered to be away from the accessible part of the windings, as described above. The leaking subconductors were blocked and the bar electrically tested and found to be down to earth. In order to minimise the work content, cost and time of repair, it was decided to disconnect this bar and the return bar from the winding, together making up one turn of a half-phase. This was carried out by rearranging link connections between the bars at each end of the winding. To nullify the effect of circulating currents around the half-phases, two conductor bars in the parallel half-phase had also to be isolated. Due to the distribution of the winding, this had to be a mirror image of the faulty bar to close the two ends of the winding without a voltage difference between the bars. To balance the three phases, the same rearrangement was made on the other half-phases.

This resulted in a balanced stator reduced to 8/9 (turns per phase). Some of the technical problems that were considered checked and shown to be acceptable, are listed below:

- Marginal changes in harmonics were expected to occur; some in fact increased, whilst others decreased, the overall effect gave no restriction on output or interference with protection, or with telephony.

- Rotor surface losses were expected to increase by some 30 to 40 kW, giving an increase in temperature rise of up to 5°C. This was not checked because of the difficulties in obtaining measurements and the calculated temperature rise was considered to be acceptable.

- Steady state stability limits were considered less restrictive, because the reactance of the machine had been lowered, but it had been decided that the terminal voltage would not be reduced *pro rata*.

- The proposed new operating voltage of 21 kV was well within the operating range of the AVR, although some minor adjustments to the AVR were carried out during recommissioning. The VAr limiter also operated satisfactorily at the reduced voltage, but was reset to the new VAr limit.

- The only setting change required on the protection was to the loss-of-excitation relay, since the machine reactance and terminal voltage had been changed. There was no significant increase in third-harmonic voltages, so additional filters were not required in the earth fault protection, but this had to be considered and checked.

- The generator transformer, which was rated at 600 MVA, 22/430 kV, +2% to −16% tapping range, had to operate at or near its tapping limit and some restrictions on lagging MVAr capability was encountered; the performance diagram is reproduced, Fig 3.83.

- Had the terminal voltage been decreased in proportion to the number of stator turns removed, this would have entailed operation at 19.6 kV. Based on the criterion of not exceeding the design core-back flux density at MCR plus 5% overvoltage, permissible operating voltages were calculated for a range of loading conditions. Using this information, in conjunction with the limitations due to the generator transformer tapping range, led to a sensible operating terminal voltage of 21 kV.

- The unit transformer was rated at 27.5 MVA, 22/11.5 kV, $+7\frac{1}{2}$% to $-7\frac{1}{2}$% tapping range, the tap position was altered two taps and the auxiliary voltages were maintained.

Not all of the considerations have been discussed above, but the unit was returned to service and operated not only at 500 MW, but flexibly, and two-shifted over

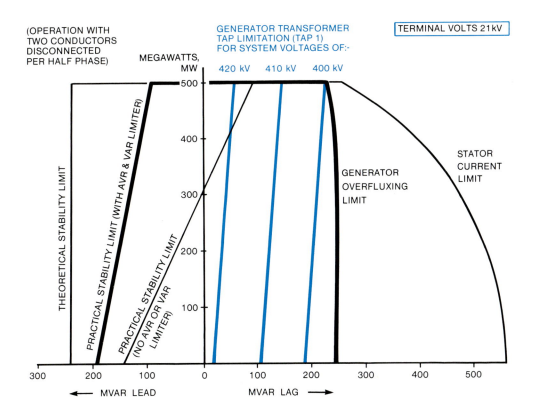

FIG. 3.83 Operating chart for generator with two conductors disconnected per half-phase

a wide operating range to the extent that no plans were made to carry out a full repair.

19 Fault conditions

Although considerable research, development and design has been done to produce safe, efficient and reliable generators, nevertheless abnormal or fault conditions can occur during the operation of these units. Protective devices connected to the various generator circuits will either cause an alarm, or trip the generator, if a fault condition occurs.

19.1 Stator earth faults

In order to minimise damage to stator windings and the core, generator stators are now earthed through a distribution transformer, loaded by a resistor, to limit the generator neutral fault currents to 10 A at nominal stator voltage. Two stages of earth fault protection are normally fitted; an instantaneous relay covering 90% of the winding, and an inverse minimum time relay set to cover 95% of the winding.

It should be noted that, with such a protection scheme, it is possible for moisture on contaminated busbar insulators to cause flashover and operation of the earth fault instantaneous relay, and for the fault-path resistance to recover and leave no trace of the fault. This condition is only normally recognised when all other parts of the circuit have been proven healthy. It is always essential with the operation of any main electrical protection on the generator unit that samples of hydrogen from the stator casing and oil from the generator transformer be obtained and analysed for hydrocarbons and other gases as an aid to fault diagnosis and location.

19.2 Stator phase-to-phase faults

The stators of large generators are normally connected to the 400 kV grid system through a step-up transformer, and the connection between the step-up transformer and the stator is normally by phase-segregated busbars. In theory, therefore, a phase-to-phase fault should not occur, so the protection of the stator is included in the generator transformer protection by means of overall biased differential protection. Operation of this protection without the operation of the stator earth fault protection would indicate either a fault within the stator casing, or an interwinding fault in the generator or unit transformer. Again, hydrogen gas and oil samples and, in the case of the transformers, operation of the Buchholz gas alarm or trip would indicate the faulty unit. It should be remembered that the generator circuit generally goes

from the HV side of the transformer to the 400 kV busbars by overhead connections, which are normally protected by their own balanced-current protection, and a fault on that part of the circuit is thus clearly indicated by a separate alarm.

Phase-to-phase faults within the stator casing are extremely damaging and the stored electromagnetic energy can give rise to damage even after the electrical circuit-breaker has cleared. Areas most at risk to phase-to-phase faults are the connections at the ends of the windings, where nylon hoses take cooling water to the windings, and the end-windings themselves. If conducting and magnetic debris is left on the winding during maintenance, this debris can work its way through the insulation, causing interstrip faults that can deteriorate to a phase-to-phase fault. Hence, great care needs to be taken to exclude the possibility of debris being left in the generator. Phase-to-phase faults on the winding within the generator can give rise to currents that will produce forces in a direction to eject conductors from the slots.

19.3 Stator interturn faults

Interturn faults within the stator are not unknown, but no protection is fitted to deal with this, in spite of the fact that each phase consists of two parallel windings. Should an interturn fault occur, a high circulating current will flow and considerable damage will occur at the point of fault, to such an extent that either an earth fault or a phase-to-phase fault develops and the appropriate protection operates to trip the generator. Causes of interturn faults can be debris on the winding, as already discussed, fatigue cracking of insulation due to looseness and vibration of end windings, moisture condensation on surfaces across end-windings adjacent to copper links connecting the conductors, and loss or contamination of the water circuit.

19.4 Negative phase sequence currents

Unbalanced loading or unbalanced faults cause negative sequence currents to flow in the rotor iron, slot wedges and damper circuits. These currents heat these parts and there is a limit to the temperature permissible on the rotor surface. With the most widely used method of large rotor construction, Fig 3.84, unacceptably high temperatures can arise with what was previously regarded as a low level of negative sequence current. Since the flexibility slits force any axially-induced rotor currents into narrow areas of the rotor, providing high localised heating of these areas and the possibility of cracks initiating from the ends of the slits. Protection is fitted to trip the generator for negative sequence currents that could damage the rotor, these relays have modelled thermal

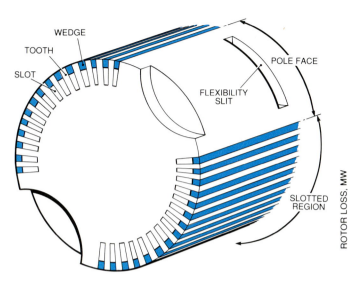

FIG. 3.84 Typical two-pole turbine-generator rotor construction

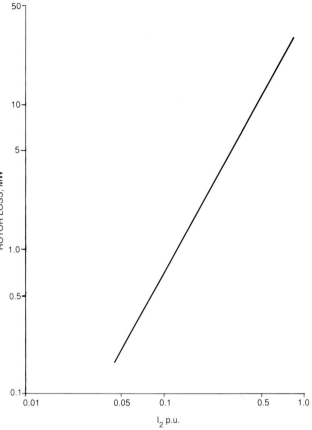

FIG. 3.85 Variation of total rotor loss with I_2

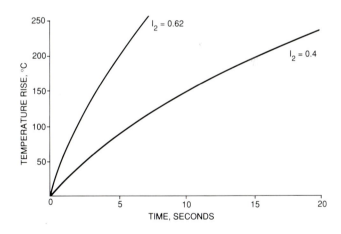

FIG. 3.86 Transient temperature response for two values of suddenly applied I_2 at end of flexible slot

characteristics that emulate the $(I_2)^2 t$ capability of the rotor. Another alarm relay is fitted for negative sequence currents below this level to warn operators of unbalanced loading conditions. It is essential that action be taken to identify and eliminate the source of unbalance; System Operations should be informed, as the source of this unbalance may be outside the control of the power station operator. Should the unbalance, for instance, be due to an open-circuit in a major interconnection then, as loading increases, the negative sequence contribution will increase and could reach a level to trip the generators. All generators connected to a common busbar will share the negative sequence current and, for identical machines, this will be an equal share, so that all generators will trip for a fault condition giving rise to currents above the trip level.

Figure 3.85 shows the rotor loss against I_2 p.u. for a typical rotor and Fig 3.86 shows the transient temperature response for two values of suddenly applied I_2 for the position at the end of a flexibility slit, both results being obtained from tests in manufacturer's works.

19.5 Loss of generator excitation

Failure of the excitation system results in the generator rotor-angle increasing, as the flux in the rotor decays, eventually losing synchronism: the generator then operates as an induction generator, as described in Section 12 of this chapter, and draws its excitation from the system. The effect of this is to depress the terminal voltage, which could cause problems with the auxiliary load connected to the generator termi-

nals. Protection is connected to the generator to cater for this condition. It usually consists of an impedance relay to monitor the impedance, as seen looking into the generator from its terminals. The impedance seen by this relay, for a generator operating without a field, is significantly different from that on-load, so it can identify this condition and trip the machine.

However, under recoverable transient conditions, the generator can enter the range of the relays, so a time delay is incorporated in the trip sequence to cater for this.

Since, with loss of excitation, magnetising current is drawn from the system, resulting in a large input of MVAr, a relay operating from a large MVAr import can also be used to trip for loss of excitation; again, the trip is time-delayed to allow recoverable transient swings to take place.

19.6 Pole-slipping

Pole-slipping occurs when a generator loses synchronism:

- Through the excitation system not being able to maintain the torque component required for synchronism, without a total loss of excitation.

- Because of system conditions, or a fault causes the generator to lose synchronism.

This will cause large fluctuations in voltage, MW, and MVAr and could cause the loss of auxiliary motors, followed by the shutdown of the unit. Such a shutdown, being non-sequenced, could cause damage to boiler plant. The effects would also be felt on consumers' plant and other generating plant. The interchange of synchronising power across the system, to and from the pole-slipping generator, could cause indiscriminate operation of transmission circuit protection and the disconnection of supplies. However, pole-slipping is rare, due to the criteria established for system operation, so pole-slipping protection is fitted only where it has been found necessary (for example, Dinorwig pumped storage where, due to its location in a site of great natural beauty, the number of transmission circuits into the site is limited).

19.7 Rotor faults

The rotor winding of a generator is fed from static or rotating rectifiers with DC at, typically, 550 V and 4500 A. The excitation circuit is insulated from earth, except for a monitoring supply connection that establishes a negative bias to earth (see volume C). An earth fault of less than approximately 20 000 Ω will bring up an earth fault alarm. It should be noted that the whole of the excitation circuit is in the monitoring circuit; the earth fault can be due to carbon dust on slipring insulation or to moisture on insulated surfaces of the excitation circuit, and the rotor winding will need to be isolated by removing the brushes to check which item of equipment has the earth fault.

However, failure of the rotor winding insulation

system can give rise to an earth fault and, although this in itself is not damaging, a second failure to earth at another point on the winding will short-circuit the rotor winding and currents will flow through the rotor forging or endrings. This current may be sufficient to cause severe damage to the forging or endrings involved. In order that high merit or strategically important generators can continue to be kept in service with a high-resistance rotor earth fault, another type of earth fault protection can be fitted which monitors the position and degradation of earth fault resistance: it can be arranged to trip either for a reduction of resistance to below a set level or for a change in earth fault position.

Similarly, the insulation between turns can be bridged by foreign material, mechanical failure of insulation or by copper dust derived from wear of the copper winding moving relative to the slot insulation. Whilst any of these mechanisms can give rise to the earth faults previously mentioned, the fault can also be clear of earth. The current at the point of the short-circuit will cause damage to the copper and overheat the insulation, possibly developing and bridging more turns. Electrically, the effect on the unit will not necessarily be noticed, since the loss in ampere-turns will be compensated by AVR action increasing the rotor current to compensate, although the open-circuit curve can indicate the effect of shorted turns, if measured accurately.

If only one set of shorted turns is present, then only one pole will be affected and, depending on the position of the shorted turns, an unbalanced magnetic pull can occur and give rise to vibration, which is dependent on rotor current. If there is a delay in a rotor-current-dependent vibration, this could be a thermal effect. The shorted turns carry a smaller proportion of current and produce less heating than other turns, giving rise to a thermal bend in the shaft and producing vibration that increases with rotor current. These vibrations can exceed commercial limits at high rotor currents and thus limit the excitation that can be applied to the generator. Obviously the rotor will need to be repaired at the earliest opportunity, but for economic or strategic reasons the unit may be required to continue running until a suitable outage occurs. Advantage can be taken of the vibration characteristic of this type of fault, that is the offset magnetic pull and thermally-induced band, in that the vibration phasor amplitude is dependent on the rotor current and follows along a straight phasor line, such that offset balancing at no-load can be applied to keep the amplitudes within commercial limits; Fig 3.87 shows a plot of such a characteristic; previous to offset balancing the cold datum was at B Fig 3.87 (a) going out to B4 at 1850 rotor amperes. Then weights were applied changing the cold datum to G Fig 3.87 (b) going out to H4 at 2470 rotor amperes, allowing a higher load.

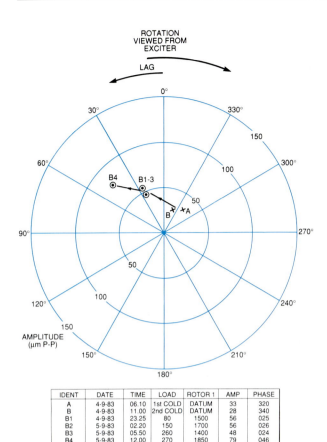

ROTATION
VIEWED FROM
EXCITER

LAG

AMPLITUDE
(μm P-P)

IDENT	DATE	TIME	LOAD	ROTOR 1	AMP	PHASE
A	4-9-83	06.10	1st COLD	DATUM	33	320
B	4-9-83	11.00	2nd COLD	DATUM	28	340
B1	4-9-83	23.25	80	1500	56	025
B2	5-9-83	02.20	150	1700	56	026
B3	5-9-83	05.50	260	1400	48	024
B4	5-9-83	12.00	270	1850	79	046

(a) Bearing 12V initial response to loading

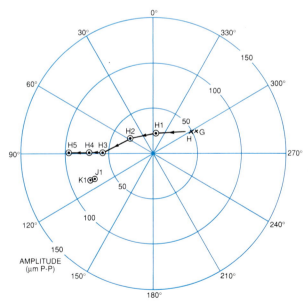

AMPLITUDE
(μm P-P)

IDENT	DATE	TIME	MW	ROTOR 1	AMP	PHASE	
G	19-9-83		COLD	DATUMS	54	299	(MEAN)
H	19-9-83				49	300	(MEAN)
H1	20-9-83	04.50	280	1600	24	352	
H2	20-9-83	06.10	440	2100	33	055	
H3	20-9-83	11.50	470	2300	56	088	
H4	20-9-83	13.55	460	2470	72	088	
H5	20-9-83	15.36	460	2400	97	088	
J1	7-10-83	10.15	470	2220	71	112	
K1	21-10-83	14.30	460	2250	74	112	

(b) Response to loading following balance

FIG. 3.87 Thermal bending of rotor shaft, giving vibration
phasor dependent on load and the effect of offset
balancing

The presence of shorted turns can be confirmed by examination of the voltage output waveform of flux coils fitted in the airgap, as shown in Fig 3.88, where the tooth harmonic waveform is seen to be different for each pole. Figure 3.88 (a) shows faults indicated on the D + F coils whilst Fig 3.88 (b), for the same rotor, shows additional shorted turns developing on the D coil.

Semiconductors are now widely used in excitation systems, either connected on the shaft or static, supplying the rotor windings through sliprings. The rectifier system consists of three-phase full-wave sections connected in parallel to give high reliability over long periods. Fuses are fitted to protect the diodes from overcurrent and resistance/capacitance networks to suppress against voltage surges. If more than two paths fail, then the bridge will fail on short-circuit, because of diode overcurrent. Rectifier bridge-arm protection will trip the main exciter field circuit-breaker and trip the unit through the low forward-power interlock. The loss-of-excitation protection should operate and trip the unit for the failure of the rectifier bridge arm.

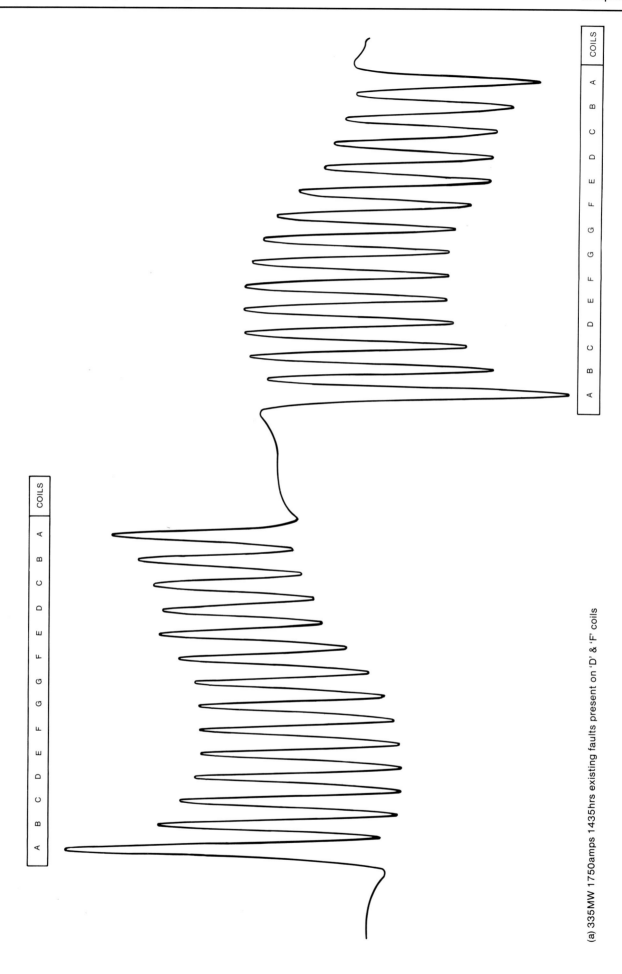

(a) 335MW 1750amps 1435hrs existing faults present on 'D' & 'F' coils

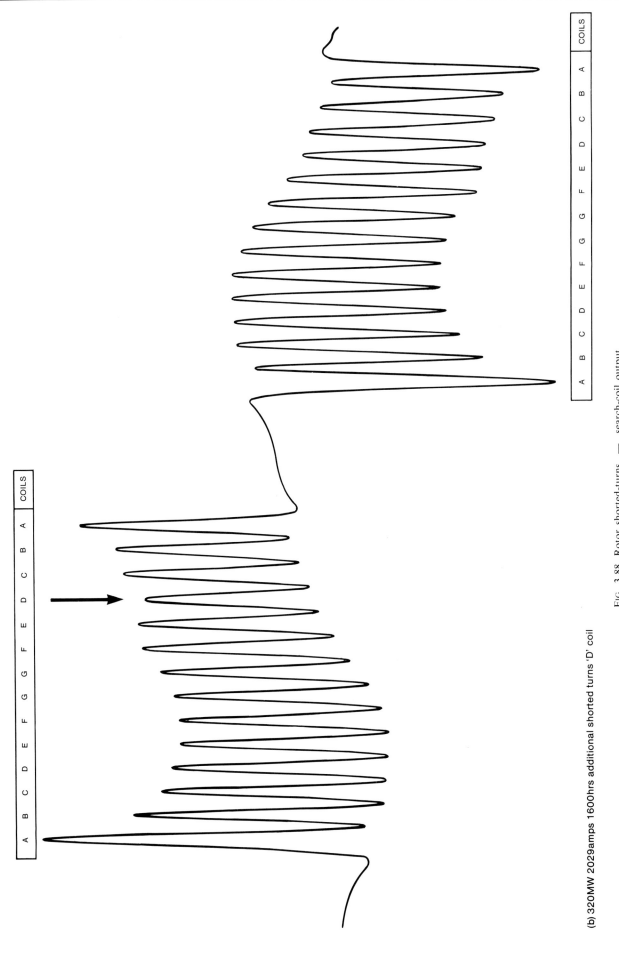

(b) 320MW 2029amps 1600hrs additional shorted turns 'D' coil

FIG. 3.88 Rotor shorted-turns — search-coil output
Faults indicated on D and F coils (a). At increased rotor current, fault on D coil worsens and more turns are shorted (b).

20   References

[1]   Park, R. H.: 'Two reaction theory of synchronous machines': Trans. Amer. IEE, Vol. 48, pp 716: 1929

[2]   Park, R. H.: 'Two reaction theory of synchronous machines II': Trans. Amer. IEE, Pt. 1, Vol. 52, pp 352: 1933

[3]   Adkins, B.: 'The general theory of electrical machines': Chapman and Hall

[4]   Adkins, B. and Bharali, P.: 'Operational impedances of turbo-generators with solid rotors': Proc. IEE, Vol. 110, No. 12: 1963

[5]   BS4296: 'Methods of test for determining synchronous machine quantities': 1968 (1984)

[6]   Shackshaft, G.: 'General purpose turbo-alternator model': Proc. IEE, Vol. 110, No. 4: 1963

[7]   Shackshaft, G. and Neilson, R.: 'Results of stability tests on an underexcited 120 MW generator': Proc. IEE, 119 (2), pp 175–188: 1972

[8]   Demello, F. P. and Concordia, C.: 'Concepts of synchronous machine stability as affected by excitation control': IEEE, Trans PAS–88, pp 316–329: 1969

[9]   Kalsi, S. S. and Adkins, B.: 'Transient Stability of power systems containing both synchronous and induction machines': Proc. IEE, 118, pp 1467: 1971

21   Additional references

Barton, S. C., Carson, C. C., Gill, R. S., Ligan, W. W. and Webb, J. L.: 'Implementation of Pyrolysate analysis of materials employing tagging compounds to locate an overheated area in a generator': IEEE, PES: Summer Meeting: 1981

Binns, K. J. and Smith, J. R.: 'Prediction of load characteristics of turbo-generators': Proc. IEE, 125(3), pp 197–202: 1978

BS2757: 'Classification of insulating material for electrical machinery and apparatus on the basis of thermal stability in service': 1956 (1984)

BS4999: 'General requirements for rotating electrical machines'

BS5000: 'Rotating electrical machines of particular types or for particular applications'

Canay, I. M.: 'Calculation of negative-sequence losses and negative-sequence resistance of turbo-generators': Trans A.IEEE, T 74, 304–2: 1974

Carson, C. C., Barton, S. C. and Echeveria, F. S.: 'Immediate warning of local overheating in electrical machines by the detection of Pyrolysis products': IEEE Summer Power Meeting TP 154 PWR: 1971

Gibbs, W. J.: 'Induction and synchronous motors with unlaminated rotors': Proc. IEE, Pt. II, 95: August 1948

Gove, R. M.: 'Geometric construction of stability limits', Proc. IEE, Vol. 112, No. 5, pp 977: 1965

Grobel, L. P. and Carson, C. C.: 'Overheating detector for gas cooled electrical machines': US Patent 342 7880: 1969

Humphries, H. J. and Fairney, W.: 'Excitation rectifier schemes for large generators': Proc. IEE, 119(6), pp 661–671: 1972

Jack, A. G. and Stoll, R. L.: 'Negative-sequence currents and losses in the solid rotor of a turbo-generator': Proc. IEE, Vol. 127(2), Pt. C, pp 53: 1980

King, E. I. and Batchelor, J. W.: 'Effects of unbalanced currents on turbo-generators': IEEE, Trans. PAS–84, pp 121–125: 1965

Lokay, H. E. and Bolger, R. L.: 'Effect of turbine-generator representation in stability studies': IEEE Trans., PAS84, No. 10; pp 933: 1965

Mason, T. H., Aylett, P. D. and Birch, F. H.: 'Turbo-generator performance under exceptional operating conditions': Proc. IEE, 106A, pp 357: 1959

Richardson, P.: 'Design and Application of large solid-rotor asynchronous generators': Proc. IEE, Vol. 105, pp 332: 1958

Salon, S. J., Shah, M. R and Montgomery, L. W.: 'Analysis and testing of negative-sequence heating of turbine-generator rotors': IEEE; PES, Winter Meeting, 81 WM 196–5: 1981

Say, M. G.: 'Alternating current machines': Pitman

Shackshaft, G.: 'Model of generator saturation for use in power system studies': Proc. IEE, Vol. 26(8) pp 759–763: 1979

Shackshaft, G.: 'New approach to the determination of synchronous machine parameters from tests': Proc. IEE, Vol. 121 (11), pp 1385: 1974

Shildneck, L. P.: 'Synchronous machine reactances': General Electric Review, Vol. 35(11), pp 560–565: 1932

Smith, J. R., Binns, K. J., Williamson, S., Backley, G. W.: Determination of saturated reactances of turbo-generators': Proc. IEE, Vol. 127(3), pp 122–128: 1980

Stephen, D. D.: 'Connecting large machines to power systems': Proc. IEE, Vol. 110, pp 1425: 1963

Stephen, D. D.: 'Effect of system voltage depression on large AC motors': Proc. IEE, Vol. 113, pp 500: 1966

Szwander, W.: 'Fundamental electrical characteristics of synchronous turbo-generators': Proc. IEE., Vol. 91, Part 2, pp 185: 1944

Takahashi, N., Kawamura, T. and Nishi, M.: 'Improvement of unbalanced current capability of large turbine-generators': IEEE, PES Summer Meeting T 74, 319–0: 1974

Wagner and Evans: 'Symmetrical components': McGraw-Hill Wagner (editor): 'Transmission and distribution reference book': Westinghouse

Waring, M. L. and Crary, S. B.: 'The operational impedances of a synchronous machine': General Electric Review, Vol. 35(11), pp 578–582: 1932

Weedy, B. M.: 'Electric power systems', 3rd Edition: Wiley: 1979.

Williamson, A. C.: 'Measurement of rotor temperature of a 500 MW turbine-generator with unbalanced loading': Proc. IEE, Vol 123(8), pp 795–803: 1976

Williamson, A. C. and Urquhart, E. B.: 'Analysis of the losses in a turbine-generator rotor caused by unbalanced loading': Proc. IEE, Vol. 123(12), pp 1325–1332: 1976

CHAPTER 4

The planning and management of work

1 Introduction

The advances in power plant design and complexity have progressively increased the number and capital cost of the components needed to keep plant operating at the highest levels of availability, flexibility and efficiency.

The maintenance requirement for individual items of plant varies considerably, depending upon the operating conditions, the importance of the item to the continuing safe operation of the generating unit, the penalty costs likely to be incurred in the event of failure and the predicted operating regime. It is therefore essential that considerable effort is devoted to determining the optimum maintenance policy for each significant item of plant or equipment in order to achieve its minimum overall lifetime cost.

Once the technical and commercial performance targets have been identified, it is necessary to ensure that work plans are based upon defined engineering needs. The Operational Engineering Division makes engineering assessments and provides guidance to stations on the work requirements deemed necessary to satisfy CEGB policies on plant integrity and performance. The Operational Engineering Department also quantifies plant risks, so that judgements can be made on the economics of specific work proposals.

When the work requirements have been determined, the role of the Resource Planning Department is to produce co-ordinated plans and identify the resources necessary to carry out the work programme. This involves detailed planning and, for major plant overhauls, requires the use of critical path network analysis techniques.

The overall objective of any maintenance management system is to optimise, within accepted operational and safety considerations, the economics of maintenance work programmes. In order to achieve this, a comprehensive work planning and scheduling system is needed, which embraces all aspects of maintenance activities. The system must be dynamic and able to react to contingencies by rescheduling jobs or redeploying resources in response to day-to-day changes, such as breakdown or resource limitations.

Most large power stations use computer systems to handle the large quantities of information now required for resource planning. These systems may be fully integrated to provide a comprehensive management control system, incorporating plant inventories, materials management, work planning, plant history and finance control systems.

Computer analysis of this data may then be used to determine the effectiveness of the maintenance policies. It also enables optimum solutions to be sought more easily than with manual systems, as statistical analysis is more readily available.

By the provision of simple reports on trends in both work bank statistics and manpower utilisation and performance, the effectiveness of station management policies can be kept under continued review.

An increasing volume of documentation has to be handled in power stations. This arises as a result of the increased complexity of power plants and the need to be able to demonstrate compliance with statutory, mandatory and other CEGB procedures. A plant history and records system must therefore be established to ensure that information can be quickly located and retrieved.

2 Corporate planning

2.1 Corporate planning principles

Corporate planning is defined as a continuous process under which an industry can set objectives and priorities to form a coherent strategy for its business as a whole, organise the actions needed to achieve its objectives and to monitor the results of its actions against those objectives.

The major purpose of corporate planning is to obtain benefits by considering the implications of individual actions, against the background of an up-to-date strategic plan covering the industry as a whole. This produces a greater sense of cohesion between the various parts of the organisation. It enables individual managers to take a more comprehensive view of the aims, objectives and future developments within the industry.

When corporate planning is properly applied as a management process, it focuses all parts of the organisation and all aspects of the management task into one comprehensive approach. Corporate planning aims to improve both the process of decision making and the process of implementing the decisions taken.

The features which distinguish corporate planning from other types of planning are as follows:

● Corporate planning is central to the management process at top management level.

● The focus of corporate planning is long range; it is pre-eminently concerned with the future performance of the industry.

● Corporate planning embraces the industry as a whole and involves detailed analysis of the organisation and its environment, as an aid to decision making.

● The process is continuous, not just the periodic production of a plan, which is then forgotten. It requires the evaluation, monitoring and adjustment of the plan, as necessary, in a systematic manner.

The purpose of any planning activity is to produce and guide action, not just to produce a plan. The corporate plan aims to set out the action required to achieve the desired objectives in accordance with the policies determined by top management. A corporate plan could therefore be considered as a set of instructions to managers, describing the role each constituent part of the organisation is expected to play to achieve corporate objectives.

The corporate plan should set out:

- Corporate purpose and objectives.

- Targets necessary to enable the achievement of these objectives.

- Strategy to be followed to achieve the targets and standards of performance, within a given timescale.

- Any assumptions underlying the strategy plan.

- Mechanism for reviewing progress and feedback.

A corporate plan is not the sum of departmental or functional plans. On the contrary, the final shape of such plans will be determined by the corporate plan itself, since this may require fundamental changes in some departmental or functional area.

One of the basic merits of corporate planning is that it attempts to achieve consistency between the overall plan and the sub-plans needed to carry it out. This unified approach helps to ensure that deficiencies in certain areas do not endanger the achievement of the corporate objectives.

The corporate plan represents a series of decisions taken at the present time, but based upon predictions of the future. The validity of the assumptions upon which the plan was made must be kept under regular review by top management, as should the progress made towards the achievement of objectives. The extent of any deviation from targets, particularly the trends, must be carefully monitored and corrective action taken; however, any change should be carried out in a systematic way and be formally approved.

The corporate planning process is therefore required to:

- Develop and specify objectives for the industry as a whole.

- Set targets for the achievement of these objectives.

- Identify and evaluate any alternative options.

- Decide the best policy from the options available, to achieve the objective within the required timescale.

- Evolve strategies for meeting them, taking into account any current or foreseeable constraints.

- Produce detailed action plans and define responsibilities.

- Identify resource requirements.

- Implement the plan, ensuring proper allocation of priorities and resources.

- Monitor and regularly review performance against the plan.

- Review and update plans in the light of progress and changing circumstances.

- Ensure that there is a feedback of information in respect of problem areas and difficulties encountered in achieving the plan.

- Initiate corrective action where required.

- Review the outcome of the plan.

2.2 Corporate planning within the Electricity Supply Industry

The following major plans are produced within the Electricity Supply Industry (ESI) in England and Wales:

- Medium term develop- — Electricity council.
  ment plan

- CEGB business plan — CEGB.

- CEGB resource plan — CEGB.

- Production formation — production formations.
  plan

- Management unit plans — power stations, transmission districts and divisional departments/ branches.

The Electricity Council is responsible for deciding the corporate objectives for the Electricity Supply Industry in England and Wales. The Electricity Council produces the Medium Term Development Plan which sets out the major aims and objectives for the industry as a whole. The plan indicates various aspects of the CEGB's responsibility for the generation of electricity and the Area Boards responsibility for distribution and sales.

The Medium Term Development Plan incorporates the predicted trends in electricity sales and fuel costs, together with aspects such as finance and pricing policy, electricity marketing, transmission and distribution requirements, environmental aspects, public and overseas relations.

Within the CEGB, members of the Board are appointed by the Secretary of State for Energy and the full-time members function as the CEGB Executive. The Executive is responsible for determining policy, setting objectives and targets, and for financial and business control. It is also responsible for the approval of the CEGB Resource Plan and the Business Plan (Fig 4.1).

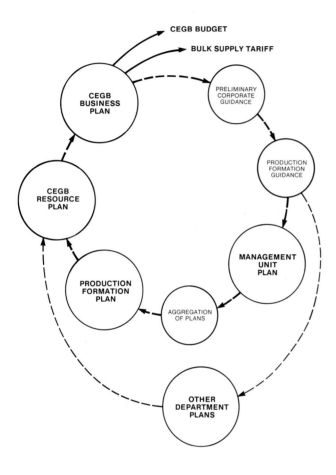

FIG. 4.1 Annual cycle of plans in the CEGB corporate planning process

Strategic planning is carried out at corporate level, the responsibility for strategic planning resting jointly with the Corporate Directors of Finance, System Planning and Strategic Studies. Their departments are responsible for ensuring that strategy plans are prepared, and that contributions from the production formation and from specialist functions within the CEGB are incorporated into the strategic planning process. The Corporate Director of Finance is responsible for the production of the CEGB business plan, which includes any major proposals, capital and revenue expenditure, manpower and changes in working capital. The Production Directorate is responsible for producing the production formation plan.

2.3   The production formation plan

The principal stages in the derivation of the pro-

duction formation plan and budget by the Production Planning Department (PPD) are as follows:

● Definition of what is required in terms of technical and commercial performance.

● The adoption of agreed strategies for the production formation.

● Ensuring that work plans to achieve the above are based upon defined engineering needs.

● Identification of the required timescales in which work programmes are to be carried out.

● Quantification of the resources necessary to undertake the work.

● Preparation of a detailed budget for financial approval.

● Detailed implementation of the budget.

● Finalisation of work plans.

The CEGB Executive issues policies, directions and guidance, which are translated via an appraisal of optional strategies into Production Directorate strategy and ultimately into guidance. This defines the technical and commercial performance required from the production formation expressed in terms of availability, performance, utilisation, operating regimes and resource levels for power stations, and equivalent factors for other management units.

The derivation of performance and resource targets and the preparation of specific guidance to each management unit requires close co-operation and collaboration by PPD with the functional departments of Finance, Personnel, Generation Services, the Operations Division, the Operational Engineering Division and the Nuclear Co-ordination Group.

Once the technical and commercial performance targets have been identified, it is necessary to ensure that work plans are based upon defined engineering needs. For example, the Operational Engineering Division (OED) carries out formal engineering reviews at each power station, while the Field Engineering Unit is responsible for this activity at each transmission district. The purpose of these reviews being to agree the programmes of work to be adopted by each location in the light of the perceived future operational requirements, engineering policies, executive guidance and commercial factors. Also, OED develops and considers the various options by which the performance requirements may be achieved and selects the most appropriate.

The timescales and the resources needed to undertake the proposed work programme are then required to be defined and incorporated in the locations management unit plan.

Unlike strategic planning, an integrated resource, work planning and budgeting process requires detailed inputs from each management unit within the organisation. It is therefore necessary to agree with location managers all the resources to be contributed by them towards the corporate objectives of the organisation. They must then provide detailed work plans indicating how they intend to deploy their resources to achieve these objectives in the most effective manner and within the planned timescale.

2.4 The management unit plan

Factors such as the predicted operating regime, operational performance targets, finance or other resource constraints are contained in policy directives and specific notes of guidance, derived from Production Directorate policies relating to the required technical and commercial performance. Following the receipt of these policy directives and specific guidance from PPD, each power station, transmission district and other management units are required to produce a 'management unit plan'.

The Operational Engineering Division (OED) produces an 'engineering status report' annually for each power station, as part of the engineering review process. This report identifies the work judged necessary by the OED to meet engineering standards and plant performance requirements for the six year plan period, and that necessary for the long term life of the plant. Similar status reports are produced by the Transmission Division for each district. The proposed engineering work programme derived from these engineering reviews provides the basis for the structured development of future plans at the locations.

When the engineering needs have been defined, the required timescales and resources necessary to carry out the work can then be quantified and incorporated into the locations Management Unit Plan. Each power station and transmission district produces a plan covering the next six year period. This represents two overhaul cycles for large fossil-fired power stations, up to three cycles for those with boiler/turbine-generator units of less than 500 MW nominal capacity and, currently, three cycles for all nuclear plants. The first three years of the plan are given in some detail, whereas the second three year period provides only an outline of the likely requirements.

The major schemes and work programmes based upon these defined engineering needs must be clearly identified in the plan. This is necessary to ensure that adequate provision is made for the total resources required for the work programme to be achieved. In this way, the overall requirements of finance, materials, plant spares, manpower and any other manufacturing or specialist services may be properly planned and coordinated. This is especially important where major

plant components, such as turbine-generator rotors, steam chests or boiler headers, are involved, as lead times are likely to be several years from the time the decision is taken to replace them.

Each location is required to submit its management unit plan, together with supporting data, in a standard format. The data schedules provide detailed statistical information relating to the future performance, resources and work programme. The schedules are divided into sections covering the following topics:

Management summary This draws together all the major performance, resource and financial aspects (apart from fuel) to present an overview of the station plan proposals.

Performance These schedules indicate a projection of the expected technical performance, an indication of the problem areas and potential for improvement. They comprise capability available, loss and thermal performance data, together with other performance issues such as plant inflexibility, reactive power, or any other limitations.

Work programme These schedules indicate the proposed work programme, revenue and capital expenditure and relate to engineering plant status reviews. They include the proposed major overhaul work programme for a period of six years ahead, other major capital or revenue schemes, proposed contracts, proposed use of national spares and the need for other working capital items, such as major plant spares.

Manpower This section indicates the proposed location and contractor manpower resource requirements and summarises the main manpower trends expected during the period covered by the plan.

Finance These schedules summarise the finance requirements from the work programme and include justification for outline scheme and major work plan proposals, revenue expenditure predictions, and the working capital needed for overhaul items, plant spares, consumable items and endurance stocks.

The provision of data relevant to future plans and work programmes from each generating station and transmission district, together with information from other corporate functions, is an essential part of the CEGB corporate planning process.

2.5 Corporate planning process timescale

Preliminary corporate guidance is issued by the Finance Department in November each year and production formation guidance is derived by the Production Planning Department and is issued to each location in January.

Following receipt of this guidance, each individual location's management unit plan is prepared and submitted to the production planning department by the end of March. The plans are analysed prior to the review meetings in June and July and a critical assessment of overall needs carried out as a basis for the production formation plan in September.

An assessment of the total CEGB resource needs are then carried out by the Finance Department, and the production formation plan, when approved by the Production Directorate, is included in the CEGB resource plan and ultimately into the Business Plan for final approval or amendment by the CEGB Executive. Once approved, the Business Plan and draft Bulk Supply Tariff are submitted to the Electricity Council; after discussion and inclusion of Government financial targets, the level of resources available to achieve the CEGB financial objectives is determined.

2.6 Dynamic planning

The planning process is designed so that the potential impact of changing conditions, both internal and external to the CEGB, can be accommodated. This is achieved by considering the effect of credible changes in the key assumptions adopted when the plan guidance was prepared, and identifying the amendments needed to deal with them. There is also a facility for modifying the plan or budget and their constituent elements, to reflect real time changes as they occur during the implementation of the plan. Where there are any implications for the CEGB's Business Plan or the CEGB Bulk Supply Tariff, any proposed amendments to the plan must be endorsed by the CEGB Executive.

The corporate planning process therefore is not just the once-off annual production of a plan, but an on-going dynamic and flexible process, requiring re-evaluation, monitoring and readjustment, as necessary, but in a systematic and carefully controlled manner.

2.7 The location work support process

The power station managers task is to establish a future work programme derived from engineering reviews and other planning criteria. The main purpose of the location work support process is to assist him to implement his work programme with the help of other functions within the organisation. The process provides a corporate view of activities which gives assurance that individual locations receive the services required, at the correct time.

The location work support process interfaces aspects of the engineering review process and the planning process, and includes the implementation of the work programme. The location work support process therefore draws together the procedures for work implementation, outage planning, strategic spares planning, plant manufacturers or major repairers plans, contract placing plans and other repairs and maintenance forward planning procedures.

Many work items have a timescale of several years from the first registration of a problem until the required work is completed. Therefore, sound planning procedures are required throughout, to ensure full commercial benefit and that the work is completed on time. A work implementation procedure provides location managers with a process by which they can plan, control and monitor the progress of each significant activity associated with specific major work items, from inception to completion. It also provides the means for the support departments to plan their workload and agree their commitment to timescales with the location manager. The work implementation procedure brings together the key dates in the engineering, sanctioning and procurement phases of each job, for all major activities, so that the adequacy of the overall timescales and resources can be assessed.

The timetable for each job covered by this procedure is dependent upon the lead time for materials, outage dates and upon the extent of engineering of the work. The power station manager is responsible for setting and agreeing dates with other functional departments in order to complete the work within the limits approved in his management unit plan and budget. Progress of the work is reviewed in plan progress meetings at the location.

2.8 Monitoring performance

Locations are required to report monthly on their activities and progress towards achieving targets set in the management unit plan. In addition, plan progress meetings involving the location, production departments, finance and procurement, are held at intervals during the year to review the progress being made in meeting the plan objectives. The station's performance in terms of plant performance, finance, resource and materials management is reviewed, together with any additional engineering work or other factors affecting the achievement of the plan objectives.

The key resource and performance parameters which underly the production planning process are incorporated into the planning module of the Integrated Production Formation Information System (IPFIS). This computerised system incorporates the collection of all key management unit plan data, thereby enabling regular monitoring of key indicators to be carried out, and provides a comprehensive analysis of the performance of each management unit.

Out-turn reporting is an essential part of the corporate planning process; hence, a view of the effectiveness of policy decisions relating to the management and operation of the power station and the achievement of targets and objectives is carried out annually, and is included in the locations annual report to the Divisional Director of Generation.

2.9 Major overhaul outline planning

The Operational Engineering Division makes engineering assessments and provides guidance to stations on the work requirements deemed necessary to satisfy CEGB policies on plant integrity and performance. The Operational Engineering Division also quantifies plant risks, so that judgements can be made on the economics of specfic work proposals at the plan review meetings.

The power station management makes proposals for work programmes, resource requirements, outage duration and placement preferences, taking into account statutory dates and any other relevant factors.

The Production Planning Department is responsible for carrying out generation outage planning and the co-ordination of overhaul outage placement throughout the CEGB. The System Operations Department, however, has the final responsibility for authorising the release of generating plant from the system at the time the outage is planned to commence.

2.9.1 Timing of major overhauls

Generating plant is normally taken out of service for major overhaul mainly during the summer months, when the system demand is low and generation replacement cost is at its minimum. If the system demand curve is inverted and allowances are made for some spare capacity to cover breakdowns and other unplanned outages, an envelope can be produced within which the generation overhaul programme must be contained (Fig 4.2).

Provided there is adequate spare generating plant available, it may be economically justified to run certain plant for the full period between statutory examinations, even though the overhaul may occur outside the normal summer overhaul period. The economic justification for this is dependent upon the difference in generation replacement costs at the time of year under consideration, and the financial gains to be made by improved availability.

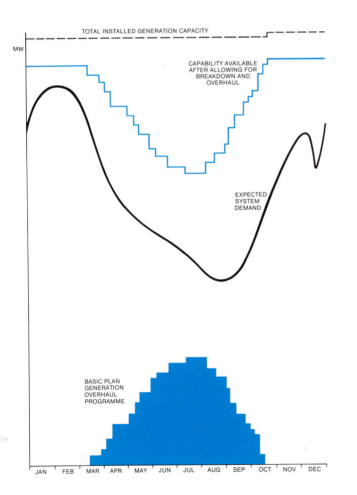

FIG. 4.2 Basic plan for generation overhauls

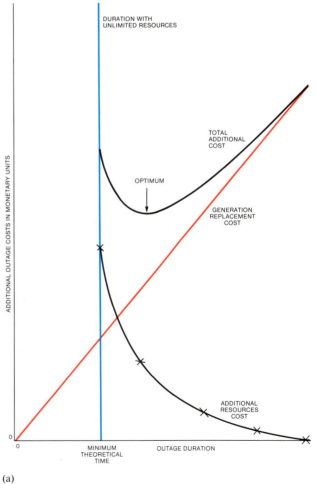

(a)

FIG. 4.3 Optimum outage duration

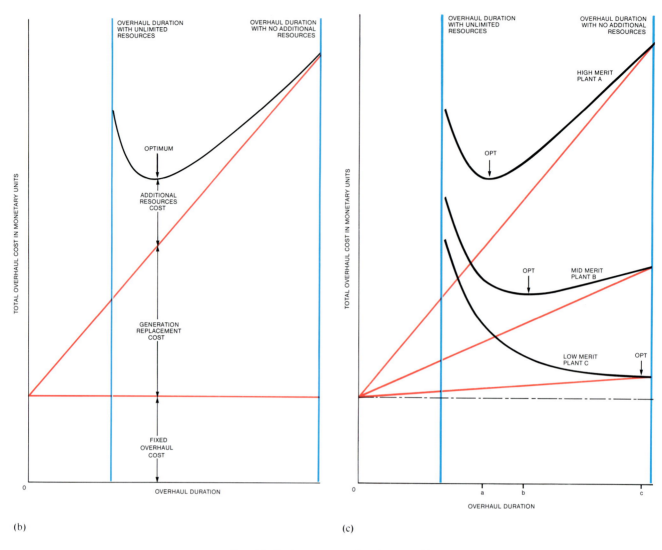

FIG. 4.3 Optimum outage duration (*cont'd*)

(b)

(c)

The generation replacement cost for the period of the overhaul depends on the predicted utilisation of that plant and its position in the merit order relative to that of the marginal plant needed to replace it. This cost can only be predicted by considering the generating system as a whole, since the timing of outages at other power stations also influences the generation replacement cost.

2.9.2 Optimum overhaul duration

The overhaul duration should be chosen so as to minimise the overall costs to the CEGB. To achieve this, the cost of additional resources, either contractors or mobile work teams which are required to reduce the outage duration, has to be offset against the saving in generation replacement costs gained by returning plant to service earlier than would otherwise be possible. This is illustrated in Fig 4.3 (a).

The total costs incurred in carrying out a major overhaul are illustrated in Fig 4.3 (b). The total cost comprises three major components:

- The fixed costs of the overhaul.
- Generation replacement costs.
- The additional resource costs required to reduce the outage duration.

The *fixed cost* includes the cost of materials, plant spares and other fixed resource costs.

The *generation replacement* cost is a function of the level of spare generating capacity available during the overhaul period and its cost of operation. For simplicity, the generation replacement cost has been indicated as being directly proportional to the duration of the outage. In fact, the generation replacement cost increases significantly if the outage extends beyond the normal summer overhaul period.

The *additional costs* incurred in obtaining the resources necessary to reduce the outage duration are obtained by considering various options. These range from carrying out the work programme entirely by location staff over a much longer period, to reducing the outage duration to the minimum practicable by utilising maximum resources.

The extent to which additional manpower resources are cost effective depends on the generation replacement cost for that particular plant. For high merit, high generation replacement cost plant, it is economic to complete the overhaul in the shortest practicable time. For lower merit, lower generation replacement cost plant which may not be required during the summer months, it may be more economic to carry out the overhaul over a longer period of time and reduce the additional manpower resource costs to a minimum. Figure 4.3 (c) illustrates a comparison of the total cost against outage duration when carrying out similar work on high, mid and low merit generating units.

For high merit plant *A*, which has the highest generation replacement cost, the minimum overall cost is obtained with the greatest practicable utilisation of additional resources, the optimum outage duration case being *a* days.

For the mid merit plant *B*, since the generation replacement cost is lower, it would not be cost justified to incur the same level of expenditure on additional resources as for the high merit plant. The optimum outage duration is now *b* days.

The high cost, low merit plant *C* has minimal generation replacement cost and will probably have a low utilisation factor. Therefore, subject to the overhaul requirements of other generators at that station, it would be more economic to minimise expenditure on additional resources and carry out the overhaul over the much longer period of *c* days. However, should it be necessary for any reason to limit the outage duration to somewhat less than this period, the additional costs will not be significantly increased, as the graph illustrates.

In addition to this analysis, an assessment of other indirect or potential costs would be taken into account including, for example, grid system security, transmission line maintenance and any other generating plant or grid system limitations.

To determine the optimum duration, the often conflicting requirements of plant availability, resource and financial limitations have to be resolved and the most appropriate solution found.

The foregoing planning must be carried out far enough ahead of the event to allow the identified costs and outage duration to be incorporated into the management unit and production formation plans.

2.10 Computer aided management systems

Substantial advances have been made in recent years in the area of computer aided management systems. From experience gained in the development and use of these systems within the CEGB, the benefits of using computers to assist managers to monitor key areas such as work planning, materials, manpower and finance resource management were quickly recognised.

In the early 1970s, mainframe computers were installed at regional centres, data being transmitted from each location in 'batches', with printed outputs being despatched overnight. Advances in computer technology and reduction in cost during the early 1980s enabled 'distributed' computers to be installed at each major power station site. These are capable of supporting each station individually, but also provide access and updating facilities to data held on mainframe computers situated at various locations within the CEGB.

More advanced systems are currently being implemented which will enable executive management, corporate functions, power station managers and other specified staff to have access, with the appropriate levels of data protection, to the complete CEGB corporate data network.

Management information systems provide a means of controlling, monitoring and supporting the various facets of power station management. Information relating to forward plans and budgets, finance and manpower resource utilisation, plant performance data and information relating to the compliance with statutory and other mandatory requirements must be readily available. This information is held on computerised systems which enable plans, budgets and resource expenditure to be correlated to provide an effective management control and information system.

In order to achieve a system which not only benefits individual power station management teams, but also provides a source of corporate management information, the system will provide a structure of common databases at all locations throughout the CEGB.

These information systems bring together data relating to total resource requirements and therefore provide an integrated approach for strategic and corporate planning purposes. There are two basic levels of computerised information systems:

- Corporate level information systems.

- Power Station level information.

The Integrated Production Formation Information System (IPFIS) is an example of one of a number of corporate level information systems. It provides the information required by the Divisional and Corporate Functions in the CEGB and consists of six basic elements:

- Integrated heat accounting, fuel and byproducts.

- Planning.

- Plant condition.

- Nuclear fuel cycle.

- Power system management and information.

- Policy and procedures.

Station computerised information systems primarily provide data to be used solely at location level and may or may not be interrelated. Their scope incorporates all the functions involved in managing a power station and includes the following systems:

- Plant coding.

- Plant inventory.

- Materials management and purchasing.

- Work planning and control.

- Manpower performance monitoring.

- Thermal performance monitoring.

- Plant availability monitoring.

- Contracts and contractor management.

- Finance and budgetary control.

2.11   Management policies

The adoption of the correct maintenance policies has such a significant influence on the cost effectiveness of both the operation and the maintenance of power station plant, that adequate consideration and attention needs to be given to the various aspects involved to ensure that only the optimum policies are chosen.

The major activities in the policy making process include:

- Formulating policies.

- Implementing the policies.

- Monitoring their effectiveness.

- Revision of policies to accommodate changing circumstances.

In addition, where specific actions are needed, the policies should clearly identify:

- What is to be done.

- Who is responsible for doing it.

- How it is to be done.

- What resources will be required.

- How progress will be monitored.

- How achievement will be recorded.

The policies relevant to the operation and maintenance of plant in a generating station are normally formulated by the management team. They must include compliance with the requirements of statutory and nuclear inspectorate regulating bodies, as well as CEGB internal directives. They must also take into account recommendations from plant manufacturers, specialist groups within the CEGB and experience obtained with similar plant at other locations.

The optimum maintenance policy should give maximum plant performance, reliability and flexibility, at minimum overall cost.

In determining maintenance policies, an endeavour is made to assess the likelihood of the occurrence of future events, about which we cannot always be certain. Predictions can only be made using the best possible information available at that time.

The policies should be reviewed at regular intervals to ensure that the initial assumptions and predictions were valid. This review is most important, otherwise decisions which may have been based upon very little firm data may be considered as being absolute and not requiring reappraisal. It is advantageous, in such instances, to fix a policy review date.

The policies should indicate the manner in which maintenance engineers and managers intend to achieve their desired objectives, by providing a structured framework of policy statements, actions and responsibilities.

Detailed consideration needs to be given to the following aspects of maintenance planning:

- The extent to which summer availability may be reduced in order to achieve maximum winter availability.

- The extent to which operational flexibility may be improved. It is necessary to identify any potential savings that may be achieved by improved flexibility and to optimise the costs with any plant modifications or increased maintenance requirements.

- Specification of the areas and the extent to which plant performance will be allowed to fall, before planning outages for remedial work.

- The operating regime for standby plant.

- The need for a comprehensive plant inventory to enable an assessment of total maintenance requirements to be made.

- Definition of the areas where maintenance requirements rely on information obtained from the use of plant condition monitoring techniques.

- The extent to which a preventive maintenance policy is to be adopted and hence the requirements for routine planned maintenance work.

- The extent to which additional manpower resources may be used during peak work-load periods, to supplement those normally based at the location.

- How the use of computer aided work planning systems may improve the effectiveness of the work planning process.

- The extent to which network analysis techniques are to be used for maintenance activities.

- The manner in which work requirements are to be recorded, specified, co-ordinated, scheduled, monitored and progressed to completion.

- Ensuring that a formal procedure is instituted for the approval of plant modifications for both nuclear and conventional plants.

- Specifying the work control and quality control procedures that are to be implemented.

- The extent of plant history records, the level of detail required and how the information is to be used to influence future policy making.

- Identification of means of ensuring that essential maintenance activities are being carried out, and the formulation of a reporting system which identifies any area where this is not being achieved.

- The extent to which computerised systems can be used to analyse the effectiveness of these maintenance policies.

The correct application of these policies will ensure that maintenance resources are utilised effectively. They ensure that the plant condition is adequately monitored, enabling maintenance effort to be directed to deal with any deterioration before a serious breakdown occurs and enable planned preventive maintenance to be carried out in the most effective manner, thereby ensuring maximum reliability, availability and efficiency.

The major aims and objectives of the policies and plans at each location must conform to the corporate planning objectives for the industry as a whole. The manner in which they are to be achieved are included in each power stations management unit plan.

3 Plant inventory system

Before any comprehensive maintenance planning system can be introduced, it is necessary to create an inventory which identifies all main plant, auxiliary plant and ancillary equipment in the power station. This is necessary to define the maintenance and spares requirements for each item of plant.

All items within the boundaries of the power station are included. Particular attention should be given to any area where the ownership and the responsibilities for operation and maintenance cross boundaries.

3.1 Plant inventory structure

The plant inventory system consists of a number of schedules, which specify in a structured manner, basic data relevant to the plant installed in the power station.

The purpose of these schedules is to ensure that reliable up-to-date information is readily available to staff responsible for the management, operation and maintenance of the power station.

The major components of the plant inventory system are illustrated in Fig 4.4. Individual inventory schedules are prepared by engineers responsible for particular items, or aspects of the plant. These schedules can then be indexed to the plant code, enabling a comprehensive plant inventory system to be created. The creation of the plant inventory system requires a substantial amount of effort. The plant code alone probably contains several thousand individual plant items. Since this provides the framework to which other information may be cross-referenced, it is important that it is created in a structured manner covering all plant and equipment on site. It is desirable to create the basic plant code at the earliest possible opportunity, preferably prior to commissioning. This has the advantage that initial plant performance data can be obtained and recorded in the inventory system, enabling performance comparisons to be made in the future.

To be of maximum benefit, the inventory must be readily accessible and easy to use. In this respect, computerisation offers considerable benefits, even for small power plants. The computer can be used to index or link directly other plant inventory information to the station plant code. This provides a comprehensive plant inventory system, available to all users who have access to a computer terminal. This has particular advantages in areas such as plant spares requirements, and instrument and control valve setpoint schedules.

Where an existing inventory is not adequate, or the structure of the coding system is such that it is not ideally suited to computer access, it may be advantageous to recompile the inventory in a more structured manner.

In the majority of instances, it is obvious which departments or sections are responsible for the maintenance of any particular plant item; there will however be some instances where this needs to be

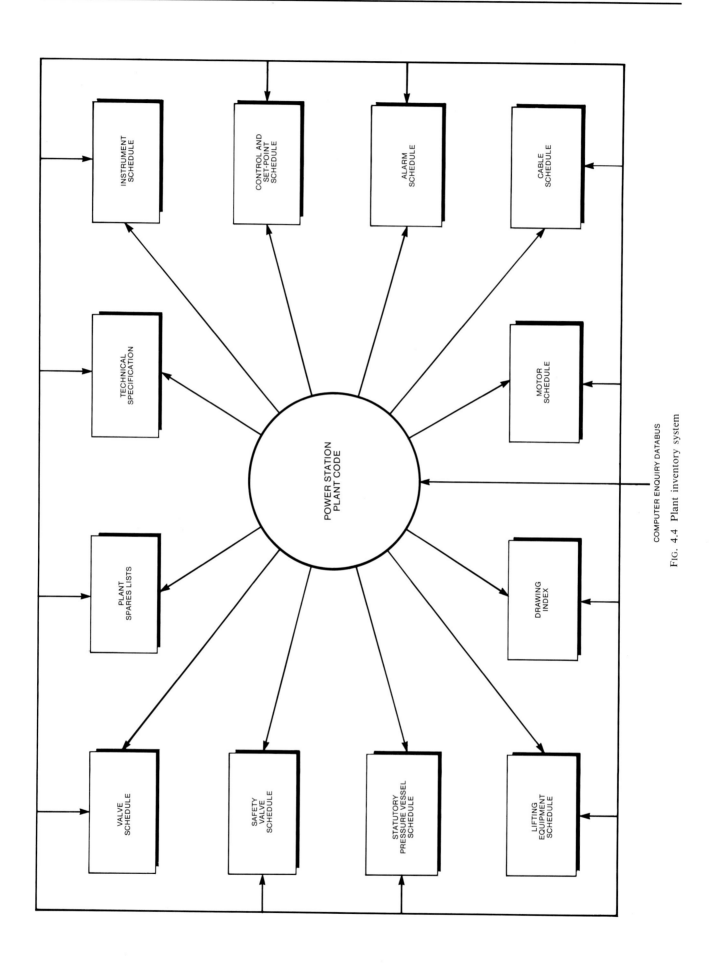

COMPUTER ENQUIRY DATABUS

FIG. 4.4 Plant inventory system

defined specifically. Typical examples are fluid or other couplings, pneumatic or hydraulic servo-systems and some protection equipment. It may be appropriate in such instances to indicate this maintenance responsibility in the plant inventory system.

3.2 Plant code

It is not practicable to record all the information relevant to a large item of plant, such as the turbine-generator, under a single plant code, as any enquiry relating to plant spares would involve searching through several thousand records. To simplify the search it is necessary to subdivide this into a number of different levels of code. How far the plant items are subdivided depends upon the complexity of the plant, the importance of the item of plant concerned and the amount of detailed information to be recorded against each individual code. For example, small hydraulic system motor pump units may be allocated a single code, whereas a boiler feed pump unit may be allocated 50 or more codes, for individual components. The complete plant code for a typical generating station with 500/660 MW units may involve in excess of 2000 plant codes. In addition, where identified separately, there may be of the order of 7000 different instruments and 3000 valves associated with these plant items, which may also have individual plant inventory records.

The primary objective of a good plant code structure is to minimise the overall time taken to find the desired information.

The major areas where frequent enquiries are made are:

● Work planning outstanding work file.

● Standard jobs file.

● Plant spares file.

The outstanding work file may contain up to 3000 records, the standard jobs file up to 10 000 records and the plant spares file up to 20 000 records. It is therefore beneficial to spend a few seconds longer finding the correct plant code than wasting several minutes scanning a substantial number of screen pages of outstanding work, standard jobs, or plant spares to find the desired information.

It is highly desirable for the plant coding system to have a hierarchical structure. This enables the user to determine the plant code in a step-by-step approach, selecting the first, then subsequently each of the lower levels of code in turn.

The CEGB Plant Reliability and Availability (PRA) code is a four-digit code having a hierarchical structure. This is now being used in power stations to provide the structure for the site plant inventory code, the titles or actual plant item descriptions being specific to the plant at that particular site.

The title or plant item description will probably need to be limited to approximately 100 characters, to limit computer file sizes. These parameters need to be established before plant inventory data collection is commenced.

The basic plant inventory therefore consists of the title or description of the plant item and an associated plant code reference number, a typical example being:

4446 Generator seal-oil coolers and filters

This particular code is used as an example in Fig 4.5 which illustrates a section of a typical power station plant coding system having a hierarchical structure. The plant code is found by selecting the appropriate plant item at each of the four levels of code.

In this example the generator seal oil coolers and filters are integral units. If they were physically separate units, or were required to be identified separately, they might be allocated individual codes at fourth-digit level. Alternatively, if no spare codes were available and the code structure allows a fifth level of coding, they could be separately identified at that level.

Figure 4.5 illustrates a typical four-figure hierarchical plant code structure, based upon the CEGB Plant Reliability and Availability Code. If used as part of a computerised plant coding system, the code is selected by a sequence of four screen menus. The final selection being at the greatest level of detail, this being the level at which spares lists and work planning information is normally entered.

When used in conjunction with computer systems, this hierarchical form of coding enables enquiries to be made at various levels within the code structure. The enquiry may be made on a wide or narrow basis. For example, all outstanding work on the turbine-generator would be found by entering 4 in the first character position of the plant code, outstanding work on the generator seal-oil coolers and filters would be found by entering code 4446 in the first four character positions.

This facility for using the computer to provide a 'masked enquiry' enables the user to widen or narrow his search, as required, to include more or less information.

Together with the unit number and the individual plant identifier (A, B, C or D), this uniquely identifies the item and provides a means of recording all information relevant to that particular item of plant.

3.3 Plant spares lists

The CEGB operates a nationally co-ordinated plant spares and materials purchasing and stock holding

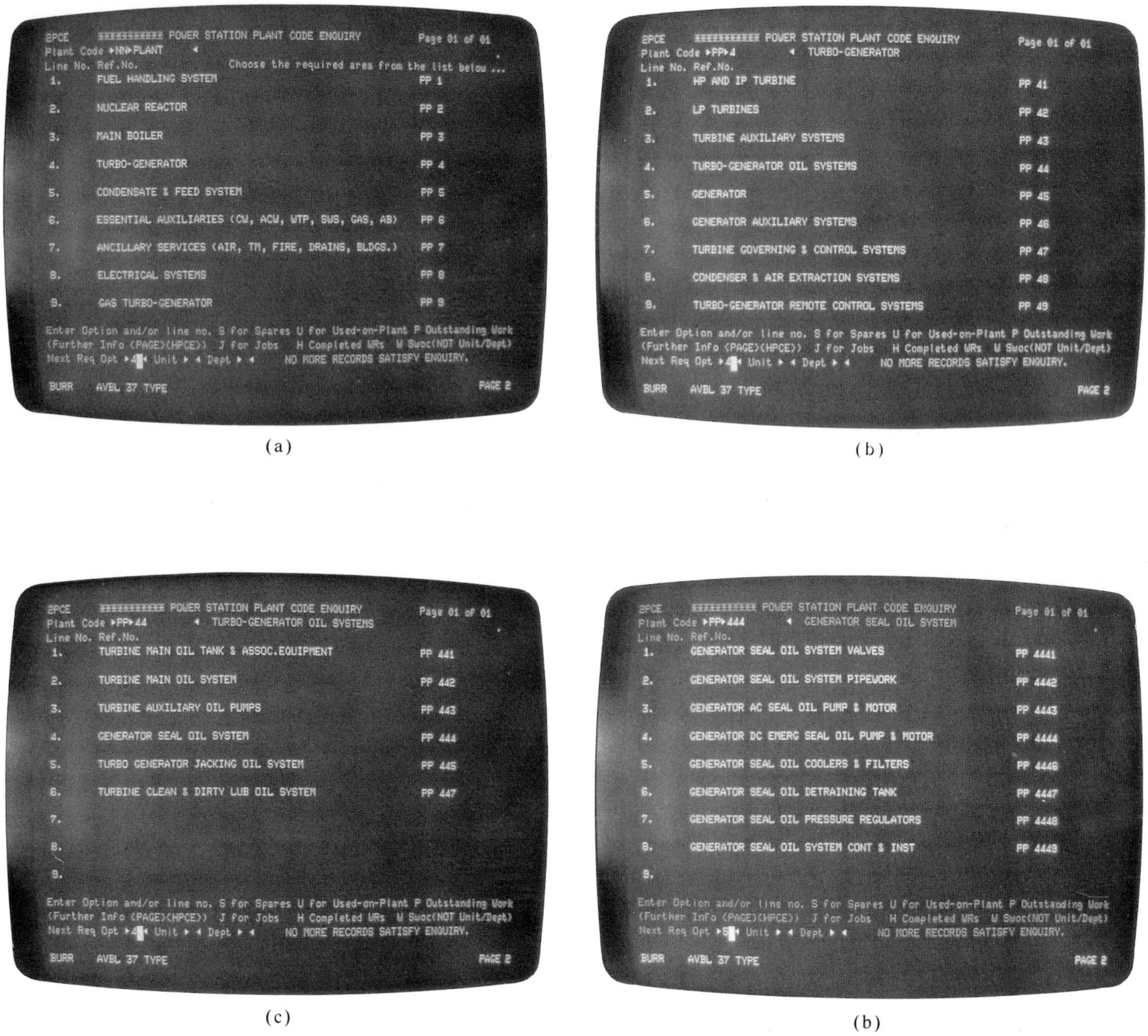

(a)

(b)

(c)

(b)

FIG. 4.5 Plant code structure

(a) First level plant code enquiry

(b) Second level plant code enquiry

(c) Third level plant code enquiry

(d) Fourth level plant code enquiry

policy. This is based upon a national commodity code system, with some limited regional and local variations to suit specific circumstances. The system requires all purchases of new equipment to be allocated a commodity code. Plant spares and consumable items are grouped into appropriate main and subgroups. A typical commodity code structure is:

Main group	Sub group	Detail	Item description
75	31	022	Element and pad pack

These commodity codes and their descriptions are available on power station computer systems to enable

spares lists to be produced for any individual plant item. These spares lists may then also be linked to the appropriate plant codes, forming one of the most important components of the plant inventory system (Fig 4.6).

3.4 Plant technical specification

A technical specification catalogue for each significant item of plant or equipment is produced as part of the plant inventory system. This should preferably be compiled prior to plant commissioning, as design data and 'as new' performance characteristics are more readily available at that time.

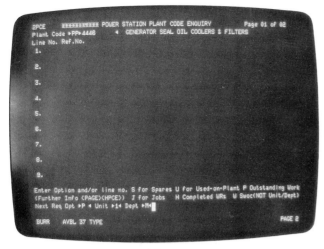

(a)

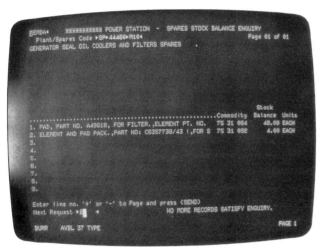

(b)

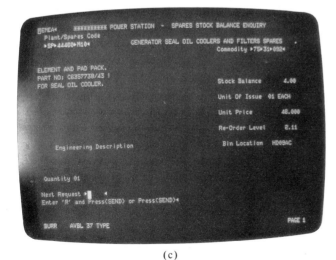

(c)

FIG. 4.6 Plant spares enquiry

(a) Once the plant code has been determined (Fig 4.5) the spares lists, or sets of lists, can be obtained for the plant item.

(b) The computer displays an abbreviated description of the item and its CEGB stores commodity code.

(c) The full description and other details can be obtained for any selected item.

During the operational life of the station, occasions arise when it is necessary to refer to this technical specification, particularly when any deviation from optimum performance is suspected.

The specification should include basic design and/ or 'first run' data, as appropriate. Examples of the type of information required, are as follows:

- Power or rated output.

- Full load current.

- Overload capacity.

- Temperature rise.

- Speed at rated output.

- Motor or hydraulic coupling slip.

- Parallel operating characteristics.

- Performance curves.

- Pump lift, or closed valve discharge pressure/head.

- Pump discharge pressure/head at rated output.

- Pump minimum suction pressure.

- Similar data for fans, as for pumps.

- Fluid or gas temperature limitations.

- Cooling system temperature limitations.

- Maximum vibration levels.

- Operating times.

3.5 Instrument schedule

Over recent years, control and instrumentation engineering has developed towards the use of modular systems, where interchangeable units can be readily fitted. These units can be overhauled, or recalibrated in clean workshop conditions, and may then be either replaced in their original location or returned to stores as serviceable spares. In the latter case, they may be used for a similar purpose, but different location, or possibly for an entirely different application.

There are, therefore, two distinct forms of data relevant to an item of instrumentation, its specific application on a plant item and the historical record of that particular instrument's previous function and reliability.

Instrument application numbers are usually specified at the design stage of the power station. The form of numbering system may relate to the item of plant on which the instrument is located:

IB1/14	Instrument boiler	Unit 1	No 14
IT2/32	Instrument turbine	Unit 2	No 32
IF3/88	Instrument feedwater	Unit 3	No 88
ICW3/24	Instrument cooling water	Unit 3	No 24
IACW4/12	Instrument aux. CW	Unit 4	No 12

The actual form of the instrument number is not important. What matters is that there is a simple numbering system that uniquely identifies the plant application of that particular instrument, to which information such as location, function and setpoint may be cross-referenced (Fig 4.7).

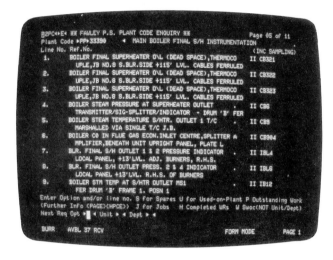

FIG. 4.7 Instrument application numbers may be linked to the plant code for the associated plant item

The instrument schedule refers only to the specific application and provides data relevant to the use of the instrument on that particular item of plant.

The instrument schedule includes data such as:

- Instrument application number.

- Instrument application title.

- Function of instrument.

- Associated control loop.

- Set point values.

- Standard of accuracy required.

- Location of instrument.

- Tapping point number

- Drawing references.

- Cable and core references.

- Manufacturer and type.

- Manufacturer's serial number.

3.6 Control and setpoint schedule

The control and setpoint schedule is simply a listing of just the instrument numbers with their descriptions and setpoints in one document, whereas the instrument schedule is comprised of individual data sheets, or computer screen records, giving all relevant data for each individual instrument.

3.7 Alarm schedule

The alarm schedule provides a listing of the alarm reference number, its description and initiating source, together with relevant cable references.

3.8 Cable schedule

This includes the cable number, its associated plant item description, cable type, size, number of cores and terminating points. This is of particular importance for fault finding and should be readily available either in hard copy form, or on a computer screen display.

3.9 Motor schedule

This schedule lists all motors used on site, their functional description, code, type, power, speed, metric or imperial frame size, and any associated equipment, such as induction or electrolyte slip regulators.

3.10 Drawing index

Although CEGB plant manuals contain a comprehensive set of plant detail drawings, it is essential for fault diagnosis purposes to have a set of drawings and schematics readily available. Drawings are usually identified by a power station prefix letter and CEGB reference number. In addition, the manufacturer's drawing reference number may also be quoted. The drawing indexing and filing systems must enable drawings to be readily obtained, located and accessed. An effective system is also needed ensure their return, on completion of the work.

3.11 Lifting equipment schedule

It is a statutory requirement to maintain a register of all cranes, hoists and lifts, together with all fixed and portable lifting appliances. This register is part of the inventory system and consists of a separate record sheet for each item of portable lifting equipment and a separate insurance certificate for each crane, hoist and lift.

Portable lifting equipment is inspected every 6 months and, to facilitate this, the equipment is usually divided into two or more separate groups, or phases, and colour coded (often red and green). Each phase is used for a period of three to six months and then the equipment is changed to another colour phase. If two groups of equipment are used, the inventory may be divided into two halves to include similar items of equipment in each group.

Portable lifting equipment also forms an important part of the inventory system and an effective method of control and recording is essential.

3.12 Statutory pressure vessel schedule

Currently it is a requirement of the United Kingdom Factories Act (Sections 33, 35 and 36) that all pressure vessels must be examined at the prescribed periods. In 1990, new regulations 'The Pressure Systems and Transportable Gas Container Regulations 1989' are to be introduced.

A schedule of pressure systems is required to be produced to ensure that all pressure vessels and pressure systems encompassed by these regulations are clearly identified.

3.13 Safety valve schedule

This is required to ensure that all safety valves located on the plant are identified and that their settings and testing requirements are specified.

Safety valves associated with statutory pressure vessels are required to conform with the requirements stated in the United Kingdom Factories Act 1961.

The safety valve schedule should include the following data:

- Set pressure — the pressure at which the safety valve commences to lift.

- Full lift pressure — the lowest pressure at which the valve will reach its maximum lift.

- Reseating pressure — the pressure at which the valve closes fully

Pressure relief valves are often fitted in sections of pipework which may be subject to heating, with subsequent expansion of the working fluid and possible risks of over-pressurisation. Typical examples are the feed side of high pressure feedheaters, or on trace-heated fuel oil lines. These pressure relief valves should also be included in the safety valve schedule and be tested periodically.

3.14 Valve schedule

The valve schedule is one of the most important sections of the plant inventory system. The operational performance of the power station is probably more reliant upon the satisfactory maintenance and operation of valves, than upon any other item. Most of the activities performed during the two-shift operation of power station plant are valve operations. Valve repairs and replacement may also represent a significant proportion of the maintenance effort and financial expenditure. It is particularly important, therefore, that valves are properly identified, labelled and are included in the plant inventory and routine maintenance systems.

The amount of data held relevant to each valve depends upon its size, duty and operational importance on the plant. For all valves, the unique application number, description, location, manufacturer, type, size and method of fixing is required. For items such as boiler stop valves, feedwater regulating valves, bled-steam and feed system isolating valves, other data, such as torque settings, maximum differential pressure and operating times should be recorded, as appropriate.

Various methods of numbering are in use within the CEGB; typical examples are:

C1/60	Condensate	Unit 1	Valve No 60
LPS2/4	Low pressure steam	Unit 2	Valve No 4
HPS 3/16	High pressure steam	Unit 3	Valve No 16
HPF4/32	High pressure feed	Unit 4	Valve No 32
FOX/4	Fuel oil	Common	Valve No 4
FO1/9	Fuel oil	Unit 1	Valve No 9
ASX/44	Auxiliary steam	Common	Valve No 44
AS1/10	Auxiliary steam	Unit 1	Valve No 10
CW1/12	Cooling water	Unit 1	Valve No 12
ACW2/96	Auxiliary cooling water	Unit 2	Valve No 96
TMX/64	Towns main	Common	Valve No 64

The particular form of numbering is not important, what is important is that there is a number which uniquely identifies each valve application in the power station (Fig 4.8).

3.15 Plant inventory summary

The plant coding and plant inventory systems provide the foundation upon which the maintenance work planning, quality and materials management systems are constructed.

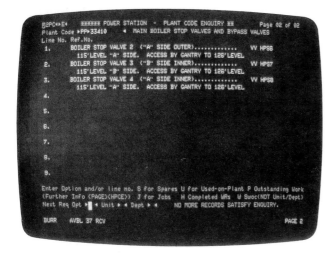

FIG. 4.8 Valve application numbers may be linked to the plant code for the associated plant item

Finance and budgetary control systems also utilise the CEGB PRA plant coding structure, and the CEGB cost code allocation system.

The form in which power station plant inventory systems are structured determines how effectively data can be utilised for the management, planning and control of maintenance activities.

4 Work identification

Once maintenance policies are determined, corporate and location planning objectives are known and a plant inventory system created, it is then possible to determine the basic maintenance work requirement. This encompasses all the work that is necessary to keep the power station operating safely, in compliance with statutory requirements, with maximum availability, reliability, flexibility and efficiency, and at the lowest overall cost. The operational role required of the power station influences the relative importance of the latter items, and also influences where the financial and manpower resources should be expended to achieve the optimum overall benefit to the CEGB.

Work requirements fall into the following categories (see Fig 4.9):

- Statutory.

- Safety.

- Routine preventive maintenance.

- Work identified from plant condition monitoring.

- Major overhaul work.

- Plant modification.

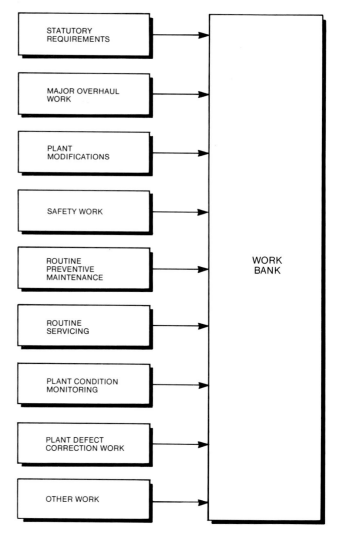

FIG. 4.9 Work bank constituents

- Plant defect corrective maintenance.

4.1 Statutory requirements

Statutory and other mandatory work represents a substantial proportion of the overall manpower resource allocation, particularly in nuclear power stations. The identification of the requirements for the statutory examination of all main and auxiliary plant is therefore essential to enable the total resource requirements to be determined and planned work programmes prepared.

All generating stations within the CEGB are currently required to operate under the Electricity Act 1957, the Factories Act 1961 and the Health and Safety at Work Act 1974. The Factories Act designates an electricity generating station as a factory.

In 1990, new regulations 'The Pressure Systems and Transportable Gas Containers Regulations 1989' were introduced. Parts of the regulations came into force in July 1990, others in January 1991, and all provisions included in the regulations come into force in July 1994.

The Health and Safety Executive have issued a guidance document for these regulations and in addition the Health and Safety Commission have produced an approved Code of Practice. Both of these documents provide guidance with respect to the new regulations, to which reference should be made. This is of particular importance since statutory regulations are continually under review, therefore, a system of ensuring compliance should be instituted and audited at appropriate intervals.

In addition to the above statutes, nuclear power stations are controlled by licence under the Nuclear Installations Act.

One of the conditions stipulated for a nuclear power station site licence, is the requirement for a maintenance schedule. This specifies the requirements for the examination, maintenance and testing of plant and its safety mechanisms. For this reason the routine maintenance requirements included in the Maintenance Schedule are referred to as 'statutory routines' and have been identified as statutory work, rather than the general category of routine preventive maintenance work.

The statutory examination of cranes, lifts, hoists, beams, runway trolleys, forklift trucks and other items of portable lifting equipment forms another important aspect of statutory inspection work. The disciplined planning and recording of these inspections is a particularly important aspect of the maintenance planning function. If items of lifting equipment are out of statutory certificate when required, costly delays to maintenance work programmes may ensue until the item is inspected by the appropriate engineer surveyor.

4.2 Safety work

This is normally considered to be work which is required to be carried out to ensure the safety of personnel, as distinct from safety to the plant. It does not normally include items which are categorised as statutory, even though this work clearly has a safety aspect. However, as statutory and safety work are both top work priority categories, the distinction is not too important.

Safety work should not include plant safety items unless it involves a direct safety-to-personnel problem. For example, steam leaks may, or may not, constitute a safety hazard, depending upon the location of the leak. If the steam leak is in an area out of reach of normal access, or it can be safely barriered off, then this may not be considered an immediate safety problem. The category of safety work therefore normally refers to the safety aspects of the working area, or its environment. It includes items, such as inadequate access, tripping hazards, obstructions, floor or walkway condition and any other aspect that could lead to hazards or danger to personnel.

4.3 Routine preventive maintenance

The major objective of a routine preventive maintenance policy is to achieve maximum plant availability, whilst reducing overall maintenance costs to a minimum.

When determining the requirements for routine maintenance, it is also important to take generating plant out-of-merit replacement costs into consideration. These may be substantial if plant has to be shutdown, or run at reduced output, whilst routine maintenance work is carried out. Conversely, costly breakdowns during high system demand periods may be avoided by carrying out routine maintenance work at periods of low system demand, such as overnight, or at weekends.

In determining what items should be considered and how far it is practical to implement a routine preventive maintenance policy, it is first necessary to identify items of plant or equipment which are particularly important to the safe and continued operation of the generating unit. Therefore items of plant or equipment should be identified which, should they fail in service, would lead to considerable consequential damage to themselves, or to other related plant. For example, the failure of the sump pump and its high level alarm in a main circulating water pump pit could result both in an expensive motor fault and the possible loss of a 500 or 660 MW generating unit. The maintenance of such equipment would therefore have a higher priority than the maintenance of similar equipment situated in less critical areas.

The primary objective of a routine preventive maintenance policy is therefore to prevent the item of plant or equipment from failing in service and thereby:

- Preventing the loss of generating capacity, particularly at periods of increased demand.

- Minimising damage to the particular plant item.

- Preventing consequential damage to other related plant items.

The same policy may not necessarily be appropriate to similar items of equipment since the importance of the item to the operation of the generating unit, and the duty and conditions under which it is required to operate, may be considerably different.

The plant operating regime also influences the routine maintenance requirements. Equipment subject to increased cyclic operation will require components to be replaced, or readjusted more frequently.

The amount of duplication of equipment, and whether it is a main or standby item, also influences the routine maintenance requirements. The nature of the routine may also be different, as in certain instances exchangeable items may be used for duplicated equipment.

The work content of each maintenance routine is determined by experience with the equipment over a period of time, from manufacturer's recommendations, or from experience at other locations within the CEGB. It is necessary to assess which components are most likely to fail and to anticipate the most probable cause of failure. In addition, the environment in which the equipment is operating needs to be considered, since dirt ingress, wet and humid conditions or high vibration adversely affect most equipment.

Setting up or reviewing a routine preventive maintenance system for an established power station where the incidence of failure is known, is more readily accomplished than for a completely new power station. In new stations it is usual, as a first step, to draw up routine maintenance schedules based upon the manufacturer's recommendations. The nature and frequency of maintenance may then be progressively improved by monitoring plant behaviour, assessing performance trends and modifying maintenance schedules to arrive at the optimum frequency and work content.

The frequency of carrying out the routine may be specified as a fixed time interval, or based upon the number of running hours, the number of operations, or main generating unit starts.

Hence, it is evident that to adopt the same maintenance routine for all similar items irrespective of their duty, may not necessarily be the optimum in terms of maintenance effectiveness, or resource utilisation.

4.4 Routine servicing work

There is an important difference in concept between routine servicing work and the more general classification of routine preventive maintenance work. Routine servicing is work which is essential to keep the plant running, or in a serviceable standby state, without which the plant item would quickly fail, often with expensive consequences.

The major areas of routine servicing work include:

● Lubrication checks.

● Oil filter changes.

● Pump gland adjustments.

● Valve gland adjustments.

Since routine servicing work usually provides one of the greatest returns in terms of improved reliability, availability and reduction in maintenance costs, it is important that this work has a high status level

in maintenance staff attitudes and the allocation of work planning priorities.

When analysing plant failures, some of the most expensive breakdowns are caused by inadequacies or omissions in carrying out the basic routine servicing requirements listed above.

4.5 Plant condition monitoring

The anticipation of failure of items of plant, or their components, by the use of condition monitoring techniques can considerably reduce total maintenance costs and avoid unplanned generating plant shutdowns (Fig 4.10).

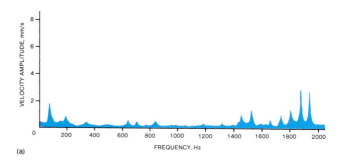

(a)

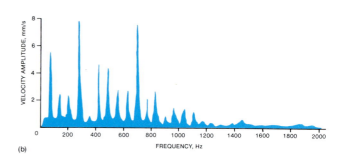

(b)

FIG. 4.10 Plant condition monitoring

(a) The chart illustrates the vibration frequency spectrum for a high pressure fuel oil pump gearbox. The peak on the left of the chart is due to output shaft vibration, indicating some slight out of balance or misalignment. The high frequency peaks on the right of the chart correspond to the vibration frequencies associated with gear tooth damage in this particular gearbox.

(b) This chart illustrates the vibration frequency spectrum of another high pressure fuel oil pump gearbox. In this example, peaks are evident in the mid-frequency range corresponding to frequencies associated with ball and roller bearing wear at the shaft speeds encountered in this particular gearbox. Note that the velocity amplitude scale values are greater than in the previous case.

The main reason for using plant condition monitoring techniques is to identify the need for corrective maintenance and, where possible, to provide information indicating the nature of the remedial work required. The monitoring technique used depends upon the plant item and the way in which it deteriorates, for example:

● Gear or bearing wear will generally result in increased noise and vibration levels. These would be monitored regularly and analysed to indicate which components are most likely to be at fault.

● Pump impeller wear would normally be determined by a deterioration in pump performance, hence pump lift or throughput at a set of standard conditions may be a convenient method of monitoring. Alternatively, heat pick-up or temperature rise across a pump will also provide a means of monitoring a deterioration in pump performance.

● Thermal performance monitoring provides a major source of data to indicate the condition of boiler and turbine-generator plant. This information is essential when determining the cost effectiveness of various maintenance work programmes.

● Oil sampling for metal particles is used to detect the rate of wear in bearings and in rotating elements, such as boiler induced draught and forced draught fan fluid couplings. By taking regular samples through a micropore filter paper, increased wear can be detected and bearings changed before components touch and cause consequential damage.

● Gas sampling has for many years been used as a method of detecting incipient electrical faults in transformers and other electrical equipment. This principle is also used to detect generator core or winding overheating by continuous monitoring of the hydrogen cooling gas.

● The chemical monitoring of steam and feedwater circuits, apart from being a preventive measure, also indicates the need for corrective maintenance. For instance, high oxygen levels in feedwater may be the result of leaking glands on pumps, or valves operating below atmospheric pressure. High solids in steam may be the result of damaged boiler drum steam separators. Increased condenser hotwell water acidity during vacuum raising may be an indicator of boiler reheater leaks.

In its broadest sense, plant condition monitoring is a quality control process covering a wide range of applications and techniques. It enables plant to be operated more reliably within defined limits and provides a means of monitoring any deterioration in performance on a periodic, or continuous basis. The method chosen and the frequency of inspection depends on the importance of the plant item, its reliability, the cost of failure and the cost of providing the monitoring system.

When determining maintenance policies, an assessment of the most likely causes of failure and their effects should be made for each important plant item. If it is decided that condition monitoring is appropriate, it is important that a procedure is set up to ensure that adequate monitoring is carried out, records are retained, trends noted, appropriate actions taken in time and the success (or failure) of the policy established for that item of plant. If failures do occur, the primary cause should be established and an examination carried out to determine what measures or improved condition monitoring techniques are required to correct the situation.

Plant condition monitoring can identify the need to carry out corrective maintenance, or suggest the opportune time to carry out auxiliary plant overhauls. As a result it can produce significant improvements in generating plant reliability, availability and cost effectiveness.

4.6 Major overhaul work

Major overhauls are normally planned to coincide with the statutory inspection periods and hence will include a significant amount of statutory work. In addition to this work, a substantial amount of other work is usually carried out during this period. This may include the examination of items, such as:

● Turbine bearings.

● Turbine shaft alignments.

● Turbine expansion arrangements.

● Turbine blading.

● Steam chests and valvegear.

● Generator windings.

● Generator excitation system.

● Boiler steam generating tubes.

● Boiler superheater and reheater tubes.

● Boiler headers and supports.

● Gas passes and ductings.

● Airheaters and precipitators.

● Boiler fans and motors.

● Steam pipe hangers.

● Feedheaters and de-aerator.

● Feed system and boiler feed pumps.

● Valve overhauls.

● Electrical switchboards and essential circuit-breakers.

The items included in this category should be those which can only be carried out during a major overhaul, or which would impose an unacceptable risk or reduction in output, if carried out at other times (Fig 4.11).

4.7 Plant modifications

It is particularly important that a formal procedure is instituted for the identification, approval and implementation of all plant modifications. This is necessary to ensure that proper consideration is given to all conceivable aspects and implications of any change to the plant design, however small.

A management system is required to ensure any modification necessary to improve plant safety, reliability and performance is properly identified and recorded. It also ensures that there is an adequate feedback of information to those groups responsible for the specification, design and manufacture of the plant item.

Plant modifications should be initiated:

- Whenever it is obvious that an item of plant or equipment, or a control system, fails to fulfil its required function.

- Whenever new data, or experience gained at other locations, suggests that the original plant design criteria are no longer valid.

- Whenever improvements in safety standards can be identified.

The formal approval of all changes to the design of the plant is necessary, whether this is a mandatory requirement by a statutory regulating body, or otherwise. The management system should ensure that the appropriate level of authorisation is given. A locally written instruction may be all that is necessary, whereas, in other cases, it may be necessary to consult specialist departments within the CEGB, or other statutory regulating bodies. For nuclear plants, all modifications must be referred to the Nuclear Installations Inspectorate for their formal approval.

4.8 Plant defect corrective maintenance

The main objective of a routine preventive maintenance policy is to reduce the incidence of plant breakdown to a point where overall costs are a minimum. This policy does not aim to eliminate plant defects entirely; hence the need for corrective main-

FIG. 4.11 Bearing dimensions and other quality control checks being carried out on an ID fan runner prior to being fitted as part of a major overhaul

(see also colour photograph between pp 434 and 435)

tenance. It assumes that defects on equipment that may affect the availability or reliability of the plant are attended to promptly, in order to reduce total costs to a minimum.

4.9 Specification of the work

Once the total maintenance work requirement has been identified, it is then necessary to produce detailed Work Specifications and Standard Work Instructions (Fig 4.12). These work instructions specify the working method, any special safety requirements, quality standards, quality check hold points, special tools or equipment, a list of essential spares which are necessary before the work can be carried out and a list of spares that may be required.

When each job has been specified, an assessment of the work content can then be made. This is necessary to enable total resource requirements to be determined and work programmes produced.

5 Work assessment

To enable the maximum amount of work to be carried out with the resources available, it is necessary that:

- The work requirements are adequately defined.

- The best working method is determined, especially where the work is of a routine nature.

- The work content is assessed.

- The work is properly planned, supervised and carried out to the required quality standards.

The second and third of these objectives require the application of work study techniques. Apart from being necessary to ensure the effective utilisation of manpower, this also enables work to be properly planned and scheduled, and provides a basis for manpower performance monitoring.

Work assessment, more commonly referred to as work study, as often thought to be a device which is intended to make people work harder. The aim of work study is to endeavour to make people work more effectively. The British Standard definition of Work Study BS3138 is 'The systematic examination of activities, in order to improve the effective use of human and material resources'.

In order to achieve this greater effectiveness, work is examined in two ways, as shown in Fig 4.13:

- Method study.
- Work measurement.

Unfortunately only the second of these, the work measurement aspect, is generally associated with work study, whereas in reality work measurement is only a small part of the whole process of studying work. In fact, increases in the effectiveness of carrying out work arise primarily from improved job methods or working practices.

5.1 Method study

The British Standard definition of method study is 'The systematic recording and critical examination of ways of doing things, in order to make improvements'.

Developing this definition further, it can be said that method study is the systematic recording and critical examination of existing and proposed ways of doing work, as a means of developing and applying easier, more effective and less costly methods of working.

The main objective is to find better ways of doing things and to contribute to improved efficiency by the avoidance of unnecessary work, delays and other forms of waste. This is achieved through improvements in:

- Layout and design of buildings, plant and the workplace.

- Working procedures.

- Use of materials, plant, equipment and manpower.

- Working environment.

- Design or specification of the product.

Method study involves a systematic approach and requires an ordered analytical manner, which may differ from what, at first sight, appears to be just plain common sense. For its successful application, method study must do three things:

- Analyse and reveal the true facts concerning the situation.

- Examine the facts critically.

- Develop the best answer possible for the given circumstances.

5.1.1 Ineffective time

Method Study improves people's effectiveness at work, by eliminating time which was previously spent ineffectively.

The total time taken to carry out a job can be considered to have four separate components, as shown in Fig 4.14:

- The *basic work*, which is the time taken for a person to do the job when it is immediately before

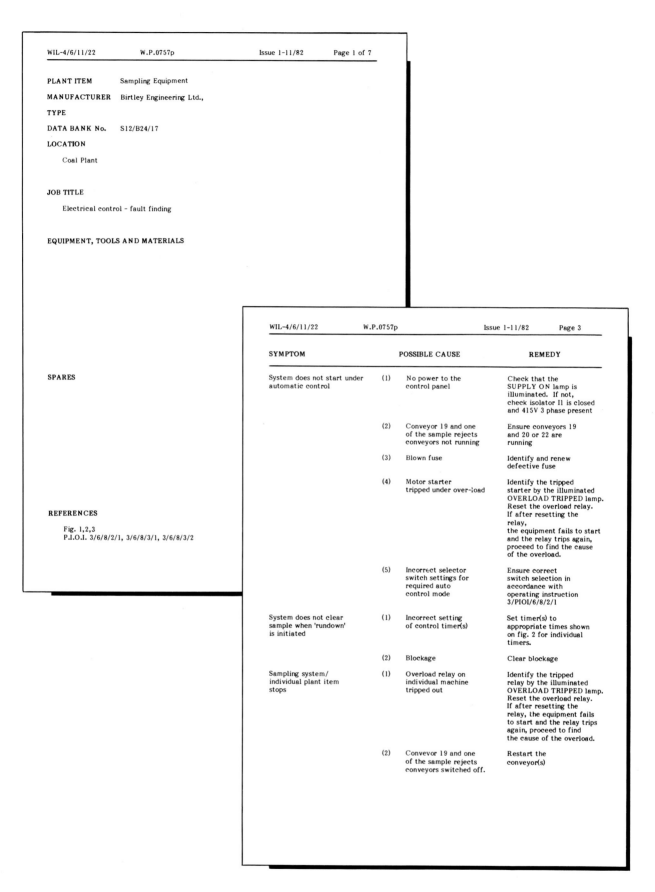

FIG. 4.12 Work specifications

Work specifications are produced for items of plant or equipment that are maintained on a regular basis. This will include the routine maintenance and overhaul of a variety of items such as pumps, motors, fans, switchgear and control and instrumentation equipment. Work specifications are also used to assist in fault finding and defect correction work.

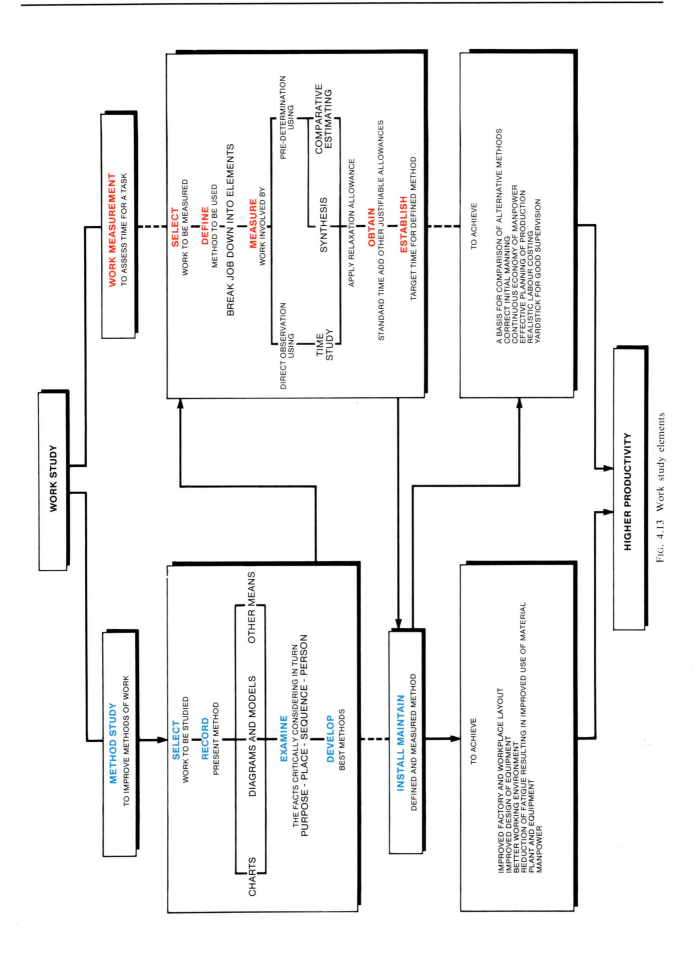

WORK STUDY

METHOD STUDY
TO IMPROVE METHODS OF WORK

SELECT
WORK TO BE STUDIED

RECORD
PRESENT METHOD

CHARTS

DIAGRAMS AND MODELS

OTHER MEANS

EXAMINE
THE FACTS CRITICALLY CONSIDERING IN TURN
PURPOSE - PLACE - SEQUENCE - PERSON

DEVELOP
BEST METHODS

INSTALL MAINTAIN
DEFINED AND MEASURED METHOD

TO ACHIEVE

IMPROVED FACTORY AND WORKPLACE LAYOUT
IMPROVED DESIGN OF EQUIPMENT
BETTER WORKING ENVIRONMENT
REDUCTION OF FATIGUE RESULTING IN IMPROVED USE OF MATERIAL
PLANT AND EQUIPMENT
MANPOWER

WORK MEASUREMENT
TO ASSESS TIME FOR A TASK

SELECT
WORK TO BE MEASURED

DEFINE
METHOD TO BE USED

BREAK JOB DOWN INTO ELEMENTS

MEASURE
WORK INVOLVED BY

PRE-DETERMINATION
USING

COMPARATIVE
ESTIMATING

SYNTHESIS

DIRECT OBSERVATION
USING

TIME
STUDY

APPLY RELAXATION ALLOWANCE

OBTAIN
STANDARD TIME ADD OTHER JUSTIFIABLE ALLOWANCES

ESTABLISH
TARGET TIME FOR DEFINED METHOD

TO ACHIEVE

A BASIS FOR COMPARISON OF ALTERNATIVE METHODS
CORRECT INITIAL MANNING
CONTINUOUS ECONOMY OF MANPOWER
EFFECTIVE PLANNING OF PRODUCTION
REALISTIC LABOUR COSTING
YARDSTICK FOR GOOD SUPERVISION

HIGHER PRODUCTIVITY

Fig. 4.13 Work study elements

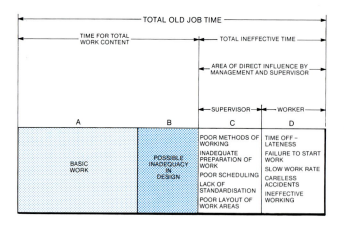

FIG. 4.14 Ineffective time

him, with all the necessary tools and equipment to hand. The basic work content of the operation is the irreducible minimum time theoretically required to complete the task under these conditions.

● Added to this basic work content are any problems which may result from inadequacies in design, construction, methods of manufacture, or other operational aspects outside the person's control.

Together, these first two components constitute the total work content of the job. The person carrying out the work cannot reduce the time necessary to complete this work requirement. The ineffective time is then added to the work content. This basically consists of:

● The *ineffective time* directly influenced by the *supervisor*.

● The *ineffective time* that lies within the control of the *worker*.

The time for the total work content, along with the total ineffective time, gives a total time for the job or operation, under existing conditions. This is generally referred to as the 'total old job time' prior to carrying out the method study.

5.1.2 Method study basic procedure

The simple flexible framework for the application of method study techniques in all circumstances is as follows:

● *Select* the work to be studied, including the limits or terms of reference for the study.

● *Record* all the relevant facts.

● *Examine* these facts critically and in sequence.

● *Develop* the most practical and effective method.

● *Install* the method.

● *Maintain* the method.

Select the work to be studied

Whilst cost savings or improvements in safety, or quality standards may be the reason for selecting the job to be studied, a wider examination of the organisation may indicate other areas of benefit, such as:

● Excessive movement of men or material.

● Bottlenecks in the system.

● Wastage of material.

● Poor equipment utilisation.

● Excessive handling.

● Poor labour utilisation.

● Areas of highly fatiguing work.

Record the relevant facts

The nature of the job to be studied determines the recording technique to be used. The most useful forms of charting for applications within the CEGB are outline and flow process charts, and multiple activity charts; where appropriate, flow diagrams, string diagrams or three-dimensional models may be used.

Outline process charts give an overall view of the process. This is the simplest form of graphical representation of the process and indicates the points at which materials are introduced and the sequence of all operations and inspections.

The flow process chart is a graphical presentation of all operations, inspections, transportations, delays and storage that occur during a given process. It illustrates the process pictorially in terms of the events as they occur in respect of either the man or the material, but not both.

Multiple activity charts provide for the charting of groups of people, or people and machines, or groups of machines plotted against a common timescale to show their interrelationship. They are used to spotlight ineffective or idle time, or poor utilisation of machines.

Flow or route diagrams are used where there is a relatively simple sequence of movements. The flow diagram can be produced by a line drawn on a plan or drawing of the area and may be in two or three dimensions. It has the disadvantage, however, that it cannot be used to indicate as many movements as a string diagram, but may suffice in many instances.

String diagrams and models can be used whenever it is desired to show the paths of operators, materials or equipment during a specified sequence of events.

After a study of the movements of persons or materials has been made and recorded, it may be illustrated on a scale drawing or three dimensional model by the use of a length of string or thread. The string diagram will then illustrate where movements are congested and where the most frequently used routes should be made as short as possible.

In addition to these charts and diagrams, several other techniques are available for special applications. These are generally used in industries which utilise production lines, where the tasks involved are more of a repetitive nature.

Examine the facts

This is the most important and critical phase of method study. The purpose of critically examining all the recorded facts, is to determine the true reasons underlying each event and to draw up a systematic list of all possible improvements. From this analysis, improved methods of handling or working may be developed. It may, of course, prove that the existing method is the most effective, as people usually try to find the easiest way of carrying out tasks, within their own sphere of influence.

When examining existing methods, certain questions need to be asked to provide background information and to establish whether the existing method is based upon sound reasoning:

● Why must it be done at all?

● How is it done — is there a better way?

● Who carries out the task — is it done by the right person?

● When is it done — is it done at the best time?

● Where is it done — would a change in place reduce movement, or permit an improved method?

The primary questions why?, how?, who?, when?, where?, are normally followed by secondary questions, such as 'Is there a better way?' which may then suggest, or lead to improved methods of carrying out the work. However, care should be taken to avoid jumping to conclusions before completing the examination of the existing method, as this may adversely influence its evaluation.

Develop the best method

The most practical, economic and effective method should be developed having due regard to all contingent circumstances; mechanical aids, equipment design, jigs and fixtures, plant layout, materials handling, planning, supervision and working conditions should all be considered (Fig 4.15).

Where a new working method is proposed, a revised process flow chart should be constructed and

critically examined to ensure sound logic and that no essential activities have been overlooked. It is particularly important to involve personnel who may be affected by the change, as this ensures that they have the opportunity to contribute to and participate in the development of any proposed new working method.

Install the method

Once the best working method has been determined, it is important that the new procedure is properly documented. This ensures that everyone involved is made fully aware of the changes and of what contribution they are expected to make.

The implementation of changes in working methods in manufacturing industries, especially those which involve production lines, require extensive planning, preparation, negotiation, procurement of new equipment and staff retraining. Within the sphere of maintenance activities at CEGB power stations, changes in working methods usually involve only a small number of staff. However, where individuals or small groups of staff are involved in changes in working methods, or procedures, it is nevertheless important that they are consulted about, made aware of, understand the reasons for, and participate in, the implementation of new working practices and methods.

Maintain the method

There is little benefit to be gained in putting effort into identifying improved working methods and implementing them if, after a period of time, practices lapse or revert back to their previous state.

After the new method has been in operation for an adequate, but not-too-lengthy period of time, it should be reviewed to ensure that it is working as intended. If unforeseen problems have been identified and variations have resulted, either the problem should be resolved, or the variation accepted and formally included in the new working method. The degree of control over variations to specified or authorised working methods and procedures will depend on the task or type of work. Quality procedures should specify the important areas where very strict adherence to specified working methods are necessary (Fig 4.16).

There are many aspects of power plant maintenance that can benefit from studies of the way in which work is carried out. Method studies can be of particular benefit when determining the optimum layout of stores and workshop areas. Substantial benefits are also to be gained by the application of method study principles by maintenance engineers in their everyday thinking and decision making. For instance, when deciding where to place the turbine LP cylinder top-half cover for diaphragm and gland repairs, there is the need to consider the often conflicting aspects of distance from the workshop area, congestion in

FIG. 4.15 Mobile lifting beam
The use of mobile lifting beams can improve the effectiveness and utilisation of the workforce during overhauls, since turbine hall cranes are often fully employed for main turbine-generator dismantling operations. Mobile lifting beams can be used for dismantling turbine valvegear components and for the removal of bearing keeps, pipework and other small fitments.

(see also colour photograph between pp 434 and 435)

the workshop area, congestion around the generating unit and minimising the restrictions that can occur to the free movement and hence use of turbine hall cranes. The optimum location for the overhaul of large motors, pumps, actuators and valves during a major overhaul also needs to be considered. Should they be removed to the workshop, or overhauled *in situ*?

Method studies on selected items can provide these answers and develop easier, more effective methods of working.

5.2 Work measurement

An assessment of the manpower resources necessary to carry out work is required in order to provide basic information to the staff responsible for the planning and control of maintenance work.

The prime reasons for measuring work are therefore to:

● Quantify the manpower resource requirements.

● Enable work to be planned to match the resources available.

● Evaluate alternative methods of doing the work.

● Provide data for manpower performance assessment.

● Provide data for optimising overall maintenance costs.

● Provide the supervisor with an assessment of the expected job duration.

● Provide basic data to assist in determining optimum maintenance policies.

The fact that work is measured imposes a discipline on all involved and helps to ensure that the basic work requirement is clearly defined and adequately specified. This should enable the standards of maintenance to be improved by increased attention to detail and hence the quality of work. Work which is well specified, the optimum working method determined, well measured and well supervised will generally lead to good man-

264

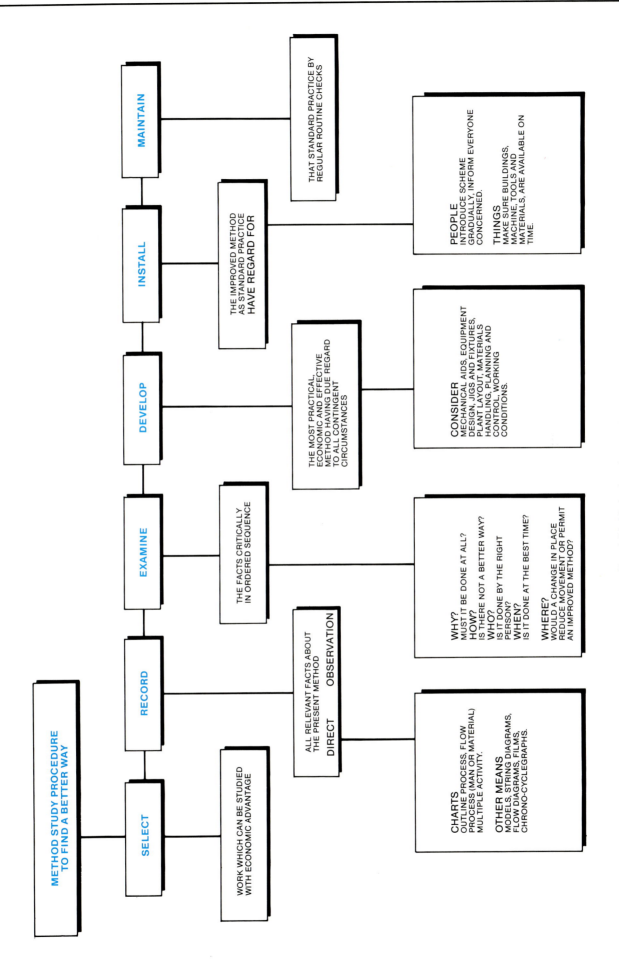

FIG. 4.16 Method study procedures

the activities of individual maintenance sections which are necessary to achieve committed work programmes within the required timescales and the resources available. It also identifies any shortfall of resources, so that various options may be explored and the appropriate corrective action taken.

The work planning and control organisation staff are responsible for the preparation and co-ordination of both long term and short term work programmes. They are also responsible for agreeing with operation and maintenance engineers, the priorities necessary to achieve these plans. To assist this process, a daily meeting is held at most locations to agree detailed plans for the next few days, and to produce outline plans for a period of one or two weeks ahead. In addition any operational problems, conflict of work priorities and lack of progress on important items are discussed, together with a regular review of outstanding work in each plant area. Separate meetings are arranged to discuss specific engineering or operational issues, longer term outages and major overhauls.

One of the most important aspects of a maintenance work planning system, however, is not how detailed the plans are, but more importantly, how effective is the planning process? It is of no benefit at all producing detailed work plans if they are inflexible and unable to accommodate uncertainties in the work programme. The planning system has to be set up in such a manner that it is dynamic and quickly able to cater for changing circumstances. Contingency plans have to be prepared to cater for uncertainties of a significant nature, otherwise the plan will become ineffective and will be discarded as soon as it fails to fit the current situation.

6.2 Computer aided work management systems

The computer system retains a record of all work requirements, whether routine maintenance, defect correction work, overhaul work, new work, or plant modifications (Fig 4.17). This information is held in a central file of data referred to as the 'work bank', which contains the power station's total maintenance work requirements, together with the state of progress of each job. This enables anyone with access to a computer terminal to make a wide variety of up-to-date enquiries, at any time of day, to establish the current state of progress of the work. This would not be possible with a paper-based system.

Computer-based systems are reducing the dependency upon paper as a means of storing and transmitting information, especially where it has to be accessed by a large number of users and has to be regularly updated. Computer systems are capable of providing rapid access to plant technical data, work specification catalogues, standard work instructions, plant spares listings, special tools and equipment listings and other supporting data. However, it is essential that this data is indexed in such a manner that the required information can be rapidly retrieved and a hard copy taken, if required.

The work bank on a large power station will contain a substantial quantity of routine maintenance and defect corrective maintenance jobs. It is therefore important that users of the system can obtain the information they require quickly by means of a number of simple input statements. This is achieved by narrowing the field of search by stating parameters such as:

- Enquiry type.
- Plant code.
- Plant unique identifier.
- Department responsible.
- Work type.

The computer searches for jobs that meet the criteria specified and displays several jobs per screen page. Clearly the more precise the selection parameters, in particular the plant code, the faster is the response time of the system and the smaller the number of records found.

Should more details of any one job be required, if this is selected the computer displays the work control card descriptions which constitute that particular job. If it is then required to examine the details of the work instruction, the plant spares requirements or any other aspect of the work control card, the relevant card can be selected and viewed on the screen. Since the work planning and materials management system are fully integrated, the plant spares and materials required, as identified by their description and CEGB commodity code, are checked and the current stores stockholding for each item specified is displayed on the computer screen.

6.2.1 Computer system structure

The computer system contains a number of different levels of data which relate to the work management process. These are:

Work request level This defines what the problem is or what is to be done. It may be raised either from the routine maintenance work scheduling process, or by the work originator entering a work request.

Standard job level A standard job defines a sequence of tasks identified on individual work control cards, which are required to complete the job in total. The job may involve one or more maintenance sections, or may define a number of smaller or manageable tasks which make up a large or complex job.

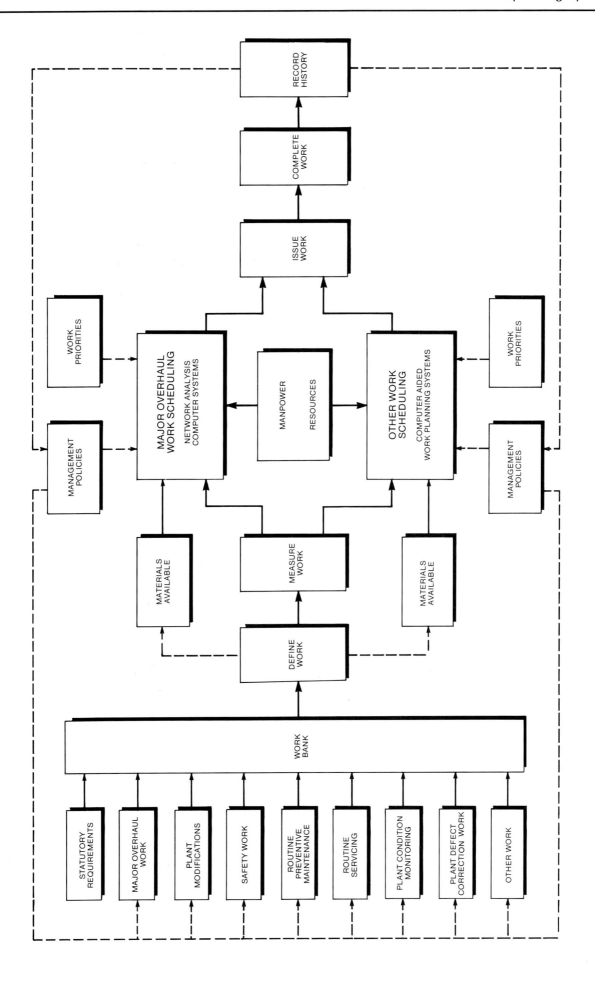

FIG. 4.17 Computer aided work management systems

Work control card level The work control card identifies the task which is required to be carried out on the specified item of plant stated on the card, by a particular maintenance section and defines the method of work. The method of work may:

- Simply refer to a CEGB standard work specification, which may then be attached to the work control card.

- Include a computer held 'standard work instruction', which can be printed directly onto the work control card without any change to the instruction.

- Include a modified version of the 'standard work instruction' held on the computer, to produce a 'temporary work instruction' applicable only to this unique work control card.

- Consist of a one-off work instruction written specifically for this task.

Figure 4.18 illustrates the interrelationships between the various levels of the work management computer system.

Standard work instruction This defines the standard working method by which a particular task is to be carried out on a specific type of plant or equipment. The same standard work instruction may be used on a number of different work control cards without creating more than one computer record. This is achieved by pointing each individual work instruction to the standard work instruction reference number. Although the computer prints the full details of this

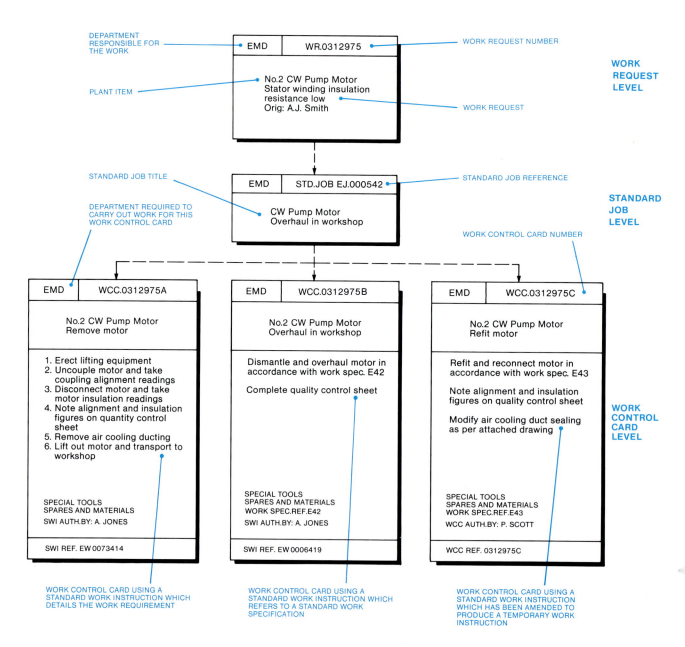

FIG. 4.18 The three main levels of a work control system, their component parts and interrelationship

standard work instruction on the work control card, it only holds the one reference copy. This standard work instruction may be used for the routine maintenance of a number of similar items of plant or equipment. Work requests for plant defect correction work may also be pre-engineered in a similar manner where a standard work instruction can be used to specify the work.

Where a standard work instruction is authorised by the maintenance engineer on a computer system, adequate safeguards must be incorporated, or a management system introduced to prevent unauthorised changes to the data. If changes are made, the computer-held authorisation should be removed and the use of the standard work instruction inhibited until it has been re-authorised by the designated engineer.

Temporary work instruction Standard work instructions may not always cover the work requirements and may, in some instances, require amendments or other additional instructions to be included to cover some specific aspect of the work. The computer system is therefore designed to allow a copy of the standard work instruction to be taken, and for this to be amended as required on the computer screen to create a one-off 'temporary work instruction'. In this case, however, the computer retains a complete copy of this work control card work instruction text, both for printing the work control card, when required, and for record purposes. Since the work content of the work instruction has now changed, it will be necessary for it to be re-authorised by the appropriate maintenance engineer.

Routine work logs Depending upon the nature, importance, frequency and history requirements of some routine work, it may be issued on individual work control cards, or on a 'routine work log'. Similarly, small tasks such as lighting repairs and other workshop or good housekeeping jobs can be controlled on a 'minor task log'. These logs do not necessarily specify the work in great detail, their function being to identify and briefly describe what is to be done, indicating references to standard working instructions, where needed, and to provide a means of controlling the work.

Controlling work in this manner can remove substantial quantities of unnecessary paperwork and will also prevent the computer history record system from being filled with large quantities of low level detail which is not required. This is also of benefit in enabling more important or pertinent information to be obtained quickly and with much less effort.

6.2.2 Work request initiation

Power station maintenance work is initiated by one of the following routes, either by raising a work request,

emergency job card, or maintenance routine for the work to be carried out by location staff, or by raising a scheme and/or contract for the work to be carried out by a contractor. In every case the requirement should be recorded on the work management computer system, otherwise the computer will not have a true comprehensive record of the total work requirement on the power station, or provide a reliable historical record of the work carried out.

Requests for work to be carried out by station resources may be entered directly into the computer system, provided the work requirement has been agreed by an engineer or supervisor. The work request originator is required to enter the unit number, the plant item identification reference and the plant code, or, alternatively, the valve or instrument application reference number (Fig 4.19 (a)). A duplication check is then carried out against the outstanding work bank file in the area defined by these parameters by the person wishing to enter the work request (Fig 4.19 (b)). Routine maintenance jobs are not displayed in this enquiry since they are not normally relevant to the originator of new work.

When the originator has assured himself that the work requirement has not already been identified, he can then call up the work request input screen (4.19 (c)). The originator is required to specify precisely the plant item, the problem, fault or new work requirement, ensuring that the plant code, valve or instrument numbers, where relevant, relate to the plant item description. It is particularly important that no ambiguity or inconsistencies arise at this stage. The description of both the plant item and the requirement should be clear and concise, especially if the information is printed on the craftsman's work control card. One of the advantages of the originator entering the details of his work requirements directly into the computer is that it helps to ensure that his requirements are not open to misinterpretation or errors by another person entering details into the computer. This information is then accessed directly by the person responsible for specifying the work.

Where a drawing or detailed description of the work required is necessary, depending upon local procedures, it is either forwarded to the works office or to the engineer responsible for pre-engineering the work, quoting the unique computer-generated work request number.

Since all new work is entered directly into the computer by the originator, there is no delay in maintenance, operations and planning engineers having access to the information (Fig 4.20). This improved accessibility of new work requirements enables priorities and work programmes to be agreed more easily and improves the co-ordination of effort.

All requests for work, once entered, are retained on the computer system. This ensures that a comprehensive record is always available for reference purposes, irrespective of whether or not the request

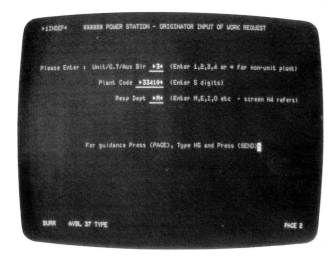

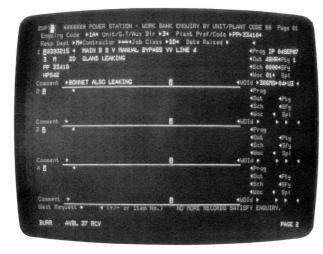

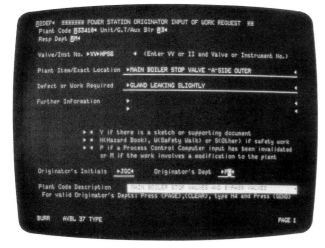

FIG. 4.19 Work request initiation

has been cancelled, either because it is a duplicate or simply that it is no longer required. In this case, once the appropriate comments have been added,

stating why it has been cancelled, and any relevant cross references have been added, the work request is transferred from the outstanding work bank file to the history file.

6.2.3 Routine maintenance scheduling methods

The scheduling of routine maintenance work can be arranged by a number of methods. These methods include:

- Fixed frequency scheduling.

- Last-done date scheduling.

- Meter file scheduling.

Fixed frequency scheduling The time intervals are all calculated from a predetermined start date. It is therefore possible to arrange that routine maintenance work on the same item, but of a different frequency, may be co-ordinated so that the routines are coincidental.

For instance, where a six-monthly routine includes the work content of a monthly routine, the scheduling of the latter item can be inhibited on this occasion. The fixed frequency scheduling method can be used to smooth out variations in the maintenance workload throughout the year, by scheduling certain routines outside the summer overhaul period.

Last-done date scheduling This method operates on the principle that the routine maintenance work is scheduled after a predetermined period of time following the completion of the last routine. This method has particular advantages for scheduling the work associated with the statutory examination of pressure vessels, and for cranes and other classes of lifting equipment.

Careful thought needs to be given to the use of this method for routine maintenance work having a time interval of greater than three months, since this may destroy any attempt to use the scheduling of routine maintenance work for annual workload smoothing. In addition, it can give rise to problems in co-ordinating routines in different departments, unless the work is grouped by means of a co-ordination number triggered by the master or most pertinent job. Alternatively, a job sequence pack can be produced which contains the complete set of maintenance routines to be carried out on that particular item, at that time.

In some instances, corrective maintenance work such as instrumentation recalibrations may be carried out by bringing forward the next routine maintenance. In this case, it may be desirable to change the date the following routine maintenance work is scheduled, so as to prevent wasted effort. Last-done date scheduling takes this into account automatically.

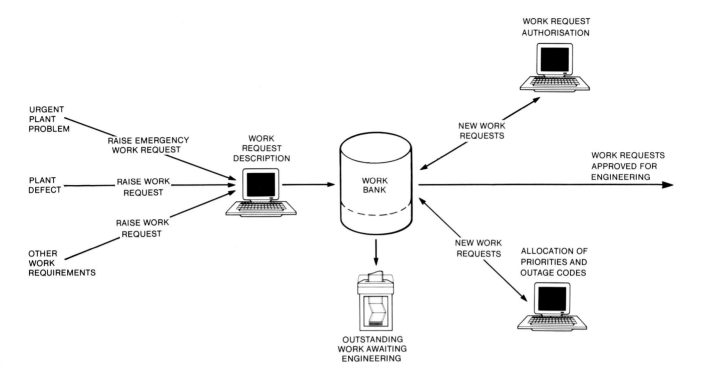

FIG. 4.20 Work request progression

Meter file scheduling This method is appropriate to items of plant which require maintenance work or examinations to be carried out after a predetermined number of running hours, starts, cycles of operation, or throughput; examples being coal milling plant running hours, gas turbine running hours or starts, circuit-breaker operations, water treatment plant throughput.

This method of initiation of routine maintenance work has considerable advantages where the generating plant has a variable running regime. Once a maintenance requirement has been established based upon one or more key parameters, the meter file can then be set to schedule the routine maintenance accordingly, provided the meter file is updated by computer data transfer, or by regular data input.

6.2.4 Pre-engineering and specifying the work

The CEGB has for many years operated a comprehensive safety system, which ensures that all CEGB staff and contractors work in such a manner that they do not endanger themselves or other personnel. More recently, this has been re-enforced by the UK Health and Safety at Work Act 1974. The CEGB safety system requires that specific instructions are given on the work instruction section of each work control card, detailing exactly what is required to be carried out when working on any item of plant or apparatus in the power station.

The existence of an effective quality assurance system enables the quality control mechanisms built into work specifications to be monitored and audited. This ensures that the required standards of quality are in fact being achieved. Where quality control check points are included in work specifications, they may be extracted and stipulated in the work instruction text of the work control card. Alternatively, they may be produced on a separate record sheet attached to the work control card, which may then be filed separately, if desired, for audit purposes (Fig 4.21).

The work control card is the document given to the craftsman identifying and defining the work he is required to carry out. On some locations this document is also referred to as a 'work order card'. It principally consists of a statement of the work requirement taken from the work request or the routine maintenance schedule, together with a work instruction specifying the working method. The work instruction may simply refer to a CEGB standard work specification document which can be attached to the work control card, or may consist of a computer-held standard work instruction, temporary work instruction, or one-off instruction prepared specifically for this task, which is printed directly onto the work control card.

The work control card should include, where appropriate, the following items:

- The main unit and individual plant identification number.

- Plant item description.

273

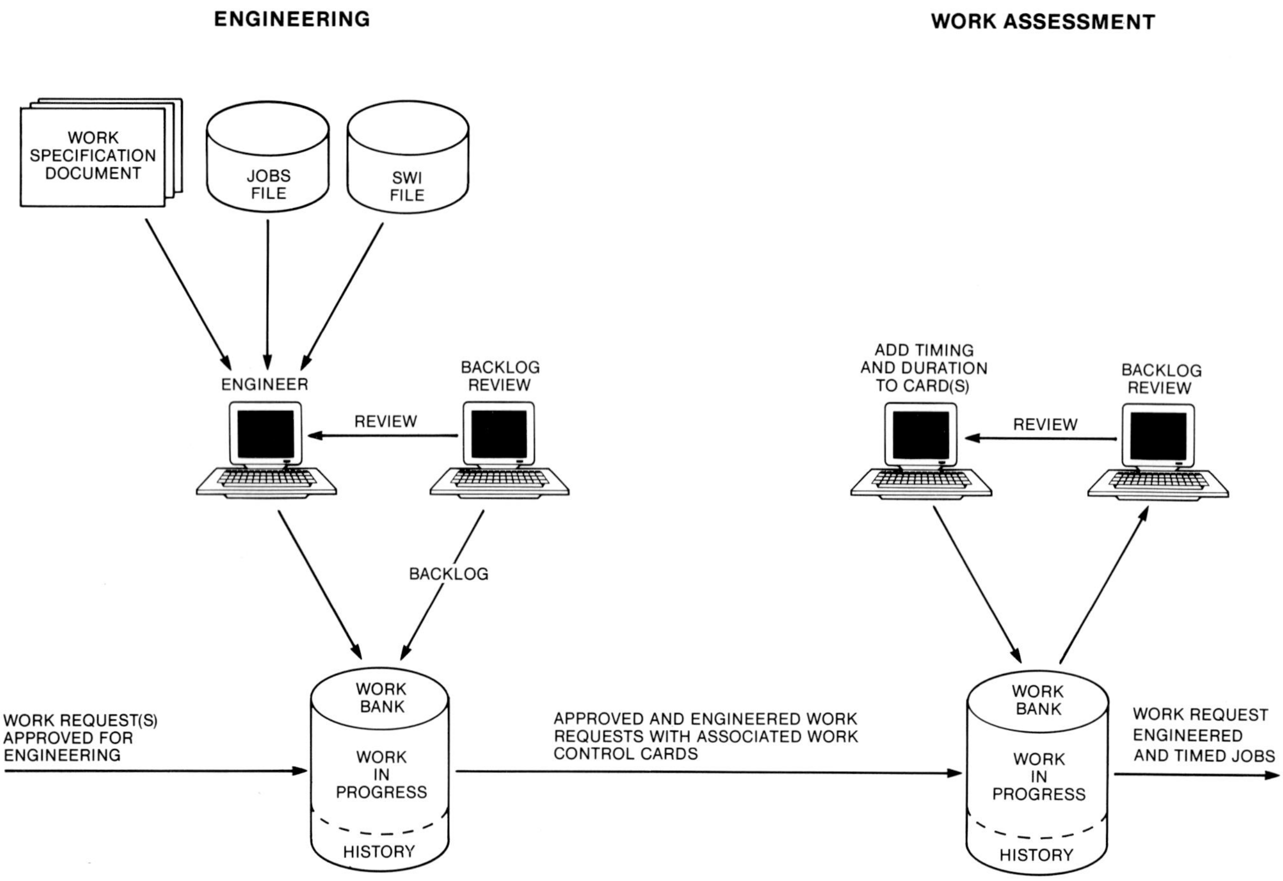

ENGINEERING

WORK ASSESSMENT

FIG. 4.21 Engineering and work assessment

- Location in the power station.

- Details of exactly what work is to be carried out and the limitations of that work.

- Details of any special safety precautions required including protective clothing.

- Work specification reference numbers.

- Drawing reference numbers.

- Cable schedule, setpoint or other data sheet references.

- Details of any special tools or equipment.

- Quality control check points.

- Details of manpower resources required.

- Planned work duration time.

- Details of supporting services, such as cranes or rigging equipment.

- A list of essential spares or materials.

- The unique work control card reference number.

- Standard work instruction reference number, when used.

- Work instruction authorisation signature or code reference.

- Manhours worked.

- Spares used.

- Plant history statement and tick-box entry.

- Quality requirements acceptance signature.

- Work completed satisfactorily acceptance signature.

Each work control card is uniquely identified by its parent work request number, followed by either one or two alpha characters.

Requests for work, once authorised, are examined by maintenance engineers and the work to be carried out is defined. Depending upon the nature and complexity of the work required, it may be pre-engineered or specified by a maintenance engineer, or if the work is of a more regular nature it may be delegated to a maintenance foreman. The objective should be to identify, define and categorise all new work requirements as early as possible to ensure that adequate time is available to properly specify what has to be done, and to order and obtain materials, spares and any special equipment that may be needed before the work can be carried out.

The work instruction should be specified to an adequate level of detail, bearing in mind the complexity (or otherwise) of the work and the craftsman's training and skills. The work instruction should be clear and concise with the limits of the work clearly identified, beyond which further instructions and permit clearance are required. The craftsman also needs to have an indication of the tolerances or quality standards necessary for the particular task in hand, which should be set to a level which is compatible with 'fitness for purpose', not necessarily the highest standard attainable at any expense.

The maintenance engineer is able to use a standard enquiry to determine all new work requiring assessment and authorisation, specification, or pre-engineering. The work request may be 'specified' by carrying out an enquiry of the standard job sequence packs, or individual standard work instructions held on the computer for that particular item of plant or equipment. Provided a standard work instruction is found that adequately specifies the work required, or can be amended to do so by creating a temporary work instruction, it may then be linked on the computer to produce the work control card, or sequence pack of cards. Should no suitable standard job or work instruction be found, a 'one-off work instruction' can be prepared on the computer for this particular task.

The specification of a high proportion of work from previously engineered standard job packs or standard work instructions can save a substantial amount of time and effort. This effort can then be re-directed towards improving standards, identification of plant spares and the identification of quality control check points on other jobs which would benefit from improved pre-engineering.

When plant defect correction work is being specified, it is an advantage to be able to access details of the previous work carried out on that particular plant item. This will identify the recurrence of any similar previous problem and enable the underlying reasons for any failure of the plant, or deficiencies in working methods or procedures to be identified. It can also provide the craftsman with valuable information relating to the expected condition of the plant.

Once the work has been specified, individual work control cards are assessed for work content and timed for work control purposes. A status marker is provided at both the work request and work control card levels to indicate the progress of the work to date. The work request status level is automatically derived from its individual work control card status levels. Once the work is specified the work request status

is therefore automatically set to 'engineered'. The work is then available in the work bank to be scheduled, depending upon its priority relative to other outstanding work, at the most suitable opportunity.

6.2.5 Work scheduling

The work scheduling process should endeavour to satisfy the total work requirements in the best possible way in the light of changing circumstances and priorities, and should enable the total resource effort to be utilised in the most effective manner.

This requires that work is scheduled in the areas where the greatest benefit can be realised, which in turn demands a knowledge of the total work available for issue in the work bank, an assessment of the relative priorities, the state of the plant and any constraints that may affect the issue of the work (Fig 4.22).

Computerised work planning systems now enable the work scheduling process to extend beyond the horizon of just the work that is readily available to hand. The facility makes it easier to plan and co-ordinate that which is required to be carried out when an item of plant is to be taken out of commission.

Computer enquiries based upon the status level of work in the work bank will indicate:

- Work requests awaiting authorisation.

- Work awaiting pre-engineering or specification.

- Work waiting to be scheduled.

- Work awaiting a suitable plant outage.

- Issued work.

- Replanned work.

- Held up work.

- Completed work.

- Cancelled work.

One of the major advantages of computer aided work planning systems is that the work may be grouped and co-ordinated before being scheduled to be carried out. Therefore, once authorised and the work priority assessed, work scheduling codes are allocated identifying any special requirements, such as shutdown codes for overnight, weekend, minor or major overhauls; or other special markers, such as plant modification, safety, scaffold required, etc.

Work may also be co-ordinated by plant item, working area, work type, or in various other categories to improve the effectiveness of carrying out and supervising the work. This enables outstanding jobs to be co-ordinated into logical groups of equipment which may be maintained at the same time. The

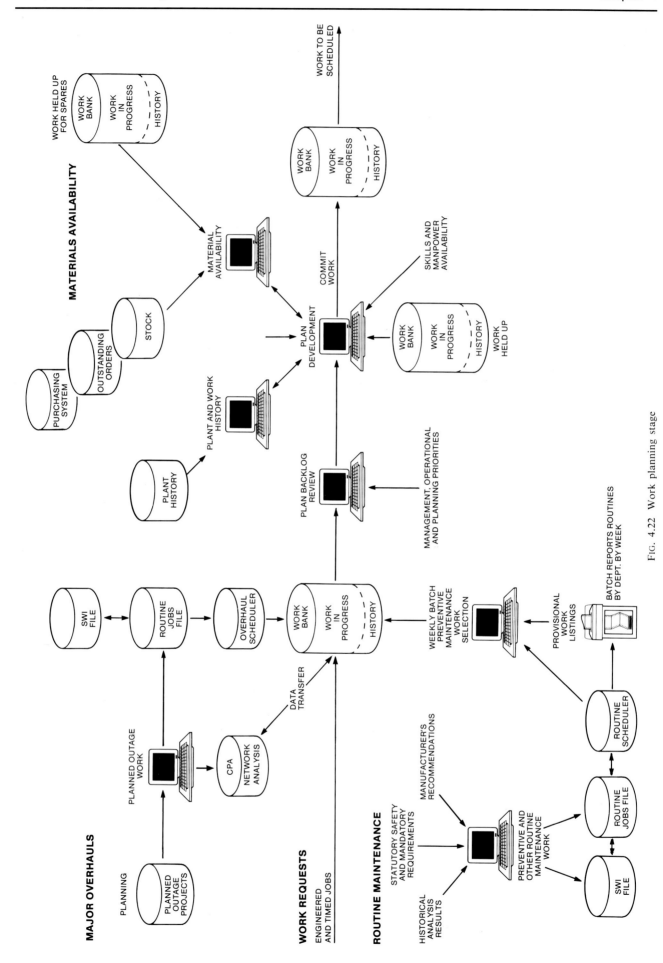

Fig. 4.22 Work planning stage

co-ordination of work between various working groups is a particularly important aspect of work scheduling, since each supervisor will tend to concentrate mainly on his own specific area of work. Systems should therefore be established to ensure that items of plant or equipment that should be maintained at the same time are co-ordinated, so as to schedule all the jobs together. Computerised maintenance work planning systems therefore include a 'co-ordination group' identification field.

The correct scheduling of work requires striking the right balance between a number of often conflicting requirements. These may include plant operational requirements, manpower resource limitations, requirement for specialist skills and the availability of spares or materials, cranes, scaffolding, rigging or other supporting services.

Work scheduling involves the matching of the work programme and the resources available to do the work and to ensure that work is planned to avoid any unnecessary interference (Fig 4.23). For the work scheduling and control process to be fully effective and to ensure that the work flows smoothly, it is necessary to know whether individual jobs are progressing according to plan, or are running ahead of or behind the expected times. The person scheduling the work must therefore be aware of the priority of new jobs, yet conscious of the need to keep the work in progress flowing smoothly. This entails, wherever possible, scheduling resources to complete work that has been commenced, rather than to divert resources to new jobs having only a slightly higher priority.

Once the work requirement has been identified and specified, before it can be scheduled it is necessary to identify all the resources required. The primary resource is manpower; however, the availability of spares, materials, special tools or equipment are all equally important, if the planned work programme is to be successfully completed with the minimum of delays. Work specifications or work instructions often contain parts lists, rather than a list of essential spares which are needed in order to carry out the job. There are some advantages to be gained by 'bonding' the latter items, before the work is scheduled and issued to the craftsman who is to carry out the work.

Where the essential spares requirements have been defined on each work control card, a detachable section or separately printed stores requisition slip may be produced, upon which those essential spares and/or materials required to carry out the work are pre-printed. These items may then be drawn from the stores upon presentation of the correct work control card at the stores counter. This process can enable the spares or materials to be collected together during the less busy periods in the stores, thereby reducing delays at the commencement of each shift or working day.

When a committed work programme has been determined, a work register is produced which indicates all the work to be carried out by that particular work team over a specific period of time. The work register may be either a handwritten or typed paper document, a computer printout, or a computer display of scheduled jobs which are to be progressed in accordance with the scheduled priority.

The committed work programme is the final programme or action plan, giving the best allocation of resources for the particular circumstances at the time the plan was prepared. Work should normally be carried out in accordance with this programme; however, instances will arise where scheduled jobs cannot proceed due to unforeseen factors or circumstances that may arise during the course of the work. However, the level of authority and conditions required to deviate from the committed work programme must be defined and controls applied to prevent abuse. It is essential that 'critical jobs' in the work programme are clearly identified on the work register, to ensure that they are not delayed.

6.2.6 Issuing the work

The craftsman who is going to perform the work needs to be properly instructed on exactly what is required. The issue of the correct documentation and the provision of the correct equipment, spares and materials, in addition to good supervision, is essential if the work is to be completed to a high standard of achievement.

When issuing work, the supervisor should endeavour to build up an experienced working team, able to cope with a wide variety of work. The value and strength of having such a team can easily be recognised, since once built, such a team can handle the most difficult tasks provided adequate job planning and preparation has been carried out. Ineffective time spent waiting at the stores, waiting for new jobs or for technical assistance are all wasteful of resources, and also demotivating to the craftsman.

The feedback and elimination of problems that have arisen during the course of the work which may have resulted from inadequate specification, non-identification of essential spares and materials, lack of job co-ordination, or any other aspect of the work is particularly important. In this way the effectiveness of the maintenance planning organisation and the motivation of the work force will be maintained at the highest possible level.

6.2.7 Completion of the work

The supervisor is responsible for satisfying himself that the work under his control has been satisfactorily completed and that all quality control checks have been carried out and the relevant documentation completed. He is responsible for signing-off the job card either as complete or, if the work is being suspended, indicating the reasons for suspension. Where

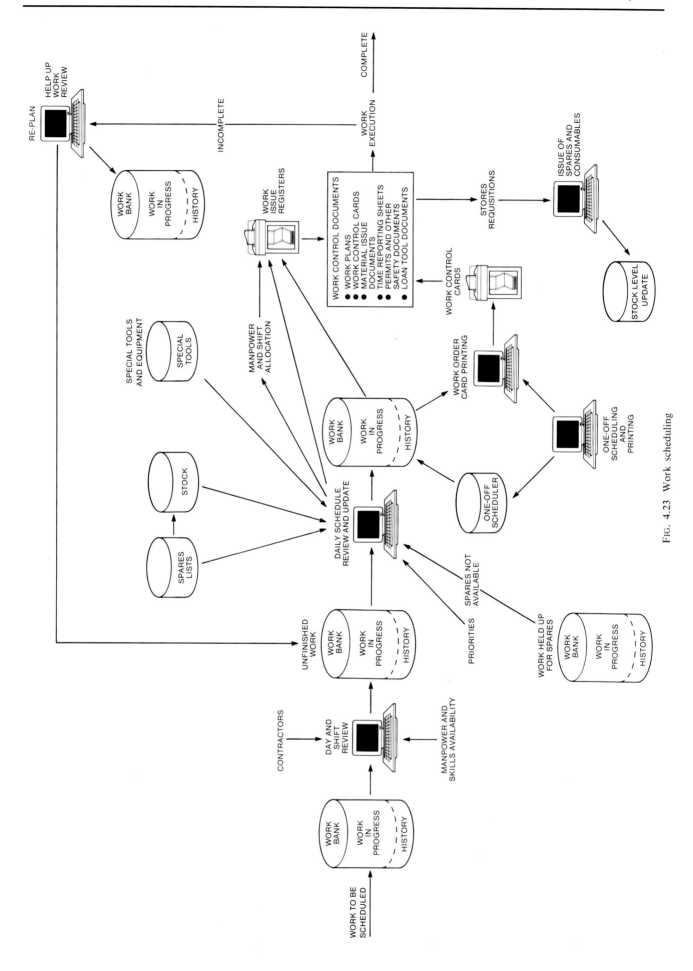

FIG. 4.23 Work scheduling

additional work is required, or only a temporary repair has been effected, a further work request should be raised for the additional work, making reference to the original job.

It is important that the work order card status is entered into the computer as soon as any change occurs, so that the computer can provide a reliable indication of the state of work in progress, held up, or completed.

6.2.8 Plant and work history recording

A history record of all work undertaken by both station and contractor resources should be retained. This is required for finance and manpower resource analysis and also to provide data which may be required to form the basis of future work programmes.

Management policies need to be established and decisions taken in respect of the way in which historical information is to be recorded and retained for future reference. This topic is covered in some detail in the 'Management of records' section of this chapter. Computer-based history systems should be used primarily for plant reliability or fault analysis, finance and manpower resource analysis, and as an indexing system for microfilmed or original work control card documents. Short text statements, such as 'overhauled pump unit fitted', may be included to provide a computer screen display history comment of the actual work carried out. However, if further information is required, reference should be made to a paper copy of the microfilm record, or the original work control card, if not microfilmed.

The plant history record should include coded details of the nature, cause and component that failed, thereby enabling a plant reliability analysis to be carried out on selected components.

The plant history system should be capable of identifying the frequent recurrence of a specific plant defect. This should then be examined in detail to ensure that the root cause of the problem is established. The failure may be found to have been caused by incorrect operation, inadequate or ineffective maintenance, or to a design or manufacturing fault. This may then require a change in operating procedures, maintenance practices or quality control methods, or a reassessment of the frequency and method of initiating routine maintenance.

If historical data is needed to determine policies in the future, it is first necessary to capture the data and then to have an efficient method of retrieving and analysing it. The maintenance of an adequate quality of information requires a discipline by craftsmen, supervisors and engineers to consistently record the required information in the first instance.

6.2.9 Work management analysis

The work management system is required to provide specific information relating to the progress of individual jobs and to provide global statistical data relating to progress, work backlogs and to changing patterns or trends. A monthly analysis of completed and outstanding work is therefore necessary to indicate trends and to provide an evaluation of the effectiveness of the utilisation of manpower. This analysis is carried out by computer and the results displayed graphically for circulation to managers and senior engineers each month. Typically, this comprises:

(a) Emergency injected work, manhours worked.

(b) Plant defect correction work, manhours worked.

(c) Routine preventive maintenance, manhours worked.

(d) Other miscellaneous routines, manhours worked.

(e) Major overhaul routine work, manhours worked.

(f) Other work categories, manhours worked.

(g) Routine maintenance work not accomplished, planned manhours.

(h) Priority 1 routine maintenance jobs not accomplished.

(i) Work request backlog, awaiting specification.

(j) Work request backlog, total.

(k) Average work request duration, annual running average, manhours.

(l) Estimated work bank backlog (j) $\times$ (k), manhours.

(m) Work bank age profile.

(n) Manpower cost analysis by plant area, as requested.

The above statistical analysis can be portrayed graphically to indicate trends, which are probably of more value than spot figures. This can be reproduced for the station in total, or for individual maintenance departments, as required (Fig 4.24). Alternatively, if predetermined default values are decided, it can be arranged that reports are only generated if the parameters are outside the required values.

There is a continuing requirement for location management to monitor achievement against the manpower performance targets stated in the power stations Management Unit Plan. A standard set of manpower utilisation and management indices are used to monitor the management and control of work in power stations. These indices provide a measure of the standard of pre-engineering, work assessment, work planning, supervisory control and manpower performance. Computerised systems enable diagnostic reports to be produced, which indicate areas where attention should be given to identifying the underlying causes of low performance work. This may be commenced simply by investigating jobs which have taken over,

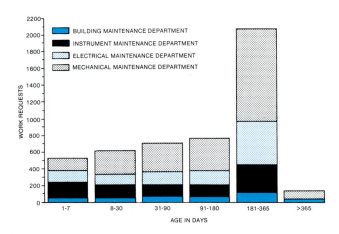

FIG. 4.24 Work bank age profile

say, 6 hours to complete and are in excess of twice the predicted time for the work. The selected jobs should then be analysed to determine:

● How frequently this job over-runs.

● Whether this was a one-off problem job, or a regularly recurring one. If the latter, urgent action is required.

● Whether the initial problem was diagnosed.

● Whether the work was adequately pre-engineered or specified.

● Whether the required spares, materials, tools and supporting services were available.

● Whether the necessary information was available to the craftsman.

● Whether the work was adequately supervised.

● Whether the working method could be improved upon in the future.

It may not be necessary to follow up all of these aspects for each job identified; in fact, time would probably not permit this. However, jobs that have gone badly astray should be analysed; particularly so, if the problems are not confined to just that particular task.

7 Major planned outages

The successful execution of generating plant major planned outages needs a substantial amount of forward planning and detailed preparation.

In the first instance it is necessary to identify clearly the main objectives and reasons for the outage and only then to take into consideration any other work requirements.

When the work programme has been decided and the critical activities identified, the minimum possible outage duration can be determined.

The total resource requirement can also be calculated and, depending upon the rate of increase of the cost of additional resources and the plant out-of-merit generation replacement cost, the optimum outage duration can be determined.

Network analysis methods will assist in identifying critical areas of work, together with an analysis of resource requirements, and can indicate how the work can be more effectively controlled and progressed throughout the duration of the outage.

It is important that effort is directed towards maximising the effectiveness of the planning, work control and monitoring systems. If the plan contains a higher level of detail than is really necessary, it is possible that the project will be frozen at the start, become unmanageable, or even discarded completely. What is more important is to consider how the plan will cater with uncertainties in the overhaul programme. Where these arise, for example, when the work to be carried out cannot be fully ascertained before the plant is opened up, the most likely options should be considered beforehand and alternative plans prepared.

Maintenance planning systems above all must be dynamic and flexible. They should be able to cater quickly with changing circumstances and therefore must rely upon a good communication system, providing accurate and up-to-date feedback of information to some central authority.

However well major plant outages are planned, it is essential that they are properly managed, coordinated, monitored and controlled. For this reason, it is usual to nominate an outage manager to carry out this task.

7.1 Major overhaul work content

The maintenance of power station plant in the CEGB is carried out in order to achieve high standards of safety, availability and efficiency.

Apart from statutory, mandatory and other safety related requirements, maintenance expenditure must be optimised to ensure that the total costs incurred are less than the expected benefit. The costs include out-of-merit generation replacement cost, contracts, materials and manpower resource costs. The benefits to be gained include:

● Improved availability of high merit, low cost generating plant and hence reduced system operating costs.

- Longer plant life and hence a reduced capital requirement for new replacement plant.

- Improved efficiency with consequent reduction in fuel consumption and hence generating costs.

These factors require to be fully considered and an optimum cost solution determined before work is committed to the overhaul programme. This is particularly important for high merit plant, where generation replacement costs are so high that any work requiring additional outage time, beyond that which is absolutely necessary, requires a very high level of cost justification.

High merit generating plant is normally required to run at full output between major overhauls. All 'shutdown' work is then carried out within the duration of the critical work, utilising additional resources, as necessary.

Factors which influence the work content of generating plant major overhauls in the CEGB are:

- Health and Safety at Work Act.

- UK Factories Acts statutory requirements.

- Nuclear Installations Inspectorate requirements.

- CEGB plant inspection requirements.

- Plant manufacturers' recommendations.

- Plant modifications.

- Assessment of remaining plant life.

- Remaining creep life of high temperature components.

- Assessment of operating hours before next overhaul.

- Assessment of the number of hot and cold cycles of operation, if limited, before next overhaul.

- Rectification of major defects.

- Restoration of thermal efficiency.

- Refurbishment programmes.

- Outage duration.

- Finance available.

- Manpower resources available.

- Availability of spares and materials.

- Manufacturers', or specialist repairers', 'in works' capability.

- Identification of future work requirements.

A particularly important aspect affecting the total work content is the degree of plant monitoring, inspection and testing that is required to be carried out.

Quality assurance procedures are required to ensure that systems are set up to monitor plant conditions regularly. Monitoring is required to ensure that plant and its components are not being subjected to conditions of a steady state or transient nature, of a severity or frequency greater than it was designed to withstand. The number and severity of these excursions should be recorded, reviewed regularly and examinations of critical areas carried out during suitable overhaul periods during the life of the component.

This process is particularly important where changes in operational requirements, such as frequent start-ups and shutdowns, or changes in operating techniques are made.

Typical examples of items of plant requiring this type of close monitoring are:

- Boiler drums and headers.

- Superheaters and reheaters.

- Boiler stop valves.

- Steam pipework welds and bends.

- Steam receivers and strainers.

- Turbine steam chests and valvegear.

- Turbine rotor shafts, discs and blading.

- Generator rotors.

- De-aerator and feedheating system vessels.

These inspections may involve ultrasonic, radiographic, magnetic or dye penetration testing methods. It may also include the removal of small sections of the component material, at representative points and at suitable periods of time during the life of the item, for detailed metallurgical examination.

Superheater headers can be fitted with 'creep pips' which are small projections attached to the outside of the header. This enables dimensional checks to be made during each major overhaul to provide an assessment of remaining creep life, or to confirm predictions made purely from temperature records.

Vibration vector analysis will provide data relating to the stability of turbine-generator rotors, shafts and blading, which may necessitate immediate shutdown of the unit, or internal examination or realignment during the major overhaul.

The purpose of these techniques is to check predictions made of the expected life of the component and to warn of changes which may require immediate, or longer term, actions to be taken.

It is important that the plant monitoring and inspection procedures are adequately documented and that quality assurance procedures are set up to audit their correct functioning. It is also essential that the appropriate functional specialists responsible for giving guidance in that particular area are formally notified of any relevant information.

When the major work areas have been defined, consideration should be given to any areas of work where specialist skills are required, or where work is to be carried out off-site at repair specialists, plant manufacturers, or at CEGB central workshops.

It is necessary to categorise the work into that which is essential and must be carried out on this particular outage and that which could be carried out at some other opportunity. The latter category should only be included if adequate resources are available and provided that this additional work is not allowed to extend the outage duration.

7.2 Overhaul planning methods

The management and control of the overhaul of a large generating unit is a complex operation having many interrelated activities, involving several groups of people. This requires the planning and preparation of the work to be carried out to an adequate level of detail, otherwise important aspects of the work are likely to be overlooked, with adverse consequences once the overhaul has commenced.

There are several ways in which this may be achieved, ranging from simple work lists, hand-drawn bar charts or magnetic planning boards, to a fully computerised system, including all activities and presenting an analysis of the duration of the work and the total resource requirement. The extent to which computer assistance is used depends upon the size of the project, the number of interrelated activities, the resources available to plan and co-ordinate the work, whether or not the plan will be used on future occasions, and the overall cost savings that can be achieved. The costs incurred by even a small delay in completing the overhaul of a large high-merit generating unit can be substantial. It is therefore necessary to ensure that adequate time and effort is spent in preparing the overhaul plan in adequate detail.

Whether or not the overhaul plan is prepared in the form of manually produced bar charts, or of computer-generated bar charts, the first step should always be to produce a logic diagram or 'activity network'. This helps to ensure that important steps are not overlooked and identifies the activities which are interdependent. The duration and, where applicable, the manpower resources necessary to carry out the work are required for each activity in the network.

Once the network has been produced, a 'time analysis' can be carried out. In this analysis, the minimum time required to carry out the plan is calculated, assuming no resource limitations. It identifies the specific sequence of jobs, or activities, which deter-

mine the minimum time required to complete the project. This sequence of activities is called the *critical path*. In addition to identifying the critical path, the time analysis also calculates the earliest and latest start and finish dates for all activities in the network.

Since manpower resources are almost certain to be limited, a *resource analysis* is then carried out. In addition there may be fixed dates when certain events, such as 'rotor in and out of works', have to be achieved, which must also be taken into account. The resource analysis allows alternative methods of resourcing to be examined and also determines the effect of any imposed changes in calendar dates. The method of calculation of 'time' and 'resource' analysis are covered in more detail later in this chapter.

For a large overhaul programme, with many activities, this process would become a complex and very time consuming task if carried out manually. For a number of years, computer programs have been used extensively by the CEGB for the computer analysis of activity networks.

The programs have mostly created activity networks in arrow format. More recently, 'precedence' networks have been used for the creation and display of overhaul activities. Later computer programs allow networks to be created in either the arrow or precedence form (Section 7.3 of this chapter).

The three main variables in an overhaul programme are the work content, time available and manpower resources available. Where a computer is used, the program allows the user to construct networks and observe the effect of change in variables such as manpower, or outage duration.

Networks having more than a few hundred activities are normally run on centrally-located mainframe computers, or at power station sites on their own distributed computers, where they are capable. Alternatively, updating and output facilities may be obtained at power station sites using either network connections to the mainframe computer, or by the transfer of data by tape or disc.

There are a number of network analysis programs now available which will run on small standalone microcomputer systems; such programs can provide a very useful aid to the maintenance work planning process where mainframe computer facilities are not available.

Some locations within the CEGB are able to run network analysis programs on their own site computer. This has the advantage that validation and time-analysis runs can be carried out on site at periods of low system utilisation. Printed outputs can then be obtained directly, eliminating the transportation delays from a centrally-located mainframe computer.

7.3 Network construction

It is particularly important that network construction is approached in a logical and accurate manner, since this provides the basis from which any subsequent analysis is derived. The network must express the logical sequence of work, or other relevant activities, as well as any interdependencies and restraints inherent in the plan.

For large projects, prior to producing a fully detailed network, it is desirable to produce an outline network indicating all the major activities. This will then form the basis for the master network for the project. Each activity on the master network may represent a single activity, such as 'steam cool turbine', or may represent a sub-project such as 'turbine valve-gear overhaul', containing many individual activities. The master network should include all important activities and sub-projects to present a complete picture of the overhaul and to indicate the logical relationships between these major activities.

Since networks may be drawn for different levels of detail, it is important to establish the degree of detail required for each particular application. The level of detail and sophistication should take into account the benefit to be gained and the resources required to produce a very detailed network and to input that data into the computer.

Accurate network construction can often best be achieved as a team effort by personnel with specific knowledge of the work requirement and the intended method of carrying it out. This also has the advantage that the team has a common interpretation of the objectives and the possible problem areas in the work programme.

The network expresses the logical sequence of activities, the interrelationships and dependency of one activity upon its preceding activities.

There are two basic methods of representing the activities in a project. They are:

● Arrow, or 'activity on arrow' networks.

● Precedence, or 'activity on node' networks.

In the arrow diagram, the activity requiring time, or time and resources, is represented by an arrow. In the precedence network, the activity is represented by a box and the arrows are used to indicate the preceding activities upon which it is dependent. Hence the precedence network does not normally require the use of 'dummy' arrows to maintain the network logic. It also incorporates the ability to have dependencies other than 'start B when A is complete' and more easily allows changes in the network logic. Arrow networks, however, were the first to be developed and widely used within the CEGB.

7.3.1 Arrow network construction

The arrow network (Fig 4.25) consists of the following elements:

● Activities.

● Events.

● Dummy activities.

An *activity* is an operation or process consuming time, and possibly resources. The activity is represented by an arrow, its length and shape have no significance for normal activity networks, although it is possible to produce arrow networks which can be time-scaled. The diagram should always be drawn from left to right, the preceding event on the left and the arrow head pointing to the succeeding event on the right.

(a) An activity represented by an arrow is always drawn between two nodes, the preceding and succeeding events

(b) Activity A must be completed before activity B can commence.

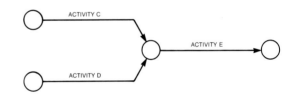

(c) Activities C and D can be carried out concurrently, but activity E cannot commence until both activities C and D are completed.

FIG. 4.25 Network series-parallel activities

An *event* is a defined state in the progress of a project after the completion of all preceding activities, but before the start of any succeeding activity. This point in time, or node in the network, is represented by a circle.

A *dummy activity* is a link in the network which is necessary to maintain the network logic. Sometimes it is necessary to indicate that a particular event cannot be achieved before another, although no specific operation occurs between the two. In this case, a dummy arrow is inserted in the network to ensure correct logic (Fig 4.26). The use of a dummy activity here prevents the easily included logic error which would be committed by drawing this network incorrectly.

There are two other forms of error that may arise when constructing a network, especially if it is a

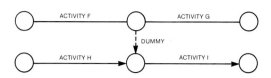

(a) The use of the dummy activity is illustrated, where activity G must follow F, but activity I must follow completion of activities H and F.

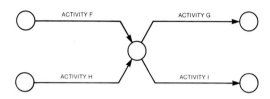

(b) In this network, activities G and I are both dependent upon the completion of activities F and H.

FIG. 4.26 Dummy activities

complicated one. They are known as 'looping' and 'dangling' (Fig 4.27).

All activities whose start is not constrained by another activity should lead out of the first event in the network, the 'project begin event'. Similarly, all activities with no following activities preferably should lead into the last event in the network, the 'project end event'.

When a network has been drawn, all the events are numbered. The numbering may be sequential or random, but no two events can have the same number. *Activities are identified within the network by their preceding and succeeding event numbers.* When

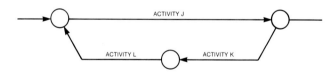

(a) Looping is a common error that can arise in the construction of a complex network. It usually arises as a result of incorrect construction, since activities should always proceed from left to right. Looping is obviously illogical as activity J precedes activities K and L, but cannot start until L is completed.

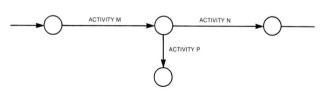

(b) Activity P has no succeeding activity. If this is in fact the case, as distinct to an error in the logic, the activity P should terminate in the end event of the network.

FIG. 4.27 Errors in network construction

284

two or more activities have the same preceding and succeeding events, then they must carry an additional identifier, or tag (Fig 4.28 (a)). This is referred to as the 'uniqueness identifier' which identifies each of the two parallel activities P and Q between events 1 and 2 in the network as 1–2a and 1–2b. An alternative method is to insert a dummy activity (Fig 4.28 (b)), but this can complicate the network if many parallel activities are involved.

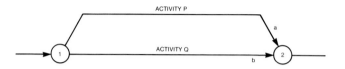

(a) Two parallel activities P and Q sharing the same preceding and succeeding events are identified by uniqueness identifiers a and b. The activities can then be defined within the network as activities 1-2a and 1-2b.

(b) A dummy activity 3-2 may be inserted in one of the parallel activities as an alternative to using the uniqueness identifier. The activities are then defined within the network as activities 1-3 and 1-2.

FIG. 4.28 Uniqueness identifiers

As a general rule, networks should be drawn on the basis of technical feasibility and should not be influenced by resource limitations. This aspect is taken into account when the resource analysis and work scheduling is carried out. Not only is it easier to construct the network ignoring resource constraints, the resulting network provides a much better basis from which to start the resource scheduling process.

7.3.2 Activity durations

Once the network has been constructed, it is necessary to identify the resource requirement and the minimum time required to complete each activity.

When preparing time estimates, it is not sufficient simply to quote the time taken from past records, since these may be distorted by some particular problem, or shortage of manpower or equipment, experienced at that time. It is important to consider each task in the light of current conditions and working methods and not to make generous allowances for contingencies.

The activity durations and resource requirements should be determined in an objective manner and should preferably be based upon method studies or other synthetic data, or derived following a critical examination of previous work programmes.

Where estimates have to be made, a more objective assessment may be achieved by splitting the task into a number of components. The more predictable aspects can then be assessed separately and an estimate made in the areas of greatest uncertainty. This is obtained by weighting the worst and best case activity durations with their probabilities, so as to obtain a more objective result.

Considering a typical problem, where the work involved in the repair of a pump cannot be determined until it is opened up for examination. From historical records, there is a 30% probability that it will require a complete overhaul requiring 28 hours' work. However, there is also a 70% probability that it will only require an internal inspection, bearing change and gland repack requiring 8 hours' work. If there are several pumps of this type to be overhauled, the resulting overhaul programme would be pessimistic if the 28 hour duration were used and optimistic if the 8 hour duration is used. A more objective assessment of the time required to overhaul the pump is: $(28 \times 0.3) + (8 \times 0.7) = 14$ hours each. The advantage of this method is that since the expected activity duration is weighted with its probability estimate, it is likely to be more accurate than by simply taking an average or mean value of the most pessimistic and optimistic times, which here would have been 18 hours.

The CEGB has for some years collated data relating to the maintenance of power station generating plant and a substantial amount of data is available from the CEGB General Maintenance Planning Standards data banks. These provide sub-networks and data sheets for maintenance work on major plant components. The data sheets indicate the master network reference, the network node references, task description, any relevant physical data, governing time or expected duration, the number of people and the standard time or manhours required to complete the task. The number of people required excludes services, such as crane drivers, storekeepers, or supervisors. Figures 4.29 and 4.30 illustrate a typical network and the associated activity work list for the removal of a steam turbine IP rotor for a large generating unit, as extracted from the CEGB General Maintenance Planning Standards data bank.

7.3.3 Precedence networks

The precedence network consists of two elements:

- Activities.

- Dependencies.

In the precedence network, the activity is represented by a box and its dependency is indicated by a line with an arrow head. The activity is identified only by an activity identifier such as A1, A2, B7 or B10 (see Fig 4.31); unlike the arrow network where the activity is identified by its preceding and succeeding events. The activity dependencies in precedence networks allow greater flexibility of network construction than do arrow networks. There are four types of dependency:

- Finish to start.

- Start to start.

- Finish to finish.

- Start to finish.

Precedence networks have the advantage that changes in network logic can be easily carried out since the identification of an activity is independent of its position in the network. When changes are necessary to an arrow network, dummy activities are usually required at interface events to maintain the network logic. In the precedence network, this is not necessary.

The difference between the two networks arises from the difference in their basic concept. Arrow networks are described in terms of nodes or events and therefore an activity takes place between two events. The activity is identified by its preceding and succeeding event numbers in the network. Precedence networks are constructed in terms of activities and dependencies. The activity is defined by its unique activity number, which does not change if its dependencies and hence position in the network are altered.

Where network analysis by computer is used in conjunction with computerised work planning systems, the precedence form has an advantage since it is not necessary for the overhaul jobs on the work planning and control computer system to carry specific network preceding and succeeding event numbers. These are, however, required for the automatic updating of an overhaul control network created in arrow form.

There are appreciable advantages to be gained in automatically updating the network analysis computer from the work planning and control computer, thereby enabling new work schedules and bar charts to be produced with the minimum of effort.

The choice of the arrow or precedence form of network is usually determined by the number of networks already existing. Additional factors to be taken into consideration are, the number and extent of changes normally encountered in the plan, the resources available to implement those changes, possibly at short notice, and the flexibility required of the work planning function.

Some contractors will produce networks for their work, if specified as a requirement of the contract. These networks may then be incorporated in the overall plan, if so desired.

Figure 4.32 illustrates part of a turbine valvegear overhaul network, produced in arrow and precedence form.

7.3.4 Standard library networks

Collecting the data and creating networks for major planned outages requires a substantial amount of time and effort by maintenance and planning staff. Since the overhaul is likely to be repeated on future occasions, often in a similar form, a significant proportion of the effort can be saved if complete sections of the network can be retained for future use.

It is therefore an advantage to break down large or complex networks into a number of smaller sub-networks, or sub-projects. These sub-networks can then be assembled or linked to create a new network for any future overhaul by nominating interface events, or adding interconnecting dummy activities. These networks are referred to as standard library networks and may be unique to any particular location, or may supplement or relate to the CEGB standard networks illustrated in Fig 4.29.

7.4 Time analysis

The purpose of carrying out the 'time analysis' is to determine the shortest possible period of time in which the overhaul programme can be completed and also to identify the critical activities. This can be carried out once the duration of each job or activity in the work programme has been determined.

Two dates or times are calculated for each event in the arrow network, the earliest possible for a given start date and the latest possible to meet the required overhaul completion. The latest date minus the earliest date indicates how much the achievement of an individual event can be delayed, without affecting the overhaul completion. Those events and activities that cannot tolerate any delay are 'critical'. They are often coloured red or distinguished by a double or heavy activity arrow on the network.

The British Standard (BS6046: Part 2: 1981) method of labelling the events in an arrow network (Fig 4.33) is with the event number entered on the left hand side of the label. The earliest event time is written in the top right hand side quadrant and the latest event time is written in the bottom right hand side quadrant. This method allows critical events to be more easily recognised, since the earliest and latest event times are the same.

The terms and concepts related to time analysis are best understood with the aid of an example. Figure 4.34 illustrates a small network in which the earliest and latest event dates have been calculated. Reference to this network will clarify some of the terms used when describing network analysis functions:

Earliest event time The earliest time by which an event can be achieved. This is determined by the longest path leading to the event. The longest path leading to event 6 is via activities 1–3 and 3–6, and has a duration of 15 time units. Since the project start time is 0 the earliest time for event 6 is therefore 15 time units.

Latest event time The latest event time by which an event must be achieved without affecting the project completion. This is determined by the longest path leading from the event. The longest path leading from event 2 is via activities 2–4, 4–6 and 6–7, and has a duration of 15 time units. Since the project finish time, as determined from the longest irreducible sequence of events is 23, the latest time for event 2 is therefore eight time units.

Slack The calculated time span within which an event must occur, without affecting the project completion. This is the difference between the latest and earliest times for the event.

Earliest start time The earliest possible time that an activity can start. This is equal to the earliest event time of its preceding event.

Earliest finish time The earliest possible time that an activity can finish. This is equal to the earliest event time, plus the activity duration.

Latest finish time The latest possible time by which an activity must finish, without affecting the project completion. This is equal to the latest time of its succeeding event.

Latest start time The latest possible time by which an activity must start, without affecting the project completion. This is equal to its latest finish time, minus its duration.

Total float The amount of time by which an activity may be delayed or extended without affecting the total project completion. This is equal to the latest event time of the succeeding event, minus the sum of the earliest event time of the preceding activity and the activity duration. Figure 4.34 shows activities 1–2, 2–4 and 4–6, all of which have a total float of three time units.

Free float The amount of time by which an activity may be delayed or extended without delaying the start of any succeeding activity. This is equal to the earliest event time of its succeeding event, minus the sum of the preceding activities earliest event time and the activity duration. As shown, activities 1–2 and 2–4 have zero free float and activity 4–6 has a free float of three time units.

UNBOX AND REMOVE IP ROTOR

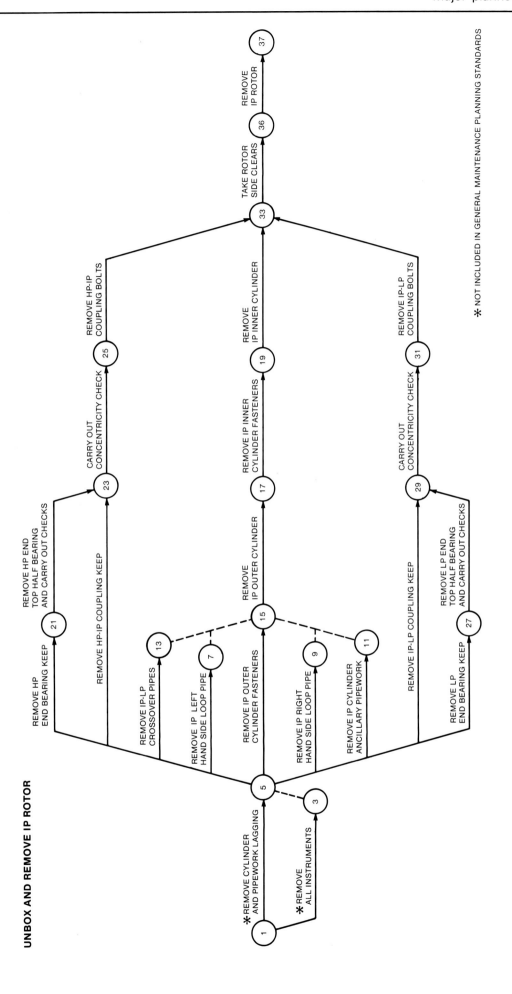

FIG. 4.29 CEGB planning standards network

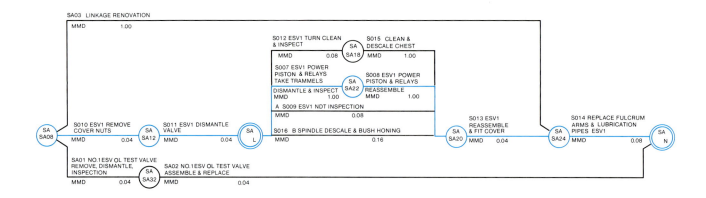

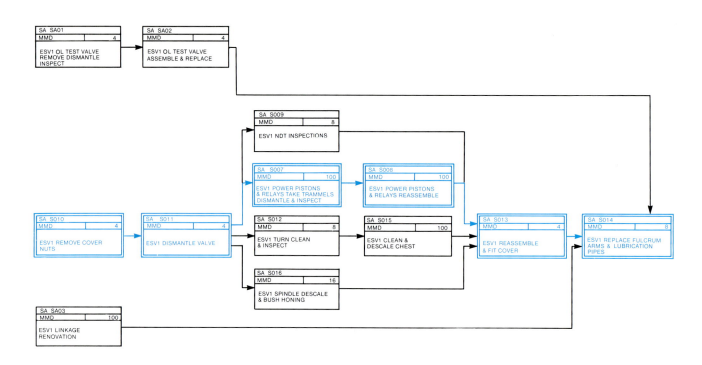

FIG. 4.32 Part of a turbine valvegear overhaul network produced in arrow and precedence form

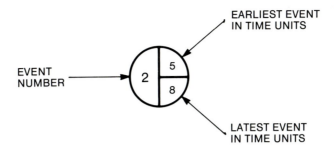

FIG. 4.33 Network event labelling

Critical path The critical path is the path through the network with zero float, or the least float in work programmes where any imposed dates make the duration longer than that of the critical path. For a network in which the critical path has zero float, all events lying on the critical path will have the same earliest and latest event times. This, together with the fact that the critical path is often highlighted in colour, or drawn with double lines, should clearly identify it within the network. In the network illustrated, the activities having the minimum float, in this case zero, are activities 1–3, 3–6, 6–7 and therefore lie on the critical path.

The time analysis results for the network illustrated in Fig 4.34 are produced in Table 4.1. The earliest start and finish times are calculated commencing from the start event. Then by working backwards from the final event, the latest finish and hence latest start times can also be calculated. The total float for each activity and any free float that may be available on certain activities can also be determined. The critical path can then be identified from the activities having zero total float.

The bar chart (Fig 4.35 (a)) illustrates the earliest start schedule from the time analysis. The total float and, where applicable, any free float that may be available, are also indicated.

The resource chart (Fig 4.35 (b)) illustrates the

mechanical fitting staff and welder resources required to meet this earliest start time analysis.

The time analysis therefore identifies:

- The critical path for the network and hence the shortest possible period in which the overhaul programme can be completed.

- The earliest and the latest start and finish times for each activity in the network.

- The total float and free float for each activity.

In addition, if actual dates are used, the time analysis will also indicate whether any imposed scheduled dates in the work programme and the proposed end date can be achieved, irrespective of resource availability.

7.5 Resource analysis

The earliest and latest times calculated in the time analysis are determined by the network logic and the activity durations. No account is taken of any restrictions arising from resource limitations. It is assumed that adequate resources will be available when required, which may not necessarily be economically justified in practice.

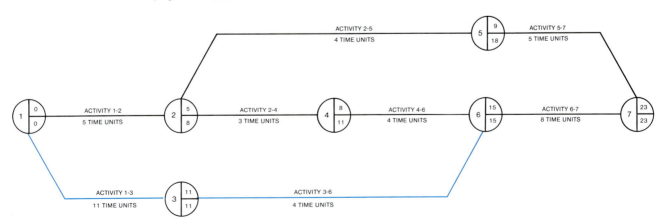

FIG. 4.34 Earliest and latest event times

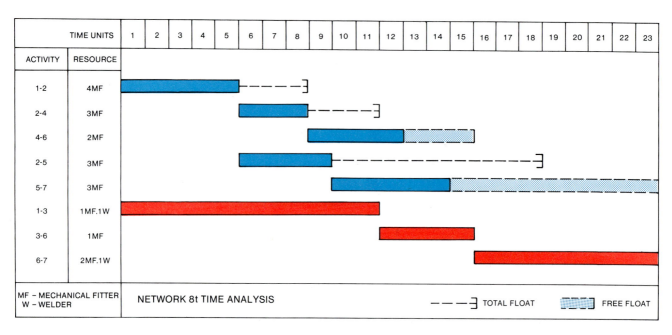

(a) Bar chart from time analysis, all activities commencing at earliest start time

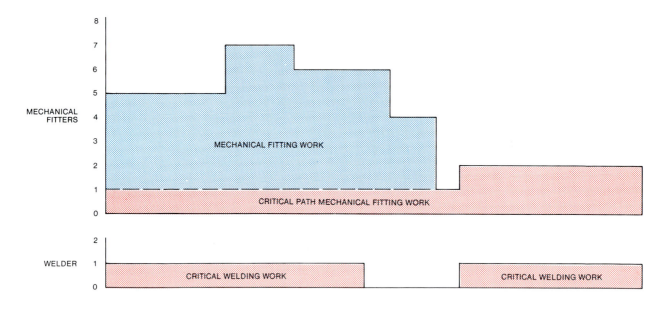

(b) Resource requirement histogram based upon time analysis, hence no resource limitation, or smoothing

FIG. 4.35 Time analysis

The resource analysis endeavours to smooth the work load, by utilising the float available in some activities to limit the peak resource requirements. Therefore, resource analysis used in conjunction with time analysis will produce work schedules that are more feasible than those generated from time analysis alone.

In carrying out the resource analysis, two new *times* and the term *remaining float* are introduced for each activity in the network:

Scheduled start time This is the earliest time when there are enough resources available to work on the activity.

TABLE 4.1
Time analysis

Activity number	Duration	Earliest start	Earliest finish	Latest start	Latest finish	Total float	Free float
1 – 2	5	0	5	3	8	3	0
1 – 3	11	0	11	0	11	0	0
2 – 4	3	5	8	8	11	3	0
2 – 5	4	5	9	14	18	9	0
3 – 6	4	11	15	11	15	0	0
4 – 6	4	8	12	11	15	3	3
5 – 7	5	9	14	18	23	9	9
6 – 7	8	15	23	15	23	0	0

Scheduled finish time This is equal to the scheduled start time, plus the activity duration. It assumes that the activity will be worked upon continuously until the work is completed.

Remaining float This is the float that remains after the resource analysis has rescheduled activities to smooth-out peak work demands. It equals the latest finish time, minus the scheduled finish time.

In the previous example, if there were only five mechanical fitters available to carry out the work, rather than the seven required to meet the time analysis earliest start schedule (Fig 4.35 (b)), then some activities would have to be delayed beyond their earliest start time. In this situation, it is preferable to delay activities which have enough float to enable them to be completed before their latest finish times. In this way, the peak work load demands are avoided and the work programme can still be completed within the critical work duration. Table 4.2 indicates the results of the resource scheduling. Figure 4.36 (a) illustrates the rescheduled bar chart and Fig 4.36 (b) the resource histogram with a limitation of five mechanical fitters.

Although both time and resource analysis can be carried out manually, for large networks it is very laborious and time consuming. Most large or complicated networks are therefore analysed by computer. Another advantage is that the rules for deciding the priority and sequencing of the scheduling of activities in a resource analysis can be varied to suit particular circumstances. Also the effect on the overhaul duration can be determined by varying or limiting the resources available. For instance, in the foregoing example; if the resources available are reduced even further to four mechanical fitters and one welder, the effect is to delay completion of the critical work and hence overall programme by six time units. The ability to vary resource and other network parameters and to obtain an assessment of the impact of those changes on the work programme, enables the outage duration and the resource requirements to be optimised and, hence, keep overall costs to a minimum.

7.5.1 Calculation of scheduled dates

The work programme can be managed on a calendar or numeric date system. In the latter system, the work listings refer to the number of days from the commencement of the project, as distinct from an actual date and time.

TABLE 4.2

Resource analysis
Resource analysis results, indicating the scheduled start and finish times and any remaining float after resource smoothing, for each activity in the network illustrated in Fig 4.34

Activity number	Resources required	Duration	Earliest start	Scheduled start	Scheduled finish	Latest finish	Remaining float
1 – 2	4 MF	5	0	0	5	8	3
1 – 3	1MF 1W	11	0	0	11	11	0
2 – 4	3MF	3	5	5	8	11	3
2 – 5	3MF	4	5	12	16	18	2
3 – 6	1MF	4	11	11	15	15	0
4 – 6	2MF	4	8	8	12	15	3
5 – 7	3MF	5	9	16	21	23	2
6 – 7	2MF 1W	8	15	15	23	23	0

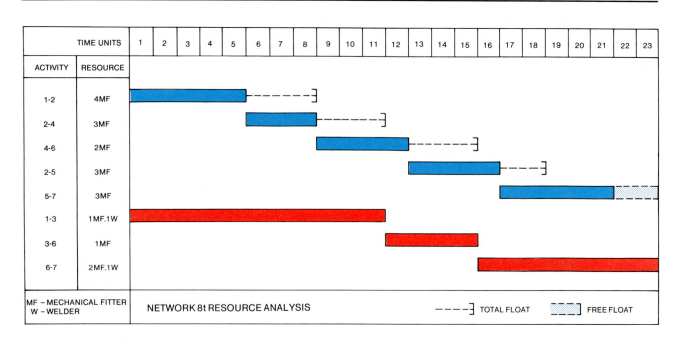

TIME UNITS		1	2	3	4	5	6	7	8	9	10	11	12	13	14	15	16	17	18	19	20	21	22	23
ACTIVITY	RESOURCE																							

MF – MECHANICAL FITTER
W – WELDER

NETWORK 8t RESOURCE ANALYSIS

----] TOTAL FLOAT   FREE FLOAT

(a) Barchart illustrating how activities have been rescheduled from the time analysis bar chart Fig.4.35(a) as a result of resource analysis

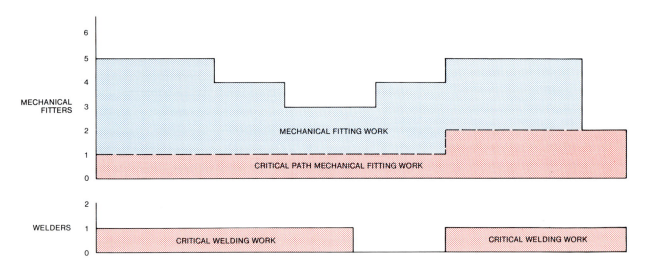

(b) Resource requirement histogram after resource analysis. Note, the work programme is completed within the duration of the critical path work and does not exceed the resource limitation of 5 mechanical fitters.

FIG. 4.36 Resource analysis

The numeric date system has the advantage if the date of commencement of the work is uncertain. The calendar date may be preferred if there are imposed dates in the work programme when certain events have to be achieved, or activities carried out. Computerised systems normally accept projects defined by either method.

Individual activities are usually defined in weeks, days and hours, or any combination thereof. What constitutes a working day and a normal working week must be defined at the time when the network is constructed.

7.5.2 Output options

Computerised network analysis systems provide a wide range of output options. These include plotted networks, activity bar charts, resource histograms, as well as a variety of standard listings in any specified order:

Network plotting With the distributed computing facilities available at large power station sites, work planning data may be transferred directly from the site to a centrally-located network analysis computer, from which printed networks may then be produced on a graphics plotter. These methods are now tending to replace the traditional method of drawing networks manually and hence save the effort involved in completely redrawing them whenever a significant change in logic or working method is made.

Output listings Printed outputs may consist of lists of events, or schedules of activities in various sequences, such as earliest or latest start or finish date, or by total float sequence. Tabular outputs are usually produced for internal use by work planning staff.

Plotted bar charts These can be produced from either the time analysis, or resource analysis data. The charts produced from the resource analysis are more practical, since account is taken of the availability of resources. Bar charts are more useful for supervisory staff who are involved in the issue, control and monitoring of the work and can be produced for selected areas of the work programme, such as boiler, turbine or generator, if so desired. They suffer from the disadvantage that the logic is not always apparent, unless used in conjunction with the corresponding activity network. However, they are generally more acceptable to users, especially as they can be produced relatively easily on a colour graphics plotter.

Histograms Resource tables and histograms enable the effects of resource limitations, or outage duration to be compared. Histograms produced by a 'time-limited' resource analysis indicate whether the resources allocated are adequate to meet the planned completion date. Histograms produced by a 'resource-limited' analysis indicate whether the proposed overhaul duration is adequate, or requires to be extended.

7.6 Network analysis summary

The technique of network analysis is a management tool which can provide a systematic and comprehensive method of defining, planning and co-ordinating work programmes. The choice and method of construction of the network should be made to enable the resulting programme to be flexible and cater for rapid revision. This requirement arises as a result of changes in working methods, unforeseen problems, or changes in key dates which frequently occur after the work programme has commenced. The use of precedence networks has an advantage in this respect, but this can be achieved in the arrow network by

the use of dummy activities. This enables the logic to be changed at points of uncertainty in the network and, if required, allows additional activities to be introduced.

Care should be taken, however, not to create unnecessarily complicated networks, with superfluous constraints or detail which can impede the flexibility and effectiveness of the work programme.

For rapid response to changes in the manner in which the work is to be carried out, or the time or resources available, computer processing of the time and resource analysis is preferable for all but the simplest of networks.

By the creation of links between the work planning and the network analysis computer systems, automatic transfer of data relating to the progress and completion of work can be achieved. This eliminates a large proportion of the time and effort spent in entering data into the network analysis computer.

Network analysis techniques can be used equally well to control generating plant major overhauls, or to provide work programmes for the repair and reinstatement of main or auxiliary plant items which are out of commission on breakdown. Plant breakdowns are often of a similar nature: hence, standard repair networks and standard jobs can often be created for the more significant items.

7.7 Outage management

Network analysis will assist in producing a plan, or programme, which enables the work to be scheduled in an efficient and logical manner. However, to control the work effectively, adequate monitoring, identification of problem areas, communication, co-ordination and control are essential. On a large generating unit major overhaul, or a major breakdown outage, this is usually achieved by appointing an outage manager.

The outage manager should be chosen at an early stage in the planning of a major overhaul and, preferably, should be released completely from all normal duties. The management skills required are those of a forward thinker, decision maker, an efficient organiser, a good co-ordinator and a capable leader, motivator and controller of the outage work team.

The outage team should be comprised of the outage manager, nominated maintenance, planning and operations engineers, site safety officer, supervisors and contractors' representatives. It is particularly important that the site safety officer is actively involved in the project, because of the intense activity and the large number of people working on site during the overhaul period.

It is one of the responsibilities of the outage manager to ensure that each work area in the project is allocated to an engineer. It is then the responsibility of that individual engineer to ensure that his areas of work are efficiently monitored and controlled in accordance with the overall plan. It is particularly important that these nominated engineers have direct access to the outage manager, who acts as a focal point, dealing rapidly with problems as they arise and thereby ensuring the smooth running of the overall project. He should not, however, attempt to take on the responsibilities of the nominated engineers, or of their line management, but should resolve problems at the appropriate level of authority.

The role, level of authority and responsibilities of the outage manager need to be clearly defined at each location, since they may vary somewhat, depending upon whether the generating plant is nuclear or fossil fuel fired, and to some degree upon the organisational structure.

The general objectives should be to draw together all the relevant people to produce an outage team working in a well co-ordinated and effective manner. In particular, it is necessary to control the project, rather than merely to monitor what is taking place. For instance, every endeavour should be made to identify potential problem areas at an early stage, thereby enabling adequate time to be given for corrective actions. Once identified, it is equally important that all decisions are based upon the most reliable, accurate and up to date information available in other related areas. Communications systems therefore have to be such that rapidly changing situations are quickly recognised and reported, so that prompt corrective actions can be taken.

7.8 Major planned outages summary

The commitment of finance and resources to the major overhaul of generating plant forms a substantial proportion of a power station's revenue budget. It is therefore necessary to identify future work requirements at the earliest possible stage, for inclusion in the forward planning process. Once approved, provision can then be made for financial and other material resources to be made available.

When preparing preliminary work lists, it is essential that all aspects of statutory, mandatory and any other requirements are examined in some detail, to ensure that no essential work is overlooked. The preliminary work list should be categorised into that work which is essential and that which may be carried out at some other opportunity.

A library of standard minimum times can be produced for major overhaul work, such as statutory examinations, boiler burner refurbishment, airheater element change, turbine valvegear overhaul, LP turbine blading repairs and generator inspection.

When first preparing the outline plan for the overhaul, once the main areas of essential work have been decided, the critical activities can be identified and the minimum outage duration determined. The op-

timum outage duration can then be determined by estimating the additional resource costs likely to be incurred and comparing these with expected savings in generation replacement costs. The additional resource costs arise from the substantial extra resources required to carry out the work in the required time. An example of the additional resource histogram for a major and minor overhaul of two of the four generating units in a 2000 MW power station is illustrated in Fig 4.37.

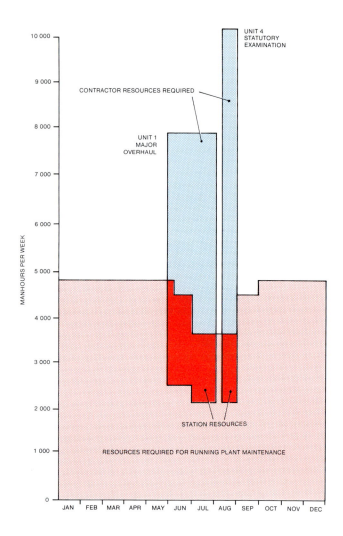

Fig. 4.37 Manpower resource requirement chart

The planning of the overhaul should be carried out to an adequate level of detail, including operational, maintenance and recommissioning activities. The planning methods used range from work listings supported by manually produced bar charts or magnetic planning boards, to fully computerised systems incorporating time and resource analysis. For the overhaul of large generating units, because of the large number of activities involved, network analysis by computer is an advantage. Networks should be created in such a way as to provide a flexible system capable of accommodating changes to the work programme with the minimum of effort. They then provide valuable assistance in the control of work as the outage progresses. If, in addition, the network analysis computer can accept data transferred directly from the work planning computer system, which may or may not be the same computer, then up to date bar charts and work schedules can be produced with the minimum of effort.

An outage manager should be appointed at an early stage in the planning of the overhaul to define the overhaul objectives, organisation and the responsibilities of key staff. The role, level of authority and responsibilities should be clearly defined. The outage manager should ensure that the responsibility for each project or plant area has been assigned to a nominated outage engineer. It is the responsibility of these engineers to ensure that the areas of work for which they are responsible are properly planned, controlled and progressed to meet the requirements of the overall plan. For these reasons the outage manager and the nominated engineer should be actively involved in the preparation of the plan, at the earliest possible stage. The use of computers for the analysis of overhaul networks can assist management in making decision at all stages in the preparation and control of the overhaul work programme. This is achieved by providing answers to a number of questions, which may be of help when determining the correct policy for a given set of circumstances. In a changing situation, it is possible to show the likely effects of alternative courses of action and so enable decisions to be made quickly, based upon an extent of knowledge which would otherwise not be available without the use of computer aided network analysis techniques.

8 Review of maintenance policies

When a power station has been operating for a number of years, various information systems will have been used to monitor the cost of operating and maintaining the plant. This will include operating experience in terms of failure statistics, running hours, usage of plant spares, failure modes and a record of lost availability. This information provides the basis for a periodic review of the effectiveness of the existing maintenance policies. The process for carrying out such a maintenance review consists of the following major activities:

● Identification of problems and major costs.

● Examination of historical data to ascertain causes.

● Investigation of the optimum or most cost effective solutions.

● Amendment of policies and implementation of the changes required.

Problems are identified by monitoring the total repairs and maintenance costs, together with the out-of-merit generation replacement costs incurred. For high merit base-load power stations, the most significant costs arising from maintenance and reliability problems will be the generation replacement costs incurred as a result of losses in station availability.

The CEGB Plant Reliability and Availability (PRA) reporting scheme is used to monitor losses in power station availability on a national basis. The losses are categorised by type, such as overhaul, breakdown, restricted output and plant code. This identifies the major problem areas and enables the station to carry out a more detailed cost analysis of its most important losses of availability.

In addition to identifying the item of plant which results in a loss of generated output or availability, the underlying cause of this restriction should be determined. It may not be possible to identify this precisely at the time the fault is first reported but, for significant items contributing to the greatest area of loss, the underlying cause should always be ascertained. This should be determined by carrying out an analysis of the exact nature and cause of failure of the individual component that is primarily responsible for the loss of capability.

The detailed causes of losses of availability should be identified from plant history records. These records may be in the form of fault and abnormal occurrence reports, plant history reports, quality assurance records, plant condition monitoring records, maintenance logs, work instruction and job card records of the work previously carried out.

The level of detail required from the plant history system should be sufficient to identify the nature of the fault, its cause and the particular component that has failed. The cause of failure may be attributable to one or more of the following reasons:

● Inadequate design to ensure fitness for purpose.

● Operational conditions outside design parameters.

● Inadequacy in manufacture or construction.

● Lack of routine servicing or preventive maintenance.

● Routine maintenance requirements incorrectly identified.

● Inadequate specification of the work to be carried out.

● Incorrect or sub-standard materials, plant spares, tools, or special equipment.

● Inadequate quality assurance systems.

The future maintenance requirements of each item of plant is carefully considered, defined and, to some extent, predetermined at the power plant design stage. This is necessary to ensure that the individual com-

- Demonstrate that plant has been commissioned, operated and maintained in accordance with the design intent and with all statutory and regulatory obligations.

- Facilitate decommissioning.

Non-permanent records are generally those records which are necessary to demonstrate the accomplishment of certain activities in accordance with some specified requirements for a limited time and are not included in the categories above.

Non-permanent records are only retained for a period of one to five years; after this period, the category should be reviewed, or the record disposed of.

9.3   Generation and acquisition of records

Since records will be generated in all phases of the life of a power station, the responsibility for the generation, acquisition and transfer of records should be with the manager of that particular phase, delegated where necessary to named staff. Where records are obtained from a plant supplier or contractor, the contract document should state the contractor's responsibility for the generation, collation and transfer of specified records. Delivery or transfer of any item of plant or equipment should not be accepted as complete, until all specified records are complete and have been received. This is equally applicable to the replacement of significant components throughout the operational lifetime of the power station.

In nuclear power stations there is a statutory obligation to produce and maintain certain records under the terms of the Site Licence. The station manager is responsible for ensuring that all records in this permanent category are updated, and for generating records which demonstrate that the plant is operated and maintained in accordance with the Site Licence.

Records should only be considered valid if they are dated and signed, or otherwise authenticated by authorised personnel. All records should be legible, complete, identifiable to the item involved and made of appropriate material to resist deterioration during the required retention period.

9.4   Indexing and schedule of records

Due to the quantity and diversity of records which are relevant to the operation of a large power station, a schedule of types of record, together with an effective indexing system, may be helpful.

The schedule should include:

- Title of the record or report.

- Item or activity to which it relates.

- Mandatory or optional record type.

- Permanent or non-permanent category.

- Nuclear related safety classification marker.

- Organisation, or person generating the record.

- Method of storing information.

- Cross reference code, or codes.

- Location of the master records.

- Location of the security duplicate copies.

Each individual record should provide only sufficient information to permit identification of both the record and its related item or activity. It should also be possible to determine whether any document or record replaces or amends any existing record to avoid reference to out-of-date information.

It is desirable that a system of classification and coding is utilised to provide a common database. Where this is not achievable, cross-indexing of records is necessary to enable rapid retrieval. The indexing should preferably be arranged so that searches for information are possible through a number of previously defined routes, such as date or plant code.

Maintenance engineers' reports relating to plant condition and maintenance work carried out are an important element of the history record system. These reports should be adequately indexed and coded, and copies made available at some central location.

9.5   Storage of historical records

A safe and secure method of storage of important records which facilitates their maintenance, preservation, protection and retrieval should be established from the time of their receipt, until their ultimate transfer or diposal. The storage of original (or master) documents and duplicate records should be at diverse physical locations. The storage facility should take account of the perishable nature of the records concerned. Consideration should therefore be given to the adverse effects of extremely high and low temperatures, humidity, dust, water or moisture ingress, fire, excess light, infestation by insects, mould, rodents and the unauthorised removal of documents.

9.5.1   Paper documentation

Paper is not the ideal storage medium for the large number of historical records relating to the design, construction, commissioning and operation of modern power plants. It is bulky, expensive to file and store, and prone to deterioration.

9.5.2 Micrographic image

Where there are large quantities of relatively static information which does not require updating, microfilming provides an efficient, cost effective, compact and secure method of record storage, occupying a fraction of the volume of an equivalent amount of information stored on paper. It has been widely used for many years and has well proven standards of reproduction and long term archival properties. The microfilm provides a true facsimile of the original document, which is of particular importance where there may be financial or legal implications.

The micrographic image may be produced in the form of an indexed 16 mm roll film, or on a microfiche, which is a postcard-sized sheet of film containing several hundred images. The simplicity of the technology ensures that there should be no difficulty in reading images. If it is only required to view the image there are many low cost viewers available for both microfiche or roll films.

The indexed roll film has an advantage that automatic retrieval based upon reel and frame number is available, providing a rapid means of accessing indexed document images. The microfiche is normally used in situations where the image is found by manually scanning the fiche. Its advantage is that it is more compact than the 16 mm roll film cassettes.

Photocopiers are available which will produce plain paper copies manually from microfiche, or automatically using the frame number for roll film cassettes.

Microfilm systems are commonly used for drawings or data sheets to provide a low cost method of storing and disseminating information.

9.5.3 Computer stored data

Computers have limited 'on-line' storage capabilities for history records. Historical data is therefore usually transferred to off-line disk storage, which can then be loaded and accessed when required for any subsequent analysis or enquiry. Where records are kept on disk, security copies are required to guard against the possibility of inadvertent data loss, or corruption of the data file.

Computer-based records will be dependent upon the compatibility of computer systems and their ability to read historical data sets produced on computer systems, possibly from many years previously. For this and other security reasons, historical records may be produced in the form of computer printouts, or computer output direct to microfilm. In the latter case, a microfilm record is produced of the information contained on the computer disk by a number of different access routes, such as unit and plant code, department, or job number.

Storage of large amounts of static historical data by magnetic disk is currently expensive, although the cost of disk storage is reducing. Disks are subject to mechanical damage and require a security back-up copy. Optical disks provide considerably more storage capacity, they are less likely to suffer mechanical damage and, since data cannot easily be erased, the data is more secure. However, a security copy is still required to guard against loss of the disk for any reason.

There are significant advantages in being able to transfer certain types of historical records from one computer system directly to another. Design and construction data may be made available and be transferred directly to the computer systems used during the operational phase of the power station. In addition, certain plant history data may be obtained from plant control computers and be transferred to planning and work control computers. Data, such as main and auxiliary plant running hours, operating cycles and remaining plant life may then be used to initiate maintenance work programmes.

Plant history data held on computer systems can be used to provide diagnostic reports which identify areas of plant with high maintenance costs, with a view to developing cost reduction schemes. Another advantage with computer analysis is that patterns may emerge which otherwise might go unnoticed.

9.5.4 Computer aided micrographics

Computers offer a convenient means of indexing and retrieving historical records. There is also considerable scope for improving history record systems by using microfilm techniques with computer links. By combining the advantages of the computer in flexible indexing and retrieval, with the good mass storage capability and the degree of authenticity obtained by microfilming original documents, a much improved system is produced.

Computers, in conjunction with micrographic storage, can be further exploited by automatically retrieving the microfilm, electronically scanning the image and transmitting the document picture to a remote computer terminal.

9.5.5 Record storage methods summary

Although it is important that methods of storage of records meet the current requirements, there is also a need to take into account possible future changes and developments in order to ensure that data recording and retrieval systems remain compatible.

Historical record system requirements are unlikely to be satisfied by computer techniques alone. Computer systems were never intended or developed for the permanent storage of large quantities of relatively static historical information. Optical disks are relatively new technology and as yet their permanent storage capability is not fully proven. Only high grade paper and microfilm records can be regarded as truly archival.

It is likely that all four storage media will play a part in the historical record storage systems of the future; high grade paper and microfilm for archiving, computers and optical disk units for short term text and graphics. Computers coupled to microfilm, optical disk and photocopying units have considerable potential in providing indexing, on-line enquiry, automatic retrieval and photocopying facilities. Systems are also available which electronically scan micrographic images of documentation or drawings and transmit them to terminals at a remote location.

9.6 Plant and work history

The CEGB operates a national Plant Reliability and Availability (PRA) history reporting system, which requires that the reasons for any shortfall in a generating station's reliability or availability is recorded and entered into a central computer. This permits an analysis of each station's performance and availability losses, thereby enabling resources to be directed to reducing the most significant areas of loss.

The PRA coding structure is also used at power stations as the basis of financial control and plant status reporting. In addition, it is used for plant inventory, plant spares, work control and plant history recording systems. Its structure is described in the 'Plant inventory' section of this chapter.

With the introduction of computer aided work planning systems within the CEGB, facilities were included for the retention and analysis of plant and work historical information. This enabled a substantial proportion of the plant history records to be obtained without a significant amount of additional data input, or the need for extensive manually produced records. Careful consideration should be given, however, to the amount of historical information in text form which is to be entered and stored in computer systems. It may be preferable to record only the basic information consistently and to cross-reference codes on the computer, referring to more detailed records held on microfilm or to the original documentation, where necessary.

Where microfilming is to provide the archival medium for work control card history records, the cassette reel and frame number should be cross-referenced to the planning and work control system computer. Then, as the result of an enquiry, the computer can produce the microfilm reel and film reference numbers, to provide a quick and efficient method of locating and, if required, producing a photocopy of the work control card information.

The need to be selective about the acquisition and storage of information depends upon the costs and benefits associated with the volumes of data involved. Modern storage and retrieval systems, such as computer-linked microfilm systems, reduce the need to be highly selective.

Plant history systems are only as reliable as the quality of the information they contain. Therefore plant and work history records must be updated consistently, since there is little point in carrying out a detailed analysis based upon unrepresentative information.

There is a continuing need to improve the systems for the collection and use of reliability data, particularly in the area of risk assessment. Detailed information is required on performance, often of small components, to effect improvements in design by a critical analysis and understanding of the factors affecting reliability and maintainability.

The plant and work history system will contain a wide variety of documentation and data, varying from comprehensive technical reports to the large quantities of mainly statistical data relating to the completion of routine maintenance work. For the latter, simply recording that the work was completed and the resources used, normally suffices. However, plant history records must provide sufficient information on plant defect or breakdown occurrences and the extent of the repairs carried out, to enable problem areas to be identified and corrective actions to be taken. It is also essential that the history system provides a comprehensive record of all work carried out on the plant, irrespective of whether the work was carried out by station or contractor resources.

History records or reports that are fragmented and available only to the engineers or departments who recorded them preclude their use for other purposes, either because their existence is not known to other people, or because their form is unsuitable. It is therefore necessary to identify the data needed to satisfy the various purposes of the plant and work history system, and to determine the types of analysis and output that are likely to be required. There is little point in recording substantial amounts of data if it cannot be retrieved in a given time, or in a form acceptable to potential users.

9.7 Management of historical records summary

The reason for creating and maintaining a power station history record system is to influence future events, policies and decision making by taking full account of previous knowledge and experience. History record systems are also needed to provide assurance and objective evidence that activities affecting quality and safety have been performed in accordance with specified requirements, and that the required quality and safety standards have been achieved and maintained. In addition, it is necessary to be able to demonstrate that the plant has been operated and maintained within statutory, mandatory, designed or subsequently approved limits, that materials used are of the appropriate quality and that staff have the required training and expertise.

It follows that the establishment and continued maintenance of an effective history record system is an essential aspect of good management. It needs a disciplined approach, with clearly defined policies specifying what information should be recorded and retained, the method of storage and its retrieval mechanism. The individual responsibilities for the establishment and maintenance of the history record system should be defined and documented in an appropriate management control procedure.

The method of storage must be carefully considered. Computer-indexed microfilming probably offers the most efficient and cost effective solution. It produces a true facsimile of the original documentation which is of particular importance where there may be legal or financial implications.

A review and audit should be carried out periodically to verify that the record system has been effectively developed and maintained, that environmental conditions are satisfactory, and that the records are properly filed and are not deteriorating.

The history record system therefore provides a means of demonstrating compliance with mandatory and other regulatory requirements. It also provides a means of monitoring and adjusting maintenance policies, thereby ensuring that the optimum maintenance strategies are adopted. In general, it provides factual information upon which sound judgements can be made, leading to sound commercial and engineering decisions in the future.

10 Additional references

British Standard BS5750: 'Quality Systems': 1979

British Standard BS5882: 'Specification for a Total Quality Assurance Programme for Nuclear Installations': 1980

British Standard BS6046: 'Use of Network Techniques in Project Management'

Part 1: Guide to the Use of Management, Planning, Review and Reporting Procedures: 1984

Part 2: Guide to the Use of Graphical and Estimating Techniques: 1981

Part 3: Guide to the Use of Computers: 1981

Part 4: Guide to Resource Analysis and Cost Control: 1981

CEGB Quality Memorandum CQA1: 'A Guide to Quality Assurance Practices for Conventional Generation and Transmission Plant and Equipment'

CEGB Quality Memorandum CQA2: 'A Guide to the Quality Assurance Practices for Nuclear Safety — Related Plant and Equipment'

HMSO: Factories Act 1961

HMSO: Nuclear Installations Act 1969

HMSO: Health and Safety at Work Act 1974

International Atomic Energy Agency Guide 50-59-QA2: 'Quality Assurance Record System'

The Pressure Systems and Transportable Gas Containers Regulations 1989

CHAPTER 5

Power plant maintenance

Contents

1   Maintenance organisation

The power station maintenance department exists to help the production function to maximise plant reliability, availability and efficiency, by determining both short and long term maintenance requirements and by carrying out the work accordingly. This includes work to comply with statutory and mandatory requirements and investigations into plant problems. The department has to make the most economic use of its available resources; this is achieved, in part, by having a level of staff (engineering, supervisory and craft) to deal with the general day-to-day steady workload, and by making alternative arrangements to cater for workload peaks, for example, by using contractors or by staff mobility between locations.

It has been traditional in power stations for the maintenance department to be separate from the production function. A Maintenance Superintendent, accountable to the Station Manager or his deputy, is responsible for all aspects of maintenance; section engineers for each of the main disciplines, mechanical, electrical and instrument, assist him in the formulation of maintenance policy and are responsible for the safe and efficient execution of the work within their domains.

A more recent approach to maintenance organisation (adopted by the CEGB during the 1980s) results from a redefinition of the three functional areas within a power station; production, engineering and resource planning. The production role now encompasses day-to-day management of staff and plant in terms of the control of work programmes for both plant operation and maintenance. The engineering role is to identify, define, and co-ordinate the engineering needs of the station in terms of commissioning, operation, maintenance and plant modification. The traditional functions of the maintenance department are thus split between the production and engineering departments. In simple terms, the production department is responsible for the execution of work, whereas the engineering department is responsible for diagnosis and work specification.

1.1   Typical staffing arrangement

Figure 5.1 illustrates a typical organisation chart for a 2000 MW coal-fired power station.

The Production Manager is responsible for the management of the maintenance resource and is assisted in this role by the Principal Engineer (Maintenance). The co-ordination of the maintenance activities in each of the main disciplines, mechanical/heavy electrical and light electrical/instrument, is carried out by the respective maintenance section engineers and includes the supervision of all craftsmen and allied trades. In addition, a services and civil section are responsible for all building and site maintenance. The organisation below section head level varies between stations, dependent upon local circumstances. The staff tree in Fig 5.1 illustrates one example, where the plant is divided into task areas and a group comprising an engineer, foreman and craftsmen assume responsibility for all maintenance within their area of plant.

A typical task area would comprise the pulverised-fuel firing equipment, boiler air and gas ducts, and

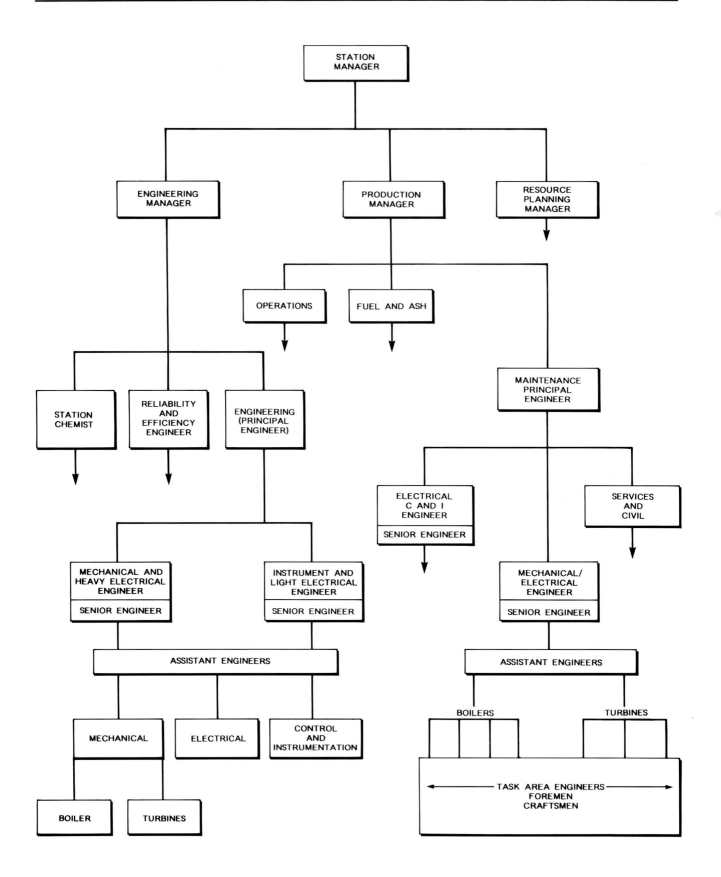

FIG. 5.1 Typical 2000 MW coal-fired power station staff tree, with task maintenance concept
in the mechanical/electrical section

associated plant. The task area maintenance system enables the group to develop as a team, gaining expertise through familiarity and providing a high degree of job satisfaction. When combined with long term job rotation to retain a degree of flexibility and allow matching of manpower to workload, such a system can greatly enhance the performance of the maintenance department.

Most maintenance staff are employed on day work, as this permits job continuity, together with more effective management and utilisation of resources. Dependent on local circumstances, a small shift-maintenance team may also be employed to deal with emergencies and jobs that require 24-hour cover.

An assessment of the average daily planned-maintenance and general repair workload, together with an allowance for emergency breakdown work, determines the number of craftsmen employed. These numbers are subject to review in the light of experience, changes in maintenance philosophy and changes in station operating regime, for example, the requirements for maintenance personnel are different for base-load, two-shifting and very-low load-factor regimes.

The engineering department contains specialist mechanical, electrical and instrument engineers, whose duties include the formulation of long term maintenance plans, advising on the frequency of planned preventive maintenance and monitoring its effectiveness, and the specification of maintenance work both for in-house and contract resources.

1.2   Staff responsibilities

The *Production Manager* has overall responsibility for the production function of a power station, including the specification of maintenance policy and the co-ordination of the operations and maintenance activities to ensure that availability and efficiency requirements are achieved. He is responsible for the recruitment and management of all maintenance staff.

The *Principal Engineer (maintenance)*, is directly responsible to the Production Manager for managing the work undertaken by the maintenance section in accordance with maintenance programmes and station policy. He co-ordinates and approves the detailed plans and budgets for the maintenance section and ensures the most effective use of resources. He is assisted by *two senior engineers* who have specific responsibilities for mechanical/heavy electrical and light electrical/instrument work, respectively. They co-ordinate the execution of work within their discipline to approve technical standards as specified by statutory legislation and the engineering branch. They are responsible for promoting and maintaining good industrial relations consistent with the policy laid down by the Production Manager. The senior engineers liaise with the engineering department on such matters as spares holdings,

routine preventive maintenance policy and the more complex technical considerations.

Assistant maintenance engineers are responsible for the work carried out by craftsmen to previously agreed work programmes and to the required quality and safety standards. They have direct charge of the workshops and are responsible for compliance with the relevant statutory regulations. They participate in the production and monitoring of work programmes on a daily basis and may be associated with, and responsible for, maintenance of a particular plant or 'task' area.

Provision of engineering expertise, information and advice is the responsibility of the engineering department. The *Principal Engineer (engineering)*, is accountable to the *Engineering Manager*, and is responsible for the co-ordination of all three engineering disciplines (mechanical, electrical, and control and instrumentation). With the assistance of the *two senior engineers*, he formulates long and short term engineering policies and specifies overhaul and development work in all three disciplines.

The senior engineers are responsible for the preparation of technical specifications and tender documents, ensuring that all relevant quality standards are included in work specifications, recommending the type and extent of spares holdings, and advising on the frequency and type of routine preventive maintenance. Each senior engineer manages a team of *assistant engineers* to provide the necessary expertise in his particular discipline. The assistant engineers may be responsible for a particular plant area in line with the 'task area' philosophy.

2   Workshops

The workshops at a power station are provided to enable maintenance work to be carried out on installed plant and equipment. The extent of the workshop facility is determined by the type of plant installed and by policy regarding maintenance of plant on-site versus maintenance at remote locations, such as manufacturers' works or central workshop facilities.

The location of the workshop complex takes account of the need to be in close proximity to main plant areas, i.e., boilers and turbines, as well as being adjacent to personnel and administrative facilities, i.e., changing rooms, offices, etc. These requirements are generally achieved by providing an island of ancillary accommodation, containing administration, welfare, workshops and stores functions, with easy access to the main building. Some stations have the workshops as an extension of the turbine hall and thus have use of the turbine hall cranes to lift heavy items direct to the shop. This arrangement has the disadvantage of subjecting workshop personnel to noise from the main plant.

2.1 Workshop layouts

The functional parts of a workshop and typical floor areas are:

- Mechanical workshop and fitting shop   1000 m²
- Electrical workshop and fitting shop   400 m²
- Instrument workshop   450 m²
- Welding shop   200 m²
- Carpenters' shop   50 m²

In addition, the workshop design is influenced by, and should take account of:

- Easy access and proximity to main plant.

- Proximity to engineering and work planning offices, and to the stores.

- The need for adequate cranage to serve the mechanical and electrical workshops.

- The segregation of hazardous areas and processes, e.g., pressure test bay, grinding bay, battery charging bay, etc.

- Safe and well defined access for personnel and vehicles.

The layout shown in Fig 5.2 illustrates an arrangement that satisfies these criteria. The locations of work benches have been omitted for the sake of clarity. The major part of the building consists of a central area with a cranage facility (typically of 25 tonne capacity), comprising the mechanical, electrical and welding workshops. In addition, it is useful for the crane to cover at least part of the heavy stores. Situated around this central area are the fitters' work areas, the carpenters' shop and the main and rigging stores.

The *mechanical workshop* is the area where repair of heavy mechanical plant items takes place. It is provided with special facilities such as a pump maintenance bay, pressure test bays, etc.; it also houses all the machine tools described in more detail in the following Section 2.2. Leading off the mechanical workshop is the *mechanical fitting shop* and accommodation for the mechanical maintenance foremen. The fitting shop houses sufficient benches, with cupboards, for the fitters and their tools, and may also contain such items as small pillar drills, bench grinders and joint-cutting tables.

The *welding shop* area is preferably walled-off from the main workshop for arc screening and noise suppression reasons, but open to the roof to allow crane access. Fume extraction hoods are provided over the benches. In order to limit the storage of oxygen and

acetylene bottles in the shop area, it is often possible to pipe gases to benches from an outside bottle-rack. It is also an advantage if the station gas-bottle store is located nearby.

The welding shop also contains fixed welding transformers, a grinder, a marking table, a heat treatment oven and a drying oven for electrodes. A profile cutter may also be provided.

The *electrical workshop* layout follows the same design as the mechanical workshop. An open area allows maintenance on large items such as motors, transformers and switchgear. An enclosed area houses the fitters' benches and it is here that the smaller jobs are carried out. Office accommodation is provided for the electrical maintenance foremen. With the machine shop adjacent, it is not necessary to provide separate machines for jobs such as commutator skimming. However, special facilities are provided for battery charging and testing of electrical apparatus. The battery charging bay is enclosed and ventilated to the outside of the building.

Maintenance of modern *instrumentation and control* equipment is largely on the basis of replacement of units/components at the plant and the repair of those components in a workshop environment. Thus a clean, quiet enclosed workshop, equipped with electronic diagnostic and test apparatus, forms the major part of the *instrument workshop complex*. Also included are offices, a general workshop for the maintenance of heavier plant items, storage areas, a calibration room, and a drawing and library facility. A small lathe, pillar drill, grinder and engraving machine are located in the general workshop. In addition to a comprehensive range of power supplies and test equipment, the clean workshop also contains purpose built test benches to allow the repair and testing of components associated with electronic boiler and governor control loops.

Further information on instrument workshop facilities is provided in Section 12.10 of this chapter.

2.2 Machine shop equipment

The range of machine shop equipment provided for a power station workshop facility will reflect the policy regarding on-site maintenance, and may also be influenced by the proximity of manufacturers' works, central workshops or local contractors. Although large lathes capable of handling turbine rotors may be installed, it is more usual for these to be located at central workshops; such installations would additionally require a larger crane capacity for the workshop.

A typical range of equipment is given in Fig 5.2. The machine tools should be arranged to minimise hazards to operators and passing pedestrians, and yet be capable of performing all tasks required of them, for example, lathes must be capable of accepting long lengths of tube without obstructing walkways,

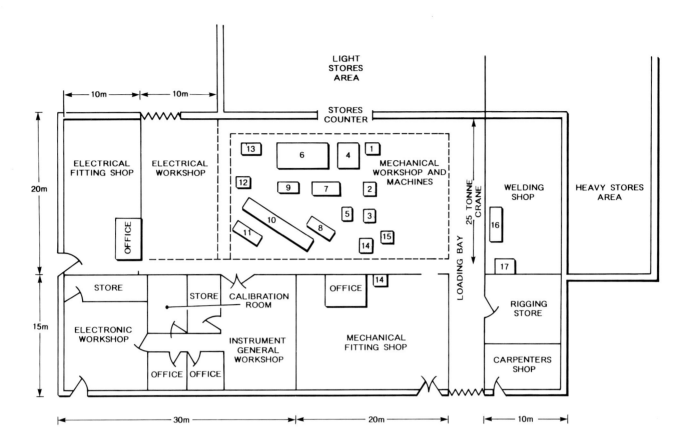

KEY TO WORKSHOP MACHINE TOOLS

ITEM	NUMBER OFF	MACHINE TOOL	SIZE OR CAPACITY	PURPOSE
1	1	THREAD CUTTING MACHINE	150mm DIAMETER PIPE	PIPE SCREWCUTTING
2	1	SURFACE GRINDER	750mm x 300mm TABLE	PRECISION GRINDING OF PLATE ETC
3	1	TOOL GRINDER	300mm WHEEL	PRECISION GRINDING OF LATHE TOOLS ETC
4	1	RADIAL DRILL	2400mm RADIUS	DRILLING BORING TAPPING ETC
5	1	DOUBLE ENDED GRINDER	150mm WHEELS	GRINDING CARBON TIPPED TOOLS
6	1	HORIZONTAL BORER	1200mm SWING	FACING VALVES FLANGES ETC
7	1	SHAPING MACHINE	600mm STROKE	GENERAL SHAPING KEYS ETC
8	1	LATHE	200mm SWING	GENERAL TURNING
9	1	MILLING MACHINE	2000mm x 400mm TABLE	SLOT CUTTING KEYWAYS SURFACING ETC
10	1	LATHE	750mm SWING	MACHINING IMPELLERS ETC
11	1	LATHE	300mm SWING	GENERAL TURNING
12	1	MECHANICAL SAW	300mm STROKE	CUTTING, PIPE BAR ETC
13	1	MARKING TABLE	2000mm x 1000mm	GENERAL PURPOSE
14	2	PEDESTAL DRILL	20mm CHUCK	HOLE DRILLING TO 20mm DIAMETER
15	1	HYDRAULIC PRESS	20 TONNE	GENERAL PURPOSE
16	1	PLATE ROLLER	2000mm PLATE	PRODUCTION OF ROLLED PLATE
17	1	GUILLOTINE	2000mm x 12mm PLATE	CUTTING PLATE

FIG. 5.2 Typical layout of a power station workshop

comprises *consumable and general engineering spares*, which are typically high turnover, low cost items not related to one particular piece of plant. They may be competitively purchased from any one of a number of suppliers and their widespread use at different locations enables large utilities such as the CEGB to negotiate national supply contracts, with very favourable purchase terms. Typical of this group of spares are such items as nuts and bolts, ball and roller bearings, jointings and packings, light bulbs, fuses and cable, etc.

The second category comprises *plant spares* that are associated with a particular item of equipment. Commercial and technical considerations often limit the source of supply to the original manufacturer. Valve spares, turbine blades and pump impellers are typical of this group. They tend to be expensive items, because of their specialist nature and limited application. Obsolescence of plant spares can require considerable redesign of components and/or expensive one-off production, unlike consumable spares, where alternatives are usually readily available. Under these circumstances, care must be taken to avoid infringing patents or copyright.

A system of coding is used to identify all spares uniquely. The CEGB uses a seven-digit commodity code which can be demonstrated by an example:

Code 23/31/010

The first two digits identify the main group to which the item belongs — 23 refers to pipes, tubes and fittings.

The second two digits identify a sub-group — 31 refers to socket weld fittings.

The last three numbers identify the item within the sub-group — 010 is a cap for 1/8 inch nominal bore pipe.

Main groups 01 to 31 refer to consumable and general engineering spares, whilst plant spares are coded in groups 34–90. In this way, the commodity code provides each item with a unique shorthand identification, which can be conveniently used with a stock control system, and is particularly suited to the latest computerised systems.

4.2 Stock levels

The Station Manager is responsible for all working capital on his site, which includes stores stock. The initial spares requirements are identified by the maintenance engineers from recommendations obtained from the plant manufacturers. Stock and reorder levels are specified after considering the manufacturer's recommendations, together with supply lead times, likely rate of usage, availability from other locations and the consequences of being out of stock. These initial requirements are then modified regularly throughout the life of the station, as dictated by experience and changing circumstances.

In practice, the high-turnover consumable items are continuously monitored by the stores staff. Often their rate of usage and low value render the use of an exact method of stock control uneconomic, especially if they are easily obtained from local suppliers. Under these circumstances, a simplified system of control is used, whereby regular visual checking by the storekeeper initiates reordering. The issue is recorded on a general issue sheet for audit purposes only and does not form any part of the stock control system. Review of the more specialised plant spares is carried out in conjunction with the maintenance engineers on a less regular basis. These reviews also serve to identify excess or redundant stock for withdrawal and disposal.

There are occasions when the maintenance engineer may require a temporary increase of stock levels to accommodate anticipated high consumption. The preparation of stores requirements for an overhaul at the planning stage enables the stores supervisor to have such material available for the programmed date without overstocking for the remainder of the year.

4.3 Central stores

The modern trend towards standardisation of plant within power stations has led to the establishment of central storage facilities in order to reduce the investment in very high value spares. Thus a power generating utility can hold one set of turbine rotors that serve as spares to perhaps six individual stations. These items are held against breakdowns, and for use during refurbishment programmes. For instance, the turbine rotors referred to above are available, either to cover the in-service failure of a unit, or as replacement for a rotor programmed to undergo refurbishment, such as renewal of eroded last-row blades.

5 Use of contractors

5.1 Determination of requirements

Section 1 of this chapter explained how the power station maintenance resource is usually sized to deal with a steady daily workload. This includes routine preventive maintenance, scheduled to even-out the workload throughout the year, together with the repair and overhaul of auxiliary plant. The non-urgent nature of much of this work enables staff to be diverted to provide a flexible and rapid response to emergencies, such as the breakdown of a main generating unit.

There are occasions of high work-load, when additional resources are necessary to expedite the work and minimise plant outages. These are often associated with statutory inspections and major refurbishment of the boiler and turbine components, and necessitate the use of contractual arrangements to procure the extra manpower on a short term basis.

The same arrangements are necessary to implement major plant modifications and to undertake capital building projects, as these are inevitably outside the capacity of the maintenance departments.

Maintenance work carried out infrequently on specialist items of plant does not provide sufficient opportunity for the maintenance department to gain the necessary expertise. Such work therefore requires the use of a specialist contractor: often only the original manufacturer has sufficient detailed knowledge. Similarly, the use of specialist contractors can extend to activities like scaffolding, which involve the provision of equipment subject to statutory inspections, the use of which is controlled by special regulations, involving trained personnel.

The above represent the circumstances requiring the use of contractors and may be summarised as follows:

- To cover peak workloads.

- To carry out major alterations or capital works.

- To provide specialist skills or services.

Once a requirement has been identified it is important to compile a detailed work specification, together with the relevant contract conditions, to enable the contractor, first, to submit a realistic price and, later, to execute the work to the satisfaction of all concerned. Adherence to laid-down procedures for contract control helps to ensure a successful outcome.

5.2 Contract control

A contract is defined as an agreement (normally in writing) between two or more parties creating obligations, which are binding in law. It is important to realise that these obligations apply to all parties and that failure to comply with contract conditions can result in severe financial penalties. The commercial risk to a company increases with the size of the contract and so administration of the higher-value contracts is usually undertaken by a separate commercial section.

Each company will have its own policy regarding personal responsibility for contracts. On a power station, a *Responsible Officer* (normally the Station Manager) is accountable for the management, execution and completion of a contract, at the date, within the cost

and to the performance specified. He is personally responsible for authorising all contract payments and contract price adjustments, and ensuring that all the necessary procedures for effective control of the contract are carried out.

The Responsible Officer appoints a *Technical Officer* to be responsible for the general control of the contract work, including the technical specification, assessment of tenders, monitoring work progress and the initiation of variations to the contract.

The foundations of a contract are the job specification and the contract conditions. Detailed attention to these at an early stage can substantially reduce the likelihood of problems arising later. The specification should clearly and unambiguously identify the full extent of the work, quoting the necessary technical standards and timescales to be achieved, together with any special quality assurance and safety requirements. The contract conditions are the responsibility of the commercial section and detail the method of dealing with matters such as unsatisfactory performance (staff or material), defects during the warranty period, liability for losses or damage, bankruptcy and delays in completion.

Supplied with these details, the tenderer is able to quote for carrying out the work; ideally, this will be a fixed lump-sum price, so that the onus is with the contractor to control his costs and complete to programme. The power station engineering involvement is then mainly in ensuring compliance with the technical specification.

However, the nature of power station plant often precludes exact knowledge of all the work to be done until access is available. Under these circumstances, the contract can be let based on 'schedule of rates', a 'bill of quantities', or 'time and materials'. Each of these puts greater responsibility on the engineer for the contract administration, if value for money is to be assured. This is illustrated by Fig 5.4.

Jointly agreed plans are an essential prerequisite for effective contract control. Both work progress and quality can be monitored against these, and any shortfall in performance quickly identified, so that corrective action can be taken.

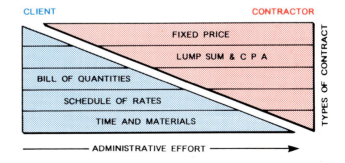

FIG. 5.4 Diagram showing comparative administrative effort between client and contractor for different types of contract

All work should be initiated by a form of written job instruction, so there is no doubt of what the contractor has been asked to do. This is particularly important for 'schedule of rates' and 'time and material' type contracts. The instructions are also useful to record work completion and can assist with identification of invoices.

6 Boiler maintenance

6.1 Milling plant

There are four principal types of mill in use on modern CEGB boiler plant. These are three medium-speed mills by Babcock and Wilcox, NEI International Combustion and PHI Engineering, and a low-speed tube ball-mill by Foster Wheeler.

These mills have 'grown up' with the increasing size of boiler units, hence the principles of maintenance are applicable to the smaller mills of the same family used on smaller boiler plant in use in various parts of the world. Small milling plant is also used in other industries, such as cement works, quarrying and waste disposal and there have been developments in these fields which have been helpful to the power generation industry.

Since mills grind coal exceedingly fine, it follows that the mill must contain wearing parts and a considerable amount of development work has gone into extending the life of these components. In the 1960s and 1970s, Nihard materials were fairly widely used, the actual alloy varying from Nihard 1 to Nihard 4. In recent years, there has been increased usage of high chrome cast-steel alloys. These materials, and others, are still being developed and different materials appear to suit different coals; it is not possible, therefore, to make a single recommendation to cover all applications.

The cost of milling plant maintenance is one of the major financial burdens in a coal-fired power station. When medium-speed mills are used, at least one spare mill is normally provided to allow maintenance to proceed with the boiler on full load. The amount of maintenance is greatly influenced by the quality of coal actually burnt, compared with design. Coal quality can vary widely during the life of a station and spare milling capacity can vary between nil and three mills. Spare mills allow milling plant maintenance to be carried out on a planned basis, usually related to running hours which are proportional to tonnes ground. Spare mill capacity also allows the user to exploit all the wearing material in the grinding components by monitoring wear at defined intervals. It is important to match the wear rate of all components, as far as possible, to op-timise costs. When circumstances dictate that there is no spare milling capacity and milling plant maintenance is debited by the cost of lost generation, then more expensive 24-hour working is used and maintenance is kept to a minimum between major boiler outages. This problem has been exacerbated by the increase of the boiler statutory period in the United Kingdom from 30 months to 38 months. When milling plant overhaul is dictated by boiler statutory periods rather than mill wear, this inevitably means discarding components which have substantial life left but not sufficient to survive until the next major outage, which again adds to costs.

The four major mill manufacturers now only produce pressure mills, so as to avoid the need for an exhauster fan. Some of the mills derive from suction mills: consequently, early smaller versions operating with exhausters can still be found. On pressure mills, particular attention must be paid to sealing arrangements, as the egress of pulverised fuel presents a major safety hazard as well as a cleaning problem.

On most milling plant, lifting and transporting heavy components for maintenance often involves problems of access compared with, say, turbines or pumps.

The mills are constrained by the bunkers above them, plus the need to feed new raw coal to the mill, and pulverised fuel (PF) away from the mill. Fortunately, in more recent designs, these difficulties are being given more consideration. Where plant constraints have prevented installation of adequate cranage, considerable use has been made of forklift trucks with special lifting brackets for the various components.

Having considered the general principles which apply to all milling plant maintenance, it is appropriate to consider the special requirements of the main different mill manufacturers.

6.1.1 Babcock mills

The Babcock E-type mill is shown in Fig 5.5. This mill has been developed over many years and is a well proven design. The mill used on the 500 MW units is the 10E model. The profile of the top and bottom rings, particularly the bottom, can be monitored to plot the wear rate but perhaps the simplest and quickest parameter to measure is ball diameter. Currently, these grinding elements are giving lives of 20 000 hours, which represents three years of generation. Even this performance has to be viewed in perspective with the overall maintenance picture. Typically, a 500 MW unit has eight E-type mills, therefore a 2000 MW station has 32 mills so, on a routine basis, mill maintenance is virtually a continuous task.

Particular, but not exclusive, areas worthy of note on E-type mills are as follows: the bedding of the bottom ring to the yoke must be checked carefully;

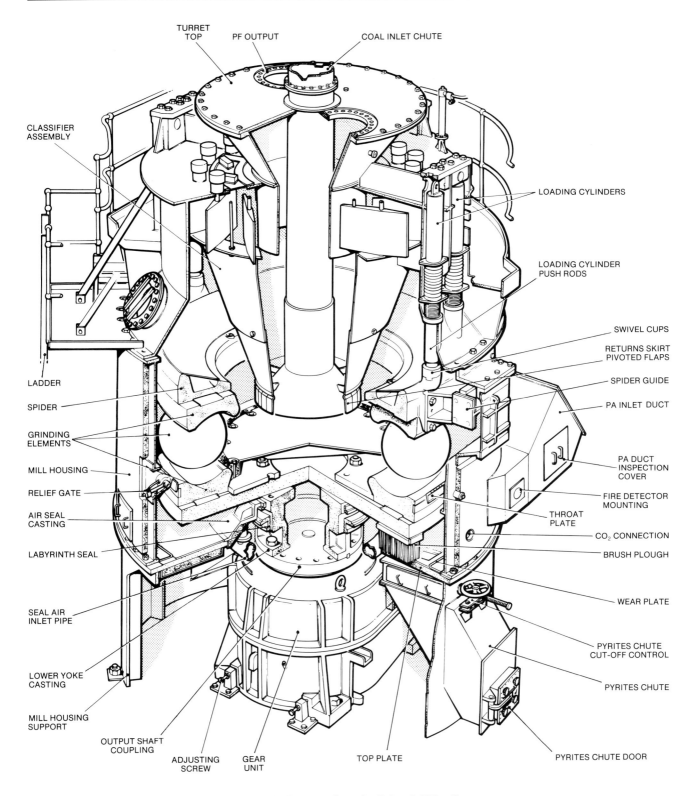

FIG. 5.5 Cutaway view of a Babcock E10 mill

any damage, machining errors or fretting can leave the ring poorly supported and cause it to crack prematurely. The top-ring alignment pads must be correctly adjusted to prevent undue movement and premature wear. Loading cylinders (or springs on early E-types) should be adjusted to give the correct ball-loading for adequate fineness and throughput. The classifier should be overhauled with the mill to ensure the fineness of the product. It is often convenient to hold spare classifiers to minimise mill downtime and allow off-site overhaul. The reject (pyrites) system has to be maintained airtight, as any

315

air leakage in this area tends to cause quite severe fires. The main shaft seal is only really accessible during a major mill overhaul and this should be kept in good order. The deterioration of the seal can be monitored on-load by roughly assessing the amount of air required to keep it packed. Any failure of the seal invariably exposes the gearbox to a measure of PF contamination, which is undesirable (gearbox maintenance is dealt with in Section 6.1.8 of this chapter).

The current E-type mill requires cranage above the mill, as it has to be stripped down from the top in the pipework, classifier, grinding elements sequence. Some early mills on 120 MW units had large doors in the side of the mill to allow forklift access, but this practice has been discontinued. Due to the cost of this large stripdown, the manufacturer's recommendation of fitting a fill-in ball halfway through the mill life is not generally followed. The throughput and quality of product do not deteriorate to an extent sufficient to justify the work involved. The mill gearbox can be slid out underneath the mill if the element loading is released. This permits workshop maintenance of the gearbox; a great improvement on the early models which had the gearbox integral with the mill body.

6.1.2 ICL mills

The majority of International Combustion Ltd (ICL) mills in service in the UK are of the Lopulco design, as shown in Fig 5.6. This has now been superseded by the Loesche mill, which is made under licence by NEI-ICL. These two mills are generally similar in that they employ a roller mounted above a segmented grinding-table. The Lopulco has three rollers while the Loesche has two. The wear rate of the grinding elements can be assessed by monitoring roll and table profiles. There is an on-going maintenance obligation to check that the gap between the roll and table is correct to ensure adequate grinding, although there is some measure of external indication on the roller assembly and by monitoring the quality of mill rejects. A typical 500 MW unit would be fitted with six Lopulco or Loesche mills. Depending on coal quality, these mills would require the roller tyres changed at 8000 hour intervals and the table segments at 16 000 hour intervals. This life span is slightly shorter than the Babcock E-type but is compensated by the reduced number of mills and the excellent 'design for maintenance' which is fundamental to the Loesche mill. The roller assemblies can be changed, as shown in the series of photographs in Fig 5.7. The mounting points and pivots for the hydraulic equipment are built into the basic structure of the mill and its foundations. The minimum of cranage is required and that cranage is outside the body of the mill. It is conceivable that a simple change of rollers could be achieved in less than a day. When the rollers are removed, this leaves two large

apertures to the mill body for work on table segments and body liners.

The principle of a grinding roll on bearings inside a mill means that it is essential to maintain a seal to prevent ingress of PF to the bearing assemblies, thus these mills are usually equipped with grinding-roll oil-temperature monitoring and the sealing-air pressure is controlled above mill inlet pressure.

On these ICL-NEI mills, the maintenance requirements of the tensioning devices, classifiers, reject system and main shaft seal are similar to that already outlined for Babcock mills, although the design details are different. Again, the gearbox can be slid out from under the mill without completely dismantling the mill.

6.1.3 Foster Wheeler tube ball-mills

The Foster Wheeler tube ball-mills, shown in Fig 5.8, are a completely different concept from the previously mentioned medium-speed spindle mills. Since the initial design concept was to provide a mill which did not require routine maintenance between boiler major overhauls, it follows that boilers do not usually have spare milling capacity. As mentioned previously, this is obviously subject to variations in coal supply but, nevertheless, lack of spare mills is inevitably a maintenance restraint. All the maintenance must be done during boiler major shutdown periods when labour and working space is at a premium. The possible exception to this is on some early 500 MW units, where it was recognised that coal feeders would not last for the life of the mill and it is possible to isolate one coal feeder at a time and sustain full-load on, say, $5\frac{1}{2}$ of the 6 mills installed. Unfortunately, in the early days of these tube ball-mills, they did not live up to their design expectations and some difficulties were encountered in the drive system and with body liners, although these problems have now mainly been overcome and the mills are generally reliable. Grinding ability is maintained by regularly adding balls to the mill to replace those worn. The manufacturers now have sufficient experience to advise on quality and mix for various fuels.

When the mill is to be overhauled, it has to be emptied of balls. The balls have to be graded where suitable for re-use and the small ones rejected. Since there are 40–50 tonnes of balls, this is a major materials-handling exercise in its own right. The body of the mill is lined with abrasion-resistant segmented liners interspaced with lifting bars to pick up the contents to encourage grinding. The mill liners normally last for several years although the lifting bars are subject to a rather higher wear rate. The maintenance engineer must be certain that these liners and lifting bars will survive until the next service before he authorises recharging the mill.

Particular areas which require attention are the classifiers, scroll assemblies and stationary casings at

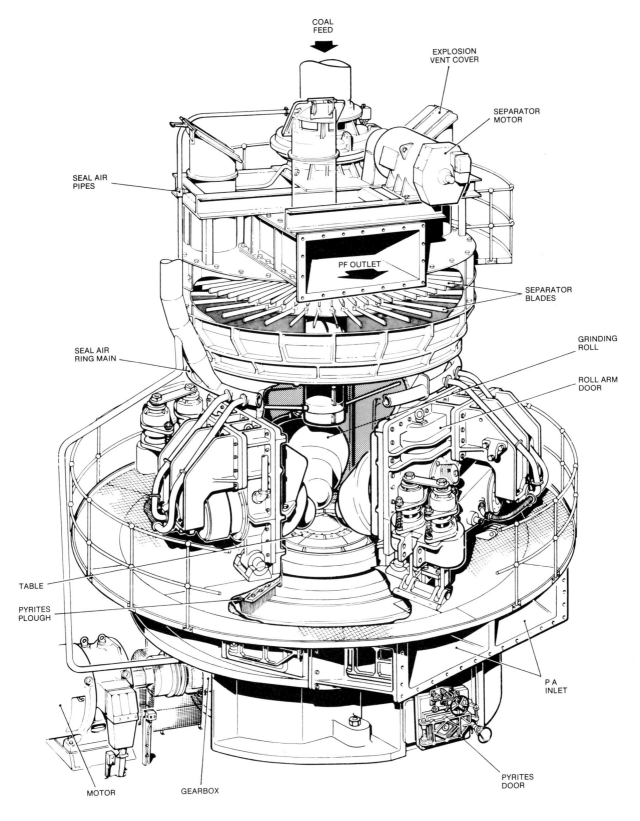

Fig. 5.6 Cutaway view of an NEI/ICL Lopulco mill

each end of the mill which are sealed to the rotating body. As on other types of mill, monitoring seal-air flow indicates the state of the seal. Any PF egress from these seals is dangerously close to the large and crucial trunnion bearing which supports the mill body. The mill drive system is physically large and transmits tremendous power. Long life can be obtained by ensuring a high standard of alignment throughout

317

FIG. 5.7 Shows the sequence of events in removing the tyre assembly on a Loesche mill
Note the provision at the design stage of built-in anchors for the hydraulic handling equipment and the easy access to the
mill body through the roller apertures (Loesche GMBH).

(see also colour photograph between pp 434 and 435)

the drive chain and adequate lubrication. The rate of
wear of the drum gear-ring and the driving pinion
can be measured by monitoring tooth profiles. Unlike
the other mills referred to, there is no reject system
from the mill as the mill consumes all it is offered.
However, as the classifier is a separate entity from
the mill, there is a classifier reject system which feeds

back to the mill.

6.1.4 PHI mills

The PHI mill is made by PHI Engineering, part of
the Deutsche Babcock group. It is a relatively-recently
developed mill and there are very few fitted in CEGB

FIG. 5.7 (cont'd) Shows the sequence of events in removing the tyre assembly on a Loesche mill
Note the provision at the design stage of built-in anchors for the hydraulic handling equipment and the easy access to the
mill body through the roller apertures (Loesche GMBH).

(see also colour photograph between pp 434 and 435)

plant. The PHI mill was developed from the German Berz mill, indeed, the PHI mills at Rugeley *B* were converted from Berz mills, but the PHI mill is now recognised in its own right. The PHI mill combines features of the ICL and Babcock mills in that three-tyre assemblies grind on a segmented table but are held in contact with the table by a top loading ring, which is loaded using springs and three wire ropes to the mill foundation.

On this mill, wear is monitored by measuring roll and table profiles and wear rates at Rugeley give approximately two years' life, using Nihard materials. Longer lives are being pursued, using high chrome steels which are currently being evaluated. Signifi-

FIG. 5.7 (cont'd)   Shows the sequence of events in removing the tyre assembly on a Loesche mill
Note the provision at the design stage of built-in anchors for the hydraulic handling equipment and the easy access to the
mill body through the roller apertures (Loesche GMBH).

(see also colour photograph between pp 434 and 435)

cantly, this type of mill achieves a slightly higher throughput of coal than its competitors, so only seven mills are fitted to a 500 MW unit and full-load is usually maintained using 6 mills (subject to coal quality). The PHI mill has the maintenance problem of roll bearings inside the PF enclosure. Obviously the temperature could be monitored, as ICL mills, but normally it is sufficient to sample the oil on a routine basis and to change it, if contaminated. The usual method of maintenance is similar to the Babcock mills, to strip down from the top, requiring extensive cranage, although PHI can offer side access for new plant with limited headroom. The maintenance requirements for tensioning devices, classifiers, reject

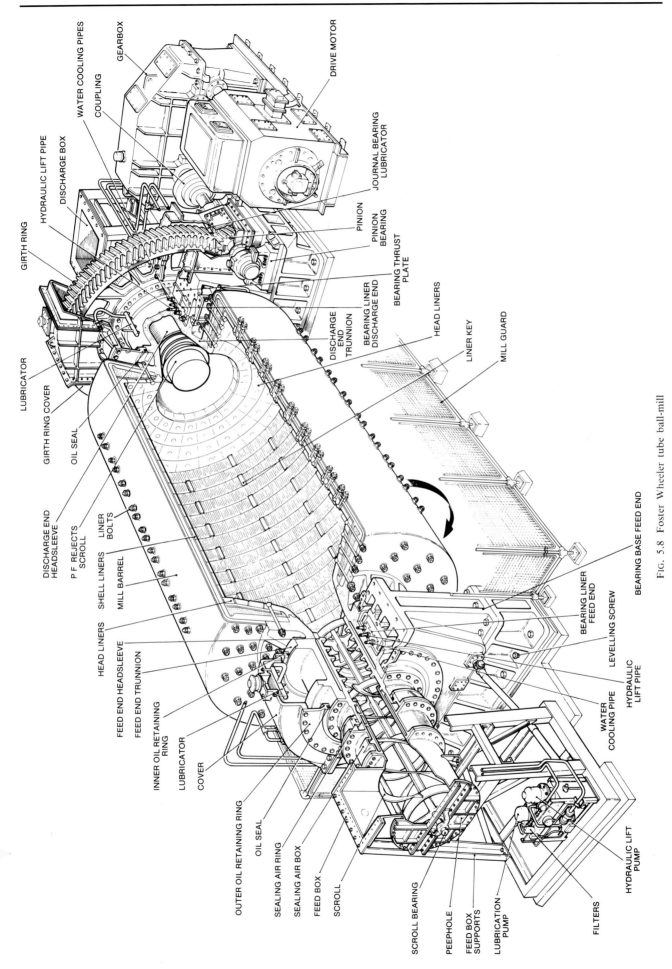

WATER COOLING PIPES
GEARBOX
COUPLING
HYDRAULIC LIFT PIPE
DISCHARGE BOX
GIRTH RING
DRIVE MOTOR
JOURNAL BEARING LUBRICATOR
PINION
PINION BEARING
BEARING THRUST PLATE
BEARING LINER DISCHARGE END
DISCHARGE END TRUNNION
HEAD LINERS
LUBRICATOR
GIRTH RING COVER
OIL SEAL
LINER KEY
MILL GUARD
DISCHARGE END HEADSLEEVE
P F REJECTS SCROLL
LINER BOLTS
SHELL LINERS
MILL BARREL
HEAD LINERS
FEED END HEADSLEEVE
FEED END TRUNNION
INNER OIL RETAINING RING
LUBRICATOR
COVER
BEARING LINER FEED END
LEVELLING SCREW
WATER COOLING PIPE
HYDRAULIC LIFT PIPE
BEARING BASE FEED END
OUTER OIL RETAINING RING
OIL SEAL
SEALING AIR RING
SEALING AIR BOX
FEED BOX
SCROLL
SCROLL BEARING
PEEPHOLE
FEED BOX SUPPORTS
LUBRICATION PUMP
FILTERS
HYDRAULIC LIFT PUMP

Fig. 5.8 Foster Wheeler tube ball-mill

321

system and main shaft seal previously mentioned for Babcock and ICL mills are also applicable to this mill.

6.1.5 Coal feeders

All the large coal-fired boiler units are equipped with some form of drag-link chain coal feeder, as the early rotating-table designs were inadequate for the larger mills on modern units. There are a number of manufacturers in the field but the maintenance philosophy is similar on all types.

The major maintenance problem on feeders, as on any coal handling plant, is the erosion of components by the passage of coal. Current designs of coal feeder are designed to withstand PF explosions which produce internal pressures of 3.5 bar. The structure should be maintained to withstand such an explosion, even though such strength is not required to fulfil the prime objective of moving coal.

The chain assembly with its crossbars are the components most exposed to wear. The chain links wear and stretch with the continued load. The range of the feeder tensioning device should allow several months of wear whilst maintaining tension. Chain links are removed periodically so that the total feeder life is commensurate with that of the mill it is feeding. It is usually convenient to carry out a major overhaul of a feeder in parallel with the mill major overhaul, so any increase in chain life would have to be by multiples of mill life rather than marginal increases. The cross-bars easily last the life of the chain; sometimes they can be recovered by hard-facing the leading edges but this is purely a matter of cost option. Manufacturers' experience is now so wide that they can offer different shapes of cross-bar for differing fuels, which can be a very important factor when long-term fuel supplies are changed. Chain sprockets, bearings and especially casing-to-bearing PF seals, and the tensioning devices are normally overhauled with the chain assembly. If these spares are held as overhauled sub-assemblies, feeder downtime can be dramatically reduced.

The pans along which the chain drags the coal are normally made of stainless or wear-resistant steel and last more than one chain lifespan. The body of the feeder itself is now protected by liners or tiled with ceramic materials to maintain the pressure containment ability already referred to. The drive trains must be correctly aligned and lubricated to ensure long life. The advent of cheaper solid state variable-speed electric motors is eliminating the need for complex variable-speed gearboxes.

6.1.6 Pulverised-fuel pipework

Since the inception of PF systems, the containment of pressure surges from PF explosions has caused problems. With the rapid growth in unit size in the 1960s, there were a number of PF explosions which initiated a considerable amount of research and design modifications. The result of this work has meant significant changes to installed plant which qualifies more as design than true maintenance but nevertheless has occupied maintenance staff at power stations. Hopefully, the modification stage is now finished and maintenance of PF pipework systems can be restricted to maintaining the parameters set by the designers.

All PF pipework systems are designed to withstand over-pressurisation and, in some cases, the effects of shock waves propagating through the system, namely:

Total shock reflection	48 bar
Sharp bends and partial shock reflection	35 bar
Straight pipes and long radius bends	24 bar
Expansion joint restraints	17 bar
Non shock plant	7 bar

The plant should be designed of ductile material with wearing liners, such as cast basalt, alumina or ceramic materials, or, alternatively, with sufficient allowance for wear to maintain the containment properties. It follows that overhaul inspections should ensure that linings are adequate to protect the main structure for the next period of service and that minimum component thicknesses are set for wearing components.

Where fittings, such as riffles or valves, are included in PF pipework, the bodies should be adequate for containment of surges, while the internals should be adequate for their purpose whether that be distribution of PF or the isolation of plant. The penalties of failure of such components can be very far reaching. Bad PF distribution can cause worn riffles, which results in poor combustion in the furnace; this is an efficiency loss and can exacerbate fireside corrosion and cause premature failure of boiler tubes. Ineffective PF dampers can, at best, prevent mill isolation, mill maintenance and, at worst, result in serious PF explosions.

It is usually sufficient to remove bends in PF pipework to inspect the bends and adjacent straights and to open riffles and valves for visual inspection. This work should cause few technical problems if it is carefully done during planned boiler shutdowns.

6.1.7 Pulverised-fuel burners

The PF burner is the termination point of the PF pipework and it is subject to a more arduous duty than the pipework since the PF inlet is always through a sharp bend: the fuel is swirled to improve combustion and work on the fuel means wear on the components. Leakage of PF from a holed burner tube would not be readily visible and could result in a windbox fire with major damage to the boiler.

In addition to the wear from PF, the burner unit is also subject to radiant heat from the furnace. A burner in-service tends to be protected by its own air supply but a burner out-of-service must be able to resist direct radiant heat. The end section of the central tube is usually tipped with heat-resistant steel. This tip and its weld is worthy of at least visual inspection on boiler overhauls. Wear or complete loss of the tip can cause poor combustion, with its consequent problems. The centre tube of the PF burner carries the oil burner and light-up equipment. This assembly is removed for workshop inspection, cleaning and overhaul on major shutdowns. The whole burner assembly must be correctly aligned to the boiler front wall. Misalignment can cause flame impingement to side or division walls and consequently accelerated 'fireside corrosion' (see Section 6.2.3 of this chapter).

The PF burner also embodies the secondary air inlet vanes. Original designs were movable to control quantity and swirl of secondary air. Such designs involve pivots and links in the fairly hostile environment of a windbox. It is possible to maintain such a system by regular overhaul but current practice is to set the blades to the optimum settings for service and to install a sleeve damper to prevent excess air when the burner is shut down. This system gives better operational control and is easier to maintain than the adjustable system. Whatever system is in use, correct setting in the load position is important.

The burner quarl serves two purposes; first, to provide a smooth outflow for the PF/air mixture and, secondly, to protect the tubes in the immediate area of the burners. The quarl must be round, concentric with the burner and true to the desired profile. This is usually achieved by a template mounted on the burner tube. The installation of quarl refractory, whether it be pre-fired tiles or rammed silicon carbide, should be rigorously controlled to the manufacturer's specification. Defects will only become apparent when the refractory falls out on-load, resulting in very bad combustion and its consequences.

6.1.8 Gearboxes

Gearboxes are used on PF plant to drive mills and coal feeders. As the same technology applies to all gearboxes, it is appropriate to consider them as a discrete plant group.

The mill drive gearboxes are among the largest in the power station; usually they have separate lubricating oil systems, with pumps and coolers. *In situ* maintenance of these components should be minimised or, if such work must be undertaken, precautions to prevent ingress of PF and dust must be observed. Similarly, contamination of lubricating oil should be avoided: the greatest risk of contamination is during filling or topping-up. The large mill gearboxes are particularly suitable candidates for plant condition monitoring by vibration analysis techniques, together with oil temperature and bearing temperature monitoring. Permanent records for every hour of operation are not necessary, providing the parameters for normal start-up and operation are known, so that abnormal variations can be noted.

When it is necessary to strip down and inspect any gearbox, the problems encountered can usually be ascribed to one of two causes:

- A short-term failure, which is usually associated with premature bearing failure.

- Long-term wear which occurs on the mating gears and is a simple matter of the transmission of power through two mating faces causing wear.

The life of a gearbox is one of the design parameters and consequently can be varied by the designer's selection of geometry, materials and finish. On a long-life continuous-process gearbox, such as the majority of large power station gearboxes, one would reasonably expect 10–15 years of working life. If the gearbox is wearing rapidly, such that a life of less than ten years is anticipated, then often the gearbox is overloaded. Frequently, overloaded gears also mean overloaded bearings which may bring about sudden failures. The exceptions to this general life expectancy are small variable-speed boxes, which have a life expectancy of about two years, although solid state electronic variable-speed control is overcoming this problem.

When a gearbox is stripped, all components should be inspected, starting with the casing. The large fabricated casings have been shown to be prone to cracking and distortion which can result in bearing failure. All shafts, gear wheels and bearings should be accurately measured for wear. Usually it is sufficient to check the wear on the mating face of a gear against the trailing face. Bearings are rather more difficult to check. Most manufacturers issue detailed instructions for their own products, giving limits on cage free-play and end-float. The feel and sound of a spinning bearing judged by an experienced craftsman and visual inspection for discoloration or damage are still very valuable tests. Generally, bearings in a large gearbox, whilst still expensive items, are a smaller percentage of the total overhaul cost than on a small gearbox. Hence, there is some merit in a policy decision to change bearings that have been in service for a number of years. Particular weak points in the large reduction gearboxes, as used in mills, are the first-stage input cartridges and the thrust bearings. The input cartridge is the input shaft and first motion spur-gear, complete with bearings. This unit is relatively small and operates at high speeds compared with later gears in the train, yet it still transmits the full power of the gearbox and therefore is highly stressed. The thrust bearing usually has to absorb the thrust of the mill which is grossly in excess of anything the gearbox could generate. Some gearboxes use roller bearings; others use tilting-pad bearings. The latter require far more care in initial set-up to achieve a good bed and oil wedge, and are far more susceptible to oil contamination in service.

When gearboxes are rebuilt, the following parameters should be accurately monitored:

- Limits and fits of shafts with their bearings and wheels.

- The alignment of mating components, which includes clearances between these components and the provision of expansion allowances, i.e., float of shafts.

- Cleanliness of the whole gearbox, particularly the less obvious points, such as oilways.

- Wherever possible, the oil system should be primed and tested prior to running the gearbox under load.

6.2 Pressure parts

The pressure parts of a boiler are the tubing, headers, boiler drum and pipework which contain the water and steam that are the fundamental operating fluids of the thermodynamic cycle. Reasonably, these components occupy a considerable amount of the maintenance engineer's time. In operation, these components are protected by avoiding large temperature differentials through the walls or gross overheating by lack of coolant. The maintenance engineer is concerned primarily with preserving the long-term integrity of these pressure parts by monitoring their degradation, and preserving or replacing, as required.

Despite the generic term 'pressure parts', it is worth noting that the operating temperature of the components is generally far more significant than the operating pressure. In addition to the long-term aspects of pressure part maintenance, the engineer is faced with short-term problems caused by pressure part failures which take the unit off-load. The engineer may well take decisions for legitimate short-term commercial gain which could be quite disastrous in the long-term. For this reason, breakdown tube repairs are considered in their own right.

6.2.1 Boiler drums

The boiler drum is usually the largest single item on the boiler, the most expensive and, should it become necessary, the most difficult to replace. During construction, the drum is usually raised before any other part of the boiler. As the drum is basically a thick-walled cylinder, it is susceptible to damage from thermal shock due to operational error. If one assumes correct operation, the structural integrity of the drum can be ensured by fairly routine inspections during overhauls, the most obvious being a detailed inspection of the supporting structure. The drum is made up of rolled plates welded together; given good initial construction and quality control, these welds should give little trouble during the life of the plant. However, their presence must be acknowledged and eventually these welds may require some detailed non-destructive tests (NDT).

All the inlet and outlet connections to and from boiler drums are welded. The detailed designs of these connections are worthy of examination. In the early days of 500 MW units, some designs included an unfused land from which cracks propagated. Clearly, modern designers would not permit such details but examples of this fault are still found in older plant. Past experience or insurance examiners may well dictate examination of all or a selection of these welds. The vast majority of these connections have site welds very close to the drum. If any of these fail, the escaping fluid may well erode the drum. Local repairs to a drum are possible but would require extensive heat treatment, making them expensive and time consuming.

The bulk of drum inspections and repairs are concerned with the fittings inside the drum. The water distribution pipework, sampling and dosing pipework, steam separators, driers and baffles are invariably bolted into the drum and there are some flanged connections. It is important that all of these items are checked for security and any indications of corrosion or erosion. It is fair to say that losing the odd bolt here or there rarely detracts from the function of the drum internals. However, failure of a number of fasteners is indicative of a fault or undue stress somewhere in the system. The most likely result of a bolt falling off in the drum is a blockage and tube leak somewhere else in the system.

Finally, the drum and its surrounding area should be inspected to ensure that it is not being locally corroded by intermittent spillage from items such as safety valve drains or aircocks.

6.2.2 Headers

Headers on modern power station boilers have attracted a considerable amount of attention in recent years. The reasons for this have been two-fold; the increases in the statutory periods between inspections for CEGB boilers and a review of plant operating lives. It is now possible to obtain statutory inspection certificates in the United Kingdom for 30 months on boilers over 21 years old and for 38 months on boilers rated 500 MW and above. These certificates, known as Exception 63 Certificates, can only be obtained after carrying out extensive examinations, particularly on the high temperature headers which may be affected by creep. In addition to these statutory requirements, the Board has revised plant lives from 100 000 hours to 200 000 hours for some boilers and detailed predictions of header life are essential to long-term forecasts.

Header creep-lives must be assessed by considering temperature, materials and dimensions. There are three increasingly more precise routes for this exercise:

(a) Based on normal station-instrument header temperature records, with correction for the operating temperature-spread along the header and using minimum specified dimensions.

(b) Based on actual header temperature, as part of a specific recording exercise with actual minimum physical dimensions.

(c) Physical and NDT inspection of headers at routine intervals of the statutory period.

Route (a) is a general approach and hence a substantial factor of safety is built into the calculations. This means that it is a common occurrence for a header to fail under these criteria. In these circumstances, it is necessary to use the actual figures of route (b). This obviously involves a lot more engineering effort and hence expense. Often one finds that the headers are actually thicker than the design, which is beneficial providing that no extremely high temperatures are found. Should the header fail to meet the creep life criteria under route (b), then route (c) must be employed. Route (c) monitors actual physical deterioration, that is the onset of creep damage.

To establish which headers are affected by creep, it is necessary to consider their material and operating temperatures. The critical temperatures are:

Carbon Steel	430°C
1% Cr $\frac{1}{2}$% Mo	505°C
$2\frac{1}{4}$% Cr 1% Mo	520°C
Austenitic Steels	620°C

Below these temperatures, allowable stress is limited by proof stress hence actual temperature is not relevant. Above these temperatures, creep strength or stress rupture criteria apply. In practice, headers operating within 20°C below these temperatures have been considered against creep criteria.

The inspection procedure for high-temperature headers is a complex procedure, involving detailed recording of information. These techniques are outlined in Chapter 4. The technical aspects of the inspection briefly consist of material composition and condition checks, dimensional checks, i.e., diameter thickness and deformation axially or circumferentially, NDT examination of welds and inspection of support systems.

The *material composition* checks are one-off in-spections, which are now part of original construction quality control. Once the correct material is proven by Metascope testing, maintenance involvement is merely to ensure that replacement header caps and welds are of the correct material. During subsequent inspections, the header material should be examined to ensure that it is not subject to corrosion or any other damage, such as adjacent steam leakage impinging on the body.

The *physical checks* involve recording of thickness, which is usually a one-off check, and the periodic diametrical checks to monitor creep. It is usual to monitor 2 diameters at 90° every 2 metres, with at least one check on each component section of header, using micrometers or dial comparators. The spots for the checks have to be carefully polished with absolute minimum metal removal and marked to ensure they can be found again in three years. The measurements should be taken to an accuracy and repeatability of ±0.025 mm. Should a lesser standard be accepted, it is possible that the onset of creep may not be discovered until significant damage has occurred. There is the need to carry out these measurements at the same low temperature on each occasion. For this reason, the checks are best carried out when the boiler has been allowed to cool naturally for a week or so. Slight differences in temperature can be accommodated by mathematical formulae.

NDT examination of welds can be split into four distinct areas, end-cap welds, butt welds, take-off branch welds and stub-tube fillet welds. The number of header welds which have actually leaked steam have been very few, the majority have been end-cap welds which have failed explosively with obvious risk to human life. Therefore, end caps should be subject to rigorous NDT, using MPI and ultrasonic techniques. Main butt-welds on fabricated headers and take-off branches should be tested using MPI and ultrasonics, remembering that while dimensional checks remain stable main butts are not likely to deteriorate, but take-off branch welds are also subject to system stresses from the take-off pipework. Stub-tube fillet welds are generally checked on a sampling basis, say 10%, using MPI until defects are discovered but concentrating on known hot areas or high system stress areas. Apart from stub-tube fillet welds, it is necessary to monitor a sample selection of circumferential and axial ligaments between stubs to ensure that these critical areas are free from cracks. Each header must be considered on its own merits, for example, using owner's or manufacturer's information on similar but older headers, the onset of measurable creep from the dimension checks, or known system stresses. Peculiar construction details, such as the use of forgings for branches or equal-diameter tees, would require special consideration.

The support systems for headers should also be checked to ensure that all parts are load bearing, particularly in multirod systems. Expansion arrangements, such as sliding supports or bearings, should receive special attention (see Section 6.3.4 of this chapter on boiler structural steelwork).

The foregoing comments have emphasised the need for inspection of high-temperature headers. It is clear from the temperature limits given that the headers on water circuits, saturated-steam inlet headers and reheat inlet headers would not be subject to the detailed repetitive measurement checks. However, the checks at construction and repair for metal composition, NDT of welds, integrity of support and corrosion in service are equally applicable to the relatively cold headers.

On statutory inspections carried out in the United Kingdom, it is normal practice for the boiler inspector to carry out an internal inspection of a selected sample of headers. This should be done in accordance with a long-term plan which allows the inspector to progress around the boiler over a period of years. Repeated inspection through the same nipple effectively shortens the nipple stub at each occasion, eventually necessitating an insert.

6.2.3 Water circuits

The water circuits are defined as the tube work of the evaporative circuits in the boiler. This is always the economiser and the furnace walls, and occasionally the walls of the rear convection pass, which are water on some designs and steam on others. Some designs include curtain walls of water tubing in the furnace, which are similar to the furnace walls but heated all round. As the saturation temperature of most modern high pressure boilers is about 350°C (depending on actual operating pressure), the water tubes are not subject to any discernible high-temperature creep problems so, from a design point of view, the water tubes are fairly straightforward, long-life items. Their lives are generally limited by our inability always to present perfect fuel and water, and to avoid built-in faults.

When the boiler is surveyed, it is usual to scaffold the furnace to inspect the interior, which obviously is mainly the furnace wall tubes. These tubes can be subject to external corrosion from the products of combustion, internal corrosion from the water, external erosion from sootblowers and grit-laden gases or PF particles. The water tubes are susceptible to welding errors, particularly at the attachments (see Section 6.2.5 of this chapter). The engineer should inspect to find any faults which may have developed during the life of the boiler, with a view to ensuring the integrity of the pressure parts in future, then carry out the necessary repairs and pressure test the finished work. In the UK, the inspection and pressure test are part of the statutory requirements for a boiler.

The most obvious risk to the tubes is corrosion of the outer tube surface on the furnace side. This is known as 'fireside corrosion': it is essentially a pro-

duct of the combustion process and is mainly dependent on the chlorine content of the coal. In the UK, the chlorine content is steadily rising overall, so this problem is growing. The corrosion rate of the furnace walls can be established by grit blasting the tubes in the furnace at 2 m intervals and ultrasonically measuring the wall thickness. The engineer can then determine a tube replacement philosophy based on running hours, corrosion rate and tube failure thickness (plus safety margin). Using a low chlorine fuel, it may well be that the corrosion rate is so low that an inspection is not required at every overhaul and a full furnace-scaffold can be avoided. At the other end of the spectrum, fireside corrosion can be so intense that new ordinary mild-steel boiler tube would not survive until the next overhaul. In this event, alternative materials can be used, such as a stainless steel/ mild steel co-extruded tube. This tubing has mild steel on the inside and a stainless outer layer, which is more resistant to the corrosive combustion atmosphere. Alternatively, an oval-shaped tube can be fitted to offer more wearing material to the corrosive atmosphere. Finally, it may be possible to obtain low-chlorine fuels, but this option is rarely available to the maintenance engineer. When considering a repair plan to combat fireside corrosion, it is essential to forecast the commercial value of all the alternatives over a long period, as the costs and lives of all the solutions vary dramatically.

Having dealt with fireside corrosion as a problem which embraces the whole furnace due to coal quality, it is worth noting that the same fireside corrosion can occur due to a maladjusted PF burner or poor mill grinding. In such an incident, if the fuel is impinging on the wall, or contains an excessive amount of coarse particles, a local reducing atmosphere will be produced which will cause severe corrosion. This is difficult to predict and is best prevented by monitoring mill performance and quality control of maintenance work.

Corrosion of tubes on the inside, normally known as 'waterside corrosion' is now virtually unknown in the CEGB: this is due to the excellent control of water chemistry exerted by the CEGB's chemists. However, the potential for failure is always there, as was demonstrated by many failures in the early 1960s when the chemistry control techniques were developed, and is still demonstrated by some power authorities in the world who have yet to employ these techniques. For the maintenance engineer, waterside corrosion is extremely difficult to deal with; the effects are likely to be widespread, necessitating large-scale tube replacement. There is no NDT technique available which will show the incidence of corrosion to determine the tube work necessary. The use of ultrasonics for finding general cracking inside a component (rather than specific cracks, such as from a keyway) is not very reliable. The results merely indicate something is amiss, rather than specific depth of cracks, etc.

When faced with chemical waterside corrosion, invariably the only choices are to replace all the tubes likely to be affected or to accept the incidental failures until all the affected tubing has failed. It is possible to cause waterside corrosion in boilers by mechanical rather than chemical phenomena. Some tubes by design run in a virtually horizontal plane; due to construction errors or movements with age, it is possible that these tubes do not drain completely when the boiler is emptied. If the boiler is stored under nitrogen all is well, but when tube work is in progress air is admitted and corrosion occurs, ultimately resulting in tube failure. In water tubes, this phenomenon can occur in ash hopper tubes and in economisers, both of which can be difficult to repair due to access problems. Inclined water tubes have been known to dry out on their upper surfaces in areas of high heat transfer, leading to rapid and severe internal corrosion. The adoption of rifled bore tubes or tubes fitted with internal swirlers has been effective in overcoming this problem. The use of superior material is another alternative.

The damage done to water-wall tubes due to erosion by sootblowers and PF particles can be accommodated in the same manner as fireside corrosion by thickness checking and by defining the affected areas. Sootblower erosion can be minimised by good sootblower maintenance, including accurate alignment (see Section 6.3.3 of this chapter), and PF erosion, which is fairly rare, by good burner and burner quarl maintenance.

The technique for replacing tubes in a conventional tangent-tube furnace-wall boiler is to fit inserts and to butt-weld the tubes together. The defective tube has to be cut out without dropping debris inside the tube, the ends prepared to the shape defined in the weld procedure, including removing magnetite film and deposits from the bore, and a new tube welded back to an approved welding procedure. The length of the tube insert has to be measured accurately to ensure that the tube will fit in the gap and produce the necessary gap at the root of the weld. When the weld is complete, it should normally be subject to NDT to ensure its fitness for purpose. The level of NDT is largely a commercial matter based on the quality required, cost of NDT and the cost of time taken to achieve it, the consequences of weld failure and the reliability of the welders. Typical CEGB practice is 100% ultrasonic testing, followed by 10% radiography. When fitting inserts into a boiler wall, the engineer should remember that the tubes are essential to the boiler structure and indeed support the scaffold inside the furnace. Isolated tube replacements cause no problems but large-scale replacements should be planned to ensure that the boiler headers and buckstays are adequately supported, particularly on rear walls where some tubes provide direct support and others form the boiler nose. The actual position of the cuts should be made to favour the welder. In particular, the closing welds will be difficult as the tubes have to be pulled out-of-line to make the welds and pushed back on completion. Membrane walls have to be made good at this stage to avoid air ingress on-load.

Economisers are situated in the relatively low temperature gas outlet from the boiler where the heat transfer is mainly by convection. The economiser tubes are contained in closely-packed nests which means access is so restricted that tubes are difficult, if not impossible, to examine physically. The low operating temperature means there are no problems of welding or materials technology. However, the tubes are prone to gas erosion, particularly if boiler fouling restricts the flow area and concentrates the grit-laden gases into certain areas at increased velocity. Sootblowing can minimise boiler fouling but can bring sootblower erosion. It is usually possible to achieve an acceptable compromise, although on some boilers finned-tube has been replaced with plain tube in restricted areas to alleviate fouling.

When it is necessary to replace tubing in an economiser, the top and bottom tubes of a bank can often be inserted *in situ*; more extensive work usually requires cutting a hole in the side of the boiler to slide the whole element out, which is obviously expensive. Normally, isolated tubes are blanked and abandoned until loss of heating surface dictates major repairs.

6.2.4 Steam circuits

The steam circuits of the boiler are the superheater and reheater tubes (which are usually in ascending order of temperature), banks of tubes in convection passes, radiant walls or platens in the furnace area and pendants at the top of the convection zone. As previously stated, the saturation temperature of large units is approximately 350°C and, as creep becomes relevant when temperature rises above 400°C, it follows that the creep life of the material in the majority of steam circuits is important when considering the requirement for replacement tubing. Steam tubing is occasionally used to form the walls of the rear enclosure. When this method of construction is used, it is invariably the first stage of superheating straight from the drum, so temperature is not a problem. It follows that such tubes can be subject to the same inspections as water tubes (Section 6.2.3 of this chapter), even though they contain steam.

When the boiler is scaffolded during overhaul, it is usual to provide access to platens and pendants to allow close inspection. The tubes are subject to corrosion by furnace gases, erosion by sootblowers, and structural damage by tube or attachment failure and fatigue. The convection banks are more difficult to examine due to their construction, they are subject to erosion by dust-laden gases and sootblowers, to a limited amount of fireside corrosion on the leading tubes in hot banks, and to attachment and support failures. As the tubes are subject to the highest temperatures in the unit, inspection should recognise the particular needs of each tube material and the necessity to join together the different materials. There is no internal corrosion of steam tubing on-load, but horizontal tubes are susceptible to localised corrosion pockets off-load when condensation does not drain away.

When the vertical tubes of platens and pendants are inspected, the most obvious damage is corrosion or erosion. The corrosion is known as 'high temperature fireside corrosion' which occurs due to a mechanism slightly different from the fireside corrosion of watertubes. In the high-temperature tubes, the corrosion occurs in two distinct bands at an angle of 30° to either side of centre on the front face of the tube in the direction of the gas flow. Leading tubes in each bank are most susceptible. Sootblower erosion obviously occurs on faces of tubes which base the blowers and is more predictable. The geometry of these tubes does not generally lend itself to grit blasting. To prepare for NDT, it is more usual to lightly grind a test area on each tube to measure wall thickness ultrasonically. On high temperature tubes, it is not sufficient to set minimum wall thickness on rupture stress criteria. When a hot tube has lost, for example, as little as 20% of its wall thickness, the stress level is increased and the creep life is dramatically reduced. Therefore, to avoid the two-stage failure of mechanical damage plus creep, it is essential to set the limits for minimum wall thicknesses carefully. If a particular area of tubing is prone to sootblower erosion, yet the sootblower is essential for boiler cleaning, it is sometimes possible to line the affected face with a stainless-steel erosion shield without any detrimental effect.

The steam elements should be examined for any signs of creep damage due to overheating; this shows as a swelling of the tube and will first be detected at the hottest part, i.e., the outlet end of a particular material section. The steam-tube elements are generally constructed around one tube (or group of tubes) which structurally supports the element, with attachments to transfer load and maintain alignment. These attachments are dealt with in Section 6.2.5 of this chapter. The vertical tubes themselves tend to swing in the gas stream and a recently emerging problem is a fatigue failure of the tube caused by this cyclic loading. The most susceptible tube is the main support tube and the most likely position of failure is adjacent to the locking mechanism. If the support tube fails on-load, it is possible for the entire element to collapse before the unit can be shut down and cooled.

When repairs to vertical steam tubes are required, inserts can be fitted in much the same way as on water walls. The structural considerations are less significant, as with only odd tubes removed the ele-

ment will remain self supporting. Even the main support tube can usually be removed on a cold element. As always, the welding should be done to an approved procedure. It is important to note that, on higher alloys of thick section, it is necessary to carry out heat treatment of the welds, as specified in BS2633. Similarly, some of the bends used in platens and pendants require elaborate heat treatment after tube manipulation, so it is essential to keep a strategic stock of formed bends.

As already mentioned, convection banks are difficult to examine in detail due to their construction. It is usually only possible to view and check the thickness of one or two tubes at the top and bottom of each bank. Fortunately, this covers the hot leading tube which is most subject to fireside corrosion and most sootblower and dust erosion occurs on the first two tubes which generally serve to protect those below them. If a tube in the centre of a bank is projecting into the gas lane, it is reasonable to assume that the tube is subject to dust erosion and some remedial action is necessary. When the convection tubes are inspected, the attachments and support provisions should also be inspected to ensure the structural stability of the element. Minor failures result in sagging tubes which trap condensation, with resultant internal corrosion. Major failures can result in a total collapse of the element.

When it is necessary to replace convection bank tubing, it is possible to insert the top one or two tubes *in situ*. More extensive repairs usually mean cutting a hole in the boiler side and sliding the affected element out. For all welds made outside the boiler, a very high standard of NDT is required, since the cost of repairing a failure in service is dramatically high, compared with the NDT costs. When an element is removed from the boiler, there is often sufficient room to allow detailed examination of the adjacent elements and all such opportunities should be utilised to the full. Even if no work is envisaged at the time of inspection, useful data for tube life forecasts can be obtained.

As the temperature of the steam through the tubes rises so the material used becomes more sophisticated, i.e., carbon steel, 1% Cr $\frac{1}{2}\%$ Mo, 2.25% Co, 1% M, and finally stainless, austenitics, etc. When welding like materials, even the higher alloys can be done successfully *in situ* and a good weld is as strong as the parent material. A transition weld between low alloy materials can be similarly made. However, a transition weld between two dissimilar high-alloy materials is a different proposition. The transition weld is usually weaker than the parent material and often requires special heat treatment and NDT. Indeed, the heat treatment invariably uses some of the material creep life. It is usual to keep transition welds outside the gas pass, or at least to protect them from the direct gas flow. Also, transition welds are made in short sections, with straightforward butts in similar metals either side. Inspection of these transition pieces is usually done on a selective basis, concentrating on those which are known to be hottest. The transition weld is examined by magnetic particle and ultrasonic NDT to identify any crack propagation and the parent material stubs are examined for any signs of swelling due to creep or overheating. When defects are found in the hot transitions or failures occur, the inspection is increased to cover all the transitions.

6.2.5 Tube attachments

A tube attachment is a general description covering any device which is attached to the pressure tubing of the boiler. The device may, for example, be a tie-back to maintain tube alignment to tie-bars and buckstays, a clip to locate a tube element or a soot-blower mounting box. The attachments are generally simple, but until recently have not received sufficient attention. Attachment failure and subsequent loss of alignment can lead to tube overheating or sootblower erosion, both leading to subsequent failure. Wall-tube misalignment can expose skin casing to the furnace with consequent damage. Poor welding can initiate cracking, again with subsequent tube failure. Therefore, all attachment welding should be carried out by certified pressure welders and given the same quality control as a butt weld; particular care should be taken to avoid unfused lands and notches at toes of the weld.

The most common attachment is the surface wall-tube to tie-bar clip. The buckstays align the tubes and the tubes support the buckstays. There are many thousand such ties in every boiler. Manufacturers offered many designs but the more successful were those that offered some flexibility in the vertical plane, such as that shown in Fig 5.9. There is invariably an erratic distribution of flow in tubes, with consequent differential expansion of adjacent tubes; this is accommodated by the cranked shape of the attachment.

All steam tubes in pendant, plates or convection elements require some mechanism to align the elements. The most recent development is to use a system of wrapper tubes to make them self-aligning. However, there are many boilers containing more traditional welded-on ties. These ties come in a variety of detailed designs, all generally known as 'T and C clips'. These deserve careful attention on boiler inspection, as they are stress concentration points and attract localised pockets of fireside corrosion.

Fins on tubes are sometimes produced as manufactured items, rather than a tube with an attachment. Where the tubes are manufactured by welding or finned on site for repairs, the welding should be to a high standard, with separate sections of fin made continuous by welding and the ends beaded. A broken fin or a notched end will cause a crack to propagate through the tube weld.

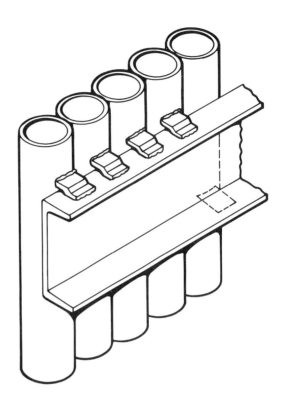

FIG. 5.9 Boiler tube tie-back assembly
The base of the clip must be flat to the tie bar. The weld to the tube must be stronger than that to the tie bar to avoid a failure pulling a piece out of the tube. Every 5th tube should be welded below the tie bar to prevent the tie bar twisting. The pattern must be maintained vertically to ensure a constant distance between ties.

6.2.6 Skin casing and insulation

The majority of large boilers in use in the CEGB have tangential wall-tube chambers which are then sealed in a mild-steel skin casing. This casing is protected from the heat of the furnace by the wall tubes. The alternative construction is a membrane wall, where each tube is solidly linked by an integral fin which seals the gap between individual tubes and is cooled by conduction to the tube so that there is no need for a separate skin. Both types of wall are then insulated to contain the heat for efficiency and for protection of staff. The insulation is invariably completed with a veneer cladding. One of the major causes of loss of boiler efficiency is air ingress through a defective casing. Whenever any tube repairs are carried out to furnace walls, it is necessary to remove and refit this insulation and casing. When it is necessary to remove casing, it is essential that the work is done very carefully to avoid inadvertently cutting the pressure-parts tubes behind. The casing should always be fully seal-welded on completion of work. If commercial pressure to generate dictates that full seal-welding is temporarily omitted, then the cost of that decision must be recognised. Skin casing can be damaged when wall tubes become detached from tube bars and spring into the furnace: thus the casing is exposed to the fire and rapidly burns through. Usually fire damage of this nature also destroys the insulation and cladding and is easy to find. The skin casing can also be damaged due to minor movements of the boiler on start-up and shutdown, particularly on two-shift operations. These minor tears in casing can be difficult to find, yet their combined effect can be of major significance. Defects in skin casing are best found prior to overhauls by the tedious but effective method of going over the entire surface area with a smoke generator, with the boiler on-load. Smoke will be sucked in through any defects.

Particular problem areas which cause air ingress are expansion joints, ash hopper doors and ash hopper seals. Although, strictly speaking, these are not actually skin casing areas, any defects will be shown up when testing the skin casing and should be dealt with accordingly.

In general terms, insulation will only need to be replaced with material as defined by the original manufacturer, unless there is a technical reason to change the material or it is necessary to protect personnel or some part of the plant, for example, structural steelwork. The CEGB design parameters are that insulation shall limit surface temperature to 55°C for metallic surfaces and 65°C for non-metallic surfaces, when accessible from a permanent working floor. Where surfaces are accessible from temporary access, they should not exceed 50°C, irrespective of surface finish. There are detailed standards available for fixing insulation which should be followed. Briefly, insulation should be adequately and evenly supported, inspection should ensure there are no voids or gaps between sections and that, where multiple layers are used, joints are overlapped. Where cladding is fitted, the finished work should be neat, tidy and functional in that it should be shaped to allow access to doors, valves, etc.

6.2.7 Safety valves

There are two main types of safety valve, using either a coil spring or torsion bars to load the valve. For the high pressure applications, such as superheaters and boiler drums, Hopkinsons offer the 'Hylif' torsion-bar valve (Fig 5.10), while Dewrance offer the 'Consolidated maxiflow' valve, which employs a coil spring. Both manufacturers offer coil spring valves for lower pressure applications, such as reheaters and auxiliary boilers. The high pressure valves are generally fully welded on the pressure side necessitating *in situ* overhaul, while the lower pressure valves are flanged and can be removed to the workshop. Accepting the differences in construction and materials for the variety of pressure and temperature applications, the same general maintenance considerations apply.

The safety valves are required to isolate effectively during normal running to avoid passing fluid to

waste, yet they must operate effectively when the set pressure is attained to collectively pass the steam generated at the maximum boiler firing rate. Maintenance is aimed at opening reliably, passing an adequate flowrate, closing and holding tight. Generally, safety valves are overhauled, or at least inspected, at every statutory inspection and a functional test is a mandatory part of that inspection. Between statutory inspections, it is occasionally necessary to relap seats to prevent fluid loss to waste. To enable this work to be carried out, both major manufacturers design their valves with provisions for locking up the spring or torsion bar loading to remove the headgear complete. However, on overhauls, it is usual to unload the bars or springs completely to allow a thorough inspection.

Boiler safety valves normally operate in a hostile environment exposed to boiler dust, exhaust gases, occasional condensate, and even weather. When the valve is to be overhauled, the initial settings should be noted; the valve is then cleaned and stripped down to its component parts. All components should be cleaned and inspected for corrosion, erosion and mechanical damage. Spindles and bores should be checked for truth and concentricity. All parts should be carefully labelled and stored. It is important not to mix parts, as high and low temperature valves can be of similar appearance but of different materials; errors could lead to premature and disastrous creep failures in the high temperature application. Generally speaking, valve seats will require lapping, or reprofiling and lapping. The seat profile and dimensions are important parameters. If the seat diameter is too large, the set pressure is reduced for a given spring load. If the seat is too wide, the sealing ability of the valve is reduced. Seats on flanged valves can obviously be reprofiled on workshop machines, while seats on all-welded valves can be reprofiled using purpose-made portable machines supplied by the valve manufacturers. These machines are usually operated by compressed air. In the event of the seat being beyond recovery, the same machines can be used to remove the old seat and produce a suitable weld profile for the seal welding of the replacement seat. When components are recovered, using welding techniques, or valves have suffered considerable mechanical damage, they should be subjected to appropriate crack detection tests to ensure that components are suitable for re-use. When the valve is rebuilt, the valve lift should be rechecked; this will ensure that the valve is capable of passing its rated capacity. The components for adjusting both the valve lift and blowdown should be set to their original positions prior to the functional test.

In the United Kingdom, the functional test of a safety valve is one of the statutory requirements for pressure vessel inspections. For reheaters and unfired vessels, this is achieved by setting the valve on a hydraulic or pneumatic test rig. The main drum and

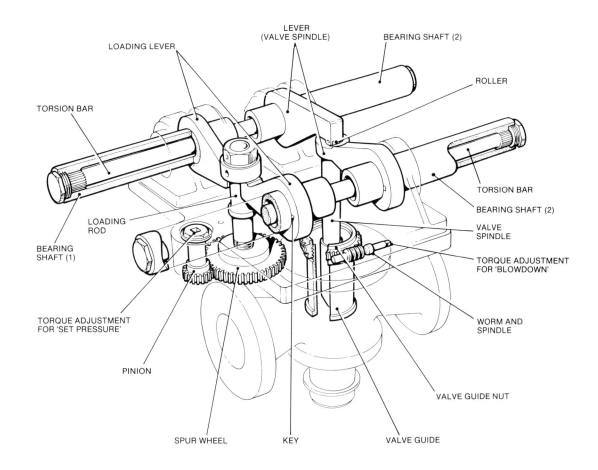

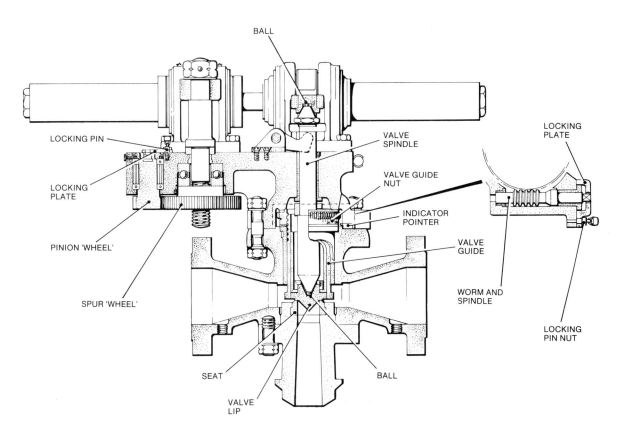

FIG. 5.10 Safety valve, 'Hylif' torsion-bar type

superheater valves have to be 'floated' on the actual boiler as an *in situ* functional test. This is obviously a very expensive operation in terms of time and fuel. To mitigate this cost, flanged valves can be initially set up on test rigs. Welded-in valves have to be set *in situ*. A recent development, known as the 'Trevitest' system, by Furmanite Engineering, allows the testing of valves using a hydraulic pack to simulate the actual fluid load. This process is not currently recognised as a substitute for a full boiler-fired test but it does serve to set the valves initially, with a consequent fuel saving. The system can also be used for functional testing of valves with the plant on-line, should this be necessary.

When the valves are maintained, due attention should be paid to the extraneous equipment associated with the valves. This includes drainage and escape pipework, exhaust silencers and, where fitted, electrical-assistance mechanisms.

6.2.8 Miscellaneous valves

The majority of valves on a boiler/turbine unit are either isolating valves or control valves. The *isolating valves* are generally required to be 'open' or 'shut' to allow or prevent fluid flow, for example, drainage during start-up which is not required on-load, or to allow isolation of sub-systems while the main plant is on-load. A typical example of a large parallel slide valve is shown in Fig 5.11. The *control valves* are required to regulate flow over a given range as part of an overall unit control system. Control valves do not always include provision for absolute isolation.

Valves are manufactured by many companies, each making a range of valves dependent on fluid, pressure, temperature, function (e.g., isolation/control, uni- or multi-directional flow). Therefore, the following comments are general and not aimed at any specific valve.

Most valves have fixed and moving sealing faces to achieve isolation. Control valves have similar mating faces in a variety of designs, such that motion of the moving face adjusts the flow rate to a specific characteristic. The moving face is generally operated by a spindle which passes through the body of the valve for external hand or powered operation. This spindle passes through a gland and there is usually a cover or base plug which allows access to the valve internals. Valve maintenance is usually concerned with the condition of glands, spindles, cover joints and sealing faces.

The selection of gland-packing material is dependent on the fluid being sealed, its temperature and its pressure. The various manufacturers are continuously developing their products and the engineer should endeavour to review new products which may minimise costs. Currently, expanded graphite preformed rings are used to pack both steam and water glands; these give a much longer service life than

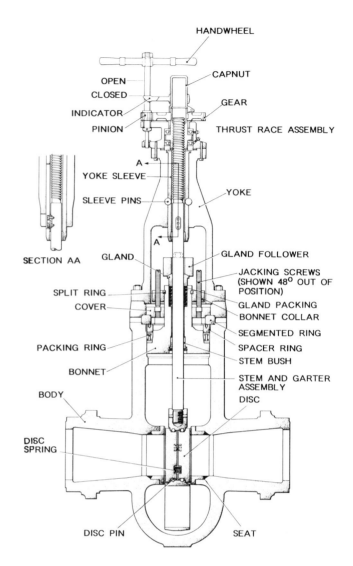

FIG. 5.11 Typical example of a large parallel-slide valve

the materials previously used. Packing manufacturers have products suitable for a variety of fluids, such as oils, air and various active chemicals. The detailed instructions of these manufacturers should be followed, particularly with respect to the need for cleanliness. Gland life is generally longer for vertical-spindle valves than horizontal-spindle valves, as the gland in the former does not take any of the weight of the spindle assembly.

When a valve is overhauled, one of the fundamental checks is to ensure that the spindle is true. The spindle has to pass through the gland and the spindle nut; whether this be associated with a handwheel or actuator is irrelevant. If the spindle is bent by any significant amount, the valve will inevitably seize. The serviceability of a bent shaft can only be established by experience as manufacturers recommend that the spindle be perfectly true. The spindle can be bent at the valve-disc end by mechanical damage to the discs, or by temperature of the fluid, or at the external end by an actuator fault or abuse. The spindle can be damaged when it is operating in a hostile environment where the exposed section of spindle passes into the gland. Scoring in this area does not detract from the function of the spindle but it means extremely short gland life. When spindles are used for high-temperature applications, they are sometimes faced with stellite or a similar alloy. Should this plating become detached, the valve operation is in jeopardy.

Cover or base plug joints come in a variety of designs. The basic joint is a simple bolted flanged joint with a suitable gasket. An improved version is a pressure-seal assembly, where the geometry of the components allows the fluid pressure to compress a seal which contains the fluid. Some designs employ a screwed-in plug which is seal welded to prevent fluid leakage up the thread. When a cover joint has been leaking in service for some time, severe erosion can take place; this can usually be cured by building-up with a suitable weld material and machining. On all these joints, cleanliness is essential and all thermal scale must be removed. The joint faces must be true and dimensionally correct, especially when joints are compressed to a predetermined amount. The actual surface finish of a joint face depends on the design of the joint and the ability of any jointing material to absorb imperfections.

Valve sealing faces are mainly metal-to-metal discs and seats, as in the parallel slide valve, although some, such as Saunders valves, employ rubber-to-metal or even plastics. Metal sealing faces have to be lapped flat and unblemished. There are several proprietary machines available, including the valve manufacturer's own products, which can be used for machine lapping of fixed seats. If the valve is beyond *in situ* lapping, then a new fixed seat is required, which usually means removing the valve for a workshop repair. The valve seats can be damaged by foreign bodies within the fluid flow, misuse of the valve, or by fluid erosion. In any event, a damaged valve seat is further damaged by fluid erosion. When valves are used for control functions, they are obviously subject to a significant pressure drop and are prone to fluid erosion. The geometry of the seats is an important feature of the valve characteristic and these faces should be very carefully examined.

When considering valve overhauls, the engineer must decide whether the valve should be removed from the line for a workshop repair or repaired *in situ*. Repairs *in situ* are always more difficult and more expensive than repairs in a workshop. Quality is more readily assured in the workshop, not least because valves can be fully tested. Obviously the cost of removing and refitting the valve must be added to the basic workshop costs but this is not the only factor. The availability of suitable labour also has to be considered. It may be possible to minimise unit outage costs by holding spare valves, so that out-of-service valves can be overhauled during the winter running periods; alternatively; it may be economic to ship valves to a specialist valve-overhaul company. The decision must be made for each type of valve dependent on all the circumstances. As a general indication, it is usually economic to workshop repair flanged valves below 150 mm nominal bore and welded-in valves below 50 mm nominal bore.

6.2.9 Main pipework

The main pipework of a boiler unit can be defined as the pipes linking the drums and headers within a boiler, the main steam pipe from the superheater outlet header to the boiler stop valves and the pipework conveying steam in both directions between the boiler and turbine. The loop pipes are between turbine valves and the actual turbines are of similar nature and should also be considered by the maintenance engineer.

Cold pipework tends to present few problems brought about by actual operation of the plant but can present problems emanating from manufacture or design. Hot pipework, which is usually of a higher alloy, often Chrome-Moly-Vanadium (CMV), is often subject to defects which grow dramatically in service due to the elevated temperature. The CEGB knowledge of defects in CMV welded joints commenced when the first excursions to elevated steam pressures and temperatures were made. One unit suffered an explosive failure of a steam loop pipe-joint very early in its life which caused a critical examination of several thousand similar joints. As a result, the difficulties encountered in welding CMV have been identified and are controllable.

The defects are broken into seven type-fault classifications:

- Transverse cracks.

- Circumferential weld cracks.

- Circumferential heat affected zone cracks.

- Slag and/or porosity.

- Lack of side wall fusion.

- Lack of root fusion.

- Other defects.

The defects originate from physical difficulties in depositing weld metal and the development of critical local metallurgical conditions. The defects can usually be prevented by meticulous adherence to the welding specification, particularly pre-weld and post-weld heat treatment. In order to avoid welding with defective material, it is usual to NDT the parent material after machining the pipe-end profile and before welding.

When the weld is complete and after a recommended period in service, which will allow minor defects to grow with temperature, the weld should be inspected to establish the status of any defects. The usual approach is to grind to a flat surface and carry out both magnetic particle inspection (MPI) and ultrasonic inspection. The defects listed above are further separated into non-planar defects occupying real volume, for example, porosity and slag, and those of planar type having length and breadth, but no significant thickness. The non-planar defects are usually innocuous unless their volume is a significant proportion of the weld or they are associated with planar defects. Planar defects can lead to premature failure by creep crack-growth. The orientation of the crack is significant: transverse cracks parallel to the pipe bore are much less important than circumferential cracking, which could ultimately lead to a complete failure of the weld and parting of the joint. Much research work has been done in the field of fracture mechanics to assess what residual life remains in a cracked component: specialist advice should be obtained if the option of running plant with cracked pipework is to be explored. With the benefit of this advice, it is often possible to defer repairs to a more commercially suitable occasion.

Hot main pipework requires a substantial amount of support since, when the pipework is at temperature, its ability to support its own weight is significantly impaired. Support systems should be examined carefully on- and off-load to ensure that they are functioning correctly; in particular, spring-type supports should be in range and not jammed at an extremity. It may be necessary to weigh pipework in the same manner as boiler structural steelwork (Section 6.3.4 of this chapter). It follows that hot pipework is very well supported when it is cold and quite significant repairs can be undertaken without additional support. Only when the pipe is actually parted is it necessary to restrain the pipework and re-establish the cold pull.

6.2.10 Attemperators

Attemperators are devices which spray a relatively small quantity of cold water into the steam flow to control steam temperature. Thermodynamically, it is more efficient to control the heat input, but realistically spray attemperators will remain and the maintenance engineer must ensure their continued availability.

The attemperator is generally contained within a section of main or reheat steam pipework of an alloy appropriate to the inlet conditions. This pipework and its associated butt welds are similar to the pipework dealt with in Section 6.2.9 of this chapter. The parts of the attemperator within the pipework are simply a feed pipe with spray nozzle, and a liner, usually of mild steel. Since the device passes cold water into superheated steam, it is obviously subject to a considerable thermal shock which can induce cracking in the long term. Should the internal fittings break up, it is unlikely that their loss will be noticed until a significant amount of damage is done. Either the main steam pipe will crack due to corrosion fatigue or, more likely, pieces of the attemperator will block superheater tubes downstream of the device, with consequent failures.

The maintenance engineer should periodically check the integrity of the fastenings of the liner and sprays and the structural integrity of the components themselves. How this is achieved varies with detailed design, but internal television inspection, NDT down the axis of fasteners and radiography can all be applied. At some point it may be necessary to completely remove, strip and examine the components. Complete removal of the attemperator is very expensive and it is generally not necessary to do this frequently. The attemperator should be selectively examined, using NDT techniques, until adverse results suggest further exploratory work on, say, one attemperator, the work being expanded or contracted as found necessary.

6.2.11 Breakdown tube repairs

Failures in the pressure parts of boilers are a major cause of loss of generating plant availability in the CEGB. Generally speaking, the direct repair costs of the failures are low compared with the additional replacement costs incurred when 500 or 660 MW units are involved. These high replacement costs, currently up to £60 000 per day, mean that it is worth spending significant sums of money to avoid tube failures. In the CEGB system, the replacement costs of smaller units are much lower and it may be economic to accept a higher rate of failure and avoid preventive maintenance cost. Such policy decisions can only be made for each electricity utility, depending on local conditions. Whatever the replacement costs, it is always economic for a maintenance engineer to develop rapid response techniques when faced with tube leaks. Some of these techniques are based on commercial pressures rather than top quality engineering procedures, with a view to effecting permanent repairs at the next planned outage. Examples of these techniques are as follows:

● Replacement of boiler wall tubes by welding from the outside of the boiler only, using 'window welding' techniques (see Fig 5.12). This avoids cleaning and scaffolding for access but gives a poor internal finish to the weld with pits for corrosion pockets. Window welds must be removed at the first opportunity.

● Blanking-off tubes at header ends or by-passing tubes external to the boiler. This technique is particularly useful in convection zones. In economisers, the gas temperature is usually low enough

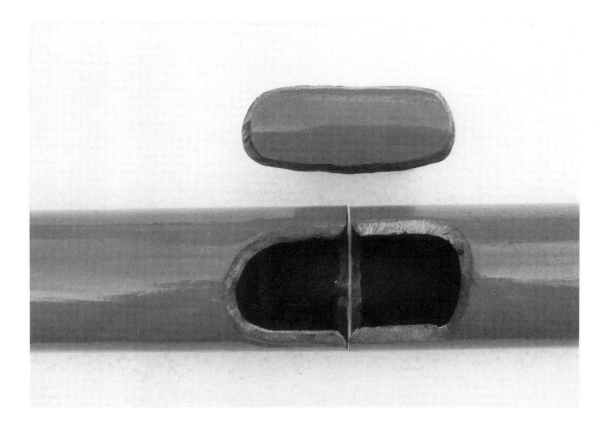

FIG. 5.12 The window welding technique

This technique allows all the welding to be completed from one side of the tube when access is limited, e.g., furnace wall tubes with no access to the furnace side can be inserted from outside the boiler. The weld preparation is made so that the root-run is put in the outside wall of the tube, working through the window. The weld is then filled to the inside of the tube. Finally, the window is closed by welding-in the blank which is made from a piece of similar tubing. This repair is purely temporary and should be removed at the first suitable opportunity.

to allow sound tubes which are by-passed to be blanked to avoid internal contamination and the tubing recovered when permanent repairs are made.

- Superheater tubes can be damaged by steam erosion from tube leaks. The engineer must be aware of the remaining creep life of such damaged tubes so that the minimum number of tubes are replaced on the breakdown and the repair completed on the next shutdown.

When the maintenance engineer plans his repair strategy, he must have the necessary materials and services available. The materials should include replacement straight tubing, formed tubing (especially bends which would require extensive heat treatment), fixings and fastenings, temporary repair parts, such as blanks or by-pass tubes, skin casing and lagging material. The services should include cleaning, lagging, scaffolding, erection and welding services. The welders will obviously require electrical supplies, consumables and detailed welding instructions for the various materials. Clearly considerable pre-planning for all potential tube leaks is necessary, preferably prior to commissioning and the first tube leak.

Whilst tube leaks provide a great commercial stimulus for the engineer to return plant to service, they are also the best source of information for planning future overhaul work. In the various foregoing sections, particularly Sections 6.2.1 to 6.2.5, fundamental inspections are outlined, but it is fairly obvious that an inspection of sufficient detail to identify every incipient failure would take a tremendously long time and consequently be very expensive. Such an inspection could easily be more expensive in manpower and downtime than the breakdown it would avoid. Therefore, when a tube leak occurs, it is important that its cause be identified so that any necessary specific inspections can be carried out, or even spares and repairs organised. Identifying the cause can be difficult, particularly where more than one factor is brought into play. In some failures, such as the superheater element illustrated in Fig 5.13, it may be difficult to find the end. Information should be gathered from as many sources as possible, including similar boilers, whether they be of similar design, firing or loading regime. The causes of tube leaks can be roughly broken down into three groups, these being fast, slow and two-stage failures.

FIG. 5.13 Typical major superheater failure
The secondary damage which occurred after the initial failure makes access difficult, takes longer to repair than the original failure and can obscure the cause of failure. On this occasion the failed tube actually parted.

Fast failures are not usually time dependent, in that the time from onset of the fault condition to final failure ranges from a few minutes to a few days, i.e., insignificant in the life of the boiler. Examples of fast failures are sootblower erosion, possibly caused by a jammed blower (Fig 5.14), tube blockages by foreign bodies or waterlogging of steam elements (Fig 5.15), and even flue-gas erosion can cause rapid failures in a very badly fouled convection pass. From the examples, it can be seen that the fast tube fail-ures occur randomly in a boiler. The action to avoid a recurrence is usually to remedy a specific fault only indirectly related to the pressure parts.

Slow failures are those which occur due to some gradual degradation over a significant period of time. They are generally a result of either a design or operational feature of the boiler. Examples of slow failures are fireside corrosion (Fig 5.16), creep (Fig 5.17), fatigue and, prior to modern water chemistry techniques, waterside corrosion (Fig 5.18). Describing these

FIG. 5.14 Typical sootblower erosion of the front joggle-tubes caused by a furnace wall sootblower passing when retracted

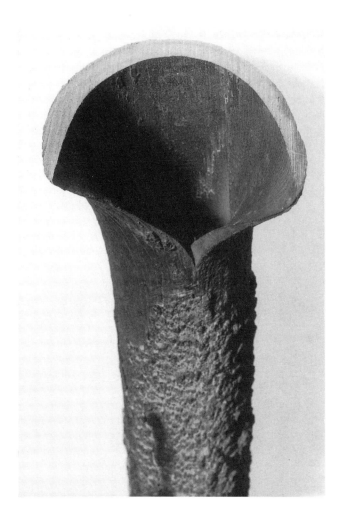

FIG. 5.16 Fireside corrosion of furnace wall tube
The material is faceted tube and was originally symmetrical
about both axes. Machining the four facets on suitable
tube allows the use of 10 mm wall thickness facing the
furnace, compared with a design of 6 mm, whilst
maintaining the tangential tube wall construction.

FIG. 5.15 Short term overheating caused
by a tube blockage
Note that the majority of the failure edge is thin-walled in
a chisel-point formation with thick walls at the extreme
end of the split. Note also the axial stretch marks on the
outside of the tube. Occasionally, score marks can be seen
on the bore where the foreign body has been ejected.
Failures of this nature can occur in minutes.

failures as 'slow' is a generalisation, as they can all be precipitated very rapidly by unusual or fault conditions, for example, fireside corrosion caused by high chlorine coals or bad combustion, and creep caused by operating above design temperatures. As a general rule, once the slow failure mechanism has matured and the first failure occurred, the rate of failure will increase with operating hours. Thus it becomes necessary to take some form of remedial action unless the boiler is close to the end of its operating life. Slow failures nearly always involve replacing large amounts of material. For example, fireside corrosion can affect a large expanse of furnace wall or superheating surface. Creep can mean the renewal of particular tubes or even whole superheating elements. Alternatively, fatigue may affect

small pieces, such as attachments. Singly these items are cheap; for an entire boiler they are expensive due to their number and complexity. These mechanisms may act together on occasions, such as a combined fireside corrosion and creep failure. Once the tubes affected by slow failure are replaced, the failures stop, at least for the period until the new tubes are life-expired by the same mechanism. If the causes can be removed, the failures will stop completely. It may be economic to combat the problem with different materials that are more resistant to the failure mechanism.

Two-stage failures occur where a defect is built into the boiler on primary design or construction and this produces a source for a secondary failure mechanism. For example, a welding defect may act as a source for crack propagation or as a corrosion pit. Once random potential two-stage failures are suspected in a boiler, there is usually little point doing anything about them. The cost and time of the necessary inspections is rarely worthwhile. The exception would be an identifiable type-fault which is repeated in a certain place over the boiler width, for example, a difficult access site weld. The majority

FIG. 5.17 Long term creep failure
This material has distended rather more than is usual in a creep failure and is a dramatic example.
Note the distention, axial/transverse scale cracking and the thick wall failure.

of two-stage failures emanate from weld defects (Fig 5.19) or constructional errors. These can be minimised by good quality assurance during construction and repair. On major overhauls, good quality assurance and careful planning are complementary and the flow of work should not be significantly impeded even by the non-destructive testing requirements. On breakdowns, the need to return plant to service may mean that non-destructive testing is delayed until the next planned outage. Two-stage failures usually mature fairly quickly after installation, so the rate of failure decreases with boiler operating hours.

FIG. 5.18 Shows the results of bad welding which caused corrosion pits to be set up downstream of the weld
Before modern boiler chemistry, this sort of failure could occur in an entirely random manner without the weld defect to
initiate it. Technically this photograph is of a two-stage failure but it does demonstrate waterside corrosion.

6.3 Boiler ancillaries

6.3.1 Airheaters

Rotary airheaters are now the only type of airheaters used in modern power plants. The older plate and tubular heaters were unsuitable for large units due to their excessive volume compared with the rotary design. There are two types of rotary design:

- The Rothemuhle, manufactured by Davidsons, which has a stationary heating element with rotating ductwork.

- The Ljungstrom by James Howden, which has stationary ductwork with a rotating heating element.

It is necessary to seal the gas and air systems in both models of airheater. Air under pressure will leak into the gas system under suction and will travel straight from the FD or PA fans to the ID fans, at great cost. On the Davidson airheater, the seals wear at a predictable rate and must therefore be replaced at a given wear point. The spring-loading mechanism and articulation joints which allow stator distortion must be maintained operable to ensure minimum leakage. The Howden airheater seals tend to move predictably in their wear rate but, as the whole element must be sealed radially, peripherally and axially, this involves more setting. On both airheaters, seal setting involves a compromise. If the seals are too loose, there is excessive leakage. If the seals are too tight, they wear more quickly and require replacement; they can wear the mating faces and can even overload the drive motor. Continual development on self-adjusting seals is taking place; some designs are within maintenance abilities although, strictly, they are modifications. A major problem of regenerative

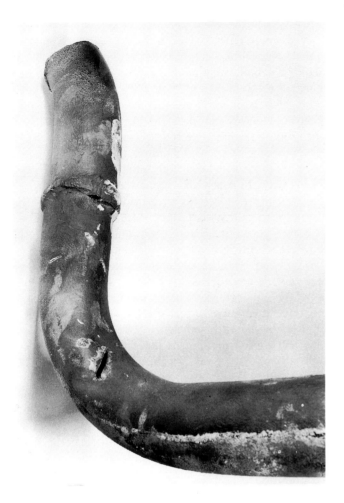

FIG. 5.19 A two-stage failure due to a weld defect
A relatively small hole in a weld has eroded an adjacent tube (not shown) which has in turn eroded the first tube in a slightly different place. It is quite conceivable for an original weld failure to erode a significant number of tubes and still be missed on inspection. If it is missed the failure will occur again. Such a problem demonstrates the importance of locating all original failures and hydraulic testing.

airheaters is corrosion of the elements, particularly at the cold end. This deterioration is long-term but must be monitored to allow forward ordering of spares. Replacement of elements allows a review of element configurations and materials to take advantage of latest advances in the technology. Replacement of an element is more an exercise in logistics than a technical challenge. The flow of elements out and in must be carefully controlled, as other boiler work will be taking place in the area. In particular, there is unlikely to be adequate floor-loading capacity in the immediate area of the airheater to store any significant number of elements temporarily.

The drive train for the airheaters is subject to the usual alignment and lubrication requirements of

gearboxes (see Section 6.1.8 of this chapter). Regular visual inspection of components should be sufficient to ensure a long life. The exception is the drive gearing for the moving duct-hoods on the Davidson airheater, which is operating in a dust-laden environment. This should be carefully inspected and wear monitored against tooth profiles.

It is not usual to provide airheaters with isolating dampers which would permit any internal on-load maintenance, since the installation and maintenance costs could not be justified. However, occasionally major breakdowns have occurred and these costs have been minimised by temporarily blanking-off the offending airheater to the same standard as surrounding ductwork to allow the boiler to operate on the remaining airheaters.

Airheaters are prone to fire damage in the event of malfunction of combustion equipment or sootblowers. Fire detection and fire fighting equipment should be tested and maintained, together with the hopper drainage facilities to avoid overloading with water. Correct maintenance of this equipment can avoid the necessity of employing the extreme tactics of the previous paragraph.

6.3.2 Fans

Most modern coal-fired units have two forced draught (FD), two induced draught (ID) and two primary air (PA) fans. The maintenance criteria for all are generally similar, although the amount of repair work required is influenced by the cleanliness of the air or flue gas. The major areas for concern are the fan runner, the shaft and bearing systems, the dampers (including flow control systems), and the casing with its associated ductwork.

Fan runners are subject to erosion, corrosion and fouling by boiler dust. Without exception, the best method of protecting a fan runner is to minimise the dust burden with good precipitation. The wear on the fan runner of a centrifugal fan is generally most severe at the leading edge of the blades and at the centreplate. Obviously, ID fans are most susceptible, but PA fans, coming after airheaters, are subject to this erosion due to airheater leakage. Wear plates and replaceable nosings are designed into fan runners to accommodate this wear. These items should be replaced before they disintegrate to avoid major damage to the runner. If a runner is subject to erosion, this can cause imbalance which can overload bearings. Ultimately, erosion can weaken a fan runner to such an extent that its structural stability is affected. On unit overhauls, the inspection of fan runners should include a check on the integrity of fabrication welds to maintain structural integrity.

Fan shafts and bearings should, in common with all such rotating machines, be kept supplied with adequate clean oil. White metal bearings should be checked for white metal adherence and correct oil clearances. The journal should be maintained true and unblemished. It is possible to hone shafts *in situ* using specialist machining contractors, if required. On large fans, separate forced-lubrication systems should not be neglected, as the costs of fan repairs due to lubrication failure would outweigh the cost of oil system maintenance. To avoid overloading bearings, the fan and motor must be correctly aligned. Flow from large double-flow centrifugal fans is usually controlled by movable inlet guide vanes (IGVs). These must be maintained free to move to ensure adequate control. On base-load plant, it is worth exercising them occasionally. On overhaul, the system should be inspected for wear and the structural integrity of each component, which usually involves a system of links and bearings. If the IGV on one side of a double-flow fan fails and closes while the other side remains open, this can impose dangerous thrust loads on the fan.

Casing and ductings should be inspected for structural integrity and cleanliness. It is particularly important that no debris is left inside the ducting, as fans may suck it in and eject it through the casing.

Major items such as fans can be maintained using machinery condition monitoring techniques. The common faults, such as runner erosion, running fouling, foundation problems, alignment errors and bearing deterioration, will show in small vibration changes. Amplitude and frequencies can be analysed to define the problems ready for remedial work on fortuitous outages.

6.3.3 Sootblowers

On a 500 or 660 MW coal-fired boiler, there are typically 100 to 120 sootblowers. These are invariably electrically-powered, using steam, air or air-with-water injection as the blowing medium. The blowers operate in an atmosphere contaminated with dirt and gases and are used intermittently. The duty of sootblowers is thus very arduous and maintenance work will usually occupy a significant proportion of the on-load maintenance effort. Should they be neglected, lack of sootblower availability can rapidly cause lack of unit availability while a major complex boiler-clean is carried out.

The choice of blowing medium influences maintenance work. The majority of 500 MW boilers have steam-blowing installations while newer plant now usually has compressed air blowers. The steam systems require high-temperature pipework, together with drainage systems for prewarming and evacuating condensate. Usually pipework is lagged to prevent excessive heat loss; even if this is not cost effective, due to the intermittent use, some lagging to ensure operator safety is necessary. Where air is the blowing medium, the pipework and valve components are simple but the cost of compressor maintenance must be considered. The large compressors required for

this duty are complicated machines in their own right. Maintenance staff will probably need special training to a standard similar to that required for turbine work.

The main types of sootblowers found on boilers are simple wallblowers, long-lance retractable blowers, semi-retractable long-lance blowers, special airheater lance or rotary blowers. These blowers always contain, and thus require the maintenance of, a valve mechanism usually known as the poppet valve to admit the fluid, a gland assembly to seal the moving lance or gun and a gearbox to transmit the drive. The lance blowers obviously have additional maintenance problems with the lance and support arrangements.

The *poppet valves* for the various types of blower are often to a similar design to allow the use of standard parts, or at least variations on a theme. Air blowers require low alloy steel valves, whereas steam blowers may require high-temperature steels. The poppet valve is usually operated by the movement of the lance or gun and this setting is crucial. If the valve operates prematurely, this can result in boiler tube erosion: if the valve opens too late or fails to open, this will expose the gun or lance to the furnace heat without it being cooled by the blowing medium. The maintenance of the poppet valve seats, integral gland and joints are subject to the same procedures as described for miscellaneous valves (see Section 6.2.8 of this chapter).

The *gland assembly* on a sootblower seals the fluid inside the gun or lance while this component moves into the boiler. The selection of the gland packing material depends mainly on the blowing medium. The early steam installations were fitted with asbestos-based packing and these are still available. Current practice is mainly to use expanded graphite preformed rings. Systems, using air as the blowing medium, tend to have polytetrafluoroethylene (PTFE) preformed rings.

The main reasons for gland-packing failures are usually external to the packing itself. The thermal cycling caused by intermittent use of blowers shortens packing life. Ash particles from the boiler tend to accumulate on the metal sealing faces of the gun or lance tubes and are dragged into the gland, causing failure of the gland packing and of the lance. Obviously, when a lance is damaged, packing life will be dramatically reduced. Finally, the gland packing is usually subject to unfortunate design loads. As the lance travels into the boiler, the forces supporting the lance are redistributed. Inevitably, a gland seal suffers vertical loads which are detrimental to the sealing ability of the packing. To minimise this effect, the gland is often contained between two sleeves to provide this vertical support. Unfortunately, these sleeves are subject to wear due to ash impregnation and eventually the load is transferred to the packing material. The development of packing materials is a continuous process and the maintenance engineer should keep abreast of developments in new and improved materials.

The *guns* on short-travel wall gunblowers or the fixed lances on static low-temperature-zone lance blowers tend to give little trouble. When blowers are overhauled, it usually is sufficient to inspect the components physically to ensure that they are still straight and intact, so that they are able to fulfil their function. These components are subject to wear from erosion by the blowing medium (more from wet steam than air, obviously) and dust erosion or slag damage from the flue-gas side of the blower. Usually these components are not exposed to high temperature damage unless a gun blower is jammed into the boiler without the blowing medium, which is rare.

The *lance assemblies or long-lance blowers* are obviously extremely vulnerable; they are long slender components operating in an extremely hot environment. Should the lance suffer a reduction or total loss of the blowing medium which also cools the lance, the lance will be destroyed in minutes. Similarly, if the lance loses its drive and does not rotate and travel, it will fail rapidly. A simultaneous failure of blowing medium and drive results in almost instantaneous destruction of the lance. The control system can be used to minimise this sort of damage. If the supply pressure is low, the blower start can be inhibited. If the initial blowing flow is low, the control system can reverse the lance and stop the blowing sequence, and the drive motor protection can be arranged to reverse the lance if it stalls on the inward stroke, for example, if the lance strikes a slag lump. The use of modern computer control systems on recent installations is particularly amenable to the inclusion of such protection systems. When lances are protected from such sudden failures, they are obviously subject to longer-term factors causing wear. Lances are supported occasionally by roller assemblies at the wallbox and, on long lances, a moving-roller support between the gland and wallbox. These supports are prone to wear by trapped ash particles, or the roller may seize and rub the lance. The lance is also subject to external corrosion and erosion as it travels in and out of the boiler. The internal face of the lance can be corroded by condensate from steam blowing, damp air in air blowers, or by condensation of flue gases escaping from the boiler. The feed tube or inner lance is susceptible to the internal corrosion mechanisms but less so than the aforementioned external mechanism.

However, the feed tube is the metal surface of the gland and can be damaged, as previously discussed. When sootblower lances are achieving long times and good availability, these long-term wear parameters can be monitored by NDT techniques during major unit outages. Sudden rupture of sootblower lance components from these causes should be prevented to avoid injury to staff.

Gearboxes used on sootblowers usually present few maintenance problems other than the sheer number of them. Failures are usually due to contamination or loss of lubricant, and occasionally overloading. When the gearboxes are overhauled, visual inspection of used parts and the observation of fitting tolerances is adequate to ensure a good repair. *In situ* repair is not recommended and special care should be taken to avoid contamination of lubricant by boiler grits.

Drive transmission mechanisms have given a number of problems over the years on a variety of designs; the CEGB have modified a number of blowers to a rack-and-pinion linear drive which has been successful, although this cannot be described as the only successful drive system.

The philosophy of sootblower maintenance should be considered by the maintenance engineer. Operational experience will show which blowers are most vital to boiler operation, so that priorities can be established. Routine maintenance which involves stripping-down rather than mere lubrication and functional checks, involves a tremendous number of man hours. Current experience suggests the most economic method of maintenance is to overhaul gun blowers on major surveys and lance blowers when they break down. To minimise outage times, components, such as gearboxes and poppet valves, should be stocked as sub-assemblies.

6.3.4 Boiler structural steelwork

During the late 1970s, there were a small number of significant structural failures on large boilers of 15 to 20 years of age, i.e., in the middle of their design lives. This prompted examination of other boilers and the result is that structural inspections to some extent are now a regular part of a boiler overhaul.

All modern boilers have their pressure parts suspended from the boiler house portal frame and expand downward in service. The structure of the boilers between this portal frame and the furnace roof is then a complex mass of support rods, headers, pipes and attachments. Below the furnace roof, there are also support arrangements for headers and hoppers, the loading of which is usually transmitted upward through the walls to the roof structure. When the majority of 500 MW boilers were built, little information on load transference was available and designs were usually based on static loads, with assumptions of transferred load. The result is that surveys both on- and off-load have revealed completely unloaded and grossly overloaded supports, with consequent damage.

In the light of this recent experience, a new boiler should be fitted with suitable strain gauge and thermocouple installations to monitor support system loadings, and sling rods should be designed with facilities for direct jacking to measure loads. At the construction stage, such embellishments are fairly cheap; on existing plant, they are very expensive.

A boiler support system should be inspected to ensure that it is not failing by overloading and progressive load migration. Only experience will determine the necessary frequency and extent of this examination. Before checking the support system, the inspector should be fully conversant with the roof expansion philosophy, tube attachments and the design of the seal systems to accommodate differential expansion, as well as a knowledge of civil engineering inspection work. The sorts of problems which may be uncovered are original design or construction defects, local overheating due to seal failure or proximity of high-temperature parts, stresses due to inadequate expansion arrangements or fouling of those arrangements, and the wasting of sling rods due to dewpoint corrosion at canopy roof level.

Where overloading is suspected, it will be necessary to weigh the sling row and its associated sling rows, using calibrated hydraulic jacks. It is not sufficient merely to prove that overloading is occurring, its source must be identified. If on-load overloading is suspected, jacking is impossible (or, at least, difficult) and strain gauges to a remote recorder may be necessary. When the overloading is completely identified, it can be accommodated either by redistributing the load or by strengthening to accommodate the boiler's natural loading.

Where expansion arrangements involve tubes sliding in refractory material or with seal bellows, these designs are prone to failure, which allows gas and dust ingress to deadspaces. Dust loading becomes an important factor and increases in ambient temperature may dictate the use of higher alloy steel. Some designs call for complex lagging details to protect the sling rods which are unlikely to be achieved in practice. From the maintenance point of view, it is more practical to use high-temperature steelwork and eliminate the complex lagging.

Boiler suspension knowledge has grown tremendously in the 1970s and 1980s and new designs will require less onerous lifetime inspection in view of this knowledge.

7 Turbine maintenance

7.1 Maintenance philosophy

Breakdowns and unscheduled outages on large modern turbine plant are expensive, both in direct repair costs and replacement costs. Maintenance is therefore directed towards obtaining sustained reliable operation between major overhauls. The scope for carrying out routine preventive maintenance during periods of operation is limited except on peripheral items, such as automatic greasing equipment, oil filters and coolers, etc. This section is therefore concerned mainly with work carried out at times of major overhaul. The extent and frequency of turbine maintenance is

governed by a number of requirements, some of which vary according to the type and manufacture of the machine. To minimise unit downtime, the work is planned for times of statutory inspection and all other requirements have to be modified to suit this timing.

Plant condition monitoring has a major influence on the extent of maintenance carried out during overhauls. It can be used to indicate the state of plant and thus give confidence for continued trouble-free operation without resorting to expensive stripdown inspections. Nevertheless, it is important to realise that condition monitoring will not necessarily warn of an incipient defect that could lead to failure on return to service.

Thus it is usual practice to undertake thorough inspections at intervals of between 6 and 9 years (dependent on experience, plant type and mode of operation), with minor overhauls between to carry out routine work on governor gear, steam admission valves, statutory inspection of vessels and welded joints, etc. Additional criteria to be considered when planning major turbine overhauls are:

- Inspection/renewal of components subject to high temperature creep (studbolts, casings, spindles).

- Work to restore turbine heat rate.

- Modifications to improve plant reliability and efficiency.

- Major refurbishment of components for plant life extension.

Turbine maintenance is covered in detail in the following Sections 7.2 to 7.6. When planning turbine overhauls, it is essential to consider the interactions between the various tasks and to cater adequately for the need to maintain a clean and tidy working environment. Clean-condition working areas should be set up whenever turbine cylinders, valves, pipework and bearings are opened. Adequate checks must be incorporated when rebuilding plant to ensure that no foreign material remains in turbine components on completion.

7.2 Turbine rotating parts

Turbine rotating parts here comprise shafts, bearings, couplings and barring gear, together with the related subjects of alignment, coupling concentricity and lubrication. Turbine blading and glands are dealt with in Section 7.4 of this chapter.

7.2.1 Shafts

At times of major stripdown, it is usual to carry out a comprehensive survey of turbine shafts. The degree and type of inspection varies between the different turbine stages. Shafts subjected to high temperatures (HP and IP reheat) are examined for creep effects as well as fatigue, whereas, on the higher-stressed LP shafts, stress corrosion and/or corrosion fatigue effects predominate. All shafts are examined at gland areas for signs of corrosion pitting and at the journal areas for wear and scoring. If necessary, the surfaces are skimmed or polished at these positions.

High temperature creep is monitored by carrying out dimensional checks of shaft diameter, together with truth checks. It may be necessary to remove scale by grit blasting, using either sharp sand or fine aluminium oxide. The prepared surfaces can then be examined for fatigue cracking (using either magnetic particle inspection (MPI) techniques or dye penetrant), particularly at changes in section, or at blade-root fixings.

Some machines have experienced stress corrosion and/or corrosion fatigue cracking of LP shafts, with the presence of condensing steam providing a sufficiently corrosive environment. The problem can be accentuated on reheat machines undertaking two-shift operation by contamination from boiler/reheater leaks during vacuum raising. Ultrasonic techniques are now available which permit inspection for cracking at vulnerable areas, such as changes of section and disc/keyway locations.

7.2.2 Journal bearings

The shafts of a turbine-generator are supported in journal bearings. These are white-metal lined, thick-wall bearings and may be either plain or spherically seated, dependent on their position and duty. Figure 5.20 shows a typical spherically seated turbine bearing, with shimmed pads for adjustment. The bearing is lined with a tin-based white metal. The bore is machined to an elliptical shape to combat any tendency for shaft whirl; this shape is achieved by machining the bore with shims in the half joint to suit the shaft journal size, as recommended by the turbine manufacturer. Once machined, there is no requirement to bed the shaft or adjust clearances, except that horn clearances may need slight adjustment.

Maintenance of turbine bearings commences with the removal of bearing covers (keeps) and top halves. As with other parts of the turbine, it is important to record the 'as found' condition, as this may assist in determining the work to be carried out and in accomplishing a successful rebuild. Readings are taken of the bearing top and side oil clearances and of the pedestal bridge gauge readings. Bearing top half clearances are taken by assembling the bearing with a piece of appropriately-sized lead wire on top of the journal and measuring the thickness to which it is compressed. Each bearing pedestal is fitted with a bridge gauge

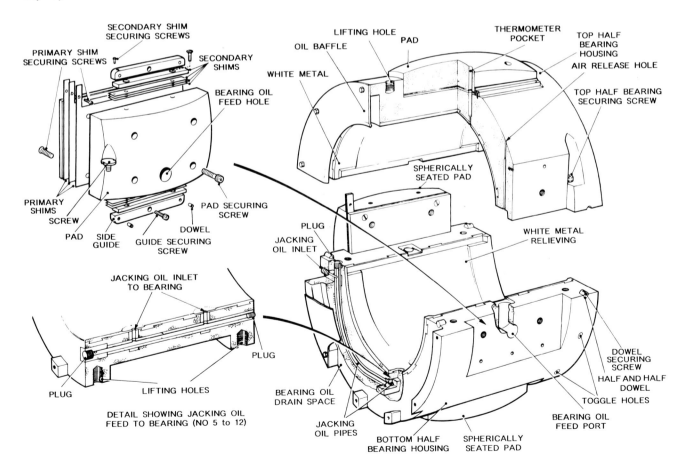

FIG. 5.20 Typical spherically-seated turbine bearing

345

for the measurement of the side and top clearances of the journal, thus determining precisely how the shaft is lying with respect to the pedestal. These datums are useful as checks of correct reassembly.

Having obtained these initial measurements, the shaft can be supported either by overhead crane or by special shaft-raising gear to permit removal of the bearing bottom half. At this stage, various inspections are carried out on the bearing.

The bearing surface is first examined for signs of deterioration. The relatively low melting point white metals undergo a rapid loss of fatigue strength with increasing temperature, so the surface should be inspected for cracking or crazing. This loss of strength may also result in extrusion or wiping under adverse loading/lubrication conditions. Although, in less severe cases, the bearing may subsequently be dressed for re-use, it is important to find the cause of the damage before doing so.

During manufacture, the white metal forms a metallurgical bond to the steel housing. This adhesion should be checked, using ultrasonic techniques; any significant deterioration of the bonding renders the bearing unfit for further service.

The surface contour must be inspected at jacking-oil holes, lubricating-oil feeds and at any anti-whirl grooves that may be included. If necessary, the metal is dressed. The condition of the anti-rotation devices, housing half-joints and spherical pads should be examined and any roughness removed with a file. Fretting of the spherical surfaces may indicate incorrect bedding of the pads to the pedestal; this can be checked using engineers' blue marking and the pad shims adjusted accordingly.

When rebuilding the bearing, all surfaces must be kept scrupulously clean. Bridge gauge readings and clearances are recorded for future reference. With spherically seated bearings, the weight of the shaft must correctly seat and align the bearing in the pedestal. The keep is then required to 'nip' the bearing in this position; this is checked by means of lead wire on the top pad with shims at the keep joints. A 0.02–0.07 mm 'nip' is generally acceptable. Before the keep is finally refitted, the pedestal is cleaned and an oil flush carried out.

7.2.3 Thrust bearings

Two main types of thrust bearing are used on turbine-generators. The *high duty tilting-pad* variety is used to absorb the residual end-thrust of the main turbine shaft system, whilst maintaining the correct axial alignment of fixed and moving blades. The bearing consists of two rings of white-metal-faces pads on either side of a shaft thrust-collar. Individual pads are able to tilt to allow formation of a stable hydrodynamic oil film. The other form of thrust bearing incorporates *white-metal thrust faces* at each end of a journal bearing. These are associated with lighter duties, typically being fitted to exciters and gearboxes.

The thrust-face metal may be grooved radially to feed oil and to aid cooling of the surface. Figures 5.21 (a) and (b) illustrate the two types.

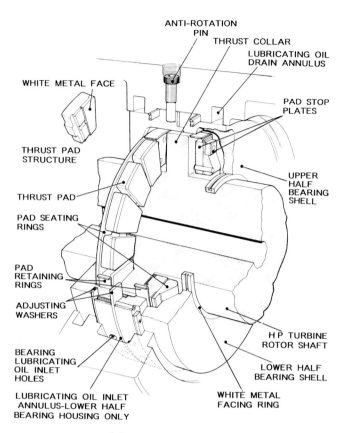

(a) High duty tilting-pad

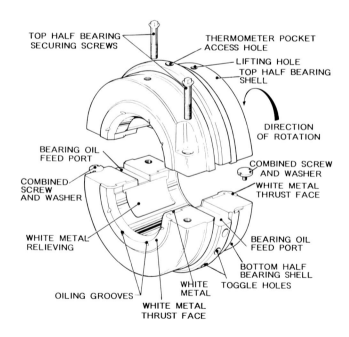

(b) White metal thrust faces

FIG. 5.21 Two types of thrust bearing used on turbine-generators

In both, it is important that the load is transmitted evenly; this requires the collar to have a flat, true surface, perpendicular to the axis of the shaft (i.e., without swash), and an equally true and flat surface of white metal. The bedding is checked using engineers' blue and the white metal scraped accordingly. The thickness of the tilting pads must be matched. In the tilting-pad bearing, the float between thrust faces is adjustable by means of shims and liners attached to the pad carriers. The liners are also used to position the turbine shafts to give the correct axial blade clearances (see Section 7.4 of this chapter).

During maintenance, the condition of the white-metal lining is inspected for wear, scoring and adhesion. To ensure freedom to tilt, the ribs or steps on the backs of pads must not be worn or damaged. All oil feed galleries must be clear before reassembly.

7.2.4 Couplings

The high power transmission associated with steam turbines necessitates the use of solid couplings, with fitted bolts, between individual shafts. Very little maintenance is involved unless the holes suffer scoring, or concentricity checks indicate that coupling hole reboring is necessary. Bushes are often fitted to couplings which enable badly scored or oversize holes to be recovered to near normal size.

Bolt-hole refurbishing is carried out with the coupling halves clamped together concentrically. The holes are bored or drilled to remove any step between the two halves. A honed finish is used to give a smooth surface. The recesses for the bolt heads and nuts must be faced perpendicular to the hole to ensure that there is no distortion of the bolts as they are tightened. The nuts must be locked, as recommended by the turbine manufacturer.

Flexible couplings, which are able to accommodate the large axial movements caused by turbine expansion, are often used to couple exciters, where power transmission duties are less onerous.

Figure 5.22 illustrates a claw-type coupling which drives through pads on each claw. These pads need checking periodically for wear or for ridging of the surface, as this may cause the coupling to resist expansion movement and overload the thrust bearing. Badly-worn pads should be renewed and bedded to the driven surfaces of the coupling muff or sleeve. This is best accomplished by means of a special jig that holds the claw and sleeve concentric: each pad is bedded in turn. The pads are adjusted so that they all contact the sleeve surfaces in this concentric condition.

7.2.5 Shaft alignment

Large, modern turbine-generators consist of up to six solidly-coupled rotors that must be aligned such that

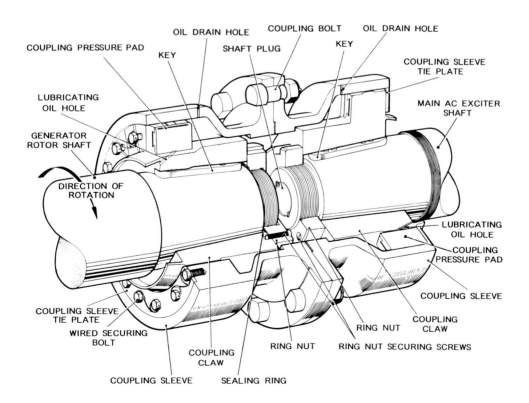

FIG. 5.22 Generator/main exciter, flexible-claw type coupling

347

the shear forces and bending moments at the couplings are zero. Each shaft exhibits elastic deformation due to its own weight and thus the perfectly aligned system forms a curve, the shaft catenary. Figure 5.23 illustrates a shaft catenary for a 660 MW six-rotor machine. The rotors retain their natural deflections at all speeds except when passing through a shaft critical. To ensure the smooth running of a machine, it is first necessary to build the shaft system to the static deflection curve and later to check the alignment periodically. Compensation, by biasing the heights of particular bearings, is built into the shaft alignment to cater for the effects of:

- Changes of bearing height due to thermal expansion.

- Large differences in journal diameter between adjacent bearings, such as may exist, for example, between the generator rotor and exciter.

- Shaft whirl associated with lightly-loaded bearings.

An initial approximate alignment of bearing pedestals and cylinders has traditionally been obtained during erection by means of a taut piano wire stretched along the axis of the machine. The various pedestals are positioned with respect to this wire, their height being adjusted in line with the expected catenary. Modern methods employ optical systems, using either precision telescopes or lasers.

Final alignment of the rotors is obtained by adjusting the bearings to give parallelism and concentricity at each pair of couplings. This procedure is followed at times of maintenance when previous condition monitoring observations, or changes of rotors or bearings, have dictated that shaft alignment should be checked. The alignment is carried out by measuring the face gaps and periphery errors between couplings. Gap measurements, using suitable gauges, are taken at the top, bottom and two side positions and a repeat set taken with both shafts rotated through 180°. The average of the two sets indicates the true parallelism of the coupling faces, eliminating any errors due to out-of-truth between coupling and shaft.

Concentricity is checked by measuring between the peripheries of the two couplings, using either dial indicators or feeler gauges, to a finger attached to

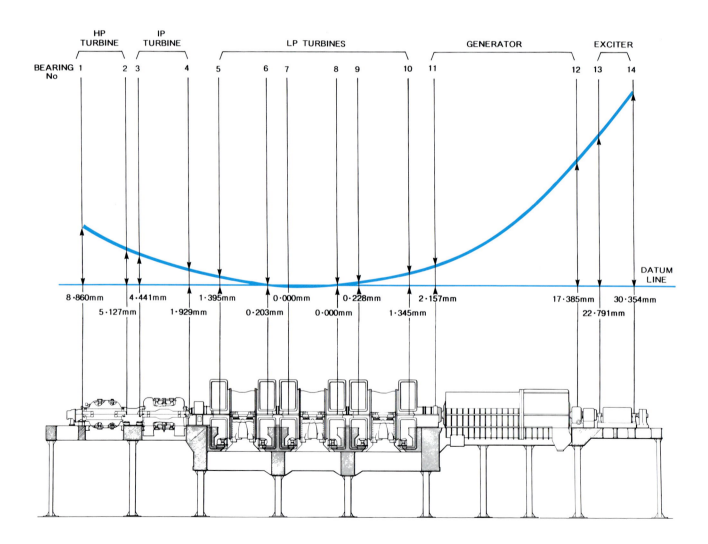

FIG. 5.23 Typical shaft catenary for a 660 MW turbine-generator, showing the vertical heights above datum of each bearing

one of the couplings. To eliminate coupling errors, both shafts are turned together and readings taken at the top, bottom and two side positions. Calculations, using ratios and similar triangles, taking account of shaft length, distance between bearings and coupling diameter, determine the adjustments required at the bearings to obtain shaft alignment. With solidly-coupled shafts, it is usual to work to very close tolerances of the order of ±0.025 mm.

Although it is possible to calculate the adjustment needed for both gap and periphery errors at the same time, the correction of gap errors can require substantial bearing movement and so it is often advantageous to correct them first. It should also be remembered that such corrections will affect blade and gland clearances and consideration should be given to the removal of cylinders to check these.

Where a major deviation from the catenary is suspected, the vertical alignment of the shafts can be checked *in situ*, using a laser and suitably designed sighting targets mounted on the shaft journals. Such a system has been developed and used successfully by the CEGB.

7.2.6 Coupling concentricity

When two turbine shafts are coupled together, it is important to ensure that they are virtually concentric with one another, i.e., no misalignment between the axes of the two shafts, and that they remain so during subsequent operation. Failure to achieve good concentricity may result in unacceptable bearing vibration set up by an out-of-balance due to the two shaft masses being on different axes. Fitted bolts are used at couplings to maintain alignment and to transmit the high torque loadings. Any error in pitch circle diameters between couplings will adversely affect concentricity. The following procedure is adopted to ensure concentricity to within 0.02 mm at adjacent bearing journals (refer to Fig 5.24):

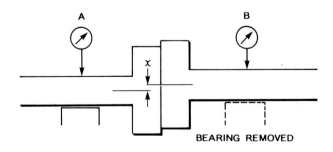

FIG. 5.24 Concentricity measurements between adjacent shafts

Misalignment between shaft axes = X
Therefore total indicator reading at B = 2X
Dial indicator at A measures any movement of the supported shaft in the bearing.

- The coupling halves are bolted together. If the concentricity is suspected or known to be in error then only a few undersize bolts are used. This enables the couplings to be moved relative to one another for correction purposes.

- One bearing bottom-half is removed and dial indicators rigidly mounted at both journals.

- The shafts are rotated and the indicators are read at 45° intervals. The indicator at the unsupported journal responds to any misalignment of the shaft axes. The indicator at the supported journal is purely for control purposes, recording any movement due to irregularities or ovality of the shaft. The difference between the indicators represents the amount of eccentricity present.

It must be recognised that any slight ovality or deformation of the journal surfaces will affect the readings when such small deflections are being observed. Plotting the net deflection versus angular position can aid analysis of the results. Theoretically, the plots should form a smooth cyclic curve that reduces to a straight line for perfect concentricity. By matching the best curve to the measured deflections, any errors due to deformation can be eliminated and the eccentricity determined.

Small errors may be corrected within the tolerances of the fitted bolts; larger errors will require the use of undersize bolts to obtain concentricity, and then the boring of the coupling holes to fit new bolts, as described in Section 7.2.4 of this chapter.

7.2.7 Lubrication system

Maintenance of lubricating oil system components consists of stripping the various regulating valves, oil pumps and their drive arrangements for inspection. There are no specific requirements other than to ensure that all items are in good order, replacing any worn parts as necessary. Oil coolers and filters must be cleaned and procedures established for flushing the oil system on completion of maintenance work to remove any contamination of the pipework, etc. This is best accomplished by installing temporary pipework at the pedestals to by-pass the bearings and thus promote flows greater than those under normal operation. The oil filter should be cleaned again after flushing.

7.3 Turbine casings and support arrangements

Operational flexibility of a turbine is largely dependent on accommodating the large thermal expansions that take place whilst, at the same time, maintaining the alignment and clearances of the rotating parts. This necessitates regular maintenance of turbine casings and their associated support arrangements.

7.3.1 Turbine casings

The opportunity to inspect and carry out work on turbine casings usually arises from the need to do other work, for example, blade or high temperature bolt inspections. Maintenance consists mainly of renewal of defective thermocouples, checking for distortion, and inspection, including non-destructive testing (NDT) of the structure.

Large modern turbines employ double-cylinder arrangements at the HP and IP stages to reduce the operating stresses and consequently the size of the cylinder flanges. This means that components, such as bolts, keys and thermocouples, operating at the highest temperature conditions are not easily accessible, without resorting to an expensive stripdown of the outer cylinder. So, when the opportunity exists, a thorough examination of all components is carried out. This includes NDT of the casings themselves, particularly at changes of section, to examine for thermal fatigue or creep cracking; inspection of inner cylinder support and retaining arrangements is also undertaken. The double-cylinder configuration necessitates special steam inlet connections incorporating a piston ring joint at the inner cylinder; the condition of these piston rings must also be checked.

After several years of operation, cylinder distortion may arise, resulting in difficulties in making leak-free joints and in obtaining gland/blading clearances, due to ovality. The degree of distortion is checked by bolting the cylinder halves together with the spindle removed and checking the bore with internal micrometers. Usually the majority of the distortion is removed by the bolting operation and the bore measurements will indicate the allowances that need to be made during the rebuild to cater for any small amount of residual distortion. In more extreme cases, machining of the joint faces and/or bores is necessary to remove the distortion effects.

Attachment of components to casing materials must be carried out to the manufacturers' approved procedures. In particular, the welding of insulation retaining pins and the fixing of thermocouples to high temperature casings needs careful attention to avoid stress-raising.

Lifting attachments are often left on cylinders during operation and therefore require careful examination before use.

LP turbines may be either single- or double-casing designs, with the fixed diaphragms located into a cast cylinder section; exhaust sections and outer cylinders are of welded and bolted construction. Inspections of the welded casing, baffles and supports should be made at times of overhaul. Bled-steam connections often incorporate expansion bellows within the condenser steam space and these also require periodic inspection for damage.

The jointing of turbine cylinders requires special care if problems of steam leakage and air ingress are to be avoided during operation. Joint faces need to be scraped clean, with particular attention being paid to areas around studs where old jointing material may accumulate.

7.3.2 Turbine support arrangements

Support arrangements associated with turbine cylinders and pedestals are important in maintaining alignment of the shaft system during expansion and contraction. Any malfunction may lead to crabbing of the pedestals with a consequential effect on shaft eccentricity. This, in turn, may result in axial and radial fouling of the fixed and moving components. Figure 5.25 illustrates typical turbine support arrangements which comprise:

- Pedestal guide keys to maintain bearing and shaft alignment.

- Bearing plates on which pedestals slide.

- Cylinder centreline guide keys to maintain cylinder alignment.

- Cylinder-palm transverse keys to allow transverse expansion and to transmit axial push-pull forces during expansion and contraction.

- Cylinder-palm bearing plates.

The first two items accommodate the largest movements and are lubricated with high temperature grease, usually from an automatic greasing facility. The higher operating temperatures of the others make lubrication difficult to achieve but, because of the smaller movements, this is not usually necessary. All keys and bearing plates require examination for excessive wear and periodic cleaning to remove corrosion products. The frequency of these inspections depends on the type of operation of the machine and maintenance experience. For instance, turbines undergoing two-shift operation, or subjected to high pipework loadings, require more frequent inspections. Some of the keys may only be accessible with the cylinder thermal insulation removed and so opportunity should be taken to examine these items at least at every major overhaul. When removing keys for examination it is important to anchor the components concerned securely, using the turbine manufacturer's approved methods. Failure to do this may result in pipework forces causing movement of cylinders, making it difficult to refit the keys involved.

Cylinder-palm transverse keys need to be examined for both excessive clearance and tightness. The push-pull action on the keys can result in apparently slight wear on the flanks being accentuated at the corners, allowing the keys to roll as the cylinder expands and contracts. The lost motion caused by rolling keys is often greater than that due to uniform excessive clearance and both types of defect must be corrected if reduction of axial clearances (leading to rubbing) is to be avoided. The unlubricated keys may also experience seizure during transmission of axial movement. This can cause cylinder distortion and abnormally high loading of the cylinder centre guide keys, leading ultimately to radial rubbing. Thus the palm keys must be cleaned and checked for freedom of movement with the minimum of keyway clearance, typically 0.07 mm.

The cylinder-palm bearing plates are usually of a bronze material supporting the steel cylinder-palms. The movements and loadings on these plates are generally small and maintenance usually only consists of removal, cleaning and refitting. Any hard scoring is dressed out and the cause ascertained (this may be due to corrosion products on the steel palm).

Cylinder centreline guide keys are also only subject to small movements and so wear should be light unless other factors, such as pipe loadings or seized transverse keys, are involved. These items often only require a feeler clearance check.

Pedestal guide keys and bearing plates experience movement equivalent to the full cylinder expansion; as their operation is essential in keeping shaft alignment, they require special attention at times of maintenance. Each key is removed in turn, with the pedestal anchored to prevent movement. The key and keyway are cleaned to remove old grease and each sized to obtain clearances. Keys having excessive clearances can lead to crabbing of the pedestals with consequential operational difficulties. When renewing keys, consideration should be given to the likely effect on alignment. If it is not apparent on which side the wear has taken place, in the absence of initial size data, a replacement key could hold the pedestal out of alignment. If necessary, therefore, alignment checks should be carried out when renewing keys, the greaseways associated with the key cleaned and the key refitted with a new charge of grease. Once all the keys have been inspected, the pedestals can be jacked to release the bearing plates. These again should be thoroughly cleaned of old grease and inspected for wear before being refitted.

7.4 Turbine blading and glands

Turbine cylinders are opened at times of major overhaul to allow maintenance inspections of the blading and glands. The need for blade examinations may also be indicated by performance monitoring; when blade damage is suspected, an enforced outage is necessary for investigation and repair. During minor overhauls, it is usual to carry out partial inspections of the easily accessible areas, such as the glands and the last row of LP blades, the latter to monitor for the effects of erosion. The sequence of operations involved in a major stripdown is as follows:

1 The cylinders are opened and the 'as found' blade and gland clearances recorded.

2 Moving and fixed blades are inspected for mechanical damage, erosion, corrosion and chemical deposits.

3 The components are cleaned and subjected to nondestructive testing.

4 Repairs and modifications are carried out, including the refurbishment of sealing fins.

5 Radial and axial clearances are adjusted.

6 The turbine rotor/cylinder is reassembled, recording the final clearance data.

Any major realignment of the shaft system should take place prior to step 5; minor alignment corrections may be carried out with the cylinder assembled.

7.4.1 Measurement of clearances

Clearances are measured at stripdown to evaluate any remedial work that may be necessary and, on reassembly, to provide a record for quality control purposes and for future reference. Axial clearances are measured at the two half-joint positions; radial clearances are taken at the sides, top and bottom positions, although the latter may be omitted if there is no other reason to remove the turbine spindle. Side clearances are measured using feeler gauges. Top and bottom radial clearances are obtained by lowering the cylinder or spindle onto appropriately-sized lead wire and measuring the compressed material with suitable dial calipers.

All clearances should be related to the turbine normal running position, for example, when checking axial clearances, the shaft system should be located against the thrust pads. Because of the support arrangements, some designs of cylinder may necessitate the bottom halves to be raised to obtain running radial clearances; information concerning this will be contained in the manufacturer's maintenance procedures.

7.4.2 Inspection, cleaning and NDT

Initial inspections are carried out at the time of stripdown, when a general impression of the condition

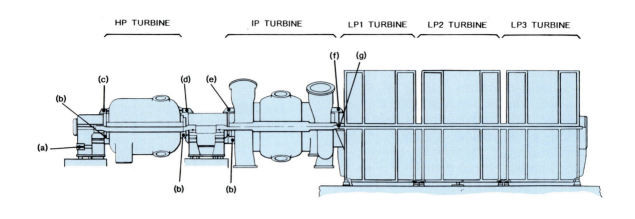

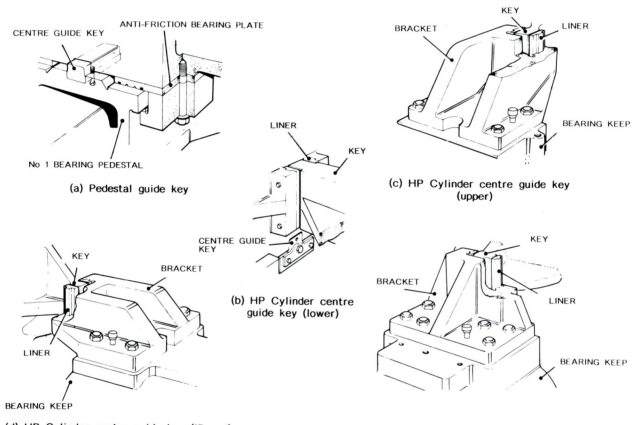

(a) Pedestal guide key

(b) HP Cylinder centre guide key (lower)

(c) HP Cylinder centre guide key (upper)

(d) HP Cylinder centre guide key (IP end)

(e) IP Cylinder centre guide key (HP end)

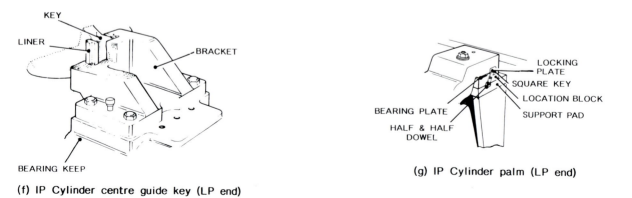

(f) IP Cylinder centre guide key (LP end)

(g) IP Cylinder palm (LP end)

FIG. 5.25 Typical turbine support arrangements

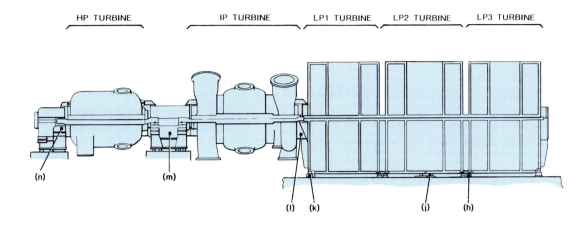

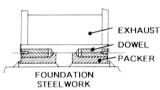

(h) LP frame supports

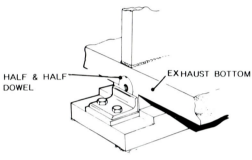

(j) LP Cylinder anchor

(k) LP frame centre guide key

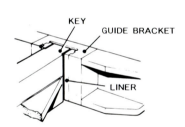

(l) IP Cylinder centre guide key to LP frame

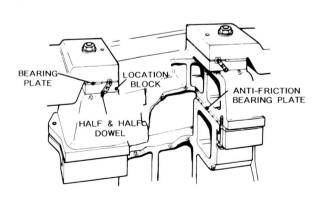

(m) HP and IP Cylinder palms

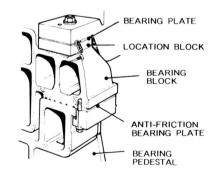

(n) HP Cylinder palm

FIG. 5.25 (*cont'd*) Typical turbine support arrangements

of the blading components becomes apparent. More detailed visual examinations are undertaken of the critical areas, in particular, blade roots and associated securing arrangements, blade support systems (such as lacing wire and shrouding) and interstage sealing arrangements. The condition of LP blade and erosion shields is also examined and the security of diaphragm retaining and anti-rotation arrangements are checked. The presence of blade deposits has a detrimental effect on turbine efficiency, due to a reduction in area or steam flow and a deterioration of surface finish and profile. The improved boiler water quality associated with modern generating units has greatly reduced the problem of blade fouling and thus the need for routine cleaning. However, at times of major overhaul or inspection, it may be necessary to remove any such deposits. Soluble accumulation is easily removed by water washing, but the more firmly-adhering insoluble material, mainly composed of silica and iron oxide, requires the use of mechanical methods for its removal.

Insoluble deposits are removed by a blasting process which, for turbine rotors and removable diaphragms, is carried out remote from the turbine in a fully-enclosed tented area. Fixed blading of bottom half cylinders is usually cleaned *in situ* for convenience, in which case special precautions are necessary to prevent the ingress of deposits and abrasives into bled-steam branches, etc. The blasting medium used depends on the severity of deposits and the surface finish required. Alumina is commonly used, being both an effective abrasive and also providing a surface finish suitable for NDT.

Detailed examination of blading is carried out to monitor for defects, particularly cracking. MPI and dye-penetrant NDT techniques are used at areas of concern; the critical areas are those that are most highly stressed and where construction joints are located. Thus LP blading (including blade roots, lacing-wire holes, ferrules and brazing, erosion shields and blade-shroud riveting) are items that receive attention. Obviously, experience of similar machines or previous failures will influence the inspection policy of the maintenance engineer.

7.4.3 Repairs and refurbishment

Work on blading often necessitates the use of specialists for the diagnosis of defects and method of repair, because of the skill and expertise required to carry out the work. In particular, since any defect on moving blading can have catastrophic consequences, expert advice should always be sought. Small surface defects, such as cracks or impact damage, can sometimes be merely dressed out by careful grinding. More serious defects may require the renewal of blade components.

This work, including lacing wire and shrouding repairs which may entail removal of blades, needs

the services of appropriately trained personnel who possess the skills and equipment required to undertake the tasks. Renewal of erosion shields also necessitates special techniques to ensure secure bonding of the shield by full penetration of the braze.

The method for re-establishing blade clearances depends on the condition of the sealing strips and the extent of adjustment necessary. If the strip is undamaged and clearances are required to be increased, then either machining of the seal areas on the spindle or machining of the fixed strip, using a boring-bar arrangement, will suffice. Where clearances are excessive or the strip has been severely damaged, the specialist operation of fitting new strip is the only option. After adjustment of clearances, all sealing fins are dressed to give a knife edge. This ensures that should a rub occur, the edges of the sealing strip will deform with the minimum production of friction heat.

Final setting of axial clearances is carried out with the cylinders assembled. The shaft is floated in the bearings (with the thrust bearing removed) until the fixed and moving blades contact. The thickness of the thrust-bearing liners or coupling spacers is then adjusted to give the designed offset from this 'contact' position.

7.4.4 Turbine glands

Maintenance associated with the modern turbine spring-back labyrinth gland is similar to that already described for blade sealing strips.

The main difference arises due to the ring of sealing strips being made up of a number of segments mounted in an adjustable carrier ring, and held in close proximity to the shaft by springs (see Fig 5.26). Should the shaft contact any of the segments, the springs allow movement, limiting any friction heating effect on the shaft.

Glands are dismantled, cleaned and inspected checking for damage to the segment fins and springs. Clearances are checked generally as described in Section 7.4.1 of this chapter, except that the segments must be wedged to prevent movement that would otherwise be allowed by the springs. The use of small jacking screws through the segment against the carrier ring is a convenient and positive method of obtaining this condition.

Damaged or excessively-worn segments must be renewed. Those showing only slight wear can be refurbished for re-use by machining their backs to allow them to spring into a smaller diameter. Subsequent operations are common to both new and refurbished components. The ends are machined to maintain the design clearance between adjacent segments. The whole ring is then assembled in the carrier, each segment is wedged or jacked to its normal running position and the assembly is machined at the sealing fins to the required shaft diameter, plus the clearance allowance. This final machining may need

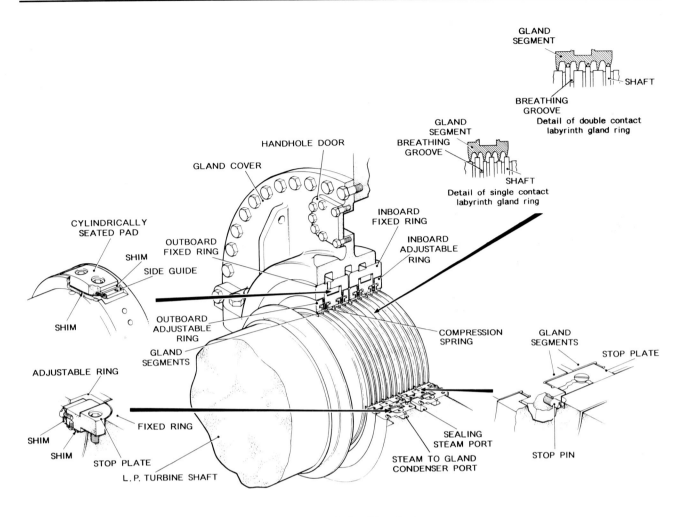

FIG. 5.26 Typical turbine gland

to allow for centralisation of the gland to the shaft, although adjustment is often possible by means of shimmed pads in a similar way to journal bearings.

Prior to final assembly, the segment fins are dressed to give a knife edge and the jacking holes are plugged. Each segment of the assembled gland is checked for freedom of movement. When the gland ring is fitted to the shaft, the axial position must be checked. Errors in this position may lead to damage when differential expansion takes place. Liners are provided to enable the axial position to be adjusted.

7.5 Turbine governing systems

Modern turbine-generators are often equipped with electronic governing systems in which steam admission valves respond to electronic speed or load signals via electrohydraulic relays. Maintenance of both mechanical and electronic systems is described, with many of the procedures being common to both. Normally maintenance is restricted to times of statutory unit overhaul, with only defects or breakdown situations

being dealt with outside these periods. The types of problem encountered with governing systems fall mainly into two distinct categories: a great deal of the system is, because of its duty, continually moving, leading to problems of excessive wear; other parts tend to run at a single position, when excessive friction becomes the main factor.

Electronic governing systems generally use a phosphate-ester fire-resistant fluid as the hydraulic operating medium. The rapid speed of response to match that of the electronic governor itself is achieved by using the fluid at pressures of up to 150 bar. The systems are therefore high pressure and low capacity; very tight clearances are necessary at the various components to reduce leakage and wear. System response and capacity is increased by the use of hydraulic accumulators. The fluid is maintained to a very high quality by means of conditioning plant, incorporating filters and vacuum chambers for the removal of moisture, particulate matter and gums. Phosphate-ester fluid has the advantage of a higher temperature capability than lubricating oil, but must be handled with care due to its toxicity.

7.5.1 Steam admission valves

The maintenance of turbine valvegear is most conveniently considered in two parts; the steam side and the operating relay equipment. Although detailed designs vary between turbine manufacturers, many of the maintenance aspects are similar. Figure 5.27 shows a typical governor valve and operating relay of a mechanical system. Figure 5.28 (a) illustrates a throttle valve of an electronic system, and Fig 5.28 (b) its electrohydraulic relay.

At times of major overhaul, it is usual to dismantle the steam valves for inspection. Valve cover nuts are removed, using procedures described in Section 7.6 of this chapter, and the valve internals withdrawn, together with any integral steam strainer baskets. All components are dismantled and cleaned for inspection, including the pilot valves of emergency stop valves, where applicable. Clearances between valve spindles and leak-off bushes are checked and any scoring or scale build-up removed by honing. When clearances are excessive, the bushes are renewed.

Valve seats are examined for signs of damage that could impair their shut-off capability, and for secure fixing within the valve body. In particular, seats that are bolted in position may come loose due to creep relaxation of the retaining screws, so these require special attention. The screws themselves must be adequately secured to prevent any loose or broken ones from entering the turbine and damaging blades. Pilot

valve assemblies need similar attention to screwed components, as well as a general check for the absence of excessive wear and for freedom of movement.

Comprehensive NDT is carried out to check items for mechanical and thermal fatigue damage. Valve seats and valve heads are checked for cracking of the seat facing material, using dye-penetrant. Valve spindles are subject to large steam velocities during operation and flow-induced vibration can ultimately cause failures. Vulnerable areas, such as changes of section, crosshead tapers and threaded portions, are checked for cracking, again using dye-penetrant. Valve body forgings are inspected for thermal fatigue or creep damage, concentrating on changes of section, other known areas of weakness and welded attachments. These checks usually take the form of ultrasonic and/or magnetic particle inspection and are particularly relevant to machines subject to thermal cycling.

If the turbine blading is to be protected from damage, cleanliness during maintenance of valvegear is of prime importance. Wooden blanks (or similar) must be available and strictly used to cover apertures where debris could enter the interconnecting pipework during operations, such as cleaning joint faces, etc. On reassembly, the valve internals must be checked as clear of all foreign material. The jointing of covers will be to the manufacturer's specification, with nuts tightened evenly and in accordance with Section 7.6

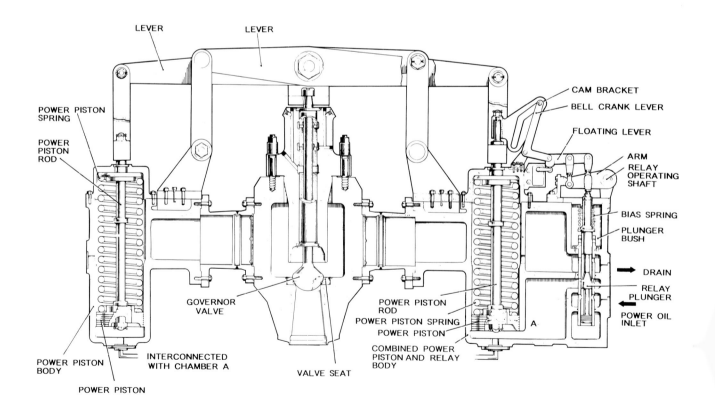

FIG. 5.27 Governor valve and operating relay, typical of a mechanical governing system

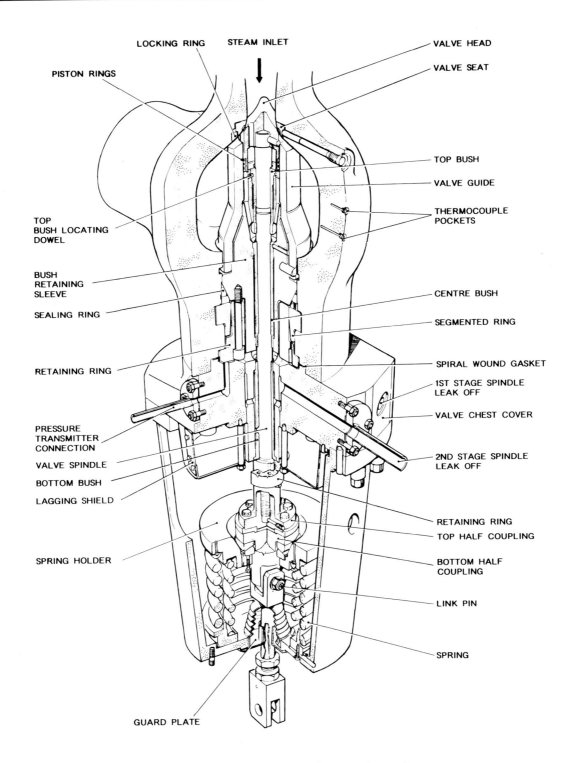

(a) Governor valve for an electronic governing system

FIG. 5.28 Governor valve and operating gear for an electronic governing system

of this chapter. Where linkages are connected to the valve spindle by a crosshead arrangement, the assembly must be securely fixed to prevent fatigue failure of the spindle. Where the spindle incorporates a taper fitting to the crosshead, this should be lapped to ensure firm contact.

The hydraulic operating relays and valve power pistons associated with mechanical governing systems tend to need less maintenance than the steam valves, especially if the operating oil is kept clean and free from moisture. However, it is usual to dismantle them at major overhauls and to inspect for wear at

357

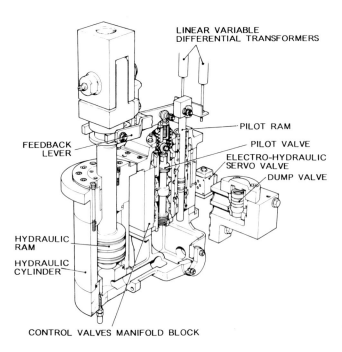

LINEAR VARIABLE
DIFFERENTIAL TRANSFORMERS

PILOT RAM

PILOT VALVE

ELECTRO-HYDRAULIC
SERVO VALVE

DUMP VALVE

FEEDBACK
LEVER

HYDRAULIC
RAM

HYDRAULIC
CYLINDER

CONTROL VALVES MANIFOLD BLOCK

(b) Operating gear for an electronic governing system

FIG. 5.28 (*cont'd*) Governor valve and operating gear
for an electronic governing system

pistons and bushes that could eventually affect their operation. Most designs of power piston incorporate strong springs requiring special care during dismantling to ensure that the stored energy is released in a controlled manner. With tight clearances between precision components, cleanliness is again an important factor. Springs and spindles are checked for corrosion, which may occur if moisture control of the oil is poor.

The pins and bushes associated with operating linkages often require considerable refurbishment due to wear resulting from the transmission of large operating forces. Some designs use PTFE-lubricated bushes, whilst others use grease lubrication from the turbine automatic greasing facility.

The servo valves and operating relays associated with electronic governing systems operate with a high quality hydraulic fluid and are therefore less likely to require regular maintenance. The high precision components are also very susceptible to damage and when maintenance is required it is usual to interchange complete modules. Therefore, maintenance requirements are determined by thorough pre-outage function checking of the system, so that any components identified as defective or suspect can be exchanged. Piston spools and sleeves are changed as a complete assembly to prevent damage to surfaces. Servo valves are removed to check for internal leakage and, if this is found to be excessive, they would be exchanged. The refurbishment of linkage pins and

bushes is similar to that described above for mechanical governor systems.

Routine maintenance of the valve servo systems is restricted to filter changes and accumulator checks. Each steam valve relay has its own filter, typically 3 μm, which should be changed at each major overhaul. Accumulator pressures are checked periodically and their bags should be changed at a frequency advised by the manufacturer, typically every 4–6 years.

Defective servo and relay valve components are usually returned to the manufacturer for refurbishment. Strict clean condition workshop facilities are required with the application of special working practices. Polished and mirror finishes are employed on components. These are very susceptible to both mechanical damage and corrosion; the latter may be initiated by corrosive fluids or even perspiration. Parts need to be scrupulously cleaned, using an approved solvent and non-linting cloth, then stored, coated with oil. Hydraulic fluid must be used on components during reassembly to prevent damage and seizure.

7.5.2 Governors

This section is restricted to the work associated with mechanical governors; the maintenance and calibration of the electronic components follows the philosophy described in Section 12 of this chapter.

A typical bobweight governor is illustrated in Fig 5.29. The centrifugal force of the spinning weights is balanced by the springs such that the weight position is proportional to the rotational speed. The lever system translates the weight position to an oil relay plunger which in turn produces an oil pressure proportional to speed. The relationship is a function of spring rates and relay design and is so arranged to give the required governing characteristic (droop). Overhaul of governors must ensure that their relationships are maintained whilst eliminating any hysteresis effects due to friction or backlash. It is often preferable to return governors to the manufacturers for major refurbishment, where the necessary special equipment and calibration rigs are available.

Wear of the linkage arrangements and oil relay sleeves are the areas usually requiring attention. Linkage bushes are usually PTFE-impregnated, allowing very tight clearances. These may be renewed, provided the necessary tools are available to allow controlled dismantling of the spring assembly. Special care is needed when pressing in the new bushes so as not to damage the bearing surface. Oil relay clearances are checked and any worn components renewed. When reassembling the governor to the drive shaft, checks are carried out to ensure that the assembly rotates concentrically.

7.5.3 General

There are many other components requiring inspection during major overhaul. These include boiler pressure

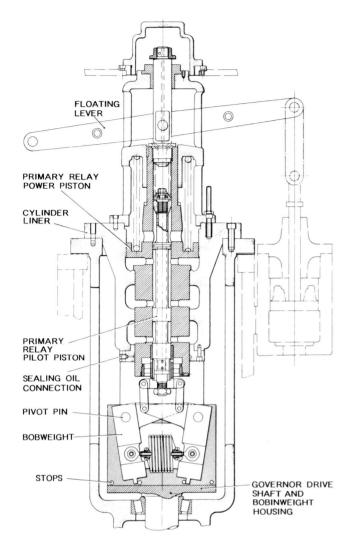

FLOATING LEVER

PRIMARY RELAY POWER PISTON

CYLINDER LINER

PRIMARY RELAY PILOT PISTON

SEALING OIL CONNECTION

PIVOT PIN

BOBWEIGHT

STOPS

GOVERNOR DRIVE SHAFT AND BOBINWEIGHT HOUSING

FIG. 5.29 Typical 'bobweight-type' speed governor and oil relay

and vacuum deloading equipment and various other ancillary oil relays. The maintenance associated with these is similar to that previously described for valve-gear relays.

The tripping system arrangements must be inspected to ensure reliable operation. The impact loads that occur when these systems operate can cause bruising of latching surfaces and excessive wear at keyways. If such damage is allowed to go unchecked, the efficiency of operation may be seriously impaired.

On completion of maintenance and before return to service, the manufacturer's recommended procedures must be followed to set the equipment for correct operation. Valve linkage settings must be adjusted and the valves operated to check that the opening characteristics are correct. A check of valve position against control oil pressure in both the opening and closing directions will indicate whether excessive friction is present.

7.6 Turbine high temperature bolts

Bolts and studs used on turbine cylinder and valve steam-chest joints, which are subject to high temperatures ($>370°C$), are made of special material having enhanced creep-resistant properties. However, at such temperatures, the fasteners have a basic life due to their limited capacity for creep strain before they crack and fail. The creep effect also results in stress relaxation from the initial value of about 300 MN/m^2, which will lead to steam leakage from the joint unless periodic retightening is carried out. The turbine manufacturer usually specifies the maximum operating hours between tightenings, typically 30 000 hours. Each tightening operation increases the average stress on the fastener and hence the rate of accumulation of creep strain. The material life is therefore reduced each time the fastener is tightened.

The materials used are either Chrome-Moly-Vanadium (CrMoV) steels or nickel-based alloys. The latter, although possessing superior basic life, have certain disadvantages. At temperatures below 538°C, they may exhibit a gradual reduction in length, with a resultant increase in stress. Therefore they should be slackened and re-tightened at intervals of about 30 000 hours to limit this stress increase. Nickel-based alloy material is also susceptible to stress corrosion, particularly in wet or condensing steam conditions. Thus any fasteners of this material suspected of operating under these conditions require more frequent examination. Lubricants containing sulphur compounds must not be used with nickel-based materials for the same reasons. All the creep-resistant materials tend to be brittle when cold and it is important not to subject them to impact forces. Temperature gradients in fasteners and joints can significantly affect material life, particularly in studs. Thermal distortion of the surrounding structure imposes bending stresses on the fastener causing abnormally high strains at the outside of the bend. In studs, these effects are concentrated at the first engaged thread which is also usually the hottest area. In addition to the accumulation of localised creep strain, cyclic straining may also lead to thermal fatigue cracking in a two-shifting regime. The maintenance of thermal insulation, as described in Section 7.7 of this chapter, is therefore very important to high temperature bolt life and joint security.

Maintenance aspects of high temperature bolts are predominantly concerned with:

● Routine tightening at recommended intervals.

● Renewal of life-expired fasteners.

● Inspection and NDT of fasteners subjected to abnormal circumstances or when approaching the limit of their basic life.

● Maintaining records of operating hours, adjusted to take account of retightenings and temperature gradients.

Fastener life depends upon the controlled application of the initial strain at each tightening operation. For CrMoV, this initial strain is usually in the range 0.12% to 0.165%; for nickel-based alloys, 0.08% to 0.11%. The percentage strain is that existing in the shank of the fastener between the first engaged threads (the effective length). Controlled application of strain ideally means direct measurement of the applied extension or application of a measured hydraulic load from which the strain can be derived. Where these methods are not possible, the strain may be derived either by measurement of the angle of rotation of the nut or by torque tightening of the fasteners whilst using an appropriate thread lubricant. Direct measurement of the extension of large bolts and studs necessitates the use of either ultrasonic instruments or special measuring equipment; the latter requires a hole through the length of the fastener. Figure 5.30 illustrates such an arrangement. It consists of a tube with a system of collets at its end which engages the bottom of the stud (or studbolt) hole. A sleeve locates at the top of the stud and the change in the relative position of sleeve and tube is a measure of the stud extension.

The actual tightening operation must be achieved without impacting the fastener and one of three methods is usually employed:

● Gradual tightening, using a hydraulic torque-wrench.

● Rotation of the nut, following extension by heat.

● Rotation of the nut, following extension by hydraulic stretching.

Prior to tightening, the joint faces are closed and any jointing material crushed by tightening the nuts sufficiently to give metal to metal contact. If direct measurement of extension is being used, the initial lengths of the fasteners in their unstressed condition will have been measured also. Extension by heat and hydraulic stretching requires the use of specially adapted studs/bolts. For heat, an electric heating element is inserted into a hole running through the length of the stud. When the applied heat has caused sufficient elongation, the nut is rotated through an angle calculated to give the required strain. This can be checked by measurement when the stud has cooled. Hydraulic stretching uses studs with a lengthened threaded portion onto which is screwed the stretching equipment. Hydraulic pressure is used to stretch the stud and the nut is then tightened. The hydraulic pressure is proportional to the stress in the stud and so the system can be calibrated to tension fasteners accurately. Figure 5.31 illustrates an arrangement for hydraulic stretching. Whichever method is used, nuts must be tightened in accordance with the manufacturer's recommended sequence. This will involve slackening each nut in turn from its torqued

FIG. 5.30 Extensometer for measuring bolt extension
The lever arrangement is used to engage the collets in the bottom of the stud hole (NEI Parsons Ltd/Hedley Purvis Ltd).

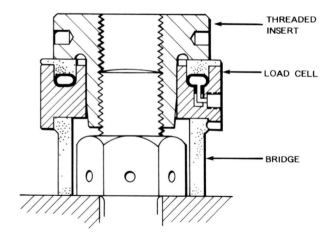

FIG. 5.31 Bolt extension tool
Hydraulic pressure at the load cell is used to stretch the bolt. The nut is then run down to mate with the surface (SPS Technologies Ltd).

condition and then applying the controlled tightening procedure. The quality management of fasteners in a power station demands the maintenance of adequate records, so that the integrity of operating joints can be confirmed and replacement of fasteners can be arranged at the appropriate time. Each joint involving high temperature bolts should be identified in the records together with data on the type, material, size, design, strain and operating temperature of the fasteners. The records are updated during the life of the joint, incorporating the appropriate life penalties associated with retightenings and/or excessive temperature gradients. Such data can to used to initiate NDT of fasteners when they reach 80% of their basic material life and planned renewal when approaching their life expiry. In the event of discovery of cracked or completely failed fasteners in a particular joint, all fasteners should be renewed and the old ones examined metallurgically to determlne the cause of failure. Premature failure may be associated with stress corrosion or thermal gradients which would require additional measures being taken to avoid a recurrence of the failure. Under such circumstances, consideration must be given to other similar joints that may be affected.

7.7 Turbine thermal insulation

Thermal insulation of turbine components is of the utmost importance in reducing temperature gradients across thick-walled forgings and along fasteners. Such gradients can cause severe distortion, leading to operational difficulties and an increased likelihood of failures due to thermal fatigue. The effects of poor thermal insulation on studbolt performance has previously been discussed in Section 7.6 of this chapter. Insulation on turbine cylinders must also minimise the differences in metal temperature between top and bottom halves at start-up and shutdown.

7.7.1 Materials

Two forms of man-made mineral fibre (MMMF) are commonly used in the insulation of turbine components, either as a spray material or in mattress form. Whichever is used, it is important that the insulation is applied to approved procedures by skilled applicators to ensure a high quality installation.

Sprayed MMMF consists of chopped rock-based fibre pre-mixed with cementation binders. It is applied with specialised spray equipment whereby the fibre/binder is mixed with water at the spray nozzle. When properly applied, it forms a continuous layer of small fibrous bundles bonded together with no joins or voids. Degradation of spray material occurs with age and is accelerated by vibration and steam/water leakage. The external coating can sometimes hide the poor condition of the spray material beneath and the taking of core samples to check material condition will assist in determining maintenance requirements.

MMMF mattress consists of rock-based fibre woven or formed into a quilt and bonded to 25 mm stainless-steel wire mesh on one side. This relatively soft and flexible arrangement is ideal for insulating complex shapes although it has the disadvantage of compressing easily and of slipping, if not adequately supported. Joints are inevitable and special techniques are needed to ensure the absence of voids.

7.7.2 Application

Insulation material is supported by means of steel studs attached to the valve chest or cylinder, usually by welding. The studs are 5–6 mm diameter, with a maximum pitch when measured between stud tips of 300 mm. Their length is such that the perpendicular distance from the tip to the surface equals the required thickness of insulation. Special arrangements are required at flange bolts to ensure the absence of voids and to maintain the insulation thickness, particularly when mattresses are used. Nuts/studs are wrapped with glass cloth and stainless steel wire is looped around them to be used later to pull the outer insulation mesh into contact with the insulation. Gaps between nuts are filled with compressed MMMF mattress so that voids do not occur when mattresses are applied. The applied thickness is dependent upon the highest casing metal temperature and is usually calculated to give an outer cold face temperature of 50°C for an ambient temperature of 30°C. Figure 5.32 shows a typical relationship between insulation thickness and hot-face temperature. The insulation thickness is built up in one, two or three layers, each retained by stainless steel wire mesh. Bottom-half cylinders have their insulation thickness increased by 25 mm to minimise temperature differences between top and bottom halves.

Spray materials should only be applied by specially trained personnel, using material that has been stored in a dry location in sealed bags. The material should not be more than 1 year old. Ideally, the process should be continuous; special attention is required to ensure good adhesion of the material to the casing and the achievement of the specified density. This may be checked by using hand pressure to see whether there is any movement of the material; the resilience will give an indication of the density. Density can also be checked by taking core samples. The final thickness is checked using a suitable probe.

Mattress is applied with the wire mesh facing outwards. The insulation must be in intimate contact with the surface and radiation paths prevented by ensuring that all joints in the lower layer are covered by mattress in the subsequent layer, the overlap being ideally at least four times the insulation thickness.

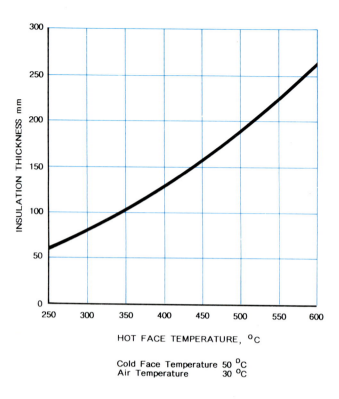

FIG. 5.32 Relationship between insulation thickness and hot face temperature

At the ends of bottom-half cylinders, the thermal insulation is tapered to finish flush with the gland housing to avoid the possibility of forming pockets which could trap oil. Casing location keys are kept free of insulation.

The insulation surface is coated with a layer of self-setting cement, 6–8 mm thick, vent holes being provided to aid drying out of the material. When thoroughly dry, the insulation system is finished with an oil- and water-resistant glass-reinforced sealant.

7.7.3 Inspection and quality control

As with all turbine work, quality assurance is of paramount importance if the reliability of plant is to be assured. Checks are required at every stage of the insulation process to ensure that approved procedures and materials are being used. The checklist shown in Fig 5.33 is an example of the documentation required to ensure this.

8 Ancillary plant maintenance

8.1 Feed pumps

Modern large generating units commonly use main boiler feed pumps of the cartridge type. These com-

prise a forged-steel barrel casing incorporating suction and discharge branches, into which is inserted an assembled cartridge consisting of shaft, impellers and diffusers. Such a pump is illustrated in Fig 5.34. Some designs take this concept further by including end covers, balance arrangements, seals and bearings in the assembly. The end covers take the form of plugs, with the joint being made and sustained by internal pressure, the plug seal being retained by a removable segmental ring. In addition to the traditional balance arrangements to cater for thrust loading, modern pumps also have thrust bearings to give a limited capability for running with low suction head.

The cartridge design, together with high rotational speeds (up to 7000 r/min), results in a compact pump with few stages; this, in turn, gives a high shaft stiffness and robust design. When major maintenance is needed, the cartridge can be quickly exchanged for a spare unit, thus minimising downtime. Refurbishment can then take place without the urgency associated with a breakdown. The specialist nature of the work, together with the equipment required for cartridge overhaul, often dictates that the task be undertaken at the manufacturer's works or at a suitable specialist facility.

As an essential feature of the generating unit, boiler feed pumps are required to be highly reliable. Consequently, much of the philosophy associated with turbine maintenance also applies to feed pumps. A condition monitoring programme is essential to give confidence in continued operation and to help to predict when both minor maintenance and major cartridge changes should be undertaken. As well as the usual pressure, temperature and vibration monitoring, the shaft running position is also checked; any substantial change indicates excessive wear of the hydraulic balance arrangements. Routine maintenance is generally associated with peripheral equipment, such as lubrication and gland-sealing systems, together with any special requirements relating to drive motors and gearboxes.

Defect maintenance associated with bearings, seals/glands, and balance arrangements can be undertaken without a complete cartridge change. It does, however, involve dismantling of the pump bearings and an essential prerequisite to any such work is a check on the bearing pedestal dowelling, together with the taking of datum measurements. These may be in the form of alignment checks and centralisation readings at the gland housings. This ensures that the pump can be rebuilt, whilst maintaining concentricity of the rotating and stationary elements.

Journal bearings are of the white-metal-lined steel bush type. They are cylindrical and it is important to ensure that they are in precise axial alignment with the shaft. This requires accurate machining, together with a bedding check with the shaft. Because of the high rotational speeds, design clearances are minimised consistent with providing sufficient oil cooling,

CONTRACT NO: .... CONTRACT START DATE (APPROX).......................

CONTRACTOR: ....

Note: Station signature indicates that agreement has been reached with the Contractor that work in the previous section has been completed to specification requirements

	Signature of Compliance	
	Station/Date	Contractor/Date
Edges of existing insulation stepped (if applicable)		
Casing cleaned		
Check that all necessary support studs are present		
Stud Sites prepared for any additional/replacement studs		
Studs/bosses attached (check for sound welds)		

STUD ARRAY CORRECT
    Correct length   spacing and coverage
    Permission to continue given by Station
    Engineer
		———————

Flange nuts/studs individually wrapped in glass cloth of correct type		
FIRST LAYER SPRAYED Compressed Density checked Check for adherence All studs visible Wire mesh in place and push-fix washers on all studs Permission to continue given by Station Engineer		———————
FINAL LAYER SPRAYED Compressed Density checked Check thickness All studs visible Stainless steel mesh in place Push-fix washers on all studs retaining the wire mesh Wire mesh fixed together with stainless steel wire (Stud tips nominally flush with the surface) Permission to continue given by Station Engineer		———————
Self-setting cement applied (vent holes provided if necessary)		
Oil and water-resistant glass reinforced sealant applied (when armouring is dry)		
ACCEPTANCE OF INSTALLED INSULATION		———————

Station Engineer's Comments

Details of defects/variations to contract requirements found during inspection and the remedial/corrective action taken before authorising continuance of the installation

FIG. 5.33 Checklist for insulation process

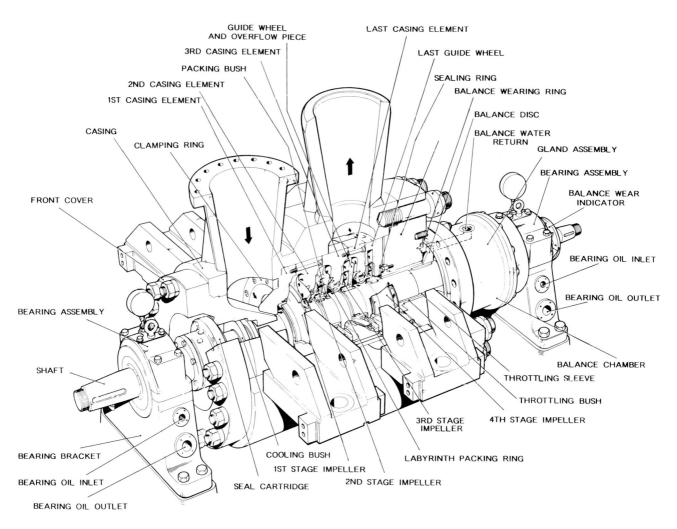

FIG. 5.34 Boiler feed pump incorporating removable cartridge

and bearings will often incorporate anti-whirl features. When thrust bearings are provided, they are of the tilting-pad design.

Two types of shaft seal are commonly used; the mechanical seal and the water-injected seal. *Water-injected seals* rely on keeping very tight clearances between the gland sleeve and the gland bush components. Water leaking out is reduced in pressure and leaked off to the de-aerator. In addition, sealing water is injected into the gland which also leaks off to the de-aerator, or to a drains tank at the outboard end of the gland. A spring-assisted stationary seal element is provided to seal under low pressure shut-down conditions. Wear of the seal components results in excessive leakage and usage of sealing water, requiring the renewal of glands, bushes and sleeves before the situation deteriorates and affects the pump bearings.

Mechanical seals use tungsten-carbide and carbon sealing faces, held in contact by spring and hydraulic pressure. They have the advantage of virtually zero leakage, with no sealing water required. Axial move-

ment allowance and sealing of the components in the seal housing is accomplished by using flurocarbon rubber O-rings. To limit the temperature of the rubber components, the seal has an external circulating water system incorporating a cooler and filter. A cooling bush is also fitted in the seal housing. The seal faces wear with time, the rate being dependent partly on the water quality. For this reason, it is important to keep the seal external circuit leak-free, as any leakage is made up from the pump barrel, which may contain abrasive particles. It is prudent to inspect the seals and renew components periodically. During the re-build of a seal, it is important to ensure correct pressure between the seal faces; this is usually set by adjusting the length of a renewable spacer sleeve. The rotating seal face should also be checked for swash during rotation; if this is excessive (>0.025 mm), it will seriously affect the performance of the seal. Figure 5.35 illustrates a typical mechanical seal for boiler feed pump duty.

Deterioration of pump performance or a major mechanical failure necessitates a cartridge change. At

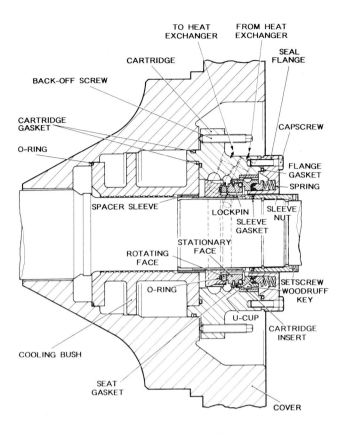

FIG. 5.35 Mechanical seal for a boiler feed pump

the same time, opportunity should be taken to examine gland/seal components and the barrel support arrangements, together with associated guide keys. Although the precise dismantling procedure varies between pumps of different manufacture, the following description highlights the main steps in a cartridge change for the pump shown in Fig 5.34. First, all auxiliary oil and water pipework is removed, making sure that all exposed ends are covered. The shaft couplings are dismantled and the bearings opened. The pedestals can now be removed and the seals dismantled. The balance chamber is unbolted and removed, giving access to the shaft nut that secures the balance disc. The nut and disc are removed together with the balance disc wear ring. The main discharge end-cover can now be released and removed. Special hydraulic stud stretching equipment is used to release the cover nuts. Removal of the suction-end cover gives access to the cartridge retaining screws. Once released, the cartridge can be pulled from the barrel onto a purpose built support frame, then lifted clear.

After stripping, all reusable components are thoroughly inspected. If the reason for changing the cartridge was a fall-off in performance, then recirculation within the pump or excessive clearances are the most likely causes. A check of the clearances at the first impeller inlet may indicate whether wear is the pro-

blem. Recirculation can be either between casing elements (interstage), in which case a complete cartridge overhaul is necessary, or across the cartridge joint face in the barrel. This joint face is carefully inspected whenever a cartridge is removed, and, if necessary, the surface machined true to the axis of the barrel.

If mechanical failure has occurred, the first impeller inlet must be checked for signs of foreign debris having passed through the pump; if so, the pump suction filters and equipment upstream of the pump must be inspected. If a heavily-worn balance disc and wearing indicate loss of hydraulic balance, the pump operation and/or loss of suction head should be investigated.

Cartridge reinstatement follows a reverse procedure to that for the stripdown. It is important to incorporate certain checks during the rebuild to ensure reliable operation. First, when the cartridge is secured in the pump barrel, the pedestals and bearings must be adjusted to centralise the rotating components within the stationary housings. This is accomplished by measuring radial clearances at the first and last stage impellers. The pedestals are dowelled at this position to ensure that, during the reinstatement of end covers, seals, etc., these too can be centralised whilst maintaining concentricity within the cartridge elements. Freedom of rotation is checked at each stage of the rebuild. Shaft sleeves and nuts are lapped true to their mating components to prevent shaft bending and water leakage under sleeves. The running position of the pump is set by adjustment of the throttle sleeve length (or positioning of the thrust bearing, if fitted) to give the position predetermined during the cartridge rebuild. Close-running surfaces perpendicular to the axis of the shaft must be checked for swash run-out, if premature wear is to be avoided. On completion of the rebuild, the alignment of the pump with its drive is checked and adjusted, as necessary.

To refurbish a feed pump cartridge, a complete stripdown is necessary with the objective of proving the mechanical integrity of all components and restoring the running clearances to design figures. Stripdown is best accomplished using a special rig that handles the cartridge in a vertical position. Each stage is dismantled in turn, using purpose built pulling and heating equipment to remove the components that are a shrink fit. NDT (dye penetrant and/or MPI) is carried out on impellers, casing elements, diffusers and shaft. Impeller wear rings are renewed and bored to suit the impeller sizes. Before reassembly, the rotating components (including the balance disc and locknut) are built up on the shaft, checked for truth and dynamically balanced. The cartridge is reassembled again in the vertical position. The end float is checked on completion of each stage to ensure each impeller takes up the same position relative to its diffuser. When assembled, the cartridge is turned horizontal and a shaft lift and drop test carried out.

Dial test-indicators are used to monitor the shaft movement, as constrained by the pump internal clearances, to validate that the rotating and fixed components of all the stages are concentric and to the design clearance.

Other items of equipment associated with feed pumps that need regular maintenance are the leak-off valves and suction strainers. Leak-off valves are usually high pressure parallel-slide valves with the pressure being dissipated across a multiple orifice unit. The usual maintenance requirements of attention to valve glands, cover joints and seat faces apply. The arduous duty associated with the control of high pressure and the velocity of water flow can cause rapid disc and seat wear, so these valves often need higher than normal maintenance.

Mechanical and magnetic strainers are commonly incorporated in feed pump suction lines. These normally only need inspection at times of major unit overhaul. The mechanical strainers consist of a rotating element and stationary backwash arrangement, or vice versa. The strainer should be dismantled to check the element for damage and/or blockage and the seals associated with the backwash system are renewed or reset. Magnetic strainer elements are removed and cleaned, using high pressure water jets.

8.2 Auxiliary pumps

8.2.1 Centrifugal pumps

The maintenance of centrifugal pumps closely follows the principles described for boiler feed pumps. Pump overhaul is concerned with the reinstatement of internal wear-ring clearances to design to overcome deficiencies in performance, together with the refurbishment of glands and bearings. In most instances, the pump will be of the horizontal-split casing design: it may be single or multistage. Often a complete rotating element is kept as a spare, so that rapid replacement is possible. It is necessary to lap the components (impellers and sleeves) when assembling rotating elements to prevent bending of the shaft as they are clamped and locked together; this is especially important with multistage pumps.

Shaft sealing may be by conventional packed glands or by mechanical seals. Both need careful consideration of materials if reliable operation is to be assured. Rubbing speeds, type of fluid, fluid pressure and temperature all influence material choice. Packed glands tend to require regular adjustment and periodic renewal; the friction involved causes shaft-sleeve wear, necessitating dismantling of the pump for renewal. Mechanical seals have the advantage of virtually no leakage and no maintenance requirements between seal changes. By the careful choice of materials, it is possible for the seal to have a life in excess of the period between pump overhauls.

When overhauling pumps containing aggressive fluids, special attention should be paid to the effects of corrosion on pump components. Protective coatings are used to protect pump bodies and these require close inspection and repair, as necessary. Coatings may be used by the maintenance engineer to combat the effects of galvanic action by seawater in circulating water pumps.

Routine preventive maintenance consists of lubrication, attention to packed glands, checking the security of holding-down arrangements, together with any special requirements of the gland-sealing system. Where clean fluid injection or gland cooling is employed, any associated strainers need periodic cleaning.

The glandless type of centrifugal pump (see Fig 5.36) is often used on condensate and boiler circulating duty. Here, there is no pressure barrier between pump and drive motor, the bearings and motor windings being immersed, lubricated and cooled by water. The pump/motor unit is mounted vertically, with the motor below the pump. This configuration minimises journal bearing loading and the pump thrust loading is partly balanced by the weight of the rotating assembly. Lightly-loaded bearings and the absence of a shaft seal result in good reliability as long as certain conditions are met. The motor cooling circuit must be filled and primed with clean demineralised water to prevent damage to the windings and bearings. Water quality is maintained by means of filtration. Leaks on the motor and cooling circuit are not permitted, as these lead to make-up from the pump, the water of which may contain contaminants and be at a temperature to cause motor winding damage. Effective neck cooling must be maintained to act as a heat barrier between the pump and motor, particularly when the pump is shut down with no internal coolant flow.

Routine maintenance is restricted to regular attention to the cooling circuit filters, together with periodic measurement of thrust wear. Often a thrust wear measurement facility is included with the pump, but otherwise a simple datum measurement can be made by removing the cooling water inlet pipe at the base of the motor.

When a unit is to be overhauled, the motor/impeller assembly is usually withdrawn from the pump casing. The motor unit is stripped in a vertical maintenance cradle, with special attention being paid to the winding bushing seals and bearings. If nylon or asbestos reinforced plastic material is used for the bearings, it must be remembered that it will absorb water and swell, thus reducing clearances. Either special allowance must be made for this, or the final machining must be carried out after a period of soaking in water. Rebuilt motor units are stored full of demineralised water with, if necessary, the addition of antifreeze.

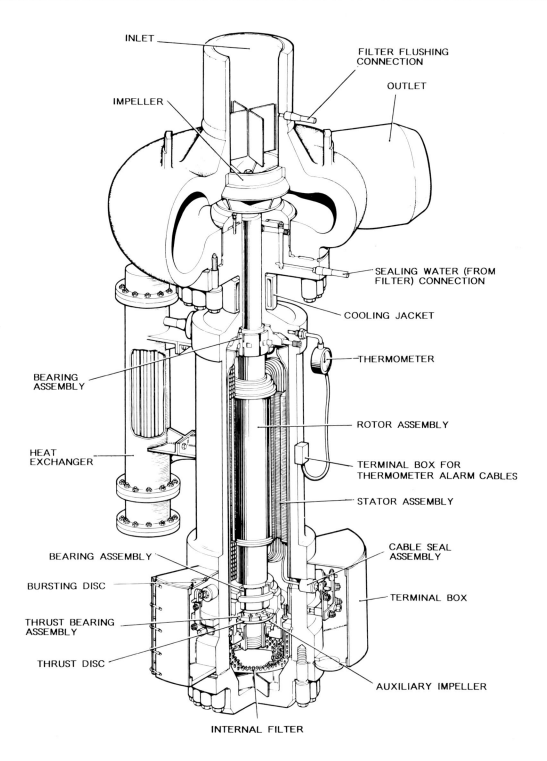

INLET

FILTER FLUSHING CONNECTION

OUTLET

IMPELLER

SEALING WATER (FROM FILTER) CONNECTION

COOLING JACKET

THERMOMETER

BEARING ASSEMBLY

ROTOR ASSEMBLY

TERMINAL BOX FOR THERMOMETER ALARM CABLES

HEAT EXCHANGER

STATOR ASSEMBLY

CABLE SEAL ASSEMBLY

BEARING ASSEMBLY

BURSTING DISC

TERMINAL BOX

THRUST BEARING ASSEMBLY

THRUST DISC

AUXILIARY IMPELLER

INTERNAL FILTER

FIG. 5.36 Glandless-type centrifugal pump

Motors are filled with water as soon as possible during their installation. If this is taking place whilst the main generating unit is in operation, any slight leakage of hot water past the isolating valves can result in damage to the motor windings. Under these circumstances, the unit should be installed with either its temporary or its permanent neck cooling in service.

8.2.2 Liquid-ring air pumps

The maintenance of liquid-ring air pumps closely follows the principles applied to other rotary pumps described earlier. However, very tight clearances are necessary to prevent leakage between suction and discharge ports within the pump. Thus, when they are overhauled the internal component surfaces must be refurbished and the pump rebuilt to maintain these clearances (see Fig 5.37). The low running speed helps the reliability of the units. Routine maintenance consists mainly of attention to bearing lubrication, and to glands to limit air ingress. Pump performance also depends on peripheral equipment, such as coolers and seal water pumps; these should receive attention whenever maintenance is carried out on the main pump. The integrity of coolers is particularly important, as contamination of the pump liquid circuit can lead to corrosion within the pumps, eventually causing seizure or a fall off in performance.

8.2.3 Positive-displacement ram pumps

These pumps are commonly used for chemical and high pressure injection duties, so the first maintenance consideration must be safety from dangerous chemicals. Precautions must be taken to protect skin and eyes, with the pump components being flushed clean as soon as possible in an approved and safe manner.

Complete overhauls include inspection of pump components (rams, glands, valves, relief valves) and drive arrangements. However, the most common cause of loss of performance is deterioration of the suction and delivery valves and, dependent upon the

fluid being pumped, these may require attention on a routine basis. At the same time, opportunity can be taken to inspect visually any drive gearing and to check the effectiveness of lubrication.

8.3 Feedheaters

Maintenance of feedheaters differs for the two main types; the direct-contact heater and the tubed heater. In addition, the high integrity bled-steam non-return and isolating valves need careful and regular maintenance in order to safeguard the turbine plant.

Direct-contact heaters are commonly used in the low pressure condensate system, up to and including the de-aerator. Maintenance outside the statutory outage periods is minimal; routine checks on the level control system are important, together with the repair of any leaks. The latter may manifest themselves as air leaks, affecting the turbine vacuum and feedwater oxygen levels. At statutory outages, the vessel shells are inspected to normal pressure vessel standards. Internal fixtures, such as waterspray nozzles, perforated trays and baffles, are inspected for security and damage. Final cleaning and inspection before return to service is important so as to avoid debris damaging extraction and feed pumps.

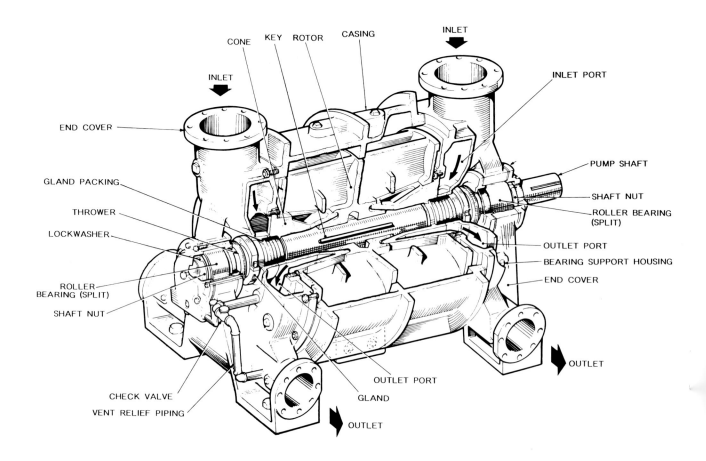

FIG. 5.37 Liquid-ring air pump

Tubed heaters are installed in both low pressure (LP) and high pressure (HP) feed systems, the design being dependent upon the duty involved. A typical HP heater is illustrated by Fig 5.38. Maintenance requirements of HP and LP heaters are similar, although higher pressures do lead to an increased workload. Apart from statutory inspections, work is mainly associated with the repair of tube leaks and rectification of any joint leakage that may occur.

There are a number of causes of tube leakage. The system of expanded tube-to-tubeplate joints often used in LP heaters can exhibit leakage due to inadequate expansion. Usually the tubes can be re-expanded to overcome this. HP heaters have steel tubes welded to the stubs of the header tubeplate and so do not suffer this problem. Tube leakage also occurs due to steam erosion, either opposite steam entry ports, where it is usual to fit a deflector plate, or at support plates and drains cooler baffle plates. Leaking tubes are plugged, one of the most successful techniques being explosive welding.

Joint leakage on tubed heaters has traditionally been a problem and is caused largely through thermal cycling, particularly in HP heaters. For this reason, many shell joints are now welded and those that still employ bolted joints use special sealing techniques. Spiral wound gaskets are used, but most success has been achieved with tapered-steel seal rings. Whichever technique is adopted, special attention to joint machining tolerances and finish is essential to ensure a reliable leak-free joint. LP heaters usually have the tubeplate sandwiched between the steam shell and waterbox with one set of bolts compressing both joints. Thus both joints should be completely remade, whenever the bolting is disturbed. Waterbox leakage on HP heaters operating with feedwater pressures in excess of 140 bar has largely been eliminated, using a combined tubeplate and waterbox of welded construction. Access into these headers is by means of an internally-jointed, pressure-seated manhole door. To permit access to the tubeplate, removable bolted division plates are used. It is important that, when replaced, they are properly jointed and the bolts secured, to prevent by-passing of the heater tubes.

Another requirement at times of maintenance is an inspection of the header and pipe welds for thermal fatigue cracking. This can result from thermal cycling of the relatively thick material needed to withstand the high feed pressures. Magnetic particle and ultrasonic NDT techniques are used.

Bled-steam isolating valves and non-return valves are inspected at major overhauls and are operated regularly during normal working to ensure that they are capable of protecting the turbine. This is particularly important with direct-contact heaters and de-aerators, where there is no physical barrier between bled-steam and feedwater. Valves should be inspected for freedom of operation and for the condition of the seating surfaces.

8.4 Condensers

There are three categories of problem associated with condensing plant that concern the maintenance engineer; tube leakage, tube fouling and corrosion. The approach to each is influenced by the plant design and the type of circulating water employed.

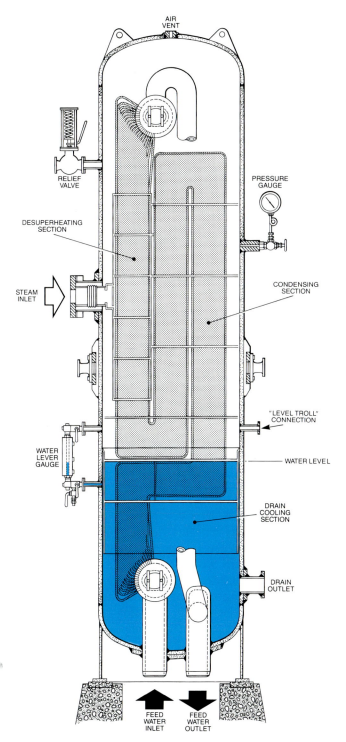

FIG. 5.38 Typical HP heater

In many power stations, condenser tube leak repairs are carried out by the operating staff; for this reason, the subject is comprehensively dealt with in Chapter 2 of this volume.

Tube fouling by debris (or seawater organisms, such as mussels) leads to tube erosion, because of localised increases in water velocity and eddies. An effective solution to this problem is to install plastic inserts at the inlet tubeplate position. These serve the dual purpose of not allowing the entry of objects large enough to become lodged in the tubes and of protecting the initial length of tube from erosion damage caused by changes in direction of water flow. Silt deposits in condenser tubes can be removed by high pressure water jetting, using nylon or wire bristle 'bullets', if necessary.

A major problem at coastal power stations is corrosion, which results in the rusting of steel components and in galvanic action between brass tubeplates and ferrous materials. Impressed-current cathodic protection is often used to protect tubeplates and requires regular monitoring to ensure correct operation. A more recent approach has been to apply a protective coating to exposed and vulnerable areas. Many coatings are used, including neoprene rubber, glass flake, and polyurethane materials. Whatever is used, it is important to inspect it regularly for damage, and to carry out repairs before deterioration of the base material occurs. Many of the materials involve the use of flammable and volatile solvents, which can be especially hazardous in confined waterboxes. Special safety precautions are therefore necessary when undertaking this work, including the provision of adequate ventilation and the use of sparkproof equipment.

A complete answer to problems associated with corrosion of materials, particularly in seawater systems, is to replace the troublesome item with one manufactured entirely of glass-reinforced plastic (GRP). The CEGB has gained experience with items as large as 500 MW condenser waterboxes together with 1.5 m diameter pipework and associated valves made of this material. The items have proved entirely satisfactory in service over a number of years with no apparent deterioration and little or no maintenance requirement.

8.5 On-line leak sealing

8.5.1 Introduction

At one time, a serious leak on a non-isolatable section of the plant called for a choice between two courses of action. The plant could either be shut down and the leak repaired or, alternatively, the leak could be allowed to remain until a convenient outage permitted repairs to be carried out. The disadvantage of the former is the cost of the outage necessary to carry out the work, in terms of plant replacement cost, water and the heat for restarting, plus the fact that the thermal cycle could well promote further leaks. The second course also involves the cost of water and heat whilst living with the leak, together with the attendant hazards of a usually deteriorating situation.

A third choice has gained increasing prominence in recent years — although the process was invented in the 1920s; this is to repair the leak with the system under normal conditions of pressure and temperature. Within the CEGB, on-line leak sealing is normally carried out by fully trained contractors' staff working to the National Safety Code of Practice GS−M1 'Repairs of leaks on charged pressurised systems by injection processes — bolted joints and valve lands'. The code does not exclude the use of CEGB staff for this work, provided that they have been properly trained and that they use material and equipment supplied by specialist manufacturers. Safety depends upon the accumulated experience and expertise that can only be obtained by persons regularly employed to carry out this work: it is not a process that should be undertaken on an occasional basis by CEGB staff.

It is essential that anyone contemplating this type of repair, first acquaints himself with the requirements of the Code of Practice referred to.

8.5.2 Technique

The technique is based on the principle of injecting a substance into the leak and joint area which is compounded so as to be plastic at the pressures applied by the injection gun but cures hard, usually by elevation of temperature caused by contact with the hot metal of the defective component.

The substance injected will normally be recommended, compounded and supplied by the specialist contractor. Various basic materials are used in the manufacture and usually a small quantity of filler is also incorporated. The key requirements are to ensure compatibility with the steam, fluid, substance or gas contained in the pipework system or vessel, also compatibility with the materials of plant components, together with considerations of toxicity or other effects on staff handling the compound. It must also have suitable properties to fill the leak and joint area effectively and to cure at the temperature conditions prevailing. In exceptional cases, on systems at lower temperatures and when a cold cure cannot be achieved, suitable means of raising the temperature may be used, provided that the temperature does not exceed the design limits for the bolts or other components. Attention should be paid to stresses induced by local temperature differences and the effect of the peak temperature to which the component is exposed.

Generally there are no special problems with an experienced contractor in this field supplying suitable compounds for steam and water at pressures and temperatures normally encountered in the CEGB. Materials for use with oil systems need more care in selection but satisfactory experience has been obtained in sealing leaks at typical pressures and temperatures found on oil-firing schemes for large boilers. Beyond these conditions, and where special oils may be involved, e.g., transformer oils, consultation will be necessary between technical staff of the contractor and appropriate specialist staff in the CEGB to ensure that a chemically compatible compound is used.

In the case of industrial gases, compounds are available and have been used satisfactorily for sealing hydrogen leaks on generators. For CO_2, no special problems arise from the leak sealing aspect, but special attention is necessary to ensure compatibility of material for nuclear reactor applications and to take account of any statutory control on modifications to plant. In the case of all other gases and chemicals, each proposed application should be carefully considered in conjunction with specialist contractors' technical staff and the appropriate staff in the CEGB technically competent to assess such an application.

8.5.3 Application

Subject to considerations discussed in the Code of Practice, the process is suitable for the repair of leaks in pipe flange and valve joints, pipeline joints, valve glands and bonnet seals, joints in pressure vessels, steamchests, heater vessel joints and transformer tank joints.

In the most common case of a leak from a flange, the process involves forming an injection moulding around the failed gasket. In systems pressurised up to 40 bar, the gap between the flanges is fitted with a peripheral seal (Fig 5.39 (a)); the sealant is then injected through adaptors (Fig 5.39 (b)) which are fitted under the flange nuts, commencing at the side opposite the leak, into the space between the peripheral seal and the failed gasket (Fig 5.39 (b) and 5.39 (c)). During injection, the compound fills all the grooves and pits in the faces of the joint and rapidly cures to form a tough, resilient, leak-tight mass. The repair can be treated as permanent but, like any gasket, its life will depend upon the extent of thermal cycling and other factors, such as the nature of the fluid in the line.

At pressures greater than 40 bar where the flange gap is less than 10 mm, the peripheral sealing clamp is impractical. In order to achieve a seal, a soft brass wire is held in the flange gap by lightly peen locking over the inside edges of the flanges. Sealing compound is then injected through holes drilled in the lightly stressed outside edge of the flange to take

suitable adaptors (Fig 5.40).

For sealing leaking flange type connections at all pressures where the gap between the flanges is greater than 10 mm, tailor-made clamps are used, each clamp being specifically designed for the flange and duty involved. The tongue of the clamp is sized to fit the gap between the flanges and injection points are provided alongside each flange bolt to ensure proper filling of the gap and bolt clearances (Fig 5.41).

In the case of leaking valve glands, adaptors are fitted at holes drilled into the stuffing box; compound is then injected to seal the leak. After injection, the follower nuts are sometimes backed-off and a further topping up injection attempted. In these circumstances, full nut-thread engagement must be maintained after backing off. Before attempting any work on valve glands, calculations must be carried out to determine the likely stresses in the follower stud material and in the wall of the stuffing box. If either of the stresses calculated exceeds 50% of the ultimate tensile strength of the material involved, then consideration must be given to reducing the injection pump pressure for that particular case.

9 Generator maintenance

Maintenance of each part of the generator can be conveniently considered under the headings of 'on-load' and 'off-load' maintenance. Whilst the on-load aspects will be seen to consist mainly of carrying out monitoring activities, their importance cannot be over emphasised, since they play a key role in the early detection of conditions which, if allowed to deteriorate, could lead to catastrophic failure.

9.1 On-load maintenance and monitoring

The degree to which this may be carried out varies with the design of machine and the extent of instrumentation built into it. Recent experience has led to a rapid increase in the monitoring techniques employed. The following paragraphs give an indication of some of these developments, as well as the more standard maintenance and monitoring activities.

9.1.1 Stator

Temperature measurement

It is now normal practice to install thermocouples throughout the stator which are used to monitor the temperatures of stator-bar insulation, core teeth, core end-plates, stator water connections and hydrogen gas temperatures. A typical 660 MW generator has about 80 thermocouples for this purpose, all of which can

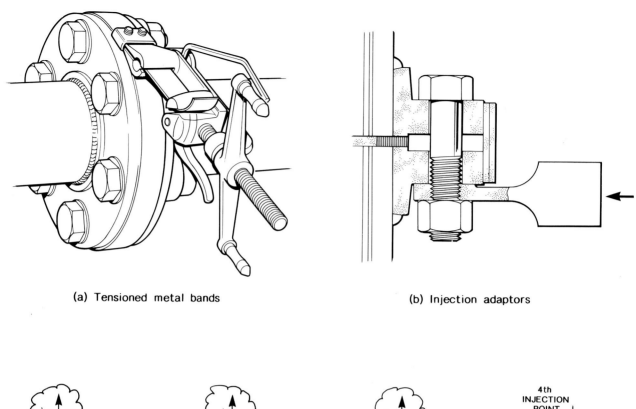

(a) Tensioned metal bands   (b) Injection adaptors

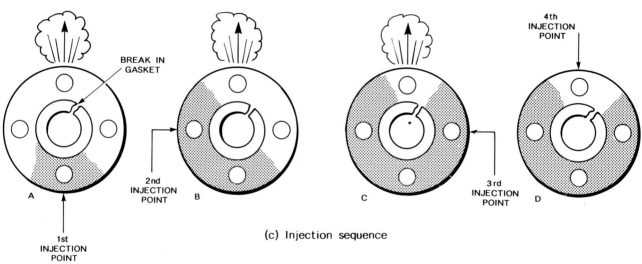

(c) Injection sequence

FIG. 5.39 On-line leak sealing

be read from the control room. Analysis of the information thus provided can help to detect deficiencies within the cooling circuit, developing faults and other temperature-related stator problems.

Endwinding vibration

A vibration transducer module 'wound' into the endwinding, which measures vibration in all planes, can be used to carry out periodic checks to detect any deterioration in the structural integrity of the endwinding and to ensure that stress levels in the conductor coils do not approach values where fatigue cracking could start.

Stator water flow

Efficient conductor cooling can only be achieved if the correct water quantity flows through each part of the stator circuit. Regular monitoring of flow indicators, where installed, can give warning of a change in the balance of flows which may demand further investigation during a shutdown. These are simple sightglass indicators with a flap in the flow, the angle of which give a rough indication of flow rate.

Hydrogen dewpoint

Regular checks of the dewpoint of the hydrogen within the frame ensure that conditions cannot exist

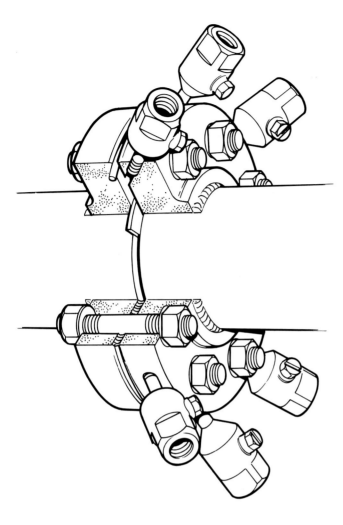

FIG. 5.40 High pressure narrow-gap flange repair

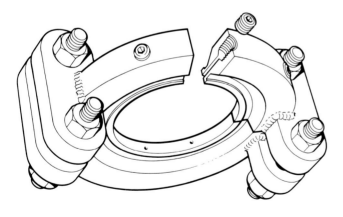

FIG. 5.41 Insert clamp

where moisture is able to condense out on the cooler parts of the machine. Deterioration in the dewpoint should lead to the following possible causes being investigated:

- Moisture in lubricating oil.

- Leaking hydrogen coolers.

- Leaking stator coolant.

- Deterioration in dryer performance.

Hydrogen entrainment into stator water

Slight hydrogen entrainment will be evident in most machines. Whilst deterioration will be detected as an alarm condition, occasional checks should be made of the volume of hydrogen detrained from the stator water system to facilitate early detection of a developing leak.

Hydrogen leakage

Evidence of increased hydrogen consumption should initiate hydrogen leakage checks over the complete machine. Points at which leakage may occur are numerous. Standard on-load and off-load check lists should be prepared, detailing every part requiring checking. The presence of hydrogen may be detected by means of a non-flammable gas detector. It is then often possible to locate the exact point of leakage by use of an aerosol foam or a soap solution. Included in the checks should be a search for hydrogen in the slipring cooling circuits and in the lubricating oil.

Core monitor

The decomposition products of overheated insulation, perhaps due to a developing fault, can be detected by means of an on-line gas sampling device known as a 'core monitor'. As well as giving an alarm, the monitor can be arranged to collect a gas sample for future analysis of entrained degradation products.

Radio frequency monitor

Corona discharge generated by developing insulation breakdown may be detected by an aerial within the stator, or by the detection of radio frequency currents in the neutral conductors. Analysing equipment is then required to enable the pick-up due to a fault to be distinguished from background noise.

9.1.2 Rotor

Vibration monitoring

The generator, in common with the turbine, should be subject to on-load and run-down vibration monitoring. The vibrations are monitored in the axial, vertical and transverse directions at each bearing. The signals are recorded on magnetic tape and subsequently analysed to give the amplitude and phase of each vibration at the fundamental and first few multiples of the fundamental frequency.

On-load measurement should be taken at regular intervals and at consistent machine conditions (load, power factor, vacuum, etc.). Any unexplained change in amplitude or phase will require further investigation, which may include arranging for a run-down plot to be obtained as early as possible.

Run-down vibration monitoring may be used initially to fingerprint the machine. The information obtainable from subsequent changes in the run-down plot as the shaft is excited through a wide range of frequencies may, after analysis, give early warning of incipient faults on the machine.

Rotor earth fault detection

A serious earth fault on the rotor should be detected by the installed earth fault protection. A periodic check of the insulation resistance of the rotor winding should be made, using the measuring circuits of the earth fault protection.

Shaft voltage measurement

Some machines are fitted with a shaft voltage alarm system. On all machines, it is advisable to take regular shaft voltage readings in order to build up a pattern of the normal values expected from the machine and hence enable an unusual trend to be recognised and investigated.

Whilst it may be necessary to seek specialist advice on the interpretation of such abnormal voltages, the following possible contributory factors should be considered:

- Change in outboard bearing insulation value.
- Shaft earth-brush problems.
- Magnetised shaft.

Flux monitor

An increasing number of machines are being fitted with a search coil in the airgap between rotor and stator which can be used to identify changes in the flux due to interturn faults. Regular monitoring of this device can thus be employed in conjunction with other lines of investigation to monitor the condition of the rotor.

Shaft earth-brush

This is often very prone to contamination which can render it ineffective. Frequent inspections of the brush are thus required to maintain its effectiveness.

Bearing pedestal insulation

Periodic checks are required on the insulation resistance of the generator outboard bearing, shaft seal ring, pipework and exciter baseplate. Failure of this insulation, due to accumulation of dirt, or by moisture ingress or by shorts due to badly placed conduit, can lead to substantial current flow through the bearings with consequential damage to bearing and journal surfaces.

Some bearings integral with the generator end plate are designed to allow on-load measurement of insulation resistance.

The insulation of pedestal bearings may be checked on-load as follows:

(a) Connect the generator rotor shaft to earth at the turbine end, using a portable copper gauze brush and lead on an insulated handle.

(b) Measure the AC voltage between shaft and earth (again using a gauze brush) at the exciter end of the rotor shaft.

(c) Short between the rotor shaft and pedestal at the exciter end and measure the voltage between pedestal and earth.

If the insulation is sound, both voltage readings will be similar. If the voltage measured in (c) is lower than that in (b), the insulation is faulty. Finally, measure the voltage between the exciter end of the shaft and the pedestal. Whilst the value depends on the relative resistance of the pedestal and bearing oil-film, the voltage measured should not normally exceed 25% of that measured in (b), if the insulation is sound.

9.1.3 Excitation system

Brushless system

The on-load maintenance of brushless excitation systems is minimal. Exciter and rectifier cooler leaks are two of the greater hazards to the system: although leakage detection is fitted, it is advisable to carry out visual checks for evidence of moisture within the cooling circuits.

Should a diode failure alarm be indicated from the protection fitted to the rotating rectifier, it is advisable to change the failed diode at the earliest opportunity to prevent a further failure overloading the remaining diodes and causing a bridge-arm failure.

Slipring systems

Trouble-free operation of the rotor slipring and exciter slipring brushgear is only achieved by regular and careful checks and maintenance, the frequency of which is set from experience on each type of machine, but should initially be carried out weekly. Unless the machine is shut down regularly, this work will often need to be undertaken on-load.

The schedule of work should include checks to determine:

- Freedom of brushes within boxes.
- Brush length.
- Individual brush current.
- Condition of brush tails.
- Evidence of sparking.
- Evidence of brush vibration.
- Spring tension.
- Evidence of ingress of oil from adjacent bearings.

Opportunity should also be taken to check the slip-ring and exciter cooling circuit, including the condition of filters.

A policy should be developed as to the maximum number of brushes which may be changed on-load within a given period. Typically, it may be decided that, to maximise evenness of wear, all the brushes in a box should be changed as soon as one reaches the minimum length, but that only one box should have brushes renewed in any one 24-hour period. This allows the brushes to bed in and share current proportionally before a further change is made. Prior to use, brushes must be profiled to the dimensions of the slipring, either by the manufacturer or by the use of a profiling rig in the workshop.

All on-load brushgear must be covered by a site Code of Practice detailing the safety precautions to be employed, including putting the machine on to manual excitation, disconnection of earth fault protection and the adoption of earth-free working by means of rubber mats, insulated tools, rubber gloves, etc.

9.2 Off-load maintenance

The following paragraphs cover work which will be carried out on every planned generator overhaul and also that which may be included as an outcome of observation of the condition of the machine whilst running or as the result of previous experience.

9.2.1 Stator internal work

The extent to which internal access is possible without removal of the rotor varies between designs of machine. On most, by removal of hydrogen coolers or access doors, it is possible to inspect the core back, endwindings, cooling connections and bushings. Removal of the rotor, of course, gives access to the whole stator.

Before any entry is made into the stator, it is imperative that an effective *clean-conditions system* is set up to prevent any debris being carried into the stator and to ensure that no tools or equipment are left in the machine upon completion of work. It is also essential that the frame is thoroughly ventilated and that atmosphere checks are carried out before entry to the machine is permitted.

The main features of a clean-conditions system should be:

- Entry to machine available at one point only.
- Point of entry to be surrounded by clean-conditions enclosure.
- Entry to enclosure to be locked, except when work in progress.
- Whilst working, a 'gatekeeper' to be positioned at the entrance to control personnel entering the area and to log 'in' and 'out' all tools and equipment.
- Entry to be restricted to approved personnel.
- All personnel to wear clean-conditions clothing (pocketless overalls, overshoes).
- Very thorough internal inspection of generator immediately prior to boxing-up.

When a stator is subjected to periods of exposure to the atmosphere, every opportunity should be taken to close the openings and to circulate the air within the casing through a dehumidifier in order to keep the winding as dry as possible.

Some machines are now fitted with off-load hydrogen circulators which are also rated to be able to circulate air. Providing the hydrogen drier is available, this system can be used as an alternative to the use of dehumidifier.

It has been shown that, providing one of the above techniques is strictly applied, the need for a dryout at the end of work on a generator can often be eliminated.

Whilst working in the endwinding region of the stator, it must be emphasised that great care be taken to ensure that stator coolant hoses are not scratched, strained or damaged in any way. It is strongly recommended that some form of protection be fitted to the hoses prior to other work being undertaken in that area.

All the above precautions, plus any others found to be necessary at a particular location, are best incorporated into a local Code of Practice on generator internal work.

Activities which may be included as part of a programme of work, where generator internal access is available, are itemised below:

- Inspection of endwinding and checking tightness of the winding and its associated support structure.

- Examination of stator bars and wedges.

- Examination of core and its support structure for evidence of hot spots, stray earth paths and general mechanical damage.

- General cleaning of any contaminated surfaces, care being taken to ensure that only approved solvents are used and that any dirt is not driven into the winding.

- Detection and identification of any debris within the stator.

- Inspection of auxiliary wiring.

- Checking of phase and neutral connections to bushings.

- Checking of all stator coolant pipework and hoses. If it is considered necessary to tighten any hoses, great care must be taken to ensure that correct procedures are followed.

- Particular investigation into stator cooling leaks, as indicated by gas-into-coolant during running. If necessary, the leaking phase may be identified by draining the winding and carrying out a vacuum drop-test on each phase.

The actual leak may be located by one of the two methods:

(a) Filling and pressurising the winding with water and inspecting for leaks. This method risks contaminating the insulation with water and is not recommended by some manufacturers.

(b) Injecting the drained winding with a tracer gas and detecting the leaks by use of a gas detector (see Section 9.3.3 of this chapter).

9.2.2 Stator external work

Stator water system

Prior to shutting the generator down, a careful survey of all external pipework, pumps, coolers, valves and filters will indicate areas needing attention during an outage period.

Once the system is isolated, an early check of the filter should be carried out to look for and identify any debris collected from the cooling circuit; such a check may well indicate other areas where further investigation is needed, e.g., particles of gasket material may indicate a deterioration of the gaskets fitted to the pipework joints.

In all work associated with the stator water system, great care must be taken to ensure that correct materials are used and that no contamination is allowed to enter the water circuit.

After any substantial work on the external circuit, the pipework should be flushed, with the winding by-passed to ensure no debris is allowed to enter the winding and hence introduce a risk of partial conductor blockage.

Hydrogen coolers

Any slight internal water leakage from coolers can create considerable difficulty in maintaining the hydrogen dewpoint at a low value. Coolers suspected of leaking should be removed, blanked and pressure-tested to confirm and identify the leak. After repair, a repeat pressure drop-test should be carried out.

Great care is required in the refitting of hydrogen coolers to ensure that a gas-tight seal is achieved with the frame.

Hydrogen driers

Evidence of loss of performance of the hydrogen driers can be investigated by means of the following checks:

- Sequence.

- Heater performance.

- Valve operation.

- Pipe obstructions.

- Desiccant contamination.

Depending on experience, some or all of the above may be included in an off-load maintenance routine.

Main connections

It is essential that the integrity of the insulation of the main connections be maintained by careful attention to the condition of the enclosure, so that ingress of moisture, oil, dirt, etc., is prevented.

The insulators themselves may need cleaning and should be regularly inspected for any evidence of damage.

The conductors will require a thorough inspection of all joints for evidence of overheating. Bolted joints should have all bolts periodically checked for tightness to the specified torque and be checked to ensure that joint resistance is at an acceptable value.

9.2.3 Rotor

The extent of work carried out on the rotor will depend on whether it has been decided to remove the rotor from the machine. It is recommended that rotors on all machines be periodically withdrawn for end-ring inspection, the frequency of such inspections being determined by the end-ring design. Apart from this, unless on- or off-load tests have clearly indicated evidence of a rotor fault, there is no value in removing the rotor as part of a regular maintenance programme.

9.2.4 Sliprings and brushgear

Maintenance will consist of:

- Removal of brushes from boxes, marking them with their position and storing them carefully to prevent chipping. Brush-box sets with one or more brushes near minimum length to be discarded.

- Disconnection and removal of brushgear enclosure.

- Slipring surfaces to be wrapped with moisture absorbent paper and polythene for protection (assuming that the surfaces are satisfactory).

- All parts of enclosure, slipring sides and shaft (NOT slipring surface), ventilation circuit and connections to be thoroughly cleaned to remove carbon dust, oil and other contamination.

- Overhaul of ventilation circuits.

- Reassembly of brushgear enclosure and *in situ* bedding in of any new brushes by use of a strip of fine abrasive cloth; check that no abrasive particles are bedded in the brush between the brush and ring. All carbon dust to be removed.

- Checks of spring tension and freedom of brushes in boxes.

Until the machine is ready to be returned to service, the brushes should remain lifted from the ring and the enclosure kept dry and warm to prevent corrosion of the rings.

The above measures are advisable on any prolonged shutdown.

From time to time, the slipring surfaces may require refurbishment to remove marks or uneven wear, or the grooves may require recutting.

Slipring grinding is commonly carried out *in situ* by one of the following methods:

- Machine on barring; rings ground using a motor-driven grindstone mounted on a sliding baseplate attached to the machine bedplate.

- Machine stationary; purpose built collar with ring of grindstones assembled around the slipring and belt-driven by a pony motor. This method is usually applied by a specialist contractor.

More substantial machining of sliprings or recutting of grooves is carried out *in situ*, using the first method with a lathe tool replacing the motor-driven grindstone.

9.2.5 Exciter and pilot exciter

Maintenance activities consist of stripping-out covers, inspection, cleaning, testing windings, checking connections and carrying out brushgear and slipring maintenance, applying principles similar to those outlined in the previous section.

9.2.6 Rectifier

When insulation-testing rectifier equipment as part of a maintenance routine, it must always be ensured that test voltages applied are low enough not to damage the diodes. Where it is necessary to apply higher voltages, the diodes must first be disconnected, removed or shorted-out.

Static rectifiers

Normal maintenance comprises a thorough clean and inspection of all components, testing of diode characteristics, fuse checks, insulation testing and check-tightening of all connections.

Rotating rectifiers

Although the activities carried out on a rotating rectifier are similar to those specified above, particular care has to be exercised with regard to the following points because of the very high centrifugal forces on the rectifier assembly when running:

- Very thorough inspection of all components to ensure mechanical integrity of assembly. A failure of a component under centrifugal forces could rapidly lead to major mechanical and electrical damage.

- If any work has been undertaken on an individual module, the module must be accurately weighed and, if necessary, have its balance weights adjusted to ensure it is within the specified module weight tolerance (typically ± 0.5 g). This is essential to maintain the balance of the complete assembly.

The complexity of some designs of rotating rectifier is such that it is advisable to set up a rigid inspection system so that each stage of maintenance and subsequent rebuild is independently checked.

Following maintenance, the performance of the measuring and alarm equipment associated with the rectifier should be tested.

9.2.7 Field switch

Normal switchgear maintenance activities should be carried out on the field switch, particular care being taken to ensure the correct performance of the operating mechanism.

The field switch overcurrent trip unit should be inspected for freedom of movement and operation of the trip contacts.

9.2.8 Automatic voltage regulator (AVR)

Off-load maintenance of the AVR consists initially of cleaning the equipment, using a soft brush and vacuum cleaner to remove all accumulations of dust, followed by a thorough examination of all components and wiring for signs of damage, overheating, loose fixings, etc.

Relays should be checked and cleaned, if necessary. Motorised potentiometers and their cam-operated switches should be checked and lubricated, followed by a check of their traverse times. Correct operation of all tripping functions must also be checked

Fault finding and setting-up the AVR are specialised tasks, for which detailed procedures need to be prepared.

9.2.9 Supervisory and protection equipment

A schedule, identifying each supervisory and protection device, should be prepared. The schedule should give reference to particular test and calibration procedures for each device and thus form the basis of a routine maintenance system which ensures that the accuracy and reliability of this equipment is maintained.

In addition to the testing of individual devices in protection equipment, overall protection system checks must be carried out at regular intervals (typically annually).

9.3 Generator testing

The ways by which the condition of a generator can be monitored on-load have already been covered. Periods of shutdown for generator maintenance present the opportunity to do certain off-load tests, which further add to the information available to enable a continuous monitoring of the condition of a machine to be carried out throughout its life.

In addition to this condition monitoring aspect, some of the tests below are essential to ensure that the machine is in a satisfactory state to be returned to service, following maintenance.

Whether or not each test is carried out during an overhaul, and the sequence of testing will depend upon total workload being undertaken.

Fault investigation and repair on the machines also requires some of the testing methods outlined below.

All generator testing has the following common requirements:

- A clearly written test procedure to ensure repeatable standards and safe working.

- A comprehensive record system to allow easy comparison of results through the years.

9.3.1 Insulation testing

Insulation resistance (IR) measurement

Insulation resistance (IR) measurement is a measure of the DC resistance of a winding to earth, or between windings. The test is carried out using a battery- or mains-powered megohmmeter. Typically, a 500 V DC test is used on rotors (care having been taken to disconnect diodes), 1 kV DC for stators up to 6.6 kV and 5 kV DC for higher voltages. To compare IR readings, the average temperature of the insulation should be assessed and the IR reading then corrected to a standard temperature (20°C). A nomogram for this purpose is given in Fig 5.42. Where insulation systems are open to the atmosphere, the IR reading can be greatly affected by humidity. Water cooled windings, when drained down, need careful blowing through to remove water from all hoses if satisfactory IR readings are to be obtained. The manifold should be connected to the guard terminal on the megohmmeter so that a true reading of winding resistance to earth is obtained.

IR tests on filled winding will give a low reading due to the number of parallel paths to earth via the water circuit. To assess whether readings so obtained are reasonable, an equivalent circuit will need to be drawn which takes account of the resistance of the water (of known conductivity) in each hose.

Polarisation index (PI) measurement

Polarisation index (PI) measurement is an extension of IR testing. The DC voltage is applied to the insulation for 10 minutes. The PI is the ratio of the IR reading obtained after ten minutes to the reading obtained after one minute. As a general rule, a value of PI greater than 2 is desirable. Low PI is usually an indication of surface contamination of the end-winding insulation by dirt or moisture, although, when

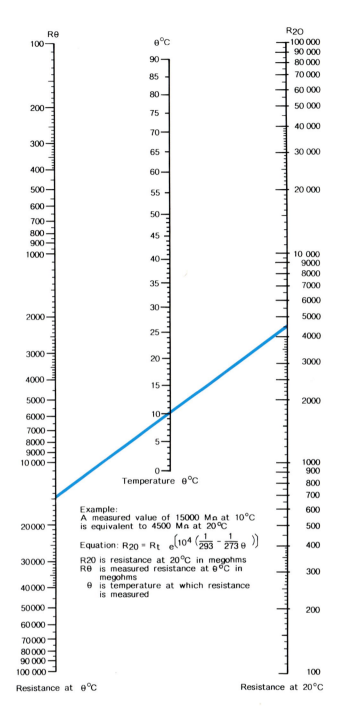

Example:
A measured value of 15000 MΩ at 10°C is equivalent to 4500 MΩ at 20°C

Equation: $R_{20} = R_t \; e^{\left(10^4 \left(\frac{1}{293} - \frac{1}{273\,\theta}\right)\right)}$

R_{20} is resistance at 20°C in megohms
R_θ is measured resistance at θ°C in megohms
θ is temperature at which resistance is measured

FIG. 5.42 Nomogram for correcting insulation resistance measurements to a standard temperature

accompanied by a low one-minute IR reading, more serious degradation of the bulk of the insulation may be indicated. This can only be determined by *Loss angle* measurement.

Loss angle testing

Loss angle testing (or tan delta testing) is a method of assessing the condition of the bulk of the insulation. As insulation deteriorates, it develops voids

which in turn cause a change in the capacitance of the winding and hence a change in the Loss angle (δ), see Fig 5.43.

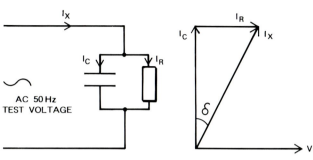

FIG. 5.43 Insulation Loss angle
The insulation may be considered as a capacitor and resistor in parallel. The ratio of the resistive leakage current to capacitive leakage current is represented by the tangent of the Loss angle, δ.

The test method is as follows:

- Check that the winding has a satisfactory IR by means of a DC test.

- Apply an AC voltage to the winding in steps of ($0.2 \times$ line voltage (V1)) from 0.2 to 1.0 V_1. (The condition of the insulation may dictate that 0.8 V_1 should not be exceeded.)

- At each step, record the capacitance and tangent of the Loss angle of the insulation, using an AC bridge.

- On completion of the test, take great care to discharge the winding.

Tan delta is plotted against line voltage. Actual results will be affected by factors such as whether or not cooling water was present in the winding, the type of gas in the generator frame and its pressure, and the amount of surface contaminant on the windings.

By comparison with other tests on the same and similar machines, assessment can be made of the condition of the insulation.

Extensive information on the assessment of insulation by IR and tan delta testing is given in CEGB Site Test Code II.

High voltage tests

High voltage DC tests, where the winding is raised to a voltage in excess of normal working voltage, are rarely carried out on a commissioned machine, the only common application being as a proof voltage test following repair, where the actual voltage used will depend upon the repair and the condition of the original insulation.

To remove the risk of fire in the event of insulation breakdown, the test should be carried out in CO_2, nitrogen or pure hydrogen.

9.3.2 Testing the stator core

When the rotor been removed from a generator, it is a wise precaution to take the opportunity to examine the stator core for evidence of damage or of core-plate insulation breakdown. Two methods are commonly used:

Ring flux test

The core is excited by means of several turns of heavy HV cable wound toroidally around the core and energised from an HV source (usually 11 or 3.3 kV). The pulsating magnetic flux produced by this simulates the rotating flux in an operating machine and induces currents in regions where the insulation between core laminations is damaged, causing hot spot (Fig 5.44). These are detected using an infra-red viewing camera. This procedure is expensive, cumbersome and time consuming.

The electromagnetic core imperfection detector (El-Cid) test

A very small flux is produced using a ring flux winding, as above, but consisting of one turn of light wire, typically energised from a 240 V 13 A supply. The El-Cid sensing head contains a pick-up coil which detects the magnetic fields produced in the air by the small fault currents flowing through the damaged region. Battery-operated electronic circuits are used to detect faults as the head is scanned along the conductor slots in the bore of the stator core.

The El-Cid method has proved to be very sensitive and is becoming the preferred method of core testing.

The El-Cid system is patented by the CEGB and is manufactured under licence by Advent Industrial Systems Ltd., who sell the equipment worldwide.

Local areas of core damage can sometimes be rectified by etching techniques or by fine grinding.

9.3.3 Stator coolant circuit testing

It is essential that the windings, hoses and manifolds within a stator frame are proved to be watertight, following work on the stator. This may be checked by a vacuum loss test, which consists of drawing a vacuum in the winding of at least 700 mbar and observing the fall over a 12-hour period. The maximum permissible fall should be specified by the manu-

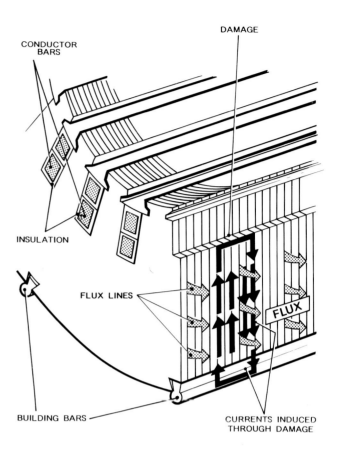

FIG. 5.44 Induced electric currents in a damaged area of the stator
The hot spots created by these currents may be detected by infra-red techniques in the ring flux test. Alternatively, the electromagnetic effects of such currents may be detected in the El-Cid test.

facturer. Typically, on a GEC 660 MW stator, the figure is 10 mbar in 12 hours. If the vacuum test is unsatisfactory, the winding should be pressurised with a mixture of air and 20% (by volume) of tracer gas. A leak search is then carried out, using an electronic gas detector.

Typical tracer gases are helium or refrigerant gases, such as Arcton or Freon. If the latter are used, precautions must be taken to ensure that the level of gas in the generator atmosphere does not exceed the 'Long-term occupational exposure limit' which is the concentration of the gas in air up to which it is safe to work for an 8-hour period. The limits for these refrigerant gases are given in Guidance Note EH/40/85, published by the Health and Safety Executive.

Another important check, following any work where the stator water coolant circuit has been broken into, is that there is a coolant flow through all coils. On hose-type machines (where the stator coolant flows in and out of the winding through translucent PTFE hoses), the most direct way of achieving this is by a 'bubble check'. The stator water system is put into service and dry filtered air injected into the inlet side of the water circuit. All hoses should be examined to check for flow, as indicated by the movement of bubbles of entrained air.

Where the manufacturer recommends that the stator winding should not be filled with water unless the frame is pressurised with gas, an alternative means of checking flow has to be devised, such as measuring the temperature rise on each conductor as the stator coolant is warmed up by pump losses.

On other designs of stator, the temperature rise test described above may be possible or, alternatively, it may only be feasible to measure total flow from each phase accurately and to compare the values with original commissioning flows.

9.3.4 Hydrogen loss test

To ensure that the frame is gas-tight, a pressure drop test should be carried out, initially with the frame pressurised with air, nitrogen or CO_2, so that leaks may be rectified without the delay of purging the frame, but finally with hydrogen.

The test may be undertaken in one of two ways:

(a) A straight pressure drop test over a fixed time period, usually at least 12 hours.

The volume of gas lost, at normal temperature and pressure (NTP), may be found by substituting in the following formula:

$$6800V/\tau \, [(P_{gs} + P_{bs})/(\theta_s + 273)] -$$

$$[(P_{gf} + P_{bf})/(\theta_f + 273)] \, m^3/day$$

where

V = volume of generator voids and associated pipework, m^3

τ = duration of test

P_{gs} = gas pressure at start of test, bar gauge

P_{gf} = gas pressure at finish of test, bar gauge

P_{bs} = barometric pressure at start of test, bar

P_{bf} = barometric pressure at finish of test, bar

θ_s = temperature at start of test, °C

θ_f = temperature at finish of test, °C

(b) By measuring a differential pressure loss over a short period of time (1–2 hours) between the generator frame and a small sealed reference box

installed for this purpose within the generator frame.

A valve arrangement is fitted to first equalise the pressure between frame and box. At the commencement of the test, an inclined gauge is inserted between the frame and box. As the frame loses pressure due to leaks, the differential pressure between the frame and box is indicated on the gauge.

The approximate volume of gas lost, in m^3/day at normal temperature and pressure (NTP) is given by the following formula:

$$\frac{24V \, \Delta p}{\tau}$$

where V = volume of generator void and associated pipework, m^3

Δp = inclined gauge reading, bar

τ = duration of test, hours

The test assumes that the temperature of the gas in the stator is at all times the same as that in the box. Because of this and the short duration of the test, this method is usually only used to give an indication of the order of a leak, method (a) being used as a more definitive test.

Where the test is carried out using air, CO_2 or N_2, the results may be referred back to H_2 by use of Graham's law.

Rate of loss of H_2 =

$$\text{Rate loss for test gas} \times \frac{\text{Density of test gas}}{\text{Density of hydrogen}}$$

The acceptability of the loss so measured must be judged against the manufacturer's recommendations and past experience.

9.3.5 Rotor winding tests

Earth faults on a rotor may be detected and measured on-load, using the installed detection equipment, and off-load by means of IR checks. Interturn faults are more difficult to detect. An accurate determination of rotor resistance may indicate an interturn fault, if the short between turns is solid.

A widely accepted method of investigating for both earth faults and interturn faults is the *recurrent surge test*. A 12 V DC square wave is injected at each end of the rotor winding in turn, through a matching resistor. The input and output waveforms for each injection are viewed on a storage oscilloscope. A fault in the winding will cause a reflected wave to travel back to the input, causing a change in the input voltage after a time proportional to the distance of the fault from the input. Superimposition of the input traces from the two ends of the winding will show any divergence between traces, thus easing the detection of the voltage change described above.

A change in the total transit time of the pulse from that of the rotor when new, i.e., a change in the electrical length, will be proportional to the number of shorted turns indicated by the test.

On slipring machines, this test may be done at speed, which gives the ability to detect those faults which are not evident at rest or which change with speed.

Figure 5.45 shows a typical trace from a recurrent surge test on a 660 MW rotor which is giving an indication of shorted turns.

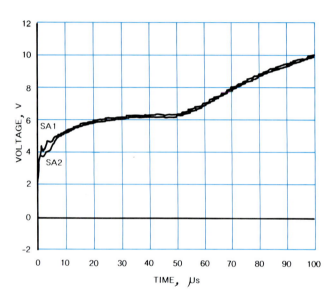

FIG. 5.45 Recurrent surge test on 660 MW rotor; superimposed traces from a digital storage oscilloscope Trace SA1 is the input voltage at end–1 of the rotor winding as the 12 V square wave is applied with end–2 open-circuited. Trace SA2 is the converse. The divergence of the traces indicates an interturn degradation of the rotor winding between 2.5% and 12.5% of the rotor winding length from SA2. The traces, when compared with those from the rotor when new, also showed that the effective rotor winding length had reduced by 5%.

10   Coal and ash handling plant

The coal, ash and dust plants in large modern power stations are the major items of no-unit plant which can halt the operation of the entire power station. Whilst a major pressure-part failure of great technical interest will only prevent one unit from generating, two minor breakdowns at strategic points in the coal

plant can starve the entire station of fuel. This statement may seem obvious but it does serve to emphasise the importance of top quality maintenance in these plant areas, which do not contain the most technically complex plant, but do convey the most vital materials for the running of the station. A typical coal handling system is shown in Fig 5.46.

The CEGB is not alone in developing technology for handling coal, ash and dust so the maintenance engineer may well look to coal mining, quarrying, civil engineering and cement industries for further information; certainly the suppliers involved in the power industry will be involved in some of these fields.

10.1   Conveyor belts

The conveyor belts are mainly used to transport coal, although they are occasionally used for cold, dry ash in some plants. Since the conveyors are carrying an abrasive cargo, it follows that they wear. In simple terms, the strength of the belts lies in the carcass while the covers are merely a protection for the carcass and are wearing items. Ideally, the belt should be replaced on a planned basis when the top cover has worn through and the carcass is about to be damaged. This aim is predictable and monitoring the cover wear allows the engineer to plan the need for a replacement belt and order accordingly. Unfortunately, the belts are prone to mechanical damage by odd pieces of tramp materials being jammed in the feed chutes in such a position that they cut the belt in two along the length of the belt. This type of breakdown can be minimised by using magnetic separators and screens, etc., but it does occasionally occur. The maintenance engineer should stock strategic lengths of belting to allow immediate replacement. CEGB practice is to use replacement belts to the British Coal Specification 158/71 Fire Resistant Conveyor Belting which was originally produced for underground mine workings. Whilst power station coal plants are mainly above ground, they are usually confined in sloping tunnels which contain both streams of conveyors and the risk of fire restricting output is high. For this reason it is also important to regularly maintain and test the fire fighting equipment on the coal plant.

The long life and reliability of coal conveyors is greatly assisted by correct routine maintenance rather than a breakdown maintenance policy, particularly as the bulk of plant can be visually inspected on load and minor defects identified and corrected before major damage is caused. When belts are in operation the coal feed should always be central, otherwise the belt will be forced off its track. Deflector plates can be fitted in chutes to ensure correct loading. Skirt plates should be inspected to ensure that they seal adequately but do not rub hard on the belt, with

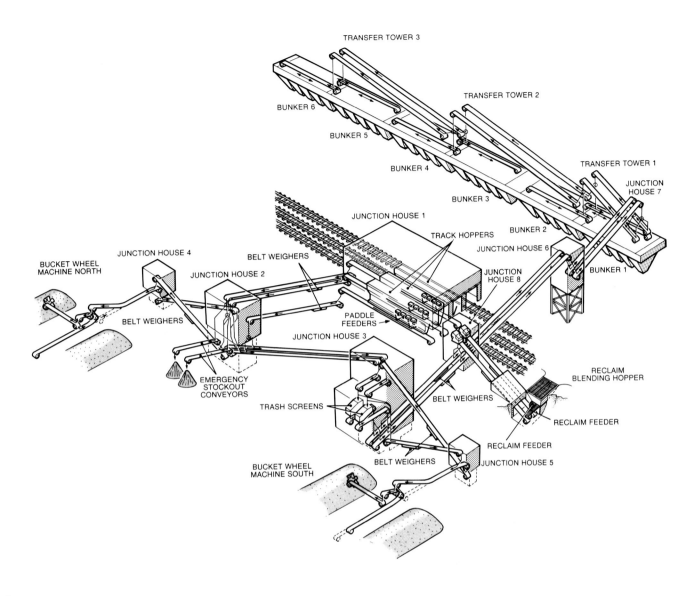

FIG. 5.46 A typical coal handling system

consequential damage. Regular attention to these features is repaid by reducing spillage. The belt should run centrally in its track with adequate support and tensioning. In particular, when a new belt of a different type to the original is fitted, it may require a different distance to change from deep trough to flat at the pulley and the belt should be adequately supported during this change. Also the tensioning device should positively tension the belt at all stages of the operation from starting to stopping, not merely full-load steady state running.

The bearing parts of the belt should be well lubricated, i.e., idler rollers, non-driving pulleys, return idlers and self-aligning idlers. Recent developments by lubrication companies include the fitting of long life pressurised containers of lubricants which last about one year. Alternatively, the bearing and roller manufacturers are offering more 'sealed for life' bearing assemblies. Whichever product is adopted, the plant should be regularly inspected and defective items rectified at suitable planned outages.

The automatic belt-clearing devices, such as scrapers or brushes, should be kept in good working order, with adequate means for disposal of their residue. If these items are allowed to fall into disrepair, the coal will merely spill in other less accessible parts of the machinery, costing more to clean and posing hazards to men and plant.

It is essential to keep conveyor plant reasonably clean: this seems a daunting task for machinery conveying coal, but a build-up of fine coal over a long period can ignite very easily. This cleaning requirement invariably conflicts with the need to guard the plant. In the UK, the Health and Safety at Work Act requires that, as far as is reasonably practicable, the conveyor is safe. In particular, but not exclusively, the dangerous areas are:

● Nips between pulleys and belts, including the gravity take-up and snub pulleys.

● Nips between the belt and conveying idlers where the upward movement of the belts is restricted, for example, by skirt plates.

● Nips between return idlers and the belt.

● Openings into chutes.

● Access ways under the conveyor (large-scale spillage or temporary obstruction may reduce this clearance).

● Power transmission machinery, such as shafts, couplings, gears, motors, etc.

● Travelling trippers with risk of trapping between tripper and fixed steelwork.

The CEGB Code of Practice for troughed belt conveyors requires that all the preceding hazards be guarded, but that access be possible for cleaning and routine lubrication with guards in position. Lubrication is easily achieved by the use of remote nipples and the pressurised canisters previously referred to, but cleaning is a problem. Some guard designs have been tried which closely follow plant contours and embody slots for inserting special cleaning tools which cannot foul the moving parts. These small guards are themselves susceptible to damage from the sheer weight of spillage and the latest designs are for large guards with safety-interlocked access for cleaning: this obviously precludes on-load cleaning but does allow a more businesslike approach. When designing such guards, the engineer should ensure that the independence of each stream of conveyors and each group of conveyors is maintained.

Normal design practice is to equip a station with two parallel 100% duty conveyor systems for the main supply and main reclaim routes. The dictates of minimising capital cost mean that changeover points between these parallel paths will be at a minimum. From the maintenance point of view, more changeover points mean greater flexibility of plant and less chance of total loss of fuel supply, but there is also the added complexity of maintaining the chute and flap systems.

As already stated, the conveyor belt is a wearing item and has to be changed periodically. This can be a difficult task because the belt is heavy and cumbersome, compared with the lifting equipment usually available on coal plants. The most effective method for changing the belt is to position the new belt somewhere under, or even over, the conveyor such

that the old belt can be cut, the old and new clipped together and the old belt run-off while the new belt is run onto the roller system. It may be necessary to fit some temporary rollers to achieve this set-up and the motive power should be carefully chosen as the normal power transmission system is far too fast and powerful for such an exercise. Occasionally an inching provision is fitted or it may be possible to fit a slow-speed pony motor, either electrically or air operated. Alternatively, a long-range winch may be fitted up to drag the belt around. If the old belt is damaged such that it cannot travel around the system, the new belt has to be positioned by repeatedly winching and clamping the belt which is a time and energy consuming method.

When the belt is in position, the two ends must be joined. In simple terms this can be achieved by clipping the belt mechanically or by vulcanising. The decision is influenced by the tendency for the belt to stretch in service. Manufacturers maintain that modern all-synthetic belts do not stretch to any significant extent and this is true compared with the older natural fibre belts. However, some stretch does occur and the need for shortening is determined by the range of the take-up unit. If shortening is required, it is usually economic to clip the belt for the initial running period and then vulcanise for normal service. Occasionally there is a need to repair mechanical damage by inserting a length of belt. Again, this should be vulcanised for long-term operation. All vulcanised joints should be made to approved standard procedures which are readily obtainable from the CEGB, manufacturers and other major belt-users throughout the world. Mechanical clipping of the belt is undesirable as it imposes severe local stresses and precludes automatic belt cleaning and close-fitting skirts. The maintenance engineer should always retain the ability to clip a belt, as this can be done in a fraction of the time taken for vulcanising: this may be important in the event of plant breakdowns.

Coal conveyors have fairly complex power transmission systems which will require attention. To avoid duplication, the reader should refer to Section 6.1.8 of this chapter which deals with the general maintenance of power transmission gearboxes.

10.2 Coal handling machines

Paddle feeders on coal plants are mechanical shovels which take coal from hoppers (whether these be rail, road or reclaim hoppers is irrelevant) and feed the coal into the adjacent coal conveyor belt. The paddle feeder (Fig 5.47) is simply a large machine running on a rail system and, in this simplified description, it is similar to a tripper car at the bunker end of the conveyor system and (extending the scale of machinery) is similar to the large stacker-reclaimer

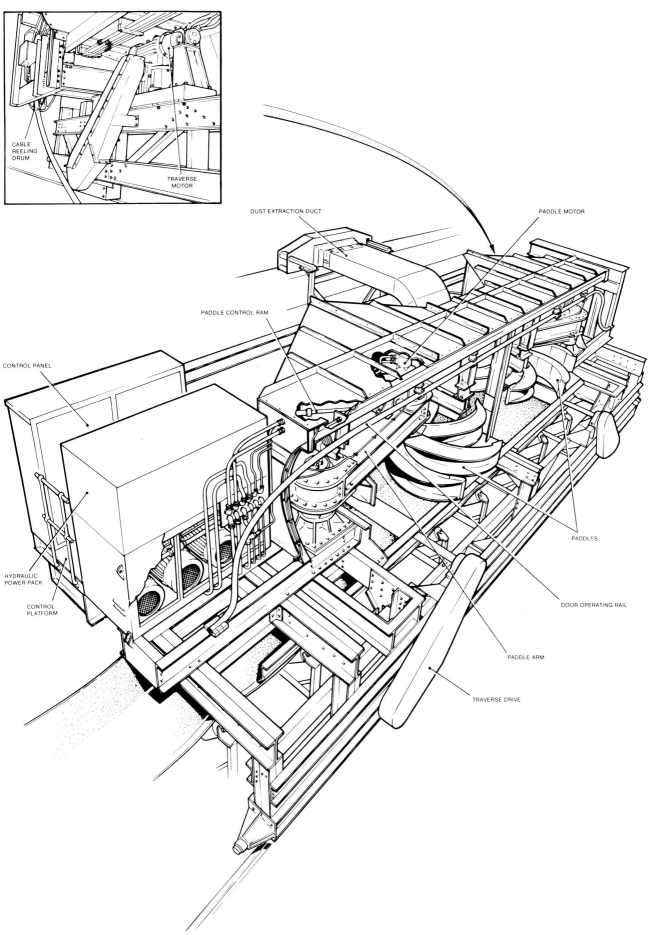

CABLE
REELING
DRUM

TRAVERSE
MOTOR

DUST EXTRACTION DUCT

PADDLE MOTOR

PADDLE CONTROL RAM

CONTROL PANEL

PADDLES

HYDRAULIC
POWER PACK

DOOR OPERATING RAIL

CONTROL
PLATFORM

PADDLE ARM

TRAVERSE DRIVE

FIG. 5.47 General arrangement of a paddle feeder

machines which are standard equipment on virtually all the major CEGB coal-fired stations.

The fundamental maintenance objective with all these major rail-mounted machines is to ensure that they remain structurally stable and on the rails. The wheels turn very slowly compared with, say, railway trains but, nevertheless, the wheel axles and bearing assemblies continually transmit all the load of the machine. Regular lubrication and a regular, albeit long-term, inspection of these traction systems and their associated braking systems on a planned basis is essential. This inspection, together with regular cleaning and a general structural inspection for overloading and corrosion, should ensure the structural stability of the machinery.

Paddle feeders and other similar machines suffer from a design disadvantage in that they must transmit tremendous power to move difficult coal deposits. Under fault conditions, this same power can inflict considerable damage upon the feeder and its associated equipment without being detected by simple electrical protection devices and so damaged paddles, hopper doors, etc., are by no means unusual. It follows that spares-stocking policies should cater for this eventuality.

The major coal handling machines are all fitted with a number of power transmission systems. As a general rule, it is preferable to carry out internal work under workshop conditions rather than *in situ*, although each job must be judged on its merits. If *in situ* repairs are carried out, positive steps are necessary to avoid contamination of gearbox internals. If a suitable maintenance bay is provided, it may be feasible to screen the machine temporarily from the running section of coal plant. For more detailed consideration of the maintenance of power transmission systems, the reader should refer to Section 6.1.8 of this chapter.

As the coal passes through the machine, it must wear some components. The components which are static relative to the body of the machine can be considered as coal chutes (see Section 10.4 of this chapter). The components which move, for example, the paddle wheel itself, require more detailed attention. The actual hub of the paddle wheel should be protected by replaceable blades on the cutting edges and erosion shields, so that the coal is not allowed to erode the expensive, heavy and accurately-made components. The wearing parts should be examined regularly and maintenance work planned to prevent damage to the basic machine. When refurbishing the paddle-wheel blades, the engineer has an opportunity to exploit the ever increasing number of wear-resistant materials on the market. When these machines are subject to major works, it is necessary to overhaul as required all the simple linkages and cam systems which are part of their overall function, for example, openers, flap dampers, etc.

This section includes reference to stacker-reclaimer machines, as they are in general terms similar to the other coal handling machines. However, there are two points worthy of attention which are specific to stacker-reclaimers. First, the stacker-reclaimers (Fig 5.48) are not naturally stable; they have to be balanced by a counterweight system. This can be achieved by a mechanical or a hydraulic control system. Whichever system is used, it must be regularly inspected and serviced in accordance with the manufacturer's instructions. Secondly, the whole stacker-reclaimer assembly stands on a very large thrust bearing. The condition of this bearing should be monitored and any necessary remedial action taken in good time. Both these factors relate to the basic stability of the stacker-reclaimer and are vital, since any accident could be devastating to personnel and plant.

10.3 Wagon unloading equipment

The majority of coal delivered to CEGB power stations is carried by permanently-coupled trains which discharge into ground hoppers whilst still moving. Other shunting operations may take place due to local track layout but the basic concept is now to discharge coal on the move, which necessitates reliable trackside equipment.

There are four types of machine: unlatching, opening, closing and latching machines. Despite their different functions, they are basically similar. The machines consist of a cam on the end of a horizontal retractable camshaft, a counterweight, a balance weight, a compressed-air operating cylinder and, on some, machines, a damper cylinder. The machines should be set to exert force within a specific range upon the wagon door. If it fails to open, the wagon overrides the lineside equipment. Setting the machines within this range is the most critical part of the overhaul and is the main parameter of a routine inspection. Hopefully, those who maintain the rail wagon are operating to the same criteria. *In situ* maintenance on this sort of equipment must be very limited and as they are easily changed and readily transportable, maintenance work is best undertaken by centralised maintenance units on a semi-production line basis.

If the power station engineer must maintain these items, the relevant areas are wear on the operating cams which are unlubricated, wear on the attachments to the main shaft which suffer shock loads, wear on the sliding faces and degradation of the compressed-air system components.

10.4 Bunkers and chutes

Coal bunkers are generally regarded as civil engineering structures while their smaller brethren, the coal

Fig. 5.48 Bucket wheel stacker-reclaiming machine
(see also colour photograph between pp 434 and 435)

chutes between conveyors, are mechanical engineering items. In both, the maintenance engineer must
preserve the structural integrity of the plant by preventing coal erosion or corrosion of the basic structure. Fortunately, the rate of coal movement per
unit surface area in coal bunkers is low compared
with chutes, so erosion is not much of a problem.
Most modern bunkers are fitted with low friction
linings. These should be inspected on major unit
outages, when the bunkers are emptied, and broken tiles replaced. Internal structural steel should
be cleaned and the protective coating repaired, as
required.

Coal chutes and diverter flaps, etc., can suffer
quite a high erosion rate, particularly in target areas.
With the coal flow rates of a 2000 MW (or more)
power station going over one line of conveyors mild
steel liner plates can wear out in a matter of days.
As already suggested, the supply of wear-resistant
materials is ever increasing and any comment can
only be a snapshot of current progress. Most recent
experience is that lining hoppers with ceramic tiles
with a high alumina content gives a lightweight but
extremely durable wearing surface. The tiles are so
light that they can even be fixed to movable items,
such as changeover flaps, etc. The tiles are glued
on with specialist adhesives and can be difficult to
remove. Looking in the long term, there is some
merit in lining the basic structure with disposable
mild-steel liners and then fixing tiles to those liners.
The detailed selection of tiles, including the method
of fixing, should be discussed with the tile manufacturer or his agent.

10.5 Ash crushers

Furnace-bottom ash from large boilers is invariable
ground at some point in the disposal route. On early
plant, up to about 120 MW units, the usual method
was to use a water sluice to convey ash to a common
pair of ash crushers. On the larger units, it is more
usual to fit pairs of ash crushers adjacent to each unit
and pump the ash away to a disposal system. From
a station operational viewpoint, the objective is merely
to produce ash particle sizes which can be readily
pumped in a water stream. From a commercial viewpoint, the crusher also produces a saleable byproduct
of consistent size. The ash crushers must wear simply
by virtue of their function of crushing ash.

The crusher has to be located in some sort of pit
in the depths of the station and is prone to flooding
whenever faults occur in the ash plant. Like other
regularly wearing plant, ash crusher running and
maintenance regimes must be planned to ensure the
continued availability of one duty crusher to each
unit and a steady flow of maintenance work. Given
that crushers are liable to damage from tramp material, for example, sootblower nozzles or scaffold

clips, the most effective maintenance policy is to
hold spare overhauled ash crushers and to change
the whole unit.

It is a fairly simple matter to establish the rate
of wear of a crusher by measurng roll tooth profile.
The crusher is worn out when the particles of ash
are so large they cause pump blockages. If optional
roll sizes are available, the larger the crushing area,
then the longer the roll will last. The casing should
be fitted with liners to prevent erosion of the basic
structure. It is unlikely that exotic materials would
be worthwhile in this application as the liner life
must match the roller life, or be reusable and match
multiples of roller life. It is possible that some bearings may be reusable, particularly on crushers with
small rollers. A decision to reuse bearings should be
made cautiously as the bearings will have suffered
considerable shock loading and will almost certainly
have been flooded with dirty grit-laden water on more
than one occasion. Modern crushers are fitted with
interlock devices to prevent access to moving parts
until the driving-motor power supply is isolated. The
integrity of the devices must be maintained at all
times.

10.6 Ash pumps

Ash pumps deliver a mixture of ash and water, usually
to a distant lagoon or disposal pit; they should not
be confused with sluice-water pumps that pump relatively clean water, which is then used to sluice ash
systems through sluiceways or ejectors. The ash pump
is designed for more regular maintenance than a
sluice-water pump due to the abrasive nature of the
ash. The pump is usually a centrifugal pump with
the impeller overhung on the end of a shaft and an
axial suction eye. The casing and impellers are made
of wear-resistant steel, usually with an adjustable
end to seal the impeller. The single gland has a shaft
sleeve to protect the precision-machined shaft and
all the bearings are in a self-contained block. Provided the bearings are correctly lubricated and the
shaft sealed to prevent ingress of dirt, the bearing
block should last for several pump bodies: hence the
design, which allows the wearing parts to be renewed
with the bearing and shaft assembly *in situ*. However, bearing life should be monitored, as there is
no merit in the loss of a pump due to bearing
failure shortly after the wearing parts are replaced.

When replacement components are considered, the
engineer should always be ready to consider the economics of longer-life materials should any be available,
bearing in mind the cost of the overhaul as well as
the spare part cost. The gland is a particularly weak
area; it must be fed with clean water, as ash ingress
to a packed gland will cause very rapid failure and
damage to the shaft sleeve, necessitating replacement
before the pump is due for overhaul.

When the pumps discharge to a distant lagoon system with different pipework lengths, it may be necessary to change the pump impeller to match the pipework flow characteristic. If the impeller is large and the pipework short, the pump will drain the supply sump and erode the pipework unnecessarily. In the event of a small impeller and long pipework run, the flow velocity will drop with a risk of pipework blockages.

10.7 Dust pumps

Dust can be conveyed as a slurry mixed with water, fluidised by mixing it with air or by pumping it as a fluid in its own right. A dust-water slurry pump is similar to the ash pump described in the previous section, although the dust pump enjoys a longer life than the ash pump as the slurry is less abrasive than ash. Air fluidisation requires fans to handle relatively-clean conveying air, as the dust itself is not pumped. Principles of fan maintenance are dealt with in Section 6.3.2 of this chapter, albeit main boiler fans are much larger than dust plant fans. When dust is pumped as a fluid in its own

right, it is usual to use a slow-speed positive-displacement pumping device. Typical examples are the screw-type dust pump (Fig 5.49) and the pneumatic transport pump (Fig 5.50), in which a measured quantity of dust is allowed to enter a hopper by gravity and is blasted out by high pressure air. These two pumps are so totally different that it is not possible to outline specific maintenance procedures.

When a dust pump contains moving parts in close contact with the dust, wear will occur which must be monitored at least until an economic life is established. Where bearings are used on either pumps or control systems, particular care should be taken to ensure that the sealing mechanism is operating correctly. When pulverised fuel ash (PFA) contaminates the lubricant, failure will occur rapidly; alternatively, oil leaks mixed with PFA cause disproportionate problems for good housekeeping. All dust pumps are susceptible to failure due to ingress of foreign bodies (invariably debris from precipitator or gas-pass repairs). Dust is regarded as waste, or at best a byproduct of the generation process, hence the avoidance of contamination does not always receive sufficient priority necessary to protect the disposal plant.

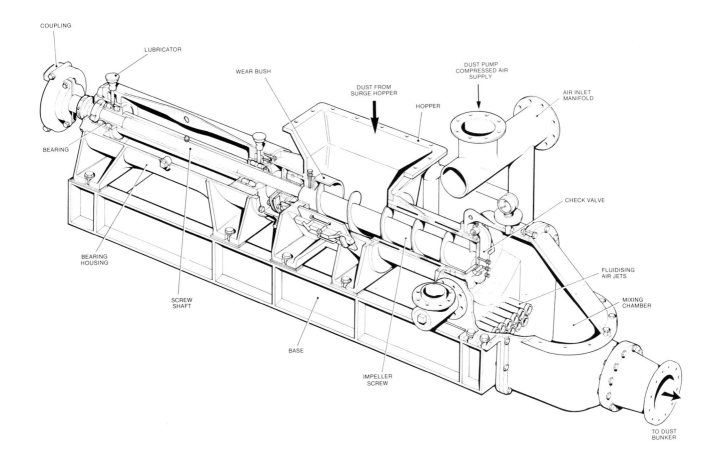

FIG. 5.49 Screw-type dust pump

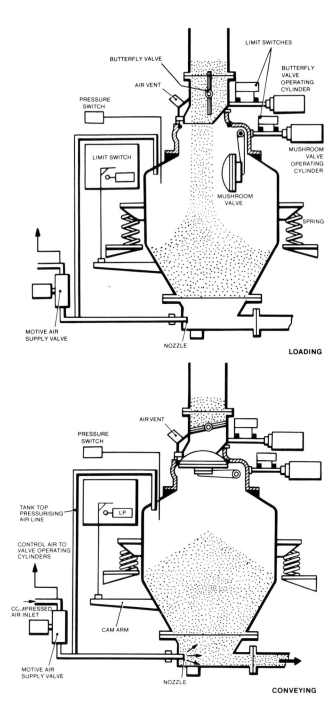

FIG. 5.50 Pressure vessel type dust pump

10.8 Ash and dust pipework

The ash pipework system is the outlet pipework from the ash pump. It can follow a short tortuous route to a grabbing pit or travel for considerable distances to lagoon systems. This pipework handles probably the most abrasive fluid mixture in the generation cycle. The erosion problem is similar to that encountered in PF pipework but without the explosion risk. Long straight sections cause little problem. They can be cast iron or, for longer life, fabricated and lined with an abrasion-resistant material, such as cast basalt. Straights tend to wear in the bottom quadrant, so pipework life can be extended by regular ultrasonic thickness-checking and by rotating the pipes. Flange joints are more difficult to make than flexible joints but cause less flow disturbance and do not offer external water traps. If a long life lined-pipework option is chosen, then mild steel pipework should be protected, or perhaps glass fibre outer casings used to avoid pipework failure due to external corrosion. If a policy decision is taken to run pipework systems to complete destruction, some measure of protection or wear monitoring should be given to areas where leakage would cause damage. Bends in ash pipework are obviously high-wear areas and it is usually worth considering lined pipework for bends. Short runs of pipework in boiler house basements are worth monitoring closely to avoid failure compared with outdoor pipework. The cost and time taken to clean spillage in basements can be prohibitive.

Dust pipework, whether wet or dry, suffers erosion but at a lower rate than the ash pipework. Again, linings prolong life and it is simply a commercial balance of life against total replacement costs. Bends are areas of weakness, particularly where plant geometry dictates a very tight radius. Invariably, that same geometry means access is difficult, so that the worn bend is expensive to change.

10.9 Dust filters

Dust-bunker filters are fitted to clean the large quantity of conveying air which transports the dust through the system, plus the relatively small amount of air displaced from the bunker. The filters are a proprietary design and the manufacturers usually provide detailed maintenance instructions. It is important that the dust filter bags are regularly cleaned by the reverse-air cleaning system.

If the bags are kept clean, they last many years. If the filter is operated until it is completely blocked, the bags cannot be recovered and replacement is very expensive. For this reason, it is well worth considering routine maintenance of the filter, which includes overhauling the reverse-air system. The reverse-air system must have means of traversing the filter-bag area and hence has moving parts operating in the dust-laden environment. These moving parts are usually relatively straightforward items, such as wheels, pins, bushes and levers. When the filter is out of service, it must be heated to prevent moisture entering the cabinet and causing the dust to solidify.

11 Electrical plant

11.1 Maintenance philosophy

The electrical plant within a power station, by the very duty it performs, must have high reliability and thus a regime of planned maintenance, rather than breakdown maintenance, must be adopted. Such maintenance can be on the basis of:

- Elapsed time.

- Running hours.

- Number of operations.

- Performance monitoring.

- Condition monitoring.

For rotating plant, the last two would seem ideal, but monitoring is usually manpower intensive and so can only be applied to certain prime items. Even the checking of running hours or number of operations would be a major exercise if carried out on the multitude of electrical plant items within a station, and thus is again only of limited application.

Elapsed time planned maintenance is therefore the most practical method for the majority of electrical plant. Considerable effort is needed to ensure that the time periods are correctly chosen and there must be a willingness to change them in the light of experience. The maintenance planning system used must therefore have a flexibility which allows such changes to be made simply as one of its main features.

11.2 Switchgear

11.2.1 High voltage switchgear

The high voltage (HV) switchgear used within the works power system on a modern power station usually operates at 11 kV or 3.3 kV and is one of the following types:

- Air circuit-breaker (Fig 5.51).

- Oil circuit-breaker.

- Vacuum circuit-breaker.

It is not intended to give a detailed description of the maintenance of each type of switchgear but rather to outline the essential aspects which must be borne in mind on any maintenance schedule.

The duty cycle of some of the switchgear is arduous and the potential fault levels that the switch-

gear may be required to interrupt are high (typically, 750 MVA on an 11 kV switchboard). The aim of all maintenance activity must therefore be not only to ensure reliable operation in normal circumstances but also that the switchgear remains capable of safe operation under fault conditions throughout its life.

Setting the periodicity of switchgear maintenance is difficult. Some devices operate many times a day and are thus subject to wear; others may only operate a few times a year and are therefore likely to give problems due to mechanical stiction. Usually an annual overhaul is planned, with an additional check if a switch has operated under fault conditions or if it performs a particularly onerous duty cycle.

All HV switchgear overhaul specifications must cover the following points (where applicable to the particular design):

- Pre-overhaul check of insulation resistances (IR) between phases and phase-to-earth, and measurement of the resistance of each phase.

- Lifting and inspection of arc chutes (air circuit-breakers).

- Removal of oil tank, replacement of oil with clean tested oil (oil circuit-breakers).

- Cleaning and inspection of moving and fixed main contact assemblies.

- Checking of contacts for wear, clearances, wipe and alignment (phase-to-phase and between phases).

- Checking of flexible connections for wear.

- Inspection and checking of 'close' and 'trip' mechanisms, carrying out all specified measurements to ensure correct operation.

- Checking the operation of isolating and earthing mechanisms.

- Examination of all auxiliary electrical equipment and overhaul of closing and tripping solenoids, contactors and anti-pump devices.

- Checking the operation and integrity of air-blowing systems.

- Checking all locking devices.

- Inspection of all insulation for cracks, discolouration, or other signs of distress.

- Checking of mechanical interlocks.

- Lubrication of moving parts.

- Final IR checks and measurement of resistance of each phase.

- Cleaning and checking cubicle — check cubicle heater.

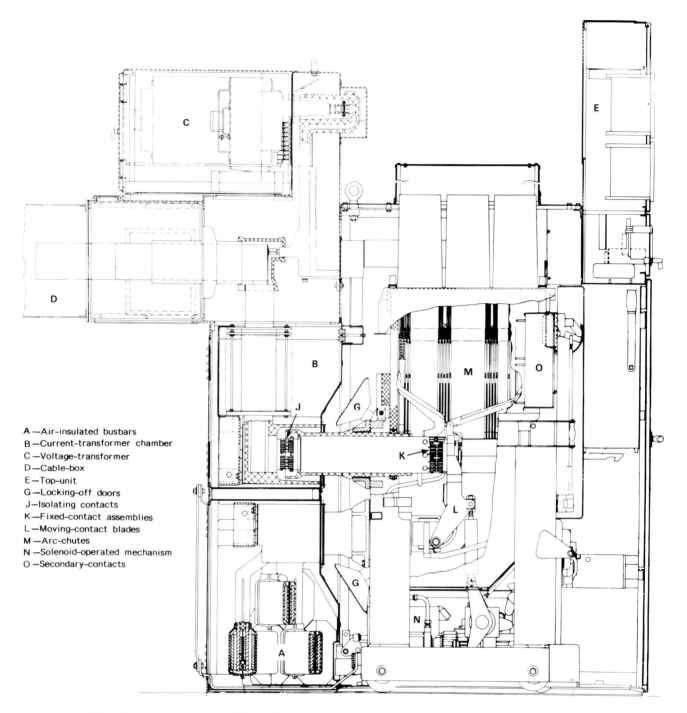

A —Air-insulated busbars
B —Current-transformer chamber
C —Voltage-transformer
D —Cable-box
E —Top-unit
G —Locking-off doors
J —Isolating contacts
K —Fixed-contact assemblies
L —Moving-contact blades
M —Arc-chutes
N —Solenoid-operated mechanism
O —Secondary-contacts

FIG. 5.51 Cross-section of 11 kV metalclad switchgear. 11 kV air circuit-breaker shown in service position.

Throughout the maintenance, all electrical checks should be recorded on a checksheet.

As experience is gained on each particular type of switchgear, particular faults and weaknesses will be discovered and additional checks will be added to the maintenance specification.

Following the overhaul of a circuit-breaker, a full function test must be carried out before returning it to service. The following checks must be included:

● Operation from all positions, taking account of circuit interlocks and group interlocks.

● Operation from each protection system.

● Alarm checks

● Trip circuit supervision checks.

In addition to the overhaul and function checking of each circuit, the switchboard itself must be over-hauled periodically. The planning of such maintenance can be difficult, since a total board outage will be required; however, since the board contains no moving parts, an overhaul every five years will usually be sufficient, with interim short outages for busbar spout cleaning and incoming circuit-breaker maintenance. Areas of work covered by a switchboard outage are:

- Cleaning of busbar spouts and checking of associated interlocks.

- Cleaning of busbars, checking of busbar joints and support arrangements.

- Busbar insulation and conductivity checks.

- Checking the security of all covers.

- Overhaul of board auxiliary wiring and supplies.

- Inspection of earthing.

Protection testing is also often carried out following HV switchgear overhaul work. This is discussed in Section 11.2.3 of this chapter.

11.2.2 Medium and low voltage switchgear

Medium and low voltage (MV and LV) switchgear is usually in a modular board arrangement, each module being the switchgear associated with a particular circuit. This is not invariable; sometimes contactors, isolators, etc., are part of a control panel which is the total control package for a plant item, e.g., a sootblower control panel or an auxiliary boiler control panel. The comments below are in the context of the component parts of a multi-circuit 415 V switchboard of modular design, but the particular maintenance details for each item apply equally to the components used in other MV and LV switchgear applications.

It must be remembered that a 415 V board in a power station is subject to a high potential fault level (about 30 MVA). There have been many instances of electrical flashovers within 415 V switchgear which have been attributed to dirt, poor connections, loose fuses and bad condition of contacts. The consequences of such faults can be serious so, whilst the overhaul and maintenance of such switchgear can appear mundane and repetitive, a high standard of work must be encouraged.

Incoming and bus-section ACBs

These are similar in many aspects to the HV ACBs previously described and do not need to be dealt with separately.

Contactors

The duty performed by the contactors in power station switchgear is often very onerous, with many

operations a day. Such duty is reflected in mechanical wear and tear and electrical contact wear; particular attention must therefore be given to these aspects when undertaking a contactor overhaul.

Maintenance instructions for a contactor must include, where applicable:

- Cleaning.

- Dressing or renewal of main and auxiliary contacts. Note that some contacts are plated and cannot be dressed successfully.

- Checking of contact alignment, wipe and spring-loading.

- Mechanical operation of armature, alignment of pole faces.

- Checking of the mechanical interlocking between reversing starters and the latching mechanism on latched starters.

- Trip mechanism inspection.

- Electrical checks on main and trip coils (resistance and insulation).

- Check tightness and condition of all connections.

- Arc chute inspection.

- Lubrication.

Isolators

The maintenance of isolators consists mainly of cleaning, checking contact condition and tightness of connections, overhaul and checking of the operating mechanism, and confirming the tightness and continuity of fuses, if fitted.

Isolators are usually interlocked with the cubicle door and often with their associated contactor. These interlocks are important safety features, which must be checked.

Control indication and protection equipment

The contactor cubicle may contain control relays, control selectors, fused supplies, mechanical and electrical indication equipment, protection relays, CTs, ammeters, etc., all of which are included in the maintenance routine. The cubicle heater is also checked.

DC starters may have a series of timed contacts to switch starting resistances. Such starting resistances are usually only short-time rated, so that the correct timing of the contact operation is important.

Function checks

At the completion of the overhaul of each circuit, a function check is performed, as outlined for HV switchgear.

Switchboard inspection and overhaul

The complete MV or LV board is isolated and over-hauled periodically. The frequency depends on the environment.

The main areas of work are the busbars, common control supplies, auxiliary wiring, security of panel fixings, and inspection of earthing.

11.2.3 Protection equipment testing

The electrical protection equipment fitted as an integral part of power station HV and MV switch-gear requires little routine maintenance apart from occasional careful cleaning and visual inspection. However, to ensure correct and reliable operation, each device needs to be tested periodically, the results being entered into a record system containing the details of the device, its settings, and a record of the tests performed on it. Additionally, it is advisable to insert a relay-setting record card in each relay case. This enables a crosscheck to be made after testing to ensure that the relay setting has not been inadvertently left at an incorrect value (Fig 5.52).

STATION			RELAY	
PANEL			C.T. RATIO	
DATE	CURRENT SETTING	TIME SETTING	INITIALS	REMARKS
....	....	....	....	....
....	....	....	....	....
....	....	....	....	....
....	....	....	....	....
....	....	....	....	....
....	....	....	....	....
....	....	....	....	....
CENTRAL ELECTRICITY GENERATING BOARD				

FIG. 5.52 Typical relay-setting record card

Two types of protection testing are normally undertaken: primary and secondary injection tests.

Primary injection tests

This test is used to check the operation of the total system associated with a protection device. Connections are made to the main primary conductors of the circuit under test and a current is injected in such a way as to check the operation of the protection system under normal and fault current conditions.

The test set used for this is a single-phase transformer, fed by a 0–415 V variable transformer on

its HV side and rated to supply about 2000 A at a few volts at its secondary terminals.

Particular test procedures need to be prepared for each type of protective device. The means of connecting the heavy current test supply into the circuit should be specified; some HV switchgear is supplied with connection devices for this purpose. Figure 5.53 shows a typical arrangement for 11 kV metalclad switchgear.

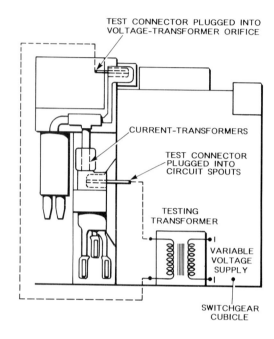

FIG. 5.53 Primary injection testing of 11 kV switchgear

Primary injection testing is time consuming and would normally only be undertaken on a few occasions during the life of the equipment, say, every 6 years, unless evidence justified more frequent testing, or a fault was suspected on a particular circuit.

Secondary injection test

This is used to check the operation of the protective relay alone and can be undertaken *in situ* or, for plug-in relays, with the relay plugged into a purpose built test rig. A test current from a secondary-injection test set is fed into the relay circuit in such a way as to check the performance of the relay over its range of operation, in accordance with a specified test procedure.

The secondary-injection test set will include:

● A low voltage single-phase variable output up to, typically, 100 A.

● A harmonic filter for use with relays sensitive to harmonic interference.

- Accurate current indication.

- A timer initiated from the test set 'start' button and stopped from the relay trip contacts.

- 0–240 V AC and DC low power variable-voltage outputs.

The frequency of secondary injection testing should initially be set at every two years and then modified as a result of the experience gained on each type of device.

11.3   Motors

The vast number of electric motors employed in a power station are so varied in size, type, application and environment that it would be foolish to stipulate a fixed periodicity of maintenance for all motors. A good basic approach is to plan regular *in situ* maintenance and an occasional full workshop overhaul for each motor, the frequency being based on a consideration of all factors influencing the wear and tear on the particular machine. As a guide, most machines would need annual *in situ* maintenance, with perhaps a triennial full workshop overhaul.

The activities comprising *in situ* maintenance are outlined below, followed by a more detailed description of the work to be included in a workshop overhaul (Fig 5.54).

The comments obviously do not apply to all machines; similarly, the complexity of each task varies between HV and MV motors, and between different designs. Sufficient information is given to identify the activities in the maintenance schedule for a particular machine.

11.3.1   Routine maintenance — *in situ*

- Generally clean down motor, check for obvious mechanical damage, check holding-down bolts, security of guards, etc. Inspect for oil leaks, coolant leaks and effectiveness of weather protection.

- *Tests*   From terminal box or switchgear, check insulation resistance of winding and supply cable; check resistance balance of windings.

- *Terminal box*   Clean terminal box internally; check connections; check insulation condition; ensure that cable glanding and earthing arrangements are satisfactory. Ensure that the terminal box is adequately sealed after maintenance.

- *Cooling circuit*   For open motors, remove covers (where possible) and blow out winding with dry compressed air.
  For enclosed motors, with air-to-air coolers, clean heat exchanger tubes. Inspect water-cooled machine heat exchangers for leaks, build-up of scale, or any other defect which may impair the cooling efficiency.

- *Windings*   It is not normally necessary to inspect the windings of an enclosed machine. Open machines should be thoroughly inspected at every

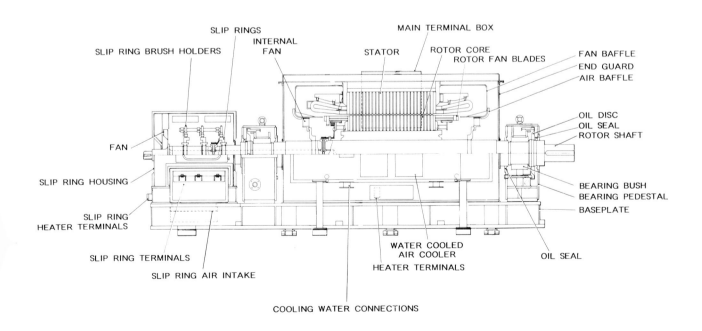

FIG. 5.54   11 kV 10 MW wound-rotor induction motor

accessible point for signs of serious contamination, winding damage, etc.

● *Bearings*

(a) Ball and roller. The *in situ* check of the bearings is made by the best practicable method. For bearings fitted with end caps, the outer cap is removed, having first supported the inner cap with guide studs. The bearing can then be inspected for scoring, roughness, cracking, metallic debris or any other sign of distress. It may be feasible to assess bearing wear by checking the lift on the shaft. Where the bearings are inaccessible, it may be possible to run the motor to listen for bearing noise, or at least to turn the shaft to try to detect any roughness or rubbing that may indicate wear. Finally, the bearing should be lubricated. Care is needed, since overgreasing can cause the bearing to overheat due to churning of the grease.

(b) Pad or sleeve bearings. Examine bearing housing for leaks from joints, shaft seals, drain plugs or level indicators, including (if possible) a check for internal oil leakage into motor. Drain the bearing oil and examine its condition, looking for signs of sludge, metallic debris or water. If unsatisfactory, further investigation of the bearing will be required. Check that bearing clearances are within tolerance by measuring shaft lift or by the gap between shaft and bearing top, as applicable.

● *Airgaps* On motors provided with access points for airgap readings, measure the airgaps, using long clean feeler gauges, and check that they are within tolerance.

Where the access points are placed at 120° around the stator three sets of three readings should be taken, turning the shaft through 120° at each set. The three readings at each position are then averaged.

Where the points are at 90° around the stator, take two sets of readings, turning the rotor through 180° between readings.

● *Coupling* Remove guards and examine for excessive play.

Excessive play in Bibby couplings requires checks of the spring and teeth for wear or damage. The grease on a Bibby coupling should also be checked and replaced, if necessary.

Excessive play in a rubber-bushed coupling needs checks for rubber-bush wear, elongation of coupling bush holes, bolt waisting and bolt tightness.

● *Sliprings* (variable-speed AC motors) The overhaul of sliprings and brushgear has already been outlined in the generator maintenance section (9.2.4 of this chapter). The principles described therein apply equally to slipring motors. On completion of maintenance, particular care must be taken to ensure that the assembly is clean and free of contamination from oil, grease or carbon dust.

● *Commutators* (DC machines and AC commutator machines) Whilst the same general comments on slipring and brushgear maintenance apply to commutator machines, particular care has to be taken to ensure that correct brush position is maintained.

A good commutator electrical surface must be achieved. The surface must be uniform in appearance and irregularities should be regarded as indicators of electrical or mechanical faults requiring further investigation.

11.3.2 Routine maintenance — full workshop overhaul

The main stages of a full motor overhaul are now described. It is assumed that the motor has been checked electrically and any evidence of faults has been noted for investigation during the overhaul:

Cleandown and strip After a thorough external clean, normal dismantling techniques are used, whereby each item is marked, where necessary, to show its position relative to other parts. All non-maintainable parts, such as covers, bearing housing, baffles, etc., should be cleaned after removal and inspected for cracks, distortion or other damage.

Rotor removal The larger the machine, the higher the probability that special equipment will be required to enable the rotor to be removed. A typical method, employing a trolley, an extension shaft (a steel tube machined such that it just slips over the protected journal surface) and an overhead beam or crane, is shown in Fig 5.55.

Stator The method of cleaning the stator depends upon the extent and nature of contamination. Common methods are:

● Dry dusting and blow-out.

● Wiping clean, using grease solvent.

● Spraying with electrical solvent.

● Washing with distilled water (followed by dryout).

When cleaning the windings of HV motors, particular care must be taken to ensure that contamination is not driven into the windings, thus becoming the focus of a future failure. Insulation resistance readings of the stator are taken before and after cleaning.

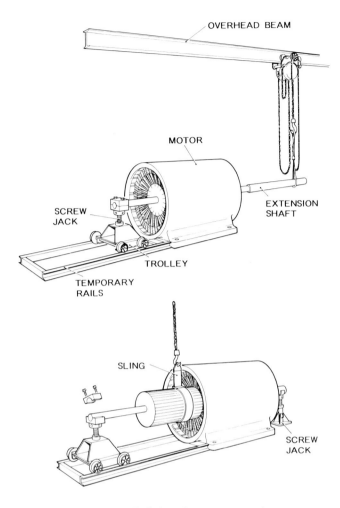

FIG. 5.55 Technique for rotor removal

A thorough inspection of the stator is now undertaken and any remedial work carried out. The following points should be covered:

- Condition of winding insulation. Is the taping satisfactory? Is revarnishing required?

- End-winding integrity. Inspect for evidence of looseness, defective lashings, bracing or spacers.

- Slot conductors. Check for any evidence of discharge, loose or damaged wedges.

- Winding connections. To cable box and star point (if accessible).

- Stator laminations. Evidence of hotspots or looseness, rubbing or scoring. Looseness of the stator pack may be indicated by evidence of brown powder from the insulation coating on the laminations.

A stator in poor overall condition may be referred to a specialist repair contractor for overhaul. General deterioration can often be reduced by thorough cleaning and a treble dipping/baking in epoxy-resin varnish, a service offered by most repairers.

Rotor Cleaning methods described above are also applicable to the rotor. The condition of the rotor is determined with regard to the following points:

- Loose, cracked or broken rotor bars.

- Cracked rotor bar-to-endring joints.

- Cracked or broken endrings.

- Blockage of ventilation ducts.

- Laminations — signs of rubbing or scoring. Evidence of hot spots.

- Rotor fan damage.

NDT techniques may be used to inspect rotor bar-to-endring joints, if cracking is suspected.

An indication of poor rotor bar-to-endring joints may be obtained by using a 'growler', an electromagnetic test instrument which produces a change of resonant note when passed over a defective bar.

Bearings

(a) Ball and roller bearings. Remove the bearing from shaft using a hydraulic bearing puller. Clean all grease from bearing and inspect for:

- Scored or worn outer journals.

- Cracked or distorted cages.

- Pitting of balls or rollers.

- Excessive play.

- Indications of slack fit between bearing and housing.

- Slack fit to shaft.

- Rough running (when spun after light oiling).

Unsatisfactory bearings should be scrapped and replaced.

To enable the bearing to be refitted to the shaft it should be expanded by heating to about 80°C using either a thermostatically controlled oil bath or a purpose made induction heater.

Great care must be taken throughout the reassembly to ensure that no dirt enters the bearing. The bearing itself should be packed with the correct grade of grease and the housing about half-filled with grease. Overgreasing will lead to churning and subsequent overheating of the bearing.

(b) Pad and sleeve bearings. The essential points on the overhaul of pad and sleeve bearings are given below; the techniques used vary with the particular design of bearing:

- Pad type bearings. Inspect for pivot wear and bearing surface scoring, cracking or overheating. Measure oil clearances by taking leads. Take care to ensure that offset-type pads are fitted in the correct relationship to the direction of rotation.

- Sleeve type bearings. Inspect the bearing surface for heavy scoring, cracking, or signs of overheating, or polishing marks where the shaft has worn the sleeve.

  Check oil pick-up rings for wear or distortion. Check bearing bore against shaft dimension in several positions. If new sleeves are being fitted, check sleeve-to-casing fit, using feeler gauges and engineers' blue.

  Check shaft-to-bearing contact arc by engineers' blue. Arc of contact should typically be 60° for high speed shafts (3000 r/min) and 100° for low speed shafts (500 r/min). Remove high spots on bearing surface. Check bearing clearance, using leads.

Slipring, brushgear, commutator Comments made under the section on *in situ* maintenance give an outline to the overhaul of these items.

Additionally, the opportunity may be taken to mount the rotor in a lathe for grinding or machining the slipring or commutator.

Cooling circuit A thorough clean and check of the cooling circuit should be carried out. Water coolers need a pressure test to ensure that they are leak-free.

Airgaps After reassembling the motor, the rotor-to-stator airgap requires checking. If necessary, the bearing positions should be adjusted to ensure that the airgap error does not exceed 10%.

The gaps between the shaft and oil seals, and between the air baffles, then also require checking and adjustment.

11.3.3 Repair following breakdown

It is common practice to employ the services of motor repair workshops, either at the motor manufacturers or at specialist repair firms, for all major repair work on electric motors.

Usually a fast turnround time can be achieved. One of the major delays on large motor repairs is the availability of the correct size of copper conductor. Consideration should be given to stocking

sufficient material within the station stores in order to avoid this delay for critical machines.

It is also feasible to stock a complete set of stator windings ready for installation by a contractor in the event of a serious fault.

Now that extensive standardisation has been achieved on 415 V machines, it is often cheaper and quicker to scrap a small machine needing a rewind and to buy a replacement ex-stock.

To ensure that a high standard of motor repair work is maintained, the CEGB has prepared a standard (No. 449903), which is specified for all repairs.

In addition to the basic repair of the motor, consideration is also given to whether it may be advantageous to:

- Uprate the motor, by winding to a higher insulation class.

- Improve the protection on particularly troublesome drives by embedding thermocouples within the stator.

- Improve damp resistance by improved sealing.

11.3.4 Motor testing

Some of the main techniques used for fault finding on a motor, or for checking its fitness for duty are outlined below. Much of the detail associated with these techniques has already been given in the generator maintenance section to which reference is made.

Insulation resistance (IR) measurement

Insulation resistance (IR) measurement is used to measure the DC resistance of a motor winding to earth, or the resistance between windings. The test is carried out using a battery- or mains-powered, or hand-operated, megohmmeter (megger). Typically, a 500 V DC test is used on 415 V machines. HV motors are usually tested at 1 kV, for all windings up to 6.6 kV, and at 5 kV for windings of 11 kV or more.

Whilst an IR is a very quick method of assessing insulation integrity, it must be realised that the value of the information obtained is limited. On open motors, IR values can be affected by atmospheric humidity. When comparing readings with previous results, a temperature correction should be made, using the nomogram given in Fig 5.42.

Polarisation index (PI) measurement

Polarisation index (PI) measurement is a method of assessing the surface contamination of insulation. The method of applying the test is outlined in Section 9.3 of this chapter. The voltages used are the same as those for IR testing, indicated above.

A PI test should be made on any machine that has been out of service for some time to determine whether energisation or testing with AC is advisable.

Figure 5.56 gives a guide to the interpretation of results obtained from IR and PI testing. If values at or below those listed in the guide are obtained after correction to 20°C, serious moisture or other contamination is indicated. Further investigation of the cause of the condition, and possibly a dryout, will be required prior to AC energisation or testing.

Line voltage of machine	Minimum acceptance value of IR of complete winding at 20°C (Megohms)			
	PI > 1.6		PI < 1.6	
kV	IR_{1min}	IR_{10min}	IR_{1min}	IR_{10min}
3.3	any value	15 MΩ	30 MΩ	any value
6.6	any value	30 MΩ	50 MΩ	any value
11.0	any value	50 MΩ	100 MΩ	any value

FIG. 5.56 Guide to determine acceptable values of insulation resistance polarisation index testing

Loss angle testing

Loss angle testing (or tan delta testing) is a method of assessing the condition of the bulk of the insulation of an HV motor. The test is made on each phase, with the other two phases earthed, where the neutral can be split, or on the complete winding, where the neutral is solid. A maximum test voltage equivalent to line voltage is recommended, although a lower maximum voltage may be used where insulation is suspect. The Loss angle test should not be applied until satisfactoy IR and PI test results have been obtained.

The test method is given in Section 9.3 of this chapter: interpretation of the results is more straightforward than for a generator, where the effects of the cooling media have to be taken into account.

Squirrel-cage rotor testing

Defects, such as broken rotor bars or high resistance bar-to-endring joints, in squirrel cage rotors can be detected on-load by detection of the current, flux or speed fluctuations caused by the fault.

A particularly sensitive device for the early identification of developing rotor faults is the Spirostrobe. The device works on the principle that the rotational period of a motor with an electrically-defective rotor bar will fluctuate slowly at twice the slip frequency. The device measures this fluctuation by processing the information from a speed pick-up on the motor shaft.

11.4 Transformers

For maintenance purposes, the power transformers in use in a modern power station can be divided into three categories:

- Fluid-filled.

- Dry air-cooled.

- Cast-in-resin air-cooled.

In each, reliability is very high and normal maintenance consists of routine cleaning, checking and testing to ensure that the electrical integrity of the transformer is sustained.

11.4.1 Fluid-filled transformers

The majority of HV and some MV power transformers are fluid-filled, the fluid acting as both an insulation and cooling medium. The fluid may be cooled by natural convection, by forced air, or by water-cooled heat exchangers.

Mineral insulating oil is the most commonly used fluid. In high risk areas, fire-resistant fluids may be used, such as silicone fluid or a phosphate-ester. Polychlorinated biphenyls (PCBs) are being phased out because of their high toxicity and non-biodegradability.

The main routine activities are outlined below and consist of regular checks on fluid and on air breather systems, together with outage checks on all other equipment. Normally, the outage checks should be yearly, but some flexibility in this may be needed to fit in with main unit outage dates.

If, at any time, it becomes necessary to open up the transformer tank, extreme care must be taken to impose clean conditions and to ensure that no foreign matter is allowed to enter the tank. Those working on an open transformer should don one-piece pocketless overalls and wear no loose items.

All tools must be accounted for and taped to the operative so that they can be recovered, if accidentally dropped. Carelessness in the application of these measures can turn a simple task into a major, time consuming and costly exercise.

Fluid

Regular checks of fluid levels must be made and recorded to ensure that an adequate level of fluid is maintained. Where available, a note of winding temperature should also be made, since the level will obviously vary with fluid temperature.

A particular danger applicable to tranformers fitted with Buchholz relays is that, if the fluid level is allowed to become too low in cold weather, it may fall below the relay during a period of light load and thus cause a spurious trip.

Fluid added to top-up a transformer must be clean and tested immediately prior to use. The top-up is best carried out off-load, with a period of time allowed after filling for air bubbles to detrain from the oil. On-load topping-up is possible on some transformers. Care must be taken to entrain a minimum of air and, for a period after filling, the Buchholz relay will need regular bleeding to remove air and to prevent spurious operation.

Fluid testing must be carried out at least annually and following a transformer trip, or gas or temperature alarm. The fluid tests check for appearance, moisture, electric strength, acidity, resistivity and dissolved gases. The results from these tests may indicate that the fluid needs changing or reconditioning. Some sites purchase their own conditioning unit through which the fluid is circulated and then returned to the transformer. Other sites use specialist contractors for this work, who come to the location with all the necessary pumping, filtration and test equipment to deal with any size of transformer.

Fluid test results may also indicate a possible incipient fault within the transformer, requiring monitoring by more frequent sampling.

The method adopted to collect the sample must be designed to ensure that the sample is truly representative of the fluid in the tank and that it is not contaminated in the collection process.

In circumstances where a fault is suspected or following a Buchholz gas alarm, gas samples should be collected from the Buchholz relay for analysis. The results can give a guide to the nature of the fault: in oil-filled transformers, these are as follows:

● Presence of one or more of ethane, methane or carbon monoxide indicates a hot spot.

● Presence of hydrogen and/or acetylene indicates the presence of arcing or discharge within the oil.

The sample may, of course, just prove to be air.

A comprehensive guide to the maintenance of insulating oil is found in BSI Code of Practice BS5730.

The tests described above are not very effective in detecting the low-temperature overheating of paper insulation in oil-filled transformers. Such overheating is known to have been a major factor in two generator transformer failures. A method has now been developed within the CEGB whereby furfuraldehyde, a product produced exclusively by thermal degradation of paper at temperatures from as low as 110°C, may be detected from the analysis of an oil sample. It seems likely that, on certain transformers, this test will be included with those already carried out on routine oil samples. Details of the test are published in CIGRE report 12.09.1984.

Breather

The inspection of the breathing equipment should be associated with regular fluid level checks. Silica-gel crystals are renewed if seen to be changing from blue to pink. At the same time, the breather oil seal is checked and, if necessary, topped-up to the correct level to prevent diffusion of moisture into the silica-gel when no breathing is taking place.

The silica-gel may be recharged by drying in an oven at about 130°C until the crystals have regained a dark blue appearance.

Transformer tank and compound

Transformer compounds must be kept free of litter and oil drains, bund walls, cables, cable ducts and fire fighting equipment examined.

A thorough clean down of the transformer tank and compound should be carried out as part of the outage work. Where spraywater protection is fitted to oil-filled transformers, it may be decided to apply solvent to the dirty areas of the tank and compound and then to test discharge the fire protection, which will in turn wash down the transformer.

Other work is to inspect the tank, radiator/heat exchangers, conservator, pipework valves, etc., for leaks, and to ensure that all valves and cocks are secure and locked.

Neutral earthing resistors

The electrolyte level should be checked and its resistance measured, using a low voltage AC supply. Where fitted, heaters should be checked for correct operation.

Bushings and connections

Bushings are cleaned and checked for leaks, chips or cracks. A fluid level check/top-up is carried out on filled bushings.

Tightness and resistance checks of all connections, including the transformer earth connection, must be made.

Off-load tapchanger

The following procedure should be carried out at an outage to clean the contacts and prevent contact overheating defects due to pyrolytic growth.

Unlock the tapchanger, note its position, operate the tapchange, switch it through its range several times, return it to its original position, and lock.

On-load tapchanger

The manufacturer's instruction must be followed for the particular design. Outage maintenance is usually confined to the diverter switch and operating mechanisms. Basic steps covered will be:

- Thorough cleaning and flushing with insulating fluid.

- Detailed inspection of all components, especially moving parts and diverter resistors.

- Contact overhaul.

- Check of all connections.

- Overhaul of operating mechanisms.

- Complete function check.

- Refill with clean, tested fluid.

Winding temperature indicators

During the off-load examination, the winding temperature indicator should be checked to ensure that the reading is correct and that the alarm operates at the specified temperature. This is carried out by removing the indicator bulb from its pocket in the top of the transformer and placing it in a temperature test oil-bath. The test unit is switched on to raise the temperature to 120°C and the reading of the indicator checked against the test unit thermometer. A test meter connected across the winding temperature alarm contacts will check the operation of the alarm against the reading of the winding temperature indicator: any necessary adjustments are made and recorded.

The indicator bulb pocket is then refilled with transformer oil and the bulb replaced. An injection test is then applied at the test links to the heating coil circuit and the current adjusted to the required test figure. The temperature rise is noted on the winding temperature indicator and compared with the calibration data.

Buchholz relay

The Buchholz relay is important in the transformer protection system and should be tested annually. This is carried out with a special test set in which a cylinder is charged with air at 2.1 bar gauge and connected to the test cock on the relay. Air from the cylinder is then admitted slowly to the relay until the gas-float contacts 'make': the volume of air required to achieve this is recorded. The surge float can then be tested by admitting air rapidly into the relay until the surge-float contacts 'make'. Contact operation is checked with a test meter, or with the secondary trip or alarm circuits alive. Care must be taken after the test to ensure that all air is bled off from the Buchholz chamber.

Pressure relief diaphragm

This is inspected during a maintenance outage for splits or leakage. A split diaphragm can allow the transformer to breath other than through the breather and thus cause unwanted moisture ingress.

Pressure relief valve

The valve is removed during an outage and checked for correct mechanical and electrical operation on a purpose built test rig.

Cooling equipment

Off-load maintenance consists of an overhaul of pumps, fans, heat exchangers, etc., using normal maintenance techniques. It is particularly important to ensure that water-cooled heat exchangers have no water-to-oil leakage path and this should be checked.

Marshalling kiosk

The following outage work should be undertaken:

- Clean and check connections.

- Overhaul control equipment and check operation.

- Check heater.

- Ensure that the weather protection is effective.

11.4.2 Dry air-cooled and cast-in-resin air-cooled transformers

The maintenance of these transformers is far simpler than for the fluid-filled type and consists of outage work where the main activities will be:

- Cleaning of transformer and compound.

- Checking of tightness and resistance of connections.

- Testing of protective devices.

11.5 Electrostatic precipitators

With the use of solid state voltage control and rectification, together with the application of modern control technology, the reliability of the electrical side of electrostatic precipitators has increased considerably over those of earlier generations. The device remains, however a piece of high voltage electrical machinery and as such, is best dealt with as a whole by the electrical maintenance staff of the power station. Figure 5.57 shows the block diagram of a typical electrostatic precipitator.

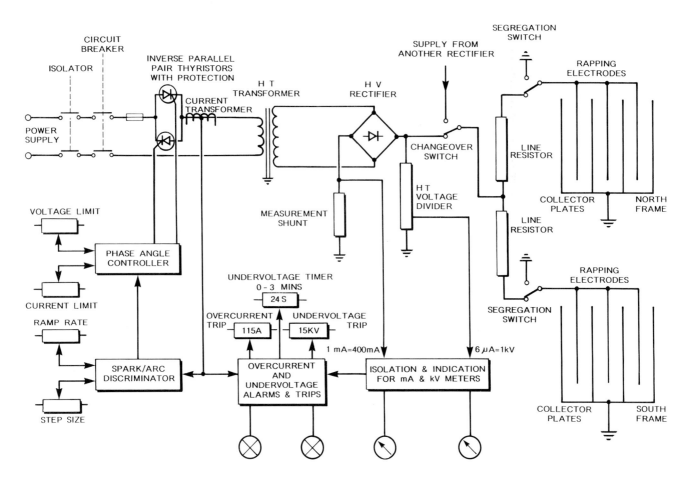

FIG. 5.57 Block diagram of a typical electrostatic precipitator

When undertaking precipitator maintenance, great care must always be taken to ensure that high voltage parts are effectively isolated and earthed before work commences. When entering the gas pass side of a precipitator, as well as ensuring that effective earths are applied, checks must also be made to ensure that there is no dangerous build-up of dust or gases.

11.5.1 Running checks

Effective precipitator operation requires the following regular work to be undertaken:

- Visual inspection for indications of internal or external flashovers. These show by violent voltage/current surges as the voltage is increased manually.

- Check for evidence of high dust levels or internal dust bridging between electrodes and plates, often indicated by high secondary current but low volts.

- Check and, if necessary, adjust the operation of automatic voltage contol circuitry, so that the precipitator is running at the maximum possible

voltage for the combined gas, dust and precipitator conditions.

- Check rapping-gear operation.

11.5.2 Outage work

Apart from short outages which may be necessary for regular lubrication of rapping drive systems, most outage work is initially carried out annually: this timescale is then modified from experience of the particular precipitator.

Internal gas-pass work

The first activity is a thorough cleandown of the electrodes and collectors, removing all dust. The method of cleaning depends on the condition of the dust. A thorough hand-rapping is often very effective; on other occasions, water washing may be required. Whilst cleaning, the pattern of dust build-up should be observed, as this can often give a good indication of possible rapping-gear problems. The following activities should then take place:

- Clean and check internal insulators.

- Inspect for, and remove, broken electrode wires.

- Replace electrode wires if the number removed is sufficient to affect precipitator performance significantly.

- Check and adjust the clearances between electrode wires and collectors.

- Check the integrity of mechanical support structure. This is particularly important, as it receives constant shocks from rapping.

- Check the gas-pass side of the rapping equipment.

Rapping drive system

Where applicable:

- Check motor electrics and overhaul motor.

- Check and lubricate gearboxes.

- Check chain drives for lubrication, freedom, tension, wear and alignment.

- Check shafts for freedom of rotation.

- Inspect all parts for mechanical failure.

- Check the tightness of all fixings.

HV distribution system

- Clean insulators.

- Inspect for evidence of arcing, poor connections, or shorts to earth.

- Check the condition of surge resistors.

- Overhaul the selector switches.

Voltage regulator, HV transformer and rectifier

- Clean, inspect and, where possible, electrically-test components.

- Where the transformer and/or rectifier are fluid-immersed, check the condition and level of the insulating fluid.

Control system

- Clean and carry out normal maintenance of contactor switchgear, relay equipment, protective devices and other control and indication equipment.

- Check calibration and operation of automatic voltage control equipment, using a test rig, if available. Check operation of protective devices.

Earthing and interlock equipment

- Clean, inspect and check for correct operation. Particular attention should be given to ensure that automatic earths operate freely and make good electrical contact.

When returning the precipitator to service following an outage, the automatic voltage controller must be set up to give optimum performance. The means of achieving this vary with the particular design, but account needs to be taken of the point at which corona discharge commences and the degree of internal arcing which can be tolerated.

11.6 Ancillary equipment

Much could be written about the maintenance of the ancillary electrical equipment within a power station, most of which performs a vital role in the operation of the main plant. The notes below are limited to the main features which need to be incorporated into the maintenance programme for these items.

11.6.1 Battery systems

Lead-acid batteries

The number of battery systems in a modern power station is such that battery cell maintenance presents a considerable workload to the maintenance department. The work is by nature very repetitive yet must be carried out conscientiously: it is probably best confined to a few operatives who are temperamentally suited to tasks of this nature.

A typical approach to the maintenance of a large 120-cell Plante-type lead-acid battery would be to carry out the following routines:

Fortnightly routine Check electrolyte levels on all cells and top-up any cells approaching or below the minimum marks with distilled water.

Note 'pilot cell' voltage, specific gravity and temperature readings. 'Pilot cells' are, say, 6 out of the 120 cells, selected so as to be evenly distributed over the length of the battery; they are used as indicators of the battery condition.

Inspect condition of connections — clean and regrease with petroleum jelly, as necessary.

Ensure that the battery room is clean, that the ventilation system is satisfactory and that the safety equipment is available.

Note the battery charger current and voltage.

Three monthly routine Similar to above, except that all cell voltage and specific gravity readings are noted.

A thorough inspection of all cells must be made and defects noted. Particular care to be taken to look for:

- Terminal and connection corrosion.

- Cells overgassing.

- Flaking of internal connections.

- Plate distortion.

- Leakage of electrolyte.

- Cracking of cell lids.

Annual work Similar to the three monthly work, but in addition:

- Remove petroleum jelly from connections.

- Check-tighten (using insulated tools), inspect and re-apply petroleum jelly.

- Clean cell casings and support framework.

The check sheet reading and notes from the routine maintenance activities should be independently inspected in order to identify any problems developing in the battery.

The state-of-charge of a cell may be assessed from the specific gravity reading. An example is as follows:

- If a typical 240 V 1300 Ah battery were fully discharged, the specific gravity would drop 90 points from 1.210 to 1.120. The rate-of-change of specific gravity with charge is, for practical purposes, linear, and thus the state-of-charge of the above battery may be determined by the points drop from 1.210 as a proportion of 90; for example, if the specific gravity is 1.180, the battery is 30/90, or one-third discharged.

For accuracy, the specific gravity readings must be corrected for any temperature variation from 15°C, as follows: ±1 point for each 1.5°C of electrolyte temperature above/below 15°C. The actual specific gravity values and points drop with charge will vary with type of cell and must be found from manufacturers' information.

It would normally be expected that, on a standby battery, all cells would indicate a near-full state of charge.

Cells which show readings out of step with the rest of the battery require investigation.

Low specific gravity which remains low and constant over a period of charging could indicate that the electrolyte has become weakened by some means,

the strength should be checked and adjusted by addition of dilute sulphuric acid of specific gravity 1.840.

Low specific gravity and low cell volts which do not respond to charging could indicate an internal cell short. It may be possible to clear a short by disturbing the electrolyte by gently blowing air into the solution.

Excessive gassing is most undesirable and can be caused by attempting to charge a near fully-charged battery at too high a rate. The excessive current charge (over and above that which the plate material can accept) decomposes the water content in the electrolyte into hydrogen and oxygen. If this is too violent, it lowers the electrolyte level, produces undesirable heat and causes active material to be scrubbed from the surface of the positive plates, which falls to the bottom as a deposit.

Nickel-cadmium alkaline batteries

These need less maintenance than lead-acid batteries. The cells are kept clean and dry, connections are kept lightly greased with petroleum jelly and are checked for tightness periodically.

The electrolyte level should be regularly checked and topped-up with pure distilled water as necessary.

The specific gravity of the electrolyte in a nickel-cadmium battery does not vary with state of charge.

Battery chargers

Solid state battery chargers are usually reliable items of plant which require very little maintenance. Nevertheless, because they play such a key role in maintaining the capacity of the vital DC supplies, they must not be neglected in the overhaul programme.

An annual inspection will include the following activities:

- Clean and inspect all components, checking for signs of overheating, loose connections, corrosion or mechanical damage.

- Overhaul switches, contactors, control relays, etc., using standard techniques.

- Check fuses.

- Check cooling fans, if fitted.

- Check operation of the 'charger fail' and 'high voltage' relays; ensure that the remote alarms are initiated.

- Carry out running checks to ensure correct operation in both 'float' and 'boost charge' selections.

11.6.2 Secure instrument supply systems

Most modern power stations have a secure instrument supply system fed from batteries via a DC/AC converter.

The DC/AC conversion is carried out by either a rotary converter or a solid state inverter.

Rotary converters

A rotating converter is basically a DC motor coupled to an AC generator, the motor being fed from a secure battery supply. Maintenance for the system thus follows the principles already laid down for these items.

A typical installation consists of three machines per unit, two running in parallel and one on standby. Correct and reliable operation of synchronising and load-sharing circuits often proves difficult to achieve: it is important that expertise is built up within the station in order to ensure maximum reliability from the particular system installed.

Inverters

Normal maintenance of an inverter system consists of an annual overhaul, which follows a very similar pattern to that already described for a solid state battery charger.

Particular features that may require regular checking are:

- Stability and accuracy of frequency and voltage over the load range.

- The automatic changeover switching (be it static or contacter switching) between running and standby inverters, or between inverters and the raw supply. It is important to ensure that any interruption to the output from the system during such switching is within the specified limits, so as to have no effect on the equipment being supplied.

11.6.3 Cabling and earthing

There is a great temptation to neglect the maintenance of cabling and earthing systems totally. Whilst this may have no adverse effect on operation for many years, such a policy will almost certainly eventually contribute to a major electrical failure. Inspections should therefore by undertaken to check for the following potential trouble spots

Cabling systems

- Debris in cableways, which could present a fire hazard or could damage cable sheaths.

- Sheath faults. Power cables may carry sufficient potential on their armouring to initiate a cable fault if the sheath breaks down and an armour-to-earth fault develops. Inspection of sheaths and, in some cases, periodic sheath insulation tests may

be considered advisable.

- Cable supports. Electrical faults can exert high forces between single core power cables even though safely cleared by protection. Cable support arrangements and cleating must be maintained in good order and inspected after faults.

- Glanding. Ensure that gland fixings are secure and that glands are correctly earthed, or insulated, according to the cable design.

- Insulation. A representative number of power cables should be tested periodically for the integrity of their insulation system to enable a continuous assessment to be made of the general condition of the cable system.

For 415 V and 3.3 kV cables, such testing will normally consist of a DC insulation resistance test between phases and between phase and earth. For 11 kV extruded solid insulation cables, CEGB Standard 095101 recommends a periodic partial discharge-voltage test be carried out. Details of this test are given in the Standard.

Earthing systems

Copper conductor earthing systems have over recent years proved to be particularly vulnerable to theft of lengths of the earthing strap, particularly from unfrequented areas of the power station. This dangerous and criminal practice can cause items of plant to be effectively unearthed. Regular checks may be necessary to ensure that the earth circuits are complete. Where theft proves to be a serious problem, consideration should be given to replacing missing copper with aluminium of equivalent resistance. If this policy is adopted, care must be taken to ensure that correct aluminium-to-copper jointing techniques are used.

More recent power stations have earthing systems constructed from aluminium cables.

The terminations of both copper and aluminium earthing systems should be periodically checked to ensure that they remain tight and free from corrosion.

The individual earth electrodes must also be checked on a planned basis to ensure that they remain an effective earth path. A common way of measuring electrode resistance in a multiple-electrode earthing system uses a comparison method, which involves disconnecting the electrode under test from the earthing system and measuring its resistance to the main earthing system, using a *null-balance earth test megger*.

A more absolute method of establishing earth-electrode resistance is given in British Standard Code of Practice CP1013.1965 'Earthing'.

11.6.4 Actuators

Periodicity of maintenance

The vast number of valve and damper actuators in a modern power station present a very onerous task to the electrical maintenance staff, especially if an unrealistic approach is taken to routine maintenance. Whilst it may be considered ideal to overhaul and test each actuator annually, it is unlikely such a target could be achieved economically. It is therefore suggested that an approach similar to the following is adopted:

- All actuators essential to safe and reliable operation of the plant to be periodically overhauled and tested, the maintenance period being initially a year but this can be modified in the light of operating experience.

- All remaining actuators to be overhauled and tested only as a result of a defect or breakdown occurring. To avoid an unacceptably high number of failures during plant running, an operational test schedule should be instituted to ensure that a periodic check operation is carried out on each of these actuators as plant conditions allow.

Maintenance activities

The particular activities which comprise a routine overhaul of an actuator depend on the design, and also on whether it has integral control gear. The following are typical for a 415 V actuator with integral control gear:

- Check for evidence of oil leakage, rectify any defects.

- Drain oil from gearcase and refill to the correct level.

- Check security of the actuator mounting bolts.

- Check cabling to the actuator, cable glanding, etc.

- Open up control gear module. Overhaul contactor, check interposing relay, pushbuttons, indicators, wiring and terminations.

- Check the motor winding IR to earth and check that the winding resistances balance.

- Check engagement of clutch.

- Engage clutch and check operation of limit and torque switches whilst winding the valve by hand. (On large drives, it may be necessary to motor the actuator to avoid excessive hand winding. This should only be done from an intermediate position and motoring stopped well short of the limit posi-

tion, the final operation to limit being carried out by hand.)

- Ensure that limits are set to correct position or torque value.

- Energise actuator, check direction of rotation, check all local controls and indicators.

- Check electrical operations to limits.

An essential feature of all actuator maintenance is to ensure that, on completion, all weatherproofing seals are in good condition and that covers are correctly fitted and fully bolted-up. Carelessness in this matter has caused many unnecessary defects due to ingress of dirt and moisture.

12 Control and instrument maintenance

12.1 The approach to maintenance

Instrument maintenance in a modern power station covers a wide range of equipment and requires a varied range of skills and techniques. In general, the emphasis is on electronics which appear in some form in most control and instrumentation equipment; however, mechanical skills are required for pressure and flow measurements and for valves, actuators and dampers in the elements of control systems. Precise skills are required for work on small mechanisms and for testing and fault finding on complex electronic circuits. An understanding of dynamics in mechanical and electrical systems is important since vibration analysis and the performance of boiler control systems are but two examples of the skills expected of the station C and I staff. Pneumatic controls are still favoured by turbine suppliers and these require a particular maintenance approach.

Control of boiler and turbine plant has become more centralised and computers are extensively used to display and control plant conditions. Operators have become much more dependent on their instrumentation now that control rooms are more remote from the plant and the number of plant attendants carrying out plant surveillance has diminished. Often, there is little or no local indication of plant conditions. Hence, it has become increasingly important that the operator is presented with a true representation of the conditions being measured.

Training plays an important part in the preparation for maintenance, since it is of utmost importance that the staff concerned understand not only the equipment but the effect that their actions may have on the plant. Fault location is often only effective if it is carried out with the plant in service, since if the plant is shut down the fault condition is difficult or impossible to simulate. This requires a high degree of personal responsibility by maintenance staff, as they quickly become aware that a mistake on their part can easily take plant out of service and could cause costly damage.

12.2 Maintenance philosophy

Since the equipment to be maintained is varied, the philosophy of maintenance must also be wide ranging. Routine maintenance has been effective and can produce good results; however, improvements in transmitter design and the stability of modern electronic components means that equipment can maintain its calibration for long periods.

The extension of statutory outage periods, particularly on boilers, means that the time interval between major outages can be up to 38 months, so that equipment, such as supervisory detectors on turbine bearings, may not be accessible for inspection or repair for similar periods. A combination of these facts and other operational constraints results in the following philosophy:

- Routine maintenance should be carried out on indication and alarm equipment during operational periods to ensure that the calibration and monitoring of alarm conditions is maintained. Periods can extend to 6–12 months, based upon the reliability of more modern equipment.

- Fault diagnostics should be built into equipment used for control and sequence, where possible, to help in identifying faults. Maintenance departments often fit equipment for this purpose themselves, where this is not provided in the original contract.

- Electronic equipment is provided on a modular basis, where possible, so that spare modules can be stocked to replace identified faulty modules. The policy is then to provide workshop test facilities to identify faulty components in modules and to carry out repairs. Also required is a means of effectively testing the repaired module before it is returned to stores, awaiting issue when the next defect arises. At peak periods of demand, assistance is available from manufacturers to carry out module repairs to ensure that a ready supply of spares is always available to keep the plant in service.

  This philosophy calls for a detailed investigation into a variety of modules that are supplied by a wide range of equipment manufacturers.

- Specialised approaches must be carefully considered for the more complex equipment which now comes within the scope of instrument maintenance. This applies to computers, electronic governors, vibra-

tion run-down monitoring equipment and other similar types of monitoring equipment. Turbine supervisory equipment also requires a different approach, due to the detectors being inaccessible for long periods of operation.

12.3 Pressure temperature and flow measurements

12.3.1 Pressure

Pressure is a prime measurement in the operation of a power station. The equipment used to measure pressure are transducers, gauges and fluid columns, each combined with indicators (see Volume F). All require calibration at some frequency to ensure that the pressure indication is accurate. Transducers vary in type from diaphragms to strain gauges. Modern materials with stable reproduction characteristics enable accuracy to be maintained over the full range over long periods of time, i.e., 6 months to 1 year. Accuracies of 0.5 to 1.0% are required for modern controls of pressure; calibration is usually carried out on the plant using substandard pressure gauges, with accuracies in the range of 0.1 to 0.25%. These gauges are checked against deadweight testers to ensure that the calibration equipment is kept up to a high standard.

Measurements of fluid levels in tanks and vessels are also important in power stations and most of these are based on some form of pressure or differential pressure measurements. Levels in boiler drums have to be measured with high static pressures of, say, 170 bar. On modern high pressure boilers, this is often achieved by two separate stainless steel diaphragms, with the deflection of each being measured electronically and subtracted to produce a difference value. Electronics are used extensively in this type of transducer and the output converted into an electrical signal in the form of a low current value for retransmission to remote indicators, computers and recorders.

Pressure measurement points in boiler gas passes which are below atmospheric conditions get blocked with ash, and these deposits require cleaning out. Most of this type of equipment can be calibrated and maintained by routine maintenance and can be taken out of service while the plant is operating. Often pressure measurements are duplicated to operators, so that a service can be carried out without interfering with the output from turbines and boilers. Typical pressure test equipment is shown in Fig 5.58.

12.3.2 Flow measurement

Flow measurements are also mainly based on some form of differential pressure measurement across an

FIG. 5.58 Pressure calibration test equipment installed in a workshop, illustrating a deadweight tester, a substandard test gauge on a comparator tester and, in the background, a high accuracy test gauge panel covering a wide range of pressures

(see also colour photograph between pp 434 and 435)

orifice plate or venturi, although some direct aerofoil types are used in fluid flows in water treatment plants at low pressures, and in some airflow measurements (see Volume F).

With pressure and flow measurements, pipework and valves are involved to take the fluids to the transducer elements. These also require periodic attention to ensure they are kept free from blockages.

12.3.3 Boiler drum level indication

A development in the measurement of boiler drum level is the Hydrastep system (see Volume F). This uses the principle of detector probes sited in a column at regular heights, which detect the difference in conductivity between steam and water and transmit this to a series of lamps on the control desk.

As the level varies in the column, the probes measure resistance, initiating a change to an electronic control unit which switches its output signal to two vertical rows of indicator lamps, depending on whether water or steam is in contact with the probe. The changing level is indicated by a change in row and colour of these lamps. Maintenance of this system consists mainly of changing probes in the columns, which can fail under the high pressure and temperature conditions. Other work involves checking the logic system used to indicate probe failure: a protection system in the control unit checks that probe faults, which could give rise to a water above steam condition, are detected and hence indicates a fault in the system.

These systems have proved much more reliable and simple to operate than the previous television and gauge glass method and have the advantage that they can be used to operate directly into boiler trip and intertripping protection schemes. They also do not suffer from a slow response to changing levels, unlike previously used manometric systems.

However, it is common practice to use the manometric column and the associated differential pressure transmitter method to produce signals for the boiler drum level control scheme. A maintenance requirement exists on these systems to check transmitter calibration: they also suffer from leaks occurring at valves and fittings.

Blowdown valves on these systems have also been a source of problems, since there are difficulties in sealing them in the closed position against the high pressures involved across the valves. Once the seating of these valves is damaged they fail to hold the pressure of the instrument lines and result in errors in the measurement of drum levels. This can result in indications being unavailable for long periods whilst boilers are in continuous service, since it is difficult to achieve satisfactory isolation to allow work to be carried out under 'in service' conditions.

12.3.4 Temperature measurements

Temperature measurements are extensively used on modern boilers and turbine systems to indicate temperatures to operators and to protect plant against overstressing conditions. The detectors are based on well established thermocouple and resistance thermometer elements. These elements are connected to various types of amplifier modules which change the operating levels from millivolts and resistance into easily transmitted current signals (see Volume F). Maintenance relates to millivolt measurement on thermocouples and resistance on thermometer detectors and the simulation of these measurements into high input impedance amplifiers.

Accuracy varies from $\pm 2\%$ on thermocouple systems to better than $\pm 0.5\%$ for resistance thermometers. Detailed calibration at regular intervals can improve the expected accuracy of these systems. Calibration equipment varies in hardware, but basically uses potentiometer principles for voltage and resistance measurements, and high stability resistance and voltage elements within the equipment for injection purposes.

Resistance elements have superseded thermocouples for main steam measurements on boiler and turbine steam lines since improvements in accuracy and repeatability enable control schemes to work to more closely defined conditions. However, failures occur in elements due to the vibration conditions they are subjected to in the steam lines. The changing of these elements, which are fitted into standard pockets, can sometimes prove difficult. Boiler metal-temperature measurements on the gas side of boilers are used to assess the life expectancies of boiler tubes. The thermocouples are attached to the tubes, using Chordal holes or metal-spraying techniques (see

Volume F), and the signals are taken outside the gas space in mineral insulated cables covered by thin metal half-tubes, called 'lap covers'.

These elements rely on the cooling of the metals by steam or water flowing in the boiler tubes to survive. Hence, it is essential that detachment into the gas space does not occur.

Maintenance employs loop resistance continuity checks and insulation measurement on the thermocouples, using low voltage insulation testers of the 'megger' type. Failures occur due to breakdown of insulation between the loop and the tube and can cause inaccuracies in indication.

Conductivity measurements are necessary to detect the continuity of the thermocouple and to detect one element from another, since a measurement scheme for a boiler can contain many elements. Replacement of an element can only be carried out under boiler outage conditions and can involve extensive scaffolding in furnaces and gas passes to gain access to the detectors where they are fitted to the tubes. Typical temperature test equipment is shown in Fig 5.59.

12.4 Turbine supervisory equipment

Turbine supervisory equipment is provided to monitor the essential moving parts of the turbine, particularly during start-up and shutdown. The parameters measured are:

- Turbine speed.
- Differential expansions.
- Eccentricity.
- Bearing vibrations.
- Valve positions.

Figure 5.60 shows the supervisory detector locations on a 660 MW turbine-generator.

Each measurement system consists of a transducer fitted at the turbine, a module or series of modules housed in a set of equipment cubicles adjacent to the turbine and the indication equipment found on the display sections of the unit control desk.

Maintenance of each system involves measurements at the turbine transducer, the calibration of the equipment modules and the remote displays. The modules can each be calibrated individually on-load and should be checked annually, using electrical and electronic simulation equipment of known accuracy. The overall loop calibration and check on the condition of the transducers can only be achieved during a lengthy unit outage when the turbine covers

FIG. 5.59 Temperature calibration equipment fitted into a portable frame in a workshop, illustrating a sandbath together with liquid fluid baths for the calibration of temperature detectors over a range of temperature from ambient to 600°C

(see also colour photograph between pp 434 and 435)

are removed and access to the turbine cylinders and bearings is possible. This can now be as long as 3 years between statutory outages. Maintenance of the differential-expansion transducers involves the mechanical displacement of the carriage supporting the transducers to check their operation and loop calibration. A check is also carried out on the insulation and continuity of all circuits, since some are subjected to steam space conditions and high temperatures. Few failures occur with more modern transducers but faults can occur with cables and connectors in vulnerable areas. Checking of *mechanical movement* is carried out using clock gauges which can measure small displacement in micrometres (μm). Transducers use magnetic circuit principles to measure shaft to casing movements. Some linear potentiometers are used in the steam spaces of LP cylinders to measure larger displacements, in the centimetre range. Transducers in the steam spaces have to be checked for their ability to seal off the internal transducer parts from the steam space.

The stainless steel bellows used to seal and protect the transducers have therefore to be sealed tightly against external pressure and vacuum conditions.

Eccentricity measurements cannot easily be simulated at the transducers, hence maintenance involves checking the airgaps between the shaft collars and the transducer faces. Some comparison checking can be carried out by fitting known spacers under the coils and checking the electrical changes of the transducers as the airgap changes. This ensures that transducers have not suffered from shorted turns in their coils.

Bearing vibration transducers can only be checked for accuracy by removing their transducers and asso-

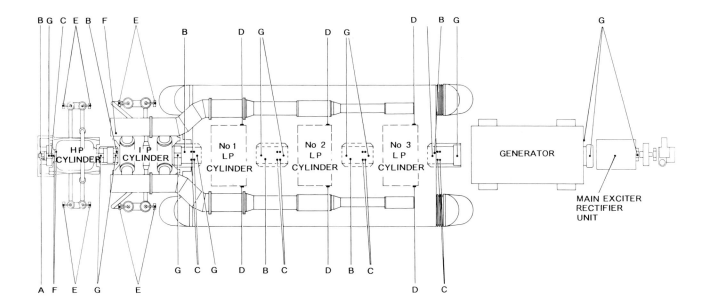

A TURBINE SPEED DETECTORS

B ROTOR ECCENTRICITY DETECTORS

C ROTOR DIFFERENTIAL EXPANSION DETECTORS

D INNER CYLINDER/CONDENSER DIFFERENTIAL EXPANSION DETECTORS

E STEAM VALVE POSITION DETECTORS

F TURBINE NON-ROTATING PART EXPANSION DETECTORS

G VIBRATION DETECTORS

FIG. 5.60 Supervisory detector locations on the main turbine-generator

ciated modules from the plant to a test centre. A limited check on the transducers can be made *in situ* by means of a light tap of the outer casing, which appears as a sharp rise and fall on the vibration indication. This proves that a channel is active but does not give any information about its measurement accuracy.

The test centre equipment must make provision for vibrating the transducer over a range of frequencies and amplitudes, to check for defects in measuring accuracy and to establish that resonances do not occur (other than the known self-resonant characteristics of the transducers). The plant modules should be connected to the transducer with which it is used to ensure a complete channel check. The calibration is checked against a standard vibration transducer which is subject to annual calibration at an approved test authority centre with defined standards. The test equipment consists of a vibrating table driven from a variable-frequency power unit. Figure 5.61 shows a typical workshop test facility.

Turbine speed is measured and used for numerous operational requirements, as well as for inputs to the electronic governor systems to control turbine speed and load.

Accuracy is important for operational requirements, such as overspeed testing, and in order to avoid critical speeds when running-up turbines towards synchronous speed. Accuracy is provided for in modern turbines by digital counting systems, which count electrical pulses derived from shaft-mounted discs. The shafts are fitted with toothed wheels so that each turn of the shaft provides many changes in the magnetic paths between the teeth and an adjacent magnetic probe.

The electrical signals produced by these probes result (after suitable shaping) in unidirectional pulsed voltages, which are counted in high accuracy digital counters. The display of shaft speed can be carried out over relatively short time intervals, i.e., 3000 pulses/second from a 60-toothed wheel fitted to a shaft rotating at 50 r/s can display 3000 r/min updated each second to an accuracy of 1 r/min. In modern turbine installations, up to four separate sys-

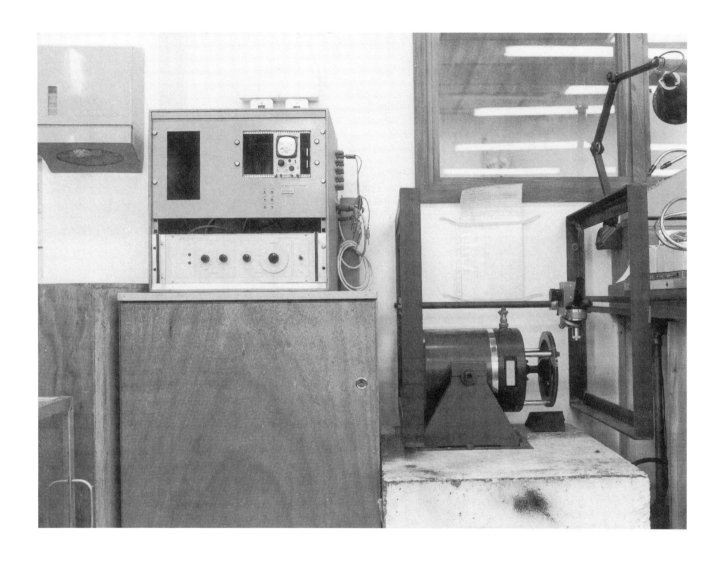

FIG. 5.61 Vibration test equipment, illustrating the variable-frequency power drive units and vibrator used to calibrate vibration transducers

tems are fitted; two provide full range speed indication, and two provide low speed measurements for barring conditions and associated interlocking and alarm facilities.

Maintenance of the speed systems involves checking the speed display and recording equipment, using frequency generators that can simulate the voltage signals received under operating conditions from the shaft-mounted transducers. Maintenance at the detector end is again hampered by the probes being fitted inside turbine covers and only accessible during certain periods. However, the outputs from the probes can be displayed on oscilloscopes whilst in service and their waveforms and voltage magnitudes can be studied for any deterioration. Continuity and insulation checks can also be carried out on the electrical parts of the probe circuit (see Fig 5.62).

12.5 Chemical analysis equipment

Chemical analysis equipment is installed on unit plant for monitoring boiler water conditions, flue gas analysis and generator hydrogen purity and dewpoint. Many more applications involve the use of portable equipment. This equipment is maintained by the Instrument Section, with the Station Chemists supplying background knowledge of the chemical aspects.

Water analysis includes conductivity, dissolved oxygen and pH measurements, as well as more complex analysis for other contaminations, such as sodium concentrations. Much of the maintenance work in this area involves the cleaning of probes and containers, and the calibration, using known solutions, and checking of the electronic units. This can be carried out effectively on a routine basis with plant in service, since downtime is minimal and the plant

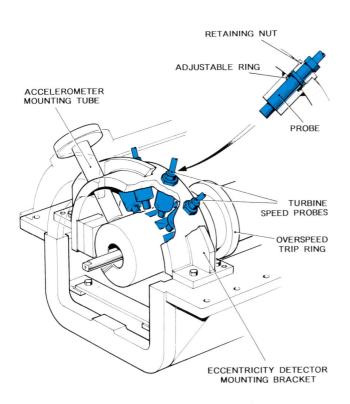

RETAINING NUT

ADJUSTABLE RING

PROBE

ACCELEROMETER
MOUNTING TUBE

TURBINE
SPEED PROBES

OVERSPEED
TRIP RING

ECCENTRICITY DETECTOR
MOUNTING BRACKET

FIG. 5.62 Turbine speed detector mounting arrangement

can operate for short periods with selected measurements not available.

Gas analysis equipment involves oxygen and CO monitoring in the main. Oxygen measurements of flue gases have progressed from paramagnetic analysers, which suffered from maintenance problems in obtaining representative samples from the gas ducts and in controlling the flow through the analyser, to chemical and zirconia cells which are able to monitor concentrations directly within the ducting and produce electrical voltage signals proportional to oxygen percentages (see Volume F).

Some manufacturers now produce equipment incorporating chemical cells which includes on-line, automatic checking of equipment calibration on a frequent basis, using known standard gases so that routine maintenance is not required. This type of equipment has operated in low percentage oxygen oil firing environments for long periods of time, requiring only minimal surveillance-type attention. This reduces the workload, which was high on gas-analysis equipment in the past.

Carbon monoxide concentration measurements employ infra-red absorption techniques, using the fact that CO gas absorbs infra-red radiation between wavelengths of 4.5 to 4.9 μm (see Volume F). Equipment development has led to a simplified system which has eliminated the need to obtain sample gas from the ducts. A heat source installed at one side of a duct produces a single beam of radiation that passes through the flue gases, any absorption being detected by a receiver installed at the opposite side of the duct. Measurements made with this single beam, using a very sensitive and stable detector, is all that is required for continuous and reliable CO monitoring.

Hydrogen dewpoint measurements for generator cooling systems require particular mention in that the available detectors can suffer from drift without this being easily detected. To avoid this problem, test equipment using a moisture generator to produce a suitable gas/moisture condition enables comparison to be made between the in-line measuring probe and equipment and an accurate mirror-operating standardising measurement system. The comparison equipment is a more accurate method of dewpoint measurement but cannot be made intrinsically safe and hence is not allowed to be used on the hydrogen system (see Fig 5.63). The method requires the removal of the moisture detector probe and associated electronic unit from the plant to the workshop for comparison with the reference equipment.

12.6 Sequence control equipment

The start-up and shutdown of auxiliary plant, such as ID and FD fans, now uses electronic sequence equipment to operate the sequences involved. Detectors measure the presence of lubricating oil pressure at bearings, and limit switches check the positions of dampers and vanes; these devices pass information to the sequence control cubicles. The cubicles contain relays, timers and switches incorporated into electronic circuitry which accepts inputs and produces output signals in a timed sequence, such that the start-up and shutdown of the auxiliary plant unit is carried out automatically.

The cubicle equipment is modular, so that faulty units can be removed and replaced when faults are detected. Diagnostic equipment is incorporated into the equipment in the form of indicating lamps which illustrate the stage of the sequence of events and can result in rapid fault correction.

Testing of modules can then be carried out in test equipment, which uses the 'faulty' module in a test sequence to assist with localising the fault. An example of this equipment is shown in Fig 5.64. In future stations, this is ideally suited to the application of a microprocessor, into which a program of the start-up procedure can be fed for each type of auxiliary plant to be controlled.

In oil-fired boiler plant and PF boilers fitted with oil overburn facilities, the burner start-up sequence equipment plays an important role in meeting plant start-up times. The burner operating equipment is subjected to the furnace temperature radiation conditions and hence can fail in service. The start-up equipment, in the form of electronic modules and associated pneumatic relays to operate the burner

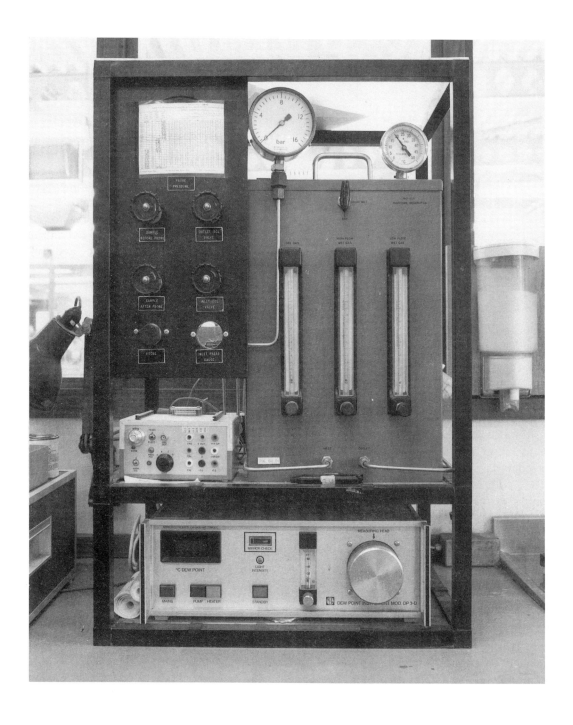

FIG. 5.63 Typical chemical analysis test equipment used for workshop test calibrations. Application; hydrogen dewpoint measurement.

(see also colour photograph between pp 434 and 435)

actuators, has to perform reliably. It has to detect failures quickly and shut down burners safely to avoid the dangerous condition where burners could be delivering oil into the furnace without ignition, being present at that burner. 'Flame eyes' to detect ignition, form part of the protection sequence system and they require frequent attention to ensure correct line-up and sensitivity.

Constant attention is required to burners and the associated sequence equipment and this imposes a heavy workload on the maintenance departments.

Logic diagrams which illustrate the order of sequences can be drawn to assist with fault diagnosis. These are then used by maintenance staff when following through sequence checking on a system. It is often found useful to provide suitable diagrams

FIG. 5.64 Test equipment used in a workshop to subject plant modules taken from auxiliary plant sequences to
a defined sequence check

(see also colour photograph between pp 434 and 435)

fitted on the inside of doors of the sequence cubicles, so that they are available when checks are being made.

12.7 Automatic control equipment

Equipment, installed to control the processes involved in a modern power plant automatically, has increased in complexity and application as the size of unit plant has grown. Computers, or equipment utilising computer equipment with fixed hardware and installed programs, are now being used to control and monitor these processes.

The control of the 660 MW units on the CEGB National Grid System calls for a rapid response to system frequency, so the units are often called upon to operate on free governor action. This has resulted in the fitting of electronic governors to control the turbine steam valves to improve the turbine-generator response (see Fig 5.65). As a result, the boiler con-

trols are required to follow a rapidly changing load pattern, and this has led to the development of fast-response control equipment. Electronics have been applied to the measurement and control facets of the system, whilst hydraulics and pneumatics are favoured in the actuator application to change valve and damper positions. Electronic systems are fitted with exchangeable modules for rapid fault correction and an exchange-type policy covers the maintenance of this part of the system. Hydraulic systems often employ fire-resistant fluids in the fire risk areas around boiler fronts and hot turbine steam chests, and these require clean condition maintenance practices. The fire-resistant fluids also need careful treatment since they present a problem of health care to maintenance staff working with them.

Boiler controls are of the well established parallel-modulating type controlling steam pressure, superheater and reheater steam temperatures, furnace pressure and feedwater controls. The system gains and measuring lags vary with load and are interactive

FIG. 5.65 Governor electronic module under test in the workshop, connected to the electronic governor test cubicle

(see also colour photograph between pp 434 and 435)

with each other; hence making it difficult to provide control over the full load range. It is typical to specify defined limits, such as the variation above and below a setpoint value which must be achieved over the specified load range of 50–100%. For feedwater controls on coal-fired boilers, the requirements are specified as ± 25 mm deviation from mid-drum level, which must be maintained upon the loss of a coal mill.

Some stations have updated their boiler control equipment by using distributed systems, in which individual control loops are each provided with a computer which are supervised by a master control processor.

Maintenance work involves two clear aspects:

- The routine maintenance of measuring equipment, control modules and actuator systems to ensure that their calibrations are correct in order to maintain the defined control limits.

- Fault detection and correction must often be carried out with plant and equipment in service, since fault conditions cannot easily be simulated off-load.

Computers are used in some installations to vary the setpoints of the respective control loops over a range of unit load, so that automatic boiler loading can be achieved.

Since boiler controls are in service over long periods, it is only at unit outage periods that a major check can be made on the control equipment. A technique of 'finger printing' the loop performance when commissioned and then comparing performance against this standard has been used. Test equipment which allows operation of all the actuators when off-load assists in improving actuator performance. Extensions of the technique, such that computers are used as a plant simulator, allow the control loops to be closed and studied off-load, and hence improve performance.

Turbine manufacturers still supply systems based on local control loops, utilising pneumatic equipment for turbine controls. This equipment requires an instrument air supply which is free from moisture and oil contamination, which necessitates a particular type of dry compressor or carbon ring compressor installation, i.e., a compressor involving no lubricants in the compression chamber which would otherwise come into contact with the air being compressed and hence contaminate the supply. Maintenance involves calibration of the controllers and actuator positioners, locating and repairing air leaks, and replacing parts which have suffered mechanical wear or dust-in-air erosion.

Instrument craftsmen have therefore to be trained in these basically mechanical crafts, as well as in the more usual electronic equipment (Fig 5.66).

12.8 Control room desks and alarms

Control room equipment has also benefited from the improvements in technology. Information is presented to Unit Operators in both conventional indicator form and on visual displays driven by computers. Analogue information appears on indicators driven from current loops, these indicators having being reduced in size to almost miniaturisation standards. The indicators and operating switch controls have been built into standard-size interchangeable modules, which can be moved around inside the fixed desk frame to provide the most acceptable arrangement of controls for the unit plant (see Fig 5.67). This allows continual development and improvement during the contract and construction phases of the station. The modules are of a defined height and of variable standard width in multiples of a fixed dimension (15 mm). Cable connections are by plug and socket fittings to allow each module to be disconnected easily without reverting to removing connectors in terminal racks. Maintenance is by replacement with identical spare modules which have been previously checked out in the workshop test environment. Very few faults occur on the desk modules since there are few active components and because miniature type indicators have been developed with exceptional reliability.

Alarms have been largely delegated to computer-based equipment and are displayed on fixed formats with flashing and audible warnings being provided. However, to safeguard against computer failures, conventional alarm facias with typically 6×4 (24-way alarm) display arrangements, are provided for about 10% of the unit alarms.

These alarms are designated 'critical alarms' in contrast with the computer-based repeat critical and non-critical designations. 'Accept' and 'reset' push-buttons form part of the facia units. The facias need little maintenance, with the exception of changing the miniature type indicator lamp bulbs, which need a special tool to remove and replace them. The alarm units which drive the facias are cubicle types, in which a 24-way alarm unit fitted with interchangeable electronic and reed relay components is connected via direct cabling to the facias. The alarm equipment cubicles are sited in equipment rooms adjacent to the control rooms. Terminals are provided to check the input alarm connections and decide whether faults are internal or external to the equipment cubicles. Alarm equipment is maintained by changing cards in the alarm cubicle units and circuit checking on external circuits to identify and clear faults (see Fig 5.68).

Recorders of the potentiometric type with conventional chart displays are provided for historical data collection. They are now based on computer technology, with step incremental motor drives, and are now extremely reliable and develop few faults.

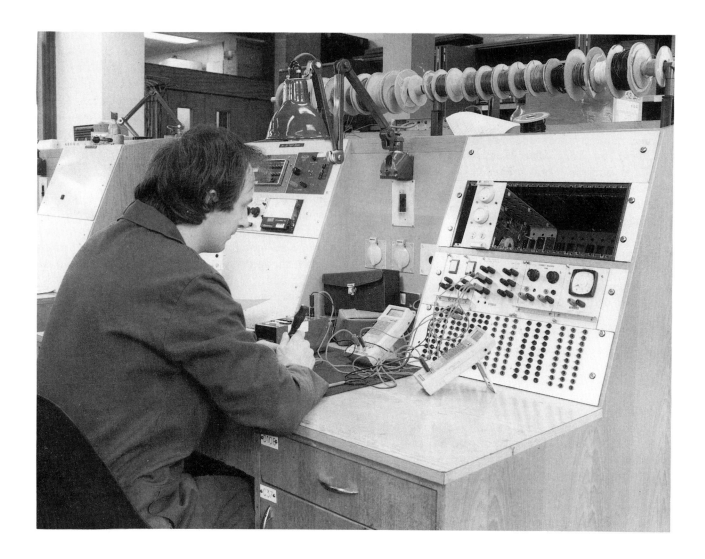

FIG. 5.66 Boiler automatic control modules under test in a workshop test bench environment

(see also colour photograph between pp 434 and 435)

They print the chart range, time and basic data, and are largely self-checking. Maintenance involves weekly chart changing and visual inspection. Historical data is also available from printer facilities which are provided by the unit computer system. Other control room displays which are common to conventional station plant relate to television displays monitoring stack emission conditions and furnace viewing of burner performance.

12.9 Computers and visual display units (VDUs)

Computers now play an important role in providing information about plant conditions to the operators. They are relied on for quick indication of alarm conditions: they also provide post-incident recording facilities, together with historical information for plant performance monitoring. Process control computer equipment is installed in clean-condition rooms directly alongside the control room to minimise cable runs to unit desk visual displays, keyboards and on-line printers (see Fig 5.69).

Maintenance involves providing stable environmental conditions, fault diagnosis and correction, and the calibration of plant equipment through the computers to the operator VDUs. Computer reliability is extremely good and few total system failures occur, but back-up facilities must be available to restore the computer systems quickly, so that unit operation is not interrupted. As more responsibility for indication and control of plant is delegated to computers, they become essential plant for operation and must be treated as such. VDUs need little maintenance other than tube replacement as the tube ages and character recognition becomes more difficult. Printer

FIG. 5.67 Sections of a unit control desk, illustrating the exchangeable modules together with the alarm facia displays and Hydrastep indications

(see also colour photograph between pp 434 and 435)

reliability has improved, but is still a high maintenance workload.

Fault detection on complex computer equipment can only be carried out utilising the diagnostic facilities designed into the system. Diagnostic tapes are provided and must be run to establish fault areas. Faults are cured largely by replacement of equipment trays or sometimes by locating faulty circuit boards.

Back-up facilities must be provided to repair faulty units since the replacement costs of some components, such as storage disks, are extremely high.

Calibration of plant inputs must be provided for at the computer terminal equipment, since checking of transducer inputs forms a significant part of computer maintenance work. Specialist staff for computer

maintenance are required and detailed training must be given to them through the computer manufacturer and software suppliers.

Updating and recording of software systems is an essential part of the computer operating regime. It is important that any reloading of programs after system shutdown is done with the most up to date disk or tapes. This means that a reliable recording system, suitably supervised, needs to be provided.

12.10 Workshop facilities

Since the philosophy of instrument maintenance relies heavily on replacement of equipment on plant

FIG. 5.68 Four 24-way alarm drive units, shown fitted into cubicle equipment, illustrating the exchangeable circuit cards
for easy fault correction

and subsequent repair in the workshop, the layout and facilities of the workshop form a critical part of the maintenance performance.

Good working conditions are essential since staff are required to do careful detailed and responsible work. Lighting and heating conditions are important, as well as space to work and suitably designed benches at which to work.

Test equipment must be extensive to cover a wide range of equipment and must be stored carefully about the workshop for easy access and rapid availability (Fig 5.70).

Workshops should be sectionalised to provide two main areas:

● The general area containing lathes, drilling machines and the heavier benches for mechanical-type work (Fig 5.71).

● A clean-conditions workshop for electronic equipment maintenance (Fig 5.72). This contains work benches incorporating built-in power supplies and basic test equipment, such as digital multimeters. Special test facilities, such as cubicles containing turbine governor test equipment, should also be within this area.

Provision of an adjoining test room for basic temperature, pressure and chemical analysis is also essential, so that much of the required testing can be kept away from the basic workshop work. A fume cubicle for corrosive materials and chemical-fume treatment should be provided, as well as facilities for dealing with mercury if this should be required, although the presence of mercury in instruments has now been largely eliminated.

FIG. 5.69 Unit computer equipment installed in clean condition rooms
The illustration shows central processor units, disk store, paper tape punch and reader, and keyboard printers.

A foreman's office adjoining the main workshops is also a desirable feature, since the supervision forms an important facet of workshop performance.

Correct storage of equipment awaiting repair is also essential, so that they can be recorded and worked on systematically to ensure that only repaired items are returned to the station stores (Fig 5.73).

Provision should be made for files of work instructions, together with the station descriptive and maintenance manuals, to be kept in the workshop, so that reference can be made to technical details concerning the equipment and work to be done.

A list of plant settings should be easily accessible for frequently-calibrated equipment.

The use of film type viewers for stores stock information enables spares requirements to be identified. The station stores must be well stocked with spare modules, components and service equipment, so that a rapid response to plant defects can be maintained.

12.11 Planning for maintenance

Effective maintenance can only be achieved as a result of good forward planning and much thought as to how work is to be carried out. Instrument and control maintenance has particular requirements because of the type of work and the conditions under which it can be carried out. It has already been emphasised that a good deal of the work is carried out by arrangement with the operations staff with plant in service, so the results of maintenance actions are reflected in plant performance. It is difficult to

421

FIG. 5.70 Workshop test equipment fitted into storage units

assess, beforehand, how long fault finding jobs will take, and hence flexibility has to be inherent in the planning function. It has become the practice to measure or estimate how long jobs take, so as to ensure that as much planned work as possible can be issued on a daily basis to each instrument mechanic. Routine maintenance systems have been developed, often utilising computers, to ensure a spread of work over a 12-month period. Routine and statutory inspections of 3–6 and 12-monthly periods then get issued through the station planning office each week and can then form the basis of about 50% of the department's weekly workload. Unlike mechanical and electrical work, C and I maintenance involves many jobs of short duration and hence a work programme for a C and I mechanic could involve up to 6 to

8 jobs in an 8-hour day. The control of this involvement and the recording of information has to be good, to make good use of the mechanic's time. This leads to the need for effective systems for stores information, job card issue, calibration-history recording and many other functions to be co-ordinated.

Station systems have developed towards a plant information system, with plant coding being developed for defect card raising, etc. However, plant coding is not a convenient way to develop a C and I system, since much of the C and I equipment is common to several plant areas. For example, it is common for best use of spares, to use one manufacturer's range of pressure transmitters across the station. For this and other reasons, it is more practical to base C and I systems on manufacturer's type codes

FIG. 5.71 Workshop area for mechanical-type bench work and machine facilities

(see also colour photograph between pp 434 and 435)

and to co-ordinate these with plant codes, where possible.

One basis for a system can be a manufacturer's six-figure code, for example:

G 01 02 3

The alphabetic prefix (G) represents the first letter of the manufacturer's name and this, together with the next two digits (01) of the numerical code, allows up to 99 different manufacturer's (whose name begins with G) to be coded under G. The next two digits (02) allow up to 99 different devices to be identified, with the final digit (3) representing 0 to 9 variations of a particular device.

This method of coding seems to give adequate cover for the range of equipment used in a large modern power station. It forms the basis of co-ordinating the numerous information requirements within the C and I Department.

12.12 Other work areas that influence C and I maintenance

12.12.1 Statutory inspection

Instrument maintenance can often only be carried out when access is given for other station maintenance work.

FIG. 5.72 Electronic workshops outlining purpose built test benches fitted with inbuilt power supply units
and basic test equipment

(see also colour photograph between pp 434 and 435)

Statutory inspections on air receivers and pressure vessels can only be carried out as a combined maintenance exercise. In these circumstances, instruments have to be removed as the first job in the programme, and calibration and repair carried out. The refitting of the instruments is then the final job to be done before return-to-service checks are carried out. This puts a strict programme requirement on the instrument maintenance section to meet defined dates and means that the work on the refitting and recommissioning is done to the required standards but in some haste, so that the plant is returned to service on time. Main unit outages for turbines and boilers also present a high workload at the end of the programme, since a substantial amount of pro-

tection and recommissioning checks have to be carried out prior to the unit's return to service. At these times, C and I staff have to be available to work with the operation staff on a continuous shift basis to ensure that the recommissioning is not held up. Most stations have now compiled recommissioning schedules which identify departmental responsibilities. A high percentage of this workload is in C and I activities.

12.12.2 Special instrumentation and test provisions

Plant investigations are frequently required to identify problem areas, perhaps as a result of plant failures

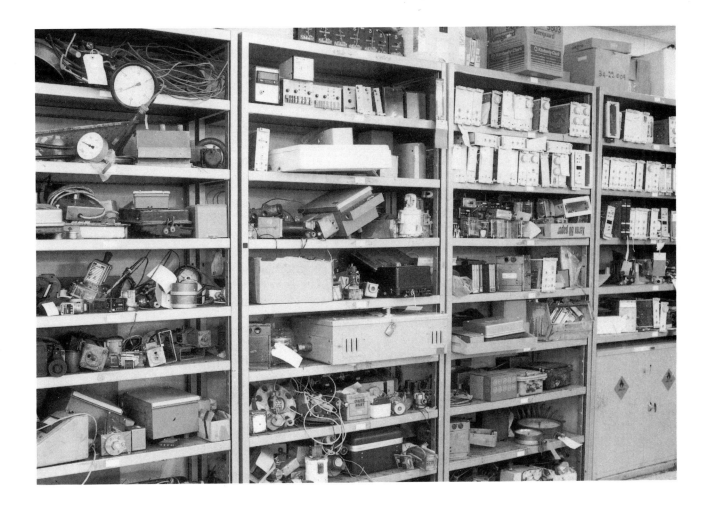

FIG. 5.73 Equipment awaiting workshop repairs, stored and labelled in workshop racks

or unit trips. In these instances, the fitting of special or additional instrumentation is often needed, which falls to the C and I section.

The provision of recording systems to investigate a sequence of events can be extensive and require a good deal of engineering and installation effort. A typical investigation necessitated the identification of the sequence of events in a governor failure which resulted in a unit trip. Several methods of approach to this problem were attempted before a conclusion was reached to use magnetic-tape recording of the parameters. These were then reproduced on conventional charts so that an analysis of the waveform and time sequences could be undertaken. Other long-term testing which can result in extensive installation work and detailed engineering is the run-down vibration monitoring of turbines to investigate the possible cracking of turbine shafts. This has led to the installation of accelerometers at the turbine bearings and the recording of up to 15 channels of information on magnetic tape. Analysis of each channel for the fundamental and harmonic magnitudes off-line then gives information relating to the condition of the turbine shaft.

13 Quality assurance

Inadequate attention to detail when carrying out maintenance work in any discipline in a power station can lead to plant failure of enormous proportions. A single unsecured ring nut in a waterbox of a 500 MW generator brought about a nine month outage costing millions of pounds in plant replacement costs. On a less dramatic scale, poorly-executed work results in wasted materials and the time, not only of the craftsman and his supervisors, but of operations staff who have to re-isolate plant and of planning staff. The guarantee of work being correctly executed, whether it be maintenance or plant operation, is known as *quality assurance* (QA).

The Executive of the CEGB re-affirmed the CEGB statement of policy on quality assurance in 1982 in the following terms:

It is the policy of the Generating Board that, for all items of power generating and transmission plant and associated items, there should be in force appropriate arrangements for providing an assurance of quality at all stages from design to de-commissioning.

Quality, like safety, is the responsibility of all employees, whatever their area of work, and it is not to be regarded as the sole responsibility of a particular person or group. The ultimate achievement of quality lies in the hands of every member of staff at a location, each of whom has his own important function to perform. It is a major task of management, however, to persuade personnel to have the correct attitude of mind towards quality, to behave in a manner which is not always 'the easiest', and to remember that prevention is usually less costly than cure.

Quality assurance, then is an essential aspect of good management; it is the principal system which combines all the administrative and practical job-orientated activities affecting the quality of a product or service. It consists of two main components:

- *Quality administration* brings together all the organisational conditions, the establishment and updating of the quality assurance programme, the systems for the preservation of records and documents, the identification and verification of corrective measures when necessary and the processes for auditing.

- *Quality control* relates to the measurements and control measures taken to prove that the physical characteristics of an item or service meet the specified requirements.

13.1 Quality administration documentation

The description of the system to establish an appropriate and acceptable level of quality at a location can usefully be embodied in three tiers of documentation.

13.1.1 Part 1

First, then, is a handbook setting out the management and policies of the quality assurance system; this document can reasonably be referred to as the *quality assurance programme* for the location.

The QA assurance programme is specific to a location and is designed to provide objective evidence that all the systematically planned actions necessary

to ensure that plant items will perform satisfactorily in service have been carried out. The QA programme requirements apply to all plant and systems. The quality system covers design, purchasing, repair and maintenance, commissioning, control of contracts, plant operation and de-commissioning. The programme itself will describe arrangements for ensuring that:

- All activities and operations are carried out in accordance with statutory and other relevant regulations and codes of practice, to ensure the safety of personnel, plant and equipment, together with the preservation of amenities.

- Handbooks are produced which set out the disciplined approach to be adopted towards all activities within the operational life of the power station. They will include descriptions of suitable approved departmental arrangements made to control the manner in which quality activities are carried out, documented and verified.

- Programmes are prepared for routine maintenance, inspection and testing of plant to meet all statutory and mandatory requirements.

- The station is staffed with suitably qualified, trained and experienced staff.

- The need for specialised advice is identified and the necessary provision arranged.

- Any proposals for engineering modifications, provision of spares and operational or maintenance procedural changes are subject to the appropriate safety and commercial assessments before authorisation is given to implement such changes.

- Operational and lifetime records are collected, stored and updated, as necessary, to ensure that retrieval and audit can easily be carried out.

An important factor in ensuring the viability and effectiveness of quality systems is the regular auditing of the arrangements made; such audits will be carried out by station staff not directly in line management of the department being audited. Auditing is a specialist task and is best carried out by a small number of suitably trained staff, who may well be members of a management services section.

13.1.2 Part 2

A handbook describing the management and procedures covering the QA assurance sections for each department or functional area is prepared. The document can be referred to as the QA manual for the relevant department. (At the time of writing the terms 'programme' and 'manual' have not been standardised and the reader may come upon the words used in the reverse sense to that shown above, hence the terms 'Part 1' and 'Part 2' have been used.) The key activities in the preparation of the Part 2 document are as follows:

- Identify key tasks to be performed:

  (a) To define the required quality.

  (b) To obtain the required quality.

  (c) To verify that the required quality has been obtained.

- Identify the management systems and procedures to be used.

- Define the responsibilities of key members of staff and produce an organisation chart.

- Identify relevent qualifications/training/experience of staff for carrying out key tasks.

- Identify interfaces with other organisations and communication arrangements.

- Identify records systems and arrangements.

- Identify audit arrangements.

13.1.3 Part 3

This comprises documents describing or identifying operations, inspections and tests (and the associated records) specific to the manufacture, construction, maintenance, repair or operation of an item of plant or equipment. Such documents are referred to as

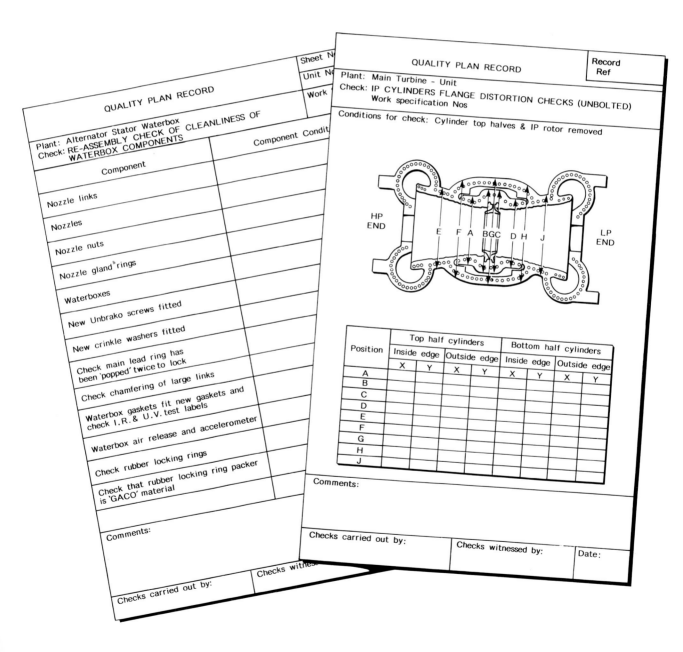

FIG. 5.74 Examples of quality plan records

quality plans; they are essentially work-orientated, as distinct from the programme and manual. They are complementary to the Part 2 document by specifying the requirements for, and the verification of, the work identified.

Working documents are essential to the systematic performance and verification of work. The contents need to be clear, concise and unambiguous, and to an approved standard. To this end, each should be checked and approved by a nominated person. Job instructions identify key checks and inspections necessary when carrying out the work, together with 'hold points' beyond which work will not progress until a foreman or supervising engineer has examined the work or checked a dimension. The completed quality plan, whether it be a single work specification or the hundreds of checksheets associated with a turbine-generator overhaul, is retained as a record of work completed and is used for comparison with the 'as found' state at future overhauls.

Examples of quality plan records relating to aspects of generator and turbine overhaul are given in Fig 5.74.

13.2 Reference documentation

13.2.1 CEGB Engineering Services Department publications

CQA1 'A guide to the quality assurance practices for conventional generation and transmission plant and equipment'

CQA2 'A guide to the quality assurance practices for nuclear safety-related plant and equipment'

13.2.2 British Standards specifications

BS4778: Glossary of terms used in quality assurance: 1979

BS4891: A guide to quality assurance: 1972

BS5750: Quality systems: 1979

Parts 1 and 4 — Specification for design, manufacture and installation.

Parts 2 and 5 — Specification for manufacture and installation.

Parts 3 and 6 — Specification for final inspection and test.

BS5760: Reliability of Systems, Equipment and Components: 1979

Part 1 — Guide to reliability programme management.

Part 3 — Guide to reliability practices: examples.

BS6143: Guide to the determination and use of quality related procedures: 1981

BSI Handbook No. 22: 1983 'Quality Assurance' contains all the above listed BS specifications.

CHAPTER 6

Safety

Contents

1 Introduction

When the supply industry was nationalised in 1948 the legal framework imposed required the industry to 'develop and maintain an efficient and economical system of electricity supply'.

The industry cannot be either efficient or economic if it does not operate safely, so irrespective of specific safety legislation this carries an implication for all staff that safety is of prime importance, and like any other supply industry activity, it must be carried out with due regard to cost and benefit to the consumer. The CEGB and its predecessors have kept safety high on the list of priorities and have sought to achieve improvements, using consultative processes with all levels of staff.

Safety activities in the Board are covered by the Health and Safety at Work Act 1974, by its enabling provisions and by various other statutory requirements (see Section 2 of this chapter). Safety policy is covered in a CEGB General Policy Statement dated April 1983, signed by the Chairman and available individually to all members of staff. This makes it clear that the Board's policy is to maintain a safe working environment for all its employees and accepts the need to conduct its undertaking in such a way that persons not employed by it are not exposed to danger by virtue of the Board's operations.

The Policy Statement goes on to make it clear that safety continues to be a major item in the consultative process with its staff and that staff have a duty to work safely within the law. The three main criteria covering safety in the CEGB are therefore as follows:

● It must be managed in accordance with Board Policy and statutory requirements.

● It is subject to the consultative process via the National Joint Co-ordinating Council and the Regional Joint Co-ordinating Council Health and Safety Committees and the Location Safety Committees.

● Training needs must be satisfied so that the designed and inherent safety of the Board's plant is fully utilised and that safe systems of work are available and adhered to at all times.

The Board operates a specialist safety service with trained staff available to give detailed technical advice to locations. Many larger locations also have a Location Safety Officer, who will be specially trained to make regular reports to location management, to carry out hazard patrols on the site, to make spot checks on lifting tackle, fire fighting equipment, etc., and to liaise with the appropriate inspector and public authorities on matters concerning safety and emergency services. In addition, he is capable of working with trade-union-appointed safety representatives and of drawing attention to specific hazards, including those of health and noise.

Finally, an additional safety service is available to the Board from the Electricity Council, which maintains a small specialised staff in various parts of the country to give advice to Area Boards as well as to the Generating Board.

2 Legal aspects

In May 1970, a Committee was set up by the Government of the day under the Chairmanship of Lord Robens, one time Chairman of the National Coal Board, to review the provision for the Safety and Health of Persons in the course of their employment.

The Committee reported in June 1972 and, as a result, the Health and Safety at Work Act reached the statute book in 1974. It provided a comprehensive and broad piece of legislation which, whilst not detailed in itself (although it fills 117 pages), is 'enabling', i.e., the Secretary of State for Employment can draw up details, regulations and codes of practice on particular aspects of health, safety and welfare. Existing legislation, for example, the Factories Acts, is not removed from the statute book until replaced by new legislation.

Readers are advised to check the current state of legislation in up-to-date literature.

Additional to the then existing legislation, the Health and Safety at Work Act set up the Health and Safety Commission and reorganised the various government inspectorates (for example, the Factory Inspectorate, Mines and Quarries Inspectorate, Alkali Inspectorate and Nuclear Installation Inspectorate) into a new body, called the Health and Safety Executive. It also provided up-dated powers and penalties for the legal enforcement of Safety and Health at Work, required the Trade Unions to appoint Safety Representatives at the place of work and required employers to consult with those safety representatives.

The act has 85 sections, a small selection of which is given: below:

Section 1: General.

Section 2: Places a general duty on employers to ensure the safety, health and welfare of their employees, and includes the need to consult with safety representatives.

Section 3: Places a general duty on employers to ensure that their activities do not cause danger to persons not in his employ.

Section 5: Requires that best 'practical means' be used to prevent or render harmless noxious matters from premises.

Section 6: Requires anyone who supplies articles for use at work to ensure (in so far as he can) that it is safe, when used in the manner recommended by the supplier. This includes the requirement that those who install plant shall install it in a way such that it can be operated safely in its normal mode.

Section 7: Employees must take responsible care to ensure that when at work they do not endanger themselves or others.

Section 8: All items pertaining to the statutory requirements of the Health and Safety at Work Act must not be misused. Other sections deal with the Establishment of the Health and Safety Commission and Executive, Enforcement, Disclosure of information and the Employment Medical Advisory Service (EMAS).

Attention is also drawn to the Offices, Shops and Railway Premises Act of 1963, to the Fire Precautions Act of 1971 and to various outstanding sections of the Factories Act which have not been amended by the Health and Safety at Work Act.

The CEGB is also governed by statute for the disposal of certain noxious refuse, which includes asbestos and nuclear waste. Regulations relevant to these include the Asbestos Regulations, Control of Pollution Act and the Disposal of Poisonous Waste Act.

In the last 10 years or so, there has been a rapid growth of safety-orientated legislation and a brief introduction, such as this, cannot in any way be regarded as comprehensive; the reader is therefore directed to the bibliography and to expert safety and legal opinion for a complete and up-to-date picture.

Health and Safety at Work, and the standards and expenditure needed to achieve it, is often qualified within the Health and Safety at Work Act by the expression 'so far as is reasonably practicable'; this phrase is not defined in the Health and Safety at Work Act, so the meaning of the expression comes from interpretations by the Courts when case law has been established. 'So far as is reasonably practicable' requires an assessment of the risks of a particular activity or environment set against the physical difficulties and expense which are required to remove or significantly reduce the risk. It should be understood that if the risks to safety or health of a particular working activity are low and the technical difficulties and financial cost of taking remedial measures are high, then it might well be considered not to be 'reasonably practicable' to rectify the situation, but the greater the degree of risk, the less the weight that can be given to the cost of avoiding measures. If the reduction of risk is insignificant in relation to the financial cost, time and trouble needed to achieve the reduction, then the obligation under the Act is met.

3 Safety rules

Early in 1982 the CEGB introduced the 4th Edition Electrical and Mechanical Safety Rules. These rules replaced existing rules, which had stood the test of time, so the 4th Edition Rules do not basically alter the way plant is isolated for work to be carried out; rather, they procedurally present an approach required under the Health and Safety at Work Act.

All the CEGB plant items are regarded as interconnected to form systems, which are required for the process of generating or transmitting electricity. In the normal operation mode, such systems are designed to be intrinsically safe and no danger arises from their routine operation carried out to defined procedures. When work or testing, other than normal operation, has to be carried out, these specific rules apply which are designed to achieve safety for the individual from the system.

In these circumstances, two types of danger are dealt with:

(a) Danger from the system, which is covered by the rules.

(b) Danger from environmental hazards which arise during the time work is in progress. Such considerations are known as *general safety*.

The 4th Edition contains eight Safety Rules: each is considered to stand indefinitely. They are expanded by a series of national safety instructions and codes of practice which can be modified from time to time to take account of changes in safety technology.

Every employee has access to the safety rule book, which contains all that is required for staff to act in a safe and proper manner on the Board's systems when out of the normal operating mode. The codes of practice are converted into local instructions for each site: these take into account the peculiarities of the plant installed and other local requirements though the code must not be weakened in any way by the conversion. Before considering the actions required to make plant safe, the following defini-

tions (culled from Part D of the 4th Edition Safety Rules) are given:

3.1 Definitions

System Safety from the system is that condition which safeguards persons working on or testing plant from the dangers which are inherent in the system.

Competent person A person who has sufficient technical knowledge and/or experience to enable him to avoid danger and who may receive, transfer and clear specified *safety documents* when so nominated.

Authorised person A person who has been nominated by the appropriate officer to carry out duties specified in writing.

Senior authorised person An authorised person nominated by an appropriate officer to carry out duties specified in writing, including the preparation, issue and cancellation of specified *safety documents*.

Control person A person who has been nominated by an appropriate officer to be responsible for controlling and co-ordinating safety activities necessary to achieve *safety from the system*.

Selected person A person qualified by technical knowledge and experience and selected by an appropriate officer to carry out tests and examinations and make recommendations regarding additional special precautions.

Limited work certificate A safety document of a format which defines the limits within which work or testing may be carried out and specifies the necessary precautions.

Permit for work A safety document of a format specifying the plant to be worked on, the work to be carried out and the actions taken to achieve *safety from the system*.

Sanction for test A safety document of a format specifying the plant to be tested, making known the conditions under which the testing is to be carried out and confirming the actions which have been taken to achieve *safety from the system*.

3.2 Safety document procedure

Plant which is required for maintenance is taken out of service with the permission of the control person, isolated and made safe in accordance with the appropriate rule, codes of practice and local instructions; a safety document is then issued to a competent person, or to an authorised person, by a senior authorised person. The safety document may be accompanied by a selected person's report which would make recommendations regarding additional safety precautions, if the senior authorised person deems it fit.

On completion of the work, the safety document is returned to the permit office or control room for clearance by the competent person or authorised person and for cancellation by the senior authorised person. The safety locks and isolation are then removed from the plant item, which is then returned to service in accordance with the operating procedures.

Staff who perform specific duties under these rules all receive initial and refresher training and are required to have suitable experience on the plant. Senior and authorised persons are authorised by examination and are only allowed to carry out their safety rule duties within the limits prescribed by a certificate awarded to them. Such authorisations are reviewed periodically and, if considered necessary, the certificates can be withdrawn until retraining and re-examination has been successfully completed.

The systems by which safety documents are issued and procedures for locking-off plant are regularly audited and any shortcomings made good.

It is estimated that upwards of 50 000 safety documents are issued in the CEGB every month. Only rarely do problems arise and safety hazards come up during the course of work. But even these few cases are too many and the Board and its staff continually seek to remove the human failure element from the scene as soon as possible, the credibility of the rules being never in doubt. This high standard of achievement can only be maintained by:

● Having correct procedures (Appendix A).

● Identifying plant properly and clearly.

● Fully trained staff applying and working to well documented and well understood rules and procedures.

● Auditing and rectifying errors.

● Retraining staff regularly.

4 Safety documentation

Safety documentation was referred to in the previous section; this section traces the system which gives rise to the issue of such a document. Slight variations in works office procedures are possible from site to site, but the content of the safety document for a specific job on a particular item of plant will always be the same.

Consider a typical electrically-driven 50% duty boiler feed pump provided on a 500 MW unit. Figures 6.1 and 6.2 show the basic layout, including valves and electrics, for such a pump set.

It is assumed that the Planning/Works Office issues work order cards and work specifications to:

(a) Overhaul the speed control electrodes.

(b) Inspect and replace, if necessary, the pressure reducing nozzle for the HP leak-off line.

Both jobs are scheduled on the 50%-duty boiler feed pump 2B. Item (a) involves an electrical fitter and mate and item (b) will involve a mechanical fitter and welder.

A typical paperwork system is detailed in Appendix A, and is initiated when the control person (who may be the senior authorised person) has given his permission for the plant to be released at programmed time for the agreed duration of the work.

The plant will be electrically and mechanically isolated, caution notices will be attached to the points of isolation and the plant will be drained down, where appropriate. All points of isolation will be secured with a padlock and chain, as required; the keys from these padlocks will be placed in a lock-out box in the permit-for-work office and locked up with a unique key, which will be issued with the safety document to the maintenance staff carrying out the work.

The safety document (permit for work) for the nozzle will be prepared by an authorised person or senior authorised person and issued by the senior authorised person to the competent person in charge of the safety aspects of the work (i.e., to the foreman or to the mechanical fitter who is the first on the job). This permit for work is illustrated in Fig 6.3.

The issuing senior authorised person has a duty to ensure that the recipient of the safety document is fully aware of its limits, knows the work to be done and knows without doubt the location of the plant item.

The work of overhauling the speed control electrodes (within the electrolyte tank) is mechanical work, but of an electrical nature, and may well be carried out by an electrical fitter, so, the senior authorised person may decide to issue a separate permit for work for these items. The procedure followed is identical and most of the isolations are the same. A second key will be turned in the original lock-out box to safeguard the isolation points for this second permit for work and is issued with the new safety document. If additional isolations are required then this second lock-out box key is labelled and placed in a second lock-out box, with the additional isolation keys, from which a securing key is obtained for the second permit.

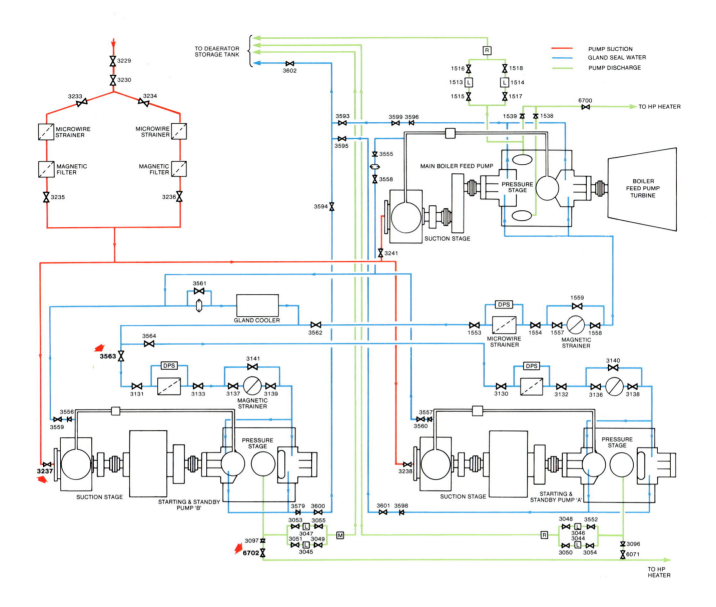

FIG. 6.1 Typical boiler feed pump system for a 500 MW unit

Both competent persons will display the permits for work in separate card safes in the workshop and the competent persons in charge of the safety aspects of the work will lock the card safe with their own personal locks.

If the work goes on beyond the day or shift of issue, and other skilled staff become involved, then the transfer procedure will be enacted in accordance with the safety rules.

On completion of the work, the competent person signs on the permit for work indicating that both men and materials have been cleared from the plant. The document is then cancelled by the senior authorised person, the plant isolations are removed and the plant recommissioned in accordance with the operating procedures.

5 Fire protection

5.1 General

There are many important aspects concerning the fire protection of plant and buildings and, of course, people.

It must never be forgotten that, whilst any given process plant will have a particular fire risk associated with it (be it low or high), this risk will be increased if people are working on or near it.

Fire risk is minimised, and the safety of personnel increased, by adequate design and siting of the plant; it is also potentially contained by such fire protection as can be properly and economically provided. The

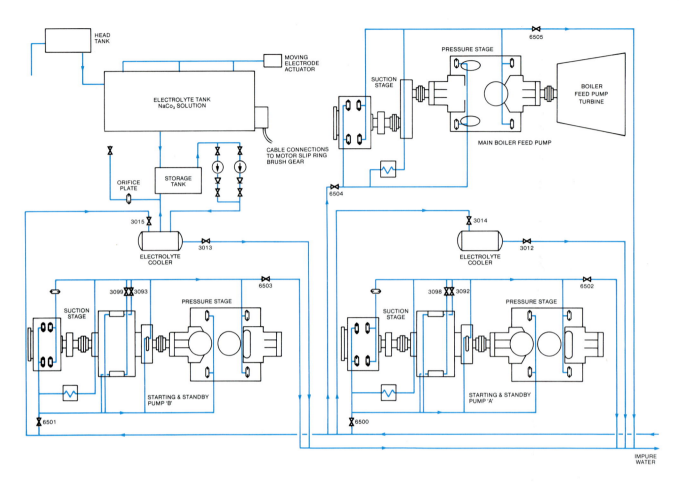

FIG. 6.2 Feed pump cooling water and speed control system

design and siting of plant must include considerations of ventilation, cable and plant segregation, the elimination of particularly high risk materials, and so on. The risks that arise when people are working on the plant increase even more if the work or personnel practices (e.g., smoking) are below an accepted standard. The risk is further increased if the fire protection system availability is reduced while work is being carried out on it (design memorandum 098/6).

A consideration of fire protection will include the following points [1, 2]:

● Acceptable basic design and maintenance of the protection system by a defined and monitored operational and maintenance procedure including realistic periodic testing of all such systems.

● Good housekeeping, and reduction of fire risk through plant cleanliness.

● Staff awareness of risk, coupled with good working practices.

● The ability to provide a manual fire fighting back-up in a timely and effective manner by the pro-

vision of fire fighting teams and equipment.

● A consideration of cost benefit arising from the provision of protective systems against the risk if they are not provided.

Good housekeeping is important for fire prevention; it requires a continuing commitment from both management and staff to ensure its attainment and the realisation of its benefits. The training of staff, maintenance planning and the regular maintenance and testing of the fire equipment, the provision and exercise of a detailed site fire plan, in liaison with the public services (Police, Fire and Ambulance), all have a part to play.

It is common CEGB practice to audit locations to ensure attainment of the Board's objectives.

5.2 Methods of fire detection for particular plant needs

Devices relying on the detection of temperature rise in a particular area or on the detection of fumes

FIG. 4.11   Bearing dimensions and other quality control checks being carried out on an ID fan runner prior to being fitted as part of a major overhaul

FIG. 4.15   Mobile lifting beam

The use of mobile lifting beams can improve the effectiveness and utilisation of the workforce during overhauls, since turbine hall cranes are often fully employed for main turbine-generator dismantling operations. Mobile lifting beams can be used for dismantling turbine valvegear components and for the removal of bearing keeps pipework and other small fitments.

Fig. 5.7   Shows the sequence of events in removing the tyre assembly on a Loesche mill
Note the provision at the design stage of built-in anchors for the hydraulic handling equipment and the easy access to the mill body through the roller apertures (Loesche GMBH).

Fig. 5.7 *(cont'd)* Shows the sequence of events in removing the tyre assembly on a Loesche mill
Note the provision at the design stage of built-in anchors for the hydraulic handling equipment and the easy access to the mill body through the roller apertures (Loesche GMBH).

FIG. 5.7 *(cont'd)* Shows the sequence of events in removing the tyre assembly on a Loesche mill
Note the provision at the design stage of built-in anchors for the hydraulic handling equipment and the easy access to the mill body through the roller apertures (Loesche GMBH).

FIG. 5.48   Bucket wheel stacker-reclaiming machine

FIG. 5.58   Pressure calibration test equipment installed in a workshop, illustrating a deadweight tester, a substandard test gauge on a comparator tester and, in the background, a high accuracy test gauge panel covering a wide range of pressures

FIG. 5.59   Temperature calibration equipment fitted into a portable frame in a workshop, illustrating a sandbath together with liquid fluid baths for the calibration of temperature detectors over a range of temperature from ambient to 600°C

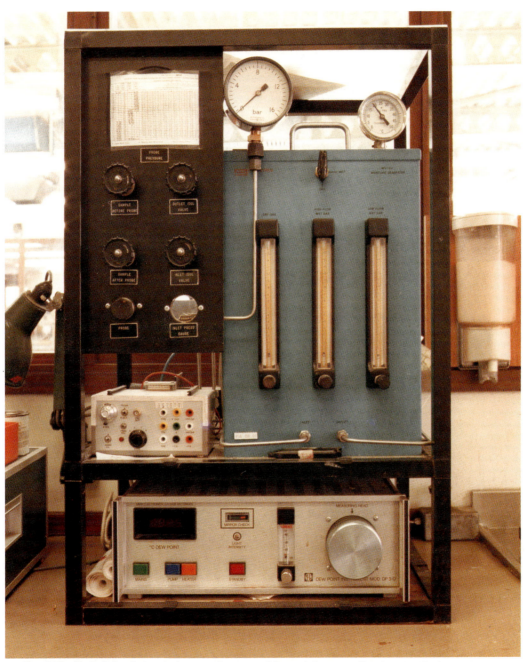

Fig. 5.63   Typical chemical analysis test equipment used for workshop test calibrations. Application; hydrogen dewpoint measurement

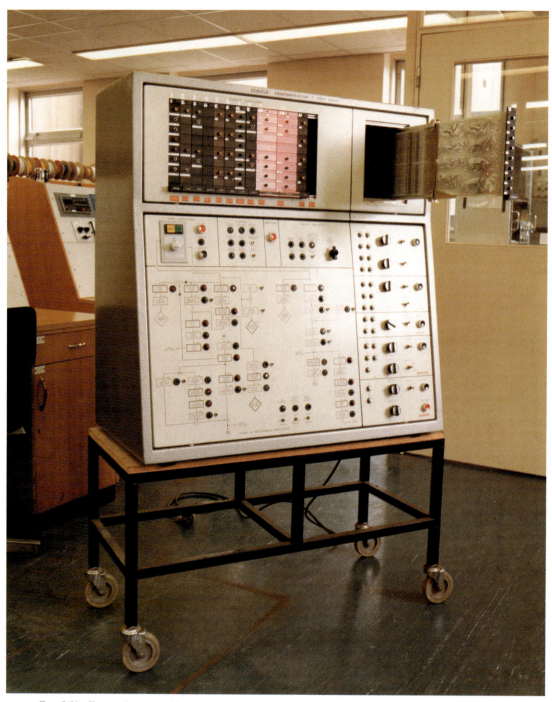

FIG. 5.64   Test equipment used in a workshop to subject plant modules taken from auxiliary plant sequences to a defined sequence check

Fig. 5.65   Governor electronic module under test in the workshop, connected to the electronic governor test cubicle

Fig. 5.66   Boiler automatic control modules under test in a workshop test bench environment

FIG. 5.67 Sections of a unit control desk, illustrating the exchangeable modules together with the alarm facia displays and Hydrastep indications

FIG. 5.71   Workshop area for mechanical-type bench work and machine facilities

FIG. 5.72   Electronic workshops outlining purpose built test benches fitted with inbuilt power supply units and basic test equipment

PERMIT FOR WORK

No. 742.

KEY SAFE | No.* 016A.

1. (i) LOCATION *BEANACRE POWER STATION*

   (ii) PLANT/APPARATUS IDENTIFICATION *2B START/STANDBY ELECTRIC FEED PUMP*

   (iii) WORK TO BE DONE *EXAMINE PRESSURE REDUCER IN HP PUMP LEAK OFF LINE IN ACCORDANCE WITH WORK ORDER CARD M1674*

2. (i) PRECAUTIONS TAKEN TO ACHIEVE SAFETY FROM THE SYSTEM: State points at which **Plant/Apparatus** has been **Isolated** and specify position(s) of **Earthing Devices** applied. State actions taken to avoid **Danger** by draining, venting, purging and containment or dissipation of stored energy.

   2B MOTOR A.C.B ISOLATED BUS BAR SHUTTERS LOCKED SHUT. VALVES 3237, 6702, 3563 AND L/O LINE ISOLATED AT No. 2 O/A LOCKED SHUT. PUMP DRAINED.

   Caution Notices have been affixed at all points of isolation

   (ii) FURTHER PRECAUTIONS TO BE TAKEN DURING THE COURSE OF WORK TO AVOID SYSTEM DERIVED HAZARDS

   I have confirmed with the **Control Person(s)** *M. SMITH*
   that the precautions in Section 2(i) have been carried out and that the **Control Person(s)** will maintain these until this Permit is cancelled. I certify that the precautions in Section 2(i) together with the precautions in Section 2(ii) are adequate to provide **Safety from the System** in respect of the work in Section 1.

   This **Permit for Work** must only be transferred under the **Personal Supervision** of a **Senior Authorised Person*** *N/A*

   Signed *M Smith* being a **Senior Authorised Person.**   Time *0700*   Date *4.3.85*

3. ISSUE: (i) **Key Safe Key** (No.)* *016A*   (ii) **Earthing Schedule*** *N/A*   (iii) **Portable Drain Earths** (No. off)* *N/A*

   (iv) **Selected Person's** Report (No.)* *N/A*   (v) **Circuit Identification** Flags (No. off)* *N/A*

   (vi) **Circuit Identification** Wristlets (No. off)* and Colours/Symbols* *N/A*

   Signed *M Smith* being the **Senior Authorised Person** responsible for the issue of this Document.   Time *0700*   Date *4.3.85.*

4. RECEIPT: I understand and accept my responsibilities under this Document and acknowledge receipt of the items in Section 3.

   Signed .... (Name (Block Letters) ....)

   being a **Competent Person** in the employ of Firm/Dept. ....   Time ....   Date ....

*N/A if Not Applicable.   RP14G   NW/R/083-1-86   April 1984

FIG. 6.3 Permit for work

and smoke or specific gases can be fitted to raise alarms at some central point, as well as triggering the appropriate zone of the fire protection system.

A brief (non-exclusive) list of detection devices is given below, and the bibliography in Section 10 of this chapter gives areas worthy of further reading:

Smoke detectors are based on ionisation chambers or photo-electric devices.

Flame detectors react to radiant energy in either infra-red or ultraviolet parts of the spectrum.

'Fire wire' consists of stainless capillary sensing elements looped together and feeding a monitoring unit.

Temperature-sensitive cabling is used to detect changes in heat level, by either analogue or digital means. Analogue is most frequently used, as the alarm temperature can be adjusted electronically.

Fusible links are solder-based and are designed to melt at a predetermined temperature, thus allowing mechanical initiation of protective equipment.

Frangible glass bulbs contain a synthetic liquid having a very low freezing point (below any 'natural' temperatures likely to be experienced) and a high coefficient of expansion. As temperatures rise, due to fire, the bulb shatters and can be used, either directly or with air as the operating medium, to operate waterspray systems.

5.3   Fire risk

The building and the plant it contains will have been designed to minimise fire risk (design memorandum 098/6). Great importance must also be attached to managing the fire risk when the building and plant are operational.

The originally-designed risk is potentially increased by:

● Additions or modifications to the plant and buildings.

● Indifferent behaviour and poor work practices of staff, e.g., poor housekeeping.

● Isolation of fixed fire fighting systems for maintenance purposes.

● Inaccessibility or obstruction of fire protection equipment.

● Poor standard of maintenance on fire protection system.

● Change in operational practice on the plant protected, for example, the changing from coal- to oil-firing, or from base-load to peak-load operation.

Details of *all* fires which occur must be logged in sufficient detail so that trends on a particular site, and between sites, may become apparent and actions initiated to reduce the incidence.

5.4   Fire fighting equipment

5.4.1   Portable plant

Portable fire fighting equipment must be provided under the Fire Precautions Act, which in this sense deals mainly with hand-held fire extinguishers. These are colour-coded, as shown in Fig 6.4. This colour-coding is vitally important, as it directs the user's attention to the type of fire on which it is safe to discharge the extinguisher. Generally, such extinguishers are activated by striking a plunger or pressing a trigger which releases the stored pressure and allows the discharge of the extinguisher's contents.

Such extinguishers are provided in banks of adequate capacity throughout plant areas, and any employee should be able to use them to contain or extinguish any small fire arising.

There is an obvious and practical requirement to report all fires, even small ones, and to make effective arrangements for discharged extinguishers to be recharged and returned to service as soon as possible. Each extinguisher should be uniquely identified and full records kept as to its whereabouts and condition.

A high reliability is required from an extinguisher since an extinguisher that fails to work when the emergency arises is useless.

Typical maintenance requirements are given below, together with periodicity, though it must be appreciated that variations will be required from one type of extinguisher to another, and from one environment to another.

5.4.2   Maintenance of fire extinguishers

Weekly: Check that all extinguishers are in position, accessible and readily usable.

Monthly: Check that all extinguishers are clean, externally undamaged, appropriate for the fire risk involved, securely fixed to the floor or wall and protected against adverse environmental conditions.

Annually: Carry out a thorough examination and maintenance of all aspects of the extinguisher, as follows:

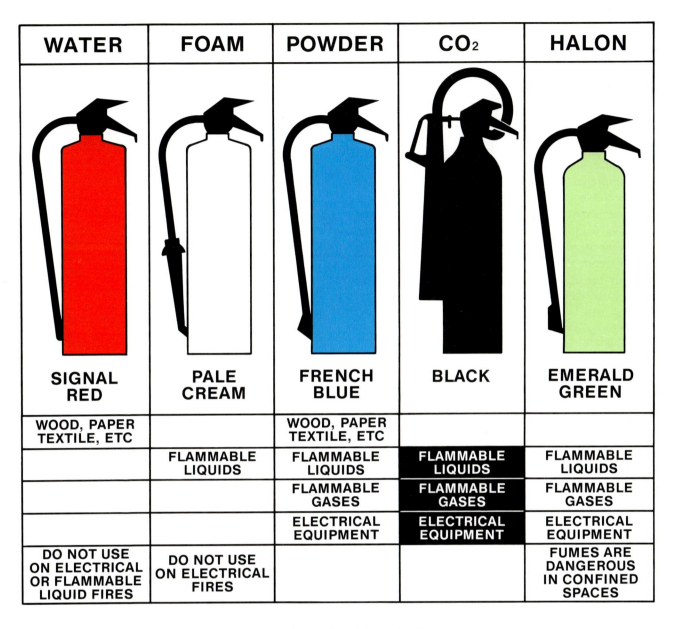

WATER	FOAM	POWDER	CO₂	HALON
SIGNAL RED	**PALE CREAM**	**FRENCH BLUE**	**BLACK**	**EMERALD GREEN**
WOOD, PAPER TEXTILE, ETC		WOOD, PAPER TEXTILE, ETC		
	FLAMMABLE LIQUIDS	FLAMMABLE LIQUIDS	FLAMMABLE LIQUIDS	FLAMMABLE LIQUIDS
		FLAMMABLE GASES	FLAMMABLE GASES	FLAMMABLE GASES
		ELECTRICAL EQUIPMENT	ELECTRICAL EQUIPMENT	ELECTRICAL EQUIPMENT
DO NOT USE ON ELECTRICAL OR FLAMMABLE LIQUID FIRES	DO NOT USE ON ELECTRICAL FIRES			FUMES ARE DANGEROUS IN CONFINED SPACES

FIG. 6.4 Colour coding of fire extinguishers

- Weigh the appliance and compare results with the previous reading.

- Empty the extinguisher and dismantle it in accordance with the correct safety procedures and examine contents, e.g., the contents of powder-filled extinguishers should be dry, free from lumps, and not mixed with any other substances.

- Examine the inside of the extinguisher body and check for the absence of corrosion.

- If it is gas-cartridge operated, weigh the cartridge and check this against the weight marked on it; if there is more than a 10% difference, replace the cartridge.

- Ensure that the vent hole in the cap is clear.

- Examine the hose nozzle and discharge tube. Do not grease or oil the operating mechanism, which should be freely operable. Examine all washers.

- Rebuild the extinguisher, using new components, where necessary.

- Refit safety clip to prevent inadvertent operation.

- Record all data and date of overhaul.

This type of maintenance procedure will give maximum availability, but the magnitude of the work must not be underestimated as a typical 2000 MW coal-fired station may hold around 1000 hand-held extinguishers of various types.

5.4.3 Fixed equipment

This equipment is provided where the fire risk is sufficiently high to warrant the cost of installation. Such equipment will always be supplemented by hand-held portable gear and may in many instances come into operation automatically (see Design Memorandum 098/6 Fire Protection of Power Station Plant).

Equipment provided under this heading will include:

- Hose reels: possibly with automatic action, for use in offices and workshops.

- Hydrant systems: dry or wet, for general use throughout plant areas.

- Sprinkler systems: used in offices, stores, turbine-generator and boiler front areas.

- CO_2 installations: used in enclosed areas, for example, switch rooms and electrical annex (building adjoining turbine hall or boiler house specifically for housing electrical apparatus), cable tunnels and gas-turbine engine cells.

- Halogen agent systems: used in computer suites, cableways and other light current electrical areas, and sometimes in gas-turbine engine cells.

- Mechanical foam installation: used in fuel oil storage protection.

All such equipment requires regular checking and maintenance. A waterspray system is considered below as a typical example:

Daily Check that all air and water pressures are normal, that all control valves are correctly set and that access to all equipment is not obstructed.

Weekly Check water levels in pressure tanks, that each fire pump starts automatically, that diesel fire-pump fuel-oil tanks are full, and log results.

Quarterly Operate all section and isolating valves over full travel. Open test valves on all control valves and check for correct operation. Examine sprinklers and detectors, the former for correct orientation, and examine detectors for signs of obstruction by paint or dirt.

Annually Deluge-test all available plant, e.g., station, generator and unit transformers, and turbine-generators and boilers, as part of the annual overhaul procedure. Check for correct operation and adequate water coverage. It is worth noting that this also gives the opportunity to check that drainage is adequate to carry away the water and affords a ready method of plant cleaning. Some plant, such as cable entries, may need additional protection before deluge-testing is carried out.

5.5 Training

To meet CEGB objectives, all staff should receive basic fire fighting training in compliance with Electricity Supply Industry Training Committee Recomendation number 33 and should know:

(a) The action to be taken in the event of a fire.

(b) How to operate an appropriate fire extinguisher and the limitations of that extinguisher.

(c) How fires are caused, and the importance of good housekeeping.

Additionally, fire team members should receive up to 40 hours' training a year in:

(a) The operation and use of all fire fighting equipment and its specific uses.

(b) Practice at fighting large fires.

(c) The use of breathing apparatus and rescue methods in a smoke-filled environment.

Fire team members should exercise with public authorities on the plant, in accordance with the location fire plan, at least once a year.

The purpose of the CEGB's fire teams is to provide a prompt and effective assault on the fire with the intention of containing it (if not extinguishing it) until the professional public authority Fire Brigade arrives.

5.6 Local and statutory requirements

5.6.1 Fire certificate

This is a statutory document issued by the Fire Authority or by the Health and Safety Executive (see Appendix B). The certificate indicates the use of the premises, the means of escape in case of fire and that the methods used to prevent the spread of fire and fumes are acceptable. Means of fighting and warning of fire are also covered.

Fire certificates are required under the Fire Precautions Act 1971 for conventional power station sites and are issued by the Fire Authority. Nuclear sites are covered by the Fire Certificates (Special Premises) Regulations 1976 and this certificate is issued by the Health and Safety Executive.

It is an offence to use a building covered by the various statutory requirements without a certificate. The issuing authority may require alterations or additions (especially to escape routes) before the certificate is issued. Any substantial changes or additions to buildings covered by a certificate must be notified to the appropriate authority.

6 Environmental hazards

Despite its sophistication (or perhaps because of it), modern industry produces many potential hazards for the people that work in it. All such hazards are regulated in one way or another, as required by Sections 2 and 6 of the Health and Safety at Work Act. Simple hazards, for example, a hole in the floor, can easily be made safe, but noise, traces of harmful gases, dusts and fibres can be much more complex and difficult to deal with.

A possible environmental hazard should be followed-up in a planned manner by asking questions such as:

● Is an undesirable substance likely to be present?

● If so, is it likely to be hazardous?

It may be necessary to organise a survey to identify the persons at risk and the timescale and physical locations when and where the risk is present. This will, if necessary, demonstrate the need for a regular survey programme and define the management and supervisory action required to minimise the risk.

The occupational exposure limits reference document in the United Kingdom is the Health and Safety Executive Guidance Note EH40, which is updated and re-issued from time to time and the reader's attention is drawn to it.

Use of this document requires an appreciation and utilisation of *control limits* and *long and short term exposure limits*. These are defined as follows:

Control limits 'These are limits which have been set, after detailed consideration of the medical evidence, to be reasonably practical for the spectrum of work.' The limits should not normally be exceeded.

Long term and short term exposure limits Substances hazardous to health may cause harmful effects, e.g., irritation of the skin, eyes or lungs after short term exposure, or by long term exposure through accumulation of substances in the body. Two types of exposure limit are listed in the Health and Safety Advisory Literature — Environmental Hygiene Series — Occupational Exposure Limits (EH40), the long term limit is concerned with the total intake over long periods, whilst the short term exposure limit is aimed at avoiding acute effects, or at least reducing the risk of their occurrence. EH40 gives both limits as time-weighted averages which are airborne concentrations averaged over a specified period of time; this is 8 hours for the long term limit and 10 minutes for the short term.

Exposure limits should be used by employers as guidance for the control of exposure to airborne contamination and will be taken into account by inspectors when assessing compliance with the various legal requirements.

The list of substances in the EH40 document contains some 900 or more entries, many of which are not used in any way in the CEGB. Some of these will only appear in laboratories and research sites of the Board, but it is as well to remember that substances are not only met by direct usage but can also be generated as a result of some other process; an example of this is ozone (O_3), which can be generated from electrical arcing and thus be present in precipitator switchrooms where mechanically-driven chopper-type rectifiers are still sometimes used to provide a unidirectional current required for the precipitator electrodes. Another example is the generation of chlorine when PVC (e.g., 'serving' on cables) is overheated.

Regulations have now come into force concerning the 'Control of Substances Hazardous to Health' (COSHH), which have been published with approved codes of practice. Additionally, the 'Introduction of New Substances Regulations 1982' places a duty on manufacturers and suppliers of new substances to declare any dangerous properties which could arise with their products. They are assessed using codes of practice, which give complex testing methods that manufacturers must use to test for potential hazards; chemicals introduced before 1982 are not required to undergo complex testing, and Section 6 of the Health and Safety at Work Act only imposes a general duty on suppliers of such substances to ensure their safety and to pass on information so that appropriate workplace safety measures can be taken.

The COSHH regulations introduce further legal requirements by placing statutory duties on employers when receiving safety information, who are then required to introduce additional occupational hazards procedures appropriate to the risk.

Some 300 substances used by the Board fall within the definitions of the COSHH regulations and potentially include such diverse items as floor cleaning materials, substances used for cleaning electrical contacts and bulk chemicals for water treatment plants.

Apart from environmental hazards caused by chemical substances, attention is also drawn to the hazards arising from dusts and fibres. The problems caused by asbestos are of particular interest within the CEGB and, although the incidence of asbestos is now significantly reduced — because of the reduction in its use and its replacement by other less harmful substances — the Asbestos Regulations still apply. It is worth noting that the Board is not only required to take adequate precautions during the removal of asbestos but special precautions also have to be taken over its disposal and these are covered in The Asbestos Regulations 1969, The Health and Safety at Work Act and The Control of Pollution (Special Waste) Regulations 1980.

Another environmental hazard which is receiving increasing attention nowadays is noise. The main standards of noise control and levels of exposure were established in the 1972 Codes of Practice for Reducing the Exposure of Employed Persons to Noise, in which it was recommended that the maximum level of exposure without ear protection was 90 dB(A) for an 8-hour day. Because of the logarithmic nature of the decibel scale, if 90 dB(A) is the maximum for an 8-hour day, then 93 dB(A) is the maximum for 4 hours, 96 dB(A) for 2 hours and 99 dB(A) for 1 hour.

The CEGB has introduced audiometry services, whereby trained staff carry out tests to detect different hearing levels during the employee's initial medical examinations and also check for loss of hearing capacity where staff have been exposed to hazardous levels of noise. Control measures for reducing noise levels at work are being followed in the design stage for new plant and equipment and, for existing plant, by the placing of baffles and the fitting of acoustic hoods. Noise levels are measured regularly around the workplace and, where noise cannot be reduced at source or by reducing staff working time in high level noise, then ear defenders of an appropriate type are required to be worn. Note that different types of ear defenders protect against different levels of noise and these must not be misused. The insertion of any form of ear plug into the ear should be carried out with care because of the medical implications of ear infection and the possible transmission of dirt into the ear.

6.1 Protective clothing

The introduction of 'so far as is reasonably practical' to reduce risks to employees implies that the risk be tackled at source and the wearing of protective clothing or devices should not be regarded as an automatic solution to a risk problem. However, some risks are inevitable and good sense requires the wearing of appropriate protective clothing, wherever the risk is likely to appear.

The Board makes available to its staff a wide range of protective clothing, with recommendations as to the use and maintenance of such clothing so that it can be effective against the implied risk. Additionally, a wide range of safety equipment is available but great care must be used to ensure that the right equipment is worn to protect against a particular risk. Such safety equipment includes various eye protectors provided under the Protection of Eyes Regulations of 1974, and ear protectors under the code of the 1972 Codes of Practice for Reducing the Exposure of Employed Persons to Noise. Other equipment includes hard hats for the protection of the head and a wide range of gloves for handling chemicals and hot substances, and complete 'hot suits' for working in very hot environments. The Board's coded standard lists include data sheets which give full recommendations on the correct usage and maintenance of all approved safety equipment.

7 Case studies

The CEGB operates to first class safety standards and procedures. They are only, however, as good as the staff who operate them. Reference has already been made to the estimated 50 000 safety documents issued per month in the CEGB and the vast majority of these proceed without incident. On very rare occasions, an incident does occur which merits special attention, especially where personal injury or plant damage has occurred. The Board requires a Panel of Inquiry to investigate such incidents and it is always found that where the Safety Rules are involved the rules are perfectly adequate, and that for some reason a human failure has occurred over the application of such rules. The remit of a Board or Panel of Inquiry set up in accordance with CEGB policy is to establish the facts giving rise to the incident and to make recommendations such that the incident can be avoided in the future.

About 8–10 Panels/year sit on safety related matters throughout the CEGB.

The paragraphs which follow give short accounts of some incidents which have occurred during recent years.

7.1 Major fire in a cable tunnel

This caused the shutdown of the four main turbine-generators, plus gas-turbine plant, at a station. Fortunately, no casualties resulted from the incident but the cost of repairs ran into several million pounds.

After extensive investigation, it was found that the cause of the fire was a leakage of lubricating oil from the turbine hall plant into the cable tunnel, and then onto the cables. Fire was then initiated by an overheating earth connection in the oil-soaked area, from a faulty plant item which was passing earth current.

Apart from detailed technical considerations on the effectiveness of fire fighting measures related to PVC covered cables, the mainpoints arising from this incident were:

- The importance of 'good housekeeping' and the containment and avoidance of oil leakage.

- The need for regular inspections of all plant areas, so that problems can be remedied in good time.

- The need to ensure that maintenance procedures are adequate to check for the integrity of earth circuits.

7.2 Fatal accident in a circulating water pumphouse

An experienced member of the Industrial Staff was employed in running telephone cabling along the wall of a circulating water pumphouse at normal ground level. At the time of the incident, he was working alone on the operating floor, which was of open-mesh construction. There were no witnesses to the accident, the employee being found lying on the floor unconscious some way from the point of work in an open area, with an aluminium ladder and his safety helmet nearby. The employee died several days later from head injuries.

It was subsequently postulated by the Panel of Inquiry that the most likely cause of the accident was that the employee was carrying the ladder on his shoulder to a point further along the proposed cable run when the end rung of the ladder caught on a 415 V crane-isolator handle which was projecting from the wall at head height. This unwitting snagging of the ladder caused him to lose his balance and fall when he hit his head on the floor. The floor was damp at the time and the casualty was wearing soft shoes, with part-worn composition soles.

Points arising from this accident were:

- That there being no medical evidence to give an alternative explanation and because floors in a CW pump house are often wet, it focuses attention on the footwear being used and it is sad to note that the casualty had a pair of new industrial shoes with non-slip soles but was not wearing them.

- That projections from walls at body height are a potential source of danger to individuals, es-

pecially in main walkway areas and that, in this case, easy alternative sites were available for this 415 V crane isolator.

7.3 Grid substation incident

An electrical fitter from a transmission section was severely burnt when he came in contact with live copper-work at 13.8 kV on the top of a transformer, where he was testing the Buchholz relay.

The incident occurred for a number of reasons, all of which arose because the staff involved failed to follow procedures associated with the Safety Rules. These failures can be summarised as follows:

- A mistake in the identification of the transformer which was repeated on the permit for work when issued.

- The position of the transformer in the electrical system was not correctly identified.

- Staff concerned did not clarify the position of the transformer by referring to the operating diagram.

- Working area demarcation was inadequate.

- The permit-for-work contents were not explained fully to the recipient of the safety document at the time of issue.

The inquiry concluded that 'the incident would have been avoided if any one of the staff involved had complied with the relevant sections of the Safety Rules and Local Management Instructions'.

7.4 Incident on 415 V switchgear

An electrical fitter sustained burns to his wrists while working on a 100 A isolator to a turbine hall crane. This isolator was part of 415 V switchgear. The work to be done involved the transposition of two phases of the three-phase connections on the outgoing side of the isolator to reverse the direction of rotation of the crane motor.

The Senior Authorised Person involved had decided (as he was entitled to) under the Safety Rules that the work could be done under his personal supervision. Having completed the changeover of the connections, the fitter was completing the tightening of the cable lugs when the open-ended spanner he was using slipped from his grasp and fell into the live busbar connection chamber below. There was a flashover, from which the fitter received burns to his wrists.

The subsequent inquiry revealed that:

- The Senior Authorised Person and the fitter were aware that the busbar connections were live but were unaware that part of these connections were not insulated.

- Staff concerned were not fully aware of the requirements of the Local Management Instructions on low- and medium-voltage working.

The main recommendation from this incident was that Senior Authorised Persons must have a proper understanding of the work to be carried out and make a full safety assessment of the work to be done.

7.5 Incident in a 400 kV substation

Maintenance work was being carried out on a 400 kV isolator, access to it being from a mobile platform, which could be manipulated by the driver from the platform in three dimensions.

The staff involved were working under a permit-for-work and the area was correctly barriered off with caution and danger notices.

Whilst manœuvring the hoist, and watched by the standby man at ground level, a flashover occurred from an adjacent section of reserve busbar to the platform. Fortunately, no injuries were sustained and the platform earthing arrangements and circuit protection operated correctly. Although various circuits tripped there was no loss of supply to the public.

The subsequent inquiry found the following:

- The section of reserve busbar from which the flashover occurred was isolated at the time the safety document was issued but for operational purposes only at the request of System Operation Department.

- Although this reserve busbar was within the platform safety clearance distance, the operational isolation was not included on the permit for work.

- During the time that the permit for work was in force, the section of the reserve busbar was recharged for operational purposes.

The moral of this incident is that plant made dead for operational purposes must be considered as live for safety purposes.

There are several themes which arise from these and other cases; they are summarised below:

- The need for strict and continuous adherence to the CEGB's Safety Rules and Local Management Instructions.

- Safety instructions must be clear, whether oral or written, and both Senior Authorised Persons and Competent Persons have an obligation to ensure full understanding of the instructions.

- The need for regular retraining of all staff involved to ensure that slipshod habits do not become accepted behaviour.

8 Management roles

Safety, like any other work activity, needs to be managed to high standards if it is to be effective [3].

It quite properly consumes resources and these need to be efficiently utilised. Safety is a difficult field to apply the discipline of cost-benefit control because personal injury, although it can be costed, also has moral and social overtones which cannot be easily costed. Safety considerations involve management in a detailed knowledge of what is happening in the activities of the workforce. At management level, detailed and consistent commitment to safety, which, provided it is genuine, will be reflected and contributed to by the workforce.

Statistics play a vital part in the management of safety. If statistics are to be properly used, however, their limitations must be respected. Sometimes the populations on which statistics are based are small and thus trends may be more useful than absolute results. Additionally, accident results have to be analysed to find out, for example, which occupations are most at risk and which plant areas give rise to the most accidents.

Typical statistics in the CEGB include:

- Accident Frequency Rate (AFR) =

$$\frac{\text{No. of lost-time accidents} \times 100\ 000}{\text{Manhours worked}}$$

- Duration Rate (DR) =

$$\frac{\text{Hours lost}}{\text{No. of accidents}}$$

- Severity Rate (SR) =

$$\frac{\text{Hours lost}}{\text{Hours worked}} \times 100\ 000$$

All these are based on accidents causing absence for more than 3 days. The figure of 100 000 is used in an attempt to make the AFR a measure of how many lost-time accidents an employee is likely to have if he spends his full working life on that site. Typically in the CEGB, these are currently around

$$AFR = 2.3$$
$$DR = 130$$
$$SR = 280$$

More detailed analysis often shows that, for instance, craftsmen (mechanical) and welders are more susceptible to accidents in the boiler house of a coalfired station. The reasons for this may include the type of work undertaken, the working environment, the attitude of mind of the staff, and so on, and can only be considered in detail and remedies found on each specific site.

8.1 Accident reporting

The statutory obligations for the employer are covered by the Health and Safety at Work Act — Notification of Accidents and Dangerous Occurrences Regulations 1980.

The Regulations require a telephone report to the enforcing authority, e.g., Health and Safety Executive Inspector, for any accident causing death or major injury (which is defined in the regulations) to an employee or to a member of the public or self-employed person in the employer's workplace. Certain defined dangerous occurrences are also reportable and include the failure of lifting gear and pressure vessels.

The telephone call must also be followed up within 7 days by the transmission of a completed Form 2508 — Report of an Accident and/or Dangerous Occurrence and Injuries Sustained. The notification may well be followed up by a site visit from the Inspector, who has right of access to the site and to related documents at any time.

8.1.1 Successful accident-reporting

The drawing-up of graphs to discover the trends and the actions necessary to prevent re-occurrence, all stem from an adequate system of reporting accidents at source. With large workforces this is considerably eased if the reporting data is computerised.

It is foolish to attempt to cover up an accident and such action may indeed break the law. In any event, only good can come from efforts made to understand the accident and to prevent it from occurring again.

8.2 CEGB safety organisations

These are covered generally in the introduction to this chapter. This is now expanded to consider the organisation on a particular site, where the location manager has overall responsibility and is accountable for the safety performance within his site. In the CEGB, he operates his safety policy in a consultative manner, whilst controlling the direction and development of that policy.

Chairmanship of the Safety Committee alternates between management and a representative of the staff. This rotation takes place yearly and it is im-

portant that irrespective of who occupies this chair the committee contains a member of management who has executive authorty to seek improvements in the safety front. Such a committee will generally meet every other month (or more often, if necessary) and is composed of staff representatives from the Local Joint Consultative Council (LJCC) and union-appointed Safety Representatives, as well as management nominees.

They consider site accidents and 'near misses', accident statistics and analysis of safety walks and plant area inspections. The committee makes recommendations to management over the resolution of specific problems. Expert staff from CEGB Headquarters and the Electricity Council can be called in to assist.

Such a committee exists in its own right and also considers health and welfare matters. It does not report to the LJCC or to the Regional Health and Safety Committee, but takes note of the actions and statistics of both bodies and keeps both informed of its own activities.

The consultative framework is also nationally-based, in that the Health and Safety Committee (HESAC), comprising Board and National Trade Union Officers (with medical and safety experts in attendance), advises the RJCC HESAC at Regional level. This latter group, comprising Regional Managers and District Trade Union Officers and Safety Officers, then advises the local committees. It is possible for a local committee member to be a member of a RJCC HESAC and thus provide an important local input to its deliberations.

Thus the location manager, operating within the Board's policy and statutory requirements, has considerable assistance available in managing the safety effort on his site. He must demonstrate his own continuing commitment to safety matters and, in so doing, carry his staff along with him and by audit and review ensure that his requirements are met.

It is noticeable that a site with a disgruntled workforce will often have high sickness levels and a high accident rate. In a well managed site, safety being one of the parameters by which the success of the site in meeting its targets is measured, the auditor will expect to find a low AFR (less than 2.0 annually) a well motivated workforce at all levels, adherence to local and safety requirements backed up by a management control which highlights any failure to do so.

9 References

[1] Requirements of Fire Precautions Act of 1971: HMSO

[2] Guide to Fire Precautions Act of 1971: HMSO

[3] Health and Safety Executive Publication OP3: 'Managing Safety' — a review of the role of management in occupational health and safety by the Accident Prevention Advisory Unit of HM Factory Inspectorate: 1981.

10 Bibliography

British Standard 5306: Fire Extinguishing Installations, Parts 1 – 5

British Standard 5423: Portable Fire Extinguishers

CEGB Design Memorandum 098/6: Fire Protection of Power Station Plant

CEGB Reference Document on Fire Precautions: CEGB 098/33 (2nd Edition)

CEGB Health and Safety at Work Act: General Policy Statement: Ref. SEC 00701305M.1: 1974

Electricity Supply Industry Training Committee Recommendation No. 33

Notification of Accidents and Dangerous Occurrences Regulations: 1980

Redgrave: Health and Safety in Factories, ISBN 406 35307 7: 'A complete Compendium of the Law Relating to Factories'.

Robens: Health and Safety at Work: 1972

TUC Handbook Health and Safety at Work: Trades Union Congress, London: 1978

TUC Handbook on Noise at Work

Appendix A
Safety and work control

Stage in process		Action required	Possible means of achievement
Initiation of work request	1	To identify 'safety' if it constitutes a main reason for rectifying a defect, carrying out a modification or doing new work.	By the inclusion of 'safety' as a category indicated by the originator. *Note:* Local Instructions for rectifying 'safety' defects are to be followed.
	2	To identify specific safety precautions, if known and thought to be necessary by the originator.	By entry of particulars on the Work Request.
		To identify plant/apparatus to be worked on.	
Routine maintenance, Inspection and testing of safety features	3	Confirmation that the inspection, testing and overhauling of safety and protective equipment is adequately covered and that the monitoring system is such that these are not omitted from work programmes as they become due.	By surveys of plant area, to ensure that existing equipment is adequately catered for and to identify possible additional safety/protective needs and, by the use of preprinted schedules, specifying the frequency for this particular type of work on safety and protective equipment.
Appraisal of a work request	4	Confirmation that:	
		(a) Reference to 'safety' in the classification of the job is correct.	By an appropriate person appraising the work request, followed by confirmation if any doubt exists.
		(b) Necessary references to general safety precautions have been made.	This to be part of the normal appraisal during the preparation of Work Order Card(s) from requests received.
Preparation and specification of work (e.g., creation of a Work Order Card, Standard Work Order Card, Work Instruction, Pre-assessed work routine, etc.)	5	Specification of safe method, correct tools and equipment, with details of SWL, etc. included as appropriate. Safety hazards/precautions highlighted. Correct identification of the work to be done.	By an appropriate person confirming the safety content of the draft specification of work prior to Departmental approval. Reappraisal of existing specifications/instructions with periodic audit by representatives of station management.
		To provide a general statement on the need to take precautions, recognising the fact that the person doing the work has a responsibility to ensure the safety of himself and others.	By including a preprinted statement on the relevant specification of work.

Appendix A

Safety and work control (cont.)

Stage in process		Action required	Possible means of achievement
Application of safety rules	6	To identify those safety rules which apply to the work. Local Instruction to be followed.	By the provision of a Local Instruction/Procedure.
		If the work is on plant/apparatus connected to the system and the Electrical and Mechanical Safety Rules apply, to state an assessment is required by an appropriate Senior Authorised Person. Specify collection, if appropriate, of a Safety Document.	
		If the work is not on plant/apparatus connected to the system this work to be confirmed by an appropriate person.	
Matching of correct Work Specification, Work Instruction or Standard Work Order Card, Pre-assessed Work Routines, etc., to the job to be done.	7	To ensure that the descriptive information is matched to the work requested.	By use of a clear, comprehensive, easily used index of the prewritten work specifications, etc.
			By providing clear information at the initiation of Work Request stage.
			By carrying out in situ job investigation where required.
			Before completing the approval stage an appropriate person should check it is correct for the work required.
Work Order Card being scheduled for issue	8	On scheduling the Work Order Card for issue the work will be confirmed as either:	
		(a) Work which is not connected to the system.	By confirmation by an appropriate person making reference to the defined system in the Local Instruction.
		(b) Work or testing to be done on plant/apparatus connected to the system and subject to the Electrical and Mechanical Safety Rules. In this case confirmation by an appropriate Senior Authorised Person on how safety from the system shall be achieved and whether a safety document is required. This requirement for confirmation shall be covered by a Local Instruction.	All requests to work on or test plant and/or apparatus connected to the system shall be submitted to an appropriate Senior Authorised Person (SAP) for assessment as to the need for, and the type of Safety Document required to achieve safety from the system.

Appendix A

Safety and work control (cont.)

Stage in process		Action required	Possible means of achievement
Prior to the commencement of job	9	For that work connected to the system, confirmation that the necessary safety document exists or how safety from the system can be achieved.	Local Instruction procedure to be followed by foremen and men who are issued with Work Order Cards or other forms of work instruction.
		For all work:	
		Confirmation that working conditions are safe.	By the use of correct terminology, plant labelling, careful work initiation and job investigation.
		To ensure work is done on the correct item.	
After completion of job	10	Clear the Safety Document if applicable. If appropriate, identify further safety documents and further work (see Action 13).	Local Instruction procedures to be followed.
	11	Correction of errors in any printed specification of work that was identified in the course of a job.	A site procedure to be set up whereby the specification of the work is identified for correction, and an appropriate person agrees the corrections.
After completion of job	12	It may be necessary in certain circumstances to identify to the Health and Safety Executive what work has been carried out on any particular item of plant.	Work Order Cards, Defect Cards, Credit Log Sheets, Work Specification Sheets and Safety Documents to be retained for five years or their contents stored in a retrieval system such as microfilm or computer files.
Jobs that are incomplete	13	Assessment by an appropriate person of the reasons for the job being left incomplete and action taken to ensure completion, i.e., ordering spares, raising additional work, etc. Confirmation by a Senior Authorised Person of safety document requirements for any further work on Plant/apparatus.	Work Order Cards to be scrutinised for written comments where appropriate by line foremen, works office staff and the engineers in that sequence. A procedure to be set up to ensure that action is taken to complete this work — and that a system for monitoring these actions is established.

Appendix A

Safety and work control (cont.)

Stage in process		Action required	Possible means of achievement
Maintenance of the Work Instruction, Specification system (e.g., the maintenance of a Work Order Card, Standard Work Order Card, Work Specification, Work Instruction, Pre-assessed Work Routines, etc.	14	To prevent the specification of work being amended without the same amendments being made to all other working and master copies in existence.	By limiting access to files of master copies and having a procedure defined for the processing of amendments, identifying persons authorised to make amendments and the means of achievement.

New Work Order Cards, etc., can be created by editing existing instructions and subject to confirmation by an appropriate person (as in Actions 5 and 8). |

Appendix B

Fire certificate under the Fire Precautions Act 1971

Address of the premises with respect)
to which this certificate is issued:)

Occupier of the premises:)
The individual occupiers of the premises are listed in)
the 'Occupiers Appendix' attached to this certificate)

Name and address of owner(s) of the premises:)

Address of the relevant building (i.e., the building)
containing the premises):)

Notified person in relation to the premises:)

The use or uses of the premises covered by this)
certificate:)

1 IT IS HEREBY CERTIFIED that:

(a) the premises described above, being premises

+ put to the use or uses designated by Order under Section 1 of the Fire Precautions Act 1971 specified above

+ used as a dwelling for which a notice under section 3 of the Fire Precautions Act 1971 is in force
   are provided with the MEANS OF ESCAPE IN CASE OF FIRE specified on the plan(s); and that

(b) the relevant building described above is provided with the MEANS (other than means for fighting fire) FOR
   SECURING THAT THE MEANS OF ESCAPE with which the premises are provided CAN BE SAFELY AND
   EFFECTIVELY USED AT ALL MATERIAL TIMES specified on the plan(s) and in Schedule 1: and that

(c) the relevant building described above is provided with the MEANS FOR FIGHTING FIRE (whether in the
   premises or affecting the means of escape) for use in case of fire by persons in the building, and with the
   MEANS FOR GIVING to persons in the premises WARNING IN CASE OF FIRE specified on the plan(s) and
   in Schedule 1: and that

(+d) the location and quantities of explosive or highly flammable material stored or used in or under the premises are
   as specified in the 'Explosive and Highly Flammable Materials Appendix' attached to this certificate.

The REQUIREMENTS IN Schedule 2 are HEREBY IMPOSED.

The plan(s), plan key and schedules attached hereto all form part of this certificate.

+ Delete as necessary

.... Signed

on behalf of, and duly authorised by,
County Council of
Fire Authority for the area in which
Date ....   the premises are situated.

CHAPTER 7

Plant performance and performance monitoring

Contents

1   Introduction

Power station running costs can be grouped into the three categories:

- Fuel.

- Materials, goods and services.

- Salaries and wages.

Figure 7.1 shows a pie-chart of the split for a medium-sized two-shifting station. Clearly, fuel is the predominant expense, accounting for about 85% of the total. For a base-load station, the proportion is even higher.

Less than half the heat in the fuel is converted to electricity and the loss to the condenser accounts for more heat than does the electrical output. The Sankey diagram (Fig 7.2), illustrates the proportions of the losses for an efficient 500 MW unit. Ideally, the losses should be kept at their optimum values, but this is not always possible in practice. Almost inevitably, plant losses are greater than the desirable minimum, so incurring extra costs. For example, if a 500 MW unit is operated continuously at an efficiency of half a percentage point worse than optimum, its extra fuel bill (assuming £2/GJ) will be about £20 000 per week. Sustained for a year, this amounts to £1 million, so it is worth going to considerable trouble and some expense to reduce the losses to the practical minimum. All staff concerned with the management, operation and maintenance of power plant have a duty to ensure that, as far as possible, generation is achieved at the lowest cost

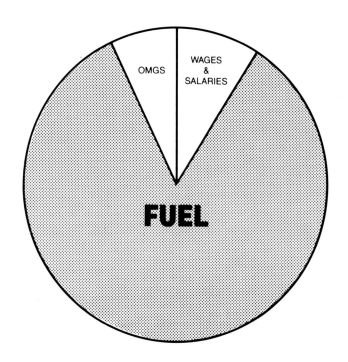

FIG. 7.1 Relative costs
Included with Other Materials, Goods and Services
(OMGS) is the cost of Major Jobs. A major job is defined
as expensive work which is not of a recurring nature.
The total cost of running this medium-size two-shift
station in 1985/86 was about £100 million. Notice
the overwhelming contribution to total cost made by
the fuel bill — over £80 million.

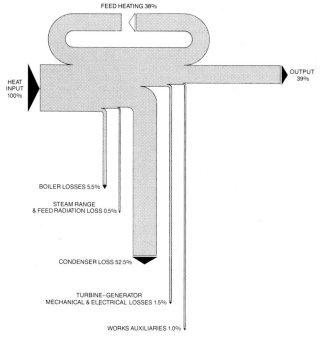

FIG. 7.2 Sankey diagram of heat flow for a 500 MW unit
Less than half the total input heat appears as useful
output. Also, more heat is dissipated in the condenser
than is used as output.

2 Ideal and practical steam cycles

2.1 Introduction

Before considering practical plant it is desirable to
study some ideal cycles, i.e. where all the components
work perfectly. For example, when steam does work
in such a cycle, it is assumed that its expansion will
be adiabatic and frictionless, so will take place at
constant entropy; there is no friction or radiation
loss from components; heat exchange is perfect, and
so on. Clearly, ideal cycles are not found in real
power plant. Nevertheless, the concept is very important as it provides a basis on which to calculate
the highest efficiency that can be achieved for particular steam conditions.

The Temperature-Entropy (T-S) diagram is particularly useful in the study of ideal cycles. The vertical
axis is scaled in temperature and the horizontal one
in specific entropy. (Fig 7.3). Thus, if boiling water
at a is heated along the boiling-water line to b,
the mean temperature of heat addition is 456.6 K
(183.5°C) and the heat per kilogram is given by 456.6
(3.7471 − 0.3530) = 1550 kJ/kg. The concept of an
area on a T-S diagram representing heat is most important and will frequently be referred to. Two
ideal cycles of particular interest are those due to
Sadi Carnot and to W. J. M. Rankine.

during the life of the station. At all CEGB power-
stations, there are some engineers whose full-time
job it is to monitor the performance of the plant,
to make recommendations for improvements and
to keep the station management fully aware of undesirable trends. In addition, each Region has specialist performance teams who provide services to
the stations as required.

Each month, station returns are submitted to Regional Headquarters detailing the efficiency achieved
and the magnitude of the losses sustained. A particularly important return is the Station Thermal
Efficiency Performance (STEP) Factor which is the
ratio of the target and actual heat consumptions. Its
derivation and use is explained later, plus a review
of the thermal performance of the main plant in a
power station.

Note that where calculations are carried out in
this chapter undue accuracy has been avoided. For
example, the absolute temperature equivalent to 0°C
is 273.15 kelvin, but the decimal places are sometimes omitted in the text because this degree of
accuracy is often spurious in practical work. Similar
reasoning is applied to other parameters such as
heat, entropy, etc.

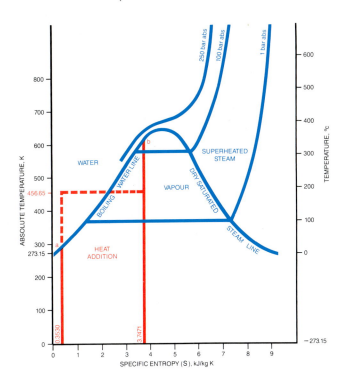

FIG. 7.3 Temperature-specific entropy (T-S) diagram
The absolute temperature scale starts from zero,
so 0° Celsius corresponds to 273.15 K. Also the specific
entropy scale starts from zero. Heat on such a diagram
is represented by an area, so it follows that the heat
content of water at 0°C is regarded as zero. At any
other condition the heat content is given by the product
of the mean absolute temperature of heat addition and
the change of specific entropy.

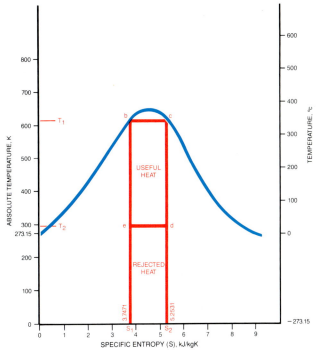

FIG. 7.4 The Carnot cycle
Steam at *b* is heated isothermally to *c*, after which
it is expanded isentropically in a heat engine to *d*. It is
then passed to a condenser where partial condensation
to *e* takes place, followed by isentropic compression to
b, thus completing the cycle. The heat given to the
cycle is equal to $T_1 (S_2 - S_1)$ and that rejected
is equal to $T_2 (S_2 - S_1)$.

2.2 The Carnot cycle

Carnot (1796–1832) developed profound abstract ideas
about thermodynamics. In particular, he conceived the
idea of a perfectly-insulated frictionless heat engine.
All of its heat is received at an upper temperature T_1
and rejected at a constant temperature T_2. Furthermore, the working fluid is assumed to do no other
work than to move the crankshaft of the engine —
no motion is given to anything else, not even to the
fluid particles. Consider its application to a steam
cycle where the expansion and compression of the
steam is isentropic (at constant entropy) and the heat
acceptance and rejection is isothermal (at constant
temperature). Such a cycle is reversible as, by reversing the sequence of events, it is possible to transfer
all the heat used back to the original conditions, as
shown on Fig 7.4.

The cycle efficiency is given by useful heat/total
heat but, useful heat = (total heat − rejected heat),
so:

$$\text{Cycle efficiency} = (\text{total heat} - \text{rejected heat})/\text{total heat}$$

$$= [T_1(S_2 - S_1) - T_2(S_2 - S_1)]/[T_1(S_1 - S_2)]$$
$$= (T_1 - T_2)/T_1$$

Note that the Carnot cycle efficiency is independent of the nature of the working substance, it
only depends upon its upper and lower absolute
temperatures.

For example, consider a machine which operates
upon the cycle illustrated in Fig 7.4. It is supplied
with dry saturated steam at 160 bar abs. and rejects
it at 29.8 mbar. From steam tables, the upper
temperature is found to be 620 K (347°C) and the
lower one 297 K (24°C). Thus, the Carnot cycle
efficiency is:

$$(620 - 297)/620 = 0.521 \text{ or } 52.1\%$$

2.3 The basic Rankine cycle

The Carnot cycle is very useful as a means of determining the ultimate performance of plant working

between particular heat input and rejection temperatures. However, the theoretical cycle upon which actual steam plant is based is that devised by Rankine. Reference to Fig 7.5 shows that the condensation of the steam is continued to completion rather than the impractical partial condensation assumed in the Carnot cycle. The work done in the machine is equal to the change of heat from c to d, the same as in the Carnot cycle, and it can be denoted as $H_c - H_d$ and is represented by the area $abcd$. The quantity of heat rejected is calculated from the heat rejection temperature and the change of entropy, and is equal to the heat at d minus the heat at a. Therefore, rejected heat is given by $(H_d - h_a)$, where H represents the total heat of steam and h the total heat of water.

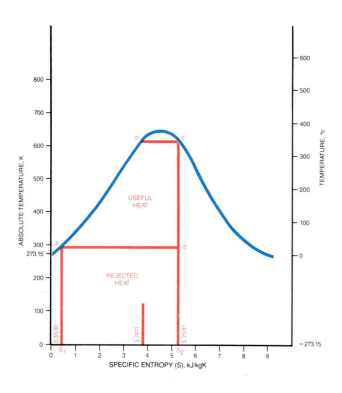

FIG. 7.5 The basic Rankine cycle
Boiling water is heated from a to b such that its state point during the change is represented by the boiling water line between the two points. Isothermal evaporation from b to c causes the steam to become dry saturated. This is followed by the frictionless adiabatic expansion shown by cd. The steam is then condensed isothermally to a.

The Rankine efficiency

$$= \text{(useful heat)/(total heat supplied)}$$

$$= \text{(total heat} - \text{rejected heat)/total heat}$$

$$= \frac{(H_c - h_a) - (H_d - h_a)}{H_c - h_a} = \frac{H_c - H_d}{H_c - h_a}$$

For example consider the conditions quoted earlier for the Carnot cycle where the input pressure was 160 bar abs. and the heat was rejected at 29.8 mbar. From Fig 7.3, the heat required to change the water state from a to b is 1550 kJ/kg. The extra heat to evaporate it is given by:

$$620 \ (5.2531 \ - \ 3.7471) \ = \ 934 \text{ kJ/kg}$$

So the total heat H_c $= \ 1550 \ + \ 934$

$= \ 2484 \text{ kJ/kg}$

The rejected heat $= \ 297 \ (5.2531 \ - \ 0.3530)$

$= \ 1455 \text{ kJ/kg}$

So the basic Rankine cycle efficiency is:

$$(2484 \ - \ 1455)/2284 \ = \ 0.414 \text{ or } 41.4\%$$

Disappointing as this result may seem, it is still significantly better than the best actual plant can achieve when operating with the stated conditions. Therefore, it is necessary to modify the basic cycle to achieve a higher efficiency. There are several ways of doing this, the first and most obvious being *superheating*.

2.4 The Rankine cycle with superheating

The modified cycle is shown in Fig 7.6. It is the same as Fig 7.5 with the addition of superheat from c to e, where e is at 570°C. From steam tables, the total heat at e is 3492 kJ/kg and at a is 101 kJ/kg.

So the total heat $= \ 3492 \ - \ 101 \ = \ 3391 \text{ kJ/kg}$

The rejected heat $= \ 297 \ (6.5463 \ - \ 0.3530)$

$= \ 1839 \text{ kJ/kg}$

Cycle efficiency $= \ (3391 \ - \ 1839)/3391$

$= \ 0.457 \text{ or } 45.7\%$

This is a significant improvement on the basic Rankine cycle. A further modification, which gives even better results, is to add *reheating*.

2.5 The Rankine cycle with reheating

From a study of Fig 7.6 it will be seen that, for a given upper steam temperature, increasing the steam pressure causes increasing wetness of the exhaust steam. Therefore, to obtain the benefits of higher steam pressure without undue exhaust wetness, it is necessary to *reheat* the steam, i.e., after partial expansion, it is reheated in the boiler. The cycle is

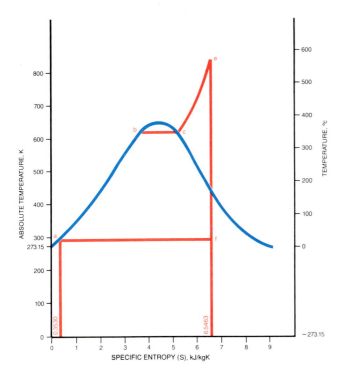

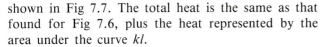

FIG. 7.6 Rankine cycle with superheat
As Fig 7.5, with the addition of superheating at constant
pressure from *c* to *e*. The quantity of extra heat is
represented by the area under the curve *ce*.

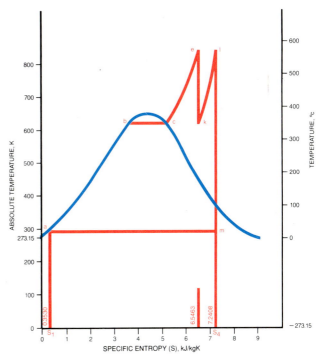

FIG. 7.7 Rankine cycle with reheat
The total heat is represented by the area S_1,
a b c e k l S_4. The rejected heat is represented
by *m a* S_1 S_4.

shown in Fig 7.7. The total heat is the same as that found for Fig 7.6, plus the heat represented by the area under the curve *kl*.

For example, suppose the expansion of the steam from inlet conditions of 160 bar abs. and 570°C continues until the pressure is 44 bar abs., at which time reheating to 570°C. takes place. Then the reheat will amount to $H_1 - H_k$. Reference to steam tables shows that this is 3600 − 3093 = 507 kJ/kg.

The heat at *k* is located in the steam tables by reference to its pressure (44 bar) and the entropy of the superheated steam (6.5463 kJ/kgK).

So the total heat	= 3391 + 507 = 3998 kJ/kg
The rejected heat	= 297 (7.2408 − 0.3530)
	= 2046 kJ/kg
The cycle efficiency	= (3898 − 2046)/3898
	= 0.475 or 47.5%

This is an improvement on the superheat cycle. Even better results can be obtained by introducing feedwater heating, usually known simply as *feedheating*. The basic advantage gained is that some steam is bled from the turbine (after doing some work) to the feedheaters where its heat is surrendered to the feedwater, thus relieving the boiler of a comparable amount

of heat exchange. Without bleeding, the steam would have continued its expansion in the turbine and then surrendered a considerable proportion of its heat to the condenser cooling water.

2.6 The Rankine cycle with superheating and feedwater heating

The cycle is illustrated in Fig 7.8. The steam, having expanded isentropically to point *g*, is then used to raise the temperature of the condensate. The result, in the example, is that the final feed temperature is raised from 24°C to 180°C. On a diagram, such as Fig 7.8 it is confusing to have both the heat surrendered by the steam AND the heat received by the water shown, as only one or the other should be considered. Consequently for calculation purposes, a modified diagram (Fig 7.9) may be more useful.

In the example, the steam at *e* is at 160 bar abs. 570°C. The condensation temperature is 24°C and the final feedwater temperature is 180°C. The heat supplied is equal to the heat at *e* minus that at *i*; from steam tables this is 3492 − 763 = 2729 kJ/kg and the heat rejected is 297 (6.5463 − 2.1393) = 1310 kJ/kg.

So the cycle efficiency = (2729 − 1310)/2729

= 0.520 or 52.0%

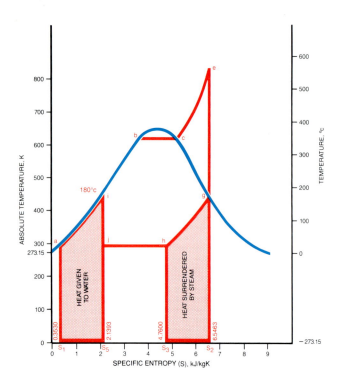

FIG. 7.8 The Rankine cycle with superheat and feedheating
Expansion of the steam takes place from *e* to *g*, at which
point feedheating commences. The heat represented by the
area under *g h* is passed, via perfect heat exchangers, to
the water side, so raising the temperature from *a* to *i*.
The lines *g h* and *i a* are parallel, so the area S_1,
a i S_5 is equal to S_3, *h g* S_2.

FIG. 7.9 The feedheating cycle used for calculations
For calculation purposes, the easiest form of the
feed-heating diagram to use is derived from Fig 7.8.
Total heat is represented by area *i b c e* S_2, S_5.
The rejected heat is represented by the area *f j* S_5, S_2.
Clearly the effect of the feedheating has been to
reduce the 'spread' of the entropy, thus increasing the
average temperature of heat addition.

2.7 The Rankine cycle with reheating and feedwater heating

Consider the previous reheating cycle, but with the addition of feed-heating to (say) 180°C, as shown in Fig 7.10. The total heat added is equal to that of the reheat cycle (Fig 7.7) minus that given to the condensate as a result of feedheating, i.e., the area S_1 *a i* S_5 which is determined from steam tables as the total heat at *i*, minus that at *a*.

Therefore heat from feedheating = 763 − 101

= 662 kJ/kg

So total heat supplied = 3898 − 662

= 3236 kJ/kg

The heat rejected = 297 (7.2408 −

2.1393)

= 1515 kJ/kg

and the cycle efficiency = (3236 − 1515)/

3236

= 0.532 or 53.2%

2.8 Improvements in Rankine cycle efficiency

It will be apparent that each modification to the basic cycle has resulted in improved efficiency and the results are shown in Table 7.1.

So, for the conditions stated the highest possible efficiency here would be 53.2%, if everything were perfect. The corresponding Carnot efficiency for an inlet temperature of 570°C (843 K) and rejection temperature of 24°C (297 K) is (843 − 297)/843 = 0.648 or 64.8%. Therefore even the best Rankine efficiency is considerably lower than that of the corresponding Carnot cycle.

One available way left to push up the Rankine efficiency a little more, whilst still retaining the inlet and rejection temperatures, is to increase the final feed temperature to (say) 250°C.

The method of calculation is now well established, so there is no necessity to explain Fig 7.11.

Total heat	= 3898 − 985 = 2913 kJ/kg
Heat rejected	= 297 (7.2408 − 2.7935)
	= 1321 kJ/kg
Cycle efficiency	= (2913 − 1321)/2913
	= 0.546 or 54.6%

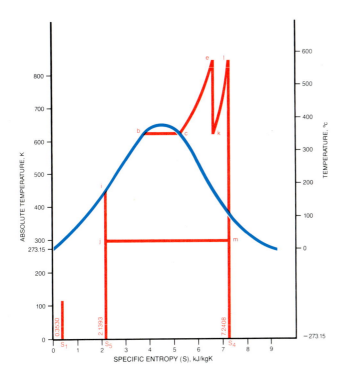

FIG. 7.10 The Rankine cycle with reheating
and feedheating
The total heat is represented by the area S_5, i b c e k l,
S_4. As with Fig 7.9, the entropy 'spread' has been
significantly reduced and so the average temperature of
heat addition has been raised.

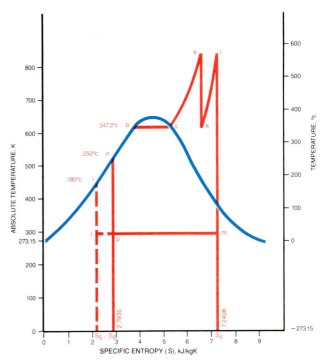

FIG. 7.11 Improved efficiency from increased final
feed temperature
By increasing the final feed from i to n, the heat
represented by the area j p S_6 S_5 is used, whereas it
would formerly have been lost to the CW system.

TABLE 7.1

Efficiency of various ideal cycles

Basic Rankine cycle	41.4%
Cycle with superheat	45.8%
Cycle with reheat	47.5%
Cycle with superheat and feedheating	52.0%
Cycle with reheating and feedheating	53.2%

This is a typical value for modern plant in the United Kingdom. The reason for the improvement from the previous reheating and feed-heating example (where the final feed temperature was 180°C) is shown in Figure 7.11. By raising the final feed temperature from 180°C to 250°C, the total heat to be supplied has been reduced by the amount represented by the area S_5inS_6 and almost all of it (area S_5jpS_6) would previously have been rejected heat.

The remaining area $jinp$ represents useful heat that is now lost from the cycle. So, increasing the feed temperature has caused the cycle efficiency to be improved from 53.2% to 54.6%. If the final feed temperature were raised to point 'b' 347.3°C, the

maximum possible in Fig 7.11, the efficiency would be increased to 55.8%. The trouble is, though, that as the final feed temperature is raised the useful heat is reduced. In the three examples considered:

Useful heat at 180°C final feed	=	1719.9 kJ/kg
Useful heat at 250°C final feed	=	1591.6 kJ/kg
Useful heat at 347.3°C final feed	=	1310.3 kJ/kg,

so the effect of increasing the feed temperature is to give increased efficiency but reduced useful heat.

2.9 The supercritical pressure cycle

Only two units which operate at supercritical pressure have been built for the CEGB, both installed at Drakelow *C* power station, with the conditions shown in Table 7.2. The corresponding ideal cycle is shown in Fig 7.12. The total heat supplied is equal to that required to produce the turbine stop valve (TSV) steam conditions, plus that for reheating.

457

TABLE 7.2

Supercritical pressure unit at Drakelow C power station

Output	375 MW
TSV pressure	242.3 bar abs
TSV temperature	593°C
Reheater pressure	50 bar abs
Reheater temperature	566°C
Back pressure	33.6 mbar
Saturation temperature	26°C
Final feed temperature	270°C

So total heat	$= 2294 + 622 = 2916$ kJ/kg
Rejected heat	$= 299 (7.1659 - 2.9260)$
	$= 1268$ kJ/kg
So, cycle efficiency	$= (2916 - 1268)/2916$
	$= 0.565$ or 56.5%

2.10 The 'equivalent' Carnot efficiency

All of the Rankine cycles considered, except the supercritical pressure cycle, have a common temperature (24°C) of heat rejection. Therefore, because the basic Rankine cycle efficiency is improved by the various modifications (superheat, reheat and feedheating), the average temperature of heat addition must increase. This follows from the theory of the Carnot cycle.

For example, consider the modified Rankine cycle which incorporates superheating, illustrated in Fig 7.6. The total heat amounts to 3391 kJ/kg. But heat is the product of temperature and entropy, so the average temperature is (heat added)/change of entropy. In the example, the change of specific entropy is 6.19 kJ/kg K. Hence, the average temperature is $3391/6.19 = 548$ K. This is illustrated in Fig 7.13.

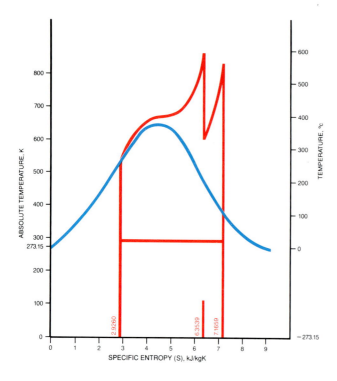

FIG. 7.12 The Rankine cycle for supercritical pressure plant
The conditions are 240 bar abs./590°C/570°C with final feed 270°C and back pressure 33.6 bar. These conditions are quite close to those employed in the only supercritical pressure plant in use in the CEGB.

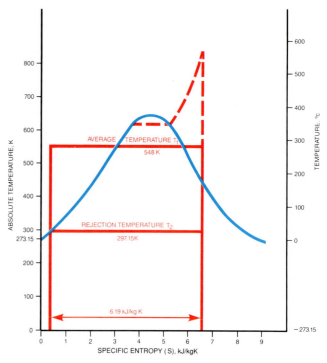

FIG. 7.13 Average temperature of heat addition
The average temperature of heat addition for the superheat cycle is 548 K. Increasing the average temperature for any of the cycles results in improved thermal efficiency.

Heat for TSV conditions	= (heat in TSV steam) −
	(heat in final feedwater)
	= 3467 − 1182 (from
	steam tables)
	= 2294 kJ/kg
Heat for reheat	= 3586 − 2964 (from
	steam tables)
	= 622 kJ/kg

In a similar manner, the average temperature of heat addition was found for each of the examples: the results are presented in Table 7.3.

TABLE 7.3
Average temperature of heat addition

Rankine cycle	Efficiency %	Average temperature K
Basic	41.4	507
Superheat	45.8	548
Reheat	47.5	566
Superheat and feedheating	52.0	619
Reheat and feedheating	53.2	634
Supercritical pressure	56.5	688

The 'equivalent' Carnot cycle efficiency, using the average temperature of heat addition as T_1, will give the same result as the Rankine efficiency. For example, consider the supercritical pressure cycle. The average temperature of heat addition is 688 K and the rejection temperature is 299 K.

The equivalent Carnot efficiency is

$$(T_1 - T_2)/T_1 = (688 - 299)/688$$
$$= 0.565 \text{ or } 56.5\%,$$

which is the same as the value determined previously in Section 2.9.

2.11 The Enthalpy-Entropy diagram

Enthalpy is another name for total heat. Although the Temperature-Entropy diagram is excellent for studying ideal cycles, it is not so good for use with practical plant. A much better diagram for this purpose was devised by Dr Mollier, having specific enthalpy for the vertical axis and specific entropy for the abscissa. A skeleton diagram is shown in Fig 7.14. Only the coloured part of the diagram is normally used.

Consider some steam expansions with reference to Fig 7.15. A perfect expansion from P_1 T_1 is isentropic, represented by a vertical line which terminates at the exhaust pressure P_2 as shown. Thus, if P_1 is 160 bar abs. and T_1 is 843 K (570°C), the specific enthalpy will be 3492 kJ/kg and the specific entropy 6.5463 kJ/kg K. If P_2 is at 44 bar abs., then its specific enthalpy will be 3095 kJ/kg.

Therefore, the maximum heat drop possible between those conditions is 3492 − 3095 = 397 kJ/kg, and this can only be achieved if the efficiency of the expansion is 100%. At lower efficiencies the heat drop achieved will be correspondingly reduced. For example, if the efficiency is 75%, the actual heat drop will be 0.75 × 397 = 298 kJ/kg, even though the exhaust pressure is still P_2. Clearly, to conform to these requirements, the expansion line must slope to the right as shown. Similarly, efficiencies of 50% and 25% are indicated. The lower the efficiency the more closely the line approaches the horizontal; at zero efficiency it is horizontal. For example, the pressure drop across the throttle valves of a throttle-governed turbine or the drop across the labyrinth shaft glands are zero-efficiency expansions, as the total heat is the same before and after the expansion takes place.

Much of the discussion can be conveniently illustrated by reference to the steam conditions from TSV to exhaust in a non-reheat turbine Fig 7.16. The pressure and temperature of the steam at the turbine stop valve (TSV) is shown at *a*. At the outlet from the throttle valves, the steam will be represented by *b*.

Even at full-load there will be some pressure-drop, but the large pressure-drop illustrated from *a* to *b* is associated with part-load operation.

At *b*, the steam is admitted to the HP cylinder where it expands at 85% efficiency and exhausts at *c*. There is a pressure drop due to friction and radiation along the HP/IP loop pipes to *d* before expansion takes place at 90% efficiency in the IP cylinder to *e*. This is followed by a pressure drop along the IP/LP crossover pipes to *f*, after which expansion takes place in the LP cylinder at an average efficiency of 80%. However, that part of the expansion above the saturation line is higher than 80% efficient, whilst that below becomes progressively worse.

The overall efficiency from inlet to outlet of the machine is given by the quotient of the heat drop from *b* to *g* and that from *b* to *h*.

$$\text{So, efficiency} = \frac{3350 - 2330}{3350 - 2150} = \frac{1020}{1200}$$
$$= 0.85 \text{ or } 85\%$$

2.12 The modern power station cycle

To complete this section, it is appropriate to look at a typical modern cycle for CEGB plant (Fig 7.17). The turbine is throttle-governed and the output is of the order of 600 MW. After doing work in the HP cylinder, some steam is bled to the boiler feed pump turbine, while the remainder is passed to the

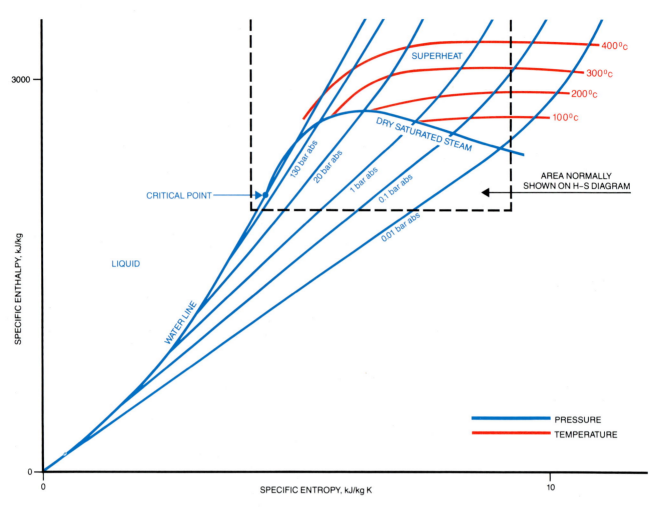

FIG. 7.14 The total heat-entropy diagram (Mollier chart)
The skeleton of a complete (H-S) diagram is shown. In practice, only the coloured part is used.

reheater where its temperature is raised to the same value as at TSV inlet. It then does work in the IP and LP cylinders. By the time the steam is passed to the condenser, it will have a wetness of about 10% and the back pressure will be about 35 mbar.

3   Fuels

3.1   Introduction

It is of great importance that the heating value and chemical composition of the fuel are established as accurately as possible, for two separate reasons:

● The incoming supplies to the station are monitored to check that the commercial quality of the fuel is in accordance with the contract to ensure that

the station gets value for money from the fuel supplier.

● To determine the characteristics of the fuel, as this information is used in performance calculations.

The two main fuels used in CEGB conventional power stations are coal and fuel oil. The properties of a particular grade of fuel oil do not vary much, so only random sampling and checking is required at the power station.

On the other hand, the bulk of fuel burned in CEGB plant is coal, and this is extremely variable. So, to achieve the above objectives at coal-fired stations entails a considerable capital outlay on sampling plant, as well as the services of specialist staff.

One of the most important characteristics of any fuel is its *calorific value* (CV), i.e., its heating power, so this will be considered first.

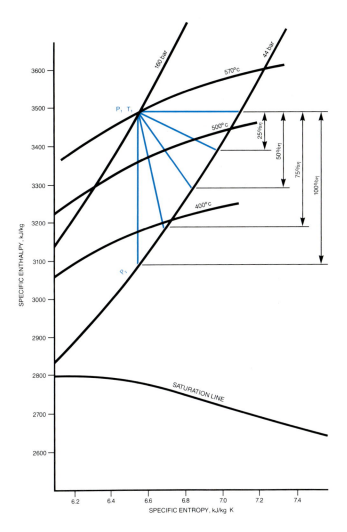

FIG. 7.15 Effect of efficiency of expansion on change of enthalpy
Maximum efficiency (100%) can only be achieved by isentropic expansion of the steam. At lower efficiencies the state line inclines to the right, resulting in higher final values of specific entropy and lower heat drop. At zero efficiency, the state line is horizontal.

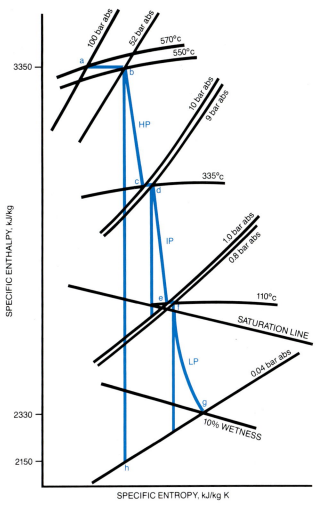

FIG. 7.16 Steam-state line for non-reheat turbine
The large pressure drop from *a* to *b* indicates that the machine is on part-load. As loading is reduced further, the isentropic heat drop *b* to *h* will become smaller.

3.2 Determination of calorific value

3.2.1 Solid and liquid fuels

The calorific value of the fuel, whether it be solid, liquid or gas, can be determined by the use of a calorimeter. The laboratory 'static' bomb type is probably the best known for solid and liquid fuels, and is illustrated in Fig 7.18. The fuel sample is placed in a strong steel shell known as the 'bomb', together with an ample supply of oxygen. It is located inside a vessel which has a water jacket of known quantity and temperature. The fuel is ignited by means of an electrically-heated platinum wire and cotton in contact with the fuel. Complete combus-

tion occurs very rapidly and the heat released raises the water temperature. Thus, the heat gained by the water, plus a few corrections (such as for the heat released by the platinum and cotton), enables the heat release of the fuel to be calculated. The 'adiabatic' version of the 'bomb' calorimeter is simpler to operate and requires less time to be spent on calculation.

They each give a calorific value which is 'gross' (because the whole heat release is determined) at constant volume and reference temperature 25°C. Therefore, such an analysis is said to be *Gross Calorific Value at Constant Volume* (GCV_v).

There are three other bases for calorific value:

- Gross CV at constant pressure (GCV_p).

- Net CV at constant pressure (NCV_p)

- Net CV at constant volume (NCV_v).

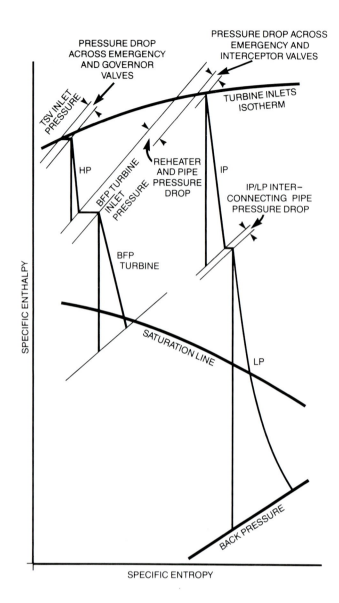

FIG. 7.17 Steam-state line for a modern turbine
The turbine incorporates reheat: a bled-steam turbine
is supplied from the HP cylinder exhaust.

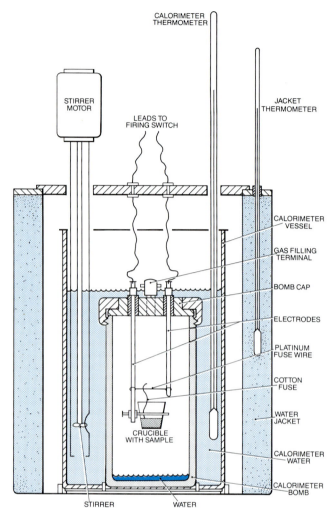

FIG. 7.18 Static calorimeter
The essential parts of the commonly used 'static' bomb
calorimeter are shown diagrammatically.

Let $\overline{H}$ = % hydrogen by weight

      $\overline{M}$ = % moisture by weight

      $\overline{A}$ = % ash by weight

      $\overline{O}$ = % oxygen by weight

The *gross CV at constant pressure* (GCV_p), is the GCV_v plus the heat equivalent of the work done by the atmosphere on the products if the fuel is burned at constant pressure.

Thus, GCV_p = GCV_v + $(6H - 0.7\overline{O})$ kJ/kg

The *net CV at constant pressure* (NCV_p), is the GCV_p minus the latent heat of evaporation at 25°C and constant water pressure.

So NCV_p =

$$GCV_v - [212.1\,\overline{H} + 24.4(\overline{M} + 0.1\,\overline{A}) + 0.7\,\overline{O}]\,kJ/kg$$

The term $0.1\,\overline{A}$ is included for coal as it is assumed that the moisture in the mineral matter is equal to one-tenth the weight of ash. This is zero for coke.

The *net CV at constant volume* (NCV_v), is equal to the GCV_v minus the latent heat of evaporation at 25°C and constant volume of the water present.

So NCV_v =

$$GCV_v - [205.9\,\overline{H} + 23.03(\overline{M} + 0.1\,\overline{A})]\,kJ/kg$$

Example

Consider a coal having the following analysis:

Moisture	16.1%	Nitrogen	1.8%
Ash	15.0%	Sulphur	1.3%
Carbon	54.1%	Oxygen	8.0%
Hydrogen	3.7%		

462

Its calorific value (GCV_v) determined in a 'bomb', is 22 200 kJ/kg.

$$GCV_p = 22\ 200 + [(6 \times 3.7) - (0.7 \times 8.0)]$$
$$= 22\ 217\ kJ/kg$$
$$NCV_p = 22\ 200 - [(212.1 \times 3.7) +$$
$$24.4\ (16.1 + 1.5) + (0.7 \times 8)]$$
$$= 20\ 980\ kJ/kg$$
$$NCV_v = 22\ 200 - [(205.9 \times 3.7) +$$
$$23.03\ (16.1 + 1.5)]$$
$$= 21\ 033\ kJ/kg$$

Until 1980, thermal efficiency in the CEGB was calculated using the gross calorific value at constant volume, after which time the net calorific value at constant pressure was used, in line with European practice. The justification for the change is that in actual boiler operation the combustion process is at approximately constant pressure. Also, the final flue-gas temperature is of the order of 130°C, so that the 'wet' products are in the form of superheated steam. Therefore, the gross heat of the fuel cannot be utilised in practice, the net heat being more representative of the real conditions.

However, the CEGB instituted one small change to the formula given earlier, in that a constant of 6 is substituted for the term $(0.7\ \overline{O})$. So for coal and coke, $NCV_p =$

$$GCV_v - [212.1\,\overline{H} + 24.4\,\overline{M} + 0.1\,\overline{A} + 6]\ kJ/kg$$

For oil, $NCV_p =$

$$GCV_v - (212.1\,\overline{H} + 24.4\,\overline{M})\ kJ/kg$$

Using NCV_p instead of GCV_v for calculation purposes increases the resultant value of efficiency significantly.

For example, if a boiler burned oil containing 12.5% hydrogen and 0.1% moisture, whose GCV_v is 45 356 kJ/kg, then:

$$NCV_p = 45\ 356 - [(212.1 \times 21.5) + (24.4 \times 0.1)]$$
$$= 42\ 702\ kJ/kg$$

Hence, the calculated boiler efficiency, using NCV_p is 45 356/42 702 = 1.06 times greater than if the GCV_v is used.

3.2.2 Gaseous fuels

Normally, the gross calorific value is determined by the continuous combustion of the gas at constant pressure, i.e., GCV_p. The following method is given in BS526 to determine the gross calorific value at constant volume. (Note The symbols *abc* are used

for convenience only in this section. They are not general substitutes.) Composition of dry gas:

Let $CO_2 = a$ % by volume
$O_2 = b$ % by volume
$CO = c$ % by volume
$H_2 = d$ % by volume
$C_x H_y = e$ % by volume
$N_2 = f$ % by volume
Total $= 100$ % by volume

where e = % by volume of hydrocarbon in the gas and y = number of hydrogen atoms in the hydrocarbon

$$GCV_v =$$

$$GCV_p - 1.044\,[(c/2) + (3d/2) + e + (ey/4)]\ kJ/m^3$$

The net calorific value at constant pressure is given by:

$$NCV_p =$$

$$GCV_p - 18.257\,[d + (ey)/2]\ kJ/m^3, \text{ and}$$

$$NCV_v =$$

$$GCV_p - (0.522c + 18.780d + 1.044e + 8.868ey)\ kJ/m^3$$

The use of these formulae can be illustrated with reference to Table 7.4.

TABLE 7.4
Values for North Sea gas

Gross calorific value	38 560 kJ/m^3
Specific gravity	0.603
Water content	nil
Ash	nil
Solids	nil
Methane (CH$_4$) by volume	93.3%
Ethane (C$_2$H$_6$) by volume	3.3%
Propane (C$_3$H$_6$) by volume	0.7%
Butane (C$_4$H$_{10}$) by volume	0.2%
Pentane (C$_5$H$_{12}$) by volume	0.5%
Carbon dioxide (CO$_2$) by volume	0.3%
Nitrogen (N$_2$) by volume	1.7%
Gas total by volume	100.0%

so c = 0

   d = 0

   e = 93.3 + 3.3 + 1.4 = 98.0

   y = (0.933 × 4) + (0.007 × 6) +

        (0.002 × 10) + (0.005 × 12)

      = 3.732 + 0.198 + 0.042 + 0.020 +

        0.060

      = 4.04

Hence:

GCV_p = 38 560 kJ/m^3

GCV_v = $38\,560 - 1.044 \left(98 + \dfrac{98 \times 4.049}{4} \right)$

      = 38 354 kJ/m^3

NCV_p = $38\,560 - 18.257 \left(\dfrac{98 \times 4.04}{2} \right)$

      = 34 946 kJ/m^3

NCV_p = 38 560 − (102.3 + 3511) + 34 947 kJ/m^3

An alternative method of obtaining the net value is to deduct 1863 kJ for each cubic metre of hydrogen (whether free or combined) from the GCV_p. Combined hydrogen is the number of molecules of hydrogen (H_2) in the compound. For example, methane (CH_4) has two.

So NCV_p = 38 560 − 1863 [(0.933 × 2) +

           (0.040 × 3) + (0.002 × 5) +

           (0.005 × 6)]

        = 38 560 − (1863 × 2.03)

        = 34 778 kJ/m^3

This differs by less than 0.5% from the earlier result.

3.3 Sampling and analysis

3.3.1 General

It is important that frequent checks on the quality of the fuel being received are carried out, and also on that being burnt. This is particularly so for coal-fired plant. For example, coal leaving a colliery in unsheeted rail wagons could have its moisture content altered significantly by the time it reached the power station during wet weather. On arrival, it may be put straight to the bunkers, or to stock, or some to each. Meanwhile, other coals may also be bunkered, so careful sampling is also necessary to determine the average quality of the coal being burned.

3.3.2 Sampling and preparation

Reliable data and fuel quality monitoring can only be achieved if there are adequate procedures for both sampling and sample preparation which minimise random and systematic errors.

Reduction of *random errors*, giving an acceptable degree of precision, can be obtained by taking the correct number of increments of the required mass, uniformly distributed over the sample unit, be it a single consignment or deliveries over a fixed period, such as a shift, week or month. The precision can be determined by methods laid down in BS1017: Part 1: 1977 'Methods for sampling coal and coke' to which reference should be made. If the sampling method is precise, the estimate of say, the ash content will always be acceptably close to the true value. Ash is the most variable characteristic of coal, whilst, for coke, it is moisture.

Even though the sampling method gives good precision, it may still give an estimate which is consistently above or below the true value. This is called *systematic error*, or *bias*, and can be caused by sampling procedures or equipment which preferentially select or reject some of the coal. For instance, the sampling scoop may reject lumps of coal above a certain size; fines may be blown away; or wet coal may be 'hung up' in the sampling chute. Since, for most coals, the ash content of the smaller particles is higher than the average, this will cause bias. Therefore, bias tests should be carried out for each procedure used, statistically comparing the ash estimate by the routine method with that from a reference sample, as detailed in BS1017. The reference sample will usually be obtained from a stopped-belt cross-section, which is not a practical method for routine sampling.

For example, bias trials were conducted on the primary stage of an automatic sampler. 29 samples were collected from the auto system, plus 29 reference samples, in accordance with BS1017. The basic calculation for bias is shown in Table 7.5.

Mean difference Z   = Z_r/n = 0.9/29 = 0.003

                        (i.e., the bias)

Variance of Z = V_Z = $1/(n-1)[(Zr)^2 - (Zr)^2/n]$

                    = $1/28\ [28.87 - (0.9^2/29)]$

                    = 1.03

Sampling precision P = $\pm t\sqrt{(V_Z/2)}$

Where t = Student's 't' = 2.05 in this case.

Precision of the estimate of bias = $\pm t\sqrt{(V_Z/n)}$

                                  = ±0.39

The bias is not significant as Z is less than 0.39.

The limits of acceptable bias are Z ±0.39.

TABLE 7.5

Bias calculation

Sample No.	Dry ash %		Difference Z_r	$(Z_r)^2$
	Ref	Auto		
1	16.5	14.7	1.8	3.24
2	14.7	15.5	− 0.8	0.64
28	12.5	12.8	− 0.3	0.09
29	13.0	13.9	− 0.9	0.81
n = 29	Totals		0.9	28.87

mass to about 100 to 250 grams, and the particle size to less than 0.2 mm. As the gross sample will be several times the required quantity, reduction is necessary.

This must also be carried out in a manner that does not introduce random or systematic errors. Some automatic samplers include one or more stages of sample crushing and weight reduction before the coal is delivered for further processing. Such preparation procedures should be subject to periodic tests for random and systematic errors, in a similar way to the collection procedures.

Systematic errors are often caused by human bias when manual sampling is carried out, so a well designed automatic sampler is preferred when large tonnages of coal are being handled (Figs 7.19 and 7.20). However, it is quite common to need samples for some specific purpose from, say, a mill coal feeder. Here, it is necessary to resort to manual operation. A typical coal feeder sampler is shown in Fig 7.21.

Having obtained a gross sample representative of the coal or coke it is then necessary to reduce its

3.3.3 Analysis

Having obtained the required samples, a *proximate analysis* is carried out, which consists of the determination of:

● Free carbon.

● Volatile matter.

● Ash.

● Moisture.

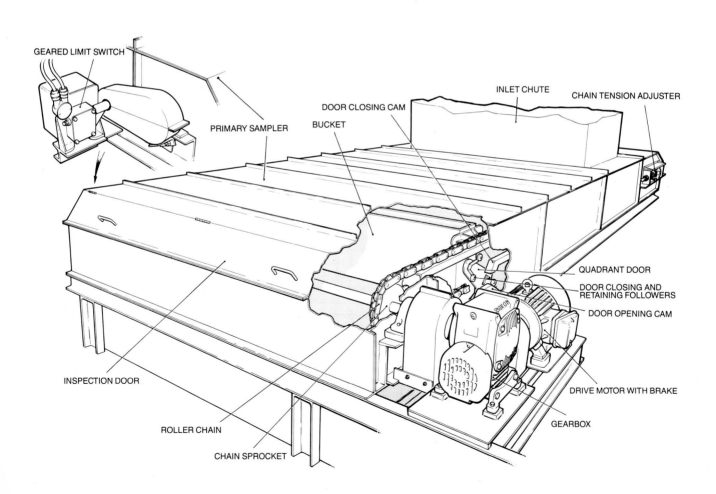

FIG. 7.19 Automatic coal sampler

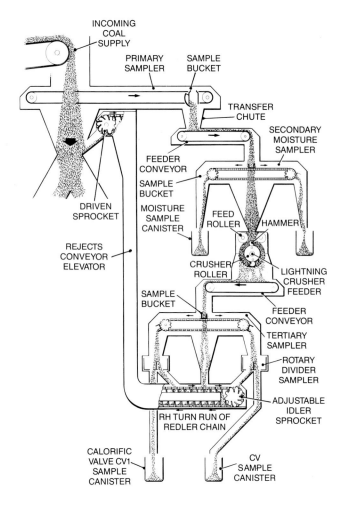

FIG. 7.20 General arrangement of coal sampling equipment
The equipment is quite high, so it is usually installed in
a junction tower. In this example, there are twin sampling
paths, each capable of giving samples for moisture
and general analysis.

The gross calorific value of the coal at constant volume (GCV_V) is also determined.

Should the chemical constituents, i.e., an *ultimate analysis* be required a long and complicated series of laboratory determinations must be carried out. Because of this, it is usual to calculate some of the ultimate from the proximate analysis. Various formulae are available to assist in this, the best known being that due to Parr.

The composition of an ultimate analysis is:

- Total carbon, $\overline{C}$.

- Hydrogen, $\overline{H}$.

- Nitrogen, $\overline{N}$.

- Sulphur, $\overline{S}$.

- Oxygen, $\overline{O}$.

- Ash, $\overline{A}$.

- Moisture, $\overline{M}$.

The Parr formulae enables the carbon and hydrogen to be calculated. The sulphur is taken as the typical percentage content for the fuel and the ash and moisture are already known from the proximate analysis.

Parr Formulae

The subscript 'Parr' indicates values calculated on a Parr basis.

$$\overline{Z} = \text{mineral matter content of the fuel burnt, \%}$$
$$= \overline{M} + 1.1\,\overline{A} + 0.1\,\overline{S}$$

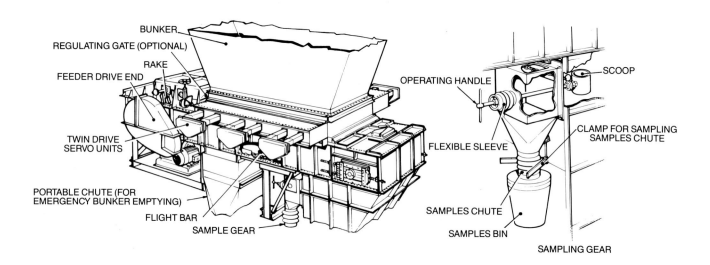

FIG. 7.21 Coal feeder hand-operated sampler
The scoop is pushed (upside down) into the falling stream of coal. Once it is in the correct position, the handle is turned
through 180° to allow the scoop to fill. It is then retracted and the sample emptied into the collecting bin.

V_{Parr} = volatile content, %

$\quad\quad\quad$ = $100 (\overline{V} - 0.1\overline{A} - 0.1\overline{S})/(100 - \overline{Z})$

GCV_{Parr} = calorific value, kJ/kg

$\quad\quad\quad$ = $100\, GCV_v/(100 - \overline{Z})$

$\overline{C}_{Parr}$ = carbon content (mineral matter free), %

$\quad\quad\quad$ = $(1.5782 \times 10^{-3}) \times$
$\quad\quad\quad\quad (GCV_{parr} - 0.2226V_{Parr} + 37.69)$

$\overline{H}_{Parr}$ = hydrogen content (mineral matter free), %

$\quad\quad\quad$ = $(0.1707 \times 10^{-3}) \times$
$\quad\quad\quad\quad (GCV_{Parr} + 0.0653V_{Parr} - 2.92)$

From these formulae, the total carbon $\overline{C}$ and hydrogen $\overline{H}$ content are derived:

$$\overline{C} = (1 - 0.01\overline{Z})(\overline{C}_{Parr} + 0.05\overline{A} - 0.5\overline{S})$$
$$\overline{H} = (1 - 0.01\overline{Z})(\overline{H}_{Parr} + 0.01\overline{A} - 0.015\overline{S})$$

The method of calculation is best shown by an example.

Consider a typical coal from the Bevercotes Colliery in Nottinghamshire with the following proximate analysis:

Moisture, $\overline{M}$	10.4%
Ash, $\overline{A}$	16.5%
Volatile matter, $\overline{V}$	28.3%
GCV_v	25 060 kJ/kg
Typical sulphur content, $\overline{S}$	1.51%

$\overline{Z}$ = $10.4 + (1.1 \times 16.5) + (0.1 \times 1.51)$
$\quad$ = 28.7%

V_{Parr} = $100 [28.3 - (0.1 \times 16.5)$
$\quad\quad\quad - (0.1 \times 1.51)]/(100 - 28.7)$
$\quad$ = 37.17%

GCV_{Parr} = $(100 \times 25\,060)/(100 - 28.7)$
$\quad$ = $35\,147$ kJ/kg

$\overline{C}_{Parr}$ = $(1.5782 \times 10^{-3}) \times$
$\quad\quad\quad (35\,147 - 0.2226 \times 37.17 + 37.69)$
$\quad$ = 84.88%

$\overline{H}_{Parr}$ = $(0.1707 \times 10^{-3}) \times$
$\quad\quad\quad (35\,147 + 0.0633 \times 37.17 - 2.92)$
$\quad$ = 5.5%

$\overline{C}$ = $[1 - 0.01 \times 28.7]\, 84.88$
$\quad\quad + (0.05 \times 16.5) - (0.5 \times 1.51)$
$\quad$ = 60.6%

$\overline{H}$ = $[1 - 0.01 \times 28.7]\, 5.5$
$\quad\quad + (0.01 \times 16.5) - (0.15 \times 1.51)$
$\quad$ = 4.1%

Hence, the ultimate analysis derived from Parr is:

Total carbon	60.6%
Hydrogen	4.1%
Sulphur	1.5% (Assumed)
Ash	16.5%
Moisture	10.4%
Oxygen and Nitrogen	6.9% (by difference)
	100.0%

NCV_{Parr} = $25\,060 - [(212.1 \times 4.1)$
$\quad\quad\quad + 24.4 (10.4 + 1.65) + 6]$ kJ/kg
$\quad$ = $25\,060 - (869.6 + 294.0 + 6)$
$\quad$ = $23\,890$ kJ/kg

For fuel oil, the following formulae may be used:

Carbon = $100 - \overline{H} - \overline{S} - 0.5$, %

Where:

Hydrogen = $(0.988 \times 10^{-3}) \times NCV_p - 29.02$, %

or

Hydrogen = $25.8 - (15 \times$ specific gravity at 15°C), %

For example, a fuel oil whose viscosity is three 100 seconds Redwood No. 1 has the following composition: Ash 0.05%; Moisture 0.25%; Sediment 0.04%; Sulphur 3.2%; Specific gravity 0.965.

So Hydrogen = $25.8 - (15 \times 0.965)$
$\quad\quad\quad$ = 11.3%

Carbon = $100 - 11.3 - 3.2 - 0.5$
$\quad\quad\quad$ = 85%

In addition to the analyses mentioned, it is common to determine properties of fuels for particular purposes. For instance, the chlorine content can have a significant effect upon the fouling propensity of coal; similarly the vanadium content of fuel oil is important because of its corrosive properties, particularly with regard to superheater metals. These, and similar characteristics of the fuels, are determined as required.

3.3.4 Dulong's formula

Sometimes the calorific value of a fuel is not known but its chemical composition is. It will already be apparent that the only components of conventional fuels which produce heat are hydrogen, carbon and sulphur. Their heat values are, in kJ/kg:

	Gross	Net
Hydrogen	143 050	121 840
Carbon	33 820	–
Sulphur	9 304	–

The formula derived by Dulong is reasonably accurate for the determination of calorific value, and is:

$$GCV_v = 33\,820\,\overline{C} + 143\,050\,(\overline{H} - \overline{O}/8)$$
$$+ 9304\,\overline{S}\ kJ/kg$$

$\overline{C}$, $\overline{H}$, $\overline{O}$, $\overline{S}$ are the proportionate parts, by weight, of the fuel. The term $(\overline{H} - \overline{O}/8)$ is a correction for the hydrogen in the fuel that combines with the oxygen to form water (see Section 4.2.2. of this chapter).

The formula is suitable for coal but not for gas. Normally, the heating value of coal so calculated, will be within about three percent of the true value.

For example, a particular coal has a measured GCV_v of 22 046 kJ/kg, and contains 3.55% hydrogen, 54.29% carbon, 1.23% sulphur and 6.99% oxygen. From Dulong's formula:

$$GCV_v = (33\,820 \times 0.5429)$$
$$+ 143\,050\left(0.0355 - \frac{0.0699}{8}\right)$$
$$+ (9304 \times 0.0123)$$
$$= 18\,361 + 3828 + 114 = 22\,303\ kJ/kg$$

The error is less than 1.2%.

4 Boiler efficiency and optimisation

4.1 General

Optimised boiler operation is a good example of the collaboration necessary between the maintenance and operations departments. The maintenance involvement includes:

- Provision of accurate instrumentation.
- Reliable and precise control of dampers, valves, etc.
- Reduction of air inleakage to a practical minimum.
- Good milling plant performance.
- Good sootblower availability.

On the operations side, the list includes such things as:

- Keeping correct air supplies at the burners.
- Using the minimum works power for the required loading.
- Maintaining correct terminal conditions.
- Taking care to avoid distortion at rotary airheaters that could cause damage to the seals.

Clearly, a considerable team effort is needed to keep boiler performance at a very high standard. Usually the performance aimed at is that achieved during the boiler acceptance tests, which is probably better than the guaranteed values.

A necessary prerequisite for the study of boiler performance is complete familiarity with combustion formulae, and this is the subject of the next section.

A knowledge of elementary chemistry is assumed.

4.2 The chemistry of combustion

4.2.1 Air and combustion

It has already been mentioned that the only combustible elements present in fuels are hydrogen, carbon and sulphur. Before oxidation (i.e., combustion) can take place, certain requirements must be met:

- The combustible substance must be at a suitable temperature.
- There must be a supply of oxygen present which is brought into intimate contact with the combustibles.
- The above two conditions must be sustained for an adequate period of time to allow combustion to be completed.

Consider each in turn.

(a) Every combustible substance has a temperature which must be attained before combustion can take place, known as its *ignition temperature*. Below this value the substance will not burn, no matter how much oxygen is present or for how long.

The ignition temperature of different combustible substances varies considerably, as can be seen from Table 7.6.

It is worth noting that the ignition temperatures of the gases given off when burning coal are much higher than that of the carbon. The gases, though, are distilled off before their ignition temperatures are attained, so the ignition temperature of the coal is regarded as that of the fixed carbon present. Once combustion has started, the heat produced will be sufficient, given the correct conditions, to produce temperatures high enough for further ignition.

(b) The oxygen involved in the combustion process is derived from the atmosphere. For present purposes, it can be assumed that air consists of 23.2% oxygen and 76.8% nitrogen, by mass. To supply 1 kg of oxygen it is necessary to provide $100/232 = 4.31$ kg of air and this contains 3.31 kg of nitrogen. Almost all the nitrogen supplied with

TABLE 7.6

Ignition temperature of some combustibles

Combustible	Approximate ignition temperature, °C
Hydrogen	610
Fixed carbon — bituminous coal	410
Fixed carbon — semi-bituminous coal	470
Fixed carbon — anthracite	500
Sulphur	245

the air or contained in the fuel leaves the boiler chemically unchanged. The small amount of nitrogen which is converted to nitrogen oxides (referred to as NO_x) is of environmental interest, but can be ignored in combustion calculations. In the combustion process, it is the combination of oxygen with hydrogen, carbon and sulphur that is important. The amount of oxygen required to burn these substance completely is of great interest, so it is desirable to be clear how it is derived.

Consider first the combustion of hydrogen. From elementary chemistry, the chemical equation is:

$$2H_2 + O_2 = 2H_2O$$

In other words, two molecules of hydrogen combine with one of oxygen to form two molecules of water.

The masses involved are proportional to the molecular weights of the elements, so the equation can be expressed as follows:

$$2\overline{H}_2 + \overline{O}_2 = 2\overline{H}_2\overline{O}$$
$$(2 \times 2) + 32 = (2 \times 18)$$

So in kilograms,

$$4 + 32 = 36, \text{ or}$$
$$1 + 8 = 9$$

Thus, one kilogram of hydrogen requires eight kilograms of oxygen to form nine kilograms of water vapour. Further, to supply eight kilograms of oxygen it is necessary to supply $8 \times 4.31 = 34.48$ kg of air and the nitrogen supplied with that air will be $8 \times 3.31 = 26.48$ kg.

Similarly, for the combustion of carbon.

$$\overline{C} + \overline{O}_2 = \overline{CO}_2$$
$$12 + 32 = 44$$
$$1 + \frac{32}{12} = \frac{44}{12}$$
$$1 + 2.67 = 3.67$$

So burning one kilogram of carbon to carbon dioxide requires 2.67 kg of oxygen, and forms 3.67 kg of carbon dioxide. The air needed is 11.49 kg and includes 8.82 kg of nitrogen.

For sulphur, $\quad \overline{S} + \overline{O}_2 = \overline{SO}_2$
$$32 + 32 = 64$$
$$1 + 1 = 2$$

So one kilogram of sulphur needs one kilogram of oxygen for complete combustion and forms two kilograms of sulphur dioxide. The air required is 4.31 kg, which contains 3.31 kg of nitrogen.

The above statements are summarised in Table 7.7 which also includes the details of a few other combustible substances which are of interest.

In the above table we are concerned only with the minimum quantity of oxygen required to give complete combustion, and this is usually called *perfect* combustion.

Solid and liquid fuel analyses are normally given in terms of mass, whereas gaseous fuels are reported in terms of volume. Table 7.8 gives the combustion data in terms of volume. The reasoning is simple, and a study of just the combustion of hydrogen makes the method clear.

Volumes can be substituted for molecules of gas, so for the complete combustion of hydrogen:

$$2H_2 + O_2 = 2H_2O$$

i.e., two volumes of hydrogen will combine with one volume of oxygen to form two volumes of water vapour. That is, 1 m³ of hydrogen requires 0.5 m³ of oxygen to form 1 m³ of water vapour. Air consists of 21% oxygen and 79% nitrogen by volume, so to supply 1 m³ of oxygen it is necessary to provide $100/21 = 4.76$ m³ of air. The nitrogen in the air will be 3.76 m³.

Table 7.8 also summarises the results for some other common gases.

(c) The time required for combustion to be completed depends upon the particle size of the combustible, provided the other conditions are met. Thus, lumps of coal take longer to burn than the same amount of pulverised coal. It should be borne in mind that combustion is a two-stage process involving:

- Physical contact of the combustible with oxygen.
- Chemical combination of the two after contact.

Once the ignition temperature has been attained the chemical process is instantaneous; but, clearly, this cannot proceed until physical contact has been made. It follows that the rate of combustion is

TABLE 7.7
Combustion data (in kg/kg of combustible)

Combustible	Molecular symbol	Theoretical		Products of combustion				
		Oxygen	Air	H_2O	CO_2	SO_2	N_2	CO
Hydrogen	H_2	8.00	34.48	9.00	–	–	26.48	–
Carbon (to CO_2)	C	2.67	11.49	–	3.67	–	8.82	–
Carbon (to CO)	C	1.33	5.75	–	–	–	4.42	2.33
Carbon monoxide	CO	0.57	2.46	–	1.57	–	1.89	–
Sulphur	S	1.00	4.31	–	–	2.00	3.31	–
Methane	CH_4	4.00	17.24	2.25	2.75	–	13.24	–

TABLE 7.8
Combustion data (in m^3/m^3 of combustible)

Combustible	Molecular symbol	Theoretical		Products of combustion		
		Oxygen	Air	H_2O	CO_2	N_2
Hydrogen	H_2	0.5	2.38	1.0	–	1.88
Carbon monoxide	CO	0.5	2.38	–	1.0	1.88
Methane	CH_4	2.0	9.58	2.0	1.0	7.52

controlled by the rate at which contact is made between the oxygen and the combustibles.

Consider a piece of coal 25 mm in diameter, at a temperature of 1000°C. For complete combustion, it needs a sphere of air one metre in diameter, arranged concentrically. The average distance of the oxygen molecules will be about 300 mm from the coal, so contact-making will be slow. If, however, the same coal is pulverised, the process is speeded up considerably. For example, if the grading is such that all the particles will pass through a 100 mesh (150 μm) sieve, then the largest particles will be about 0.15 mm diameter. The air required for combustion will be contained in a sphere only 5 mm in diameter. The reduction in size of the coal has reduced the average distance of the oxygen from the fuel considerably and so the rate of contact-making (and hence combustion) is much faster.

4.2.2 Combustion formulae and excess air

The minimum quantities of oxygen required to burn the combustible substances completely were derived in Section 4.2.1 of this chapter. For example, hydrogen requires eight times its own weight of oxygen, carbon 2.67 times and sulphur its own weight of oxygen.

Thus, the theoretical oxygen to burn a fuel is given by: $(2.67 \ \overline{C} + 8 \ \overline{H} + \overline{S})$ kg/kg fuel. But the fuel itself usually contains oxygen, which is available to combine with combustible matter in the same way as the oxygen in the air. It is assumed that all the oxygen in the fuel combines with one-eighth of its own weight of hydrogen, so the hydrogen left will be $(H - \overline{O}/8)$, so the theoretical oxygen becomes $[2.67 \ \overline{C} + 8 \ (\overline{H} - \overline{O}/8) + \overline{S}]$ kg/kg fuel. The air required is 4.31 times the oxygen, so the theoretical air = $4.31 \ [2.67 \ \overline{C} + 8(\overline{H} - \overline{O}/8) + \overline{S}]$ kg/kg fuel. This theoretical air is sometimes referred to as *stoichiometric* air.

Consider a coal which has the following analysis:

Carbon	55.5%	Hydrogen	3.8%
Sulphur	1.6%	Oxygen	7.4%

The theoretical air required is:

$$4.31 \ [(2.67 \times 0.555) + 8 \ (0.038 - 0.074/8) + 0.016] \text{ kg/kg fuel}$$

$$= \ 4.31 \ (1.48 + 0.23 + 0.016) \text{ kg/kg fuel}$$

$$= \ 7.4 \text{ kg/kg fuel}$$

As a rule of thumb guide, coal requires about 3.27 kg of theoretical air per 10 000 kJ of heat release. Oil needs about 3.19 kg and natural gas about 3.2 kg.

However, in a practical furnace, it is impossible to obtain complete combustion of any fuel with only the theoretical quantity of combustion air, as it would require every molecule of fuel and oxygen to be in exactly the right place at the right time, so what would actually happen is that some of the fuel would have an ample supply of oxygen and the rest not enough. Consequently, carbon monoxide and hydrocarbons would appear in the flue gas and the combustion would be inefficient, because heat release from the fuel would be incomplete.

To ensure that all the combustibles meet with a sufficient supply of oxygen, excess air is supplied to the boiler. This enables all the available carbon to burn to carbon dioxide, hydrogen to water vapour and sulphur to sulphur dioxide, thus releasing the maximum possible heat. On the other hand, the more excess air supplied the greater will be the flow of flue gas to the chimney, carrying considerable quantities of heat with it, as well as diluting the heat from the combustion process.

Thus, although excess air is necessary, every effort is made to keep it as small as possible. Normal quantities are:

- Pulverised coal   20%

- Fuel oil   2%

- Natural gas   8%

Consider some salient details of combustion of the main fuels (Practical combustion and firing equipment is dealt with in Chapter 2 of this Volume.)

Coal

For bituminous coals, the desirable grading is in the range of 70%–80%, through a 200 mesh sieve (75 μm). If the air supplies are correctly adjusted and the temperatures are acceptable, complete combustion should be achieved with 20% excess air, or slightly less.

The primary factors which contribute to efficient combustion are:

- Acceptable PF grading.

- Adequate coal drying.

- Acceptable air/PF temperature (usually about 70°C) at the mill outlet.

- Correct proportions of primary and secondary air supplies.

- Correct admission of the air supplies for combustion.

A station with four 500 MW units will burn about 20 000 tonnes of coal per day on full-load and almost 200 000 tonnes of air will be supplied to the boilers.

Oil

The quantity of excess air used when burning oil is critical for more reasons than just efficient combustion. The heavy residual fuel-oils burned in power stations may have a high sulphur content. During combustion most of the sulphur burns to sulphur dioxide, but some of it forms sulphur trioxide (SO_3), which combines with water to produce sulphuric acid:
$$SO_3 + H_2O = H_2SO_4.$$

This presents a very real danger of severe corrosion at the air heaters, ducts and the top of the stack, particularly if the flue gas temperature frequently falls below its dewpoint. The dewpoint is lowered as the oxygen (i.e., the excess air) in the flue gas is reduced and also as the sulphur trioxide concentration is reduced (Fig 7.22). So the excess air should be kept as low as possible.

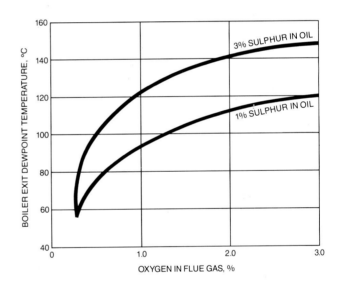

FIG. 7.22 Relationships between excess air and dewpoint with oil-firing
As the sulphur content of the fuel increases so does the dewpoint. Similarly the dewpoint increases as excess air is increased. These constraints place an economic upper limit on the permissible quantity of excess air.

A typical airheater gas outlet temperature on a modern boiler is about 130°C, which corresponds to the dewpoint when there is 10 PPM of SO_3 in the flue gas. The oxygen level is kept at about 0.5%, corresponding to about 2% of excess air.

Reducing the oxygen level enhances the possibility of creating acid smuts, so the desirable range of excess air is very small in practice, the lower limit being set by consideration of acid-smut formation, and the upper limit by the dewpoint.

The essential requirements for good combustion of fuel oil are:

- The elimination of tramp air ingress, particularly at the burners.

- Air dampers, oil burners, etc., to be kept in good condition.

- Provision for simultaneously altering the oil and air flows (to prevent a mismatch during load changes).

- Correct tip pressure and viscosity (oil/steam atomisation gives a better 'turndown' ratio than pressure atomisers).

- Correct combustion air supplies.

Natural gas

The flame temperature when burning natural gas is similar to that for pulverised fuel, i.e., about 1500°C. The air/gas mixing process starts at the burner. Due to the type of nozzle (Fig 7.23), some unburnt natural gas is preheated to a high temperature before coming into contact with oxygen at the vortex, and *pyrolisis* takes place, i.e., the gas is subjected to a 'cracking' process. This causes the formation of carbon particles which are later burnt, thus producing a flame of increased luminosity, the flame being yellow rather than blue. (The emissivity is 0.60, compared with 0.75 for a pulverised-coal flame.)

At Hams Hall *C* power station, the corner-fired pulverised-fuel (PF) boilers were converted to burn natural gas, alone or with any proportion of coal.

The conversion took place in the early 1970s and a combined natural-gas/PF corner-burner box is shown in Fig 7.24. The rate of flame propagation of natural gas is about 0.36 m/s, very slow compared with pulverised coal, which is 10 m/s. Also the limits of flammability of natural gas are about 5–15% of gas in air. Despite these factors, a turndown ratio of 4.4 to 1 was achieved.

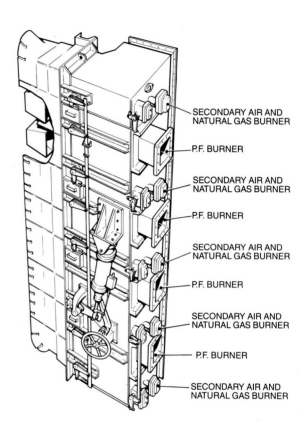

FIG. 7.24 Arrangement of natural gas/pulverised fuel corner box
The box illustrated was used on the Hams Hall combined natural-gas/pulverised-fuel fired boilers.

The theoretical air requirement is about 16.5 kg per kg of gas and the required excess air is about 8.0%. A typical operating CO_2 value at the boiler outlet is 10.8% (i.e., 2% oxygen). Because of the extra water vapour in the flue gas, compared with PF firing, the chimney plume is more dense and white.

4.2.3 Example of calculation of air required and products of combustion

Consider a fuel oil with the following analysis, burnt with 2% excess air:

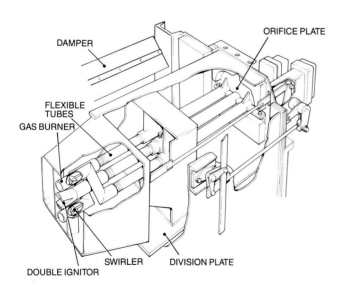

FIG. 7.23 Detail of natural gas burners
The burner illustrated is the type used on the boilers at Hams Hall power station.

Carbon	84.0%	Hydrogen	12.7%
Oxygen	1.2%	Sulphur	0.4%
Nitrogen	1.7%		

Calculate the theoretical and actual air required, and the products of combustion.

The theoretical air requirement and the resultant products of combustion are shown in Table 7.9 derived from information given in Table 7.7. For example, consider the carbon in the fuel (84%). From Table 7.7, 0.84 kg of carbon needs $0.84 \times 2.67 = 2.243$ kg of oxygen and will form $0.84 \times 3.67 = 3.083$ kg of carbon dioxide, and so on.

From the table, it is seen that one kilogram of the fuel will require 3.251 kg of oxygen and 13.996 kg of air. This is the theoretical requirement, and in practice two percent excess air is also to be supplied. Therefore, there will be:

$0.02 \times 3.251 = 0.065$ kg of oxygen extra

$0.02 \times 13.996 = 0.280$ kg of air extra

$0.02 \times 10.762 = 0.215$ kg of nitrogen extra

So the final products of combustion will be as shown in Table 7.10.

The percentage by weight of the products is found by dividing the weight of each by the total and multiplying the answer by 100.

Clearly it is also possible to work backwards. Suppose the percentage of dry products is known from a flue gas analysis (Table 7.11) and it is required to determine the quantity of dry gas. The total dry gas per kilogram of fuel is equal to the product of the dry gas per kilogram of carbon and the amount of carbon per kilogram of fuel. But all the carbon burned must appear in the flue gas. So,

Dry gas per kilogram carbon =

$$\frac{\text{total dry gas per kilogram fuel}}{\text{carbon per kilogram fuel}}$$

Also, the carbon can only appear in the flue gas as either CO_2 or CO. It has been shown already that 12 kilograms of carbon will form 44 kilograms of carbon dioxide. Thus, every kilogram of carbon dioxide in the flue gas has been formed from 0.273 kg of carbon.

Similarly, for every 28 kilograms of carbon monoxide in the flue gas there will have been 12 kilograms of carbon, so every kilogram of carbon monoxide has been formed from 0.429 kilograms of carbon.

Hence, the carbon per kg of fuel =

$$0.273 \, CO_2 + 0.429 \, CO \text{ so,}$$

dry gas per kilogram carbon =

$$\frac{CO_2 + CO + SO_2 \, N_2 + O_2}{0.273 \, CO_2 + 0.429 \, CO} \text{ kg}$$

All the products may be expressed as a % or as a weight. Therefore, for this example using percentages,

dry gas per kilogram carbon =

$$\frac{21.8 + 0 + 0.1 + 77.6 + 0.5}{0.273 \times 21.8} = 16.80 \text{ kg}$$

The fuel contains 84.0% carbon, so the dry gas per kilogram of fuel is $0.84 \times 16.803 = 14.11$ kg, the same as found earlier, allowing for the 'rounding' of values.

If the total gas weight (instead of the dry gas weight) is required, this is merely the sum of the dry gas weight and the weight of moisture present. In this

TABLE 7.9

Theoretical air and products of combustion

	kg/kg	Theoretical		Products of combustion				
	Fuel	Oxygen	Air	H_2O	CO_2	SO_2	N_2	O_2
Hydrogen	0.127	1.016	4.379	1.143			3.363	
Carbon	0.840	2.243	9.652		3.083		7.409	
Sulphur	0.004	0.004	0.017			0.008	0.013	
Oxygen	0.012							0.012
Nitrogen	0.017						0.17	
Totals		3.263	14.048	1.143	3.083	0.008	10.802	0.012
less oxygen in oil		0.012	0.052*				0.040*	0.012
		3.251	13.996	1.143	3.083	0.008	10.762	0

*These are the air and nitrogen equivalents of the oxygen in the fuel oil.

473

TABLE 7.10

Air supply and products of combustion (in kg/kg of fuel)

		Excess air	
		0%	2%
Air supply	Oxygen	3.250	3.320
	Air	14.000	14.280
Products	H_2O	1.143	1.143
	CO_2	3.083	3.083
	SO_2	0.008	0.008
	N_2	10.760	10.980
	O_2	0.000	0.060

case, the moisture is equal to nine times the hydrogen in the fuel, i.e., $9 \times 0.127 = 1.14$ kg/kg fuel.

So the total gas weight $= 14.11 + 1.14 = 15.25$ kg/kg fuel.

Furthermore, all of the fuel in this case will appear in the final products, so the air supplied will be $15.25 - 1.0 = 14.25$ kg per kilogram of fuel.

In practice though, the flue-gas analysis is normally carried out on a volumetric rather than a weight basis. Therefore, the previous formula for dry gas per kilogram of carbon must be modified by multiplying each term by its relative density; that is by its molecular weight relative to hydrogen.

So the dry gas volume per kilogram of carbon burned =

$$\frac{22CO_2 + 14CO + 16O_2 + 14N_2}{6CO_2 + 6CO}$$

$$= \frac{11CO_2 + 8O_2 + 7(CO + N_2)}{3(CO2 + CO)}$$

where CO_2, O_2, etc., are percentages by *volume*. Notice that the sulphur has been ignored as it is normally so small.

To complete this section on combustion formulae, it is necessary to look at the method of converting a weight analysis to one by volume. This is done by dividing each dry product by its molecular weight,

because the density of any gas is proportional to its molecular weight.

Consider the results for the dry products obtained in Table 7.11. They are on a weight basis and the information is reproduced in Table 7.12, along with the conversion to a volume basis.

4.2.4 Determination of optimum air

The flue gas produced by burning any fuel completely but with only the theoretical quantity of air, will contain a characteristic percentage of carbon dioxide. For example, if pure carbon were so burned, then all the carbon would be oxidised by combining with all of the oxygen present. Hence, the oxygen in the combustion air would become carbon dioxide in the flue gas, so the volume of CO_2 would be 21%. Adding excess air will cause oxygen to be present in the flue gas and the quantity of CO_2 will be reduced because of dilution by the extra oxygen and nitrogen. In fact,

excess air $=$ (Theoretical CO_2%/Actual CO_2%) $- 1$

The theoretical CO_2 is the value obtained when the fuel is burned completely without excess air.

Table 7.13 lists some common values.

So, if coal is burned with 15% CO_2, then the excess air quantity will be:

$$\frac{18.6}{15.0} - 1 = 0.24 \text{ or } 24\%.$$

Fuels containing hydrogen or hydrocarbon gases have a lower theoretical CO_2% than carbon. This is because the oxygen supplied to burn the hydrogen forms water, leaving the nitrogen to dilute the flue gases formed from the carbon.

It is now standard practice to use zirconia oxygen analysers instead of CO_2 instruments, with the considerable advantage that there is no necessity to withdraw a gas sample from the boiler for analysis. Instead, the zirconia probe operates by direct insertion into the gas stream, and its electrical output is a function of the oxygen in the flue gas. Undoubtedly, these instruments are a considerable advance on all

TABLE 7.11

Products of combustion (% by weight)

	H_2O	CO_2	SO_2	N_2	O_2	Total
Wet products	1.143	3.083	0.008	10.98	0.06	15.27
%	7.5	20.2	0.1	71.8	0.4	100.0
Dry products		3.083	0.008	10.98	0.06	14.13
%		21.8	0.1	77.7	0.4	100.0

TABLE 7.12
Conversion of weight basis to volume basis

Gas	CO_2	N_2	O_2	Total
Dry product by weight, kg	3.083	10.98	0.060	
Molecular weight	44.000	28.00	32.000	
Dry product by volume	0.070	0.39	0.002	0.462
Volume %	15.200	84.40	0.400	100.000

TABLE 7.13
Approximate theoretical CO_2 for various fuels

Bituminous coal	18.6%
Fuel oil	15.3%
Natural gas	11.8%

previous types. Figure 7.25 shows a typical modern analyser.

With the advent of improved O_2 instruments, operating staff are able to relate accurately the oxygen level to excess air, given by excess air = $O_2\%/(21 - O_2\%)$.

For example, if an oil-fired boiler is operated with a 0.5% oxygen level in the flue gas, then the excess air is $0.5/(21 - 0.5) = 0.025$ or 2.5%.

Clearly, it is possible to relate $CO_2\%$, $O_2\%$ and excess air to each other and this has been done in Fig 7.26. In power station work, only excess air values up to, say, 30% are considered, so Fig 7.27 illustrates the interrelationship of the three main classes of fuel for this limited range of excess air.

Before reliable CO meters were available, the optimum value of excess air for the combustion of a particular fuel was determined by carrying out a series of tests for boiler losses. At a particular boiler load, say 100%, four or five tests were carried out with different amounts of excess air. The total boiler loss was calculated for each test and the results plotted. The excess air corresponding to minimum boiler loss could then be established. The procedure was repeated at various loadings, after which a graph of optimum $O_2\%$ or $CO_2\%$ versus boiler-load was prepared for the use of the operations staff (Fig 7.28).

However, the point of minimum boiler loss (i.e., maximum efficiency) always coincides with the portion of the CO curve just before the knee. With the advent of excellent CO monitors, values less than 50 PPM can be measured, so a combination of O_2 and CO monitors are now used to optimise excess air. At excess air values above optimum, the CO is less than 100 PPM on a coal-fired boiler. As the excess air is reduced to just below optimum there is a dramatic rise in the CO value, so this provides a very easy means of establishing the desired air quantity. The $O_2\%$ corresponding to the optimum air required is noted, and this is used as the desired value for normal operation at that load. It has been established that only single-point sampling is required if the CO measurement is made after the ID fan, so this is the usual location. Figure 7.29 shows a typical CO monitor.

The sampling time of the O_2 and CO analysers is usually different, so allowance should be made for this, as well as for their different probe locations in the gas path.

A typical scatter diagram of CO versus O_2 is shown in Fig 7.30. This display of CO and O_2 values, which is an advance on the normal two-pen chart often used, has been developed at CEGB Midlands Region Scientific Services Department, and incorporates a visual display unit (VDU).

4.3 Boiler efficiency

This can be measured by either:

- The direct method.
- The losses (or indirect) method.

The *direct* method is the old established way of determining boiler efficiency, and consists of measuring the quantity of heat given to the steam and the quantity of heat given to the boiler in a period of say, one hour. It was in common use within the CEGB until the late 1950s after which it was gradually replaced.

It is now customary to use the *losses* method, as it offers the following advantages:

- No coal weighers are required, so a potential source of coal hold-up is eliminated.
- Accuracy of values is less critical than with the direct method, e.g., 1% error in measurement will result in only about 0.1% error in the answer.

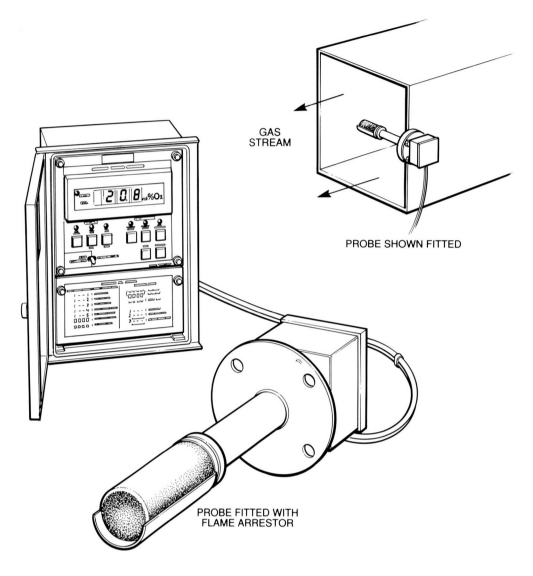

FIG. 7.25 Zirconia oxygen analyser
The advantage of this type of analyser is that it dispenses with the necessity to withdraw a gas sample for analysis external
to the boiler. The accuracy and reliability are very good.

● Not only is the efficiency obtained but also the magnitude of the individual losses.

● The calorific value of the fuel is still required, but any error in its determination has a smaller effect on the final answer.

The basis of the method is that all of the heat supplied to the boiler must become either heat in the steam or be lost from the system.

So, 100% heat input = useful heat + losses. Therefore, if the losses are known, the efficiency is easily determined.

Table 7.14 lists typical values of those losses which are usually calculated.

The percentage is obtained by dividing the loss in kilojoules per kilogram of fuel by the calorific value per kilogram of fuel. Clearly, it is important to specify which calorific value is used. In some countries GCV_v is still used, but for most of the world (and within the CEGB) NCV_p is used. Just how great is the effect on the losses of using different calorific values can be seen in Table 7.14.

Each of the losses will now be considered in turn.

4.4 Combustion losses

4.4.1 Dry flue gas

When the flue gas leaves the airheaters, it does no further useful work. However, it is at a considerably higher temperature than ambient, so it carries

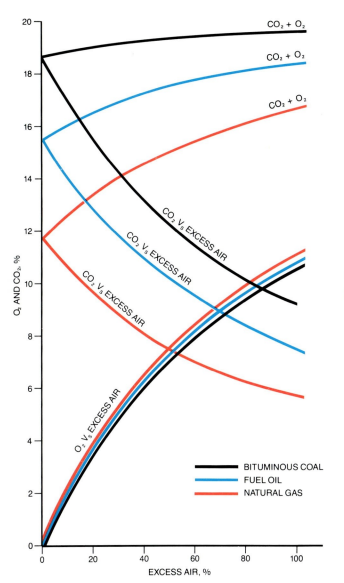

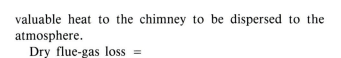

FIG. 7.26 Relationship between CO_2%, O_2%
and excess air
The three main conventional fuels are illustrated. The
normal amount of excess air used is 20% for pulverised
fuel, 2% for oil and 8% for natural gas.

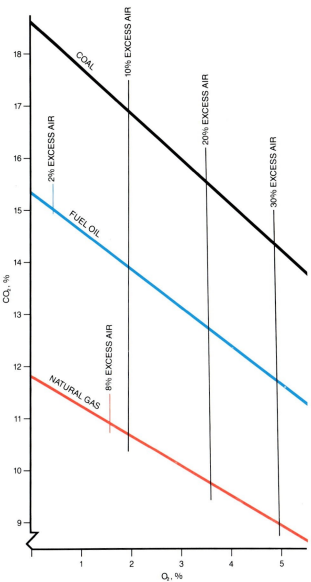

FIG. 7.27 Relationship between O_2 and CO_2 over a
limited range of excess air
The illustration shows the relationship for the three
main fuels for a limited range of excess air, and the
approximate excess air required for satisfactory
combustion of each fuel.

valuable heat to the chimney to be dispersed to the atmosphere.

Dry flue-gas loss =

$$\frac{100}{12\,(CO_2 + CO)} \left(\frac{\overline{C}}{100} + \frac{\overline{S}}{267} - C\ in\ A \right)$$

$$\times\ 30.6 \times (t_1 - t_2)\ kJ/kg\ fuel$$

Where CO_2, CO = % by volume

$\overline{C}$ = carbon content per unit mass of fuel, kg/kg

$\overline{S}$ = sulphur content per unit mass of fuel, kg/kg

C in A = combustible in ash per unit mass of fuel, kg/kg

30.6 = average kilogram molecular specific heat of gases, kJ/(kg mol K)

The term $\overline{S}/267$ is only used if the flue gas analysis is carried out by means of an Orsat apparatus, as the carbon dioxide value will include the sulphur dioxide. The denominator is derived from the ratio of the atomic weights of carbon and sulphur, i.e., $32/12 = 2.67$.

477

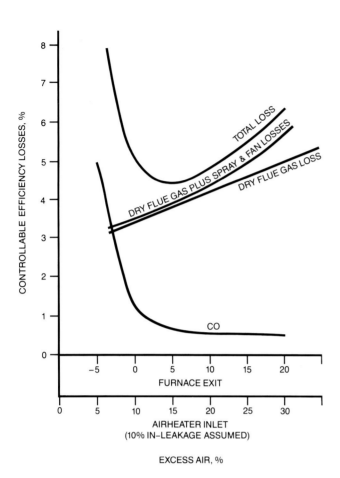

FIG. 7.28 Controllable combustion efficiency loss
for a PF boiler
Note that the point of minimum boiler loss coincides
with the knee of the CO curve.

Should the flue gas contain hydrocarbons, a slightly modified formula is required (see BS2885: 1974), but the above is adequate for power station work, where the aim is to have complete combustion of the fuel.

4.4.2 Wet flue gas

This results from steam in the flue gas carrying heat from the boiler to atmosphere. The steam is derived from two sources:

● Moisture in the fuel.

● Moisture formed from the combustion of the hydrogen in the fuel.

The first of these is obvious. For every kilogram of moisture in, say coal that is burned, heat is required to raise it to boiling temperature, to evaporate it and then superheat it to the airheater outlet gas temperature. Notice that the loss is not affected by what

happens to its temperature *within* the boiler — it is only the initial and final temperatures that matter.

The second item concerns the combustion of hydrogen.

Whatever the hydrogen content of the fuel it will, upon combustion, produce nine times into own weight of moisture. It will also release a considerable amount of heat in the process, as shown in Table 7.15. Carbon values are also given.

In fact, the heat released by the combustion of hydrogen is over four times that released by burning carbon to carbon dioxide. This is why fuels with higher proportions of hydrogen have higher calorific values, as shown in Table 7.16. For solid and liquid fuels:

Wet flue gas loss =

$$\frac{\overline{M} + 9\overline{H}}{100} \, [4.2 \, (25 - t_2) + 2442 +$$

$$1.88 \, (t_1 - 25)] \text{ kJ/kg fuel}$$

For gas, the moisture in the fuel is already in the vapour form, so:

Wet flue gas loss = $\frac{\overline{M}}{100} \, [1.88 \, (t_1 - t_2)] +$

$$\frac{9\overline{H}}{100} \, [4.2 \, (25 - t_2) + 2442 + 1.88 \, (t_1 - 25)]$$

kJ/kg fuel

4.4.3 Sensible heat in water vapour

This is only used for the NCV_P heat balance.

Sensible heat loss = Wet flue gas loss −
   $(GCV_v - NCV_p)$ kJ/kg fuel

4.4.4 Combustible-in-ash

Reference to Table 7.15 shows that for every kilogram of carbon which is mixed with the furnace-bottom ash, pulverised-fuel ash, etc., there is a potential heat loss of 33 820 kJ. Therefore, it is necessary to collect samples of the ash and determine their combustible content (which is regarded as carbon).

So combustible-in-ash loss = $i \overline{A} \times 33\,820$ kJ/kg fuel

where i = combustible matter % by weight
      $\overline{A}$ = mass of dry ash, kg/kg fuel

For a PF boiler, the ratio of furnace-bottom ash to pulverised-fuel ash is normally taken as 20:80, in the absence of a specific determination.

4.4.5 Radiation and unaccounted heat

The 'radiation' component usually amounts to about 50% of the loss. The remainder includes various losses which are inconvenient to measure, too small

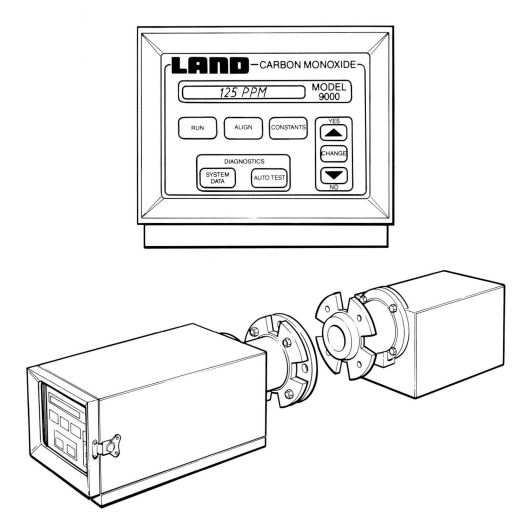

FIG. 7.29 Carbon monoxide analyser
CO analysers are capable of detecting very small quantities of CO in flue gas. Some types analyse the gas *in situ*
whilst others analyse the gas externally, away from the gas duct.

to assess separately, or impossible to measure. These include:

- Moisture supplied with the combustion air.

- Heat retained by the ash.

- Hydrocarbons in the fuel gas.

- Incomplete combustion of carbon.

An approximation for the percentage radiation and unaccounted loss for an indoor boiler is given by:

$$\log_n L_R = 1.88 - 0.4238 \log_n M_s$$

where L_R = radiation and unaccounted loss, %
M_s = specific boiler capacity, kg/s

For outdoor boilers, the value of L_R is multiplied by 1.5.

For example, at 100% loading (M_s = 561 kg/s), the loss from the indoor boiler of a 660 MW unit will be:

$$\log_n L_R = 1.88 - 0.4238 \log_n 561 = -0.8025$$
$$L_R = 0.4\%$$

At part loads, the loss is given by $100\, L_R/(\%\ \text{load})$, so that the loss for the 660 MW unit boiler at 80% MCR will be $(100 \times 0.4)/80 = 0.6\%$.

By proceeding in this manner, a series of curves of radiation and unaccounted loss can be drawn for a range of loadings and size of boiler (Fig 7.31).

4.5 Example of boiler losses calculation

The following information relates to a pulverised-fuel boiler test carried out on a 550 MW unit at full load, the steam flow being 472 kg/s.

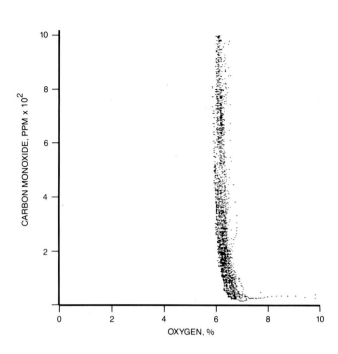

FIG. 7.30 Scatter diagram of CO versus O_2
Note the dramatic change of CO when the O_2 reaches the
critical value. Here, the value is about 6.5% O_2, a high
figure which indicates that there is significant air ingress
after the combustion zone.

TABLE 7.14

Typical values of the normally calculated boiler losses
for a PF-fired boiler

Loss	GCV_v %	NCV_p %
Dry flue gas	4.00	4.30
Wet flue gas	5.25	–
Sensible heat in water vapour	–	0.75
Combustible in ash	0.25	0.25
Radiation and unaccounted	0.45	0.46
Total loss	9.95	5.76
Boiler efficiency		
= (100% − losses)	90.05	94.24

TABLE 7.15

Heat values of carbon and hydrogen in kJ/kg

	GCV_v	NCV_p
Hydrogen	143 050	121 840
Carbon to CO	10 200	–
Carbon to CO_2	33 820	–

Coal analysis:

Carbon 55.5% Hydrogen 3.8% Sulphur 1.6%

Ash 16.0% Moisture 14.0% Nitrogen 1.7%

Oxygen 7.4%

TABLE 7.16

Typical hydrogen content of fuels and heating values
in kJ/kg

Fuel	Hydrogen %	GCV_v	NCV_p
Coal	4.5	22 000	20 800
Fuel oil	12.0	44 000	41 500
Natural gas	24.0	54 000	48 800

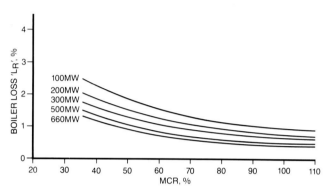

FIG. 7.31 Basic boiler radiation and unmeasured loss
The value obtained from the figure is multiplied by 1.5
for outdoor boilers.

GCV_v = 22 725 kJ/kg, determined from a bomb-calorimeter test.

O_2 at airheater outlets (average)	= 4.6%
CO_2 at airheater outlets (average) (from Fig 7.26)	= 14.4%
Combustible in pulverised-fuel ash (PFA), (average)	= 1.0%
Combustible in furnace-bottom ash (FBA), (average)	= 0.3%
Airheater gas-outlet temperature (mean)	= 121°C
Air temperature at FD inlet duct (mean)	= 32°C
CO in flue gas	= negligible

Calculation of losses

NCV_p = 22 725 − [(212.1 × 3.8) +
24.4 (14.0 + 1.6) + 6] kJ/kg fuel

= 22 725 − [806 + 380.6 + 6]

= 21 532 kJ/kg fuel

Combustible-in-ash (assuming 20% FBA and 80% PFA):

FBA $= 0.2 \times 16.0 = 3.2\%$ of coal

PFA $= 0.8 \times 16.0 = 12.8\%$ of coal

Carbon in FBA $= 0.003 \times 0.032 = 0.000096,$

$$\simeq 0.0001 \text{ kg/kg coal.}$$

Carbon in PFA $= 0.01 \times 0.128 = 0.00128,$

$$\simeq 0.0013 \text{ kg/kg coal.}$$

Dry flue gas loss $= \dfrac{100}{12 \times 14.4} \left[\left(\dfrac{55.5}{100} - 0.0014 \right) \right]$

$$30.6 \ (121 - 32)$$

$$= (0.579)(0.554)(30.6)(89)$$

$$= 874 \text{ kJ/kg coal}$$

$$= \dfrac{874 \times 100}{21\ 532} = 4.1\%$$

Wet flue gas loss $= \dfrac{14 + (9 \times 3.8)}{100} \ [4.2 \ (25 - 32)$

$$+ \ 2442 + 1.88 \ (121 - 25)]$$

$$= 0.482 \ (-29.4 + 2442 + 180.48)$$

$$= 1250 \text{ kJ/kg coal}$$

Sensible heat loss $= 1250 - (22\ 725 - 21\ 532)$

$$= 57 \text{ kJ/kg coal} = 0.26\%$$

Combustible-in-ash loss

Loss to PFA $= 0.00128 \ (33\ 820) = 43.3 \text{ kJ/kg coal}$

Loss to FBA $= 0.0001 \ (33\ 820) = 3.4 \text{ kJ/kg coal}$

Therefore total loss $= 47 \text{ kJ/kg coal} = 0.2\%$

Radiation and unaccounted loss $=$

$\log_n L_R = 1.88 - 0.4238 \log_n 472 = 1.88 - 2.6093$

$\log_n L_R = -0.729$

$$L_R = 0.5\%$$

4.6 Accuracy of losses determination

Although the boiler efficiency in the previous section was calculated as 94.9% (Table 7.17), it is pertinent to consider how accurate the result is likely to be, as every measurement used to determine the losses contains an element of inaccuracy. To illustrate this, consider the determination of gas temperature in a

TABLE 7.17
Heat balance on NCV_p basis

	kJ/kg	%
Dry flue gas	874	4.1
Sensible heat	57	0.3
Combustible in ash	47	0.2
Radiation and unaccounted		0.5
Total loss		5.1
So boiler efficiency		94.9%

large duct. The temperature pattern is far from uniform, as shown by the typical example of isotherm distribution in Fig 7.32. The average temperature is 261°C, but the chance of sensing that temperature with a single probe is slim. Therefore, where stratification may be present, some form of averaging device must be used to reduce the measurement error.

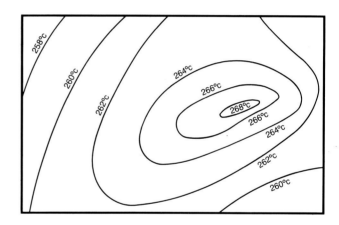

FIG. 7.32 Temperature distribution in a large duct
There is usually a considerable variation of gas temperature in ducts, a typical example of which is shown. To obtain an acceptable representative value, it is necessary to carry out a traverse of the duct.

Similarly, for high grade tests, such as acceptance tests, two separate fuel analyses are required for the determination of calorific value. The results should agree within 350 kJ/kg and the mean is used for the test calculations.

Table 7.18 has been prepared, using data from Section 4.5 of this chapter.

Consider next the effect of positive errors on the losses to demonstrate the method. (Negative errors could be calculated in the same way.) There are two types of absolute accuracy determined in Table 7.18: engineering values (such as °C) and percentages. In the case of a percentage, the effect on an individual

TABLE 7.18

Accuracy of measurement of boiler losses (Note: RSS = root sum of squares)

Measurement of estimate	Value	Source of error	Accuracy	
			Relative ± %	Absolute ±
Flue gas				
O_2	4.6%	Sampling	2.5	
		Analysis	5.0	
		Total (RSS)	5.6	$0.056 \times 4.6 = 0.26\%$
Final gas temp	121°C	Sample and measurement		5°C
Air inlet temp	32°C	Sample and measurement		1.5°C
Specific heat	1.0 kJ/kg	Approximation	2	$0.02 \times 1.0 = 0.02$ kJ/kg
Ash				
Carbon in PFA	1.0%	Sampling	20	
		Analysis	10	
		Total (RSS)	22.4	$0.224 \times 1 = 0.22\%$
Carbon in FBA	0.3%	(as above)	22.4	$0.224 \times 0.3 = 0.07\%$
Proportion PFA	0.8	Estimate		0.1
Fuel analysis				
Carbon	55.5%	Sampling	5	
		Analysis	1	
		Total (RSS)	5.1	$0.051 \times 55.5 = 2.83\%$
Hydrogen	3.8%	Sampling	5	
		Analysis	5	
		Total (RSS)	7.1	$0.071 \times 3.8 = 0.27\%$
Moisture	14.0%	Sampling	5	
		Analysis	5	
		Total (RSS)	7.1	$0.071 \times 14.0 = 0.99\%$
Ash	16.0%	Sampling	5	
		Analysis	2.5	
		Total (RSS)	2.6	$0.056 \times 16.0 = 0.90\%$
NCV_p	21 530 kJ/kg	Sampling	0.5	
		Analysis	1.0	
		Total (RSS)	1.12	$0.012 \times 21530 = 258$ kJ/kg

(Footnote: The author is indebted to Mr. A. P. Salt for his assistance in compiling the perturbation analysis)

loss is found by substitution into the appropriate loss formula. As an example, consider the effect of O_2 accuracy on the dry flue gas loss: the absolute accuracy was found to be 0.26% in Table 7.18, and the O_2 determined during the test was 4.6%. Therefore, it is required to know what the loss would have been if the O_2 had been 4.6 + 0.26 = 4.86%.

This is the equivalent of 14.14% of CO_2. Substituting this into the loss equation gives a loss of 889.6 kJ/kg, compared to the original value of 874 kJ/kg, so the effect of the error would be to increase the loss by 15.6 kJ/kg.

Where an error is expressed in engineering units, the method is simpler. Consider the final gas temperature. The absolute accuracy is 5°C and the dry flue gas loss depends upon the difference in temperature between the air and gas, i.e., 121 − 32 = 89°C. Hence the effect of a five degree error will be to alter the originally calculated loss of 874 kJ/kg by 5/89 × 874 = 49.1 kJ/kg.

The two methods described enable the accuracies of all the losses to be determined, as listed in Table 7.19.

Therefore, the total loss could be 5.1 + 0.35 = 5.4% with the positive errors of measurement used

TABLE 7.19

Effect of accuracy on losses

Measurement	Absolute accuracy	Dry flue gas	Sensible heat in water vapour				Effect on total loss
			Moisture	GCV − NCV	Net loss	C in A	
O_2	0.26%	15.6					15.6
Final temp	5°C	49.1			4.4		53.5
Air temp	1.5°C	− 14.7			− 3.2		−17.9
Specific heat	0.02 kJ/kg	17.5					17.5
Carbon in PFA	0.22%					9.8	9.8
Carbon in FBA	0.07%					0.8	0.8
Proportion of PFA	0.1					3.9	3.9
Carbon	2.83%	44.0					44.0
Hydrogen	0.27%		63.0	− 57.3	5.7		5.7
Moisture	0.99%		25.7	− 24.2	1.5		1.5
Ash	0.90%					2.6	2.6
Combined (RSS)		± 71.9			± 8.0	± 10.9	± 75.0
Percentage of NCV (NCV accuracy 1.12%)		± 0.33 − 0.05			± 0.04 − 0.00	± 0.05 − 0.00	± 0.35 − 0.06
All combined (RSS)		± 0.34			± 0.04	± 0.05	± 0.35

for the computation and the efficiency would be 94.9 ±0.35%.

Notice that the dry flue gas loss is the main one and, in fact, there is not a great deal of difference if the remaining losses are ignored.

Also, observe that when adding the component parts of an individual loss (i.e., vertical addition in Table 7.19) the root sum of squares is used, whereas when adding horizontally to obtain the effect of individual uncertainties on the total loss algebraic addition is used.

In the context of an acceptance test, the accuracy of the estimate of radiation and unaccounted loss is irrelevant as it has been agreed by the parties concerned beforehand. It can be included in the uncertainty estimate of the boiler efficiency if the *overall* accuracy is required for any purpose.

4.7 Variation of boiler efficiency with fuel quality

Direct comparison of test results for a particular boiler is often complicated by the fact that the quality of the fuel burned is different from one test series to another. For example, the fuel used during acceptance testing may not be available later, even though it is desirable to relate the subsequent results to those obtained on acceptance. Therefore, it is necessary to calculate what the test results would have been had a standard fuel been burned.

Consider the boiler loss calculation given in Section 4.5 of this chapter. Suppose it is required to refer the results to a 'standard' fuel, the 'test' and 'standard' analyses being:

	Test, %	Standard, %
Carbon	55.5	59.2
Hydrogen	3.8	4.0
Sulphur	1.6	2.0
Ash	16.0	16.5
Moisture	14.0	11.8
Nitrogen	1.7	1.5
Oxygen	7.4	5.0
Volatile matter	28.0	28.3
GCV_v	22 725 kJ/kg	25 060 kJ/kg
NCV_p	21 532 kJ/kg	−

Let subscripts 's' and 't' refer to standard and test fuels respectively. So the NCV_p of standard fuel

$$= 25\,060 - [(212.1 \times 4) + 24.4(11.8 + 1.65) + 6]$$
$$= 23\,877 \text{ kJ/kg}$$

The unburned carbon loss (L_u) is given by:

$$L_{us} = L_{ut} (A_s/A_t) \times (CV_t/CV_s) + k \text{ \%}$$

$k = 0$, unless the volatile matter of either the test or standard fuels is less than 17%. If the volatile matter is less than 17% then:

$$k = 0.013 (\overline{A}_s/\overline{A}_t) \times (CV_t/CV_s)$$
$$[\exp(0.255 \, \overline{C}_s/\overline{H}_s) - \exp(0.255 \, \overline{C}_t/\overline{H}_t)]$$

The test combustible-in-ash loss is 0.2%,

So $\quad L_{us} = 0.2 \times \dfrac{16.5}{16.0} \times \dfrac{21\,532}{23\,877} = 0.19\%$

The dry flue-gas loss (L_d) is given by:

$$L_{ds} = L_{dt} (CV_t/CV_s) \times (C_{es}/C_{et})$$

Where: C_e = equivalent carbon in the fuel

$$= [\overline{C} - L_u(NCV_p/33\,820) + 0.375\overline{S}]\%$$

So: $\quad C_{es} = 59.2 - \left(0.19 \times \dfrac{23\,877}{33\,820}\right) +$

$\qquad (0.375 \times 2.0)$

$\qquad = 59.2 - 0.135 + 0.750 = 59.8\%$

$\qquad C_{et} = 55.5 - \left(0.2 \times \dfrac{21\,532}{33\,820}\right) +$

$\qquad (0.375 \times 1.6)$

$\qquad = 55.5 - 0.128 + 0.6 = 56.0$

The test dry flue-gas loss is 4.1%

So: $\quad L_{ds} = 4.1 \times \dfrac{21\,532}{23\,877} \times \dfrac{59.8}{56.0} = 3.9\%$

The sensible heat loss (L_s) is given by:

$$L_{ss} = L_{st} \times (\overline{M}_s + 9\overline{H}_s/\overline{M}_t + 9\overline{H}_t) \times$$
$$(NCV_p)_t/(NCV_p)_s$$

$$= 0.3 \times \left[\dfrac{11.8 + (9 \times 4.0)}{14.0 + (9 \times 3.8)}\right] \times$$

$$\left[\dfrac{21\,532}{23\,877}\right]$$

$$= 0.3 \times 0.992 \times 0.902$$

$$= 0.27\%$$

Therefore the losses are:

	Test, %	Standard, %
Dry flue gas	4.1	3.9
Sensible heat	0.3	0.27
Combustible-in-ash	0.2	0.19
Radiation and unaccounted	0.5	0.5
	5.1	4.8
Efficiency	94.9%	95.2%

4.8 Tramp air

One major cause of reduced boiler efficiency is air inleakage, usually known as *tramp* air. This particularly affects coal-fired boilers, and even more so if they are fired by suction mills. The air inleakage may be before or after the combustion zone. Typical locations of inleakage before the combustion zone include:

● Suction milling plant.

● Ash hopper water seals.

● Ash hopper doors ajar or ill-fitting.

● Attemperating air to mill dampers open or passing.

The last is not strictly air ingress, but is undesirable because it has by-passed the airheater, in common with all tramp air. The operator usually maintains optimum air at the combustion chamber by trimming the FD fans until the O_2/CO meters indicate the desirable conditions. With significant tramp air present, the FD fans are, therefore, reduced from normal. As a result, the airflow through the airheater is reduced and so is the heat abstracted from the hot flue gas. Hence, the airheater gas outlet temperature rises, so reducing the boiler efficiency.

Tramp air also enters the boiler after the combustion chamber. Inleakage manifests itself as extra gas at the airheater gas inlet, causing the oxygen reading to be high. The inleakage from a hole of a given size is more significant the nearer it is to the ID fan suction, so it is also necessary to check for inleakage after the airheaters. This is done by using a portable oxygen analyser, such as the Servomex model illustrated in Fig 7.33.

Typical locations for inleakage after the combustion chamber and before the airheater gas inlet are:

● Defective duct expansion joints.

● Holed boiler casing.

● Inspection or access doors open or ill-fitting.

● Defective boiler-roof seals.

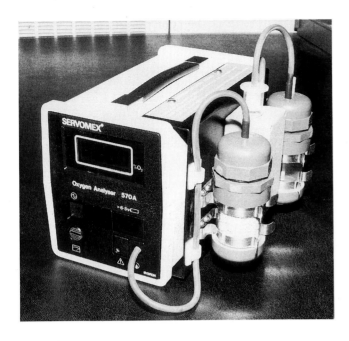

FIG. 7.33 Portable oxygen analyser
This analyser has its own filter unit and pump, so is
completely self-contained. There is a digital display of the
oxygen content.

The method of evaluating the quantity of tramp air is
quite simple, although laborious. It consists of determining the total heat surrendered by the flue gas at the
airheater and the heat accepted per kilogram of air at
the airheater. The quantity of air is found by dividing
the total heat by the heat per kilogram of air.

The optimum quantity of air required at the combustion zone is about 9 kg per kg of coal and this
should be maintained by reference to the CO meter.
Therefore, if the calculation shows only, say, 7 kg
leaving the airheater it can be inferred that 2 kg of
tramp air is leaking in before the combustion zone.
Similarly, if the air equivalent of the gas entering
the airheater is 10 kg per kg of coal then there is
1 kg/kg coal inleakage between the combustion zone
and the airheater.

The method is best illustrated by an example. Consider
a 500 MW PF boiler with the following conditions:

(1) $\overline{H} = 3.8\% \ \overline{M} = 17.4\% \ \overline{A} = 13\%$ Fuel flow =
58 kg/s (estimated)

(2) Conditions at airheater (A/H):

a	Air inlet temperature	=	29°C
b	Air outlet temperature	=	292°C
c	Gas inlet temperature	=	327°C
d	Gas outlet temperature	=	134°C
e	Gas inlet O_2	=	4.3%
f	Gas outlet O_2	=	5.5%

(3) Theoretical air required = 7.5 kg/kg fuel
(from methods already explained).

(4) Airheater log mean temperature difference
(LMTD) = $(\theta_1 - \theta_2)/\log_n(\theta_1/\theta_2)$

Where $\theta_1 = c - b = 35°C$, from (2) above
$\theta_2 = d - a = 105°C$
LMTD = 63.7°C

(5) Airheater seal leakage   = $100(f - e)/(21 - f)$
= 7.7%

(6) Excess air at A/H gas inlet = $e/(21 - e) = 0.26$

(7) Air required for gas inlet condition
= $[1.0 + (6)] \times (3)$
= 9.45 kg/kg fuel

(8) Excess air at A/H gas outlet
= $f/(21 - f) = 0.35$

(9) Air required for gas outlet condition
= $[1.0 + (8)] \times (3)$
= 10.1 kg/kg fuel

(10) Leakage air = (9) − (7) = 0.67 kg/kg fuel

(11) Flue gas at A/H inlet   = $(7) + \left(1.0 - \dfrac{A}{100}\right)$
= 10.3 kg/kg fuel

(12) Heat to leakage air   = $(d - a) \times (10) \times 1.01$
= 71 kJ/kg fuel

(13) Wet flue gas   = $(\overline{M} + 9\overline{H})/100$
= 0.52 kg/kg fuel

(14) Heat from wet flue gas   = $(c - d) \times 2.03 \times (13)$
= 203.7 kJ/kg fuel

(15) Heat from dry flue gas   = $(c - d) \times [(11) - (13)]$
= 1887.5 kJ/kg fuel

(16) Total heat from flue gas   = (14) + (15)
= 2091.2 kJ/kg fuel

(17) Heat to combustion air   = (16) − (12)
= 2020 kJ/kg fuel

(18) Heat gained by air   = $[(2b) - (2a)] \times 1.01$
= 265.6 kJ/kg air

(19) Combustion air   = (17)/(18)
= 7.6 kg/kg fuel at
A/H outlet

(20) Air at A/H inlet   = (19) + (10)
= 8.27 kg/kg fuel

(21) Tramp air from A/H air out to gas in

$$= (7) - (19)$$
$$= 1.85 \text{ kg/kg fuel}$$

(22) Tramp air, including A/H leakage

$$= (21) + (10)$$
$$= 2.52 \text{ kg/kg fuel}$$

(23) Item (21) as % of theoretical air

$$= [(21) \times 100]/(3)$$
$$= 24.7\%$$

(24) Item (22) as % of total air $= [(22) \times 100]/(20)$
$$= 30.5\%$$

(25) (Heat to air)/°C LMTD $= (17)/(4)$
$$= 31.7 \text{ kJ/kg fuel °C}$$

(26) Fuel throughput 58 kg/s

(27) Air at A/H inlet $= (20) \times (26)$
$$= 480 \text{ kg/s}$$

(28) Specific volume of air $= 0.7736[(2a) + 273]/273$
$$= 0.856 \text{ m}^3/\text{kg}$$

(29) Volume flow of air to A/H $= (28) \times (27)$
$$= 411 \text{ m}^3/\text{s}$$

(30) Flue gas at A/H outlet $= (11) + (10)$
$$= 11.0 \text{ kg/kg fuel}$$

(31) Flue gas flow at A/H outlet

$$= (30) \times (26)$$
$$= 638 \text{ kg/s}$$

(32) Specific volume of gas $0.76 \times [(2d) + 273]/273$
$$= 1.13 \text{ m}^3/\text{kg}$$

(33) Volume flow of gas at A/H outlet

$$= (31) \times (32)$$
$$= 720.9 \text{ m}^3/\text{s}$$

(34) Heat to combustion air $= (17) \times (26)$
$$= 117\,160 \text{ kW}$$

(35) Heat per °C LMTD $= (34)/(4)$
$$= 1839 \text{ kW}$$

The salient points are:

● The A/H leakage is 7.7%. This is good for rotary airheaters.

● The air leaving the A/H is 7.6 kg/kg fuel, so there is about 1.4 kg/kg fuel inleakage before the combustion zone.

● The air equivalent of the A/H gas inlet is 9.45 kg/kg fuel, therefore, there is about 0.45 kg air inleakage per kg fuel from the combustion zone to the airheater inlet.

● The A/H LMTD is 63.7°C. Comparison with the clean value will indicate how dirty the airheater is.

● The air and gas flow to and from the airheater respectively are 411 m³/s and 721 m³/s.

● The total heat transferred to the combustion air is 117 MW.

Normally such an investigation would be carried out for each individual airheater.

If a series of tests is carried out keeping the same airheater gas inlet $O_2\%$, but at different combustion chamber suctions, a graph can be plotted, such as Fig 7.34. This will give an indication of the necessary reduction of tramp air for a given airheater outlet temperature to be achieved.

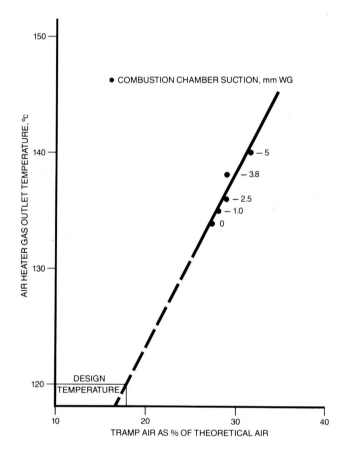

FIG. 7.34 Tramp air investigation on a 500 MW boiler
If the $O_2\%$ is kept constant for a range of furnace pressures, the points will all be in a straight line. Thus, the reduction of airheater gas outlet temperature for a given reduction of tramp air can be determined.

It should be established that the temperatures and O_2 values used are truly representative of the conditions at the airheaters. This is normally done by using averaging devices, or by carrying out a comprehensive traverse of the duct beforehand to establish the measuring locations.

4.9 Dilution effects of tramp air

Consider a flue gas analysis, without dilution, of CO_2 14.4%; H_2O 6.1%; SO_2 0.2%; N_2 75.9%; O_2 3.4%; CO 100 PPM. How will these values change with, say, 10% dilution by tramp air?

Air contains 21% oxygen and 79% nitrogen by volume, so 10% dilution = 0.1 (21% O_2 + 79% N_2) = 2.1% O_2 + 7.9% N_2.

Therefore 10% dilution will result in:

			New %	Change relative to original analysis, %
CO_2	14.4/1.1	=	13.09%	9.1
H_2O	6.1/1.1	=	5.55%	9.0
SO_2	0.2/1.1	=	0.18%	10.0
N_2	(75.9 + 7.9)/1.1	=	76.18%	−9.5
O_2	(3.4 + 2.1)/1.1	=	5.00%	−56.2
CO^2	100/1.1	=	90.91 PPM	9.1
			(100%)	

Hence, the effect of dilution is to cause a change of approximately the same percentage for all of the constituent gases *except oxygen and nitrogen*. For oxygen, the error is considerably in excess of the dilution, and this should be borne in mind when using O_2 analysers.

4.10 Computer-aided boiler optimisation

With modern sensing equipment temperatures, pressures, flows, gas analyses, etc., all produce electrical signals. Therefore, one can assemble equipment which will considerably ease the work of optimising boiler performance. Such aids are dealt with in detail in Volume F, but here it may be appropriate to look at one example, plus a mathematical model.

First, consider the equipment assembled by CEGB Midlands Region Production Services Department. It is called ADAPT (*A*utomatic *D*ata *A*nalysis for *Pl*ant *T*esting). It is a computer-based on-line system for monitoring boiler (or turbine) performance and has facilities for scanning, checking and recording

the test information. The basic layout is shown in Fig 7.35. It consists of several data acquisition cabinets located at convenient points around the plant to which the sensor outputs are fed. Each cabinet contains up to three scanners, a digital voltmeter and an extender. The purpose of the extender is to convert the data and control information to 'bit-serial' form so that it can be transmitted along a coaxial cable up to one kilometre long. The controller is a computer with a keyboard, graphics screen and printer.

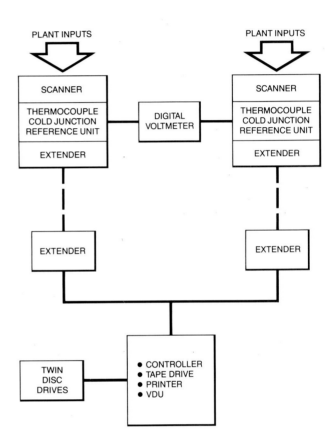

FIG. 7.35 Diagrammatic layout of ADAPT installation
The data acquisition cabinets can be up to one kilometre away from the controller, interconnected by means of a simple coaxial cable.

The tape drives provide storage for programs and data. The disks will hold 240 scans of data.

The results are calculated automatically and presented graphically, enabling selected parameters to be monitored continuously for several hours. Figure 7.36 shows a typical trace.

The second aid is a mathematical boiler model called POSTMAN devised by the CEGB Scientific Services Department.

The requirements of the model are:

(a) The equations used to describe the significant modes of energy transfer should have the simplest

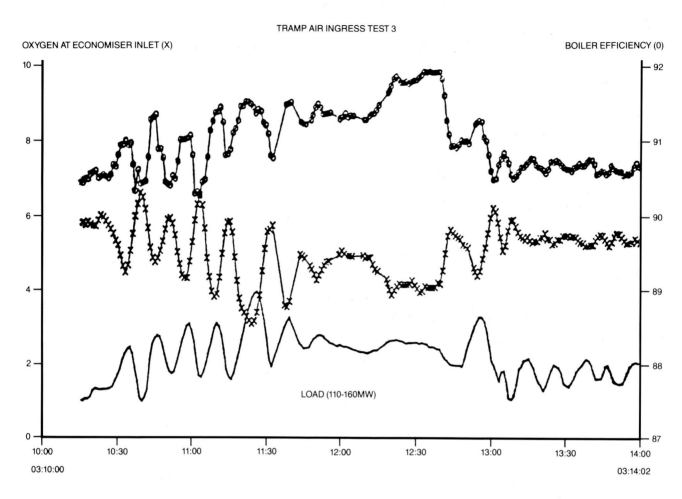

TRAMP AIR INGRESS TEST 3

OXYGEN AT ECONOMISER INLET (X)   BOILER EFFICIENCY (0)

LOAD (110-160MW)

FIG. 7.36 Representation of an ADAPT printout
Various parameters can be selected to be continuously recorded. The effects of changes of plant operating conditions
are easily seen.

possible form, based upon physical principles, whilst retaining the ability to predict plant behaviour within the accuracy of plant measurements.

(b) The constants used in the model should not need adjustment to permit adequate representation of the boiler at different operating conditions, such as load variation.

(c) The model should be constructed so that unknowns, such as the state of boiler fouling, can be evaluated from practical plant measurements.

(d) A corollary of the above requirements is that model representations should be good enough to allow valid judgements about the effects of changes in boiler operation and design on plant performance.

Figure 7.37 shows how the POSTMAN modules have been assembled to model a boiler at Cottam power station. The major uncertainty in the construction is the distribution of tramp air. However, this can reasonably be assumed to be before the combustion chamber, at the economiser exit and between the airheater outlet and ID fan inlet. Figure 7.38 shows some of the plant correlations which were used.

Considerable information can be obtained from such a model, including optimisation of sootblowing, effect of changes of heat transfer surface in the boiler, optimisation of operating parameters and so on.

5 Turbine performance

5.1 The ideal turbine stage

The efficiency with which heat in the steam is converted to mechanical work at the turbine moving

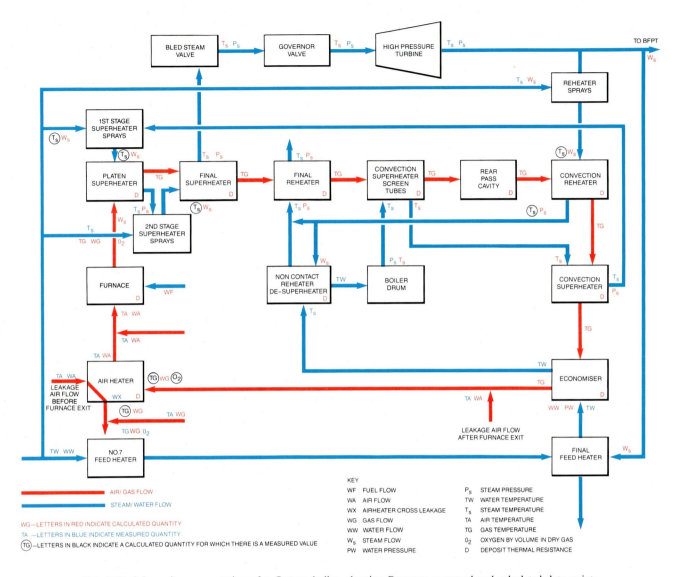

FIG. 7.37 Schematic representation of a Cottam boiler, showing Postman measured and calculated data points
Cottam power station has 500 MW units and early work on Postman was carried out there. The required data points for the formulation of the successful mathematical model are shown.

blades is primarily dependent upon the blade speed/steam speed ratio.

Consider an ideal turbine stage in which steam is admitted to the moving blades in the direction of blade motion and leaves in the opposite direction. It can be shown that the efficiency (η) is given by: $\eta = 4 [(u/V_i) - (u/V_i)^2]$, where u/V_i = blade speed/steam speed ratio.

Using this formula, the co-ordinates shown in Table 7.20 can be obtained. These are plotted to give the well known curve for an ideal turbine stage shown in Fig 7.39. The maximum efficiency occurs when the blade speed is half the steam speed. V_o is the blade outlet steam velocity and it can be shown that $V_o = V_i - 2u$. At maximum efficiency, the outlet velocity is zero and the entire velocity energy of the steam is utilised in doing work on the blades.

Notice from Fig 7.39 how rapidly the efficiency changes as the u/V_i ratio is altered. The u/V_i ratio is the design parameter which has the most influence on the efficiency of turbines.

5.2 Practical turbines

5.2.1 General considerations

Practical machines may be classified into two broad categories:

● Impulse.

● Impulse reaction (usually referred to as 'reaction').

489

the steam expands only in the fixed nozzles and its consequent increase in kinetic energy is utilised to drive the moving blades.

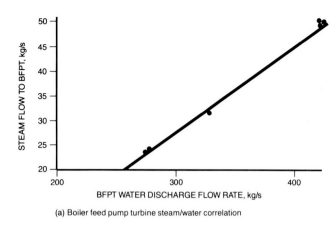

(a) Boiler feed pump turbine steam/water correlation

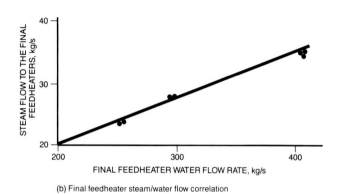

(b) Final feedheater steam/water flow correlation

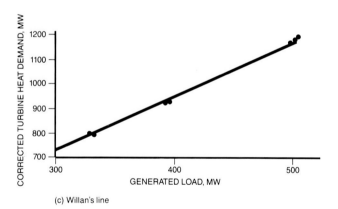

(c) Willan's line

FIG. 7.38 Some of the relationships used in formulating the Postman mathematical model: derived from series-2 heat rate data

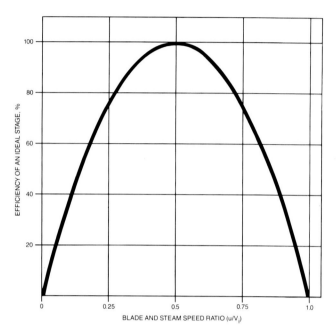

FIG. 7.39 Relationship between (blade speed/steam speed) ratio and efficiency of an ideal stage
The curve reaches 100% efficiency when the blade speed is one-half the steam speed. Variations of the blade and steam speed ratio have a marked effect upon the theoretically attainable efficiency.

In the *impulse-reaction* turbine, the steam pressure is reduced in passing across both the fixed and the moving blades as they are each formed into nozzles. Thus, the motive power of the moving blades is derived partly from the reaction of the jet of steam leaving the moving nozzles and partly by utilising the kinetic energy of the steam flowing across the moving nozzles.

Turbine-generators must run at a speed (n,r/s) determined by the frequency (f) of the system to which they are connected, given by n = f/p, where p is the number of pole-pairs.

The speed of a turbine directly driving a 2-pole, 50 Hz generator must be 3000 r/min. For 60 Hz, the necessary speed is 3600 r/min.

In *impulse* machines, there is a pressure drop of the steam across the fixed blades only — in other words,

TABLE 7.20

Stage efficiency versus u/V$_i$ ratio

u/V$_i$	0	0.1	0.2	0.3	0.4	0.5	0.6	0.7	0.8	0.9	1.0
Efficiency	0	0.36	0.64	0.84	0.96	1.0	0.96	0.84	0.64	0.36	0

5.2.2 Efficiency and nozzle angle

Consider impulse blading. The steam leaves the nozzle at an angle α, so the blade inlet angle θ will be as shown in Fig 7.40. The steam velocity can be regarded as having two components at right angles. One is the *velocity of flow* V_{fi} which causes the steam to flow along the turbine. The other is the component that does work on the blades, and is known as the *velocity of whirl* V_{wi}. Similar considerations apply at the outlet of the moving blades. For convenience, the two vector diagrams are usually combined such as shown in Fig 7.41.

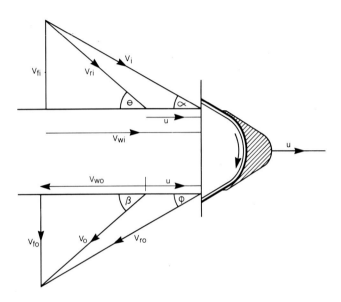

FIG. 7.40 Velocity diagram for turbine blade
The steam is required to glide onto the moving blade, and to achieve this it is necessary to have a nozzle angle of α. θ and ϕ are the blade inlet and outlet angles and β is the angle of the steam at outlet. The blade speed is u.

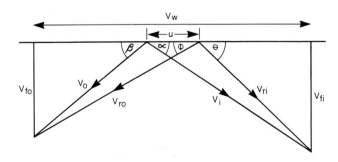

FIG. 7.41 Combined velocity diagram
The blade inlet and outlet velocity diagrams are usually combined as shown. The velocity of whirl acts in the same direction as the blade motion. The velocity of flow is at right angles to it and so causes the steam to flow through the turbine.

A simple calculation shows that the efficiency of such a stage is represented by a formula similar to that for an ideal stage, where $\cos \alpha = 1$ for the ideal turbine.

$$\eta = 4[(u/V_i)\cos \alpha - (u/V_i)^2]$$

Table 7.21 shows how η varies for different nozzle angles and (u/V_i) ratios.

Some of the information from Table 7.21 is plotted in Fig 7.42. Notice how the efficiency decreases as the nozzle angle increases; also how the efficiency changes for a given blade angle as the (u/V_i) ratio is varied. Designers try to achieve the maximum practical efficiency.

The theoretical maximum efficiency of an impulse row occurs when $(u/V_i) = \cos(\alpha/2)$. When the effects of friction are considered, the maximum attainable efficiency is reduced, although the general form of the curve is much the same. Figure 7.43 shows a typical curve for a row of blades with and without friction.

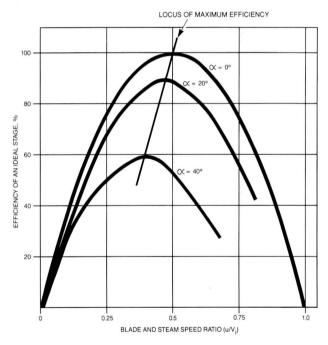

FIG. 7.42 Relationships of (blade speed/steam speed) ratio and nozzle angle
The maximum attainable efficiency decreases as the nozzle angle increases. The top curve ($\alpha = 0°$) is the ideal blade stage and cannot be used in practice.

5.2.3 Basic turbine types

These are dealt with in more detail in Volume C.

Impulse turbines can be sub-divided into the following main types:

TABLE 7.21
Efficiency for various nozzle angles

Nozzle angle α, degrees	Blade speed/steam speed						
	0	0.2	0.4	0.5	0.6	0.8	1.0
0	0	0.64	0.96	1.0	0.96	0.64	0
10	0	0.63	0.94	0.97	0.92	0.59	—
20	0	0.59	0.86	0.88	0.82	0.45	—
30	0	0.53	0.75	0.73	0.64	0.21	—
40	0	0.45	0.59	0.53	0.40	—	—

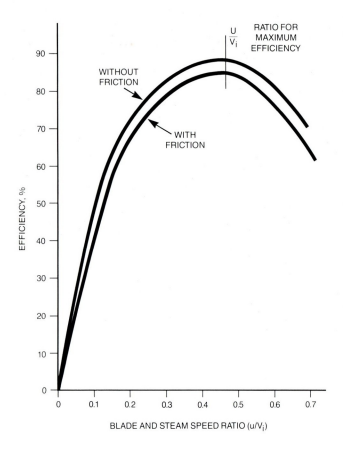

FIG. 7.43 Efficiency/u/V_i curves
The upper curve shows typical values of efficiency and blade speed/steam speed for a blade without friction. The effect of friction is to lower the maximum efficiency but this still occurs at the original u/V_i ratio.

- Simple impulse.
- Velocity-compounded.
- Pressure-compounded.
- Pressure-velocity compounded.

In all turbines, it is important to realise that the steam *glides* onto the moving blades and work is done by the change of direction of the steam as it passes over the blades.

The simple impulse turbine consists of one row of nozzles followed by one row of moving blades, as shown in Fig 7.44. The steam expands from inlet to outlet pressure across the set of nozzles, with the

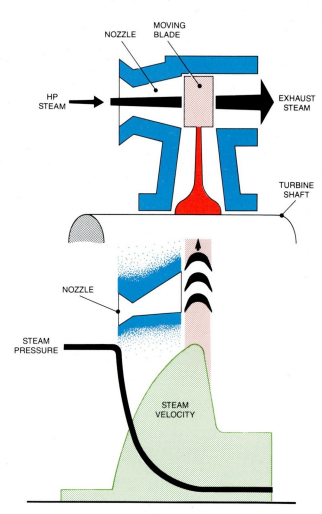

FIG. 7.44 Simple impulse turbine
This consists of one set of nozzles and one wheel. Blade speeds can be very high.

result that the steam velocity at the nozzle exit is very high. This high velocity steam is guided onto the moving blades without shock or impact. Any energy remaining in the steam when it leaves the blades is lost. This type of turbine usually runs at extremely high speeds, often over 30 000 r/min.

The velocity-compounded turbine has one set of nozzles at the inlet followed by several rows of moving blades, interspersed with fixed guide blades as shown in Fig 7.45. The guide blades merely alter the direction of the steam to make it suitable for admission to the next row of moving blades. The after-nozzle steam pressure is constant through the remainder of the turbine.

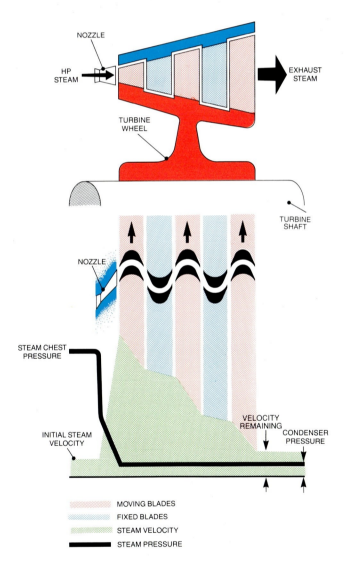

FIG. 7.45 Velocity-compounded turbine
The pressure is reduced to back pressure at the inlet nozzles. The resultant high steam velocity is used to drive the subsequent moving blades.

In the pressure-compounded turbine there are rows of fixed nozzles followed by moving blades. A row of nozzles and its following row of moving blades is known as a turbine stage. Thus, this is really a series of simple-impulse turbines. The steam pressure is reduced at each set of nozzles and the increased velocity resulting from each pressure drop is utilised in the following row of moving blades, as shown in Fig 7.46.

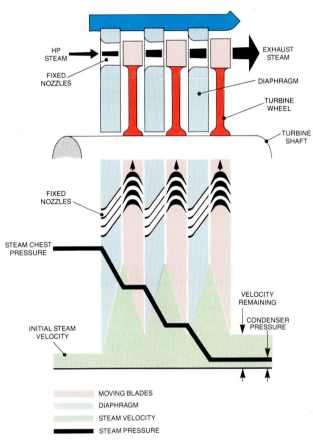

FIG. 7.46 Pressure-compounded turbine
This is really a series of simple-impulse turbines. As the pressure drop at each stage is only a part of the total available, the blade speed is much lower than in the simple turbine.

The steam velocity at each stage is rather low, so the blade velocities can also be low thus preventing excessive steam friction loss. The type is sometimes referred to as a *Rateau* turbine.

The pressure-velocity compounded turbine is illustrated in Fig 7.47. It is the equivalent of two velocity-compounded machines in series. The steam pressure is reduced at the inlet nozzles to some intermediate value and reduced to back pressure at the second set of nozzles.

493

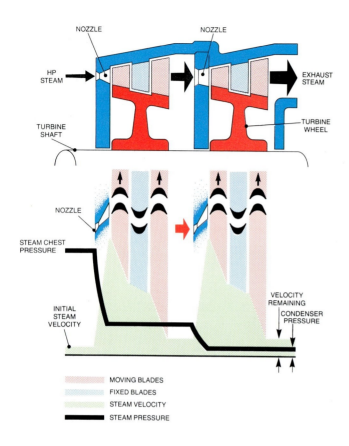

FIG. 7.47 Pressure-velocity compounded turbine
This is essentially two velocity-compounded turbines in series. The pressure reduction at each of the cylinder inlet nozzle belts causes the steam velocity to increase; this velocity energy is used to drive the turbine.

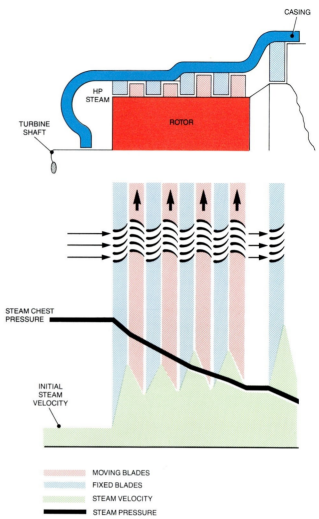

FIG. 7.48 Reaction turbine
All the moving and fixed blades are nozzles, so the pressure falls throughout the turbine. The reaction of the steam due to the pressure reduction, plus the utilisation of impulse energy, provides the turbine driving force.

The impulse-reaction turbine type was first built by Sir Charles A. Parsons and the general layout is shown in Fig 7.48. As the steam passes through the turbine, its pressure is reduced while passing across both the fixed and the moving blades. Thus, every moving row has a differential pressure acting upon it causing a cumulative thrust in the direction of steam flow. To counteract this, *balance pistons* (or *dummy pistons*) are provided. In addition, the pressure drop across every row means that the clearance between the tips of the blades and the rotor or cylinder must be kept very small to prevent undue steam leakage from stage to stage.

It is common for practical machines to incorporate features from more than one type. For example, a reaction turbine may incorporate a velocity stage at inlet. Similarly, a pressure-compounded impulse machine may have low pressure blades which have a considerable degree of reaction. There is little to choose in terms of cost and efficiency between the best impulse and reaction types, which is why they are both in common use.

From these introductory notes about turbines, a few pointers to achieving good performance will be apparent:

● Wherever there is a pressure drop across a row of fixed or moving blades, the clearances must be kept as small as possible.

● Steam conditions at the turbine should be kept at the design values, particularly at high loads.

● Blades and nozzles should be smooth and undamaged to prevent loss due to steam friction.

● Back pressure must be kept as low as practicable to obtain the maximum utilisation of heat in the steam.

494

5.3 Turbine-generator efficiency and heat rate

The aim of the turbine designer is to obtain a performance as near to the ideal as possible. In practice, about 85% of the theoretical value is attained. Thus, if the ideal cycle efficiency is 53% and the turbine-generator (T/G) design efficiency is 85% then:

$$\text{Actual T/G efficiency} = \text{ideal efficiency} \times \text{design efficiency}$$
$$= 0.53 \times 0.85 = 0.45 \text{ or } 45\%$$

However, when considering the thermal performance of turbines, it is more common to refer to its *heat rate* than to its efficiency.

Heat rate is defined as the quantity of heat supplied to a turbine to give one kWh output at a specified loading.

It follows that the units of heat rate are kJ/kWh. As both efficiency and heat rate are a measure of the thermal performance of the plant, they are inter-related.

$$\text{Heat rate} = 3600/(\text{actual efficiency})$$
$$\text{Actual efficiency} = 3600/(\text{heat rate})$$

Thus, if a 500 MW turbine-generator has a guaranteed heat rate at full-load of 7940 kJ/kWh, then its actual efficiency is $3600/7940 = 0.453$ or 45.3%.

Both the efficiency and the heat rate of a machine vary with load, so it is most important that the loading is specified when quoting performance.

The *design efficiency* of the 500 MW machine mentioned earlier will be (assuming the terminal conditions in Fig 7.10, which gave an ideal efficiency of 53.2%):

$$\text{Design efficiency} = \frac{\text{actual efficiency}}{\text{ideal efficiency}}$$
$$= \frac{0.453}{0.532} = 0.852 \text{ or } 85.2\%$$

The expressions are summarised below:

- Ideal cycle efficiency   53.2%
- Design efficiency   85.2%
- Actual efficiency   45.3%
- Heat race   7940 kJ/kWh

Of these, heat rate is of the greatest concern to most power station staff. They are particularly interested in the present actual heat rate compared to that achieved during acceptance testing.

5.4 Practical heat rates of turbine-generators

Clearly, it is important to specify exactly what a given performance figure refers to. As far as turbine-generators are concerned, the main definitions are:

- *Guaranteed heat rate* This is the value which the manufacturer is contractually obliged to achieve at a specified loading and with specified terminal conditions at the machine.

- *Acceptance test heat rate* This is the value obtained from contractual tests on the plant. They are carried out under the direction of the manufacturer and witnessed by representatives of the customer. Normally the results, when corrected to the specified conditions, will be better than guarantee.

- *'As-run' heat rate* This is the result obtained from a machine with terminal conditions other than specified. For example, it is often not possible to test with the back pressure at its specified value, or the turbine stop valve or reheat conditions may be different from those specified. Such deviations may raise or lower the 'as-run' heat rate compared with the 'acceptance' value.

- *'Generated' basis heat rate* The output basis selected will affect the heat rate result considerably. For contractual tests the generated output is normally specified.

- *'Sent-out' heat rate* The sent-out output is the generated output less the electrical power used by the unit auxiliaries. For routine CEGB purposes the 'sent-out' output is used, as the object is to determine the heat required by the turbine for a given output to the transmission system.

Whatever the test on a turbine-generator, it is usually carried out in special circumstances. For example, during the testing there will normally be no blowdown or sootblowing of the boiler; neither will there be any ashing-out, no make-up will be admitted to the unit and every effort will be made to keep all conditions as constant as possible and close to design.

Thus, the test results are normally better than those obtained under normal operating conditions.

5.5 Turbine heat consumption tests

5.5.1 Use made of test results

As soon as possible after commissioning, the manufacturer is required to demonstrate the performance of the machine by means of *acceptance tests*. After this, the machine is tested as a routine at least once every two years for the remainder of its life. Although classed as routine, the tests are conducted with considerable care by the station staff and are witnessed by independent CEGB test engineers.

The purpose of the heat consumption tests is to establish a *Willans line* for the machine. For example, suppose a series of tests on a 500 MW machine gave the results shown in Table 7.22.

be a slight scatter of the points due to testing tolerances, so the line of best fit is determined by means of regression analysis. The Willans line for the results given in Table 7.22 is shown in Fig 7.49. The heat consumption at zero load is shown where the Willans line intercepts the Y axis at 252.7 GJ/h. This is 6.28% of the full-load heat and values of about 7% are usual for throttle-governed machines.

TABLE 7.22

Heat consumption at various outputs
The heat consumption is determined at least twice, at about the same loading if possible, to ensure that the results are repeatable

Output	MW	515.682	512.460	288.429	285.300	371.426
Heat consumption	GJ/h	4149.300	4126.400	2438.000	2415.800	3026.400

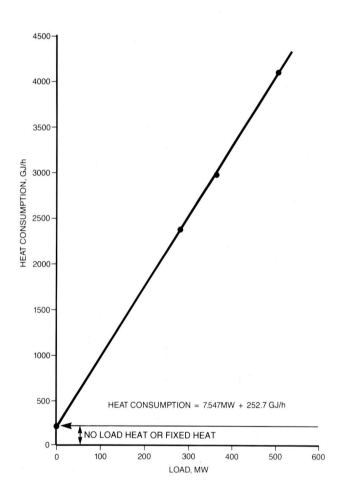

FIG. 7.49 Willans line
There is a straight-line relationship between the output and heat consumption of a turbine. The heat consumption at zero load is called the *fixed heat*, and the slope of the line is called the *incremental heat*.

When the heat consumption and loadings are plotted they will ideally lie on a straight line called the Willans line. In practice, though, there will always

The no-load value for nozzle-governed machines is of the order of 3.5% of full-load heat.

The slope of the line is called the 'incremental heat' and it is constant for all loads, in this case 7.547 GJ/MW.

Heat consumption = (incremental heat × load) + fixed heat

For example, at 450 MW load, the heat consumption will be:

(7.547 × 450) + 252.7 GJ/h = 3649 GJ/h

The corresponding heat rate is: $(3649 \times 10^6)/450\ 000$ = 8109 kJ/kWh and the efficiency is: 3600/8109 = 0.444 or 44.4%.

Table 7.23 was prepared by performing the same calculations for a series of loadings.

The information obtained from tests such as these is useful for various purposes, the most important of which is its contribution to the Order of Merit tables.

5.5.2 Method of testing

Heat consumption tests are carried out in accordance with CEGB Site Test Code No. 2 'Steam Turbine-Generator Heat Rate Tests' and BS752, to which reference should be made for further information. Care must be taken before testing to ensure that the plant is in good order. For example, that there is no tube leakage at the feedheaters, isolating valves do not pass, leakage from the system is as insignificant as possible, etc. Also, the test instrumentation must be of 'test' accuracy, preferably calibrated immediately before the tests are carried out.

TABLE 7.23
Variation of turbine-generator performance with loading

Load, MW	500	450	400	350	300
Heat consumption, GJ/h	4026	3649	3272	2897	2517
Heat rate, kJ/kWg	8052	8109	8179	8269	8389
Efficiency, %	44.7	44.4	44.0	43.6	42.9

The unit should be kept at a steady loading for a time before testing commences to ensure that everything is 'heat-soaked'. The tests themselves are usually of one hour duration, during which time every effort should be made to keep fluctuations of the terminal conditions to a minimum.

The basic information required for the heat consumption is very simple:

$$\left[\frac{\begin{array}{c} M_s\,(H_1 - h_f) + M_{R/H}(H_3 - H_2) + \\ M_{is}\,(h_f - h_{is}) + M_{ir}\,(H_3 - h_{ir}) \end{array}}{P_g} \right] \times 3600$$

where M_s = steam flow to HP turbine stop valves, kg/s

H_1 = specific enthalpy of steam at HP turbine stop valves, kJ/kg

h_f = specific enthalpy of final feedwater, kJ/kg

$M_{R/H}$ = steam flow from HP exhaust to reheater, kg/s

H_2 = specific enthalpy of steam at HP turbine exhaust, kJ/kg

H_3 = specific enthalpy of steam at IP turbine stop valves, kJ/kg

M_{is} = spraywater flow to superheater, kg/s

h_{is} = specific enthalpy of spraywater to superheater, kJ/kg

M_{ir} = spraywater flow to reheater, kg/s

h_{ir} = specific enthalpy of spraywater to reheater, kJ/kg

P_g = power at generator terminals, kWh

The various specific enthalpies are easily obtained from careful measurement of the appropriate temperatures and pressures and reference to steam tables. However, the various flows are much more difficult to measure at the required locations to the degree of accuracy required. This is such a problem that it is usual to measure the water flow at a location such as the de-aerator outlet and make suitable adjustments to derive the TSV steam flow and the reheater steam flow. The superheat spray and reheat spray-water flows are usually measured directly. Figure 7.50 shows a diagrammatic layout of a 500 MW unit and the subsidiary flows are indicated. Notice that the main flow measurement is at the de-aerator outlet (413.047 kg/s) and it is necessary to add the effects of various subsidiary flows to arrive at the TSV flow of 409.883 kg/s and the HP exhaust flow to reheater of 330.46 kg/s.

The heat added to the feedwater in the boiler ($H_1 - h_f$) was calculated and found to be 2400.8 kJ/kg, and that in the reheater ($H_3 - H_2$) 465.4 kJ/kg. The superheater spray ($h_f - h_{is}$) is 511.1 kJ/kg and the reheater spray ($H_3 - h_{ir}$) is 3021.7 kJ/kg.

Therefore the various heats are:

(a) In boiler
= 2400.8 × main steam flow
= 2400.8 × 409.883
= 984 047.1 kJ/s

(b) In reheater
= 465.4 × HP exhaust flow to R/H
= 465.4 × 330.46
= 153 796.0 kJ/s

(c) To superheater spray
= 511.1 × spray flow
= 511.1 × 5.0
= 2555.5 kJ/s

(d) To reheater spray
= 3021.7 × spray flow
= 3021.7 × 4.028
= 12 171.4 kJ/s

(e) Total heat consumption
= 1 152 570.0 kJ/s

(f) So total heat consumption
= 1152570.0 × 10^{-6} × 3600
= 4149.2520 GJ/h

(g) Output at generator terminals
= 515 682 kW for one hour

(h) 'As-run' heat rate
= [(f)/(g)] × 10^6 = 8046.1 kJ/kWh

497

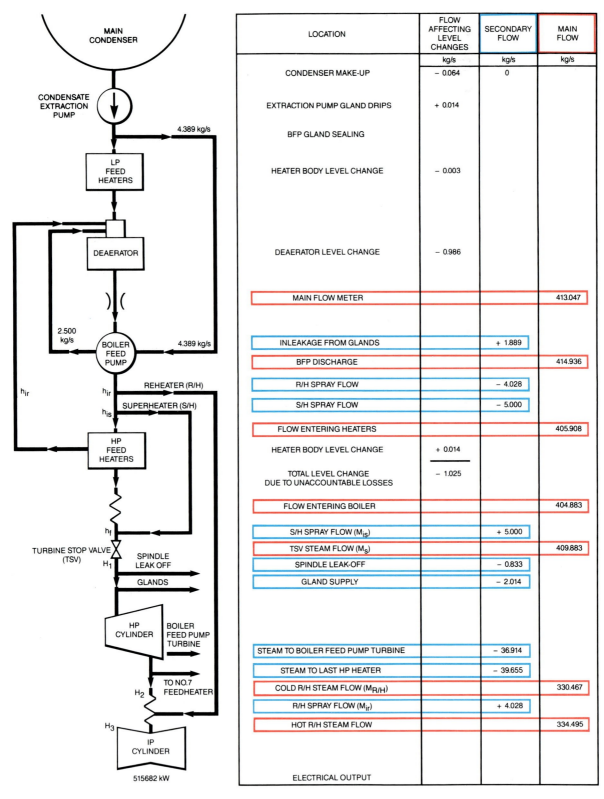

FIG. 7.50 Turbine flow diagram
The main flow measurement is at the de-aerator outlet. Therefore the subsidiary flows must be measured to arrive at the required flow values at the turbine.

For purposes of comparison with tests at other times, it is necessary to refer the performance to standard conditions of steam temperature and pressure at the TSV and IP inlet; back-pressure; final feed temperature, and so on.

This is done by applying corrections to the 'as-run' results, examples of typical corrections being shown in Fig 7.51, to determine the 'corrected' heat rate.

5.5.3 Interpretation of results

When the corrected heat rate is calculated, it may be that it is worse than expected. If so, the next requirement is to determine what has caused the deviation. The heat rate result itself is no help, except to indicate that there is a problem. It is therefore necessary to carry out further investigations to determine probable causes of the trouble.

Such extra tests will include the determination of:

- HP and IP cylinder efficiency.

- Turbine pressure survey.

- Main shaft-gland leakage rate.

- Deposition on turbine blades and blade roughness.

- Condenser performance.

- Feedheater performance.

Normally the cylinder efficiencies are determined every six months, the turbine pressure surveys every month and the gland leakage as required. Condenser and feedheater performance are considered separately in later sections of this chapter.

5.6 Cylinder efficiency tests

5.6.1 Method and effect on heat rate

Although called cylinder efficiency tests, this simple method is equally applicable to any points on the turbine whose steam enthalpy can be accurately determined, i.e., provided the steam is not wet. Thus, it can be applied to two bled-steam take-offs from a cylinder, for example.

The method is illustrated in Fig 7.52. Consider the HP cylinder. The steam conditions at the TSV are pressure P_1 and temperature T_1, from which the specific enthalpy can be determined. It is a throttle-governed machine, and so the specific enthalpy at the after-throttle pressure P_2 will be the same as that at the TSV. If the steam could be utilised at 100% efficiency, the expansion would be isentropic, and so would be represented by a vertical line extending to the HP cylinder exhaust pressure P_3. In practice,

the expansion is less than 100% efficient and so the actual exhaust conditions are determined by the temperature T_3 and pressure P_3.

The cylinder efficiency is given by:

(actual heat drop)/(isentropic heat drop).

Similar considerations apply to the IP cylinder.

It is of great importance that the pressures and temperatures used are accurate, as minor deviations can result in significant errors in the result. For example, Table 7.24 shows the calculated cylinder efficiency for a 1% change in selected parameters used for cylinder efficiency determination.

Typical values for the cylinder efficiencies of large output machines are:

HP cylinder 89% This is because the blades are short, particularly at the early stages, with consequent high tip-losses.

IP cylinder 92% The blades are longer and so the tip losses are reduced and the velocities are moderate.

LP cylinder 80% The blades are very long, velocities are very high, there is wetness in the steam at the last few stages and the steam flow path is highly flared.

Deviations of cylinder efficiency change the heat rate of the machine to a greater or lesser degree, depending upon which cylinder is affected. As a general guide, the following may be used to assess the change of heat rate for one percentage point change of cylinder efficiency:

HP cylinder	= 0.2% heat rate
IP cylinder	= 0.25% heat rate
IP/LP cylinder	= 0.75% heat rate
LP cylinder	= 0.5% heat rate

For example, if tests indicated that the HP cylinder efficiency is three percentage points worse and the IP/LP cylinder is two percentage points worse, then the effect on the heat rate of the machine will be, approximately, $(0.2 \times 3) + (0.75 \times 2) = 2.1\%$.

5.6.2 Effect of loading

It has already been mentioned that the efficiency of blading is primarily determined by the u/V_i ratio, i.e., the blade speed/steam speed ratio. Clearly, the blade speed will remain the same whatever the loading on the machine. What is possibly not so clear is that the steam speed will also remain about the same

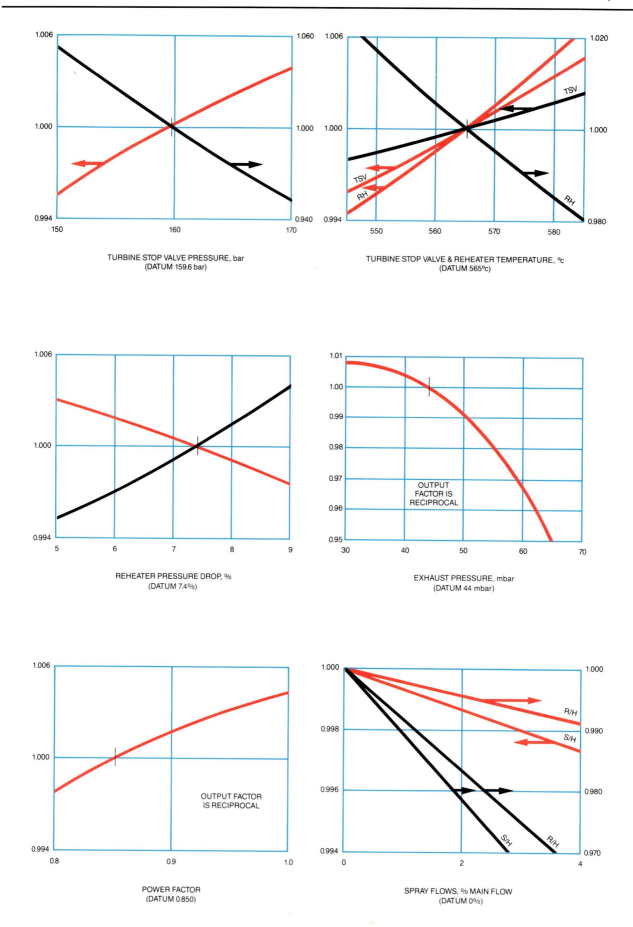

FIG. 7.51 Heat rate correction curves

The correction curves shown are typical examples for a large modern unit. The ordinates are multipliers to be applied to the test heat-rate and output to give the heat rate and output which would be obtained under datum conditions.

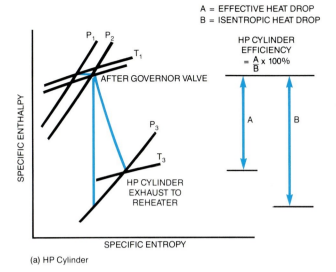

(a) HP Cylinder

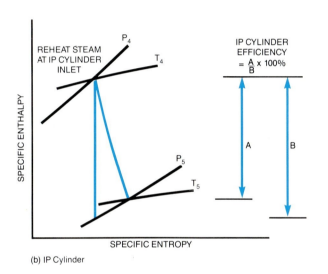

(b) IP Cylinder

FIG. 7.52 Cylinder efficiency
Provided the steam is not wet, the pressure and temperature can be used to locate the state point of the steam on a Mollier Chart. The efficiency is the ratio of the isentropic and actual enthalpy drops.

whatever the loading. Thus the cylinder efficiency will be fairly constant for a considerable range of loads.

The reason for the almost constant steam velocity is that, as steam flow is reduced so, also, is the pressure. Therefore the volume flow is unaltered. For example, consider the conditions at the HP exhaust of a 500 MW turbine given in Table 7.25.

The implication is that the loading on the machine is not important when carrying out cylinder efficiency checks, provided it is not extremely low. Similarly, the normal variations of terminal conditions (pressure, temperature, etc.) have no effect when comparing one test with another. An exception to the above is the first HP cylinder stage on a nozzle-governed turbine, as this *is* affected by loading.

5.6.3 Interpretation of results

If the results show a deterioration in cylinder efficiency, it is important that care is taken in identifying the cause, as it is not necessarily indicative of blading defects. For example, Fig 7.53 shows the layout of a particular double-cased HP cylinder. Table 7.26 shows the (simplified) information obtained from a 120 MW turbine-generator cylinder in good condition and also when a poor result was obtained (about 82% is a reasonable design efficiency for such a cylinder).

Notice that the poor result is primarily due to the HP cylinder exhaust temperature being high — but why? The clue is that, for the good result, the pressure at the exhaust was only slightly lower than at the bled-steam take-off. Thus, there was a small steam flow toward the cylinder exhaust pipeline from the bled-steam pipe take-off. In the poor result, on the other hand, there must be a higher steam flow from the bled-steam take-off towards the exhaust pipe, and this can only occur because of inner to outer cylinder leakage, probably at the inlet shaft gland. The resultant high-temperature steam flow will mix with the flow leaving the last row of blades, thus raising the temperature above normal, and producing the poor cylinder efficiency.

As a second example, consider the double-flow double-casing HP cylinder of a large turbine illustrated in Fig 7.54. The steam flows through the inner casing to the primary exhaust, after which it flows inside the outer casing to the last three stages of blading to the cylinder exhaust. The design and test data for the cylinder is listed in Table 7.27. Clearly there is a significant departure of efficiency from design and it is necessary to determine the cause.

The inlet to primary exhaust efficiency is slightly worse than design, mainly because the primary exhaust temperature is slightly high.

This is almost certainly because of either blade damage or worn blading seals at the first five stages.

Much more serious, though, is the very poor performance of the primary exhaust to cylinder exhaust. It is barely credible that this could be due to blade wear or damage as it would need to be so bad. Much more probable is excessive leakage from the centre gland, allowing leakage of high quality steam from the inner cylinder to the outer. Such steam would mix with the main flow and pass through the last three stages. Unfortunately, the steam temperature at the inlet to the last three stages cannot be measured, so it is *assumed* to be the same as at the primary exhaust. If it could be measured, though, it would probably be found to be of the order of 455°C. Thus, the actual conditions will be similar to those illustrated in Fig 7.55. On test, the steam temperature at point 4 is a little high, thus accounting for the marginally-worse efficiency from the inlet to the primary exhaust. Meanwhile, because of the steam leakage, the

TABLE 7.24

Effect of errors of measurement on cylinder efficiency calculations

Basic conditions:

TSV	158	bar/566°C
ATV	150	bar
Cylinder exhaust	42.1	bar/368°C
HP cylinder efficiency	89.4	%
IP before valve	39.0	bar/565.4°C
IP inlet	38.5	bar
IP exhaust	3.6	bar/245.1°C
IP cylinder efficiency	91.6	%

+1% error in the following parameters:		Cylinder efficiency %	
TSV temperature		92.4	
TSV pressure		89.0	
ATV pressure	gives	88.8	cf 89.4
HP exhaust temperature		87.0	
HP exhaust pressure		90.2	

		Cylinder efficiency %	
IP before valve temperature		92.7	
IP before valve pressure		91.6	
IP inlet pressure	gives	91.3	cf 91.6
IP exhaust temperature		90.8	
IP exhaust pressure		91.9	

Alternatively, +1°C temperature error		Cylinder efficiency %
At TSV		89.9
At HP cylinder exhaust	gives	88.7
At IP before valve		91.8
At IP cylinder exhaust		91.3

TABLE 7.25

Constant steam volume at various loads

Load, MW	500	400	300	200
Pressure, bar (a)	42.6	34.1	25.6	17.0
Temperature, °C	368.8	362.0	352.0	346.0
Enthalpy, kJ/kg	3136	3136	3136	3136
Specific volume, l/kg	64.1	79.0	109.0	160.0
Flow, kg/s	320.4	256.3	192.2	128.2
Volume flow, m³/s	20.5	20.3	20.9	20.5

Notice how little the volume flow changes with load. The steam passage area is constant so the steam velocity must also be constant.

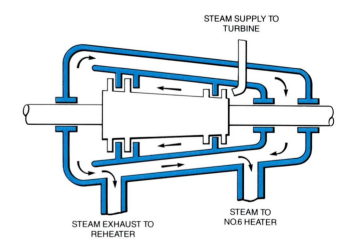

FIG. 7.53 HP cylinder efficiency
Diagram of the 120 MW HP cylinder quoted
in the example.

actual temperature of the steam entering the last three stages is probably about that represented by point 6, in which case the blading efficiency of the last three stages is satisfactory.

TABLE 7.26

HP cylinder conditions for a 120 MW turbine-generator

	Good result	Poor result
Load, MW	120.0	120.0
TSV pressure, bar	105.0	105.0
TSV temperature, °C	538.0	538.0
ATV pressure, bar	100.0	100.0
Bled-steam pressure, bar	29.3	29.8
Bled-steam temperature, °C	385.0	387.0
HP exhaust pressure, bar	29.1	29.1
HP exhaust temperature, °C	369.0	379.0
Cylinder efficiency, %	80.1	74.0

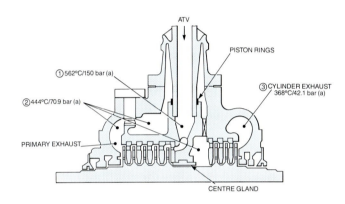

FIG. 7.54 Steam path and steam conditions
in the HP cylinder
The full-load steam conditions are illustrated. Notice
that steam leakage from the inlet stage will by-pass
the first five stages.

5.7 Turbine pressure survey

5.7.1 Introduction

The pressure survey is a graphical method of obtaining an indication of turbine internal faults. The method is basically very simple, consisting merely of comparing pressures obtained from the various available points along the turbine with reference data. It is important that the pressure indications are themselves accurate.

First, construct a graph for the machine being considered. Provide a suitable pressure scale from zero up to at least the turbine stop valve pressure, as shown in Fig 7.56, and draw an arbitrary sloping line from any convenient pressure above that of the TSV to zero in the bottom right-hand corner. Using design data or, preferably, acceptance-test data, mark

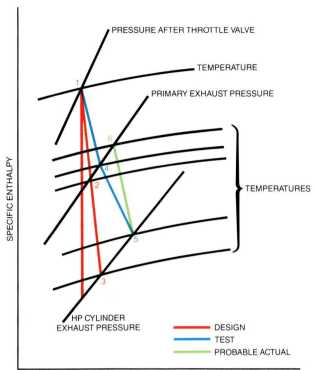

FIG. 7.55 H-S diagram of steam conditions in HP cylinder
The 'apparent' state line is from (1)(4)(5). However, it
is likely that there is significant steam leakage at the
centre gland or some similar location, causing the steam
temperature before stage 6 to be elevated to point (6)
on the diagram.

TABLE 7.27

Turbine HP cylinder results

	Design	Test
Load, % MCR	100.0	100.0
TSV pressure, bar	158.0	158.0
TSV temperature, °C	566.0	566.0
ATV pressure, bar	150.0	150.2
Primary exhaust pressure, bar	70.9	71.5
Primary exhaust temperature, °C	444.0	447.0
Cylinder exhaust pressure, bar	42.1	42.9
Cylinder exhaust temperature, °C	368.0	379.0
Inlet to primary exhaust efficiency, %	86.6	84.6
Primary exhaust to cylinder exhaust efficiency, %	91.7	80.4
Inlet to cylinder exhaust efficiency, %	89.4	83.8

points on this sloping line at the various pressure tappings, and label them. This information should preferably refer to 100% MCR, although other loadings will do. This is the basic reference diagram. Should reference lines be required for other loads,

503

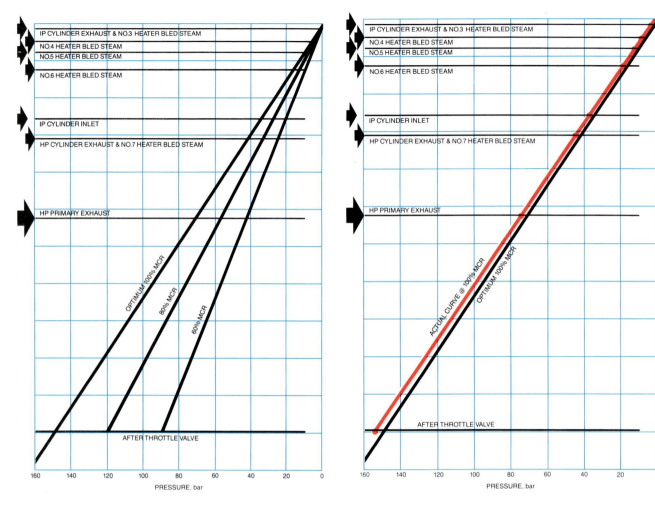

FIG. 7.56 Basic turbine-pressure survey
The absolute stage pressures are proportional to load.

FIG. 7.57 Turbine-pressure survey — interstage wear
The effect of wear is to elevate the pressures throughout
the turbine. If the interstage wear is significant, then
the main shaft glands are probably also worn.

this is done by directly proportioning the stage pressures. For example, if the 'after throttle valve' pressure at 100% MCR is 150 bar, then at 80% MCR load it will be 0.8 × 150 = 120 bar, and so on. Figure 7.56 shows the reference lines for 100%, 80% and 60% MCR.

5.7.2 Application of the method

Having constructed the diagram, it is now necessary to obtain a set of pressure readings from the machine at a steady known load, say 100% MCR. If the turbine is in exactly the same mechanical condition as it was when the reference data was obtained, then the pressures will all plot on the reference line.

On the other hand, suppose there has been general wear at the internal seals. Then the stage pressures will all be higher than normal, as shown in Fig 7.57.

Another possible defect inside the turbine is that the interstage seals are satisfactory but some blades

are damaged, or there is some other restriction to flow. The upstream pressures from the restriction will be higher than normal, whilst those following will be about normal, as illustrated in Fig 7.58.

Of course, the probability is that if there is a defect in a machine, it will probably not be there in isolation — in other words it is quite possible that, if there is seal wear, there will also be a restriction to flow. It is important that all the feedheaters are in service during pressure surveys — otherwise the steam flows through the machine will not be consistent.

As an example of a pressure survey, consider Fig 7.59. At the time the loading was 75% MCR and the pressures were as indicated. It is immediately apparent that there is general internal seal wear throughout the machine. In addition, there is a restriction to flow between the IP cylinder inlet and the No 6 bled-steam take-off point. The restriction is probably not general between the two points, but at one particular stage. In this case, as the first IP stage is in-

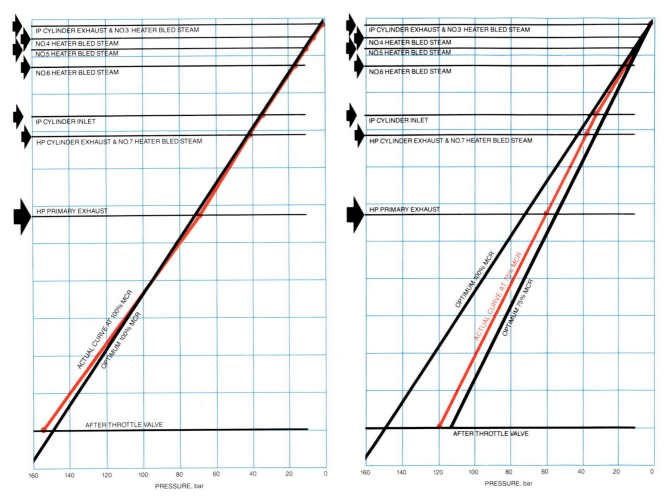

FIG. 7.58 Turbine-pressure survey — restriction to flow
A kink in the line is indicative of a restriction to steam flow somewhere between the extremities of the kink. In the example illustrated, it is between the 'after throttle valves' and the HP primary exhaust.

FIG. 7.59 Turbine-pressure survey — practical case
This is a typical example where interstage wear and restriction to flow are both present.

cluded, it is probably due to damage at the first stage blades. This is because welding slag, broken hacksaw blades, etc., can get into the boiler and/or reheater tubes during boiler repairs if adequate quality control is not applied. When the boiler is returned to service, this debris is carried forward to the turbine. Usually there are steam strainers to intercept the debris, but even so, quite large pieces can get past them and cause damage to the first moving row of blades that they encounter — here, the first IP row.

Should a pressure survey indicate internal gland wear, it follows that there is a very good chance that the main shaft glands have also been damaged.

5.8 Main shaft gland-leakage rate

Even when the main shaft glands are in good condition, they pass significant quantities of steam. Should the gland be worn, the quantity increases consider-

ably. It is, therefore, important to keep a check on the state of the main turbine glands. The method of doing this depends upon the gland leak-off system in use.

A common system for smaller machines is to pipe the gland leak-off steam to suitable bled-steam lines as in Fig 7.60. The first-stage HP cylinder gland leak-off is passed through the bled-steam pipe to No 4 HP feedheater, whilst the second-stage leak-off is piped to the No 1 LP bled-steam line. An increase in gland leak-off flow raises the steam temperature at the heater so this should be regularly monitored to give early warning of gland damage.

A second common arrangement is to pass the leak-off flows into common exhaust lines. A typical arrangement is shown in Fig 7.61 and the total second-stage flow can be measured at the drain from the gland condenser. An increase in the total flow raises the temperature of the condensate across the gland condenser. The first stage leak-off joins No. 2 heater bled-steam line.

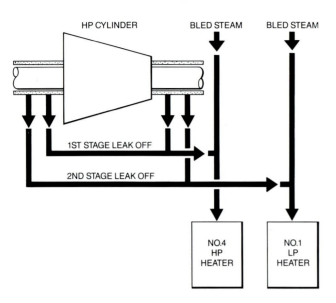

FIG. 7.60 Cylinder gland leak-off arrangements
It is common, particularly on smaller units, to pipe the
gland leak-off steam to suitable feedheater bled-steam lines.

By measuring the temperature of the individual supplies at each junction, it is possible to calculate the individual flows. Consider the arrangement shown in Fig 7.62 where three leak-offs are piped into a common manifold.

The total flow to the gland condenser can be measured at the drain, or else by carrying out a heat balance across the heat exchanger. Let this flow be F_1 and the other flows F_2, F_3, etc. Similarly, the heats are h_1, h_2, h_3, etc.

$$
\begin{aligned}
\text{Now } F_1 h_1 &= F_2 h_2 + F_3 h_3 \\
&= F_2 h_2 + (F_1 - F_2) h_3 \\
&= F_1 h_3 - F_2 h_3 + F_2 h_2 \\
&= F_1 h_3 - F_2 (h_3 - h_2) \\
&= F_1 h_3 + F_2 (h_2 - h_3) \\
\text{So } F_2 &= F_1 (h_1 - h_3)/(h_2 - h_3)
\end{aligned}
$$

So, if flow F_1 and heats h_1, h_2, h_3 are known, then flow F_2 can be calculated.

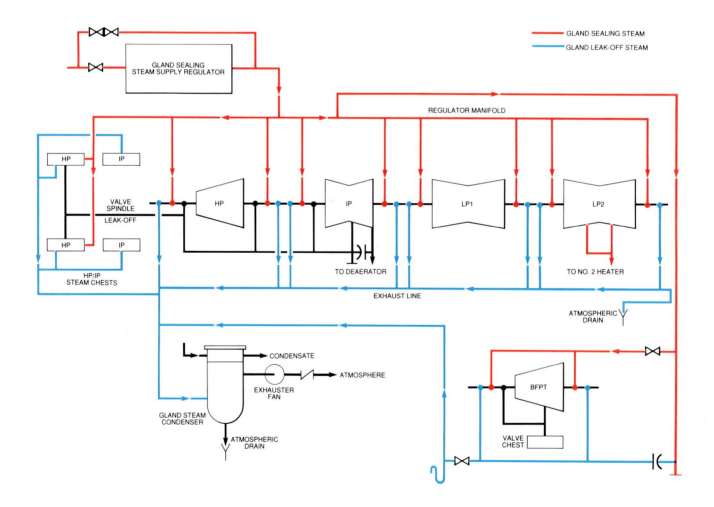

FIG. 7.61 Typical gland system of a large unit
The leak-off steam passes to a common exhauster line which is maintained at a suitable low pressure by the exhauster fan.
The heat in the exhausted gland steam is surrendered to the condensate at the gland condenser.

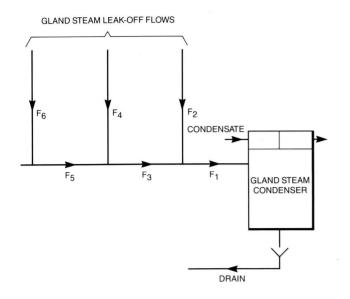

GLAND STEAM LEAK-OFF FLOWS

FIG. 7.62 Steam flow in a gland system
This is a portion of a typical modern gland leak-off
system. All the flows can be established from the total
flow and temperature data.

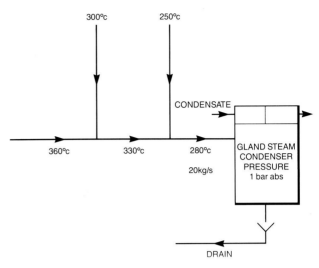

FIG. 7.63 Steam flow in a gland system
Example of the determination of individual steam flows.

Flow F_3 is then given by $F_1 - F_2$. Similar reasoning can be applied to flows F_3 F_4 and F_5 and so on.

For example, consider the values given in Fig 7.63. If the exhaust line is kept at about 1 bar absolute, the various heats are found from steam tables to be:

h_1 3034 kJ/kg h_4 3074 kJ/kg
h_2 2974 kJ/kg h_5 3196 kJ/kg
h_3 3155 kJ/kg

$$So\ flow\ F_2 = 20\,(3034 - 3155)/(2974 - 3155)$$
$$= (-2420)/(-181)$$
$$= 13.4\ kg/s$$
$$F_3 = 20 - 13.4$$
$$= 6.6\ kg/s$$
$$F_4 = F_3\,(h_3 - h_5)/(h_4 - h_5)$$
$$= 6.6\,(3155 - 3196)/(3074 - 31396)$$
$$= (-270.6)/(-122)$$
$$= 2.2\ kg/s$$
$$F_5 = F_3 - F_4 = 6.6 - 2.2$$
$$= 4\ kg/s$$

5.9 Deposition on turbine blades and blade roughness

With the high quality of water in modern boiler/turbine systems, the probability of deposition due to impurities is very low. Modern water treatment plants provide a product with virtually no undesirable contaminants. Note that the higher the operating pressure, the lower the permissible contaminants. On supercritical pressure plant, not only is the make-up water free of contaminants but, after each circuit of the boiler/turbine the condensate has picked up so much contamination from the plant that it must be 'polished' — that is, passed through a demineralisation plant when it leaves the extraction pump.

Thus, not only is it necessary to ensure that the water is acceptable, but also that the plant itself is not a source of contamination. Whilst plant is opened for maintenance (e.g., of a feed pump or a feedheater), dust may be deposited inside it in sufficient quantity to increase the silica content of the water significantly when it is returned to service. Similarly, a reheater tube leak can cause contamination by allowing dust and gases to enter the hole during light-loading periods, so leading to water contamination. Therefore, adequate quality control must be exercised to minimise contamination whenever the plant is opened for work or inspection.

Should contamination occur the most probable cause is silica. The solubility of silica in steam is very low at low pressures, but increases rapidly at higher values. At the critical pressure (221 bar) the solubility is unity, so for every kilogram of contamination in the water there will be a kilogram of silica carried over to the turbine in the steam. When they reach the appropriate condensation temperature, silica compounds will be deposited on the blades and they may be quite difficult to remove. Usually the appropriate condensation temperatures are reached at the last few rows of IP blades and the first few rows of LP blades.

The deposition, if allowed to accumulate, will gradually reduce the area of the steam passages, which will cause the preceding stage pressures to increase for a given loading. Therefore, it is appropriate to monitor the pressure at the LP cylinder inlet. A gradual rise of this pressure at a given loading probably indicates silica deposition on the LP blades.

Another common cause of loss of performance is that of blade roughness. Various researchers have tried to correlate a given degree of roughness with a corresponding reduction of efficiency. One factor to emerge from such studies is that apparently slight blade roughening is sufficient to cause a significant change of performance.

One researcher, Mr. V. T. Forster, produced a paper entitled 'Performance loss of modern steam-turbine plant due to surface roughness', showing that a given degree of roughness affects the heat rate more if it is present in the HP cylinder, followed by the IP cylinder, and least at the LP cylinder. Figure 7.64 shows the effect if present throughout the machine. The equivalent emery grade is used to give a better idea of what a given roughness is, and the ordinate indicates the deterioration of heat rate. A general roughness of 50 μm (just under 400 grade emery) lowers the stage efficiency of the HP cylinder of a 200 MW unit by about 9%, the IP by 7% and the LP by 3.5%.

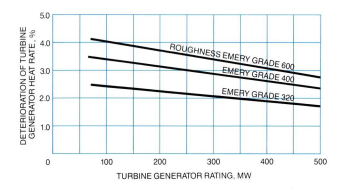

FIG. 7.64 Effect of turbine internal roughness
Roughness affects small machines more than large ones.
The greater the roughness, the greater the effect.

The corresponding values for a 500 MW machine are about 6%, 5% and 2%, which illustrates that a given roughness affects a large machine less than a smaller one.

Every time a turbine is opened, the opportunity should be taken to inspect the blading carefully, bearing in mind that roughness of only 10 μm (if present throughout the turbine) is sufficient to cause the heat rate of a 200 MW machine to worsen by 1%.

5.10   Variation of turbine-generator operating parameters

5.10.1   Types of governing

There are three types of steam governing in common use:

● Throttle.

● Nozzle.

● By-pass.

Throttle governing is used almost exclusively on CEGB machines, but the other types, particularly nozzle governing, are in use in some other parts of the world.

By-pass governing is the least common, so only a brief description will be given of its operation before considering the other two methods in more detail. The principle is that steam is supplied to the turbine via two throttle valves *A* and *B*. Steam from valve *A* is admitted to the first stage of the turbine and, at maximum steam flow, is sufficient to sustain the economic load of the machine, usually 80% of maximum. At loads above the economic rating, valve *B* admits steam to the turbine several stages down the turbine, i.e., the first few stages are by-passed by the steam from valve *B*. When both valves are fully open, the maximum steam flow is admitted to the turbine and the maximum rating of the machine is attained.

5.10.2   Throttle governing

In modern, large machines the economic rating coincides with the maximum rating so by-passing is not required. The steam flow is controlled by the degree of opening of the throttle valves which are located in the HP cylinder steam chest. Obviously the upstream steam pressure at the throttle valves will be constant, irrespective of the turbine loading, but the pressure after the throttle valves will reduce as the valve opening reduces. This will cause the available energy per kilogram of steam due to expansion to be reduced, as well as the mass flow of steam. This is illustrated for a non-reheat turbine in Fig 7.65. The turbine stop valve conditions are 105 bar, 565°C. At full-load, the loss of steam pressure due to the throttling effect of the turbine stop and throttle valves is about 5%. Hence the after throttle valve (ATV) pressure is, say, 100 bar absolute. The steam does work in the turbine by expansion until 40 mbar back pressure is reached, the state-point of the steam being represented by the line *AB*, representing 80% efficiency.

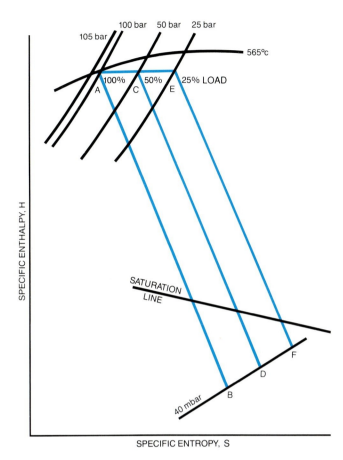

FIG. 7.65 Throttle-governed turbine
Steam state-point lines for various loadings.
Note, that the slope of the lines (i.e., the internal
efficiency) is unaffected by loading.

This reduction of heat drop, coupled with the reduced flow of steam, causes the turbine output to be reduced as the throttle valves close. Notice that the slope of the expansion line (Fig 7.65) remains the same for all loads, so the *efficiency* of the expansion remains the same, which is what one would expect from the earlier discussion about cylinder efficiency. On the other hand, it should be realised that throttling incurs losses as the steam pressure and temperature are adversely affected.

5.10.3 Nozzle governing

Here the first-stage nozzles are divided into a number of groups, sometimes more than ten. The steam admitted to each group is controlled by a valve, and each valve is opened in turn as loading is increased as shown in Fig 7.66. Consider a four-valve arrangement where the first valve only is in service from no-load to quarter-load, and at quarter load it is fully-open. This is followed by the second to half-load, and then valve three to three-quarter load. Finally, the last permits maximum loading when fully open. The only valve operating in the throttling mode is the last one to be brought into service, all the previous ones being wide open, so the losses due to throttling are not so great as with the previous type, so operation is more efficient.

At half-load, the ATV pressure will be half of that at full-load, as shown at *C*. The expansion of the steam is shown from *C* to *D*. Similarly, for 25% load, the ATV pressure will be a quarter of full-load and the expansion *EF*. Notice that the specific enthalpy of the inlet steam remains unaltered for the range of loads. However, the available energy is reduced significantly as the loading falls, as shown in Table 7.28.

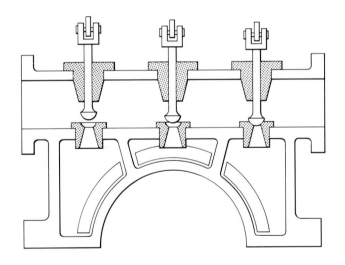

FIG. 7.66 Turbine nozzle-governing
The illustration shows three valves, and the
operating mechanism causes them to open sequentially.
When the first valve is fully open, the second commences
to open, and so on.

TABLE 7.28

*Heat drop per kg of steam at various loads
(at 80% efficiency)*

Load	Enthalpy kJ/kg		Heat drop kJ/kg
	At inlet	At exhaust	
100%	3537	2343	1194
50%	3537	2419	1118
25%	3537	2495	1042

This is illustrated in Fig 7.67, where the conditions are similar to those in Fig 7.65 except for the governing used. At full-load, the expansion is from *A* to *B*, similar to the previous example. However,

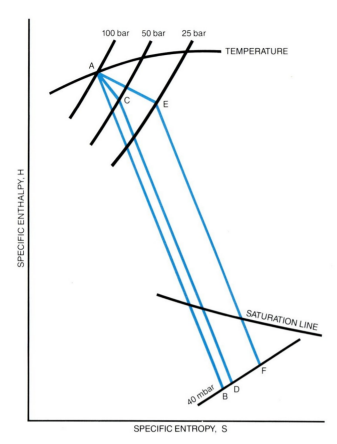

FIG. 7.67 Nozzle-governed turbine
The part-load pressures are identical to the
throttle-governed case but the utilisation of the heat
in the steam is greater at part-loads, so this type of
governing is desirable for turbines which will operate for
substantial periods at part-loads.

at 50% load, only two valves are open, but again
a pressure of 50 bar is required. This is achieved
more efficiently than before, so the line for two-
nozzle operation is shown from *A* to *C*. Thereafter
the expansion is from *C* to *D*. Similarly one nozzle
operation is shown from *A* to *E*, and thence from
E to *F*.

So nozzle governing differs from throttle governing
as follows:

- The turbine efficiency is higher at part-loads
  because of reduced throttling loss.

- The valve control gear is more complicated.

- The internal efficiency of throttle and nozzle
  governed machines can be the same (represented
  by the slope of the expansion lines on the H-S
  diagram) after the wheel case.

- Because of their improved part-load efficiency,
  nozzle-governed machines are preferable where the
  loading regime involves prolonged operation under
  these conditions.

5.10.4 Variation of TSV pressure

First consider a non-reheat machine whose after-
throttle-valve (ATV) steam conditions are 100 bar/
570°C, represented by *A* in Fig 7.68. The expansion
line is indicated by the sloping line *AB* to a back
pressure of 40 mbar. If the turbine stop valve pressure rises then, for a given throttle valve opening,
the ATV pressure will also rise. Suppose the ATV
pressure became 110 bar (with an unchanged temperature) then the expansion line would be the line
CD. In a similar manner, if the pressure fell to
90 bar the expansion would be represented by *EF*.
Obviously, varying the turbine inlet pressure causes:

- The heat drop of the steam to vary, higher pressures causing increased heat drop.

- Increasing exhaust wetness at higher pressures.

- Increased turbine output for a given valve opening
  such that, for example, 10% increase in pressure
  will produce about 10% extra output.

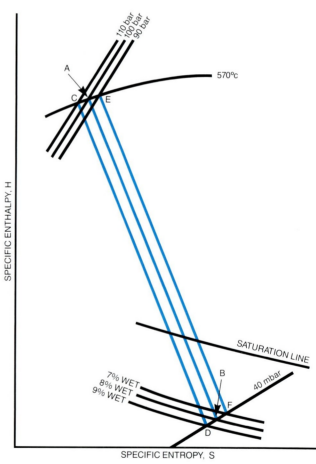

FIG. 7.68 Effect of TSV pressure variations
Increased TSV pressure on a non-reheat machine
will result in increased availability of heat in
the steam, and vice versa.

However, machines of reasonable output incorporate reheat, so attention will now be given to this, the more common type. The stage pressures are all proportional to flow and so is the reheater pressure. For example, Fig 7.69 represents the conditions of the ideal cycle at full load. ATV pressure is 160 bar and the expansion line is from A to B, where B is at the HP cylinder exhaust pressure of 40 bar. Ignoring, for the moment, considerations of pressure drops in the cold and hot reheat pipes, the inlet pressure to the IP turbine will also be 40 bar, and the expansion line will be as shown from C to D for the IP/LP cylinders.

to H. Hence, the HP cylinder enthalpy drop remains about the same, whilst the IP/LP enthalpy drop becomes smaller as load is reduced, even though there is a slight increase in the IP inlet specific enthalpy. Therefore, the efficiency of the steam utilisation is reduced by part-load operation.

Next, consider the effects of the pressure drops in the cold and hot reheat lines for a practical turbine. The conditions are shown in Fig 7.70. From the inlet pressure of, say, 160 bar the full-load ATV pressure will be 150 bar, shown by the horizontal line AB. Expansion in the HP cylinder is represented by the line BC. There is then a pressure drop in the cold

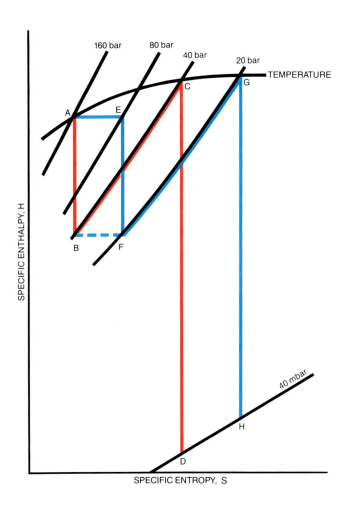

FIG. 7.69 Part-load operation of reheat turbine
The isentropic heat drop at full load is AB, plus CD. At half-load, it is EF, plus GH. Note that part-load operation only affects the heat drop in the IP and LP cylinder.

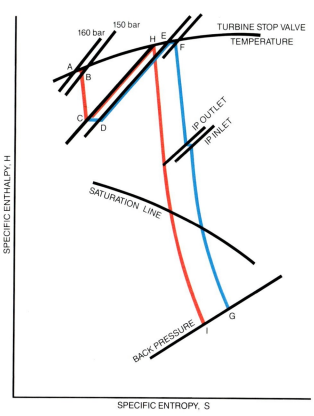

FIG. 7.70 Practical expansion of steam in turbine
Pressure drops have a significant effect on the quantity of heat which is available in the steam. Practical conditions are represented by ABCDEFG where, for example, CD represents the pressure drop in the cold reheat pipe. If the pipe friction could be eliminated, the expansion would be ABCHI and the heat available would be increased.

If the output is reduced to half-load, the expansion lines change. The inlet and outlet HP cylinder specific enthalpy remains the same, but the pressures are now reduced in proportion to the load. The IP inlet pressure will be 20 bar and the expansion from G

reheat pipe, CD, after which the steam is heated in the reheater, represented by DE. The pressure at E is slightly lower than at D because of the pressure drop in reheater. There is a further pressure drop as the steam passes through the hot reheat pipe, represented by the horizontal line EF. There is a slight loss of heat from the pipe as well as a drop of pressure, but the heat loss is so small in relation to the total that it is not discernible on the diagram. Expansion in the IP/LP turbines is represented by the line *FG*. Note the considerable change caused by the various pressure drops.

If there had been none, the expansion line would have been drawn as *ABCHI*, clearly a much better utilisation of the steam. Of course, in practice there must always be a pressure drop and so the penalty must be accepted. Nevertheless, it should be fully realised that the consequences of an excessive pressure drop can be very serious for the efficient performance of the unit. Commonly, the reheater design pressure-drop is about 7% of the HP exhaust pressure at full-load.

5.10.5 Sliding pressure control

The principle of sliding control of pressure is best illustrated by reference to a throttle-governed non-reheat machine. The appropriate diagram is shown in Fig 7.71. The TSV steam pressure is 105 bar and the temperature 570°C. If the turbine is operating on part-load, say 75% MCR, then the ATV pressure will be about 75 bar. In normal operation, this would be achieved by straightforward throttling at constant enthalpy, represented by the line *AB*. Expansion in the turbine is represented by the line *BC*. If, instead of throttling the steam, the throttle valves are left open and the boiler pressure reduced to give the required ATV pressure whilst the full steam temperature is maintained, there will be some benefit. Point *D* represents the new TSV pressure and *E* the ATV pressure with the throttle valves wide open.

Clearly, the specific enthalpy of the steam is greater at *E* than it is at *B*. On the other hand, the exhaust enthalpy will also be higher, but not by as much as at the ATV. Hence, there is a net gain of available energy of the steam. Incidentally, the final wetness of the steam is reduced, which is an added bonus. Because of these factors, control as described is quite common at part loads. However, it is important that full consideration is given to possible adverse effects before adopting the technique. For example, should steam flow to the turbine be interrupted, but firing continue, then there could be a possibly dangerous delay whilst the boiler pressure increases to a value high enough to lift the safety valves. Severe damage due to overheated tubes could result. Therefore, advice from the boiler manufacturers should always be sought before proceeding.

It should be mentioned that nozzle-governed machines, when operating at part-load, will normally have all the valves in use wide open. Therefore, under these conditions any pressure increase will cause the available energy of the steam to increase and, hence, the efficiency will improve. Consequently sliding pressure control is not applicable to this type of machine.

512

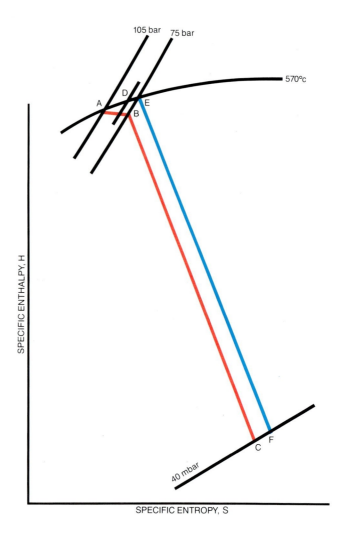

FIG. 7.71 Sliding pressure control
Extra heat is made available by reducing the boiler pressure, maintaining full steam temperature and opening the throttle wide.

5.10.6 Variation of steam temperature

First consider a throttle-governed non-reheat turbine, as represented in Fig 7.72. The optimum ATV conditions are 100 bar, 560°C and the steam state line is shown from *A* to *B*. If the pressure is kept constant, but the temperature increased to 570°C, it is clear that the specific enthalpy of the steam will be increased to point *C*. The expansion line will be *CD*, which is parallel to *AB* (i.e., the efficiency of the expansion is the same), but the final exhaust steam wetness will be less. On the other hand, a reduction in temperature will reduce the inlet steam enthalpy and increase the exhaust-steam wetness. Normally, about 12% wetness is regarded as the tolerable maximum.

Therefore, it is important that the operators keep the inlet steam temperature at its optimum value if possible. Letting it fall below optimum will cause:

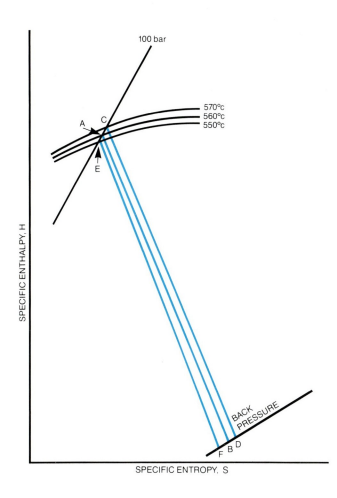

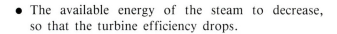

FIG. 7.72 Effect of change of TSV temperature on a
non-reheat turbine
Because of the divergence of the pressure lines, the
available heat increases as the temperature increases.

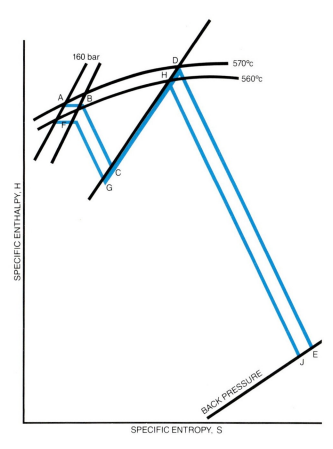

FIG. 7.73 Effect of change of TSV and reheat temperature
At 160 bar/570°C, the expansion is shown by *ABCDE*.
Should the TSV temperature fall to 560°C, the HP
cylinder inlet and exhaust pressures will remain the same,
but the expansion will be *FGDE*. If the reheat temperature
is also affected, the expansion will be *FGHJ*. Because the
pressure lines diverge, the reduction of temperature results
in a reduction of available heat.

- The available energy of the steam to decrease,
  so that the turbine efficiency drops.

- The wetness of the steam at the turbine exhaust
  to increase.

- The efficiency of the expansion to remain the
  same (i.e., parallel expansion lines on the diagram).

Next, consider a reheat turbine. Assume the TSV
optimum conditions are 160 bar at 570°C, represented
by point *A* in Fig 7.73 and the ATV conditions are
shown at *B*. Expansion of the steam in the HP
cylinder is represented by the line *BC*. This is followed by reheating to point *D* after which expansion
in the IP/LP cylinder is shown by the line *DE*.
If, now, the TSV temperature is lowered, such that
the ATV conditions become *F*, then the new HP
cylinder expansion line will be *FG*, the exhaust pressure being the same as before, but with the temperature lower. Also, the available energy of the

steam is reduced because of the convergence of the
pressure lines with reduced specific entropy. For a
constant reheat temperature, it is clear that extra
heat is required in the reheater to compensate for
the changed conditions in the HP cylinder. The expansion from *D* to *E* remains, of course, provided
that the reheater outlet temperature is unchanged.

If the reheat temperature is reduced, though, while
the TSV temperature is maintained at optimum, then
the expansion would be represented by *ABCHJ*, resulting in the following:

- The available energy in the reheat steam will be
  less, so reducing the turbine efficiency.

- The wetness of the steam at the exhaust will be
  increased.

- The heat supplied in the reheater will be reduced.

- The expansion of the steam in the HP cylinder is
  not affected.

As an example of the effect of changes of temperature at the inlet to the HP and the IP cylinders, consider Table 7.29. The expansions are considered to be isentropic and the HP and IP pressures 160 and 40 bar. The back pressure is 40 mbar.

6 Condenser performance

6.1 Introduction

After the steam has surrendered its useful heat to the turbine, it passes to the condenser. The work obtained by the turbine from the steam will increase as the back pressure is reduced, so it is always desirable to operate at the minimum economic back pressure, i.e. the condensate temperature should be as low as possible. If the condensing surface were infinite, the condensing temperature would equal the temperature of the inlet cooling water (CW). However, there is a practical limit and in practice the average temperature of the condensate is about 15°C above the inlet temperature, but even so the size of condensing plant is considerable. For example, on a 660 MW unit the condenser may have 20 000 × 25 mm dia. tubes, each 20 metres long. The reason for such a massive heat-transfer surface is apparent when it is realised that for a generator output of 660 MW about 780 MW of energy will be surrendered to the cooling water — see Fig 7.74.

Even a very small worsening of the back pressure is very expensive in terms of the extra heat required for a given output. To illustrate this for a 2000 MW station, if the back pressure worsened by just 2 mbar, the resulting extra fuel cost would be about £250 000 per year.

In fact, condenser performance is undoubtedly the most important operating parameter on a unit, so the factors which worsen back-pressure must be clearly

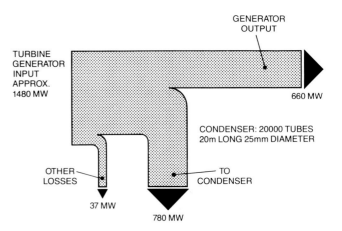

FIG. 7.74 Heat-energy Sankey diagram of heat flow
The heat given to the condenser cooling water is greater than that at the generator terminals. The condenser loss is much greater than all the other losses combined.

recognised, so that effective remedial measures can be taken once they are detected.

6.2 Factors affecting the performance of condensers

6.2.1 Determination of deviations
Condenser design is based upon authoritative documents which lay down the basic guidelines. Typical documents are:

● 'Recommended practice for the design of surface type steam condensing plant', publication No. 222, British Electrical and Allied Manufacturers Association.

● 'Standards for steam surface condensers' by the Heat Exchange Institute, New York.

TABLE 7.29
Effect of inlet temperature changes

ATV temperature °C	570.0	560.0	580.0	570.0	570.0
Reheater temperature °C	570.0	570.0	570.0	560.0	580.0
Exhaust temperature °C	29.0	29.0	29.0	29.0	29.0
Exhaust wetness %	14.7	14.7	14.7	15.1	14.4
Heat drop in HP cylinder, kJ/kg	421.8	414.8	428.3	421.8	421.8
Heat drop from reheater, kJ/kg	1407.4	1407.4	1407.4	1394.3	1422.9
Total heat drop, kJ/kg	1829.2	1822.2	1835.7	1816.1	1844.7

Note that the effect of reducing the ATV temperature to 560°C is significant with regard to the total heat drop, even though it only affects the HP cylinder heat drop and does not affect the wetness.

The manufacturer supplies various performance curves for the condensing plant and these should be verified by tests as soon as possible. There are two particularly important relationships which should be established — the CW temperature rise with load and the terminal temperature difference with load. Typical values are shown on Fig 7.75. A knowledge of these optimum values is basic to many condenser investigations. The importance of the CW temperature rise is obvious, but perhaps a word about the terminal temperature difference (TTD) is required. To make heat flow from the steam to the cooling water it is necessary to have a temperature gradient, with the steam temperature higher than that of the cooling water. Excellent heat transfer only needs a small gradient whereas poor heat transfer requires a large one, so TTD is a measure of the effectiveness of the heat transfer. For a given CW inlet temperature, it follows that the back pressure in the condenser also depends in part upon the TTD, so it should be regularly monitored when a unit is on-load and every effort made to keep it at optimum. Factors affecting the back pressure include:

• Variation of CW inlet temperature.

• Variation of CW quantity.

• Interference with heat transfer.

The effects of these factors can be illustrated by an example with the help of a diagram devised by the author (Fig 7.76).

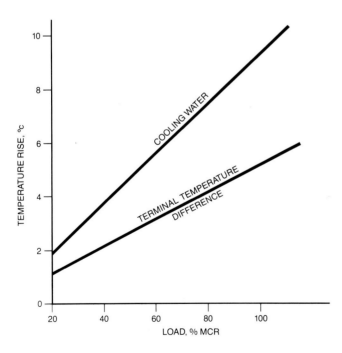

FIG. 7.75 Cooling water temperature rise and terminal temperature difference
The values shown are fairly typical for a modern machine.

The optimum and actual conditions for a condenser are listed in Table 7.30.

TABLE 7.30
Optimum and actual condenser conditions at full-load

Parameter	Optimum	Actual
CW inlet temperature, °C	16.5	20.0
CW outlet temperature, °C	25.0	26.0
CW temperature rise, °C	8.5	6.0
Saturated steam temperature corresponding to back pressure, °C	30.5	36.0
Terminal temperature difference, °C	5.5	10.0
Back pressure, mbar	43.7	59.4

To determine the contribution of each main factor to the total deviation of back pressure from optimum, proceed as follows:

Contribution due to high CW inlet temperature   Plot a line vertically upwards from the actual CW inlet temperature (20°C) to the optimum CW rise line, and then horizontally to the optimum TTD. From this point, drop a vertical to intercept the saturation temperature line at 34°C, with a corresponding back pressure of 53.2 mbar. So the back pressure deviation due to the CW inlet temperature alone is 53.2 − 43.7 = 9.5 mbar.

Contribution due to incorrect CW flow   The CW temperature rise is less than optimum, so there must be a higher than optimum CW flow. To determine the effect of this plot a line from the actual CW inlet temperature to the actual CW rise, thence to the optimum TTD and, finally, downward to cut the saturated steam temperature line at 31.5°C. The equivalent back pressure is 46.2 mbar. Therefore the pressure deviation due to the high flow is 46.2 − 53.2 = −7.0 mbar. The high flow has improved the back pressure by 7 mbar.

Contribution due to heat transfer   The TTD is a measure of interference with heat transfer. Here, the TTD is considerably higher than optimum, so it is worsening the back pressure. There is no necessity to plot the lines as in the previous cases, as they would merely connect all the 'actual' values and the resultant back pressure would be 59.4 mbar which is already known. So, to find the deviation due to the high TTD, deduct 46.2 mbar from 59.4 mbar, giving a deviation of 13.2 mbar.

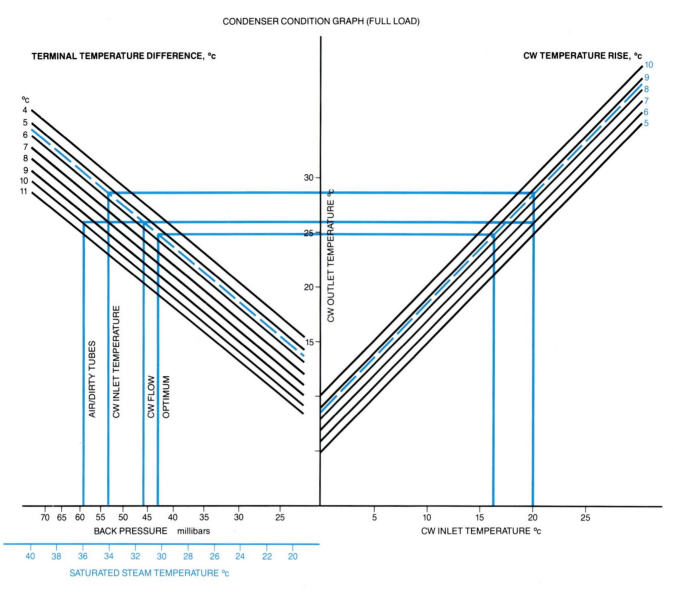

CONDENSER CONDITION GRAPH (FULL LOAD)

TERMINAL TEMPERATURE DIFFERENCE, °c

CW TEMPERATURE RISE, °c

FIG. 7.76 Optimisation of condenser performance
Using the diagram, the total back pressure deviation can be resolved into its component parts.

So

deviation due to CW inlet temperature = 9.5 mbar

deviation due to CW flow = −7.0 mbar

deviation due to poor TTD = 13.2 mbar

The user of the diagram is, in effect, adding a series of temperatures. For example, the back pressure due to high CW inlet temperature is found by adding the actual CW inlet, optimum CW rise and optimum TTD temperatures, i.e., back pressure due to CW inlet = 20.0 + 8.5 + 5.5 = 34°C ≡ (53.2 mbar). Similarly the back pressure due to CW flow will be: 20.0 + 6.0 + 5.5 = 31.5°C ≡ (46.2 mbar).

The above calculations are quite adequate for shift-to-shift monitoring by the operations staff, but more refined calculations are required for detailed performance investigations, as will be shown later.

6.2.2 Notes on deviations

Usually the loss due to high CW inlet temperature must be accepted as it is dictated by the river or sea temperature, over which there is no control from the station. On a station with cooling towers, the irrigation systems should be kept in good order and the total complement of towers kept in service except in freezing weather (when some towers can be taken out of service to raise the CW temperature to inhibit icing). In mixed cooling tower and river systems, the maximum permitted quantity of river water should be used at all times, as it is cooler than re-circulated tower water. Loss of back pressure due to the CW flow being too low should be avoided, where possible. Common causes of this include: insufficient pumps in service; CW outlet valves not open wide enough; condenser inlet tube plates fouled with debris. The last condition can be inferred if the CW temperature rise remains high, even with wide-open CW outlet valves, assuming the rest of the possibilities have been eliminated. At coastal stations there may be mussels on the tube plates. On the other hand, it is the practice in some stations to operate with wide-open CW outlet valves, in which case the temperature rise will usually be less than optimum. The benefit of this is that the back pressure is improved, but at the expense of increased pumping power. The obvious compromise is to ensure that CW pump operation is optimised, i.e., only use extra CW pump power when the benefit is greater than the cost of the extra power. Care must be taken to ensure that the velocity of the cooling water through the condenser tubes does not exceed the design limit of about 2 m/s. Excessive velocity leads to tube erosion, resulting in cooling water leakage into the steam spaces.

The last of the main deviations is that due to high TTD. As already stated, this is an indication of interference with the heat transfer across the tubes. The steam temperature is almost the same over the whole surface of the tubes (except in the air-cooling section), whereas the temperature gradient between the steam and the inlet CW starts high, becoming progressively smaller until it becomes the TTD at the CW outlet (Fig 7.77).

The average temperature gradient is normally assumed to be the log mean temperature difference and this is a measure of the mean driving force for the heat transfer. The log mean temperature difference (LMTD) is usually expressed as:

$$LMTD = (\theta_1 - \theta_2)/\log_n(\theta_1/\theta_2)$$

where θ_1 = steam temperature – CW inlet temperature (i.e., the initial temperature difference)

θ_2 = steam temperature – CW outlet temperature (i.e., the terminal temperature difference)

Figure 7.78 shows a nomogram from which the LMTD can be obtained. $(\theta_1 - \theta_2)$ is the CW temperature rise, so an alternative form of the LMTD expression is (CW temperature rise)/[$\log_n (\theta_1/\theta_2)$].

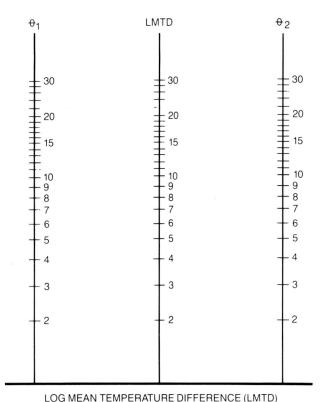

LOG MEAN TEMPERATURE DIFFERENCE (LMTD)

FIG. 7.78 Nomogram of log mean temperature difference
The inlet temperature difference is θ_1 and the terminal temperature difference is θ_2. A straight line joining the two values crosses the middle line at the LMTD.

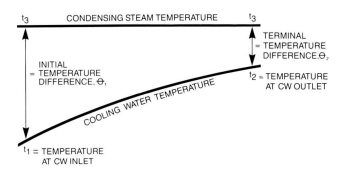

FIG. 7.77 Condenser temperature diagram
The condensing steam temperature is constant, but the CW temperature rises continuously from the condenser inlet to the outlet. The average temperature difference between the steam and the cooling water is taken as the LMTD.

The cleanliness of condenser tubes can be affected by deposits on the inside from mud, slime and scale. On the steamside, there may be air accumulation or scale. Heat transfer will be discussed in greater detail later. For now, it should be noted that, when condenser tubes are cleaned, the only part of the total deviation of back pressure that will be improved is that due to 'dirty tubes'. It has no effect whatever on deviations due to air, CW temperature, etc. So the improvement due to cleaning the tubes may not be as great as expected.

6.2.3 Shift monitoring

For shift-monitoring purposes the required curves are: optimum CW rise versus load; optimum TTD versus load; and curves of target back pressure for various CW inlet temperatures versus load. The first two were illustrated earlier, and Fig 7.79 shows the third.

Once per shift, during a period of steady loading, the back pressure on each running unit should be monitored to determine any back pressure deviations and to institute remedial measures, where possible. As with so many of these routine calculations, the task is simplified by using a standard form, such as the one shown in Table 7.31. A typical calculation for a 500 MW unit has been included.

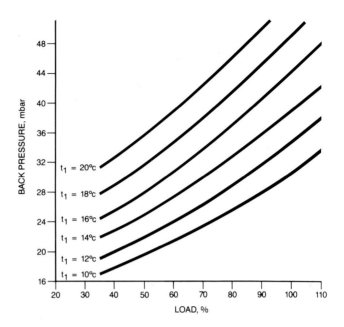

FIG. 7.79 Curves of optimum back-pressure versus load, for various CW inlet temperatures

TABLE 7.31
Spot back-pressure check

		Morning	Evening	Night
	Unit No.	Date		
	Shift			
	Load, MW	491.1		
1	Actual back-pressure, mbar	52.3		
2	Saturated steam temperature, °C	33.7		
3	CW inlet temperature, °C	17.9		
4	CW outlet temperature, °C	26.8		
5	Exhaust steam temperature, °C	33.7		
6	Condensate temperature, °C	34.9		
7	Air suction temperature, °C	24.9		
8	CW outlet valve position, % open	55		
9	Target back-pressure, mbar	48.4		
10	Optimum CW rise, °C	9.0		
11	Optimum TTD, °C	5.2		
12	Back-pressure due to CW inlet ≡ (3) + (10) + (11), mbar	47.8		
13	Back-pressure due to CW flow ≡ (4) + (11), mbar	47.5		
14	Variation due to CW inlet temp = (12) − (9), mbar	− 0.6		
15	Variation due to CW flow = (13) − (12), mbar	− 0.3		
16	Variation due to air/ dirty tubes = (1) − (13), mbar	4.8		
17	BP variation = (1) − (12), mbar	4.5		
18	No. of air pumps in service			

The air suction temperature is required because it is an indication of the quantity of air ingress to the condenser. For example, if there were no air present at all, then the temperature of the contents of the air-pump suction pipe would be about 4.5°C lower than the condenser saturated-steam temperature. Air leaking into the condenser causes the air suction temperature to be still lower than the saturated steam temperature, so this is a very easy way of ascertaining if there is significant ingress. Of course, it is only an indication — it does not enable the actual quantity of inleakage to be determined. The number of air pumps in service is noted merely to serve as a reminder that possibly an air pump can be taken out of service. The fact that air pumps are normally rated at 50% duty does not mean that it is necessary to have two in service. If the air inleakage is small it will be possible to run with only one, so saving some works power.

The following conclusions may be reached:

CW inlet temperature For sea or river stations the temperature must be accepted, as the operators have no control over it. At cooling tower stations, check the items listed earlier.

CW temperature rise This is due to CW flow. If there is insufficient flow, the first thing to do is increase the opening of the condenser CW outlet valves. If the rise is unaffected, check the number of CW pumps in service, the condenser priming and (if necessary) the effectiveness of the cooling tower irrigation system. If the rise is still high, the probability is that the tube plates are partially obstructed with debris. This is most likely at river stations in the autumn, due to leaves and twigs, etc., being carried to the condenser from the river. The river screens must be kept in good order. At a coastal station considerable areas of the tube plates can be rendered inoperative because of fish or mussels becoming lodged on the plates. Again, the effectiveness of the CW screens should be checked.

High TTD The effects of internal tube deposits and the effect of air blanketing on the outside of the tubes are indistinguishable in their effect on the TTD. However, as mentioned earlier, if there is significant air ingress, it will cause the temperature of the air suction to become lower, since Dalton's law of partial pressures applies. The total pressure in the pipe (very nearly that of the condenser) is made up of the sum of the partial pressures of any air present and that of the water vapour, but the temperature of the mixture is due only to the water vapour. Consequently, as the vapour quantity is reduced, so is the temperature.

In the example (Table 7.31) the air-suction temperature (7) is considerably below the saturated steam temperature (2), so clearly there is significant air ingress. This can be confirmed by flow measurement as discussed in Section 6.4 of this chapter. Urgent action should be taken to locate the sources of ingress and rectify them. Then a further set of readings can be taken. Any deviation ascribed to 'air/dirty tubes' will now be due to dirty tubes, as the air ingress is known to be acceptable. A decision can then be made as to the desirability of cleaning the tubes. Meanwhile it should be ascertained that any tube cleaning such as chlorination or a mechanical on-load cleaning device is working properly.

6.3 Heat transfer across condenser tubes

The temperature gradient required to transfer a given quantity of heat across the condenser tubes is determined by the total resistance to heat flow and consists of the following main components:

(a) Resistance due to the condensation film on the steamside of the tube.

(b) The effect of any air blanketing on the steamside of the tubes. Air is such an excellent insulator that only a minute film on the tubes will cause a serious resistance to heat flow.

(c) Deposition, such as copper oxide, on the steamside of the tubes.

(d) The resistance of the tube material itself.

(e) Deposition on the inside of the tubes due to scale, slime or dirt.

(f) The stagnant water film adjacent to the inside of the tubes.

Consider each item in turn:

(a) Clearly there will be a water film on the outside of the tubes from the condensation of the steam. The resistance to heat flow will depend upon factors such as the mode of condensation (dropwise or film) and the depth of liquid. Such factors are beyond the control of the operators of the plant as they are inherent in the design, so the situation must be accepted for any given installation.

(b) The effect of air ingress is often very serious. Indeed, the main factor causing poor performance of condensers is the degree of air blanketing of the tubes. A small inleakage is inevitable, and is normally removed by the air pumps. However, should the air be able to accumulate it will interfere with the steam getting to the tubes, increase the temperature gradient and so worsen the back pressure.

The position of air blanketing pockets can often be inferred from measurement of the cooling water temperature distribution at the outlet tube plate. For this, a thermocouple grid can be installed near the tube plate and the connections brought out through a suitable gland in the waterrbox. About one thermocouple per hundred tubes will suffice. An abnormally low temperature rise is indicative of air blanketing somewhere along the condenser on the path of the affected tubes. A more precise indication of the position of the blanketing can then be made by inserting a small probe into the steam space to measure the temperature and pressure within the tube nests, as shown in Fig 7.80.

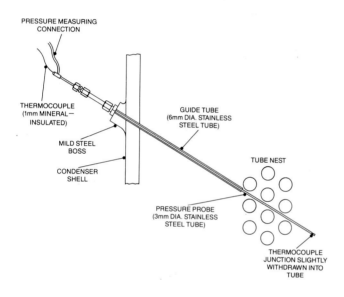

FIG. 7.80 Air blanketing surveys
The illustration shows the method of inserting the probe into the condenser tube nest to measure the steam pressure and temperature at the location.

Assuming that the steam is everywhere saturated, the air mass fraction can be found from:

$$M_a/(M_a + M_s) = (p - p_s)/(p - 0379p_s)$$

where M_a = mass flowrate of air, kg/s

M_s = mass flowrate of steam, kg/s

p = measured pressure, mbar

p_s = saturation pressure of steam at temperature t°C, mbar

t = measured temperature (air and steam are at the same temperature), °C

Air accumulation within a bundle of tubes is cooled by them until it is some 10°C below the temperature of the condensing steam. For example, using this technique an air pocket was located in a 500 MW unit in the central region of the condenser tube nest. The conversion of a small number of condenser tubes to air suction tubes subsequently reduced the size of the air pocket and significantly improved the back pressure. Improvements such as this on several large units has saved the CEGB an estimated £500 000 per year in reduced fuel costs.

(c) It is not always realised that depositions can occur on the outside of condenser tubes. Usually they are composed of oil or oxides of copper or iron, and the effect can be significant. For example, on one 500 MW unit the deposit was sufficient to worsen the back pressure by about 10 mbar. The only approved way of restoring the condition of the tubes is to soak them in a cleansing solution of Ethylene Diamine Tetra Acetic Acid (EDTA) and oxalic acid at pH4.

This is quite expensive, but the improved back pressure makes it abundantly economic.

The tubes most affected by steamside deposits are located at the top of the condenser, so these should be carefully examined during surveys.

(d) The material of the tubes does not constitute a significant resistance to heat flow.

The usual tube materials include 70–30 brass; Admiralty brass; aluminium brass; 90–10 copper nickel and titanium. The thermal conductivities vary between 0.0173 and 0.1255 kW/m K, the lowest being titanium, but the normally used metals are over 0.1 kW/m K. In any case, as far as the station staff are concerned, they are obliged to accept the material supplied.

(e) On the inside of the tubes, the most common cause of poor heat transfer is mud and slime. Basically the problem is due to algae in the cooling water. These are jelly-like organisms which find the conditions in the tubes hospitable, so they tend to settle there. They form a suitable bonding agent for particles of mud and, in a short time, the internal tube coating will seriously affect the condenser performance. The solution is to ensure that the environment is made inhospitable by intermittently dosing the condenser CW with chlorine. Care must be taken, though, to ensure that strict control is maintained over the free chlorine concentration at the condenser outlet, as it can adversely affect the fish population in rivers or near the outfall in the sea. Also, there are economic reasons for using the minimum amount of chlorination.

The chlorine is often stored on site in bulk-storage tanks containing several tonnes of the chemical, so a burst tank could release enough chlorine gas to be a hazard. Therefore, in the CEGB, a change will be made before 1990 to using sodium hypochlorite solution. This is a more expensive way of injecting chlorine, but is much safer to store in bulk. When the hypochlorite is used on the plant it acts in the same way as chlorine, so the basic tube cleaning process is the same. The hypochlorite solution falls in strength in storage, losing about 20% of available chlorine in four weeks at 20°C. An alternative is to use electrochlorination plant: this has been installed at several CEGB coastal power stations. In this, sodium hypochloride is formed from the electrolysis of seawater.

In recent years, a serious cause of internal tube fouling has appeared — scale formation due to phosphate in the CW of inland stations. Since World War II, there has been a huge increase in the number of domestic washing machines in use, particularly in North America and Europe. The washing powders used in these machines contain phosphates and much of it is eventually discharged into rivers via sewage plants. The phosphate is not regarded as a contaminant by the sewerage authorities, so they make no provision for its removal. Consequently, if the water is later used as CW for power stations the phosphate, under suitable conditions, will adhere as a tenacious scale to the inside of the condenser tubes, posing a serious problem for the power industry. Scale formation can be prevented by controlling the pH of the CW to 7.5 by dosing it with sulphuric acid. However, this is expensive and can lead to acid attack on the concrete culverts and cooling towers. Alternatively, the scale can be allowed to form and be periodically removed by chemical cleaning. The quantity of scale may be surprisingly high. For example, on a 350 MW unit over a tonne of scale could be present on the tubes.

(f) The thickness of the stagnant water film on the inside of the tubes is dependent upon two factors: the temperature and velocity of the water. The temperature of the water causes the heat transmission rate to vary as the fourth root of CW temperature and the square root of the CW velocity, results propounded in a classic paper read before the Institution of Mechanical Engineers in 1934 by Guy and Winstanley*. Although their findings are not completely accepted by some authorities, there is no doubt that the results obtained by using their work are at least very good guides. The effects of temperature and velocity can be determined from Figs 7.81 and 7.82.

---

* 'Some Factors in the Design of Surface Condensing Plant' by H. L. Guy and E. Winstanley.

For example, calculate the effect of changing the CW inlet temperature to 8.4°C for a condenser with the following design conditions:

CW inlet temperature	16.5°C
Back pressure	38.6 mbar
Saturation temperature corresponding to back pressure	28.3°C
Initial temperature difference	11.8°C

The drop in the CW inlet temperature is $16.5 - 8.4 = 8.1°C$, so from Fig 7.81 the factor $F_1 - 1.074$, so the new saturation temperature is ($F_1 \theta_1 +$ new CW inlet temperature), where θ_1 = initial temperature difference (design value).

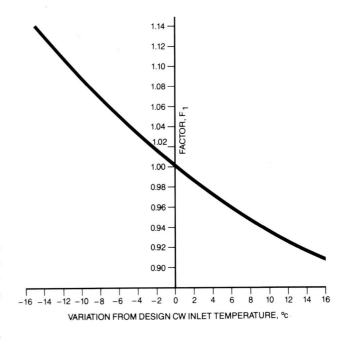

FIG 7.81 Factor F_1 for variation from design CW inlet temperature

The new saturation temperature $= (1.074 \times 11.8) + 8.4 = 21.1$ and equivalent saturation pressure is 25 mbar, so the cooler circulating water has improved the back pressure by 13.6 mbar. If only the reduction of CW temperature is considered, the saturation temperature would have been $28.3 - 8.1 = 20.2°C$. The cooler water has increased the thickness of the stagnant water film inside the condenser tubes, so that the initial temperature difference has gone from 11.8 to $1.074 \times 11.8 = 12.7°C$, an extra 0.9°C. Hence, the primary and secondary effects of a change of CW temperature work in opposite directions.

In a similar manner, the effects of a change of velocity can be determined. Suppose, in the example above there had also been a change of CW flow (which would give a corresponding change of CW velocity) from, say, 95% of design flow to 90%. The factor F_2 is obtained from Fig 7.82. Proceeding as before, the new saturation temperature is obtained from ($F_2 \theta_1$ + new CW inlet temperature). The factor for 95% flow is 1.04 and for 90% flow is 1.08. So $F_2 = 1.08/1.04 = 1.04$, the saturation temperature is $(1.04 \times 11.8) + 8.4 = 20.7°C$ and the equivalent saturation pressure is 24.4 mbar.

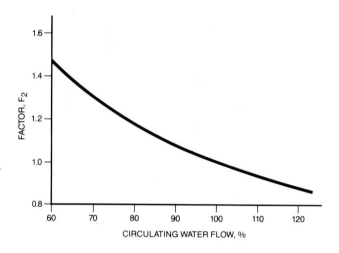

FIG. 7.82 Factor F_2 for variation of CW flow

This is also a convenient place to mention a third parameter, i.e., variation of steam flow to the condenser. Suppose, for the same design conditions as already considered, that the steam load changes from 100% to 90%. The appropriate factor, F_3, is obtained from Fig 7.83.

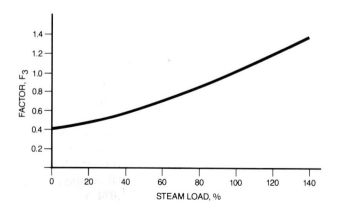

FIG. 7.83 Factor F_3 for variation of steam load

Then $F_3 = 0.92$, so the new saturation temperature is ($F_3 \theta_1$ + new CW inlet temperature), i.e., $(0.92 \times 11.8) + 8.4 = 19.3°C$ and the corresponding saturation pressure is 22.4 mbar, a reduction of 16.2 mbar from design.

Should the combined effect of all three changes (CW temperature, CW flow and steam load) be required, the procedure is the same, except the product of the factors is used, giving the new saturation temperature as ($F_1 \times F_2 \times F2 \times \theta_1$) + new CW inlet temperature.

In the example, new saturation temperature = $(1.074 \times 1.04 \times 11.8) + 8.4 = 12.1 + 8.4 = 20.5°C$. The corresponding back pressure is 24.1 mbar.

As an example of the use of the above method, consider a 2000 MW station having five cooling towers to supply its CW needs. The full-load design data for the four 500 MW units is:

CW inlet temperature	20°C
CW outlet temperature	29°C
CW temperature rise	9°C
CW flow rate	15.61 m³/s
Back pressure	50.8 mbar
Saturation temperature	33.2°C
TTD	4.2°C
Initial temperature rise	13.2°C
Dry-bulb air temperature	10°C
Relative humidity	70%

What would the back pressure become if one cooling tower is taken out of service, with the steam load to the condensers unchanged, assuming all four units remain on load?

For this it is necessary to refer to a universal performance chart for the cooling towers, Fig 7.84. From the dry bulb temperature of 10°C and 70% relative humidity, plot a horizontal line to the transfer line. The cooling range is the same as the CW temperature rise (9°C) and the recooled water temperature is the same as the CW inlet temperature (20°C). Therefore, the line from the transfer point is as shown for the design conditions. Taking a cooling tower out of service reduces the quantity of CW flowing to 4/5 of the original flow and so the CW temperature rise becomes $5/4 \times 9° = 11.25°C$, as the rise is inversely proportional to the flow. Reference to the performance chart shows that for the same cooling-tower water-load and a cooling range of 11.25°C the recooled water temperature becomes 20.8°C, so that this is the new CW inlet temperature to the condensers. Hence, the CW inlet temperature goes from 20° to 20.8°C, and the factor $F_1 = 0.994$. The CW flow is reduced to 80% of the design flow, so the factor $F_2 = 1.18$.

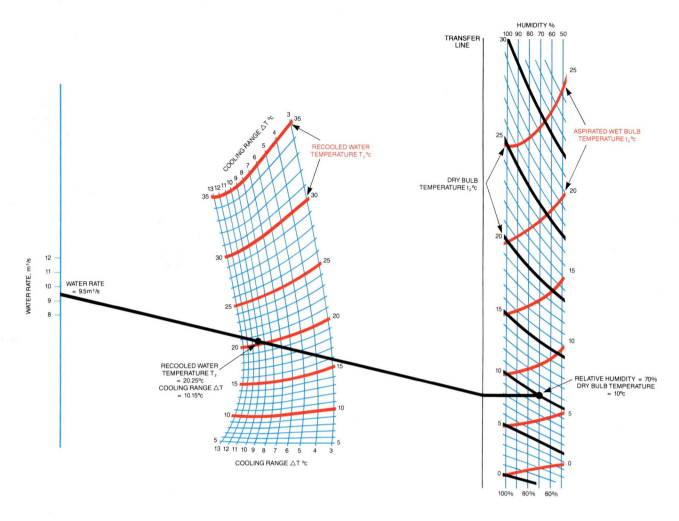

FIG. 7.84 Universal performance chart for a natural-draught cooling tower
Large-scale diagrams are used when carrying out cooling tower investigations.

The steam load to the condensers remains unchanged so the factor F_3 is unity. Thus, the new saturation temperature

$$= (0.994 \times 1.18 \times 1.0 \times 13.2) + 20.8$$
$$= 15.5 + 20.8$$
$$= 36.3°C$$

The equivalent saturation pressure is 60.4 mbar, so taking the cooling tower out of service will cause the back pressure to change from 50.8 to 60.4 mbar.

6.4 Air ingress

6.4.1 Air pumps

As has already been mentioned, air is an excellent insulator. Should tube blanketing occur, the back pressure will be significantly worsened. Obviously, adequate air-removal capacity is highly desirable. The BEAMA document already referred to suggests the dry-air ratings given in Fig 7.85 for various steam flows. Thus a 500 MW unit would have a dry-air removal capacity of about 140 kg/h. Usually three $\times$ 50% duty air pumps are provided. Sometimes it is found that the air extraction on a unit is satisfactory at high unit loads, but is not so good at lower loads. One possible cause is that the volume of air being offered to the pumps at low loads is more than they can cope with at low pressure, as the specific volume of the air is too great. Another possible cause is that there is air ingress at a part of the turbine that is pressurised at higher loads.

Even if the air pumps can cope, it is desirable to regularly measure the quantity of air being handled. Several methods are available for doing this; they are listed below (see also Volume F). The preferred position for measurement is at the air pump discharge, because measurement at the air pump suction

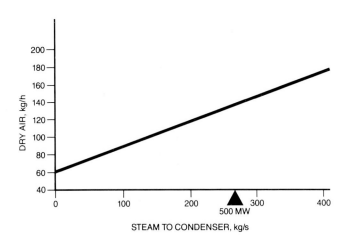

FIG. 7.85 Recommended air-removal capacity
of modern plant
A typical 500 MW unit would have a removal capacity
of about 140 kg/h of dry air.

requires the determination of the steam/air ratio and is made difficult by the low density of the steam/air mixture.

Airflow measurement at the air pump discharge will include any air leakage between the condenser and the point of measurement.

6.4.2 Variable-area flowmeters

These consist of a float which is free to move in a slightly tapered vertical tube. Air passing upwards through the tube causes the float to rise until the drag and buoyancy forces just balance the weight of the float. The tapered tube is usually glass with a nearly linear scale on which the position of the float indicates the rate of flow. The flowmeters are mounted downstream of individual air pumps; straight runs of pipe before or after the device are not required.

6.4.3 Pitot tubes

A total pressure tube used with a wall-static pressure tapping or, alternatively, a combined pitot-static tube may be used to traverse the air outlet pipe. However, the differential pressure produced is very small, and it is necessary either to employ precision measuring devices or to increase the air velocity by reducing the cross-sectional area. (Air flowing at 20 m/s at ambient conditions has a velocity head of only 24 mm water gauge.)

6.4.4 Anemometers

Rotating-vane or hot-wire anemometers may be used to measure the airflow. Both types of instrument

should be calibrated, preferably using the steam/air ratio, temperature and pressure expected during the investigations. To improve accuracy, the flow may be constricted to increase the air velocity and a correction for air density may be applied.

6.4.5 Orifice or nozzle boxes

The box contains two standard orifice plates or nozzles (one for high flow, one for low flow), a typical example of which is shown in Fig 7.86. It provides a reliable method of measuring total airflow at the discharge of the air pump. The airflow is passed through the box to atmosphere by a suitable arrangement of valves and pipework and the temperature and pressure in the box are measured. The complete box should be calibrated before a test. These devices may be permanently installed and should maintain their calibration, so long as the orifice plates are not damaged.

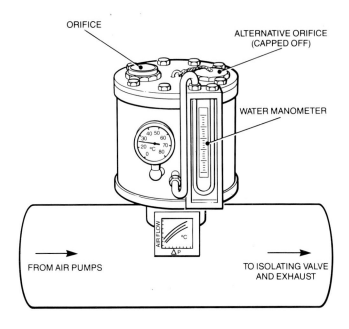

FIG. 7.86 Orifice box for air flow measurement
The assembly can be permanently installed and will
maintain its calibration provided the orifice
plates are not damaged.

6.4.6 Vortex-shedding meters

Two types of vortex-shedding meter can be used to measure airflow. The first is an insertion type. For this, the pipe is traversed to determine the velocity profile giving a correction factor, defined as the ratio of the weighted average velocity to the centreline velocity. The insertion meter can then be mounted on the pipe centreline and the reading multiplied by the correction factor.

Alternatively, from the velocity traverse, a dia-metral position can be selected which gives a velocity equal to the weighted mean velocity and the meter can be mounted at this point. A typical location for this 'mean velocity point' is at a quarter of the radius from the pipe wall.

The second is a full-flow type in which the bluff body extends across the diameter of the pipe.

Both types of meter have an accuracy of about ±1%, are reasonably linear over a wide flow range, and may be used to give a continuous indication of flow from the air pumps.

6.4.7 Extrapolation

Additional air is admitted into the condenser at known rates and measurements are of the total pressure and the wet-bulb temperature in the air suction pipe. The air partial-pressure is then calculated for each flow rate by subtracting the steam saturation pressure corresponding to the wet-bulb temperature from the total pressure.

The flow rates of the admitted air are plotted against air partial-pressure and a straight line is drawn through the points and extrapolated to intersect the air leakage axis (drawn at zero partial-pressure) to determine the initial air inleakage. The quantity of admitted air is known by passing it through a nozzle across which the pressure drop exceeds the critical pressure ratio and whose flow rate is known. Con-structional details of suitable nozzles are given in BEAMA publication 222.

As an example of the method, consider the fol-lowing:

On a 500 MW unit, air is admitted via three nozzles, each rated at 45.4 kg/h. The calculated air leakage is derived as shown in Table 7.32.

The same result is obtained in Fig 7.87 by the extrapolation of the air partial-pressures for the ad-mitted air conditions.

6.4.8 Air pump characteristics

It is possible to use the air pump characteristics to assess the airflow by measuring the air pump suction pressure and (for a hydraulic pump) the water temperature. This method is useful for day-to-day monitoring purposes but can only be used for ac-ceptance tests if the current performance is known.

6.4.9 Air density

One of the above methods should be satisfactory for any air measurements associated with condensers. To complete this section on air and air measurement, it is appropriate to mention the determination of

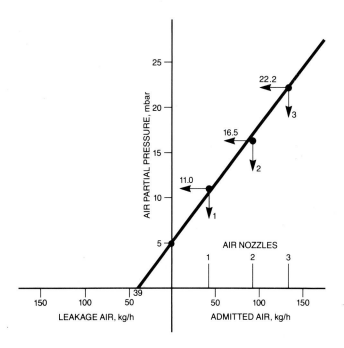

FIG. 7.87 Air inleakage by extrapolation
The conditions are plotted for 0, 1, 2 and 3 nozzles in use.
The line of best fit is drawn and extended to cut the
x-axis, which gives the normal air inleakage.

air density, which can be calculated from measure-ments of its pressure and temperature. For the pres-sure measurement, wall static-tappings are provided in the air extraction pipes, as close to the condenser shell as practicable, and also at the air pump (or ejector) end of the pipes. The air suction temperature is measured in the air suction pipework close to the points selected for pressure measurement. To obtain a quick response, it is advantageous to insert the temperature-sensing element directly into the pipe through a suitable gland.

The relevant formula is:

$$\varrho = [(p - p_s)/2.87 \, (\theta + 273.15)] + (1/v_s)$$

where ϱ = density of the mixture kg/m^3 at pressure p and temperature θ

p = total pressure, mbar

p_s = saturation pressure, of steam mixture at temperature θ, mbar

v_s = specific volume of saturated steam at temperature θ, m^3/kg

θ = temperature of the mixture, °C

Thus the air density in a suction pipe at a total pressure of 40.0 mbar and temperature 27°C is:

Condenser air leakage

Nozzles in use	n		0	1	2	3
Generated load		MW		490		
Admitted air	Q_{in}	kg/h	0	45.4	90.8	136.2
Suction main total pressure	P_t	mbar	40.4	45.4	49.5	53.8
Suction main temperature		°C	26.9	26.4	25.7	25.0
Vapour partial pressure	P_s	mbar	35.4	34.4	33.0	31.6
Air partial pressure	P_a	mbar	5.0	11.0	16.5	22.2
$\delta P_a = P_{an} - P_{ao}$	δP_a	mbar	0	6.0	11.5	17.2
Air leakage = $Q_{in}P_{ao}/\delta P_a$	Q_a	kg/h	0	37.8	39.5	39.6
Average Q_a for n = 1, 2, 3		kg/h		39.0		

$$\varrho = \frac{40.0 - 35.64}{2.87 \, (27.0 + 273.15)} + \frac{1}{38.8129}$$

$$= \frac{4.36}{861.43} + 0.0258 = 0.031 \text{ kg/m}^3$$

6.5 On-load condenser measurements

6.5.1 CW inlet temperature

The CW inlet temperature is reasonably uniform across the supply pipe. Therefore, a single measuring point in each CW inlet pipe normally gives the required accuracy. If thermometer pockets are used, they should project at least 150 mm into the pipes and should be partially filled with oil to improve the thermal conductivity.

6.5.2 CW outlet temperature

The cooling water temperature at the condenser outlet is stratified because of spatial variations of the rate of condensation within the condenser. It is therefore necessary either to measure the temperature after sufficient length of pipe to ensure adequate mixing has taken place or to provide an averaging temperature measuring system such as:

● A continuous sample which may be withdrawn using either cantilevered or diametral multi-hole probes. The cantilevered type is preferred because, by the use of a gland and valve system, it can be inserted or removed while the unit is on-load (Fig 7.88).

● Thermistors located at area-weighted radial positions.

● Long resistance elements positioned on chordal paths as shown in Fig 7.89.

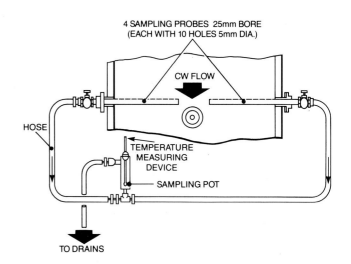

FIG. 7.88 Cantilevered probes arrangement for CW temperature determination
An advantage of this arrangement is that the probes can be removed, if necessary, whilst the unit is on-load.

The mean water temperature may be weighted by the chordal path length and by the mean path water velocity, if known.

The probes are subjected to considerable buffeting by the water, so care must be taken to ensure that they are adequately protected and supported. In systems arranged for syphonic operation, the condenser CW outlet may be below atmospheric pressure. When using multi-hole probes with this arrangement, it may be necessary to incorporate a suitable extraction device to ensure an adequate water flow over the temperature sensing elements.

6.5.3 Condensate outlet temperature

Condensate outlet temperature is measured at the exit from the condenser shell, irrespective of whether

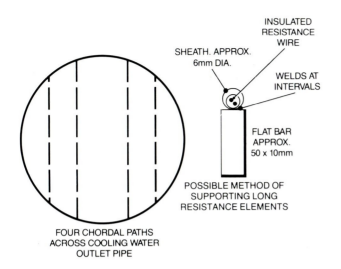

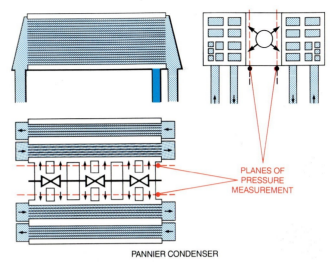

FIG. 7.89 Chordal paths for resistance elements
Long resistance elements must be very well supported
in the CW pipe because of the considerable buffeting
by the water. For an acceptable average CW outlet
temperature, it may be necessary to have, say, four long
elements mounted inside the pipe along the chordal paths,
as indicated in the illustration.

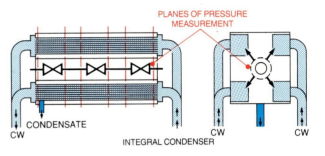

FIG. 7.90 Condenser inlet steam pressure-measuring planes
The measuring planes for pannier and integral
condensers are as shown.

the hotwell is integral with, or external to, the condenser.

6.5.4 Condenser pressure

The most important pressure is that at the condenser steam inlet.

The same plane should be used for measuring both condenser pressure and turbine back pressure. With underslung condensers, the plane can be readily identified; but for other types the pressure should be measured at the plane or planes as near as practicable to the condenser tube nests, as shown in Fig 7.90.

Considerable variations in static pressure occur across the condenser inlet so that it is necessary to make provision for measuring the mean by providing numerous points for pressure sampling. Flush wall-tappings are preferred and the holes should be 10 mm diameter where possible, but never less than 6 mm. They should be drilled normal to the wall, with burrs removed and a wide area around the hole cleaned and coated with anti-corrosive paint. Each condenser inlet duct should have wall tappings on all four sides, or on two opposite sides if access to all sides is impossible. The tappings are distributed as evenly as possible and there should be at least eight for each LP cylinder.

Should the flow at the tappings not be parallel to the wall, measurement errors may occur. Two ways to overcome this are:

- Fit guide plates about 300 mm square parallel to the wall surface containing the pressure-sensing holes at a distance of about 50 mm.

- Install an array of pressure-sensing tubes consisting of a series of closed tubes with rounded ends facing the steam flow and parallel to the expected flow direction with a series of small holes drilled around each tube about three tube diameters from the closed end.

Commonly the back pressure is displayed in the control room on an instrument such as a Kenotometer or a Vacumeter, as shown in Fig 7.91. They are very good for this purpose, but suffer from two disadvantages:

- The instrument must be zeroed by adjusting the scale, judging the correct position by eye. This is a subjective assessment which may be in error.

- The instrument is connected to the condenser tappings by small-bore pipework. The pipe should be run such that moisture cannot accumulate in it, but if some supports should fail it may permit the pipe to dip and form a 'valley'. This allows

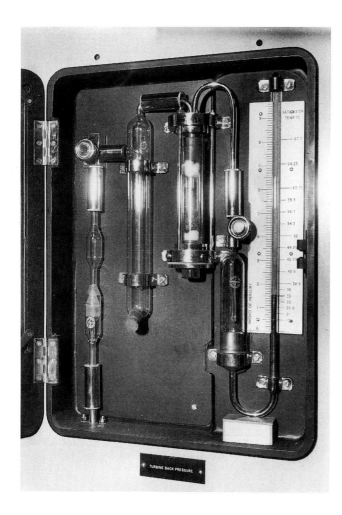

FIG. 7.91 Modified back pressure indicator
Mechanical indicators are used extensively within the
CEGB. In the illustration the right-hand scale has been
modified to show the equivalent saturation temperature
of the back pressure.

moisture to collect and the instrument will give
a wrong indication. Alternatively, air may get into
the pipe at a defective joint and again wrong
indication will be given.

An alternative to these instruments is to install
back pressure transducers near the condenser, connected electrically to a digital readout in the Control
Room.

6.5.5 Cooling water flow

General

The CW flow is very high indeed on large units.
For example, a typical value for a 500 MW unit is

about 15.6 m^3/s (56 160 m^3/h). This renders accurate measurement by direct methods difficult.

A rough indication of the best methods to determine the flow is, in descending order of preference:

- Condenser heat balance.

- Multi-path ultrasonic flowmeter.

- Bend flowmeter.

- Segmental orifice plate.

- Insertion meter.

- Differential pressure between condenser waterboxes.

Other methods are also described in CEGB Site Test
Code 3 but here only these will discussed.

Condenser heat balance

The cooling water flow is derived from the condenser
heat load and the cooling water temperature rise as
follows:

$$q_{cw} = \frac{P}{c_p \, (\theta_2 - \theta_1)}$$

Where q_{cw} = cooling water mass flow rate, kg/s

P = condenser heat load, kW

c_p = specific heat, kJ/kg K
  = [4.207 − 0.65S − 0.001(θ − S^2) +
    2.5(10^{-4}Sθ + 9)(10^{-6} θ^2)]

θ_1 = cooling water inlet temperature, °C

θ_2 = cooling water outlet temperature, °C

S = salinity, %

Alternatively, the volumetric flow can be derived
from:

$$Q = \frac{P}{\varrho \, c_p \, (\theta_2 - \theta_1)}$$

Where Q = cooling water volume flow rate, m^3/s

ϱ = water density in kg/m^3
  = (999.905 + 7.955S − 0.00614) ×

$$\left[\theta^2 - \left(7 - \frac{11}{4} S \right) \theta \right]$$

In each of these expressions it is necessary to know
P, the condenser heat load. This is done as follows,
due allowance being made for the generator mechanical and electrical efficiencies.

$$P = P_g (HR/3600) - (10^4/\eta_m \, \eta_g)$$

Where P = condenser heat load, kW

P_g = electrical output at generator terminals, kW

$\overline{HR}$ = uncorrected heat rate at the generator terminals, kJ/kWh

η_m = mechanical efficiency of the turbine-generator (typically 99.5% for large machines), %

η_g = generator electrical efficiency (typically 99.0% for large generators), %

When an electric-motor-driven feed pump is used, the value of P obtained from the above formula should be increased by:

$$P_{fp} \, (\eta_{fp}/100)$$

Where P_{fp} = power to feed pump motor, kW

η_{fp} = electrical efficiency of feed pump motor, %

The heat rate used in the formula should preferably be one recently determined, corrected for load and other conditions. Values of the density and specific heat of water are given in Site Test Code No. 3 'Performance of surface-type steam condensers'.

For example, determine the CW flow rate for a 500 MW machine with a full-load heat rate of 7884 kJ/kWh; $\theta_1 = 20°C$; $\theta_2 = 28°C$.

Then P = 500 000 (7884/3600) − (10^4/99.5 × 99)

= 500 000 (2.19 − 1.02)

= 585 000 kW

$$q_{cw} = \frac{585\ 000}{4.188\ (28 - 20)} = 17\ 460 \text{ kg/s,}$$

where 4.188 is the specific heat of water of salinity 0% at a mean temperature of 24°C. (For river water, the salinity is taken as zero but for sea and estuary water it must be determined. Here it has been assumed that river water is used.)

Alternatively, Q = 585 000/4177(28 − 20) = 17.5 m³/s

Where 4177 is the value of ϱc_p at 24°C from Site Test Code 3.

Multi-path ultrasonic flowmeter

Of the various ultrasonic techniques available for flow measurement, the preferred method uses the principle of measuring the transit time of acoustic energy travelling in both directions between two transducers located on a path angled across the pipe. If a long length of straight pipe is available upstream of the flow measurement position a single diametral-pitch system should be adequate, as the velocity profile can be predicted with sufficient accuracy to give a satisfactory correction factor. If, however, a distorted velocity profile is thought to be present due to the effect of upstream bends, etc., a multi-path system may be necessary. By using four chordal paths on an angled acoustic plane, it should be possible to achieve accuracies of ±1%, even with a poor velocity profile.

Bend flowmeter

The bend flowmeter utilises the difference in pressure between the inside and outside of a sharp bend or elbow in the cooling water system with the tappings located at the 45° position.

The differential is typically 70 mbar (0.7 m wg) and, like most flowmeters, it is sensitive to upstream flow conditions. If possible, the bend chosen should be at the end of a long straight section of pipework and well downstream of any major flow disturbance such as a partially-open valve or a swirl-producing device.

The flow rate is obtained from:

$$G = K \sqrt{h}$$

Where G = volumetric flow, m³/s

h = differential head, metres of water

K = calibration constant

The value of K is obtained from a calibration using, for example, the radioactive-isotope dilution technique, as it is not possible to predict the performance with sufficient accuracy from purely geometric considerations. It is estimated that the uncertainty of flow measurement using this device is of the order of ±2.5%.

Segmental orifice plate

This is cheap to manufacture and install, and is not prone to blockage by silt. Care should be taken to select a suitable location for the segmental orifice plate with at least ten diameters of straight pipe on the upstream and three on the downstream sides.

The flow rate is determined from the conventional flow formula given in BS1042: Part 1 using an apparent orifice diameter, given by d = 2000 $\sqrt{(a/\pi)}$.

Where d = apparent orifice diameter, mm

      a = open area of plate, m^2

The flow rate formula is:

    $q = 0.01252 \, C \, Z \, E \, d^2 \, \sqrt{(h\varrho)}$

Where q = mass flow rate, kg/h

    C = discharge coefficient

    E = velocity of approach factor

    d = orifice diameter, mm

    h = differential pressure at orifice, mm wg

    ϱ = density of fluid, kg/m^3

    Z = correction factor for pipe size and Reynolds number

The value of the discharge coefficient should be determined by calibration, possibly using the radioactive-isotope dilution method.

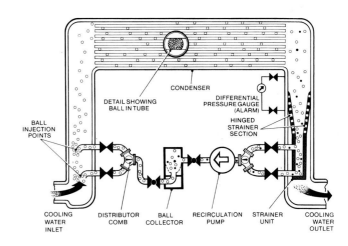

FIG. 7.92 On-load mechanical condenser cleaning
Spongy balls are re-circulated through the tubes as shown. The balls can be of various degrees of abrasiveness for different conditions of dirtiness. When such equipment is in use, the cleanliness factor of the tubes is assumed to be 0.95.

Differential pressure between waterboxes

This can be used for comparative flow measurement purposes. A differential-pressure transducer or a mercury manometer is connected to pressure tappings on the inlet and outlet cooling water pipes or waterboxes. Of course this is not an absolute method, so calibration will be necessary to relate the differential pressure and water flow rate. This can be done conveniently during a turbine test series using the condenser heat balance method.

The general form of the relationship is:

$$Q = K \, (\delta p/\varrho)^x$$

Where Q = volume flow rate, m^3/s

    K = constant

    δp = differential pressure, mbar

    ϱ = water density, kg/m^3

    x = exponent, usually between 0.5 and 0.57

The method can only be used with complete confidence for a few days after the calibration, because of possible changes of the resistance to flow due to deposits in the tubes or tube plates. However, if trouble from these sources is felt to be remote, the method provides a very convenient approximation of the CW flow. For example, if the condenser has on-load mechanical cleaning, such as illustrated in Fig 7.92, the tubes should remain relatively free from deposits of slime dirt and scale. The principle of operation of the cleaning method is that spongy balls (of various grades of abrasion, as required) are

circulated with the CW, scouring the tubes as they pass through.

Example. The conditions noted during two tests were:

$q_1 = 11.52$ m^3/s; $\delta P_1 = 2006$ mbar; $\varrho_1 = 998$ kg/m^3

$q_2 = 5.76$ m^3/s; $\delta P_2 = 501$ mbar; $\varrho_2 = 998$ kg/m^3

Find the constants from;

$$\log_n q = \log_n K + x \log_n (\delta p)/\varrho$$

Flow 1, $\log_n 11.52 = \log_n K + x \log_n 2.01$

     $2.444 = \log_n K + x \, 0.698$

Flow 2, $\log_n 5.76 = \log_n K + x \log_n 0.502$

     $1.751 = \log_n K + x(-0.689)$

Flow (1) − (2) $0.693 = 1.387x$

     $x = 0.5$

Subtitute in flow (1) $2.444 = \log_n K + 0.349$

     $\log_n K = 2.095$

     $K = 8.13$

     So $q = 8.13 \, (\delta P/\varrho)^{0.5}$ m^3/s

6.6 Heat transfer in condensers

6.6.1 Calculation of heat-transfer coefficient

The heat-transfer coefficient (U) is defined as the average rate of heat transfer from the steam to the cooling water per unit area per degree of logarithmic mean temperature difference.

$$U = P/A\theta$$

Where P = condenser heat load, kW

A = tube surface area, m^2

θ = LMTD, °C

So, for a typical condenser having the following design data, $P = 588\,430$ kW; $A = 27\,871.9$; $\theta = 8.92$°C, the design heat-transfer coefficient will be:

$$U_d = 588\,430/27\,871 \times 8.92 = 2.367 \text{ kW/m}^2 \text{ K}$$

The design value will normally be lower than the best attainable because the designer will make some allowance for tube fouling plus a factor to ensure that the specified performance is achieved.

The reference value U_r that ought to be attainable, assuming design conditions for number of tubes, heat load and CW flow, is found from:

$$U_r = 2.52\,(v_d)^{0.4}\,f(\theta_1)\,F_m \text{ kW/m}^2 \text{ K}$$

Where v_d = design CW flow velocity, m/s

F_m = tube material factor, from Table 7.33

θ_1 = CW inlet temperature, °C

$f(\theta_1) = 0.716 + 0.0236\theta_1 - 0.00031(\theta_1)^2$

So, for a condenser whose relevant data is:

v_d = 1.826 m/s;

F_m = 1.0 (for 70/30 brass, 25.4 mm OD);

θ_1 = 19°C;

the reference value is:

$$U_r = 2.52 \times (1.826)^{0.4} \times 1.0525 \times 1.0$$
$$= 3.372 \text{ kW/m}^2 \text{ K}$$

In normal operation, the condenser conditions will usually be different from design, and this affects the heat-transfer coefficient. Not only the CW temperature, cleanliness, etc., but also the number of tubes in service may be different, because some may be plugged. This will have an effect upon the CW flow, as shown in Fig 7.93. The cleanliness factor is assumed to be 0.9 for normal operation or 0.95 if on-load cleaning is used.

The determination of the heat-transfer coefficient from test is shown in Table 7.34.

The test heat-transfer coefficient obtained in Table 7.34 is the 'as-run' value. It is of interest to know what it would be if the test had been carried out at the design conditions of heat load and CW flowrate, and the method of doing this is shown in Table 7.35. Finally, it is necessary to correct the heat-transfer coefficient to standard conditions, if it is desired to compare the performance of one condenser with another of a different size or type. The standard conditions are:

Cooling water velocity	2 m/s
Cooling water inlet temperature	15°C
Tube material	Admiralty brass
Tube outside diameter	25.4 mm
Tube wall thickness	1.22 mm
Tube cleanliness factor	0.9
Heat load	Corresponding to 20 kW/m² of design surface area

TABLE 7.33

Tube material correction factors

Tube material	Composition, %	Thermal conductivity, kW/mK	Correction factor F_m			
			Tube outside diameter, mm			
			25.4	19.0	25.4	19.0
			Wall thickness			
			1.22, mm		0.71, mm	
70–30 brass	70Cu 30Zn	0.1255	1.005	1.016	1.033	1.051
Admiralty brass	70Cu 29Zn 1Sn	0.1089	1.000	1.011	1.030	1.048
Aluminium brass (Yorcalbro)	76Cu 22Zn 2Al	0.1006	0.997	1.008	1.029	1.046
90–10 copper-nickel	90Cu 10Ni	0.0481	0.957	0.967	1.004	1.020
ICI Cupro-Nickel	68Cu 30Ni 1Fe 1Mn	0.0273	0.904	0.912	0.970	0.985
Yorcoron	66Cu 30Ni 2Fe 2Mn	0.0254	0.896	0.903	0.965	0.979
Titanium	Commercially pure	0.0173	0.842	0.848	0.928	0.942

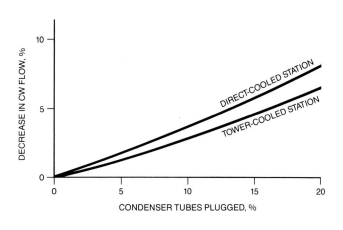

FIG. 7.93 Effect of plugged condenser tubes on CW flow
Note that even having 20% of the tubes plugged only
decreases the CW flow by about 7%.

$$U_{cl} = (23/6)(1.32/20)0.4 \text{ kW/m}^2 \text{ K}$$

Substituting values from Table 7.34 gives:

$$U_{cl} = 2.04 \text{ kW/m}^2 \text{ K}$$

To summarise the various heat transfer coefficients:

$$U_d = 2.367 \text{ kW/m}^2\text{K (design)}$$
$$U_r = 3.372 \text{ kW/m}^2\text{K (reference)}$$
$$U_t = 1.97 \text{ kW/m}^2\text{K (test)}$$
$$U_c = 1.98 \text{ kW/m}^2\text{K (corrected)}$$
$$U_{cl} = 2.04 \text{ kW/m}^2\text{K (corrected to standard condition)}$$

The heat-transfer coefficient corrected to standard conditions (U_{cl}) is determined from:

The reference heat-transfer coefficient and condenser pressure can be calculated for various inlet cooling water temperatures, and plotted for reference on graphs such as Fig 7.94. Also, the design coefficient and condenser pressure are shown. If the results of several tests are joined, such as the full lines in the

TABLE 7.34
Determination of test heat-transfer coefficient

	Description	Derivation	Units	Value
1	No. of CW passes			1
2	No. of tubes (design)			20 648
3	No. of tubes (test)			20 304
4	Tube outside diameter		mm	25.4
5	Tube wall thickness		mm	1.22
6	Tube material factor	Table 7.33		1.000
7	Tube surface area (design)		m^2	27 871
8	CW flow area in tubes	$\left(\frac{3}{1}\right)\frac{\pi}{4}\left[\frac{(4) - 2(5)}{10^6}\right]^2 \times 10^6$	m^2	8.407
9	Salinity (test)		%	0
10	Tube surface area (test)	(7)(3)/(2)	m^2	27 407
11	Electrical load		kW	491 195
12	CW inlet temp (test)		°C	17.9
13	CW outlet temp (test)		°C	26.8
14	CW temperature rise	(13) − (12)	°C	8.9
15	Condenser pressure (test)		mbar	52.34
16	Volumetric heat capacity	CEGB Site Test Code 3 or Formula	kJ/m^3K	4179
17	Turbine heat rate	At test conditions	kJ/kWh	7890
18	Condenser heat load (test)	[(17)/3600 − 1.017] (11)	kW	576 990
19	CW flow rate (test)	(18)/(16)(14)	m^3/s	15.5
20	CW velocity (test)	(19)/(8)	m/s	1.84
21	Condensing temperature	From (15)	°C	33.71
22	LMTD	$\dfrac{(14)}{\log_n[(21) - (12)/[(21) - (13)]]}$	°C	10.8
23	Test heat-transfer coefficient U_t	(18)/(10)(22)	kW/m^2K	1.95

TABLE 7.35

Correction to design heat-load and CW flowrate

	Description	Derivation	Units	Value
24	Condenser heat-load (design)		kW	588 430
25	CW flow rate (design)		m^3/s	15.6
26	Percentage loss of tubes	$[1 - (3)/(2)]$ 100	%	1.67
27	Reduction in CW flow	Fig 7.93	%	0.6
28	CW flow rate (design) for modified number of tubes in test	$(25) [1 - (27)/100]$	m^3/s	15.51
29	CW flow velocity (design) modified for number of tubes in test	$(28)/(8)$	m/s	1.84
30	Test heat transfer coefficient corrected to design conditions	$(23)[(29)/(20)]^{0.4}$	kW/m^2K	1.95
31	Test LMTD corrected to design conditions	$(24)/(10)(30)$	°C	11.0
32	CW temperature rise (design) modified for number of tubes in test	$(24)/(28)(16)$	°C	9.08
33	Exponential	exp $[(32)/(31)]$		2.283
34	Corrected condensing temperature	$(12) + [(32)(33)]/[(33) - 1]$	°C	34.1
35	Corrected condenser pressure	From (34)	mbar	53.5

figure, it will be possible to assess the effectiveness of the condenser and possibly obtain a guide to contributory causes of trouble. For example, if there is a fall-off in performance at low CW temperatures, such that the pressure curve diverges upwards from the reference curve, then this indicates either high air leakage or inadequate air venting. Furthermore, as condenser pressures are reduced through improved cleaning, correspondingly lower air-pump pressures will be required to prevent air blanketing. In some stations, auxiliary steam ejectors have been installed as they can attain lower back-pressure than pumps, which are generally cooled by condenser cooling water.

Sometimes the heat-transfer coefficients are good (indicating good operating conditions) but there is not very good absolute performance. This is possibly due to rather low tube surface area or design CW flow rate.

A measure of the effectiveness of a condenser is its *performance factor* (F_p). This is the ratio of the test heat-transfer coefficient (corrected to specified operating conditions) to the heat-transfer coefficient derived from the guaranteed condenser pressure and the specified operating conditions, i.e., $F_p = U_c/U_d$.

Thus, in the above example $F_p = 1.98/2.367 = 0.84$.

Computer programs are available for calculating routine condenser performance results.

6.6.2 Tube cleanliness

Definition

The tube cleanliness factor (F_c) is the ratio of the average heat-transfer coefficient of tubes in the condenser to that of a new, acid-cleaned tube.

Obviously, this is a very important parameter, particularly with tubes which are prone to fouling, either on the outside or inside.

On-load determination of cleanliness factor

Complete details of the method are given in Site Test Code 3. Sample tubes in groups of three or four should be selected with approximately one group per 2000 tubes. They must be representative of conditions throughout the condenser, avoiding areas (such as the air cooling section) where the CW temperature is likely to be low. One tube in each group should be selected as the clean tube, and water supplies arranged as shown in Fig 7.95. Water flow through the 'fouled' tubes should be maintained at approximately the same velocity as that in the remaining tubes, and the 'clean' tubes isolated. When the cleanliness factor test is carried out, CW is passed through both the clean and fouled tubes. The common inlet and individual outlet temperatures are measured using thermometers or thermocouples.

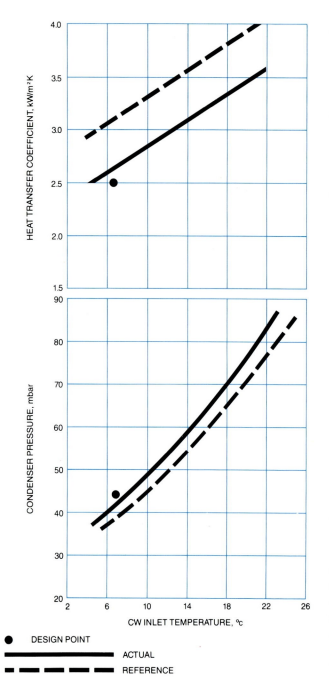

FIG. 7.94 Good condenser performance
As expected the design point is significantly worse than the reference performance. However, the actual performance is better than design over the whole CW temperature range.

● DESIGN POINT

━━━━━━ ACTUAL

▄▄ ▄▄ ▄▄ REFERENCE

The steam temperature at the group must also be determined, and this is done by recirculating cooling water through the clean tubes until the rise in temperature along the tube is zero. The temperature is then that of the steam in the vicinity of the tube. The cleanliness factor is calculated as shown in the example given in Table 7.36.

Off-load determination of cleanliness factor

About one in every 2000 tubes should be withdrawn from the condenser to give a good representation of the operating conditions. Lengths suitable for mounting in the testing apparatus are cut from the middle and ends of each sample tube. Care must be taken when removing and handling the tubes that the tube fouling is not affected.

The thermal resistance of the sample tubes is measured and compared with the thermal resistance of some of the tubes after acid cleaning. The tube lengths are mounted in a test rig in which steam is made to condense on the tube surface by water passing through the tubes.

The thermal resistance of the tubes is calculated as:

$$R = \frac{A}{q \times C_p \times \log_n (\theta_s - \theta_1)/(\theta_s - \theta_2)}$$

where R = thermal resistance, m^2K/kW
  q = cooling water mass flow rate, kg/s
  C_p = specific heat of cooling water, kJ/kg K
  A = outside area of tube, m^2
  θ_s = condensing temperature of the steam, °C
  θ_1 = temperature of water entering tube, °C
  θ_2 = temperature of water leaving tube, °C

To establish a relationship between thermal resistance and cooling water velocity, these measurements are repeated at different flow rates. The resistance 'R' is then plotted against $V^{-0.8}$, where V is the cooling water flow velocity in the sample tube in m/s and a best straight line fit is drawn through the points as shown in Fig 7.95 (b). An appropriate weighting may be devised if the tubes were not removed from uniformly distributed locations. At least two tubes should be thoroughly acid cleaned and the measurements repeated and plotted similarly. The resistance due to fouling, R_f, is then read off as the vertical separation between the lines for fouled and clean tubes.

The procedure to obtain the cleanliness factor is detailed in Site Test Code 3.

6.7 Acceptance tests

The efficient performance of the condensing plant is so crucial to the performance of the turbine that it is highly desirable to carry out exhaustive acceptance tests on the plant and routine tests thereafter. The full procedure for both single-pressure and multipressure condensers is detailed in Site Test Code 3, and reference has already been made to some of its provisions. It specifies procedures for testing condensers and for calculating and reporting the results in

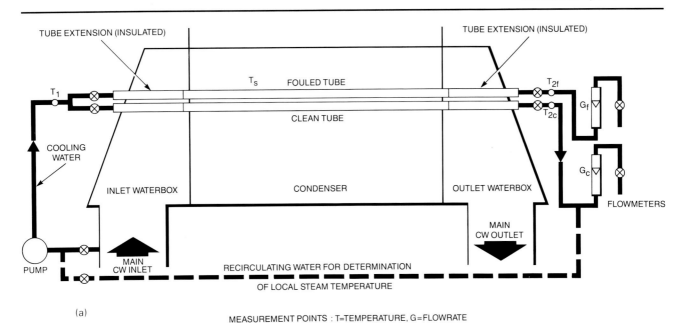

(a)   MEASUREMENT POINTS : T=TEMPERATURE, G=FLOWRATE

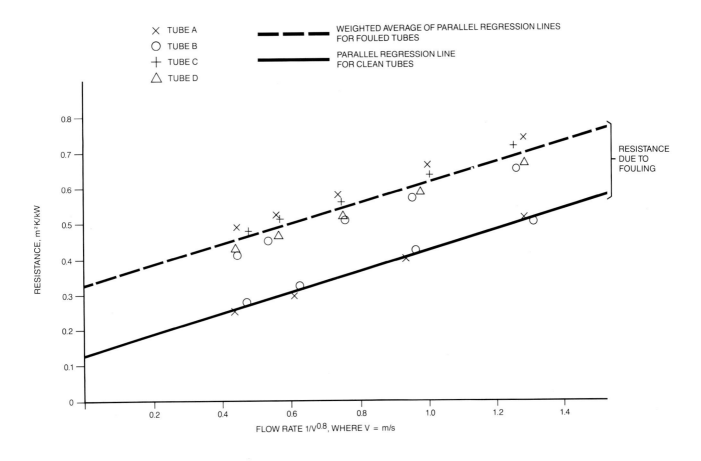

(b)

FIG. 7.95 Determination of cleanliness factor, (a) on-load and (b) off-load
The on-load arrangement enables a comparison to be made between the cleanliness of the clean
(reference) tubes and those in normal service.
The off-load method involves the determination of clean and fouled tube thermal resistance at various water flow rates.
The best straight lines are plotted and the vertical distance between them is the resistance due to fouling.

TABLE 7.36

Determination of cleanliness factor for a single pressure condenser

Test measurements

Symbol	Description	Units	Tube group position				
			V	W	X	Y	Z
θ_s	Steam temperature (see Note 1)	°C	33.0	33.5	32.5	33.1	32.9
θ_{w1}	CW inlet temperature	°C	17.91	17.91	17.91	17.91	17.91
θ_{w2fA}	CW outlet temperatures	°C	27.73	28.01	27.27	27.33	27.42
θ_{w2fB}	from fouled tubes	°C	27.70	28.03	27.22	27.35	27.45
θ_{w2fC}		°C	27.68	27.98	27.25	27.28	27.43
θ_{w2c}	CW outlet temperatures from clean tube	°C	28.61	28.97	28.38	28.34	28.43
Q_f	CW flow rate to test tubes	1/s	0.777	0.774	0.778	0.776	0.775

Calculation

Symbol	Description	Units	Tube group position				
			V	W	X	Y	Z
θ_{w2f}	Mean CW outlet temp. from fouled tubes	°C	27.70	28.01	27.25	27.32	27.43
F_{cg}	Tube group cleanliness factor (see Note 2)		0.847	0.844	0.808	0.833	0.833
U_g	Tube group heat-transfer coefficient (see Note 3)	kW/m^2K	2.517	2.501	2.462	2.321	2.419
F_{cgc}	Tube group cleanliness factor, corrected to test heat-transfer coefficient = $1 - (U_t/U_g)(1 - F_{cg})$		0.871	0.868	0.835	0.848	0.854
F_c	Mean cleanliness factor = $1/n \sum_1 F_{cgc}$ (see Note 4)		0.855				

Supplementary data

Symbol	Description/derivation	Units	Value
ϱC_p	Heat capacity of cooling water	kJ/m^3K	4179
A_t	Surface area of single tube = tube design total surface area/design number of tubes	m^2	1.350
U_t	Test heat-transfer coefficient	kW/m^2K	2.114
n	Number of tube groups		5

Notes:

1 Local steam temperature at position of tubes being tested. If this is not measured, it should be estimated from the steam temperature variation through the tube nest.

2 $$\dfrac{\log_n [(\theta_s - \theta_{w1})/(\theta_s - \theta_{w2F})]}{\log_n [(\theta_s - \theta_{w1})/(\theta_s - \theta_{w2c})]}$$

3 $$\dfrac{(10^{-3} Q_f \varrho c_p)\log_n [(\theta_s - \theta_{w1})/(\theta_s - \theta_{w2f})]}{A_t}$$

4 An appropriate weighting may be devised if the groups of tubes being tested are not distributed uniformly throughout the condenser.

order that the condenser performance may be checked against contractual obligations. It also details how different condensers may be readily compared and the operational performance of condensing plant assessed.

Before an acceptance test takes place, meetings are held between the interested parties to discuss such things as:

- Scope and objectives of the test.

- State of the plant.

- Preparation of the plant for testing.

- Programme of tests.

- Allocation of responsibilities.

- Method to be used for determining condenser heat load (i.e., using turbine heat rate or heat absorbed by the cooling water).

- Test instrumentation and calibration.

The various points at which measurements are required are shown in Fig 7.96. The Test Code lists the required types, locations and accuracy of the test instruments.

The duration of the test should be at least thirty minutes, but will normally extend over the full period of a concurrent turbine-generator test. The CW inlet temperature should be within the limits shown in Fig 7.97, unless validated corrections are available for other temperatures.

Because of the small number of measurements required, sophisticated data acquisition systems are usually not necessary. On the other hand, they may be justified on time and manpower grounds. If so, a 50-channel system should suffice.

Before the test condenser pressure can be compared with the guaranteed pressure, it must be corrected if the following test conditions differ from those specified:

- CW velocity.

- CW inlet temperature.

- Condenser heat-load.

- Tube cleanliness.

- Tube surface area.

- Tube material.

The corrected test condenser pressure is compared with the guaranteed value to establish whether the guaranteed performance has been achieved.

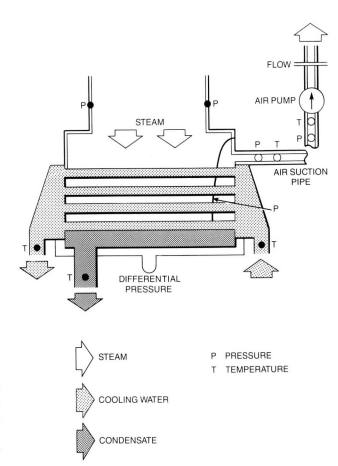

FIG. 7.96 Points of measurement for condenser testing
The desirable locations for measurements are shown.

Table 7.37 shows the information required from formal condenser tests and the required accuracy of the values.

7 Feedwater heater performance

7.1 Introduction

During the period 1950 to 1965, there was a marked advance of large power station plant conditions. In the United Kingdom, steam pressures and temperatures increased from a little over 60 bar/450°C to 160 bar/566°C (even higher on the supercritical pressure plant), and the unit size from 30 to 660 MW. Part of the resulting improvement in thermal performance is due to advances in the feed-heating systems. Not only has the performance of individual heaters improved, but the number in use on a unit has also gone up significantly. Figure 7.98 indicates

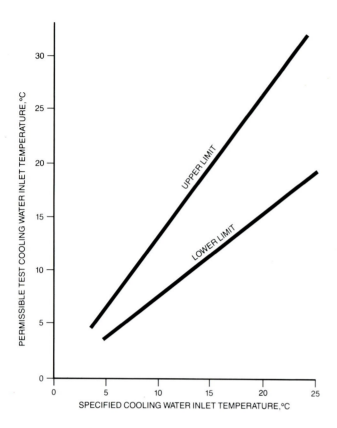

FIG. 7.97 Permissible limits of cooling-water
inlet temperature
The maximum allowable divergence from specified
CW temperature is shown.

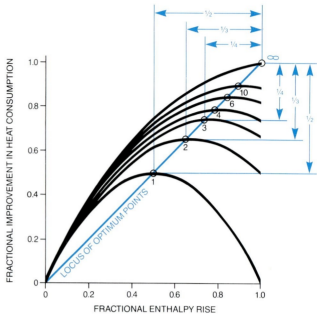

NOTE: THE FIGURES ON THE CURVES REFER TO THE NUMBER
OF HEATERS

FIG. 7.98 Curves showing improvement in heat
consumption with feed heating range and
number of heaters
The first heater is the most beneficial. Above ten heaters
there is little practical improvement. (The figures on the
curves refer to the number of heaters.)

that the more heaters used, the higher the gain in thermal performance. It follows the law of diminishing returns, though, so the greatest benefit is obtained with one heater and each succeeding one gives a reducing benefit. By the time ten heaters are considered, the further improvement by having an infinite number is not very great, so the economically justified maximum in practice is about ten. Individual evaluation is based upon the annual saving in fuel costs by having a particular complement of heaters balanced against the annual fixed charges on heater investment for minimum generating cost.

It is now common practice to incorporate a bled-steam-driven turbine feed pump in large units (Fig 7.99). The bled-steam turbine is used to supply steam to some of the HP feedheaters as well as to drive the pump, and there is some thermal benefit to the cycle from such an arrangement. Also, it is now quite common to have all the low pressure heaters in a feed train of the direct contact type. These and other developments will be discussed.

7.2   Typical system layouts

Before discussing particular heater configurations, it

is interesting to note the number of heaters used on some UK and USA plants (shown in Table 7.38).

The HP heaters are now commonly arranged in two parallel banks. Figure 7.100 illustrates a layout with all the heaters, i.e., LP and ancillary heaters (except the de-aerator) of the non-contact type. The evaporator has now been removed from the installation. A few points of interest are:

● This is a supercritical pressure unit, so all the condensate is polished when it leaves the primary extraction pump.

● The stator and hydrogen coolers, etc., provide the initial condensate heating.

● The bled-steam turbine supplies steam to all the HP heaters except No 8.

● It is a split pumping system.

The second illustration (Fig 7.101) is for a 660 MW unit. Unlike the previous example, the LP heaters

TABLE 7.37

Test summary for a single pressure condenser

Station: Unit:

Manufacturing details

Manufacturer:		Unit Size:	500 MW
Date of commissioning:		Number of tubes:	20 648
Number of condensing compartments:	3	Length between tubeplates:	18.3 m
Number of exhaust flows:	6	Tube outside diameter:	25.4 mm
Number of shells:	1	Tube wall thickness:	1.22 mm
Number of CW passes:	1	Total surface area:	27 871 m^2
Number of CW flows:	4	Tube material:	Admiralty Brass

Specified Conditions and Principal Test Measurements

Description	Units	Specified	Test	Corrected to design conditions
Date				
Time				
Number of tubes in service (see Note 1)		20 648*	20 304	
Cleanliness factor		0.9*	0.86	
Electrical load	kW	500 000	491 130	
Heat load	kW	588 430	576 990	
Steam flow	kg/s	266.26*	261.08	
CW inlet temperature	°C	19.0*	17.9	
CW outlet temperature	°C	28.0	26.8	
CW temperature rise	°C	9.0	8.9	
CW flow rate	m^3/s	15.61*	15.44	
CW velocity	m/s	1.83	1.84	
Condenser pressure	mbar	50.8*	52.3	54.1 ±0.72
Deviation from guaranteed pressure	mbar			+3.3
Steamside pressure drop	mbar		4.0	4.0
Saturation temperature	°C	33.2	33.7	34.3
Terminal temperature difference	°C	5.2	6.9	6.3
Logarithmic mean temperature difference	°C	8.9	10.0	9.4
Heat transfer coefficient	kW/m^{2}K	2.37	1.97	2.08
Performance factor, F_p				0.88
Heat transfer coefficient corrected to standard conditions	kW/m^{2}K			1.87
CW pressure drop between waterboxes (see Note 2)	bar		0.39	0.40
Condensate temperature	°C		34.9	
Condensate subcooling (see Note 3)	°C		+1.2	
Air flow rate	kg/h	120	175	
Air pump suction pressure	mbar		40.2	
Air pump suction temperature	°C		24.9	
Air suction depression	°C		4.2	
No. of air pumps in service		2*	3	
Condensate oxygen level	μg/kg			
Remarks: (see Note 4)				

* Value on which guarantee is based

Method of cleaning: Bulletting
Condenser heat load calculated from turbine-generator load assuming a heat rate of 7884 kJ/kWh.

Notes on the test summary:

1 The reasons for any difference from the specified number of tubes should be given.

2 The correction to the design CW flowrate assumes that the pressure drop is proportional to (flowrate)$^{1.75}$.

3 A positive value denotes condensate temperatures above the saturation temperature corresponding to the condenser pressure.

4 The remarks should include the method used to clean the condenser tubes and whether the test heat-load was calculated from turbine-generator heat rate or from direct measurement of cooling water flowrate.

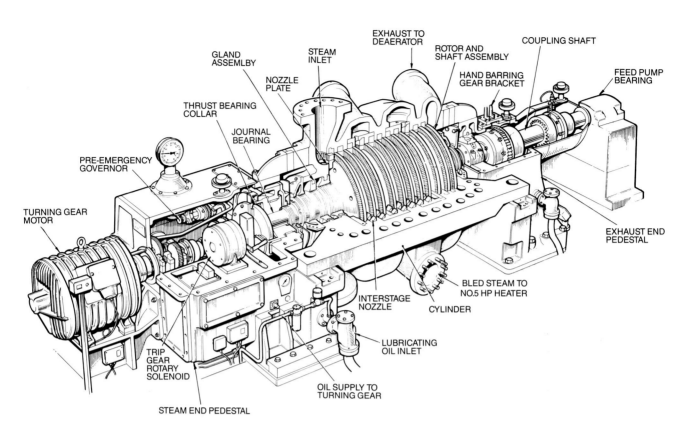

FIG. 7.99 Boiler feed pump turbine
It is now common to drive the main boiler feed-pump by a bled-steam turbine. This arrangement lends itself to supplying
selected feedheaters from the steam-driven turbine, thus improving the cycle efficiency.

TABLE 7.38
Typical numbers of feedheaters in use

Output,	Steam at TSV		Reheat temperature,	Number of heaters	Final feed temperature,	Country
MW	bar	°C	°C		°C	
375	241.3	593	566	8	266	UK
500	158.6	566	566	8	252	UK
600	158.6	566	566	9	278	UK
660	158.6	566	566	8	252	UK
800	158.6	538	538	8	254	USA
860	172.4	538	538	7	254	USA

are all of the direct-contact type and there are no booster feed pumps. Thus, the HP heaters have to withstand feedwater pressure of the order of 210 bar.

7.3 Heat balance

7.3.1 Direct-contact heater

In a direct-contact (DC) heater, as illustrated in Fig 7.102, the heat balance is simple. The condensate temperature is raised to boiling point by heat addition from steam supplied to the heater. The terminal temperature difference is zero because the heater body is maintained at the pressure of the condensing steam. For example, consider a de-aerator which operates at a pressure of 4 bar absolute. If the steam temperature is 240°C, then each kilogram of steam will surrender its superheat and latent heat, i.e., 206 + 2133 = 2339 kJ, so the total heat given to the condensate will be 2339 × steam flow (kg/s).

Heat surrendered by steam = heat received by condensate, so

$$q_s \times (H_s - h_d) = q_f \times (h_{fo} - h_{fi})$$

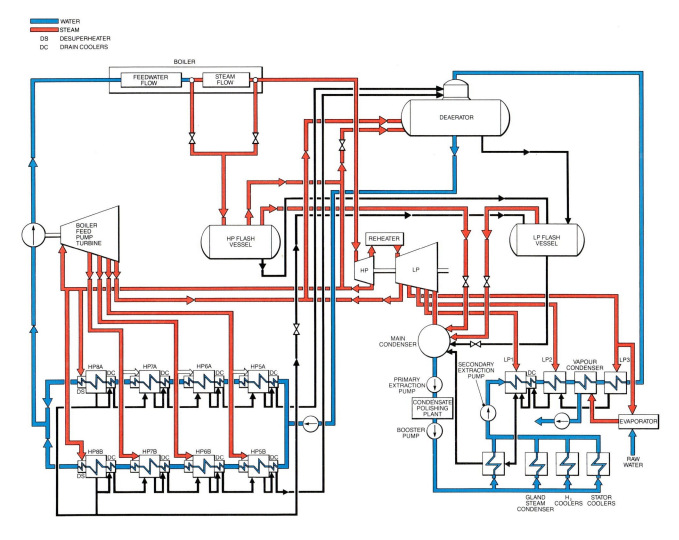

FIG. 7.100 Diagrammatic arrangement of a feedheater system for a 375 MW unit
The system shown is associated with a unit operating at supercritical pressure conditions.

where q_s = steam quantity, kg/s

H_s = enthalpy of steam supply, kJ/kg

h_d = enthalpy of condensed steam, kJ/kg

q_f = condensate quantity at inlet, kg/s

h_{fo} = condensate enthalpy at heater outlet, kJ/kg

h_{fi} = condensate enthalpy at heater inlet, kJ/kg

Substituting values from Fig 7.102 gives:

$$q_s (2944 - 605) = 400 (605 - 419)$$

$$\text{so } q_s = 31.8 \text{ kg/s}$$

and condensate flow from de-aerator = (400 + 31.8) = 431.8 kg/s

7.3.2 Non-contact heater

The same procedure is used for the heat balance across a non-contact heater. Here, the steam and water pressures in the heater are different and the condensed steam does not mix with the feed. The conditions depicted in Fig 7.103 could apply, where the heat surrendered by the steam equals the increased feed-water heat.

$$q_s (3098 - 1155) = 295 (1170 - 987)$$

$$q_s = 27.8 \text{ kg/s}$$

So the heater would need to be supplied with about 28 kg/s of steam to achieve the stated rise in feed-water temperature. About 30°C is the normal order of feed rise per heater except for the highest pressure heater, where it is economic to have a higher value. The terminal temperature difference (TTD) is the

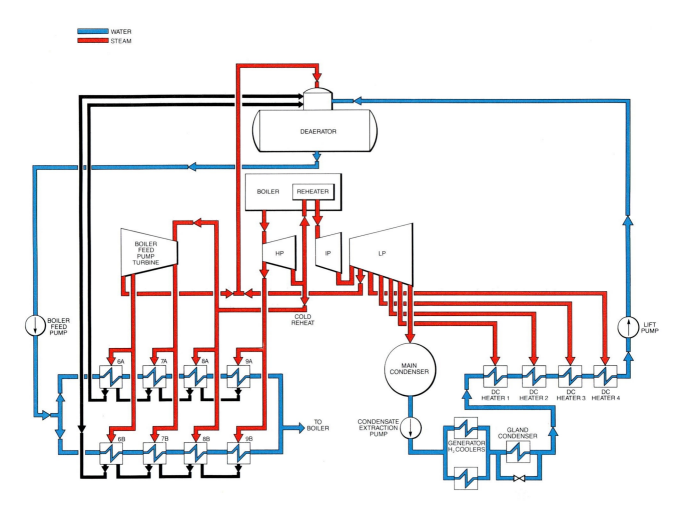

FIG. 7.101 Diagrammatic arrangement of a 660 MW unit feedheating system
All of the LP feed train consists of direct-contact feedheaters. The HP heaters must be capable of withstanding
full feed pump pressure.

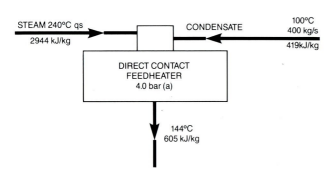

FIG. 7.102 Direct-contact feedheater heat balance
The water in the vessel is raised to saturation temperature
by the steam supplied.

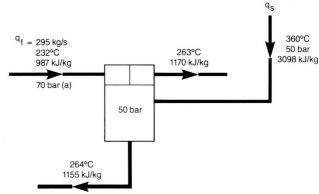

FIG. 7.103 Non-contact feedheater heat balance
The feedwater and steam are kept separate. If there is no
drains cooling section, the drain temperature corresponds
to the saturation temperature of the steam.

condensing steam temperature minus the outlet feed temperature, here (264 − 263) = 1°C. For a plain condensing heater, a typical TTD is 6°C, but in heaters with significant desuperheating TTDs of zero or small negative values are possible. The TTD is a

very important pointer to the effectiveness of the thermal performance of the heater. With modern

HP heaters, it is normal to incorporate a drains cooling section as well as a desuperheating section as can be seen in Fig 7.104.

The condensed steam, or drain water as it is now termed, falls to the bottom of the heater and submerges the section of tubes containing the incoming feedwater. Thus, some sensible heat is transferred to the incoming feed from the drain water. The thermal performance of a drain cooling section is judged by the difference in temperature between the feedwater entering and the drain leaving, known as the drain TTD: a typical value is about 8°C.

Figure 7.105 illustrates the different arrangements of heat transfer in common use.

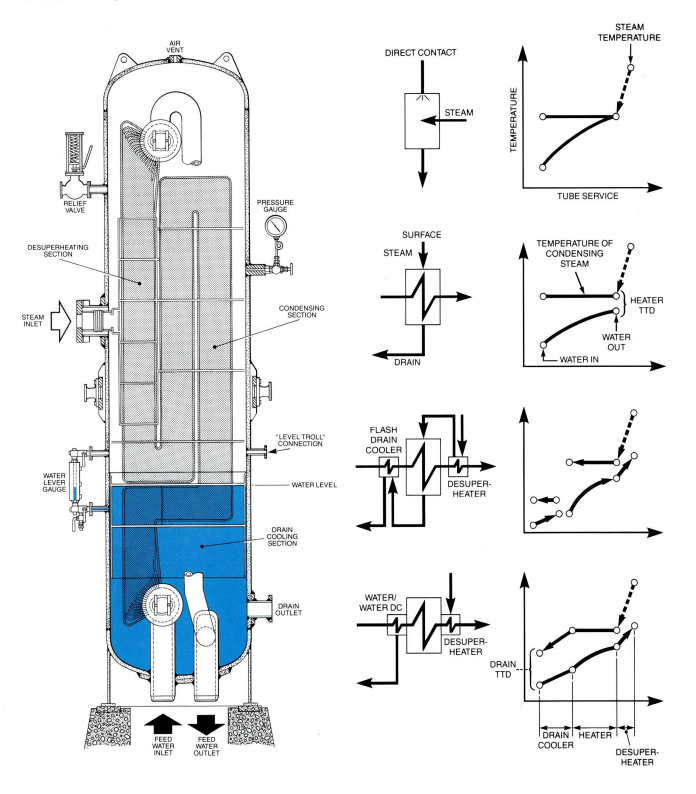

FIG. 7.104 High pressure feedheater
High pressure feedheater with integral desuperheating and drains cooling sections.

FIG. 7.105 Heat exchange in various types of heater
The diagrams illustrate the general temperature changes associated with various types of heater.

Heat transfer to the water is primarily achieved by the surrender of the latent heat of the steam, i.e., by condensation. This applies even if the steam initially contains a considerable degree of superheat. To illustrate the contributions made by the desuperheating, condensing and drains cooling sections consider the last heater in the train on a large unit.

The inlet steam conditions are 50 bar absolute/360°C and the drain outlet temperature is 240°C. Referring to steam tables:

Heat surrendered by desuperheating $= 3098 - 2794$
$$= 304 \text{ kJ/kg}$$

Heat surrendered by condensation $= 2794 - 1154$
$$= 1640 \text{ kJ/kg}$$

Heat surrendered by drains cooling $= 1154 - 1038$
$$= 116 \text{ kJ/kg}$$

So the heat transfer due to condensation is over five times as great as that due to desuperheating and over fourteen times that due to drains cooling. Note that the appropriate steam pressure from which the saturation temperature is determined is that in the condensing section of the heater. This is not the same as the pressure of the steam before the heater, because of the loss sustained in the desuperheating section.

7.3.3 Cascade drains

The drainage from the highest-pressure heater is normally passed to the next lower pressure heater

flashbox, and so on. The drainage from the HP heaters is passed to the de-aerator, while the cascaded low pressure heater drainage is passed to the condenser, usually via a drains cooler. An orifice plate is located at the entrance to each flashbox to ensure that flashing takes place in the box and not in the drain line. The heat from the drains must be taken into account when carrying out a heater heat-balance. Consider Fig 7.106. The drainage flow from heater No. 6 will be the same as the flow of steam, and the temperature of the drain water will be at saturation, assuming that there is no drains cooling, so that No. 6 heater steam flow, q_{s6}, is given by:

$$q_{s6} (3191 - 990) = 104 (991 - 810)$$
$$q_{s6} = 8.6 \text{ kg/s}$$

The drain water flow from No. 6 heater will also be 8.6 kg/s and the heat content 990 kJ/kg.

At the No. 5 heater flashbox, the pressure is 14 bar absolute so the reduction of pressure of the drain water will cause flashing to take place. The quantity of flash steam is calculated from the expression:

Flash steam
$$= \frac{\text{(inlet drain heat)} - \text{(outlet drain heat)}}{\text{latent heat at flashbox pressure}}$$

Here, flashbox pressure is 14 bar, so the latent heat is 1958 kJ/kg, and the quantity of flash steam is $(990 - 830)/1958 = 0.082$ kg, so each kilogram of drain water entering the flashbox will produce 0.082 kg

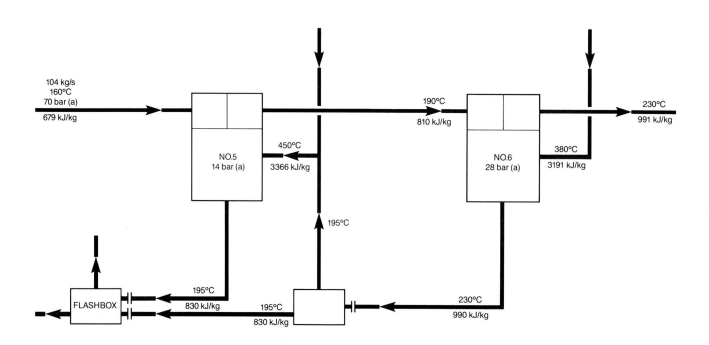

FIG. 7.106 Non-contact heaters — heat balance
The determination of the bled-steam supplies to No. 6 and 5 heaters is explained in the text.

of flash steam (i.e., 8.2%) with a heat content of $(1958 + 830) = 2788$ kJ/kg. The drain water from the box will be at saturation temperature (as will the flash steam), and is passed to the next box downstream where the flashing process is repeated, but at the conditions prevailing in No. 4 heater. Meanwhile, No. 5 heater is supplied with heat from two sources (bled-steam and flash steam), whose combined heat produces steam at 450°C with a heat content of 3366 kJ/kg.

The steam flow to No. 5 heater (Fig 7.106) is:

$$q_{s5} \ (3366 - 830) = 104 \ (810 - 679)$$

$$\text{So } q_{s5} = 5.4 \text{ kg/s}$$

and this includes $0.082 \times 8.6 = 0.7$ kg/s flash steam at 2788 kJ/kg. The heat in the bled-steam must be equal to the difference of the enthalpies of the steam-to-heater and that from the flash steam, so the bled steam heat is:

$$\frac{(5.4 \times 3366) - (0.7 \times 2788)}{(5.4 - 0.7)} = 3452 \text{ kJ/kg}$$

Therefore, the bled-steam temperature is 490°C approximately (from steam tables).

A heat balance for every heater in a train is a tedious but simple task. Additional complications are often present in the form of various leak-offs which combine with the bled-steam. For example, often the gland leak-off from the main turbine is discharged into appropriate bled-steam lines.

7.4 Deterioration of heater performance

7.4.1 General

It is not sufficient to just check the final feed temperature to determine whether the heaters are operating satisfactorily because trouble at one heater can be compensated for, in terms of feed temperature, by subsequent heaters doing more work. This will cause the steam flows in the main- or bled-steam turbine to alter, depending upon the source of the heater steam, so altering the work done. In other words, the cycle efficiency is altered. Therefore, it is necessary to check individual heaters from time to time, paying particular attention to the drain and terminal temperature differences. If necessary, a full heat-balance should be carried out for every heater. As a general rule, a heater investigation should be carried out at least once, and preferably twice, per year on each running unit.

Where deterioration of performance occurs it is normally due to one or more of the following causes:

- Air accumulation.

- Steamside fouling.

- Waterside fouling.

- Drainage defects.

Each will be considered in turn.

7.4.2 Air accumulation

Air is a superb thermal insulator, so it is most undesirable in heat exchangers. The body of each heater is vented to the condenser so that any air which gets in does not accumulate. When a unit is on-load, it is only possible for air to get into heaters which operate at less than atmospheric pressure. However, air gets into all of them when the unit is off-load. Clearly, the vents on the sub-atmospheric heaters should be permanently open, whilst those on the rest should at least be opened for long enough to vent them thoroughly when the unit is brought on-load, and periodically thereafter. Normally, vents pass about 0.5% of the steam, so opening them represents a thermal loss to the system.

The location of the vents is important and, if continuous problems with air accumulation are encountered, it may be that they are incorrectly located. Basically, they should be at the ends of the heat exchanger remote from the steam inlet, as shown in Fig 7.107. The steam is admitted towards the middle of the heater and flows upwards and downwards, so vents are required at the top of the body and also just above the water level in the drain cooler.

In normal operation, air ingress can occur at a heater itself or at connected plant, such as a bled-steam linejoint. Once air accumulation occurs, there are various manifestations of it, including:

- Reduced heater drain-water temperature.

- Increased TTD.

- Possible elevation of steam-to-heater temperature.

- Reduced temperature rise of feedwater or condensate.

A particular example of air ingress occurred on a low pressure heater. The normal values are shown in Fig 7.108, with those at the time of the air accumulation in brackets.

The cause of the trouble was due to considerable air ingress at joints on the bled-steam line in the vicinity of the turbine. The mechanism of the trouble is of interest, as one could easily conclude from an

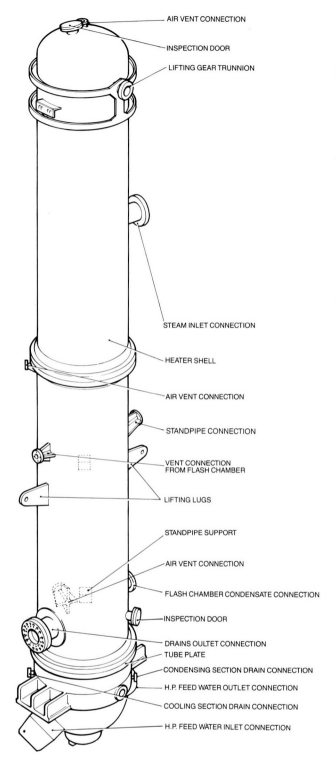

FIG. 7.107 High pressure feedheater — positions
of air vents
As the steam is admitted at about the middle of the
heater, it is necessary to provide air vents both at the top
of the heater and towards the bottom.

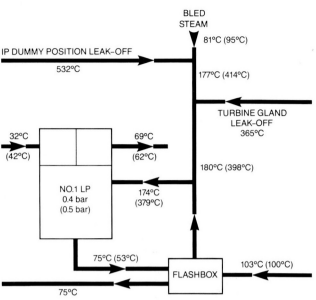

FIG. 7.108 Effect of air accumulation in heater
The optimum conditions are shown, plus those after a
short period of operation (shown in brackets). Note the
excessive steam temperature at the body of the heater.

from the turbine glands is determined both by the
condition of the glands and by the pressure difference
between the gland inlet and the bled-steam line. The
pressure in the bled-steam line was only about
0.1 bar worse than normal and the pressures at the
dummy and turbine glands were normal. Also, a heat
balance showed that the steam flow quantity was normal in both leak-off lines. As the drain was considerably under-cooled, it was a clear indication of
air accumulation in the heater. What had happened
was that the heat transfer had become much worse
across the heater nest, as indicated by the poor condensate temperature rise. Thus the quantity of heating steam required was reduced so much that it was
almost all provided by the gland leak-offs. The slight
rise in bled-steam pressure was enough to reduce the
flow of bled steam, and so the dummy-piston leak-off
steam was not being cooled as much by the bled-
steam as before, resulting in the observed high temperature steam at the heater.

7.4.3 Steamside fouling

Up to the 1950s, a common material for HP heater
tubes was 70/30 cupro-nickel. However, it become
clear after some years' experience that this material
is unsuitable because it exfoliates. As a result, the
tube surface material flakes off like dead skin, the
space between the tubes becomes blocked with debris (Fig 7.109) and heat transfer is progressively
affected. The mechanism of the process is that oxygen
comes into contact with a hot cupro-nickel surface

inspection of the diagram that the basic problem
was badly worn IP dummy-piston glands. The rate
of steam flow from the dummy-piston glands and

FIG. 7.109 Photograph of exfoliation debris
Severe exfoliation has halved the thickness of the
70/30 Cu/Ni tubes and a considerable quantity of
exfoliated material is lodged between the tubes.

and cuprous oxide is formed. If heating is continued, even if there is no free oxygen present, combined oxygen may diffuse into the alloy forming a sub-scale of nickel oxide in a matrix of copper. The effects of exfoliation includes:

- Progressive increase of TTD.

- Drain temperature unaffected.

- Reduced feed temperature rise.

- Eventual tube failures due to mechanical weakening.

- Accumulation of debris in the heater shell.

In one case of exfoliation, the TTD rose gradually from 4°C to 14°C in seven years, as shown in

Fig 7.110. The only solution to the problem was to re-tube the heater nest.

Alternative materials are now used including;

- 90/10 cupro-nickel.

- Monel.

- Mild steel.

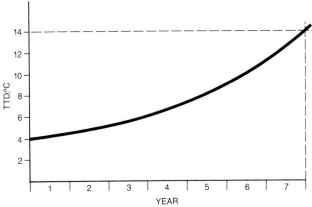

FIG. 7.110 Effect of exfoliation on TTD
The rise of TTD at the heater is gradual, but can be considerable over a period of years.

The last is in common use on modern plant because it is cheaper than the others and fastening the tubes to the tube plate is facilitated. Rusting is not a problem as the operating conditions inside the heater are oxygen-free. However, when the heater is out of service, it is important that there is no standing water at tubes such as those in the drains cooling section, to obviate corrosion. Therefore, automatic emptying of the steamside of the heater is required on unit shutdown.

7.4.4 Waterside fouling

Probably the most common cause of waterside fouling is oil. For example, oil from a leaking bearing at the turbine low pressure cylinder could get into the turbine via an imperfectly-sealed LP gland. It is carried to the feedheaters with the condensate and feed water. Deposition then occurs in the HP heaters, affecting them all, but being worst at the highest pressure heater. The thermal manifestations of the trouble are similar to those for exfoliation, except that the onset of increasing TTD is usually sudden and the rate of deterioration is rapid.

When tubes are only slightly fouled, they can be cleaned by chemical means. If the fouling is severe, high pressure water jetting may be required.

Should oil contamination be suspected, it is possible to analyse the feedwater with equipment such as the HORIBA OCMA 220, which will detect minute quantities of oil in feed to less than 0.2 mg/kg of water.

7.4.5 Drainage defects

Apart from obvious problems such as passing valves, the usual troubles are due to:

- Damaged flashbox internals.

- Reduced orifice opening.

- Enlarged orifice opening.

- Heater drain pump defects.

A typical flashbox has an orifice plate in the inlet pipe followed by a diffuser, the function of which is to direct the flashing drains downward onto the surface of the water reservoir in the box. Should the diffuser disintegrate, the pieces could cause obstructions in the pipes, resulting in flooding. Also the flashing steam/water would quickly cause erosion of the flash-box internals. Consequently it is good practice to make the diffuser of high quality material, such as stainless steel.

Another problem with drains systems is that of reduced orifice size. This can be caused by careless maintenance (where an incorrect size of plate is installed) or by a build-up of deposit. Figure 7.111 allows a photograph of an orifice fouled with copper oxide. The diameter of the hole was 19 mm, but after cleaning it was 42 mm. The effect of fouling is to cause the drains to back-up to the previous heater flashbox, causing it to flood and some water droplets will be carried with the flash steam to the junction with the bled steam pipe. Heat is then abstracted from the bled-steam to evaporate the water drops, and consequently the steam-to-heater temperature will be low.

For example, Fig 7.112 shows the conditions at the heaters on a 120 MW unit in normal operation and, in brackets, when an orifice was fouled. The No. 4 steam-to-heater temperature was so seriously reduced, that it was at saturation temperature. Hence, the steam entering the heater was wet instead of having a considerable degree of superheat.

The converse problem to fouled orifice plates is that of eroded plates. The effect of the enlarged orifice is to reduce the differential pressure from inlet to outlet. The bled-steam pressures at the upstream and downstream heaters are not altered by the orifice plate, so the reduced differential is achieved by lowering the water level in the upstream heater.

It is now quite common to have all the drainage from one heater passed to the body of the next, and so on throughout the train (Fig 7.113).

FIG. 7.111 Effect of oxide deposit on drain-line orifice plate
The *top* photograph shows the fouling of the plate as it was when removed from the train and that on the *bottom* shows the same plate after partial cleaning. Note the coin on the plate for size comparison.

Such an arrangement will be associated with heater systems having drain cooling. The water level in the heaters is maintained by the configuration of the drain pipework or by level-actuated drain valves.

All the drain water is passed to each downstream heater instead of just the flash steam as with the previous arrangement.

7.4.6 Heater drain pump defects

It is common practice to have the high pressure heater drains cascaded to a heater drains pump from which the drains are passed to the de-aerator (Fig 7.114). The water in the de-aerator flashbox is at saturation

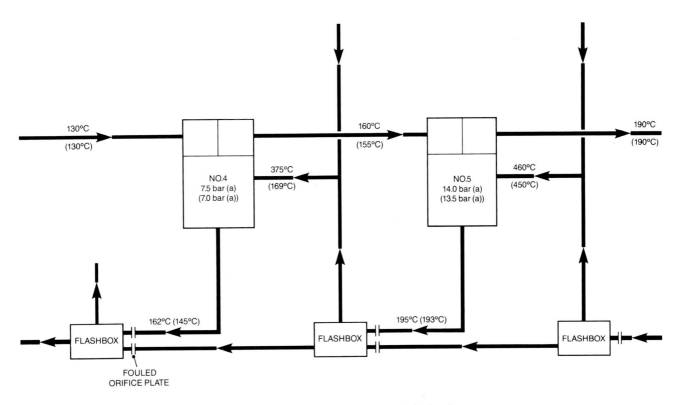

FIG. 7.112 Abnormal operation of the feed train
The steam temperature entering No. 4 heater is very low because of the fouled orifice in the drain line.

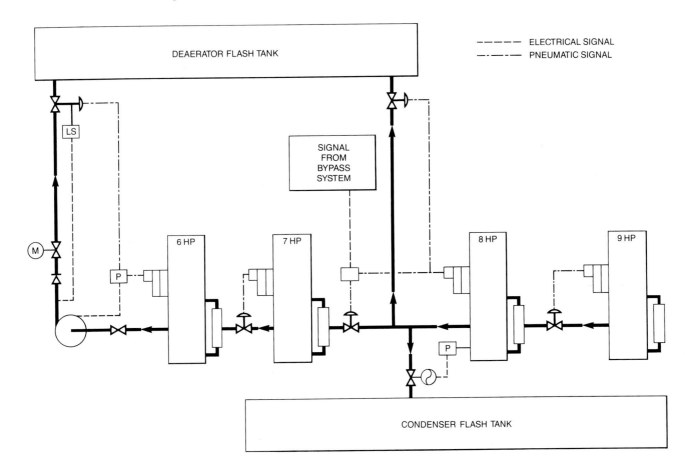

FIG. 7.113 Heater cascade drains arrangement for a 660 MW unit
On modern units with heater drains cooling, it is common to cascade the drains directly from heater to heater
via water level control valves.

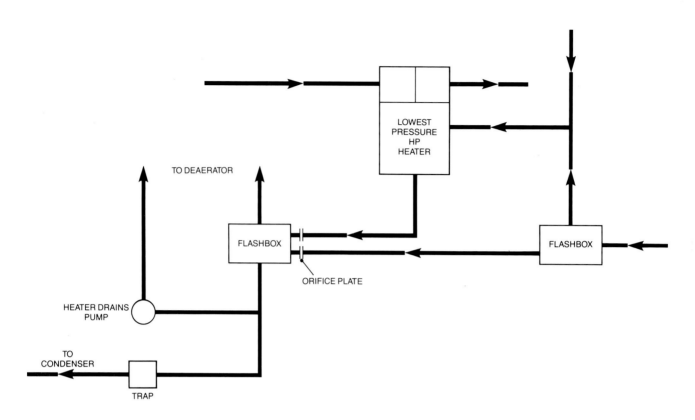

FIG. 7.114 Heater drains pump installation
The static suction head at the pump is often quite low, so the pump should not have its suction restricted in any way,
such as by throttling the suction valve.

temperature, so the head of water at the pump suction must be sufficient to prevent flashing taking place in the pump body, so interfering with pumping. The static head of water at the pump suction may only be of the order of a metre, so friction and other losses must be kept to the practical minimum. Consider the layout shown in Fig 7.115. The water in the flashbox is at the saturation temperature corresponding to the flashbox pressure, i.e., 159°C at 6 bar absolute pressure. If there were no friction loss in the pump suction system, the water entering the pump would be 2 m (about 0.2 bar) above saturation pressure. In practice, though, it is impossible to have friction-free water flow. In fact the various losses can be significant and are shown in Fig 7.115.

Suppose that all of these amount to 1.2 m friction loss, then the remaining margin to prevent the water flashing is only 0.8 m (about 0.08 bar). It is obvious that a small increase in the friction loss could be sufficient to cause flashing inside the pumps. This, in turn, will cause the delivery to be severely curtailed and the trap on the emergency line to the condenser to open, so lowering the cycle efficiency. It is, therefore, important that nothing is done to increase the suction resistance unnecessarily. Also, the pump should be kept in good repair, as a worn pump may not be able to handle the total quantity of water offered to it. Should the installation be prone to flashing even

under favourable circumstances, it will be worth considering physically lowering the pump so that the suction head is increased.

7.5 Effect of heater fouling

It has already been demonstrated that fouling always causes the affected heater TTD to increase because the outlet feed temperature will be reduced. Therefore, when the feed enters the next heater it will be colder than normal and so increase the steam consumption at that heater. For example, consider the heaters in Fig 7.116. The clean operating conditions are as shown. Suppose, now, the No. 1 heater became fouled such that the TTD rose to 10°C. The saturation temperature of the steam would be unaffected, so the condensate outlet temperature would become 60°C. The feed outlet temperature at No. 2 heater would still be, say, 100°C. Therefore, the steam flow (q_{s2}) to No. 2 heater would be:

$$q_{s2} (2745 - 439) = 400 (420 - 252)$$
$$q_{s2} = 29.1 \text{ kg/s}$$

and that to No. 1 would be:

$$q_{s1} (2645 - 318) + 29.1 (439 - 318)$$
$$= 400 (252 - 211)$$
$$q_{s1} = 5.5 \text{ kg/s}$$

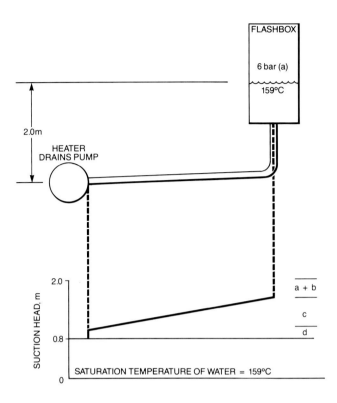

FLASHBOX

6 bar (a)

159°C

2.0m

HEATER
DRAINS PUMP

2.0

a + b

c

d

SUCTION HEAD, m

0.8

SATURATION TEMPERATURE OF WATER = 159°C

0

FIG. 7.115 Heater drains pump suction conditions
The losses incurred by delivering water to the pump are
shown. If the pressure drop is excessive, the water will
flash inside the suction pipe, resulting in poor delivery:

a Represents getting the water from the flashbox
into the supply pipe

b Represents accelerating the water from rest to its
velocity in the pipe

c Represents the supply pipe friction loss

d Represents the loss due to getting the water
into the pump via the pump inlet passages

mined by the quantity of bled-steam that can get to the heater. If, for argument, the condensate outlet temperature remained at 100°C, the bled-steam flow q_{s2} would be:

$$q_{s2} (2745 - 439) = 400 (420 - 211)$$
$$q_{s2} = 36 \text{ kg/s}$$

This is a 66% increase on the optimum flow and the steam velocity entering the heater will be increased by the same amount. This would pose a serious risk of damage to the heater, so the manufacturers normally recommend that the bled-steam supply be throttled to give only the design condensate temperature rise across the heater.

An exception to this reasoning occurs when the final heater is taken out of service, such as for emergency generation. There are no further feedheaters in the system so the feedwater temperature is reduced by the amount of the normal heater temperature pick-up. The colder feedwater, therefore, has to have extra heat given to it in the boiler. Meanwhile, the steam that would normally have been bled can now do work in the remainder of the turbine and so significantly increase the unit output, although at the expense of lower thermal efficiency. For example, consider a 500 MW turbine-generator which has a full-load heat rate of 7850 kJ/kWh. By taking the highest-pressure heater out of service, the output will rise to, say, 520 MW at a heat rate of 8000 kJ/kWh, so the change in turbine heat consumption is given by:

Heat consumption at 500 MW

$$= 500 \times 7850 \times 10^{-3}$$
$$= 3925 \text{ GJ/h}$$

Heat consumption at 520 MW

$$= 520 \times 8000 \times 10^{-3}$$
$$= 4160 \text{ GJ/h}$$

Extra heat consumption $= 235$ GJ/h

Hence the incremental heat rate of the last 20 MW of output is $(235 \times 10^6)/20\ 000 = 11\ 750$ kJ/kWh, so it is very expensive generation.

So the steam flow to No. 1 heater has been reduced from 13.1 to 5.5 kg/s and that to No. 2 has increased from 21.8 to 29.1 kg/s. The increased flow to No. 2 heater could cause mechanical damage to the heater because of the increased steam velocity and mass flow. In this example, the velocity and mass flow will increase by 33% above normal.

As a general guide, the turbine-generator heat rate will be affected by 0.07 percent for 1°C change in the TTD of an HP heater for a unit with 159 bar/566°C/566°C turbine inlet conditions.

7.6   Effect of heater out of service

If a heater is taken out of service, the effect on the heater train can be quite considerable. For example, consider taking No. 1 heater out of service (Fig 7.116). The feed entering the second heater will only be 50°C and the condensate outlet temperature will be deter-

As a general guide to the effect of having a feedheater out of service, the following efficiency reductions may be used:

TSV pressure	Up to 100 bar	Over 100 bar
LP heater	0.5%	0.5%
Last HP heater	1.3%	1.5%

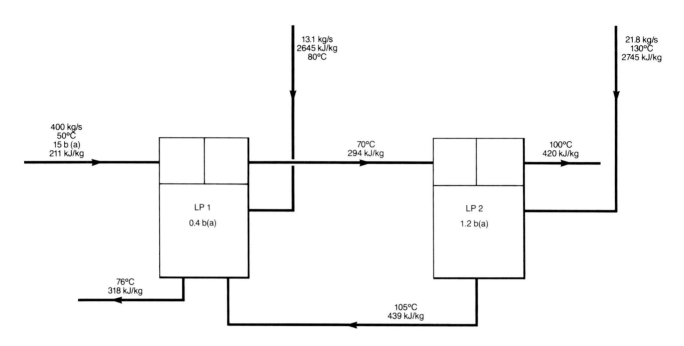

FIG. 7.116 Effect of heater fouling

Should a heater become fouled, the TTD will increase and the feed outlet temperature will decrease. Hence, the feed entering the next heater will have lower than normal temperature, so the demand for bled-steam at that heater will go up.

On large modern units the HP heaters are usually arranged in two banks; Fig 7.117 shows the layout for a 660 MW unit. Should one bank of heaters trip, the remaining one can carry 70% of MCR feed flow, causing a reduction of final feed temperature from 278°C to 233°C in the system shown. It is also possible to have only two heaters out of service in either bank. Thus, for example, if 8A and 9A heaters are by-passed, the final feed temperature will be 242°C.

It may also be mentioned that water-sealed feed pump glands may involve by-passing the HP heaters. Consider a split pumping installation in which the boiler feed pump glands are supplied with water from the booster pump.

If the values are as shown in Fig 7.118, the effect of the gland sealing will be to add 2 kg/s of water at 137°C to the main feed flow. This will cause the combined water heat to become $[(2 \times 593) + (320 \times 1086)]/322 = 1083$ kJ/kg, so the equivalent temperature is 249.6°C.

Here, the temperature change is negligible. If, however, the glands permit significant inleakage of water to the pump, the resulting reduction of the feed temperature could be several degrees. Mechanical gland-sealing is now common, eliminating this problem.

One final cause of by-passing should also be mentioned, where a heater by-pass valve is supposed to be shut but is, in fact, passing. This condition is easily detected by noting both the final heater feed outlet temperature and comparing it with the feed temperature after the junction with the by-pass line.

7.7 Monitoring the feed system

The basic requirement for effective monitoring is to have comprehensive information on the optimum operation of the system. Acceptance test results, plus design data regarding the feed system, should be adequate. Having established what the plant ought to do (from design data) and what it actually achieved under acceptance test conditions, it is then necessary to ensure that any deterioration in performance is detected quickly. Usually this requires a feed heater survey to be carried out every six months or so in normal circumstances. Meanwhile, the operation staff is expected to keep a routine check of the performance on a shift-to-shift basis, noting such items as the final feed temperature, heater TTDs and steam-to-heater temperatures.

For a heater survey, more comprehensive measurements are required, preferably using test-standard instruments. The information obtained should permit the calculation of a heat balance for each heater and the determination of ancillary flows, such as leak-offs which discharge into bled-steam lines.

As an example of a feedheater survey, consider the information given in Fig 7.119. The values without

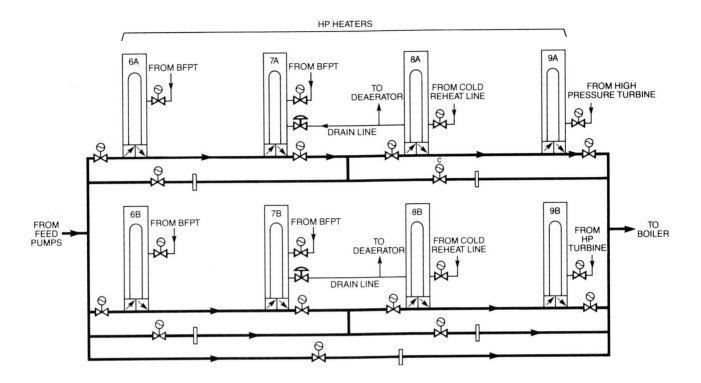

FIG. 7.117   HP heater by-pass system
The diagram shows a typical arrangement for a large modern unit.

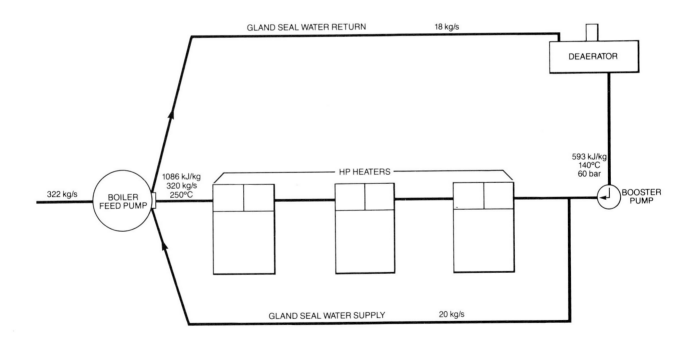

FIG. 7.118 Effect of by-passing the HP heaters
The water which has by-passed the heaters is colder than the main feed. Should an appreciable quantity of colder water enter the feed system, it could lower the feed temperature significantly.

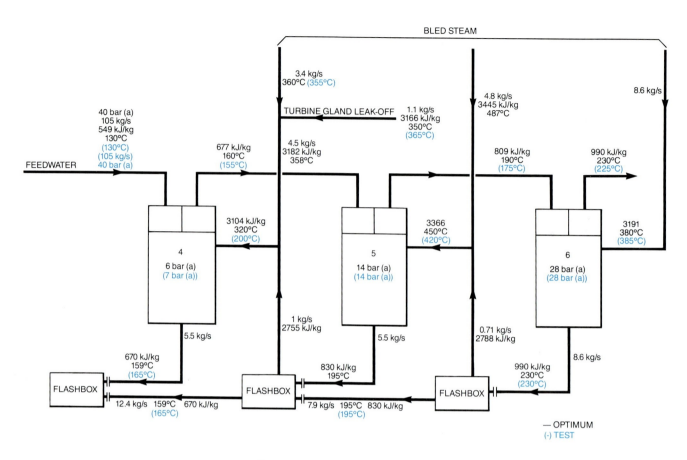

FIG. 7.119 Heater-train investigation
The test and optimum conditions are shown. From the given information, various defects can be determined.

brackets refer to optimum conditions, and those in brackets are test values. The heats corresponding to the various temperatures are established from steam tables and the method of calculation has already been described, so no further description is necessary.

Test results:

No 6. heater

Steam to No. 6 heater $q_{s6} = \dfrac{105 \ (967 \ - \ 738)}{3202 \ - \ 990}$

$= 10.9$ kg/s

No. 5 heater

Flash steam $= \dfrac{990 \ - \ 830}{1957.7} = 0.082$ kg/kg drain

$= 0.082 \times 10.9 = 0.9$ kg/s

Heat in flash steam $= 2788$ kJ/kg

Steam to No. 5 heater $q_{s5} = \dfrac{105 \ (738 \ - \ 655)}{3301 \ - \ 830}$

$= 3.5$ kg/s

Bled-steam flow $= 3.5 - 0.9 = 2.6$ kg/s

Heat in bled-steam

$= \dfrac{(3.5 \times 3301) \ - \ (0.9 \times 2788)}{2.6} = 3479$ kJ/kg

(equivalent to 502°C)

No. 4 heater

Flash steam $= \dfrac{830 \ - \ 697}{2065} = 0.064$ kg/kg drain

$= 0.064 \times (3.5 + 10.0) = 0.9$ kg/s

Heat in flash steam $= 2762$ kJ/kg

Steam to No. 4 heater $q_{s4} = 105 \ (655 \ - \ 549)$

$= 5.2$ kg/s $2844 \ - \ 697$

Gland leak-off +

BS $\dfrac{(5.2 \times 2844) \ - \ (0.9 \times 2762)}{4.3} = 2861$ kJ/kg

(equivalent to 208°C)

554

This is impossible, as the gland leak-off and bled-steam temperatures are over 350°C. Therefore, there must be water droplets in the flash steam from No. 4 flashbox, which is why the temperature of the steam to No. 4 heater is so low. This, in its turn, indicates that there is some obstruction to flow from No. 4 flashbox to No. 3 flashbox.

Further analysis of results

Terminal temperature differences, °C

Heater number	4	5	6
Optimum	−1	5	0
Test	10	20	5
Variation	11	15	5

Steam flows to heaters, kg/s

Heater number	4	5	6
Optimum	4.5	4.8	8.6
Test	5.2	3.5	10.9
Variation	0.7	−1.3	2.3

Steam temperatures at heaters, °C

Heater number	4	5	6
Optimum	320	450	380
Test	200	420	385
Variation	−120	−30	5

There appear to be several defects, such as:

(a) Every TTD is high so there is reduced heat transfer at every heater, possibly due to waterside deposits (e.g., oil contamination).

(b) The steam flow to No. 6 heater is significantly high, but this is probably due to the reduced inlet feed temperature. This, in turn, is due to the very poor heat transfer in No. 5 heater, as evidenced by the high TTD. Therefore, the tube material should be checked, as it seems that there may be exfoliation present. The heater tubes should have been made of 90/10 cupro-nickel, but a 70/30 cupro-nickel nest from another unit may have been put in at some time.

The appropriate recommendations were made for inspections of the system at the first opportunity. When they were carried out it was found that:

- The waterside of the tubes had oil deposition.

- No. 5 heater tube nest was made of 70/30 Cupro-Nickel and was exfoliated.

- The No. 4 flashbox to No. 3 flashbox orifice was smaller than design.

- The No. 5 flashbox to No. 4 flashbox orifice was eroded.

7.8 High oxygen regime

Low-alloy and mild steels rely on the layer of iron oxide, in the form of magnetite, for protection from corrosion by the feedwater. However, at temperatures between 140°C and 200°C, the magnetite layer is more porous than it would be if it formed at higher temperature. Since magnetite has a slight solubility, fast-flowing turbulent water in this temperature range can remove the protective magnetite layer, exposing the bare steel. A further magnetite layer will then form and subsequently be washed away, and so on.

By this mechanism the surface is eroded and the metal becomes thinner. For mild-steel high pressure feedheater tubing, the combination of temperature, fast (sometimes turbulent) flow and thin tube walls create a situation where so-called erosion/corrosion can occur, which leads to perforation of the metal at bends or bifurcations.

To deal with this problem, 20–150 micrograms per litre of oxygen is injected into the feedwater, preferably at a point just before the HP heater train. This oxygen causes the iron oxide to be deposited as haematite. For the haematite to be an effective protective layer, it is essential that the feedwater has a very low anion content, such as chloride and sulphate. This is indicated by a low 'after cation' conductivity, usually below 20 μS/m being specified. This feedwater quality is only achieved by good condensate polishing, so this feedwater treatment is restricted to plants where total condensate polishing is available, such as the supercritical pressure plant at Drakelow *C* power station.

8 Electrostatic precipitation

8.1 Introduction

The design and construction of dust-arresting plant is dealt with in Volume B of this series. Here we are only concerned with the measurement of the performance of the plant and a discussion of the factors which contribute to poor performance. Modern plants are designed to attain a collecting efficiency of at least 99.3% of the total weight of solids in the gas flow (about 80 percent or more of the ash content of the coal). However, the legal requirement does not state the desired efficiency of the plant, but instead requires the 'presumptive standard' to be achieved by the best practical means. By agreement between the CEGB and HM Industrial Air Pollution Inspectorate, the presumptive standard has been established as 0.115 gm/m^3 of flue gas at standard conditions of 1 bar (wet) at 12% CO_2. This is a very stringent condition and can only be attained by keeping the boiler and dust-arresting plant in excellent condition.

The basic formula for collecting efficiency, η%, due to Deutsch, is 100 [1 − exp(−k)], where k is a factor. Figure 7.120 shows the relationship between k and η. Basically the efficiency is dependent upon the total surface area of the collecting plates and the *effective migration velocity* (EMV) of the charged particles. It can be shown that EMV is given by 1000 k/F, mm/s, where F = (total collector-plate area m^2)/ (total gas volume flow, m^3/s).

For example, the required collector-plate area to produce an effective migration velocity of 100 mm/s at 99.3% precipitator efficiency with a gas flow of 550 m^3/s is found as follows:

At 99.3% efficiency, k = 5 (from Fig 7.120)

$$F = 1000k/EMV = 5000/100 = 50 \ m^2/m^3/s$$

Collector plate area $= 50 \times 550$
$$= 27 \ 500 \ m^2$$

Figure 7.121 shows the arrangement of a typical precipitator pass.

8.2   Causes of poor performance

There are many causes of poor performance. The most common are:

- Excess gas volume.

- Poor gas distribution.

- Tracking and air inleakage.

- Electrode breakage.

- Ash resistivity.

- Particle size.

- Electrical conditions.

- Over-full hoppers.

- Defective collector plates.

Consider each in turn.

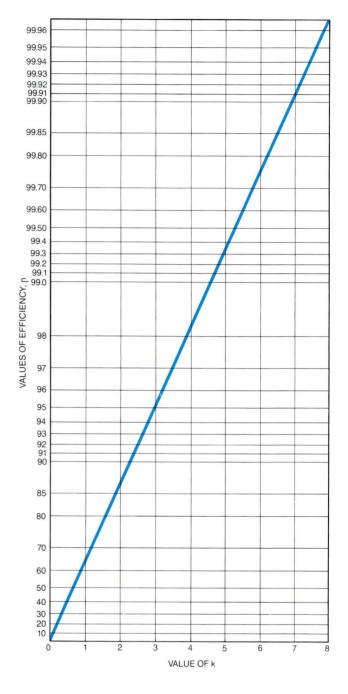

FIG. 7.120 Graph of precipitator efficiency versus factor k
The graphs are used in various electrostatic
precipitator studies.

8.2.1   Excess gas volume

If the gas volume is high, the gas velocity through the precipitator will be high so that the time available for the charged particles to migrate to the collector plates is reduced, possibly to the extent that the particles pass out of the working zone before they can be captured. Therefore the efficiency is affected adversely. The normal gas velocity is about 2 m/s, but this may be increased by:

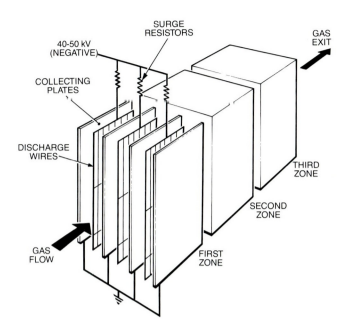

FIG. 7.121 Electrostatic precipitator
Schematic arrangement of wire grids and plates in an electrostatic precipitator. Solid particles in the turbulent gas flow become electrically-charged in the high-field regions near the wires and are then attracted to the collecting plates. The amassed layer is dislodged periodically by mechanical rapping. Even distribution of rapping forces over the plate surfaces is important to minimise the local thickness of residual dust, which may initiate 'flashover' and upset the electrical stability of a whole zone. Uniform gas-flow distribution is similarly important for good precipitator performance.

- Tramp air inleakage at the boiler.

- High airheater gas outlet temperature causing increased volume flow.

- Poor condition of rotary airheater seals.

- Operating the boiler with too much oxygen (i.e., high excess air).

In the previous example, if the gas flow increased from 550 m³/s to 600 m³/s then F becomes 27 500/600 = 45.8 and k decreases to 45.8 × 100/1000 = 4.58.

From Fig 7.120, the corresponding efficiency of collection is about 99.0%.

8.2.2 Poor gas distribution

It could be that the total gas flow is satisfactory but the spatial distribution is poor, causing good performance where the gas velocity is low and for it to be poor where the velocity is high. So the procedure,

when investigating gas flow, is first to ensure that the total flow is correct and then to determine the gas distribution. Table 7.39 shows the results of an investigation of the distribution in a large duct.

TABLE 7.39
Gas velocity distribution in duct   m/s

Position	1	2	3	4	5	6
A	4.5	9	6	2.4	1.4	1.3
B	6.0	9	4	2.0	1.5	0.9
C	7.0	4.5	3	3.0	2.0	0.6
D	2.5	2.5	0.6	1.3	1.0	0.6

The magnitude of the maldistribution of the velocities in the table is illustrated by Fig 7.122. High velocities are present in the top left hand corner of the duct cross-section, reaching 9 m/s, whilst in the diagonally-opposite corner the flow is very gentle. So, even if the total gas flow is correct, the precipitation will generally be poor. This is because, although the velocity is acceptable for about a half of the duct, the remainder is very high, so the good performance of the one half will be more than cancelled by the poor performance of the other.

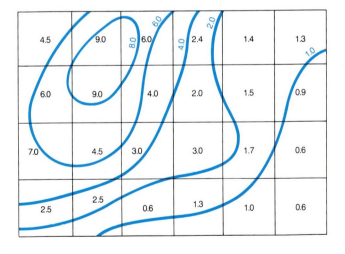

FIG. 7.122 Maldistribution of gas velocity in duct
A surprisingly poor distribution of velocity is common, particularly in the very large ducts associated with high output units.

To be completely acceptable, all the traverse point velocities should be within 25% of the average for the duct. In the example, this means that the velocities should all be within the range 2.4 to 4.0 m/s, but in fact the range is much greater than this.

8.2.3 Tracking and air inleakage

This is a common cause of trouble because precipitator structures are physically large and have numerous access doors, etc., at which leakage can occur. If the inleakage is sufficiently high, a draught of cool (and possibly moist) air will pass over the insulators, steady bars, etc. As the voltage inside a precipitator is very high (40–50 kV), tracking paths may result. Therefore, it is important that access-door seals, rapping-rod covers, expansion joints and other ingress points are kept in good condition and are airtight.

Apart from the above hazards, there are the additional disadvantages that inleakage increases the mass flow of the gas and lowers its temperature.

8.2.4 Poor rapping

In the usual precipitator arrangement the wires are negatively charged and the collector plates are earthed. In the immediate vicinity of the wires the electrostatic field is so strong that corona discharge occurs. The corona ionises the flue gases and the entrained solid particles acquire negative ions, causing them to migrate towards the relatively-positive collecting plates. A few particles acquire positive ions and these migrate towards the discharge electrodes.

The dust which adheres to the wires and collecting plates is periodically removed by rapping. The force of the rappers should be between 30 and 70g, and sometimes two-tier (two-position) rapping is used. The timing of the operation of the rapping gear is important as damage can be done to the plant if it is too frequent. Also, every time rapping takes place some of the dislodged dust is re-entrained, so reducing the collection performance. On the other hand, if the interval between raps is too great, the dust build-up on the wires and plates wlll be sufficient to interfere with the operation of the plant and this also will reduce its performance. Efforts should be made to determine the optimum timing. Typical times for a 3-zone pass are:

Inlet	— every 10 minutes
Middle	— every 20 minutes
Outlet	— every 2 hours
Discharge electrodes	— every 15 minutes

Upon inspection, the collector plate dust-layer should be less than 1 mm for high resistance dust and 3 mm for dust from coal with high sulphur. The discharge electrodes should only have a very light deposit.

8.2.5 Electrode breakage

The discharge electrodes consist of relatively thin wires which may be of flat, circular, star or square cross-section. In addition, they may be barbed or plain, twisted or straight and rigidly or loosely fastened, besides being made from a range of materials. Whichever arrangement is adopted, the essential requirement is that the wires should have excellent reliability. Out of the huge number in a zone, it only needs one to break to cause short-circuiting and electrical instability generally, which may seriously affect the performance of the plant.

Spark erosion is a common cause of wire failure, particularly near the top fixing. Figure 7.123 shows a photograph of various electrodes.

The 'unbreakable' type of wire is becoming popular because of its inherent reliability. One method of limiting the effect on performance by breakage is to divide a complete zone into four sub-zones, so that a broken wire only affects a quarter of the zone.

8.2.6 Ash resistivity

An important factor in precipitator performance is the electrical conductivity of the ash itself. Poor combustion can result in a higher than normal carbon content in the ash. This causes the particles to have a low resistivity which may cause them to be difficult to collect, as they easily surrender their electric charge on reaching the collector plate. This means that they can be readily re-entrained by the gas, particularly during rapping. On the other hand, if the resistivity is too high, the dust will accumulate on the collector plates and form an insulating layer, so preventing subsequent particles from surrendering their charge. Consequently, the charge on the surface layer will repel incoming dust, causing re-entrainment. In addition, a very high voltage-gradient will be built up across the dust layer which could result in flashover. Dust with high resistivity accumulating on the discharge electrodes will suppress the corona discharge, seriously lowering the performance of the plant.

The resistivity of the ash is a function of the surface layers of sulphuric acid, salts and moisture, particularly the acid. If the coal contains about 2% sulphur, the resulting sulphuric-acid deposition creates a desirable level of dust resistivity. If the sulphur content of the coal is very low, a highly-resistive ash may be produced which is very difficult to precipitate. One remedy for this is to inject small quantities of suitable additives, such an ammonium sulphate, into the flue gas.

The desirable range of resistivity is between 10^6 and 10^{12} ohm-cm. Should the normal fuel supplies be low in sulphur and a change to alternative supplies be uneconomic, another possible remedy besides injection is to install 'pulsed energisation'. In this, high frequency pulses are added to the rectified wave form. The resulting high ion-density corona acts along the length of the discharge electrode instead of just at discrete locations, so the dust particles are more easily precipitated. The improved performance of the plant

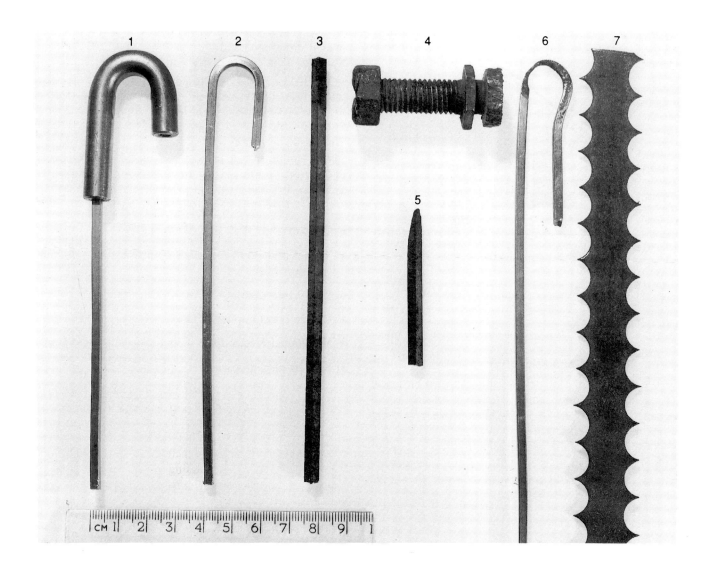

FIG. 7.123 Electrostatic precipitator electrodes and fixings
The photograph shows:
1 New 3 mm square-section wire with sheathed supporting loop.
2 New 3 mm square-section wire with supporting loop.
3 Used star-section wire.
4 Supporting fixture.
5 Used star-section wire with spark-eroded end which led to breakage.
6 3 mm square-section wire with considerable spark-erosion at supporting loop.
7 'Unbreakable' wire, 2 mm thick and 23 mm across 'spikes'.

is called the *enhancement factor H*, given by (pulsed migration velocity)/(unpulsed migration velocity). Values of H of two or even more are possible, so the effect on plant performance can be considerable. Figure 7.124 shows the order of improvement that can be gained by using pulse energisation.

8.2.7 Particle size

Particle sizing has an important bearing upon the efficiency of precipitation. Up to approximately 20 μm, the electrical and the drag forces combine to give a deposition velocity which increases in roughly direct proportion to the size of the particles. Collection is almost 100% efficient for sizes over 20 μm, the only problem being that of slight re-entrainment. A typical dust-size spectrum indicates that about half of the ash particles are over 20 μm and will almost all be captured. Dust less than 20 μm is more difficult to capture, so the stack emission will contain a preponderance of fine particles.

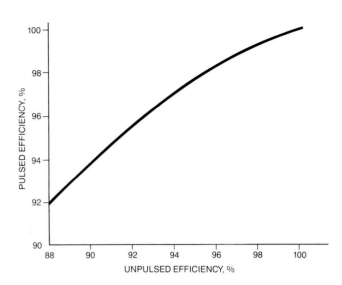

FIG. 7.124 Improvement in efficiency with pulsed
energisation ($H = 1.5$)

8.2.8 Electrical conditions

Figure 7.125 shows the relationship between electrical power input and precipitator efficiency. As one would expect, increased power results in improved performance, provided the power is useful. For example, flashover and tracking will give high power consumption, but much of it is wasted. To find the demarcation between useful and wasted power, a simple

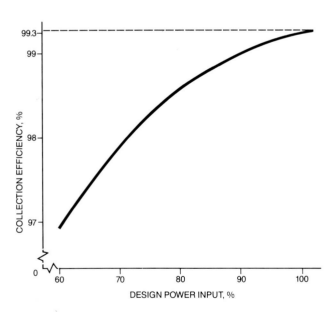

FIG. 7.125 Typical relationship between power input
and precipitator efficiency
Provided the power is usefully employed, the performance
of the precipitator improves as the power is increased.

graph is determined from tests on the plant. Starting at a low value the current is increased by suitable increments, noting the corresponding voltage. A typical diagram is shown in Fig 7.126. Notice that the current and voltage both increase until the voltage is about 45 kV. Thereafter increasing current is accompanied by reducing voltage. The knee of the curve determines the demarcation between stable and unstable operation, the highest stable current here being a little over 200 mA. The automatic voltage control always tries to operate the plant with the highest stable power input which can be maintained. The design power input to a large precipitator is considerable, being of the order of 500 kW.

Causes of electrical instability include:

● Tracking, possibly due to air ingress.

● Arcing and poor connections, possibly at sub-zone isolation selectors.

● Ineffective rapping.

● Overfull dust hoppers, causing bridging of electrodes.

● Misaligned electrodes.

● Broken discharge electrode wires.

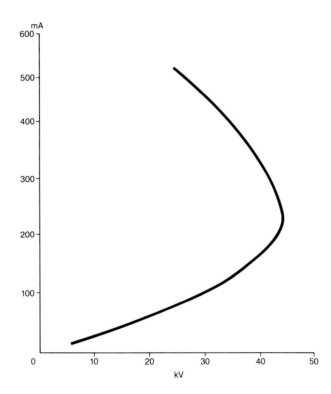

FIG. 7.126 Stable and unstable operation
As the voltage is increased from a low value, there
is also an increased current flow. There is a maximum
voltage, and any attempt to increase it results in an
increased current, but a reduced voltage. The knee of
the curve represents the boundary between stable and
unstable electrical operation.

If dust and grit hoppers are allowed to become overfull, particles will be blown off the top of the pile. This causes the dust burden of the gas to be higher than normal at the precipitator, and even if the plant is operating at its design efficiency, it will still cause the outlet dust burden to increase. However, attempts to empty overfull hoppers must NEVER include removing outlet pipework, as explosions and fires can easily result. Therefore, hoppers should be emptied before becoming full.

Overfull precipitator hoppers can also cause mechanical damage to the electrodes, and unstable electrical conditions.

8.2.9 Defective collector plates

The plates are mechanically strong, but can become defective if the gas temperature falls below the dew-point, allowing deposition of dilute acid. This can occur in discrete areas if cold air inleakage takes place. Alternatively, general corrosion can occur if the flue gas itself has a low temperature, possibly due to massive air ingress at defective seals on the ducts between the airheater and the precipitator, or if the ducting is corroded and holed. Frequent cold-starting of the plant can also lead to low-temperature gas at the precipitators.

8.3 Dust monitors

Four types of monitor are in general use in the CEGB:

- SEROP (South Eastern Region Optical Probe).

- NORDUST.

- CERL (Central Electricity Research Laboratories).

- Erwin Sick.

The four instruments are illustrated in Fig 7.127. Each has advantages and disadvantages. The Clean

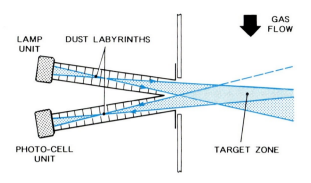

(b)

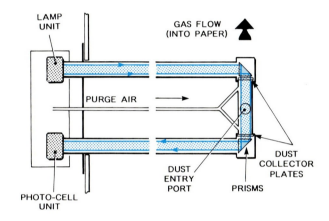

(c)

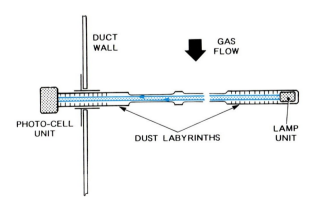

(a)

Fig. 7.127 Flue gas dust monitors

(a) The South Eastern Region's optical probe (SEROP), monitors flue gas dust by the obscuration of a light beam. The 1.5 metre-long probe is semi-portable and can be withdrawn from the duct for zero resetting. Like other instruments depending on light transmission or reflection by the flue gas, its response increases strongly with decreasing particle size and it is therefore particularly sensitive to smoke.

(b) Developed by the North Western Region, the NORDUST monitor is mounted wholly outside the flue gas duct and relies on back-scattering from a light beam directed at the gas flow within. The angle between the source and receiver tubes avoids the problem of direct reflection from the opposite wall. Standard reflector cards are used for calibration and zero setting.

(c) Larger dust particles are preferentially sampled and detected with this instrument, devised by CERL. Particles with sufficient inertia can negotiate the entry port and settle on the glass collector plates, whose optical obscuration is integrated over 15-minute intervals before an air purge re-establishes the zero conditions.

facturer and the customer. The efficiency is given by 1 − [Slip/(Dust burden at inlet)] where the *slip* is the dust burden of the gas leaving the plant. Therefore, it is necessary to sample the gas at inlet and outlet of the plant to determine the volumes and dust burdens. The gas ducts of modern, high output plant are very large so it is necessary to obtain samples from many points across the area of the duct — at least 24. Also, the sampling must be isokinetic, i.e., for the correct quantity of dust to be sampled, the velocity of the sample gas should be the same as that of the main gas flow in the vicinity.

If the sampling rate is too low or too high, then too little or too much dust will enter the probe compared to the average concentraton in the stream of gas at that point.

The time required to carry out a test can be several hours and correction curves should be agreed beforehand for items such as:

● Sulphur content of the fuel (Fig 7.130).

● Inlet dust burden.

● Gas flow rate (Fig 7.131).

● Dust surging.

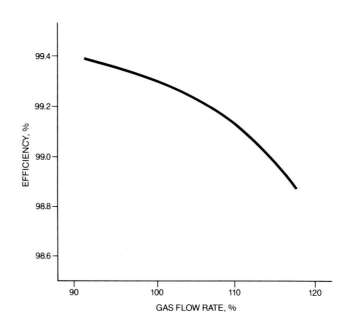

FIG. 7.131 Variation of precipitator efficiency
with gas flow
Higher than design gas flow increases the gas velocity
through the precipitator. This reduces the treatment time,
adversely affecting the precipitator performance.

Typical correction curves are shown for two of the items and precise definitions are required of the relevant data, for example, whether the sulphur is 'total' or 'combustible' and whether on a 'dry' or

'as-received' basis. Also, the contractor must declare his satisfaction with the condition of his plant and all the associated plant (including the boiler) before acceptance testing, and also with the test conditions after testing but before the results are determined.

Routine tests are carried out as required, but at least once per year. These are undertaken by specialist CEGB test teams using Mark II dust samplers. Figure 7.132 shows the schematic arrangement. The sampler heads can be fitted with a paper thimble or a glass-fibre filter. If the coarse dust burden is high, a small cyclone sampling head followed by a glass-wool filter can be used.

A calculation for a test on a 500 MW plant is shown below, the test averages and other data being:

CO_2 = 14%; $\sqrt{\text{(Velocity head)}}$ = $\sqrt{h_v}$ = 3.62;

Gas temperature = 143°C (416 K);

Barometer = 1016 mbar; moisture in flue gas = 7.0%;

Static head = − 100 mw wg;

Dust collected = 5.680 g; duration of test = 3 hours (10 800 s)

Nozzle area = 0.00046 m²

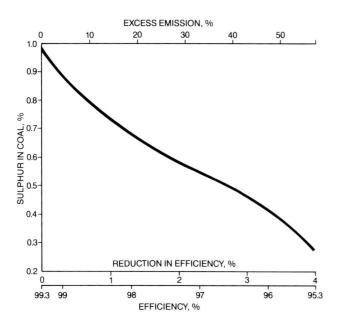

FIG. 7.130 Variation of precipitator efficiency
with sulphur in coal

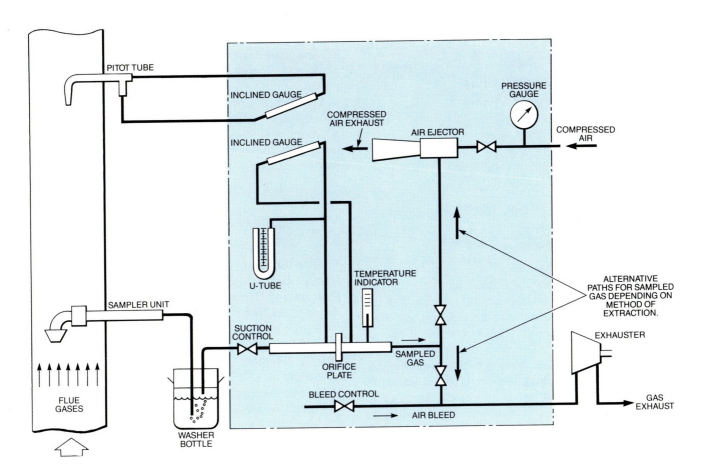

FIG. 7.132 Schematic diagram of Mark II dust sampler
This type of equipment is commonly used within the CEGB for precipitator testing.

	Dry		(1-Moisture)		Wet		Molecular weight		Relative weight
CO_2	0.14	×	0.93	=	0.13	×	44	=	5.72
O_2	0.05	×	0.93	=	0.05	×	32	=	1.60
N_2	0.81	×	0.93	=	0.75	×	28	=	21.00
H_2O	–				0.07	×	18	=	1.26
	1.00				1.00				29.58

Density ϱ of flue gas (at 0°C, 1 bar)

$$= \frac{\text{Relative weight}}{22.4} = \frac{29.58}{22.4}$$

$$= 1.32 \text{ kg/m}^3$$

Density correction for duct conditions

$$= \frac{273}{T} \times \frac{(10\,363 + h_s)}{10\,363} \times \frac{B}{1013}$$

Density ϱ at duct conditions

$$= 1.32 \times \frac{273}{416} \times \frac{10\,263}{10\,363} \times \frac{1016}{1013}$$

$$= 0.860 \text{ kg/m}^3$$

$$\sqrt{\varrho} = 0.927$$

Velocity of gas in duct $= 4.43 \sqrt{(h_v/\varrho)}$

$$= 4.43 \times 3.62/0.927$$

$$= 17.3 \text{ m/s}$$

Sampled volume at duct $=$ velocity $\times$ nozzle area $\times$ time

$$= 17.3 \times 0.00046 \times 10\,800$$

$$= 86.0 \text{ m}^3$$

565

Sampled volume at standard conditions of 15°C, 1 bar, 12% CO_2 (wet)

$$= \text{Vol} \times \frac{288}{T} \times \frac{10\,363 + h_s}{10\,363} \times \frac{B}{1013} \times \frac{12}{CO_2\%}$$

$$= 86.0 \times \frac{288}{416} \times \frac{10\,263}{10\,363} \times \frac{1016}{1013} \times \frac{12}{14}$$

$$= 50.7 \text{ m}^3$$

Dust burden at standard conditions

$$= \frac{\text{dust collected}}{\text{sample volume}}$$

$$= \frac{5.68}{50.7} = 0.112 \text{ g/m}^3$$

Suppose that a similar calculation has been carried out for the precipitator inlet conditions and that the inlet dust burden was found to be 14.012 g/m^3, then the efficiency = 1 − (0.112/14.012) = 0.992 or 99.2%.

If the routine test is merely to confirm that the plant is operating within the presumptive standard, it is only necessary to test at the precipitator outlet. On the other hand, if the efficiency of collection is required, it is necessary to test at the precipitator inlet and outlet simultaneously.

9 Cooling towers

9.1 Introduction

About the only large items of plant on a power station that are not regularly tested are the cooling towers. Acceptance or performance tests are normally carried out as soon as possible after completion (possibly only on a representative tower), but there is no requirement to carry out further tests. This is for convenience, not because the plant will not deteriorate. On the contrary, it is common for a variety of faults to be present, including:

● Damaged or blocked sprayers on the irrigation system.

● Damaged packings, possibly due to an excessive weight of ice during cold weather.

● Overflowing inlet culverts at high loads due to obstruction to flow in the irrigation system.

● Excessive mud deposition in the pond.

● Damaged de-icing sprayer system.

● Pond trash-screens partially blocked or damaged.

Consequently, it is necessary to inspect the towers periodically to ensure that they are in good condition. If the recooled water temperature is lowered by less than even one degree by the rectification of defects, it probably abundantly justifies the effort put in.

The CEGB now design their own cooling towers and employ contractors to build them, so they do not require acceptance testing, although performance tests to a comparable standard are still carried out for information purposes.

9.2 Testing

The performance of a cooling tower is determined from consideration of its actual re-cooled water temperature and that which should have been attained at the test conditions according to the design. BS44485: 1969 details methods of design, testing, etc., and consists of four parts, plus an addendum. However, only Part 2, 'Methods of test and acceptance testing' is of interest here. It details the instrumentation required and typical accuracies.

Within the CEGB, the performance is referred to a Performance Chart of the type shown in Fig 7.84. It can be seen that for a given wet and dry bulb temperature, water flowrate and cooling range, the design recooled temperature can be determined corresponding to the test conditions. The difference between this and the test value gives a measure of the cooling tower performance.

For example, for the conditions shown on the performance chart the recooled water temperature should be 20.25°C. If the re-cooled temperature obtained on test is 19.8°C, then the tower performance is 0.45°C better than design.

A tower is usually designed to handle about 110% CW flow without overflow, and so this is the normal test loading.

9.3 Test measurements

9.3.1 CW flow

If it is possible to operate the tower in conjunction with just one turbine, the flow can be derived from a condenser heat balance. This and other possible methods are detailed in the section on condensers and in Site Test Code 3. It should be remembered that the quantity of flow to be measured can be considerable, for example, the CW flowrate for a 500 MW unit is over 15.5 m^3/s. Reference should be made to BS3680 'Methods of measurement of liquid flow in open channels', which consists of 24 parts (numbered from 1 to 10C). Some of interest are:

- Dilution method (Part 2).

- Stream flow measurement (Part 3).

- Current meters (Part 8).

Also reference should be made to BS5857 'Methods for measurement of fluid flow in closed conduits using tracers'. Tracer methods are particularly suitable for cooling water flow measurement because of their sensitivity, accuracy and minimum permanent head loss. However, they only measure the average flow over a short period of time, so they are normally used to calibrate other flow measuring devices. They fall into three main categories:

- Constant rate injection.

- Transit time.

- Integration (sudden injection).

As an example, consider the first of these. The method is based on injecting the tracer into the cooling water for a few minutes at an accurately-measured constant rate. A series of samples is extracted from the system at a point where the tracer has become completely mixed with the cooling water. The mass flow rate is calculated from:

$$q_{cw} = q_t \, C_1/C_2$$

where q_{cw} = cooling water mass flow rate, kg/s

q_t = mass flow rate of injected tracer, kg/s

C_1 = concentration of injected tracer, kg/kg

C_2 = concentration of tracer at downstream position during the 'plateau' period of constant concentration, kg/kg

The method is illustrated in Fig 7.133.

Other possible types of measurement include velocity/area methods such as:

- Ultrasonic flowmeters.

- Insertion meters.

- Current meters.

- Pitot tubes.

Current meters have been used successfully for many years. The speed of rotation of the propeller is determined by the water velocity in the vicinity. A

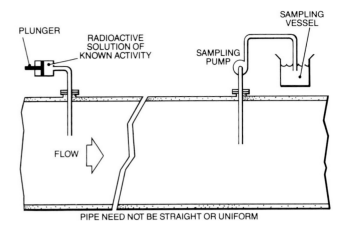

FIG. 7.133 Radio-isotope dilution method
of flow measurement
The accurate assessment of large CW flows presents many difficulties. The radio-isotope method is widely used.

number of meters are arranged on a frame as shown in Fig 7.134: the speeds of rotation are determined by electrical means and transmitted to a display unit. The meters should be regularly calibrated and, if kept in good condition, will give excellent results. They can be used in open culverts or pipes. It is necessary to use a large number of meters (over 50) to obtain an acceptable average flow measurement in open culverts. In circular pipes, measurements are taken on two diameters at right angles at about 17 points. The

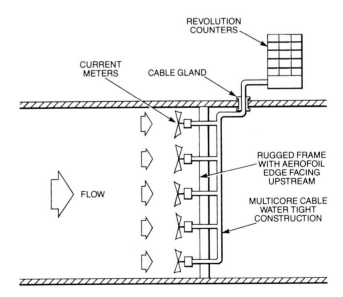

FIG. 7.134 Application of current meters to the
measurement of water flow
The method requires the construction of a substantial grid upon which the meters are mounted. It is suitable for test purposes, but not as a permanent installation.

cooling water flow is calculated from the appropriate duct cross-sectional area and the average velocity, after allowance has been made for the area occupied by the meters themselves. Results should have an accuracy better than ±1%. This method is particularly suited to cooling-tower testing, where open culverts are available.

Pitot-tube traverses have been successfully carried out on CW pipes of more than two metres diameter. BS1042: Part 2A: 1973 'Pitot tubes, class A accuracy' gives details for measurement in circular pipes. Aerodynamically-shaped tubes are normally necessary to withstand the turbulence levels in large pipes, although cylindrical tubes are satisfactory for short insertions. Sometimes a total pressure tube and wall static pressure tappings may be more convenient. A gland and valve assembly should be used so that the pitot tube can be inserted and removed without draining the CW system. The general arrangement is shown in Fig 7.135. The CW flow is determined from consideration of the average velocity from the local velocities at the log-linear traverse point positions.

The final group is the differential pressure category and includes:

- Bend flowmeter.

- Segmental orifice plate.

- Differential pressure between water boxes.

- Weirs and venturi flumes.

A typical bend flowmeter installation is shown in Fig 7.136. It utilises the pressure difference between the inside and outside of a sharp bend in the cooling water system.

For circular pipes, a curvature ratio of R/r of 3 will give a convenient differential head for flow velocities of about 2 m/s or more, where R is the mean radius of bend and r is the pipe radius.

Weirs and flumes are convenient for measuring flows through open culverts. However, they are not particularly common because of their high initial cost and also they impose a permanent head loss on the

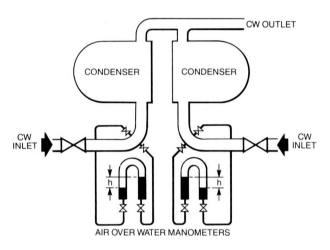

(a) Layout for condenser inlet bend flow calibration

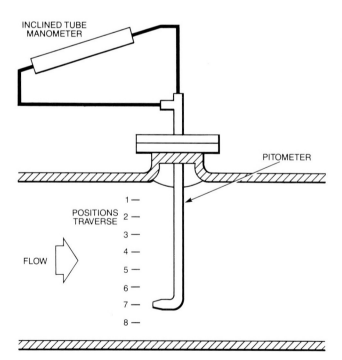

FIG. 7.135 Pitometer used for water-velocity traverse
The method of use is shown diagrammatically.

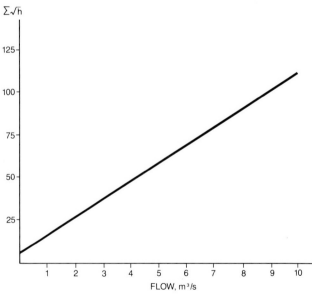

(b) Calibration curve for condenser inlet bends

FIG. 7.136 Condenser inlet-bend flow measurement
A typical installation arrangement and
calibration curve is shown.

cooling water system for the life of the plant. The general arrangements are shown in Fig 7.137.

9.3.2 CW temperatures

Measurement of the tower inlet CW temperature usually presents little difficulty, as normally there is very little stratification present. Some CW can be bled from the tower supply culvert and passed around the measuring device (which can be a calibrated mercury-in-glass thermometer or a resistance element) and this can be accepted as an average value.

On the other hand, there is normally significant temperature stratification present at the recooled water flow. Thus, multi-position sampling is necessary at the outlet culvert to obtain an acceptable average recooled temperature.

9.3.3 Atmospheric conditions

The wet and dry bulb air-temperatures are measured using Assman aspirated psychrometers. They are positioned on the windward side of the cooling tower at about 1.2 m above ground level. The wind speed is also determined, as there is a maximum value stipulated in the performance specification.

9.4 Environment

The cooling towers are usually the most striking structures at large inland power station sites. Typically, a 2000 MW site will have about eight natural draught towers. There is always non-boiling evaporation in natural draught towers — in fact, this is the primary mechanism by which the CW is cooled. A typical loss due to evaporation is about 1–1.5% of the total water cooled. Figure 7.138 shows the order of evaporative loss in m³/GJ rejected. For example, if the atmospheric dry-bulb temperature is 12.8°C and

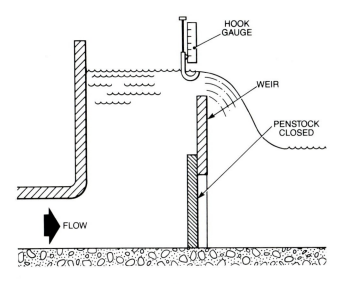

(a) Weir method

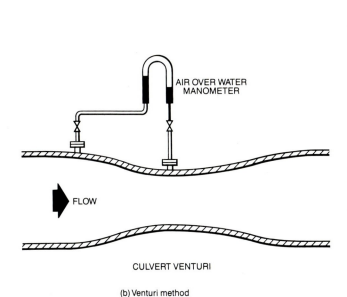

CULVERT VENTURI

(b) Venturi method

FIG. 7.137 Weir (a) and venturi (b) flume methods of flow measurement
As these methods impose a permanent pressure drop in the system, they are not in common use.

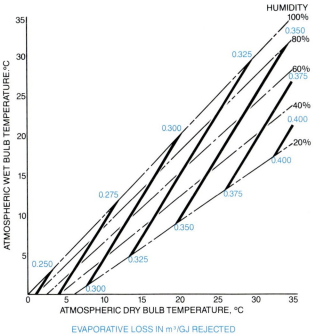

EVAPORATIVE LOSS IN m³/GJ REJECTED

FIG. 7.138 Evaporative loss from natural draught cooling towers
The chart is used to estimate the evaporative loss in m³/GJ of heat rejected.

the wet-bulb temperature is 7.8°C then, from the chart, the evaporative loss is 0.306 m³/GJ of heat rejected.

The make-up water required by a natural draught tower system is about 3.5% of the cooling water flow; this is to make good the losses due to evaporation and purge. In some countries, though, water is so scarce that it is desirable to dispense with make-up. This can be achieved by using dry cooling towers, similar to the one which was installed in the early 1960s at Rugeley power station. It uses the atmosphere as a sink for the turbine exhaust heat by a combination of jet condenser, closed water-circuit, air-cooled heat exchangers and a cooling tower. There are 648 coolers, all of aluminium multi-tube louvred-plate construction, each containing 6 rows of 40 tubes. The general arrangement is shown in Fig 7.139. Another development has been that of the assisted-draught cooling tower. As the name suggests, the draught is assisted by large fans located in the base of the tower. Plans are in hand for one such tower to provide the total cooling requirement of a 2000 MW station.

However, the majority of towers in use are still of the comparatively small natural draught type. Where mixed cooling (or river cooling) is used, it is important that the temperature at the CW outfall does not exceed the value stipulated in the water authority consent, typically 30°C. In mixed cooling installations this frequently requires the cooling towers to be used for controlling the river temperature and so appreciable operating costs are incurred for environmental requirements. Heat dissipation is the most important environmental problem posed by power stations.

10 Pump testing

10.1 Cooling water (CW) pumps

CW pump tests are carried out on-site for several reasons, including:

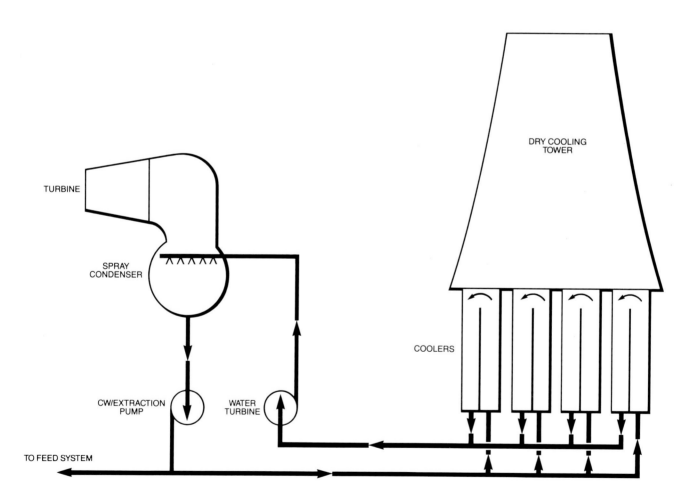

FIG. 7.139 Dry cooling tower system
This is a diagrammatic representation of the Rugeley power station installation, incorporating a jet condenser.

- The configuration of the CW system, particularly in the vicinity of the pump inlet, affects the performance of the pump.

- Modern large CW pumps are too big for existing works testing facilities in many cases.

- Many CW pumps are erected on site in volutes which have been cast in concrete, so these cannot be tested at the manufacturer's works.

Therefore, site testing is often necessary for contractual reasons and so CEGB Site Test Code No 6 'Site performance testing of Condenser circulating water pumps' has been prepared. It outlines the objectives and scope of the tests, procedures, instrumentation and other data. Reference should also be made to BS599: 1977 'Methods of testing pumps'. The manufacturer will have guaranteed the following parameters at rated speed:

- Pump generated total head at rated flow, as installed on site.

- Pump efficiency at rated flow, as installed on site.

In addition, the head characteristic of the pump must be stable and the pump must be capable of running satisfactorily in parallel with other pumps on the same CW system. The site tests may be extended to include assessments of the system resistance and the performance of ancillary plant. The pump throughput must be varied over a considerable range, so it is usual to test the plant before the first unit is commissioned. If it is necessary to test a pump with its associated turbine-generating plant running, special provisions will have to be made to supply the condenser CW flow required for turbine operation and this will probably limit the range of testing possible.

The instrumentation used should preferably be calibrated before the tests commence.

10.1.1 Measurements

Inlet head

This is taken to be the open surface of water in the pump inlet chamber, downstream of the screens. Normally the level will not remain constant, particularly at coastal stations, so the level must be determined during the test. The inlet head is the difference between the surface of the water and the highest point of the inlet edges of the pump impeller.

Discharge head

This should be measured two pipe diameters downstream of the pump diffuser outlet, if possible. It is measured by means of four static-pressure tappings arranged on two pipe diameters at right angles to each other. They are all connected via a manifold to the pressure measuring device, such as a calibrated standard test pressure gauge, a mercury manometer or a calibrated pressure transducer. Where possible, velocity-head traverse at rated flow will be carried out at the discharge head measuring point to obtain the mean measuring point, to obtain the mean axial velocity-head. Alternatively, the uniform velocity can be calculated by dividing the measured volume flow by the cross-sectional area of the duct.

The velocity head calculated by this method is multiplied by the factor 1.1 to obtain the velocity head for test purposes.

The pump discharge head at the diffuser outlet is obtained by adding a head loss (agreed between the parties) from the diffuser to the measuring point to the total head obtained at the measuring point.

Flow rate

This is determined by an acceptable method, such as radio-isotope dilution, current meters, etc., at any suitable point in the system. Specialist CEGB teams carry out this work. Reference should be made to BS3680: Part 2C: 1967 'Radio Isotope Techniques'.

Pump power input

The power consumption of the driving motor is measured in accordance with agreed practice between the parties concerned. The two-wattmeter method, using precision-grade wattmeters or watt-hour meters, should be acceptable. Voltmeters and ammeters are usually also installed to enable the motor efficiency to be calculated and to check the circuit power factor. In addition, the motor efficiency curve determined by works tests will be utilised.

The pump input power is determined using the motor electrical input power, the calculated motor efficiency and the specified gearbox efficiency, if applicable. The metering is often a considerable distance away from the pump, so a suitable I^2R loss for the cable is agreed between the parties concerned.

Speed and frequency

These are measured with approved instruments of class A standard in BS599. Preferably they should be of the digital type.

Density of CW

The specific gravity and temperature of the CW is determined for each test. Particular care is necessary at stations using estuary water as the salinity may vary during a test, so affecting the density.

10.1.2 Test procedure

The object of the tests is to obtain the values of flow rate, pump generated total head, speed, motor power consumption and system frequency. The test flow rates are usually about 0, 20, 40, 60, 80 and 100% of rated flow, obtained by throttling the pump discharge valve as required.

Some 20 tests are carried out within about ±5 percent of rated flow. Normally each test lasts about 30 minutes, but care must be taken during the closed-valve test to see that no overheating occurs.

On tidal stations, the tests are carried out when the CW outfall sill is flooded. This ensures that tidal variations have equal effects on both the pump suction and discharge heads.

10.1.3 Pump affinity laws

Tests are normally carried out when the speed of rotation of the pump is different from the design speed. The necessary corrections are in accordance with the affinity laws which state:

$$\text{Flow rate } (\alpha \text{ speed})$$
$$\text{Head } (\alpha \text{ speed})^2$$
$$\text{Pump power } (\alpha \text{ speed})^3$$
$$(\text{Flow rate})^2 \alpha \text{ head}$$

10.1.4 Numerical example

Consider a pump associated with a 500 MW unit. The pump design details are as follows:

Rated flow 15.9 m³/s;	Rated head 21.3 m;
Rated motor speed 992 r/min;	Efficiency 89%;
Gearbox ratio 5.1;	CW specific gravity 1.025; density 1023.7 kg/m³;

Pipe diameter at point of discharge head measurement 2.74 m

The results of one test are given in Table 7.40.

A total of about 20 tests will be carried out, from which the 'as-run' results can be plotted such as those shown in Fig 7.140.

10.2 Boiler feed pumps

Although site acceptance tests were not a contractual requirement for the feed pumps of the early 500 MW units, provision has been made to fit a torquemeter in some instances. Testing is carried out on site under the non-standard conditions mentioned in BS599: 1966 'Methods of testing pumps', and also CEGB Site Test Code No. 1 'Site performance testing of Boiler feed pumps'.

If the pump balance water flow is returned to the de-aerator, it represents an external loss and the flow must be measured. Main feed pumps should be tested during the following main plant tests:

- Combined condenser test, generator heat run and preliminary heat rate test.
- Official guarantee heat rate test.
- Overload test on plant specified to have overload capacity (by shutting steam off last heater).
- Part-load test with steam-driven feed pump in operation.

The electrically-driven feed pumps are also tested during the main plant test programme, but tests on all machines are not required.

10.2.1 Measurements

Feedwater flow

The pump discharge flow is the quantity passing to the boiler and does not include gland leakage or balance water. The flow is measured as stipulated for turbine tests.

Pump inlet pressure

This should be taken at a point as near as is reasonable to the pump inlet flanges.

Test-quality gauges or pressure transducers are used to measure the pressure. The accuracy should be better than ±0.5% and they should be calibrated before the testing starts.

Pump discharge pressure

Gauges or pressure transducers are acceptable, but the accuracy should be better than ±0.1%. Whichever is used, the instrument must be calibrated before the test.

Temperatures

The pump inlet and outlet water temperature should be measured close to the appropriate pressure tapping. The accuracy of the measuring devices should be ±0.5°C and that of the differential temperature device between inlet and outlet 0.01°C.

Speed

A typical instrument for speed measurement is an electronic counter with a suitable pick-up head pro-

TABLE 7.40

CW pump test results

		Derivation	Units	Test 1
	Duration		min	30
	Nominal % rated flow		%	20
	Pump head			
1	Inlet head above datum	Water level	m	15.28
2	Discharge pressure (manometer datum 0.975 m above reference datum)	Manometer (mercury)	m Hg	3.194
3	Mercury temperature		°C	13.8
4	Density of mercury		kg/m^3	13 561
5	Discharge pressure head	(2)[(4)/(9) − 1] + 1.0	m	41.1
	Flow			
6	Mean discharge flow	Current meters	m^3/s	2.88
	Other measurements			
7	Pump/motor shaft speed	Test instruments	r/min	996.8
8	CW temperature	Thermometers	°C	13.4
9	CW density	CW samples	kg/m^3	999.426
10	Opening of outlet valve		mm	360
11	Bearing temperatures (inner/outer)	Permanent instruments	°C	54/57
12	Motor temperatures	Permanent instruments	°C	54
	Electrical measurements			
13	Corrected incremental reading of kWh meters	Test instruments	kWh	1346.7
14	Duration of measurement		min	30
15	Average power	[(13) × 60]/(14)	kW	2693.4
16	Supply volts	Test instruments	kV	11.06
17	Supply current	Test instruments	A	204.6
18	Supply frequency	Test instruments	Hz	50.047
19	Supply power factor	Test instruments		0.687
20	Cable resistance per phase	Manufacturer	Ω	0.0104
21	Estimated I^{2}R loss	3[(17)^2 × (20)]/1000	kW	1.3
22	Power to motor terminals	(20) − (21)	kW	2692.1
	Motor efficiency			
23	No-load loss at normal volts	Works tests	kW	55
24	Windage and friction loss	Works tests	kW	30
25	Iron loss	(23) − (24)	kW	25
26	Stator winding line-to-line resistance at 75°C	Works tests	Ω	0.172
27	Stator current	Item (17)	A	204.6
28	Stator winding loss	$\dfrac{[(27)^2 \times (26) \times 1.5]}{1000}$	kW	10.8
29	Rotor input	(22) − [(25) + (28)]	kW	2656.3
30	Rotor slip	From items (7) and (18)	%	0.41
31	Rotor loss	(29) × [(30)/100]	kW	10.9
32	Rotor output	(29) − (31)	kW	2645.4
33	Gross motor output	(32) − (24)	kW	2615.4

TABLE 7.40 (cont'd)

CW pump test results

		Derivation	Units	Test 1
	Motor efficiency			
34	Stray losses	$0.5\% \times$ (33) (BSS)	kW	13.1
35	Net motor output	(33) $-$ (34)	kW	2602.3
36	Motor efficiency	[(35) $\times$ 100] / (22)	%	96.66
	Pump power input			
37	Gearbox efficiency	Works tests	%	97.8
38	Pump power input	(35) $\times$ [(37)/100]	kW	2545.0
39	Input corrected to design density of s.g. $\times$ 998.75	$\dfrac{[(38) \times \text{design density}]}{(9)}$	kW	2606.9
	Pump generated head			
40	Inlet head above datum	From (1)	m	15.28
41	Discharge head above datum	From (5)	m	41.1
42	Velocity head	$\dfrac{1.1[(6)/\text{Area flow}]^2}{2 \times 9.81}$	m	0.012
43	Head loss between discharge and measuring point	Agreed before tests	m	0.01
44	Total discharge head	(41) $+$ (42) $+$ (43)	m	41.122
45	Generated total head	(44) $-$ (40)	m	25.84
	CW flow			
46	Mean flow	From (6)	m^3/s	2.88
47	Mean velocity of flow	(46)/area of flow	m/s	0.49
48	Flow (as % of rated flow)	[(46) $\times$ 100]/15.9	%	18.10
	Water power			
49	Water power	[(46) $\times$ (45) $\times$ (9) $\times$ 9.81]/10^3	kW	729.6
50	Water power corrected to design density	$\dfrac{[(49) \times 1023.7]}{9}$	kW	747.3
	Pump efficiency			
51	Pump efficiency	[(50) $\times$ 100]/(39)	%	28.7
	Corrections for speed			
52	CW flow rate	$\dfrac{[(48) \times \text{rated speed}]}{7}$	%	18.01
53	Pump generated total head	$\dfrac{[(45) \times (\text{rated speed})^2]}{(7)^2}$	m	25.59
54	Pump power	$\dfrac{[(39) \times (\text{rated speed})^3]}{(7)^3}$	kW	2569

ducing several pulses per revolution of the pump shaft. The required accuracy is $\pm 0.1\%$.

Torque

The driving torque applied to the pump coupling should be measured by device capable of an accuracy of $\pm 0.25\%$. Because of the temperature effect on the stiffness of the torsion shaft, its temperature must be assessed during the tests.

If there is no provision for a torquemeter, then the pump power inlet can be calculated from a heat balance on the driving turbine, provided:

• The BFP turbine exhaust steam is superheated.

• An accurate steam-flow device is used to measure the steam supply to the BFP turbine.

For electrically-driven pumps, the input power is obtained by metering the electrical supply and allowing

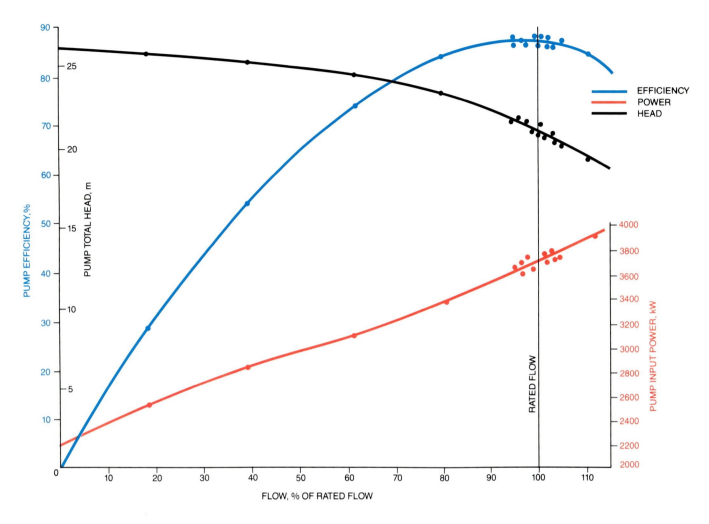

FIG. 7.140 Plot of CW pump test results
The curves are drawn from the test results obtained over the range.

for all the energy losses between the point of metering and the pump coupling. The supply frequency is also required.

Gearbox losses

When a gearbox is fitted, its losses should not be included in the pump power input. The mechanical efficiency of the gearbox is determined from curves obtained from works tests.

10.2.2 Numerical example

A main boiler feed pump has the following guaranteed performance at 100% rated flow:

> Total head generated 2210 m;
> turbine speed 4300 r/min;

> Pump efficiency 83.0%;
> volumetric flow rate 476.8 l/s

Preliminary; Guarantee; Overload and Part-load operation tests are carried out. As an example of the method the data for the preliminary test are given in Table 7.41.

When the test series is complete, curves of head versus flow rate can be plotted as shown in Fig 7.141. The effect of varying the speed is to raise or lower the head/quantity curve. Figure 7.141 shows the pump characteristic curves for a range of speeds.

11 Auxiliary plant power consumption

The power consumption of auxiliary plant is considerable, particularly for the large pump drives, such as CW, boiler feed and booster pumps. For example, the auxiliary power consumption in a station with four 500 MW units may reach 70 MW when the steam-driven boiler feed pumps are in use, and up to 120 MW when they are not. Therefore, it is most important

TABLE 7.41
Preliminary test on 100% boiler feed pump

		Derivation	Units	Preliminary test
1	Generator load		MW	512.7
	Corrected test readings			
2	Pump discharge flow	Test meter	kg/s	412
3	Suction pressure	Test pressure gauge	bar abs	6.9
4	Suction temperature	Test instrument	°C	138.3
5	Discharge pressure	Test instrument	bar abs	199.9
6	Discharge temperature	Test instrument	°C	141.1
7	Pump shaft speed	Test instrument	r/min	4055
8	Driving torque	Test torque meter	kNm	24.5
	Inlet velocity head			
9	Specific volume	Steam tables	m^3/kg	0.001 078
10	Entropy	Steam tables	kJ/kgK	1.7216
11	Velocity (pipe ID = 406 mm)	$\dfrac{4}{\pi}\dfrac{(2)\times(9)}{(0.406)^2}$	m/s	3.43
12	Velocity head	$(11)^2/2\times9.81$	m	0.60
	Discharge			
13	Specific volume	Steam tables	m^3/kg	0.001 069
14	Velocity (pipe ID = 356 mm)	$\dfrac{4}{\pi}\dfrac{(2)\times(13)}{(0.356)^2}$	m/s	4.4
15	Velocity head	$(14)^2/(2\times9.81)$	m	0.99
	Volumetric flow rate			
16	Specific volume	From (5) and (10), using steam tables	m^3/kg	0.001 07
17	Mean isentropic specific volume	(9) + (16)/2	m^3/kg	0.001 07
18	Pump volumetric flow rate	(2) × (17)	m^3/s	0.441
	Pump input power			
19	Power transmitted by torque meter	(7) × (8) × $(2\pi/60)$	kW	10 405
	Water power			
20	Pressure rise across pump	(5) − (3)	bar	193
21	Difference in velocity head	(15) − (12)	m	0.39
22	Pump total pressure rise	$\dfrac{(20)\times(17)\times10^5}{9.81}$ + (21)	m	2105.5
23	Water power	(18) × (20) × 10^2	kW	8511
	Pump efficiency			
24	Pump efficiency	(23) × 100/(19)	%	81.8
	Comparison with guarantee			
25	Correction factor for speed and volumetric flow rate	$\sqrt{(2210/(22))}$	C_f	1.0245
26	Corrected speed	(25) × (7)	r/min	4154

TABLE 7.41 (cont'd)

Preliminary test on 100% boiler feed pump

		Derivation	Units	Preliminary test
	Comparison with guarantee			
27	Corrected volume flow rate	(25) × (18)	m^3/s	0.4518
28	Guaranteed speed at (27)	From guarantee	r/min	4259
29	Guaranteed efficiency at (26) by linear interpolation	From guarantee	%	82.7%
30	Comparison test speed with guarantee	[(26) × 100/(28)] − 100	%	− 2.46
31	Comparison test efficiency with guarantee	[(24) × 100/(29)] − 100	%	− 1.09
	Test results corrected to 4300 r/min			
32	Flow rate	(18) × 4300/(7)	m^3/s	0.4676
33	Total head	$(22) × 4300^2/(7)^2$	m	2367.6
34	Pump input power	$(19) × 4300^3/(7)^3$	kW	12 407
35	Mass flow rate	(32)/(17)	kg/s	437.0
36	Water power	$(32) × (33) × 9.81/(17)10^3$	kW	10 150
37	Pump efficiency	(36) × 100/(34)	%	81.81

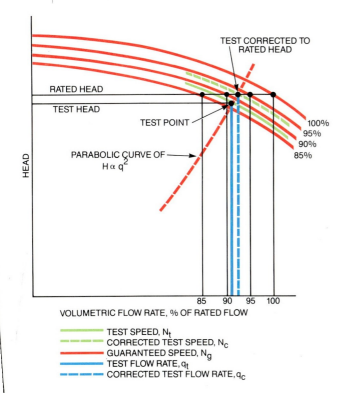

FIG. 7.141 Application of pump affinity laws
An illustration of how the test flow rate (q_t) and the test speed (N_t) are corrected by the affinity laws to their values at rated head. q_c is corrected flow rate and N_c is corrected speed.

also the following whose motor consumption is smaller, but still significant:

● Induced draught fans.

● Forced draught fans.

● Primary air and exhauster fans.

● Gas recirculating fans.

● Pulverised-fuel mills.

● Boiler circulation pumps.

The remaining drives in a power station are generally of modest size, but even so the opportunity should not be missed to introduce economies of operation wherever possible. However, the main attention is directed to those drives with large power consumption. For example, if a unit is taken off-load, the boiler feed pump should be shut down as soon as possible (along with all the other auxiliaries which are not required). Whilst shut down, the boiler should be 'tight', i.e., dampers should be a good fit, to prevent a draught through the gas passes. Also, steam leakage must be kept to a minimum, otherwise considerable extra firing is needed to bring the unit back on-load. Within the CEGB, every unit has off-load energy tests to establish exactly how much energy various types of start require and these are used by System Operations Branch in determining the economic operation of units.

It is particularly important that pumping power is not wasted if a whole station is shut down, say, overnight. It is common to keep a CW pump running

to reduce works power wherever possible. The drives mentioned have large power consumption; there are

during the off-load period to keep the CW system primed, at a tremendous cost in works power consumption. A modern CW pump motor consumes about 4.5 MW on full-load and, even if operated with the discharge severely throttled, the consumption only falls to about 2.5 MW. So, attention should be given to keeping the system primed without running a pump or, at worst, to only run the pump for short periods to top-up the system. If this is not possible, then perhaps a small off-load CW pump should be installed to keep the system primed during off-load periods to minimise power consumption.

Economic operation of the CW pumps is also necessary when the plant is on-load. Clearly, at least one pump must be run. The economic question is 'When is it worthwhile to run an extra pump?' and, in broad terms, the answer is 'when the benefit of running the pump is equal to, or greater than, the cost of running it'. The method of determining the break-even point is best illustrated by an example.

A 500 MW unit CW pump motor consumes 4500 kW on full-load, i.e., 16.2 GJ/h at a system efficiency of, say, 33%. Therefore, the heat input to the boiler to provide the pump power is 16.2/0.33 = 49.1 GJ/h. Also, the turbine heat consumption at 44 mbar back pressure on full-load is 5983 GJ/h, so assuming a boiler efficiency of 95% the heat input to the boiler will be 6298 GJ/h. If the load remains the same but the back pressure is changed, then the boiler heat will change, as shown in Fig 7.142. For example, at 44 mbar the heat will be 6298 GJ/h, whereas at 53 mbar it will be 6361 GJ/h, and so on. Thus, if the initial back pressure is 67 mbar (6550 GJ/h) the heat consumption would have to improve by at least 49.1 GJ/h to make the operation of an additional pump economic, so the back pressure would need to be better than 62 mbar.

Hence, a table of critical back pressure changes can be prepared, as shown in Table 7.42. By constructing a graph of the critical back pressure change, the operating staff can easily determine whether the operation of a CW pump is justified (Fig 7.143).

In general, the scope for reducing works power is considerable — in almost every station *some* saving is possible and in many it is substantial, bearing in mind that we are referring to *economic* savings of power. For example, in some stations the condenser CW discharge valves are kept wide open to improve the back pressure, but at the expense of increased CW pump power consumption. For this to be economic depends upon whether the benefit of the increased generator output (due to the reduced back pressure) is greater than the cost of supplying the extra CW pumping power. To illustrate this, consider the machine whose back pressure correction curve is shown in Fig 7.144. As in the example of the economic operation of an extra CW pump considered earlier, the equivalent boiler input heat is also shown.

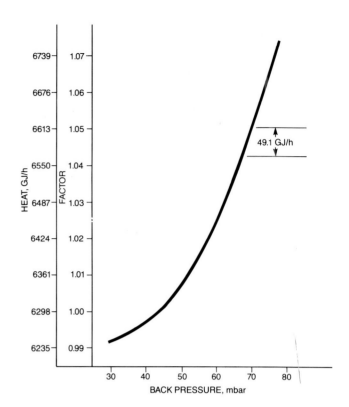

FIG. 7.142 Variation of boiler input heat at full-load with back pressure
The curve is used to determine the back pressure change needed to justify running an extra CW pump.

TABLE 7.42
Critical back pressure change for economic operation of an extra CW pump

Initial back pressure, mbar	40	50	60	70
Final back pressure, mbar	29.4	42	56	67

Suppose the extra CW pumping power required for fully-open valve operation is 2 MW (7.2 GJ/h) and the efficiency of supply is 33%, then the equivalent boiler input heat will be 7.2/0.33 = 21.8 GJ/h. Alternatively, it can be argued that every unit of extra works power consumed at a station results in a comparable reduction of power sent out to the grid system. This power must be made good by a station lower down the Merit table, and so will probably be produced at a sent-out efficiency of something less than 30%. Here we will assume that the efficiency of the supply to the motor terminals is that of the station itself, i.e., 33%. Then fully-open valve operation will only be economic if the back pressure improvement causes a reduction of at least 21.8 GJ/h of boiler input heat. For example, if the back pressure is 67 mbar (6550 GJ/h boiler input heat) when

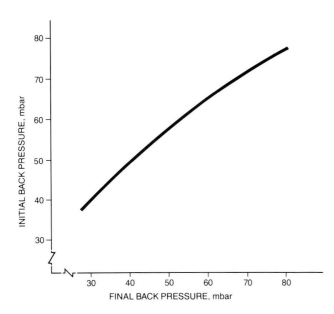

FIG. 7.143 Critical back pressure change for economic operation of a CW pump

12 Plant efficiency during normal operation

12.1 General

So far the discussion has been mainly about the performance of plant under test conditions. In normal operation, the plant is not continuously at test standard — for example, sootblowing, blowdown, load changes, shutdown and start-up all affect the performance of the plant. Clearly, the 'running' efficiency of a station will be significantly lower than the test values, and is determined from the expression:

$$\text{Station efficiency} = \frac{\text{electricity sent-out in period}}{\text{total heat of fuel burned in period}}$$

For example, consider a station which has two 500 MW oil-fired units. The following results were obtained for a particular seven-day period:

		Unit 1	Unit 2
Average sent-out load	MW	400	460
Running hours in period		130	168
Heat supplied from fuel	kJ/kg	45 356	
Quantity of fuel burned	tonnes	15 000	16 150

$$\begin{aligned}\text{Total heat to boilers} &= 45\ 356\ (15\ 000 + 16\ 150)\ 10^3 \\ &= 45\ 356 \times 31.15 \times 10^6 \text{ kJ} \\ &= 1413 \text{ TJ in seven days}\end{aligned}$$

$$\begin{aligned}\text{Electricity sent out} &= [(400 \times 130) + (460 \times 168)] \times 10^{-3} \\ &= 129.3 \text{ GWh} \\ &= 129.3 \times 3600 \text{ GJ} \\ &= 0.465 \text{ TJ}\end{aligned}$$

$$\begin{aligned}\text{So station efficiency} &= 0.465/1.4130 \\ &= 0.329 \text{ or } 32.9\% \text{ sent-out}\end{aligned}$$

the discharge valve is throttled to give the optimum CW rise, the back pressure must reduce to at least 65 mbar for the fully-open valve operation to be economic.

Co-ordinates can be determined (as shown in Table 7.43) and the information used to construct a graph (Fig 7.145). Variable-speed drives, such as to boiler feed pumps and induced draught fans, should always be run at the minimum speed which gives the desired result. From the affinity laws, the power consumption varies as the cube of the speed so, if the speed is 10% too high, the power consumption will be 33% high and the extra speed is very expensive.

Similarly, where auxiliaries are operated in parallel, it may be possible under certain conditions to obtain the required output by operating with one less auxiliary. It should be remembered that operating with, say, two pumps in parallel will not give twice the output of one pump on its own, as can be seen from Fig 7.146. The actual increase depends upon the system and pump characteristics. If one of the pumps is in service on its own the output is Q_1 (about 65% flow) and if a second (identical) pump is run in parallel, the flow becomes Q_2 (87% flow). The extra works power demand of the second pump only produces an additional flow of 22 percentage points.

Included in this result is all the heat used by the station for lighting, heating, maintenance, boiler blowdown, etc. It is not usually realised how substantial the lighting load is in a power station and significant savings may be possible. Normal station operation is dealt with in detail in the next section under the heading 'STEP factor'.

TABLE 7.43

Critical back pressure change for operating with fully-open CW discharge

Initial back pressure, mbar	40	50	60	67
Final back pressure, mbar	34	47	58	65

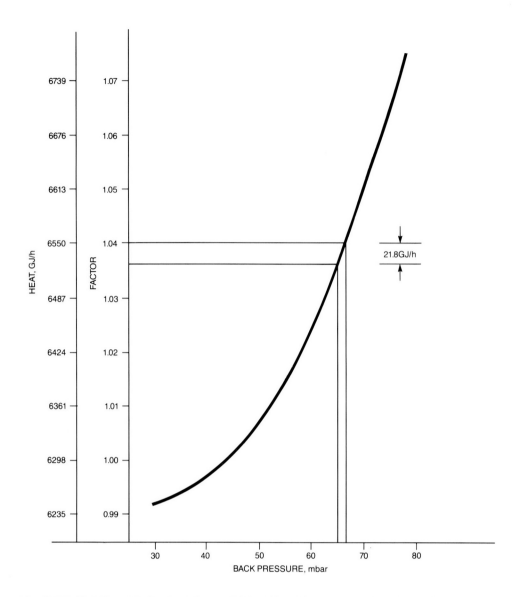

FIG. 7.144 Variation of boiler input heat at full-load for wide-open operation of CW discharge valves
The information is used to derive the critical change diagram in Fig 7.145 for operation with the
CW discharge valves fully-open.

12.2 Turbine-generator normal running performance

12.2.1 Method utilising steam flow measurement

If it were possible to have a continuous indication of the efficiency of individual units during normal operation, it would be of considerable benefit both to the station staff as well as the system operations engineers. As has already been mentioned, the main difficulty in being able to do this is that of determining directly the steam flow at the HP and IP turbines. However, promising results have been obtained from tests at Fiddlers Ferry power station, which may provide continuous on-line efficiency (or

heat rate) of an acceptable accuracy for system loading requirements and station information.

The basic as-run heat rate expression is:

Heat rate =

$$\frac{3600\,[M_s\,(H_1 - h_f) + M_{R/H}\,(H_3 - H_2) + M_{is}\,(h_f - h_{is}) + M_{ir}\,(H_3 - h_{ir})]}{P_g}$$

where M_s = steam flow to turbine stop valves

H_1 = specific enthalpy of steam at TSV

h_f = specific enthalpy of steam at TSV

$M_{R/H}$ = steam flow from HP exhaust to reheater

H_2 = specific enthalpy of steam at HP exhaust

H_3 = specific enthalpy of steam on IP turbine stop valves

M_{is} = spraywater flow to superheater

h_{is} = specific enthalpy of spraywater to superheater

M_{ir} = spraywater flow to reheater

h_{ir} = specific enthalpy of spraywater to reheater

P_g = electrical power at generator terminals

The various specific heats are readily obtainable from pressures and temperatures at the required positions. The values M_{is} and M_{ir} are of relatively small magnitude and so some error in measurement can be tolerated without it having a very significant effect upon the result. The real problem is to determine the magnitude of M_s and $M_{R/H}$. The method adopted at Fiddlers Ferry is shown in Fig 7.147. Pitot sensors

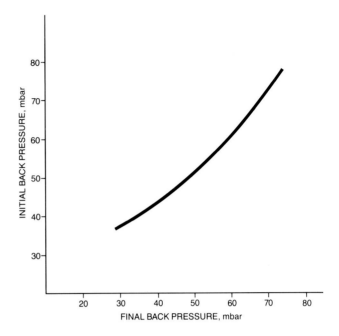

FIG. 7.145 Critical back pressure change to justify operation with CW discharge valves fully open

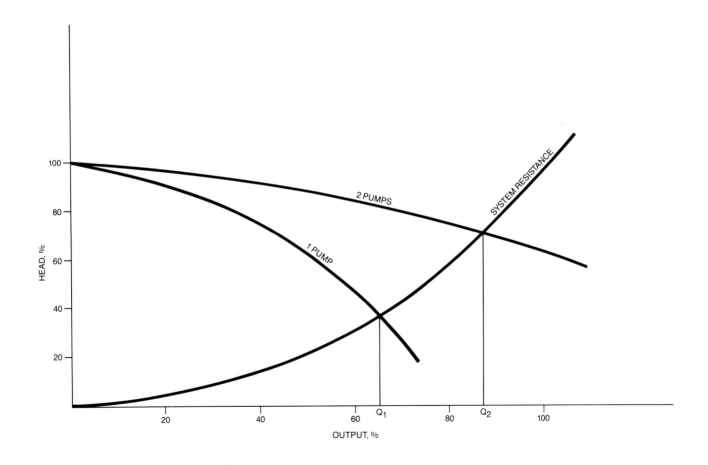

FIG. 7.146 Parallel operation of two pumps
The extra output and head which is obtained by running two pumps in parallel is largely determined by the system resistance. The one-pump flow is Q_1 and the two-pump flow is Q_2.

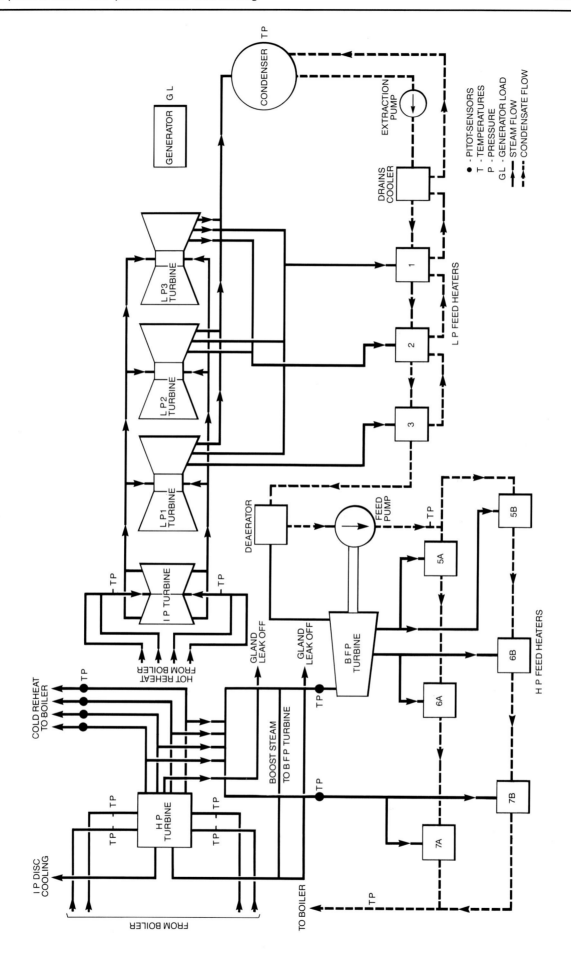

FIG. 7.147 Simplified drawing of turbine and feed system, showing location of test measurements
Note that there is no necessity to measure steam flow before the HP cylinder.

are located in the cold reheat lines, steam to boiler feed pump turbine and steam to No. 7 HP heater. By this means, the steam flow from the HP cylinder exhaust to the reheater is accurately measured, as also is that to the BFP turbine and heater. The summation of these, plus the gland leak-off, gives the steam flow leaving the HP cylinder. To determine the flow into the HP cylinder, it is necessary to establish the order of flow to the IP disc cooling and to the gland leak-off from the HP cylinder reversal, both small quantities that can be estimated with sufficient accuracy for the purpose. Notice that there are no flow sensors in the steam line before the HP cylinder, where they would be subjected to arduous operating conditions with a consequential risk of failure.

The pitot sensors used are two different round-type averaging devices, both of which can be installed across the diameter of the steam pipe to measure average steam velocity. The two types shown in Fig 7.148 are:

- 'Tekprobes'.

- 'Gendeal' probes.

The first is based upon a design which incorporates three averaging chambers for the upstream velocity head. The resultant average is derived from the higher and lower ports in independent chambers. The downstream pressure, i.e., the static pressure, is taken in the 'dead zone' at the pipe periphery.

Five of these probes were installed on the HP cylinder exhaust, i.e., four in the reheat pipes and one in the line to No. 7 feedheater.

The Gendeal probe was installed in the feed pump turbine steam pipe.

The data logging system on each unit at Fiddlers Ferry power station consists of a 'Fluke' 2280A data logger, using a remotely-mounted 2281A extender chassis to receive input signals from the plant. The 2280A is controlled by a CMB Pet 8096 computer which has a disk system attached and a touch-sensitive display terminal. A schematic layout is shown in Fig 7.149.

Output from the logger is read by the computer at ten-minute intervals and stored on floppy disk. The information is used for a variety of functions, including STEP data and condenser performance monitoring.

The results obtained so far indicate that the accuracy of the on-line equipment is within 1.2% of the results obtained by conventional turbine testing. Therefore, it is possible that selected stations could have an installation such as this and their performance be continuously relayed to System Control for Merit-Order loading, using such an installation.

12.2.2 'Thermal' method

The rate of heat input to a turbine is equal to the

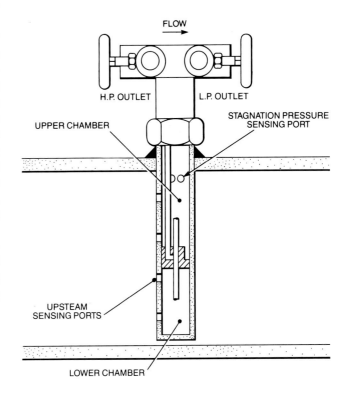

(a) 'Tekprobe' pitot-sensor

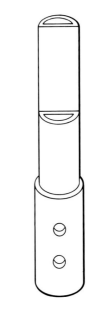

(b) Cut-away section of 'gendeal' probe

FIG. 7.148 Probes for steam flow measurement

(a) 'Tekprobe' pitot sensor. This type of probe is used for steam flow measurement in the reheater pipework.

(b) 'Gendeal' probe. This type of probe is used for steam flow measurement at the bled-steam turbine.

electrical power output plus the rate of heat rejected, i.e., rate of heat input = electrical output + heat to condenser + ancillary losses (Fig 7.150). Thus for a typical machine: 100% = 44.0% + 55.0% + 1.0%.

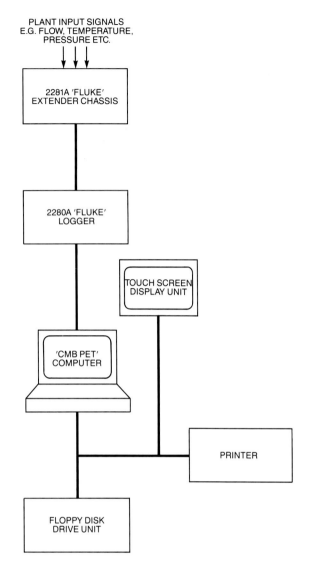

PLANT INPUT SIGNALS
E.G. FLOW, TEMPERATURE,
PRESSURE ETC.

2281A 'FLUKE'
EXTENDER CHASSIS

2280A 'FLUKE'
LOGGER

TOUCH SCREEN
DISPLAY UNIT

'CMB PET'
COMPUTER

PRINTER

FLOPPY DISK
DRIVE UNIT

FIG. 7.149 Schematic layout of the station performance data acquisition system in use at Fiddlers Ferry power station.

along a number of chordal paths on an angled acoustic plane across the pipe. A particular advantage of this method is that it compensates continuously for any changes in the velocity of sound in water, so it is not sensitive to variations in water purity, although air bubbles in the water do have an effect.

The name 'Thermal' is derived from *T*urbine *HE*at *R*ate *M*onitoring by *A*ssessment of *L*osses.

The flow and temperature signals are passed to data acquisition and computing equipment. Typical printout data from the computer comprises:

- Date/time.

- Load, MW.

- Heat rejected, MW.

- Turbine-generator efficiency.

- Back pressure, mbar.

The information is printed at pre-set time intervals. On test it has been shown that the turbine-generator as-run thermal efficiency stayed within about ± 0.2 percentage points at nominal full-load on a 500 MW machine.

Various changes in the operation of the plant can be investigated by the 'Thermal' method, such as:

- Determination of the Willans line.

- Effect on efficiency of back pressure.

- Effect of operation without main boiler feed pump.

- Effect of load swings.

- Sliding pressure operation.

The electrical power can be measured to about ± 0.2 per cent and the ancillary losses to about ± 20 per cent, but as the ancillary losses are small in magnitude the effect on the final result is acceptable. The remaining loss to be measured is the major one, heat to condenser, determined by measuring the CW flow and its temperature rise. This is quite difficult to do, but by using extended platinum resistance thermometers reaching across the pipe on multiple chordal paths, accuracies of temperature measurement better than $\pm 0.1°C$ can be achieved. The flow measurement is more difficult as the CW pipes are so large, and it is also desirable to avoid head loss when measuring the flow. This can be accomplished to an accuracy better than ± 1 per cent by using an ultrasonic 'time of flight' method.

It involves the measurement of travel times of acoustic energy transmitted forwards and backwards

Thus, at stations where the method is possible (usually determined by the CW pipework configuration to each unit), it provides an easy means of continuously monitoring the thermal performance of the turbine-generator. If condenser pressure is measured it will also provide detailed information about the condenser performance.

12.3 Boiler normal-running performance

As well as continuous turbine efficiency determination, the boiler losses can also be continuously measured, except for three parameters:

- Calorific value of the fuel.

- Moisture content of the fuel.

- Combustible in ash.

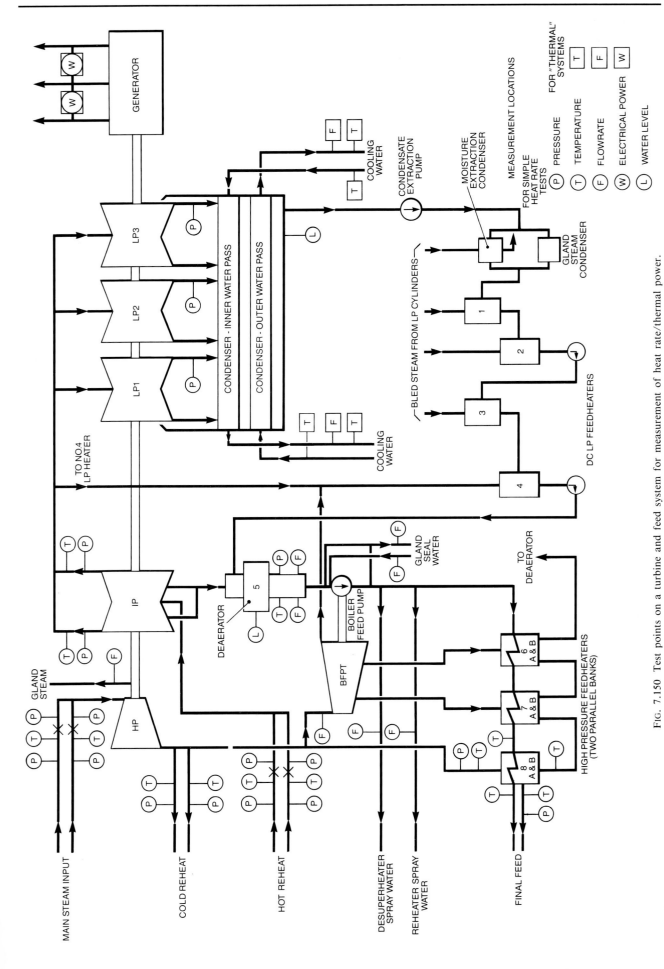

FIG. 7.150 Test points on a turbine and feed system for measurement of heat rate/thermal power.

Good estimates can be made of these items, based upon recent results, so the boiler efficiency can be approximately determined, but it would be much better if reasonably accurate current values were available. Work on this is in hand.

Similarly, the advent of microprocessors and the development of sophisticated measuring equipment and computers has enabled new boiler control strategies to be introduced to improve normal running efficiency. This aspect is dealt with in detail elsewhere, so only a few observations will be made here regarding this very important topic. For example, consider the safe operation of milling plant. Various constraints are imposed upon the plant, including:

- Maximum mill output.

- Maximum PF/air throughput that the exhausters or primary air fans can handle.

- Burner flame stability.

- Available attemperation of hot air to mill.

- Mill drying capacity.

- Flammability limit.

- Pipework erosion.

- Transport velocity.

Consider each item in turn.

12.3.1   Mill maximum output

The design capacity of the mill should be achievable under all conditions of operation, while maintaining the PF grading within the desired range. For example, for bituminous coal the range is about 70% to 80% through a 200-mesh sieve, so the mill should be capable of achieving its rated capacity with a grading of at least 70% through this size of sieve, even when due for a major overhaul.

12.3.2   Exhauster or primary air fan capacity

Clearly, inadequate exhauster or primary air fan capacity will limit output, and it will be necessary to increase the speed of the fan or increase the size of the impeller.

Plant tests indicate that a relationship of the form 'air flow plus coal flow equals a constant', i.e., $W_a + W_f$ = constant, adequately represents the performance over the normal range of air and coal flows used on conventional plant.

12.3.3   Burner flame stability

For a stable flame, the air and fuel supplies must be kept within a rather narrow operational band. The band varies from one boiler to another with loading and with the type of fuel being burned. However, for the practical purpose of controlling milling plant, it can be assumed that a pulverised-coal fire becomes unstable at loadings of less than 50% of the maximum coal throughput, unless supported by an oil fire.

12.3.4   Available attemperation of hot air to mill

When the raw coal is fairly dry, the hot air may dry it and yet still be too hot when leaving the mill, i.e., the air/PF mixture leaving the mill will be in excess of 70°C. It is then necessary to attemperate the hot air at the mill inlet to control the outlet temperature to its desired value, so the maximum available flow of attemperating air imposes a constraint on the minimum quantity of coal which can be dried.

12.3.5   Mill drying capability

The fundamental drying equation is approximately:

$$W_f = \frac{A_o}{1 + \alpha} \times \frac{\theta_i - \theta_o}{M_c} \times \frac{C_p}{h_{fg}} \times \left(\frac{1 - M_{pf}}{1 - M_{pf}/M_c} \right)$$

where W_f = fuel flow

A_o = air flow at mill outlet

A_i = air flow at mill inlet

α = leakage fraction (including elutriating air)
= $(A_o - A_i)/A_i$

θ_i = temperature of hot air to mill, °C

θ_o = temperature of air/coal mixture at mill, °C

$\overline{M}_c$ = moisture in coal (total moisture)

$\overline{M}_{pf}$ = moisture in pulverised fuel

C_p = specific heat of air

h_{fg} = heat of evaporation

For a given moisture in the coal,

$$W_f = A_o/(1 + \alpha) \times (\theta_1 - \theta) \times \text{constant},$$

which permits the effect of improved conditions to be assessed with reference to datum values. Let the improved conditions be designated by the suffix 1.

$$\text{Then } F = \frac{W_f 1}{W_f} = \frac{A_o 1}{A_o} \times \frac{(1 + \alpha)}{(1 + \alpha)1} \times \frac{(\theta_i - \theta_o)1}{(\theta 1 - \theta_o)}$$

where F is a factor for the improved drying capacity, given by $F = f_A \times f_\alpha \times f_t$.

Figure 7.151 shows typical factors for the effect on drying capacity for a particular tube ball-mill. 20°C increase of θ_i, or 10% increase in mill outlet airflow, or 0.1 reduction in inleakage all have an equivalent effect on drying capacity, and each offers about 10% increase in fuel flow.

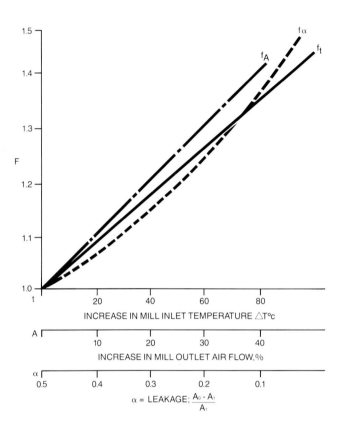

FIG. 7.151 Effect of variation of drying-capacity parameters
The effect of changing the mill inlet temperature, the mill outlet airflow and the leakage is illustrated; the datum conditions are $t_1 = 300°C$; mill outlet airflow $A_0 = 21$ kg/s and leakage $\alpha = 0.5$.

Drying capability is often the most serious of all the constraints on mill operation.

12.3.6 Flammability limit

This depends upon the air/fuel ratio. The limiting value is regarded as 5:1, and a higher ratio may produce an explosive mixture. The preferred limit is 3:1.

12.3.7 Pipework erosion

Erosion is proportional to the (air/fuel velocity)$^{2.5}$, approximately. Consequently, operation with an ex-

cessive air flow reduces the life of the plant, and in particular increases the frequency of PF pipework replacement. A range of airflows is required to accommodate the drying requirements and to exploit the coal storage in the mill (for control of load), but it is possible to achieve these within the range of 1.3 to 1.5 times the minimum transport velocity.

If the air/fuel velocity is doubled, erosion increases by $2^{2.5} = 5.7$, so doubling the velocity will enhance pipework erosion by almost six times. The velocity should be kept to the minimum which is economical and satisfactory, usually about 1.5 times that required to prevent the fall-out of pulverised fuel from suspension.

12.3.8 Transport velocity

The minimum permissible air/PF velocity is that at which the fall-out of PF from suspension is prevented. It depends upon the size and geometry of the pipework and allowance is made for distribution factors, such as multiple burners. Investigation has shown that the theoretical minimum velocity to prevent fall-out is often too low in practice. The minimum speed of air to transport pulverised fuel with a grading of 70% through a 200-mesh sieve, is about 20 m/s. Care must be exercised to ensure that this is achieved in that part of the pipe run where the lowest velocity can be expected.

12.4 The mill operating window

It is a fairly simple matter to quantify all of the constraints for a particular plant. The boundaries for safe operation can then be plotted for operation of the mill in terms of the coal and air flows delivered to the burners, known as the 'operating window'. A typical construction is shown in Fig 7.152. Automatic mill control is set to ensure that the mill operating point is within the window for all loads except during start-up and shutdown, when operation outside the window is inevitable. Effective control of milling plant performance is a major factor in improved boiler efficiency.

As an alternative to having the 'operating window' diagram axes scaled in terms of 'fuel flow' and 'air flow', it is sometimes found to be more convenient to use 'fuel flow + air flow' and 'air flow'. This arrangement is illustrated in Fig 7.153.

12.5 Coal stock accountancy

12.5.1 General

The value of the coal stock at a large power station is several million pounds. Thus, the commercial

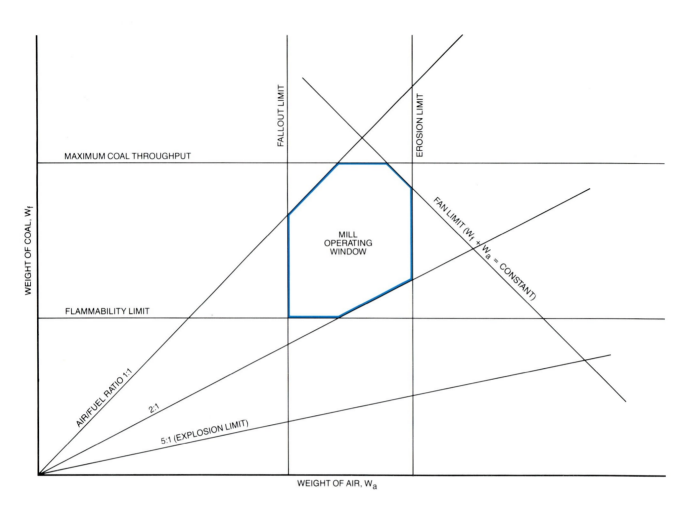

FIG. 7.152 Mill operating window
The desirable region of operation for a mill can be represented by a number of constraints. The resultant operating window defines the limits for safe and efficient operation. Mill automatic control is arranged to ensure that the operation point is always inside the window during normal working.

management of resources makes it vital that this very important item is accurately accounted for. Also, if there is a mismatch between the quantity of coal on the ground and that recorded in the books, this reflects on the apparent performance of the plant. Should more coal be burnt in a given period than is accounted for in the books, then the plant will have an enhanced apparent thermal performance but the coal stock accountancy will be in error. If this continues for a substantial period, the discrepancies both of thermal performance and stock accountancy will be considerable.

Therefore, checks and counter checks are built into the accountancy system. The basic procedure is as follows:

● The thermal performance of the plant is determined by consideration of 'book' heat burned and electricity sent out.

● The losses incurred by the plant are calculated and used to determine the plant performance by the

'losses' method.

● The coal stock is surveyed at regular intervals to determine the actual quantity of coal on the stock for comparison with the book value.

● A coal stock heat account is kept.

The first two of these items are considered in the next section. It is the coal stock accountancy with which we are concerned here.

12.5.2 Volume surveys

The volume of the stock is determined at regular intervals (say, every two months) by means of an aerial survey. These aerial surveys are supplemented from time to time by physical surveys to ensure compatability of the results. Meanwhile, the average specific density of the coal on the stock is determined, say, once per year. From a knowledge of the volume

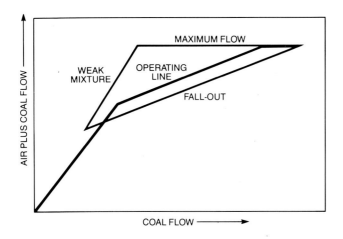

FIG. 7.153 Mill operating constraints
The desirable region of operation for a mill can
be represented by a number of constraints. At High
Marnham power station three are relevant, the upper
constraint representing the maximum available mill
throughput, whilst to the left a weak mixture or a high
air-to-fuel ratio is encountered, which leads to unsafe
operation. Below the triangle the airflow is too low to
keep the pulverised fuel entrained, leading to fall-out and
eventually to fires in the pipework to the burners. The new
control system ensures that for different demanded coal
flows, the mill is operated along a line which remains
within this triangle except during start-up and shutdown,
when operation outside is inevitable.

and the specific density of the stock, the coal quantity
can be determined.

Two items of information are required before the
aerial survey:

- A detailed base-plan of the stocking area, showing
  ground levels.

- The identification and location of agreed fixed
  control points. These should be permanent fixtures
  which are easily visible on aerial photographs,
  such as pylon bases. The height of each control
  point is accurately determined with reference to
  ordnance-survey bench-marks.

The 'fly-over' photographs are taken from a height
of about 400 m. Stereoscopic pairs of photographs
are taken with 60% overlap and printed on glass,
so that distortion is minimised. The stereoscopic images of the coal heaps can be scanned by a special
optical system from which a series of spot heights
can be measured, to an accuracy better than 3 mm.

The camera focal length is 15 cm, so the scale
of the photographs is 1:(40 000/15), i.e., 1:2666,
assuming they are taken from 400 m. The plotting
error is within ± 30 μm, so the full-scale error is
within ± 30 μm $\times 2666 = \pm 80$ mm. The spot heights
are used to compile a contour map of the coal heaps,

and usually between 1000 and 4000 are needed per
heap of coal. As each height can be ± 80 mm in
error, the resulting total error will be between $\pm 80/\sqrt{1000}$ and $\pm 80/\sqrt{4000}$ mm, i.e., between ± 2.5 mm
and ± 1.3 mm, so there is very little loss of accuracy
from this cause.

A much more significant error could arise from
setting-up the photographs with reference to the control points in the first place. Once again, the possible
error is ± 80 mm for each point, so the resulting
error could be appreciable for stocks of low vertical
height. The error is less than $\pm 2\%$ for stock over 4
metres high.

12.5.3 Density determination

The density is determined by lowering a radioactive
source into holes in the coal heap and determining the
'count' from the source. The density of the coal to
produce the given count can be determined from previous calibration tests.

A large number of readings is required to give an
acceptable average density. The stock is drilled in
several places (usually four), each hole extending to
the bottom of the heap and readings are obtained
from every hole at each metre of depth. The average
density is easily determined and so this, coupled with
the volume result, enable the weight of coal to be
calculated.

The result of the survey is usually reported in the
following manner:

- Aerial survey   951 624 tonnes
- Book stock   964 230 tonnes
- Difference   (−)12 606 tonnes
- % Difference $-1.3\% = [(12606 \times 100)/964\,230]$

Usually the total error is from ± 2 to $\pm 7.5\%$, the
larger error being associated with coal stocks made
up of several small heaps.

12.5.4 Heat accountancy

Even though the coal stock aerial survey shows good
agreement between the ground and book stocks, it
does not necessarily follow that all is well. There may
still be a discrepancy between the actual average stock
calorific value and the 'book' average value. This can
be brought about in various ways. For example, consider one of the simplest causes of inaccuracy. Coal
is delivered to a station with an invoiced calorific
value of, say, 22 000 kJ/kg and 10 000 tonnes of it
is put to stock. The coal stock average heat (after
this coal addition) is 22 500 kJ/kg, according to the
book figure. Over the next few days the *same* coal
is reclaimed from the stock, but now its calorific value
is assumed to be that of the stock average. Therefore,

the plant is credited with burning 10 000 tonnes of coal at 22 500 kJ/kg, whereas it has actually burned 10 000 tonnes at 22 000 kJ/kg, a total of 5 GJ less. Therefore, the thermal efficiency of the plant will seem worse than it actually is. If it is too bad to be credible it may be that the apparent performance is improved by reducing the weight of coal credited to the plant, but in this case the aerial survey will reveal any significant discrepancy which results.

Another source of heat discrepancy is if the invoiced calorific value is wrong. As all calculations of heat consumption must be based upon the invoiced values, this clearly leads to complications with the heat accountancy at the station.

Therefore, it is most desirable to sample the coal deliveries for quality and also the coal sent to the bunkers so that discrepancies can be rectified at the earliest opportunity. Only coal is a problem fuel. Oil and natural gas each have fairly consistent quality and so less rigorous accountancy procedures suffice.

In 1987 integrated heat accountancy was introduced in an attempt to improve the accuracy of performance data and to reduce the degree of distortion in the merit order and thermal efficiency returns.

The evaluation of the heat account is thus subject to more closely defined rules and therefore ensures a higher degree of conformity than was previously the case. A document entitled 'Integrated Heat Accountancy' was issued in July 1987 which gives full details of the new accountancy procedure.

13   STEP factor, heat balance and merit order heat rate*

13.1   Introduction

Because of the varying effects of external factors, notably loading regime, fuel quality and cooling water temperature on intrinsic plant efficiency, it is difficult to assess the reported thermal performance of a power station without comparing it with an equitable target or par performance level over the period of assessment. In the CEGB the method adopted is to compare the performance of each plant (and station) with its own best achievable performance for the operating conditions which pertained during the period. This is similar to a golfer comparing his play to his handicap. The full method also permits meaningful inter-station comparisons.

It is known as *The assessment of Station Thermal Efficiency Performance*, or *The STEP Scheme*, for short.

---

* The author is indebted to Mr. N. D. Clack, formerly of the Production Services Branch, CEGB Midlands Region, for his extensive contribution to this section.

The STEP method has been refined from time to time since its inception in April 1962 although the basic philosophy has remained unchanged, and this will be a continuing process as plant and monitoring technology improves. The method as currently applied is described in the following pages.

The *STEP factor* is the ratio of the target heat rate (i.e., the target heat consumption per unit of electricity supplied) to the actual heat rate achieved.

$$\text{So STEP factor} = \frac{\text{Target heat rate in period}}{\text{Actual heat rate in period}}$$

expressed as a percentage.

It is quite easy to compute the *actual* quantity of heat used in a given period, as this is merely the weight of fuel burnt multiplied by its calorific value. The more difficult problem is to determine the *target* heat consumption. The method adopted is to start from the heat equivalent of the electricity generated and work backwards to determine the necessary heat to be supplied to the boilers. To this value is added the heat used for various off-load operations, giving the target station heat consumption.

The *target electricity supplied* is obtained by subtracting the target works electricity from the electricity generated by each unit, and summing the values for all the units in the station. Reference to the guide given below should make the procedure clear.

13.1.1   Guide to the derivation of station STEP factor

Item	How derived
(a) Actual electricity generated.	From generator metering.
(b) Incremental turbine heat consumption in period	By applying Willans line data to (a).
(c) Total turbine heat input in period at standard terminal conditions	(b) plus turbine fixed heat consumption (Willans line fixed heat times number of hours run in period).
(d) Total turbine heat input in period at prevailing CW inlet temperature	(c) modified for effect of actual CW temperature on back pressure and turbine heat consumption.
(e) Target boiler on-load heat output in period	(d) plus allowances for on-load make-up loss and range radiation loss.
(f) Target boiler heat input.	(e)/(1 − boiler losses)

Item (f) is the target on-load heat consumption for the unit considered. This is repeated for each unit in the station, and their sum gives the station target on-load heat consumption.

Item	How derived
(g) Station target heat consumption	(f) for all units plus target heat consumption during off-load periods, start-ups, post-maintenance testing and the use of auxiliary plant.
(h) Target electricity	Total generation minus target electrical auxiliary consumption for on-load and handling, plus allowance for post-maintenance testing.
(i) Station target sent-out heat rate.	(g) divided by (h)
(j) Station actual sent-out	Station total heat consumption in period divided by the actual electricity supplied.
(k) Station STEP factor	(i)/(j) $\times$ 100%.

It should be noted that where allowances are given they are related to the individual design characteristics of the plant and are derived from test data. No allowance is given for ageing, as the effects of age can be significantly affected by the amount of maintenance the plant receives.

13.2 Derivation of STEP factor

Note: For compatibility with the STEP scheme data, the symbols used therein are also used in this section.

13.2.1 General

Section 13.1 showed the basic requirements of the derivation of STEP factor. The means by which the information is derived will now be considered in more detail. Figure 7.154, shows in diagrammatic form, the way the individual items of information are used in the calculation. Although it looks formidable at first glance, it merely lists (in more detail) the stages considered in the preceding section. It is recommended that reference is made to Fig 7.154 in conjunction with reading the detailed explanation which follows. Also, it is necessary to refer to a number of forms which are in three categories:

- Registered data (prefixed R).

- Monthly operating data (prefixed M).

- Calculated data (prefixed C).

The *registered data* is basic plant information, which changes only when required to conform with modifications or to the STEP method; this data must, wherever possible, be based on Official Test results. Tables 7.44, 7.45, 7.46 and 7.47 (Forms STEP 3A, 3B, 3C and 3D) contain typical data for a station with four $\times$ 500 MW units. For example, item R20 lists the incremental heat rate of turbines 1 and 3 as 7778 kJ/kwh, and turbines 2 and 4 as 7663 kJ/kWh.

The *monthly operating data* refers to the actual performance during a monthly review period. The relevant forms are shown in Tables 7.48, 7.49 and 7.50 (Forms STEP 4A, 4B and 4C). For example, item M5 (in Table 7.48) shows the on-load works electricity used in the period for each of the four units and the total used.

Information from the registered data and the monthly operating data forms is used to calculate various items of information, as detailed in the STEP factor calculation key (Table 7.51). For example, item C1 gives the calculation to derive the turbine fixed heat consumption as R19 (M9C − M10C) $\times$ (R21 $\times$ M10C). In other words, item 19 of the registered data is multiplied by items 9 minus 10 from the monthly operating data forms. To this is added the product of registered data item 21 and item 10 of the monthly operating data. It should also be mentioned that the monthly operating data can refer to an individual unit or the station, so to differentiate between them the item number is followed by a letter. Thus item M2A signifies the generated output of one of the units and M2B refers to the total generated output of all of the units.

To summarise, the forms used in the calculation of STEP factor are:

Table	Reference	Type of data
7.44	STEP 3A	Registered data — station summary
7.45	STEP 3B	Registered data — tabular data — turbine-generators
7.46	STEP 3C	Registered data — tabular data — boilers
7.47	STEP 3D	Registered data — tabular data miscellaneous
7.48	STEP 4A	Monthly operating data sheet 1
7.49	STEP 4B	Monthly operating data sheet 2
7.50	STEP 4C	Monthly operating data sheet 3
7.51		STEP factor calculation key

Note: These tables are assembled at the back of this chapter.

13.2.2 Example of STEP calculation

The model used is a station with four $\times$ 500 MW units.

Details of the plant are given in Tables 7.44 to 7.50.

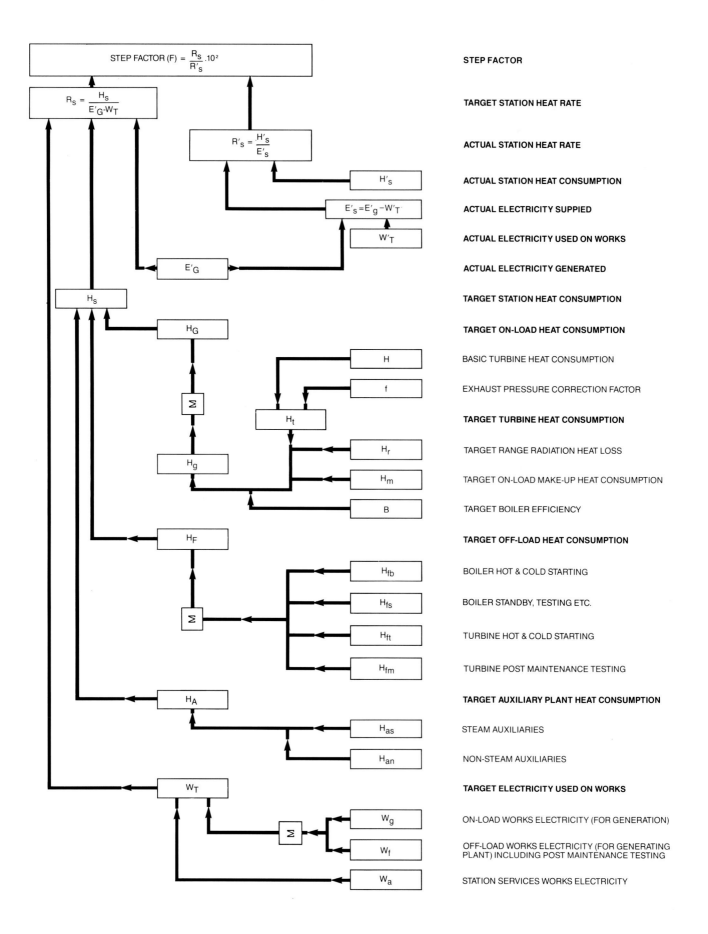

FIG. 7.154 Block diagram of step factor calculation method

The method of calculation is given in Table 7.51 and the results are listed in Table 7.52 (STEP factor calculation summary (STEP 5)).

The main items used in the calculation are:

Total target on-load heat = 8 622 938 GJ

Total off-load target heat consumption allowance = 34 269 GJ

Station target auxiliary plant heat consumption = 13 999 GJ

Total station target heat consumption = 8 671 206 GJ

Total actual electricity generated = 953.470 GWh

Total target works electricity = 49.448 GWh

So, target electricity supplied = 904.022 GWh

Target station heat rate = 8 671 206/904.022 = 9592 kJ/kWh

Actual station heat rate = 9828 kJ/kWh

STEP factor = [9592/9828] 100 = 97.60%

The derivation of some tables and factors used in the calculation varies from plant to plant and is fully explained in Operation Memorandum No. 37 'The Assessment of Station Thermal Efficiency Performance (The STEP Schemes)', issue No. 2, April 1981, published by the CEGB.

13.3 Heat balance and losses analysis

13.3.1 General

Higher management is concerned mainly with overall results: for this purpose the STEP factor for a station or group of stations is usually sufficient. At lower levels of management, more detail is required for investigative work and this is provided by the STEP factor losses analysis. It identifies the reasons for changes in performance and expresses them in both heat and monetary values.

Ideally, [STEP factor %] + [Σ STEP losses %] = 100%

In drawing up a heat balance sheet, some discretion is necessary with regard to the way in which the total loss is sub-divided, the number of items considered, and the units in which the results are stated. At present, a variety of methods is used, as devised by the different operating areas of the CEGB, but the example given may be regarded as typical. The national losses analysis scheme currently being developed will be based broadly on this approach, but will also deal

with the heat discrepancies arising from differences between 'heat as invoiced by supplier' and 'heat received as determined by the CEGB', as required by the new concept of Integrated Heat Accountancy. The method used in the example is given in detail in the CEGB Document MRS 8, issued April 1982. It requires the calculation of full target and actual heat balances from which the deviations from target are derived.

The target heat balance is taken from the STEP factor calculation. Many losses can be determined, reducing the unaccounted loss to a small value. The unaccounted loss comprises:

- Any discrepancy between invoiced heat and the CEGB determination of 'as-received' heat. The invoiced heat is used to derive the station actual heat rate.

- Any adjustment made to the heat supplied to boilers to compensate for differences between the 'book' stock and the actual stock on the ground, where applicable.

- Residual unaccounted heat.

Provision is also made to sub-divide some of the losses into their component elements to assist with investigative work.

A considerable amount of additional information about the plant is required for loss determination than is used in the STEP factor calculation. This extra data is designated RL (for registered loss data), based on manufacturer's data or reliable test results; ML (for monthly operating data); CT or CA for detailed calculations relating to target or actual values as appropriate, plus a losses breakdown summary designated 'G', as listed in the following summary.

Table	Reference	Type of data
7.53	RL	Additional registered data
7.54	ML	Additional monthly operating data
7.55	CT	Target heat balance
7.56	CA	Actual heat balance
7.57	CA2	Actual heat balance
7.58	CA3	Actual heat balance
7.59	G	Losses breakdown summary

Note: These tables are assembled at the back of this chapter.

The method of using the forms is similar to that in the STEP scheme. Although the detailed calculations are set out for manual computation, in practice they are performed on station computing equipment.

The national heat balance and losses analysis scheme is being programmed for mainframe computer calculation as part of the STEP suite of programs. As computer programs do not round-off values to the stated number of decimal points, a manual computation using the specified precision may reveal minor differences from a computerised calculation.

13.3.2 Example of calculation

The information applies to the same four × 500 MW station considered earlier.

It will be noticed that, on the calculation sheets, some of the information is derived from the earlier STEP calculations. For example, item CT 303 (Table 7.55) is derived from C15 (Table 7.52) × R 216 (Table 7.53), which gives 163 612 × 0.303 = 49 574 GJ. Proceeding in this fashion the calculations can all be made, so enabling the STEP Factor Losses Breakdown Summary (Table 7.59) to be completed.

The 'unaccounted heat consumption' (item 29 in the summary) is the difference between 100% and the sum of the STEP factor, total accounted heat consumption (item 26) and works electricity equivalent heat consumption percentages. In the example this is:

$$100\% - (97.6\% + 0.98\% + 0.11\%) = 1.31\%$$

At the bottom of the form, the last five lines refer to data calculated for the previous month under the heading 'As-received Heat Basis Monitoring for Previous Month'. This part of the analysis will soon be superseded by the introduction throughout the CEGB of the concept of Integrated Heat Accountancy.

Because the 'as-received' calorific value of the coal is determined by a central laboratory rather than at the stations, the full information is usually not available for inclusion in the current month's summary. The discrepancy between the 'invoiced' calorific value (from which the STEP results are calculated) and the 'as-received' calorific value may be significant, so it is important that its contribution to item 29 is determined, even though it is in retrospect.

The parallel problem of keeping the actual coal stock heat compatible with the 'book' coal stock heat is dealt with, where necessary, by adjusting the heat of the coal sent to bunkers from stock. This is accounted for as item 29 (b). The difference between item 29 and items 29 (a) and (b) gives the 'Residual Unaccounted Heat' 29 (c), which is an indication of the errors involved in the data used to calculate the STEP factor and losses. The residual unaccounted heat would be zero if all the data were completely accurate. However, in practice the information is often derived from averaged panel instrumentation, so inevitably some discrepancy will occur. Provided the residual unaccounted loss is less than ±0.5% it is not regarded as significant.

At the very bottom of the form the thermal efficiency and STEP factor for the *previous* month are shown, using the invoiced and as-received calorific values, giving a comparison of the commercial and technical performance of the plant.

13.3.3 Use of the losses information

A study of Table 7.59 will quickly reveal the significance of the individual losses one to another. For example, the loss due to turbine heat consumption is by far the highest. Turbine heat consumption is determined regularly (at least every two years), so that the current heat requirements for a given loading regime are known reasonably accurately. In this example the plant has been on very high loads for a considerable time, i.e., on base load, with a few off-load periods for plant repairs for such things as boiler tube failures. It is several years since the cylinders were opened during survey periods, so general wear and tear is causing the high STEP loss. Further investigation is necessary (cylinder efficiencies, etc.) to determine the actual cause; the loss analysis data enables an economic assessment to be made by comparing the cost of rectification with the excess fuel costs which will be incurred if no improvement is carried out.

In the example, the remaining losses are small, and this is due to the effort and attention devoted to containing them. However, the dry flue gas loss, whilst comparatively small, is 0.3%, so is worthy of attention — particularly as the associated excess fuel cost is over £54 000 in just one month.

Some of the losses are negative — in other words they represent better than optimum performance and so result in a cash saving. For example, the unburned carbon loss is −0.11%, a saving of £20 380 in the month compared to target expenditure.

Notice that the cost of the total accounted excess heat loss (i.e., the accounted heat consumption above target) is £432 051 for the month. At this level, over a year this would amount to about £5 million for this one station. Clearly the cost of sustaining losses over and above optimum can be very expensive indeed and there is a powerful incentive to devote resources to identifying and reducing them.

13.4 Merit order heat rate calculations

13.4.1 Background

To enable demand for electricity to be met at minimum cost to the system, it is necessary to have a procedure which will provide up-to-date information on generating plant heat rates. These are then used with other relevant data, such as individual station fuel heat costs, transmission losses, etc., to produce an order of loading based on the cheapest available production costs.

A simplified version of the method is given by the equation:

$$\text{Unit heat rate} \times \text{unit heat cost}$$
$$= \text{unit merit cost (p/kWh)}$$

13.4.2. Merit order heat rates (MOHR)

In this connection, the STEP scheme provides only the plant heat rate data to the System Operation Department, in the manner necessary for them to carry out their function of *Generation Ordering And Loading*, for which they have devised the *GOAL* program. The procedures for determining the appropriate heat rate data are set out fully in Memorandum 0(S)78 Supplement (1976) 'The Assessment of Merit Order Heat Rate Parameters for STEP Stations', but an abbreviated guide to the requirements is given below.

Whilst the Merit Order Heat Rate (MOHR) calculations rely very much on the STEP scheme data, some additional fixed and monthly operational data are required. There are three categories, prefixed *R*, *M* and *D*. The forms are:

Table	Reference	Type of data
7.60	STEP 3E	Registered data, prefixed *R*
7.61	STEP 3F	Latest turbine Willans line data, prefixed *R*
7.62	STEP 3G	Additional station registered data,
7.63	STEP 4E	Monthly operating data, prefixed *M*
7.64	STEP 4F	Monthly operating data, prefixed *M*
7.65	Sheet 1	Standard data for daily calculations, prefixed *D*
7.66	Sheet 2	Daily information, prefixed *D*
7.67	Sheet 3	Monthly form for each Unit

Note: These tables are assembled at the back of the chapter.

The MOHR calculations are carried out each month in conjunction with the STEP factor calculations as part of the STEP suite of programs.

System operation user requirements fall broadly into two categories:

(a) For daily plant scheduling and loading.

(b) For fuel and load allocation studies.

(a) *For daily plant scheduling and loading*, they need values for the unit sent-out hourly fixed heat consumption, the incremental heat rate and the capability available which properly reflect the expected performance of the plant for the period when the data will be used (normally the following day). Each month the computer calculates basic or 'normal' fixed rate and incremental heat rate values which take account of recent operational proficiency, but assume target turbine exhaust pressure and that all feedheating plant plus bled-steam-driven feed pump (where applicable) are fully available. These values are adjusted daily at each station for actual turbine exhaust pressure plus any existing feed train abnormalities (Tables 7.65 and 7.66), and are then reported to the Grid Control Centre, together with the expected capability of the unit for the following day. This information enables System Operation to produce abbreviated plant loading schedules for use by the loading engineers.

(b) *For fuel and load allocation studies*, they need the unit sent-out hourly fixed heat consumptions and incremental heat rates which properly reflect the MCR on-load heat rates over the period of the allocation study. The heat rate data must take account of recent operational proficiency (or expected operational proficiency for longer term studies), with appropriate adjustments to allow for forecast conditions, e.g., CW inlet temperature, but no allowance is made for feed train abnormalities or for the use of start and standby electric feed pumps. These heat rates are passed directly to System Operation.

Merit order ranking schedules produced from these data are distributed at appropriate intervals. Although they do not give the merit heat rates used for actual daily order of loading, they do give a good indication of the daily loading positions. They are used for load allocation studies and may also be used for deriving system costs, including generation replacement and out-of-merit generation costs.

Much of the input data is common to both the requirements in (a) and (b) above, including:

● Latest test turbine heat results.

● Deterioration from latest test to current estimate of turbine performance (where applicable).

● STEP target values for boiler efficiency, make-up and range radiation losses, and off-load heat.

To smooth any marked variations in monthly station operating performance, which are usually for reasons other than variation in technical performance, the factors for operational proficiency are the averages of the last four months' operation, weighted by total heat for generation.

In order to produce heat rate data in the form required by System Operation, i.e., sent-out fixed heat consumption and incremental heat rate for each of the requirements, the hourly heat consumptions and target units supplied at 100% and 60% MCR loads are calculated, which enables the slope of the Willans line (i.e., the incremental heat rate) and, hence, the fixed heat consumption to be determined for each unit.

13.4.3 Notes on completion of Tables 7.65–7.67 for station reporting of daily merit order heat rates

Data Sheet 1 (a) — standard data (Table 7.65)

Data Items (1) and (8) are the normal sent-out values of fixed heat consumption and incremental heat rates. These values are recalculated monthly on the Regional computer, and a copy of the output giving the updated values is sent to the Station. The other data items are the corrections to the normal values to be applied when a specified abnormal condition occurs. These are only revised when more accurate data becomes available.

Note: For simplicity, the size of the changes in the normal values for a particular abnormal condition is assumed to be equal for each set of the same size in a station, although this may not be strictly true in practice. Negative changes to fixed heat consumption values indicate that the effect of the abnormality is less severe at part load than at MCR.

Data sheet 2 (b) — daily predictions for following day (Table 7.66)

Items (20) and (21) are actual readings taken under steady load conditions near the time the form was completed. Item (20) is shown to the nearest MW, and item (21) to the nearest millibar. Items (23) to (28) are predictions of the state of the plant expected to obtain during the following day: only a simple YES/NO entry is required. The conditions listed should each be treated exclusively; for example, if all HP heaters are out of commission on a Unit, then an entry is required for that condition only, and any involving individual HP heaters should be left blank.

Sheet 2 (c) — daily calculations

Items (29) and (37) are the latest updated values from the monthly computer output sent to the Station. The changes for predicted abnormal conditions should be entered only for the appropriate abnormality on the appropriate set, the values being taken from the

standard data on Sheet 1. Only the calculated merit fixed heat consumptions, incremental heat rates and plant capabilities, items (52), (53), and (55) should be telephoned to the Grid Control centre at the time agreed with Grid Control. Item (52) is given to one decimal place and items (53) and (55) to the nearest whole number. Values should also be reported for sets out of commission, based on the expected condition and capability of the set when returned to service.

Monthly form — sheet 3 (Table 7.67)

One of these forms is required for each turbine on a monthly basis.

The totals from this working sheet are used in the completion of the Station Monthly Operating Data return, form STEP 4E, for use in computing the next month's normal incremental heat rate and fixed heat consumption values (items (1) and (8) on sheet 1, (Table 7.65)).

13.5 Conclusions

Even though the information given about STEP losses, etc., is only a guide, the value of the scheme should be apparent. The considerable advantages accruing from the information derived from the scheme are many, but a few of the main ones are:

● Its use in formulating merit-order information.

● Detailed analysis of the thermal performance of the plant.

● Indication of losses in order of importance and cost.

● Help in determining the most cost effective remedial work on the plant.

● Help in preparation of economic appraisals for thermal performance improvement work.

● Indication of best use of resources for thermal performance improvement.

● Comparison of performance of different stations.

Thus, although a lot of preliminary work is required to get this sort of scheme started, there is a considerable benefit to be gained from so doing. As greater sophistication of plant parameter measurement is introduced, so the accuracy of the derived information is improved, resulting in lower unaccounted losses and better accuracy of the split of accounted losses.

The main outstanding problem at some stations is to accurately determine the heat consumed. So long as there are discrepancies between the 'invoiced' and the 'as-received' heat then there must be some doubt about the accuracy of the STEP results, as they depend entirely upon heat calculations. Therefore it is to be hoped that consultation between the CEGB and British Coal will result in some procedure which will eliminate the problem.

The success of the various thermal efficiency improvement schemes within the CEGB may be judged by the following results (source, CEGB Annual Report and Accounts, 1986/87):

System thermal efficiency — all plant	33.81%
System thermal efficiency — fossil plant	35.08%
Annual coal bill	£3437 million

Coal and oil fired stations with the highest thermal efficiency are:

	%
Rugeley B	37.92
Ratcliffe-on-Soar	37.30
Drax	37.09
Cottam	36.37
Ferrybridge C	36.30
Fawley	36.23
West Burton	36.11
Eggborough	36.10
Ironbridge	35.76
Littlebrook D	35.71

Hopefully, the trend of improving system thermal efficiency will continue, resulting in lower comparative costs and, therefore, increasing demand for electricity.

14 Additional references

14.1 Steam cycles

Kearton, W. J.: 'Steam turbine theory and practice': Pitman

Lewitt, E. H.: 'Thermodynamics applied to heat engines'

Sorenson, H. A.: 'Principles of thermodynamics': Holt, Rinehart and Winston

14.2 Fuel

Brame, I. S. S. and King, J. G.: 'Fuel — Solid Liquid and Gaseous': Edward Arnold

BS1017: 'Methods for Sampling Coal and Coke', (2 parts): 1977

Francis, W.: 'Coal — Its Formation and Composition': Edward Arnold

Francis, W.: 'Boiler House and Power Station Chemistry'

Francis W.: 'Fuels and Fuel Technology' (2 volumes): Pergamon

14.3 Boilers

BS2885: 'Code for Acceptance Tests on Stationary Steam Generators of the Power Station Type': 1974

Francis, W.: 'Fuels and Fuel Technology' (2 volumes): Pergamon

Hart, A. B. and Lawn, C. J.: 'Combustion of Coal and Oil in Power Station Boilers': CEGB Research No. 5,: August 1977

Raask, E.: 'Mineral Impurities in Coal Combustion': Hemisphere Publishing: 1985

Thurlow, C. C.: 'The Combustion of Coal in Fluidised Beds': Proc. IMechE 192, No. 15

14.4 Turbines

BS752: 'Test Code for Acceptance of Steam Turbines': 1974

CEGB Site Test Code No. 2: 'Steam Turbine-Generator Heat Rate Tests'

CEGB Report No. FF/79/86: 'Assessment of direct Steam Flow Measurements to Obtain Turbine Heat Rate at Fiddlers Ferry Power Station': 1986

Church, E. F.: 'Steam Turbines': McGraw-Hill

Cotton, K. C. and Schofield, P.: 'Analysis of changes in the performance characteristics of steam turbines': Journal of Engineering for Power: April 1971

Salisbury, J. K.: 'Steam Turbines and Their Cycles': John Wiley

Salisbury, K.: 'A new performance criterion for steam turbine regenerative cycles': Journal of Engineering for Power: October 1959

Salisbury, K.: 'Power performance monitoring': ASME Paper No 60-WA-222

Sochaczewski, Z. W., Clay, C. A. E. and Morris, J. A.: 'Development of a Turbine-Generator Thermal Performance Monitoring System': Proc. IMechE, Vol 195, No 31: 1981

Welch, C. P. and Hoffman, H. J.: 'The application of the deviation concept in turbine cycle monitoring': Journal of Engineering for Power: October 1961

14.5 Condensers

BEAMA; 'Recommended Practice for the Design of Surface Steam Condensing Plant': Publication 222

CEGB Site Test Code No. 3: 'Performance of Surface Type Steam Condensers': 1981

Drummond G.: 'Steam Side Pressure Gradients in Surface Condensers': Proc. IMechE 186, No. 10/72: 1977

Guy, H. L. and Winstanley, E.: 'Some Factors in the Design of Surface Condensing Plant': Proc. IMechE: C 1935

Rowe, M. and Moore, M. J.: 'Improvement in Thermal Performance of Steam Condensers by Attention to Air Removal Systems': CERL: April 1980

Silver, R. S: 'An Approach to a General Theory of Surface Condensers': Proc. IMechE 178, No. 14: 1963/64

'Standards for Steam Surface Condensers': Heat Exchange Institute, New York: 1970

14.6 Feedheaters

Clemmer, A. B. and Lemezie, S.: 'Selection and Design of Closed Feed Water Heaters': Proc ASME: 1965

Haywood, R. W.: 'Generalised Analysis of the Regenerative Cycle': Proc. IMechE, 161: 1949

Weir, C. D.: 'A Generalised Analysis of Regenerative Feed Heating Trains': Proc. IMechE 179, Part 1 No. 15: 1964/65

14.7 Dust-arresting plant

Barrimore, M.: 'Energy and environment': (Sixty-fourth Thomas Hawksley Lecture), Proc. IMechE 192 No 18: 1978

BS893: 'Methods for the Measurement of the concentration of Particulate Materials in Ducts Carrying gases': 1979

Hawksley, Badzioch and Blackett: 'Measurement of Solids in Flue Gases': The British Coal Utilisation Research Association: 1961

Jones, A. R. and Sarjeant, M.: 'Reducing Dust Emissions from Oil-fired Power Stations': CEGB Research p 40–48: July 1986

Richards, D. J. W., Jones, W. S. and Laxton, J. W.: 'Monitoring and Control of Flue Gases': CEGB Research: August 1977

Seriven, R. A. and Howells, G.: 'Stack Emissions and the Environment': CEGB Research p 28–40: August 1977

14.8 Pumps

BS599: 'Methods of Testing Pumps': 1977

BS5316: 'Acceptance Tests for Centrifugal Mixed Flow and Axial Pumps' (2 parts): 1977

CEGB Site Test Code No. 1: 'Boiler Feed Pumps'

CEGB Site Test Code No. 6 'Circulating Water Pumps'

Hilliar, H.: 'Suction Supply Conditions for Pumping Installations': Transactions of Institute of Marine Engineers Vol LXIV No 11: 1952

14.9 STEP scheme

'The Assessment of Station Thermal Efficiency Performance (The STEP Scheme)': Operation Memorandum No 37, 1981

'The Assessment of Merit Order Heat Rate Parameters for STEP Stations': 0(S)78 Supplement, CEGB: 1976

'Station Thermal Performance Losses Analysis': Document MRS8, CEGB Midlands Region: 1981

14.10 General

CEGB Site Test Code No. 4: 'Turbine Governors'

CEGB: 'Steam Tables in SI Units': 1970

Cotton, K. C. and Wescott, J. C.: 'Methods for Measuring Steam Turbine Generator Performance': ASME No 60/Wa/139

Fenton, K.: 'Thermal Efficiency and Power Production': 1966

Gill, A. B.: 'Power Plant Performance': Butterworths: 1984

Hart, A. B.: 'The Chemistry of Power Station Emissions': Chemistry Division, Central Electricity Research Laboratories: 1981

Horsman, H. S.: 'Testing of Boilers and Turbo-alternators in Power Stations of the Central Electricity Authority': Proc. IMechE, Vol 170 No. 18: 1956

Lyle, O.: 'The Efficient Use of Steam': HMSO: 1947

McCrae, R. U.: 'The Influence of the Modern Boiler Feed Pump Turbine on Modern Reheater Steam Cycles': Paper to South African IMechE, Johannesburg: 1965

Sherry, A.: 'Power Station Optimisation': Journal of Institute of Fuel: November 1961

Spiers, H. M.: 'Technical Data on Fuel': British National Committee, World Power Conference: 1962

Waddington, J. and Maples, G. C.: 'The Control of Large Coal and Oil Fired Generating Units': CEGB Research No. 14: February 1983

TABLE 7.44

CEGB STEP registered data — station summary

FORM STEP 3A

Net C.V. Basis

STATION .... CODE .... REVISION DATE ....

R	Station or unit	Site or group No.		Station	1-4	
1	Type of steam supply			Unit		
2	Type of make-up supply			Demin		
3	Type of circulating water supply			Tower (river)		
4	Specified CW inlet temperature	°C		19.0		
	Target on-load works power					
5	– without CW pumps	kW		66840	16710	
6	– with standard CW pump power	kW		88040	22010	
7	– with reduced CW pump power	kW		88040	22010	
8	Target works power for station services	kW		4000		
9	Target works power for fuel and ash handling	kWh/t		5.6		
	Cooling tower pumps (mixed-cooled plant only)					
	– pump numbers (by group)					
10	– pump consumption	kW				
11	Target on-load feed water make-up heat	GJ/h		329.2	82.3	
12	Load dependent auxiliary heat consumption	%		0.1	0.1	
13	Time dependent auxiliary heat consumption	GJ/h		8.0		
14	Steam and feed range radiation loss	GJ/h		25.2	6.3	
	Turbines	Site number			1.3	2.4
15	Specified MCR capacity	MW			500.0	500.0
16	Practical MCR capacity	MW	1948.0		487.0	487.0
17	Practical NER capcity	MW			—	—
18	Practical power factor	lag			0.97	0.97
19	Fixed heat consumption (economic load range)	GJ/h			292.3	288.0
20	Incremental heat rate (economic load range)	kJ/kWh			7778	7663
21	Fixed heat consumption (overload range)	GJ/h			—	—
22	Incremental heat rate (overload range)	kJ/kWh			—	—
23	Heat rate at practical MCR	kJ/kWh			8378	8255
24	Target works electricity per cold start	kWh			35044	35044
25	Target heat consumption per cold start	GJ			292.3	288.0
	Boilers	Site or group No.		1-4		
26	Type of firing			PF		
27	Specified MCR capacity	kg/s		428.0		
28	Practical MCR capacity	kg/s		428.0		
29	Target Work electricity per cold start	kWh		97378		
30	Heat consumption per cold start	GJ		3750		
31	Natural cooling time constant			0.0588		
32	Natural cooling time for 2% residual heat	h		51		
	Registered fuel					
33	Net calorific value	kJ/kg		23340		
34	Moisture content	%		11.38		
35	Ash content	%		16.23		
36	Volatile matter content	%		28.53		
37	Carbon content	%		59.06		
38	Hydrogen content	%		3.93		
39	Sulphur content	%		1.64		

NOTE The load-variable items of registered data are given separately in the graphical relationships R40 to R49 on Forms STEP 3B, 3C and 3D. (Tables 7.45, 7.46, 7.47)

TABLE 7.45

CEGB STEP tabular data — turbine-generators

CEGB STEP TABULAR DATA – TURBO-GENERATORS   FORM STEP 3B

STATION .... CODE .... REVISION DATE...............................

ITEM R40 – TARGET EXHAUST PRESSURE | FOR TURBINE GROUP | SETS 1 TO 4 |

Fc = 0.950

CW inlet Temperature °C	Turbine load – per cent of specified MCR								
	30	40	50	60	70	80	90	100	110
	Target exhaust pressures – m bar								
10	21.86	22.94	24.17	25.64	27.30	29.21	31.40	33.93	36.70
12	24.35	25.50	26.81	28.37	30.14	32.17	34.49	37.16	40.10
14	27.15	28.37	29.78	31.44	33.33	35.49	37.96	40.81	43.94
16	30.28	31.60	33.11	34.89	36.91	39.22	41.87	44.91	48.24
18	33.79	35.20	36.83	38.74	40.91	43.40	46.23	49.49	53.06
20	37.69	39.22	40.97	43.04	45.38	48.05	51.10	54.60	58.44
22	42.03	43.69	45.58	47.82	50.34	53.23	56.51	60.28	64.41
24	46.85	48.65	50.70	53.12	55.85	58.97	62.51	66.59	71.03
26	52.19	54.14	56.37	58.99	61.95	65.33	69.16	73.56	78.37
28	58.10	60.22	62.64	65.48	68.69	72.35	76.51	81.27	86.46
30	64.62	66.93	69.56	72.65	76.13	80.10	84.61	89.77	95.39

ITEM R41 – EXHAUST PRESSURE CORRECTION FACTOR | FOR TURBINE GROUP | SETS 1 TO 4 |

Turbine exhaust pressure m bar	Turbine load – per cent of specified MCR								
	30	40	50	60	70	80	90	100	110
	Heat consumption correction factor for exhaust pressure								
15									
20	0.9734	0.9686	0.9725	0.9727					
25	0.9838	0.9783	0.9800	0.9797	0.9810	0.9828	0.9833		
30	0.9939	0.9886	0.9886	0.9865	0.9868	0.9866	0.9866	0.9868	0.9869
35	1.0038	0.9977	0.9954	0.9934	0.9920	0.9906	0.9899	0.9891	0.9884
40	1.0142	1.0081	1.0032	1.0004	0.9976	0.9949	0.9932	0.9916	0.9900
45	1.0245	1.0170	1.0121	1.0073	1.0034	1.0005	0.9975	0.9947	0.9919
50	1.0349	1.0267	1.0212	1.0157	1.0109	1.0070	1.0031	0.9992	0.9952
55	1.0452	1.0366	1.0303	1.0240	1.0184	1.0137	1.0091	1.0048	1.0005
60	1.0555	1.0469	1.0394	1.0324	1.0260	1.0203	1.0151	1.0105	1.0058
65	1.0659	1.0568	1.0485	1.0411	1.0335	1.0270	1.0212	1.0162	1.0112
70	1.0762	1.0665	1.0576	1.0496	1.0410	1.0336	1.0273	1.0219	1.0165
75			1.0666	1.0558	1.0486	1.0403	1.0333	1.0269	1.0205
80				1.0620	1.0561	1.0469	1.0394	1.0319	1.0245
85					1.0636	1.0536	1.0454	1.0369	1.0285
90							1.0515	1.0420	1.0325
95								1.0470	1.0364
100									

TABLE 7.46

CEGB STEP tabular data — boilers

FORM STEP 3C

CEGB STEP TABULAR DATA - BOILERS

STATION .... CODE .... REVISION DATE ....

DATA FOR BOILER GROUP 1 (BOILERS 1 TO 4)

ITEM R42 – UNBURNT CARBON LOSS

Boiler load – per cent of practical MCR		40	50	60	70	80	90	100	110
Unburnt Carbon loss –	%	0.11	0.14	0.18	0.21	0.30	0.42	0.58	0.79

ITEM R43 – DRY FLUE GAS LOSS FACTOR

Boiler load – per cent of practical MCR		40	50	60	70	80	90	100	110
Dry flue gas loss factor	kJ/kg	2561	2340	2114	1841	1702	1744	1882	2074

ITEM R44 – HEAT LOST PER KILOGRAM MOISTURE IN FLUE GAS

Boiler load – per cent of practical MCR		40	50	60	70	80	90	100	110
Heat/kg moisture in flue gas –	kJ	223	215	208	203	203	212	224	247

ITEM R45 – RADIATION AND OTHER HEAT LOSSES

Boiler load – per cent of practical MCR		40	50	60	70	80	90	100	110
Radiation and other losses –	%	1.32	1.05	0.87	0.75	0.65	0.59	0.53	0.47

TABLE 7.47

CEGB STEP tabular data — miscellaneous

CEGB STEP TABULAR DATA - MISCELLANEOUS

FORM STEP 3D

STATION...CODE...REVISION DATE.......................................

ITEM R46 – ON-LOAD MAKE-UP HEAT LOSS

		PLANT GROUP 1 (Units 1-4)				
Unit (or station) running load – % of MCR capacity		10	15	20	30	40	50
On-load make up heat loss –	GJ/h	82.3	82.3	82.3	82.3	82.3	82.3

Unit (or station) running load – % of MCR capacity		60	70	80	90	100	110
On-load make up heat loss –	GJ/h	82.3	82.3	82.3	82.3	82.3	82.3

ITEM R47 – BOILER OFF-LOAD HEAT LOSS   BOILER GROUP 1 | BOILERS 1 TO 4

Boiler time off-load –	Hours	1	3	5	7	10	15
Boiler off-load heat loss –	GJ	214	607	955	1265	1667	2198

Boiler time off-load –	Hours	25	40	51			
Boiler off-load heat loss –	GJ	2887	3393	3573			

ITEM R48 – ON-LOAD WORKS POWER   PLANT GROUP 1 – UNITS 1-4

Unit (or station) running load – % MCR		10	15	20	30	40	50
On-load works power – standard value –	kW	14525	14755	14965	15385	16085	17170
On-load works power – reduced value –	kW	14525	14755	14965	15385	16085	17170

Unit (or station) running load – % MCR		60	70	80	90	100	110
On-load works power – standard value –	kW	18510	19410	20340	21195	22010	22845
On-load works power – reduced value –	kW	18510	19410	20340	21195	22010	22845

ITEM R49 – BOILER OFF-LOAD WORKS POWER   BOILER GROUP 1 | BOILERS 1 TO 4

Boiler time off-load –	Hours	1	3	5	7	10	15
Boiler off-load works power –	kWh	6735	14000	20420	26860	35630	48800

Boiler time off-load –	Hours	25	40	60	80	120	
Boiler off-load works power –	kWh	69820	87530	97378			

TABLE 7.48

CEGB STEP scheme — monthly operating data sheet 1

STATION		CODE	MONTH	YEAR	STEP
				M	4A

CEGB –
STEP SCHEME – MONTHLY OPERATING DATA

CARD TYPE	M1 SET NUMBER	M2 ELECTRICITY GENERATED TOTAL GWh	M3 ABOVE N.E.R. GWh	M4	M5 ON-LOAD WORKS ELECTRICITY GWh	M6 ELECTRICITY SENT OUT DURING RUNNING HRS. GWh	M7 CW INLET TEMPERATURE °C
10	15 17	23 30	33 40	44 50	54 60	63 70	75 78
A	1	296.290	0.		13.986		16.2
	2	202.080	0.		14.493		17.9
	3	174.670	0.		8.416		18.2
	4	280.430	0.		13.534		16.8
B	STATION	953.470	0.		50.429	903.041	17.1

CARD TYPE	M8	M9 TIME GENERATING TOTAL – h	M10 ABOVE NER – h	M11 NUMBER OF STARTS HOT	M12 COLD	M13 POST – MAINTENANCE TESTING WORKS ELECTRICITY – GWh	M14 TIME – h
10	15 17	23 27	33 37	44 47	54 57	65 70	74 77
C	1	608.	0.	0.	1.	0.	0.
	2	460.	0.	1.	3.	0.	0.
	3	353.	0.	1.	1.	0.	0.
	4	567.	0.	0.	3.	0.	0.
D	STATION	672.		2.	8.	0.	0.

M84 STATION MAXIMUM DEMAND – MWso	M85 EXPORT TO GRID GWh	M86 RUNNING PLANT LOAD FACTOR – %
1975	903.041	98.8

REV. APR. 1981
R2003

603

TABLE 7.49

CEGB STEP scheme — monthly operating sheet data 2

STATION	CODE	MONTH	YEAR	STEP
		M		4B

CEGB –
STEP SCHEME – MONTHLY OPERATING DATA

CARD TYPE	M15 DECLARED NET CAPABILITY MW (AVERAGE)	M16 AVERAGE CAPABILITY AVAILABLE – MW ALL HOURS	M17 WEEKDAYS NAT.PEAK	M18 WORKS ELECTRICITY TOTAL – GWh	M19 ELECTRICITY SUPPLIED GWh	M20 HOURS IN REVIEW PERIOD	M21 AVERAGE LOAD–% OF DECLARED NET CAPABILITY
10	13 17	23 27	33 37	43 50	52 60	63 67	74 78
E	1 8 4 0 ·	1 3 2 1 ·	1 3 3 9 ·	5 0 · 4 2 9	9 0 3 · 0 4 1	6 7 2 ·	7 3 · 0

CARD TYPE	M22 HEAT FOR I.C. SETS GJ	M23 PULVERISED FUEL BURNT – TONNES	M24 TOTAL FUEL BURNT TONNES	M25 HEAT FOR GENERATION GJ	M26 STATION HEAT RATE (SUPPLIED) kJ/kWh	M27 ACTUAL THERMAL EFFICIENCY – %	M28 STEP FACTOR %
10	14 20	23 30	33 40	41 50	52 57	65 69	74 79
F	0 ·	3 7 7 8 8 7 ·	3 7 8 9 3 7 ·	8 8 7 4 9 6 9 ·	9 8 2 8 ·	3 6 · 6 3	9 7 · 6 0

CARD TYPE	M29 COAL AND COKE	M30 OTHER SOLID FUEL	M31 MAIN OIL	M32 GAS	M33	M34 LIGHTING-UP AND FLAME STABILISING OIL	M35 TOTAL FUEL BURNT
10	12 20	23 30	33 40	43 50	53 60	63 70 71	80
T	3 7 7 8 8 7 ·	·	1 8 6 ·	·	·	8 6 4 ·	3 7 8 9 3 7 ·
H	8 8 3 2 6 5 4 ·	·	7 4 9 6 ·	·	·	3 4 8 1 9 ·	8 8 7 4 9 6 9 ·
1	kJ/kg 2 3 3 7 4 ·	kJ/kg ·	kJ/kg 4 0 3 0 0 ·	kJ/kg ·	kJ/kg ·	kJ/kg 4 0 3 0 0 ·	kJ/kg 2 3 4 2 1 ·
2	·	·	·	·	·		

CARD TYPE	M36 NON-INDUSTRIAL STAFF	M37 INDUSTRIAL STAFF ON OPERATION	M38 INDUSTRIAL STAFF ON R&M	M39 TOTAL STAFF	M61 ELECTRICITY FOR TESTING & C.T. PUMPS GWh	M62 AVERAGE CAPABILITY AVAILABILITY - MW ALL HOURS	M63 WEEKDAYS NAT PEAK
10	14 17	24 27	34 37	43 47	54 60	63 67	73 77
W	1 0 8 ·	1 9 3 ·	2 9 3 ·	5 9 4 ·	0 · 0	1 3 1 1 ·	1 3 3 9 ·

PRIMARY FUEL HEAT RATIO (MIXED-FIRED PLANT ONLY)

CARD TYPE	M64/S STATION	M64/1 UNIT OR BOILER GROUP No. 1	M64/2 UNIT OR BOILER GROUP No. 2	M64/3 UNIT OR BOILER GROUP No. 3	M64/4 UNIT OR BOILER GROUP No. 4	M64/5 UNIT OR BOILER GROUP No. 5	M64/6 UNIT OR BOILER GROUP No. 6
10	16 20	26 30	36 40	46 50	56 60	66 70	76 80
V	·	·	·	·	·	·	·

REV. APR. 1981
R2004

T<small>ABLE</small> 7.50

CEGB STEP scheme — monthly operating data sheet 3

STATION	CODE	MONTH	YEAR	STEP
		M		4C

CEGB –
STEP SCHEME – MONTHLY OPERATING DATA

CARD TYPE	M40 BOILER (OR GROUP) NUMBER	M41 TIME STEAMED h	M42 STEAM RAISED TONNES	M43 NUMBER OF OFF-LOAD PERIODS CLASSIFIED BY HOURS IN PERIOD 0-10h	M44 11-30h	M45 31-(TIME R32) h	M46 No. OF TIMES LIT UP FOR MAINTAINING RANGE PRESSURE
10	15 · 17	23 · 27	32 · 40	44 · 47	54 · 57	64 · 67	74 · 77
G	1 ·	608 ·	907744 ·	0 ·	0 ·	0 ·	0 ·
	2 ·	460 ·	616860 ·	1 ·	0 ·	0 ·	0 ·
	3 ·	353 ·	534795 ·	1 ·	0 ·	0 ·	0 ·
	4 ·	567 ·	859005 ·	0 ·	0 ·	0 ·	0 ·

CARD TYPE	M47 BOILER (OR GROUP) NUMBER	M48 TIME ON HOT STANDBY h	M49 NUMBER OF COLD STARTS	M50 TOTAL TIME OFF LOAD CLASSIFIED BY HOURS IN PERIOD 0-10 h	M51 11-30 h	M52 31-(TIME R32) h	M53 TIME ON TEST h
10	15 · 17	23 · 27	33 · 37	43 · 47	53 · 57	63 · 67	73 · 77
K	1 ·	0 ·	1 ·	0 ·	0 ·	0 ·	0 ·
	2 ·	0 ·	3 ·	1 ·	0 ·	0 ·	0 ·
	3 ·	0 ·	1 ·	6 ·	0 ·	0 ·	0 ·
	4 ·	0 ·	3 ·	0 ·	0 ·	0 ·	0 ·

FIRING CARD TYPE	M54 CALORIFIC VALUE kJ/kg	M55 MOISTURE CONTENT %	M56 ASH CONTENT %	M57 VOLATILE CONTENT %	M58 CARBON CONTENT %	M59 HYDROGEN CONTENT %	M60 SULPHUR CONTENT %
10	12 17	25 29	35 39	45 49	55 59	65 69	75 79
S	·	·	·	·	·	·	·
P	23374 ·	11·55	16·34	26·98	59·63	3·92	1·6
O	·	·	·	·	·	·	·
M	·	·	·	·	·	·	·
Z	·	·	·	·	·	·	·

TABLE 7.51

CEGB STEP factor calculation key

(a) Unit plant

Item codes: C = Calculation, R = Registered data, M = Monthly data

Item No.	Item	Key
C	TURBINE GENERATORS	
1	Fixed heat consumption	R19 (M9C-M10C) + (R21 × M10C)
2	Incremental heat consumption	R20 (M2A − M3A) + (R22 × M3A)
3	Basic turbine heat consumption	C1 + C2
4	Average turbine running load	$\dfrac{M2A \times 10^5}{R15 \times M9C}$
5	Target turbine exhaust pressure	Read graph R40 at C4 and M7A
6	Exhaust pressure correction factor	Read graph R41 at C4 and C5
7	Target turbine heat consumption	C3 × C6
	BOILERS	
8	Average boiler running load	$\dfrac{M42 \times 27.8}{R28 \times M41}$
9	Target unburnt carbon loss (B_u)	* Read standard value $B_{u(t)}$ from graph R42 at C8: $\oslash$ then, Bu = $B_{u(t)} \times \dfrac{M56 \times R33}{R35 \times M54}$ + correction X where X is used for low volatile fuel only (i.e. R36 or M57 < 17%) then: X = 0.013 $\left[\dfrac{M56 \times R33}{R35 \times M54}\right]\left[\exp\ (0.225\ \dfrac{M58}{M59}) - \exp\ (0.225\ \dfrac{R37}{R38}\right]$
10	Target dry flue gas loss (B_d)	* Read factor K from graph R43 at C8: $\oslash$ then, B_d = K $\left[\dfrac{M58 + (0.375 \times M60)}{M54} - \dfrac{C9}{33820}\right]$
11	Target moisture in flue gas loss	$\oslash$ $\dfrac{M55 + (9 \times M59)}{M54}$ × (value from graph R44 at C8)
12	Target radiation and other losses	Read graph R45 at C8
13	Target boiler efficiency	100 − C9 − C10 − C11 − C12
	UNITS	
14	Unit running load	$\dfrac{M2A \times 10^5}{R16 \times M9C}$
15	Target on-load make-up heat consumption	* M9C × (value from graph R46 at C14)
16	Steam and feed range radiation loss	R14 × M9C
17	Target on-load main plant heat consumption	(C7 + C15 + C16) × $\dfrac{10^2}{C13}$
18	Target auxiliary plant heat consumption	* R12 × C17 × 10^{-2} + (R13 × M20)
	OFF-LOAD HEAT ALLOWANCES	
19	Boiler cold starts	R30 × M49
20	Boiler hot starts	(M43 x H_1) + (M44 x H_2) + (M45 x H_3) where H_1, H_2 and H_3 are read from graph R47 at values: $\dfrac{M50}{M43}$, $\dfrac{M51}{M44}$ and $\dfrac{M52}{M45}$
21	Boiler standby and testing	(M48 + M53) x (value from graph R47 at 1 hour)
22	Turbine starts and testing	1.13 |R25 (M12C + ½M11C) + (R19 × M14C)|
23	Total off-load heat allowance	C19 + C20 + C21 + C22 |Station value = (C19 + C20 + C21 + C22)|
	ELECTRICITY USED ON WORKS	
24	Target on-load works electricity	* |w_s − 0.1 (R4 − M7A) (w_s − w_r)| × M9C × 10^{-6} where w_s and w_r are read from graph R48 at C14, note: when M7A > R4, put (R4 − M7A) = 0
25	Off-load works electricity − boiler and cold starts	* R29 × M49 × 10^{-6}
26	Off-load works electricity − boiler hot starts, standby and testing	* |(M43 × W_1) + (M44 × W_2) + (M45 × W_3) + (M48 + M53) W_4| × 10^{-6} where W_1, W_2, W_3 and W_4 are read from graph R49 at values: $\dfrac{M50, M51, M52}{M43\ M44\ M45}$ and 1 hour
27	Off-load works electricity − turbine starts and testing	R24 (½M11C + M12C) × 10^{-6} + M13C
28	Target station services works electricity	$\oslash$ R8 × M20 × 10^{-6} + R9 x M35T x 10^{-6} + M61 $\overleftarrow{\qquad}$ W$_{as}$ $\overrightarrow{\qquad}$$\overleftarrow{\qquad}$ W$_{af}$ $\overrightarrow{\qquad}$ C28 components: C28a (on-load) = $W_{as} \left(\dfrac{M9D}{M20E}\right)$ + M61 C28b (off-load) = $W_{as} \dfrac{M20 - M9D}{M20}$ C28c (fuel handling) = W_{af}
29	Target station works electricity consumption	C24 + C25 + C26 + C27 + C28
	STEP FACTOR	
30	Target station heat consumption	C17 + C18 + C23 + M22
31	Target station heat rate supplied	$\dfrac{C30}{M2B - C29}$
32	Actual station heat rate supplied	$\dfrac{M35H}{M19E}$
33	STEP factor	$\dfrac{C31 \times 10^2}{C32}$

Footnote: For mixed fired plant (i.e. that with duplicate data), items marked * require weighting by heat mixture and items marked $\oslash$ require weighting by weight mixture; see notes on adjacent page.

TABLE 7.52

CEGB STEP factor calculation summary

STATION		PERIOD		M	

C			dec	Individual unit, turbine, boiler, boiler group values				
	Turbine - generators	No.		STATION	1	2	3	4
1	Fixed heat consumption	GJ	0		177718	132480	103182	163296
2	Incremental heat consumption	GJ	0		2304544	1548539	1358583	2148935
3	Basic turbine heat consumption	GJ	0	7937277	2482262	1681019	1461765	2312231
4	Average turbine running load	%	1		97.5	87.9	99.0	98.9
5	Target turbine exhaust pressure	m bar	2		44.59	45.41	49.66	46.40
6	Exhaust pressure correction factor		4		0.9951	0.9986	0.9993	0.9963
7	Target turbine heat consumption	GJ	0	7913183	2470099	1678666	1460742	2303676
	Boilers and boiler groups	No.						
8	Average boiler (or group) running load	%	1		97.0	87.1	98.4	98.4
9	Target unburnt carbon loss	%	2		0.53	0.39	0.56	0.56
10	Target dry flue gas loss	%	2		4.71	4.44	4.76	4.76
11	Target moisture in flue gas loss	%	2		0.44	0.42	0.44	0.44
12	Target radiation and other losses	%	2		0.55	0.61	0.54	0.54
13	Target boiler efficiency	%	2	93.81	93.76	94.14	93.70	93.70
	Unit or station	No.						
14	Unit (or station) running load	%	1		100.1	90.2	101.6	101.6
15	Target on-load make-up heat consumption	GJ	0	163612	50038	37858	29052	46664
16	Steam and feed range radiation heat loss	GJ	0	12524	3830	2898	2224	3572
17	Target on-load main plant heat consumption	GJ		8622938	2691857	1826391	1592403	2512287
18	Target auxiliary plant heat consumption	GJ	0					
	Off-load heat allowances			13999	–	–	–	–
19	– Boiler cold starts	GJ	0		3750	11250	3750	11250
20	– Boiler hot starts	GJ	0		0	214	1115	0
21	– Boiler standby and testing	GJ	0		0	0	0	0
22	– Turbine starts and testing	GJ	0		330	1139	495	976
23	Total off-load heat allowance	GJ	0	34269	4080	12603	5360	12226
24	Target unit (or station) on-load works electricity	GWh	3	43.512	13.385	9.757	7.817	12.553
	Target unit (or station) off-load works electricity							
25	– Boiler cold starts	GWh	3	0.778	0.097	0.292	0.097	0.292
26	– Boiler hot starts, standby and testing	GWh	3	0.031	0.000	0.007	0.024	0.000
27	– Turbine starts and testing	GWh	3	0.316	0.035	0.123	0.053	0.105
28	Target station services works electricity	GWh	3	4.810				
29	Target station works electricity consumption	GWh	3	49.448				
30	Target station heat consumption	GJ	0	8671206				
31	Target station heat rate (supplied)	kJ/kWh	0	9592				
32	Actual station heat rate (supplied)	kJ/kWh	0	9828				
33	Station thermal efficiency performance factor	%	2	97.60				

TABLE 7.53

CEGB STEP factor — additional registered data

(a)

POWER STATION

Latest Test Turbine Willans Line Data		Unit 1	Unit 2	Unit 3	Unit 4
R128	Fixed Heat Consumption, GJ/h	292.3	288.0	284.7	287.6
R129	Incremental Heat Rate, kJ/kWh	7778	7663	7802	7864

RL	Description	Derivation/Constant
201	TSV temperature correction	Table 7.53(b)
202	TSV pressure correction	Table
203	Reheat temperature correction	Table
204	Reheat pressure drop correction	Table
205	Specified Final Water temperature	Table
206	Final Feed Water Temperature correction	Table
210	Target CO_2 at Air Heater Outlet	Table 7.53(c)
211	Target Gas Temperature Rise	Table 7.53(d)
212	Standard CW Flow	15.58M³/s
213	Target Sootblowing Heat Loss – All Units	24.97 GJ/h
214	Target Blowdown Heat Loss – All Units	29.52 GJ/h
215	Target Leakage Heat Loss – All Units	27.85 GJ/h
216	Target Sootblowing Fraction of Make-up	0.303
217	Target Blowdown Fraction of Make-up	0.359
218	Target Leakage Fraction of Make-up	0.338
219	Target Leakage Rate	15.264 Te/h
220	Target Sootblowing Rate	8.172 Te/h
221	Target Blowdown Rate	18.288 Te/h
222	Ratio PF Ash/Furnace Bottom Ash	4.0
223	STEP Turbine Generator Heat Rate Correction for Superheater Sprays	1.0068
224	STEP Turbine Generator Heat Rate Correction for Reheat Sprays	1.0030
225	% Turbine Generator Heat Rate Correction for 1% Superheat Spray	0.054
226	% Turbine Generator Heat Rate Correction for 1% Reheat Spray	0.110

(b)

RL		
201	TSV steam temperature	
	for t < 565°C	$f = 1 + |(565 - t) \times 0.000245|$
	for t > 565°C	$f = 1 - |(t - 565) \times 0.000175|$
202	TSV steam pressure	
	f = 0.00007 per 3.45 bars	
203	Reheat steam temperature	
	for t < 565°C	$f = 1 + |(565 - t) \times 0.000345|$
	for t > 565°C	$f = 1 - |(t - 565) \times 0.000267|$
204	Reheat steam pressure drop	
	for % pressure drop < 7.3%	$f = 1 - |(7.3 - p) \times 0.000831|$
	for % pressure drop < 7.3%	$f = 1 + |(p - 7.3) \times 0.00104|$
205	Specified final feedwater temperature (t_s)	
	For running load (L) 80% to 100% MCR	
	$t_s = 252°C - |(100 - L) \times 0.65°C|$	
206	Final feedwater temperature correction (f)	
	t_s = specified	t = actual
	for t_s < t	$f = 1 - |(t - t_s) \times 0.000284|$
	for t_s > t	$f = 1 + |(t_s - t) \times 0.000284|$

TABLE 7.53 *(cont'd)*
CEGB STEP factor — additional registered data

(c)

RL210 TARGET CO$_2$% IN FLUE GAS

% Boiler Load C8	%CO$_2$
40	10.3
45	10.9
50	11.5
55	12.0
60	12.5
61	12.6
62	12.7
63	12.8
64	12.9
65	12.9
66	13.0
67	13.1
68	13.2
69	13.2
70	13.3
71	13.4
72	13.4
73	13.5
74	13.6
75	13.6
76	13.7
77	13.7
78	13.7
79	13.8
80	13.8
81	13.9
82	13.9
83	13.9
84	13.9
85	14.0
86	14.0
87	14.0
88	14.0
89	14.0
90	14.0
91	14.0
92	14.0
93	14.0
94	14.0
95	13.9
96	13.9
97	13.9
98	13.9
99	13.9
100	13.8

(d)

RL 211 TARGET GAS TEMPERATURE RISE °C

% Boiler Load C8	°C
40	101.4
45	99.8
50	98.6
55	97.4
60	96.4
61	96.2
62	96.1
63	95.8
64	95.7
65	95.6
66	95.6
67	95.6
68	95.5
69	95.6
70	95.6
71	95.6
72	95.6
73	95.7
74	95.7
75	95.8
76	95.9
77	96.1
78	96.3
79	96.5
80	96.8
81	97.2
82	97.4
83	97.8
84	98.2
85	98.6
86	99.1
87	99.7
88	100.2
89	100.6
90	101.3
91	101.9
92	102.5
93	103.3
94	103.9
95	104.6
96	105.4
97	106.1
98	106.9
99	107.7
100	108.3
101	109.1
102	109.9
103	110.6

TABLE 7.54

Monthly heat balance and loss analysis — additional monthly operating data

STATION...

PERIOD....................M..................

Item No. ML	Description	Derivation	D.P.	Units	Station Values	Unit No. 1	Unit No. 2	Unit No. 3	Unit No. 4
201	Average turbine load	C4	1	%	–	97.5	87.9	99.0	98.9
202	TSV Steam Temperature		0	°C	566	567	567	567	565
203	TSV Steam Pressure		1	bar	157.3	164.0	145.0	159.0	158.0
204	Turbine Reheat Steam Temperature		0	°C	567	570	568	568	564
205	Turbine Reheat Pressure Drop %		2	%	7.30	7.30	7.30	7.30	7.30
206	Target Exhaust Pressure	C5	2	mbar	46.22	44.59	45.41	49.66	46.40
207	Actual Exhaust Pressure	M106	2	mbar	47.35	45.44	51.17	51.97	43.73
208	Specified Final Feed Water Temperature		0	°C	250	251	245	251	251
209	Actual Final Feed Water Temperature (While all HP heaters in service)		0	°C	246	239	242	254	252
210 +	Actual Superheater Spray quantity		0	kg/s	–	53	47	53	53
211	Actual reheat spray quantity		0	kg/s	–	11	10	11	11
212	Actual CW outlet temperature		1	°C	–	25.6	26.2	28.1	26.5
213	Time bled steam air heater in service		0	h	–	–	–	–	–
214	Steam used for sootblowing		0	tonne	16247	4969	3759	2885	4634
215	Blowdown quantity		0	tonne	36356	11119	8412	6456	10369
216	Water for boiler filling		0	tonne	–	–	–	–	–
217	Actual make-up quantity (total)		0	tonne	84292	26074	16661	17674	23883
218	Heat Content/kg of sootblowing steam		0	kJ/kg	–	3120	3120	3120	3120
219	Heat Content/kg of blowdown		0	kJ/kg	–	1677	1677	1677	1677
220	Raw water temperature		1	°C	15.0	–	–	–	–
221	Fraction of make-up through evaporators		2	–	–	–	–	–	–
222	Fraction of leakage as steam		2	–	–	0.33	0.33	0.33	0.33
223	Heat loss/kg of leakage as steam	Hs – (4.2 × ML 220)	0	kJ/kg	–	3417	3417	3417	3417
224	Heat loss/kg of leakage as water	hw – (4.2 × ML 220)	0	kJ/kg	–	1031	1031	1031	1031
225	Steam and feed range insulation factor		3	–	–	1.000	1.000	1.000	1.000
226	Heat supplied to auxiliary boilers		0	GJ	5376	–	–	–	–
227	O$_2$ at air heater outlet		1	%	–	–	–	–	–
228	CO$_2$ at air heater outlet		1	%	13.3	12.7	12.8	13.6	14.2
229	CO at air heater outlet		4	%	0.0060	0.0060	0.0060	0.0060	0.0060
230	Final flue gas temperature at air heater outlet		0	°C	134	137	128	133	135
231	Air temperature at bled steam air heater outlet		0	°C	–	–	–	–	–
232	Air temperature at FD fan inlet		0	°C	33	33	33	36	32
233	Carbon in fly ash		1	%	1.8	2.2	2.2	1.4	1.4
234	Carbon in furnace bottom ash		1	%	0.9	1.1	1.1	0.7	0.7
235	Total duration of feed heater outages (excluding main BFPT outages, etc.)	M103/1-6 as req.	0	hrs	12 –	0	0	0	12
236	Actual heat for boiler cold starts		0	GJ	30000	3750	11250	3750	11250
237	Actual heat for boiler hot starts		0	GJ	1329	0	214	1115	0
238	Actual heat for boiler standby and testing		0	GJ	0	0	0	0	0
239	Weight of fuel burnt per unit		0	tonne	–	–	–	–	–
240	Proportion of (make-up) water recovered		2	–	1.00	1.00	1.00	1.00	1.00
241	Sensible heat in distillate leaving evaporator		0	kJ/kg	–	–	–	–	–
242	Cost of heat in fuel		5	£/GJ	2.0248	–	–	–	–

+ Only required where spray is taken upstream of the final feedheaters.

TABLE 7.55

Monthly heat balance and loss analysis — target heat balance

MONTHLY HEAT BALANCE AND LOSS ANALYSIS: – TARGET HEAT BALANCE

Item No. CT	Description	Derivation	d.p.	Units	Station	No. Unit 1	No. Unit 2	No. Unit 3	No. Unit 4
	STATION TARGET HEAT BALANCE				STATION..			PERIOD	M
301	Basic turbine heat consumption	C3	0	GJ	7937277	2482262	1681019	1461765	2312231
302	Target exhaust pressure heat allowance	C3 (C6-1)	0	GJ	– 24095	- 12163	– 2353	- 1023	– 8555
303	Target heat for sootblowing	C15 × RL216	0	GJ	49574	15162	11471	8803	14139
304	Target heat for blowdown	C15 × RL217	0	GJ	58737	17964	13591	10430	16752
305	Target heat for leakage	C15 × RL218	0	GJ	55301	16913	12796	9820	15772
306	Target heat for evaporator inefficiency	C15 – (CT303 + CT304 + CT305)	0	GJ		—	—	—	—
307									
308	Target sundry heat	C15 + C16	0	GJ	176136	53868	40756	31276	50236
309	Target boiler output heat	C7 + CT 308	0	GJ	8089318	2523967	1719422	1492018	2353912
310	Target unburnt carbon heat loss	$C9 \times C17 \times 10^{-2}$	0	GJ	44376	14267	7123	8917	14069
311 +	Target dry flue gas heat loss	$C10 \times C17 \times 10^{-2}$	0	GJ	403531	127056	81092	75798	119585
312	Target sensible heat in vapour loss	$C11 \times C17 \times 10^{-2}$	0	GJ	37576	11844	7671	7007	11054
313	Target radiation and other heat loss	$C12 \times C17 \times 10^{-2}$	0	GJ	48112	14805	11141	8599	13566
314	Target total boiler heat loss	$\sum$ (CT310 to CT313)	0	GJ	533594	167972	107027	100321	158274
315	Target heat consumption – steam auxiliaries	$R12 \times \sum C17 \times 10^{-2}$ (Station value only)	0	GJ	8623	—	—	—	—
316	Target heat consumption – auxiliary boilers	R13 × M20 (Station value only)	0	GJ	5376	—	—	—	—
317	Target heat consumption – non-steam sets	M22 (Station value only)	0	GJ	0	—	—	—	—
318	Target heat for boiler cold starts	C19	0	GJ	30000	3750	11250	3750	11250
319	Target heat for boiler hot starts	C20	0	GJ	1329	0	214	1115	0
320	Target heat for boiler standby and testing	C21	0	GJ	0	0	0	0	0
321	Target heat for turbine starts and testing	C22	0	GJ	2940	330	1139	495	976
322	Total target off-load heat	C23	0	GJ	34269	4080	12603	5360	12226
323	Total target accounted heat	CT309 + CT314 + CT322	0	GJ	8657181	2696019	1839051	1597699	2524412

Notes:
+ If $\sum$ (C9 to C13 $\neq$ 100.00% adjust item C10 by ± 0.01% as appropriate

T<small>ABLE</small> 7.56

Monthly heat balance and loss analysis — actual heat balance (preliminary calculations)

STATION ACTUAL HEAT BALANCE (1) – PRELIMINARY CALCULATIONS

STATION................   PERIOD................M........

Item No. CA	Description	Derivation	DP	Units	Station Values	Unit No. 1	Unit No. 2	Unit No. 3	Unit No. 4
351	(a) Bled Steam Air Heater (BSAH) – Willington 'B' only Heat to BSAH/kg fuel burnt	$\dfrac{305\,(ML231 - ML232)}{21 - ML227}$	0	kJ/kg	–		–	–	–
352	Total heat to BSAH	$CA351 \times ML239 \times 10^{-3}$	0	GJ	–		–	–	–
353	BSAH heat supplied by Turbine/hour	$CA352 \div M9C$	2	GJ/h	–		–	–	–
354	Specified BSAH heat from turbine	RL208 at C4	2	GJ/h	–		–	–	–
355	Excess BSAH heat from turbine	CA353 – CA354	2	GJ/h	–		–	–	–
356	Correction for excess BSAH duty	From RL207 at CA355	4	–	–	–	–	–	–
357	(b) Superheat and Reheat Spray – where applicable Heat rate correction for ML210	$\dfrac{RL225 \times ML210 \times M41G}{M42G \times 0.278}$	4	–	–	0.0069	0.0068	0.0068	0.0068
358	Heat rate correction for ML211	$\dfrac{RL226 \times ML211 \times M41G}{M42G \times 0.278}$				0.0029	0.0030	0.0029	0.0029
359	Excess superheat spray correction factor	$(1 + CA357) \div RL223$	4	–	–	1.0001	1.0000	1.0000	1.0000
360	Excess reheat spray correction factor	$(1 + CA358) \div RL224$	4	–	–	0.9999	1.0000	0.9999	0.9999
361	(c) Exhaust Pressure loss due to CW flow variation Actual CW temperature rise	ML212 – M7A	1	°C	–	9.4	8.3	9.9	9.7
362	Specified CW temperature rise	R40 at load ML201	1	°C	–	9.8	9.0	9.9	9.9
363	CW flow ratio	CA362 ÷ CA361	4	–	–	1.0407	1.0844	1.0007	1.0205
364	CW flow correction factor	$0.2774 + 0.7228\,\dfrac{CA362}{CA363}$	4	–	–	0.9719	0.9440	0.9997	0.9857
365 *	Saturation temp. for target exhaust pressure $= 9.44\,(\log_e ML206) + 1.07\,(\log_e ML206)^2 - 20.4$	See description	1	°C	–	30.9	31.2	32.8	31.6
366	Saturation temp. corrected for CW flow	$CA364\,(CA365 - M7A) + M7A$	1	°C	–	30.5	30.5	32.8	31.4
367	Exhaust pressure corresponding to corrected saturation temperature $= 3.3864\,\exp\left[0.6 + 0.71\left(\dfrac{CA366}{10} - 0.0224\left(\dfrac{CA366}{10}\right)^2\right)\right]$	See description	2	mbar	–	43.60	43.57	49.72	45.90
368	Turbine heat rate correction factor for CW flow corrected exhaust pressure	R41 at load ML201 and exhaust pressure CA367	4	–	–	0.9944	0.9968	0.9993	0.9958

* If a programmable calculator is not available, these may be read from steam tables.

TABLE 7.57

Monthly heat balance and loss analysis — actual heat balance (main calculations) sheet 1

STATION ACTUAL HEAT BALANCE (2) – MAIN CALCULATIONS

STATION / PERIOD / M...

Item No. CA	Description	Derivation	DP	Units	Station	Unit No. 1	Unit No. 2	Unit No. 3	Unit No. 4
371	Turbine and Feed Train Losses Standard turbine heat consumption	R128 × M9C + R129 × M2A (= C201)	0	GJ	7994926	2482262	1681019	1463274	2368371
372	Turbine deterioration (acc. to latest test)	CA371 - C3	0	GJ	57649	0	0	1509	56140
373	Turbine deterioration (latest Test to current)	CA371 (M107 - 1)	0	GJ	0	0	0	0	0
374	Correction for TSV temperature	RL201 at ML202	4	–	–	0.9996	0.9996	0.9996	1.0000
375	Correction for TSV pressure	RL202 at ML203	4	–	–	0	0	0	0
376	Correction for reheat temperature	RL203 at ML204	4	–	–	0.9985	0.9991	0.9991	1.0003
377	Correction for reheat pressure drop	RL204 at ML205	4	–	–	1.0000	1.0000	1.0000	1.0000
378									
379	Correction for final feedwater temperature	RL206 at (ML209 - ML208)	4	–	–	1.0034	1.0009	0.9992	0.9998
380									
381	Correction for actual exhaust pressure	R41 at ML207 & ML201	4	–	–	0.9958	1.0054	1.0018	0.9942
382	Turbine heat consumption – current est.	CA371 + CA373	0	GJ	7994926	2482262	1681019	1463274	2368371
383	Loss/gain due to TSV temperature	CA382 (CA374-1)	0	GJ	-2250	-993	-672	-585	0
384	Loss/gain due to TSV pressure	CA382 (CA375-1)	0	GJ	0	0	0	0	0
385	Loss/gain due to reheat temperature	CA382 (CA376-1)	0	GJ	-5842	-3723	-1513	-1317	711
386	Loss/gain due to reheat pressure drop	CA382 (CA377-1)	0	GJ	0	0	0	0	0
387	Loss/gain due to final feedwater temperature	$\dfrac{\text{CA382 (M9C-ML235) (CA379-1)}}{\text{M9C}}$	0	GJ	8318	8440	1513	-1171	-464
388	Loss/gain due to HP heater outages	CA382 (CA356-1)	0	GJ	531	0	0	0	531
389	Loss/gain due to BSAH duty	CA382 (CA381-1)	0	GJ	–	–	–	–	–
390	Loss/gain due to actual exhaust pressure	CA382 + $\sum$(CA383 to CA390)	0	GJ	-12451	-10426	9078	2634	-13767
391	Actual turbine heat consumption	CA382 (C6-1)	0	GJ	7983232	2475560	1689425	1462835	2355412
392	Loss/gain due to CW temperature	CA382 (CA368-C6)	0	GJ	-24303	-12163	-2353	-1024	-8763
393	Loss/gain due to CW flow	CA390 - CA392 - CA393	0	GJ	-5948	-1758	-3026	0	-1184
394	Loss/gain due to condenser fouling etc.	C16 × ML225	0	GJ	17800	3475	14457	3658	-3790
395	Sundry Losses Steam and feed range loss	ML236	0	GJ	12524	3830	2898	2224	3572
396	Actual heat for boiler cold starts	ML237	0	GJ	30000	3750	11250	3750	11250
397	Actual heat for boiler hot starts	ML238	0	GJ	1329	0	214	1115	0
398	Actual heat for boiler standby and testing	$\sum$(CA396 to CA398)	0	GJ	0	0	0	0	0
399	Total boiler off-load heat consumption	CA399 × 0.25 (1-ML240)	0	GJ	31329	3750	11464	4865	11250
400	Water loss to boiler starts, etc.	ML217-ML214-ML215-ML216-CA400	0	tonne	0	0	0	0	0
401	Actual leakage quantity	CA401 (ML224 + ML222 (ML223-ML224)) $/ 10^3$	0	tonne	31689	9986	4490	8333	8880
402	Heat loss to leakage	ML214 (ML218-4.2 ML220) $\times 10^{-3}$	0	GJ	57623	18158	8165	15153	16147
403	Heat loss to sootblowing	ML215 (ML219-4.2 ML220) $\times 10^{-3}$	0	GJ	49666	15190	11491	8819	14166
404	Heat loss to blowdown	ML221 × 2.5 (ML241-4.2 ML220) $\times 10^{-4}$	0	GJ	58679	17946	13577	10420	16736
405	Heat loss to evaporator inefficiency	CA395 + $\sum$(CA402 to CA405)	0	GJ	0	0	0	0	0
406	Total sundry losses	CA391 + CA406	0	GJ	178492	55142	36131	36616	50621
407	Heat supplied from main boiler		0	GJ	8161724	2530684	1725556	1499451	2406033

TABLE 7.58

Monthly heat balance and loss analysis — actual heat balance (main calculations) sheet 2

STATION ACTUAL HEAT BALANCE (3) – MAIN CALCULATIONS

Item No.CA	Description	Derivation	D.P.	Units	Station	Unit No.1	Unit No.2	Unit No.3	Unit No.4
	Boiler losses								
408	Unburnt carbon/kg of fuel burnt	$\frac{M56(RL222 \times ML233 + ML234)}{100[100(RL222+1) - (RL222 \times ML233 + ML234)]}$	4	kg/kg	—	0.0033	0.0033	0.0021	0.0021
409	Unburnt carbon loss/gain	$(CA408 \times 33820) \div M54$	5	kJ/kJ	0.00394	0.00478	0.00478	0.00302	0.00302
410	Weight of flue gas/kg of carbon burnt	$255 \div (ML228 + ML229)$	2	kg/kg	—	20.07	19.91	18.74	17.95
411	Equivalent carbon burnt	$\frac{M58}{100} + \frac{M60}{267} - CA408$	4	kg/kg	—	0.5990	0.5990	0.6002	0.6002
412	Dry flue gas loss/gain	$\frac{CA410 \times CA411 (ML230 - ML232)}{M54}$	5	kJ/kJ	0.04941	0.05349	0.04848	0.04668	0.04748
413	Unburnt gas loss/gain	$\frac{23800 \times ML229 \times CA411}{(ML228 + ML229) M54}$	5	kJ/kJ	0.00028	0.00029	0.00029	0.00027	0.00026
414	Vapour heat detriment	$1.88ML230 + 58 - 4.2ML232$	1	kJ/kg	—	177.0	160.0	156.8	177.4
415	Sensible heat in vapour loss/gain	$\frac{CA414 (M55 + 9 \times M59)}{M54 \times 100}$	5	kJ/kJ	0.00340	0.00355	0.00321	0.00314	0.00355
416	Boiler effy. losses less radiation, etc.	$100(CA409 + CA412 + CA413 + CA415)$	3	%	—	6.210	5.675	5.311	5.431
417	Boiler efficiency	$\frac{(100 - CA416) CA407}{CA407 + CT313}$	3	%	93.745	93.245	93.720	94.149	94.039
418	Boiler input on load	$\frac{CA407}{CA417 \times 10^{-2}}$	0	GJ	—	2714016	1841182	1592636	2558548
419	Unburnt carbon heat loss	$CA418 \times CA409$	0	GJ	34311	12973	8801	4810	7727
420	Dry flue gas heat loss	$CA418 \times CA412$	0	GJ	430258	145173	89261	74344	121480
421	Unburnt gas heat loss	$CA418 \times CA413$	0	GJ	2416	787	534	430	665
422	Vapour sensible heat loss	$CA418 \times CA415$	0	GJ	29629	9635	5910	5001	9083
423	CO$_2$ factor	$(RL210 \text{ at load } C8 \div ML228) - 1$	4	0	—	0.0913	0.0938	0.0171	−0.0259
424									
425	Excess air heat loss	$CA420 \times CA423$	0	GJ	19748	13260	8368	1268	−3148
426	Gas temperature rise heat loss	$(CA420 - CT311) - CA425$	0	GJ	6979	4859	−199	−2722	5043
427	**Total boiler heat losses**	$\Sigma(CA419 \text{ to } CA422) + CT313$	0	GJ	544725	183373	115647	93184	152521
	Other losses								
428	Actual heat for turbine starts and testing	$\frac{100 \times R128 (M12C + M11C + M14C)}{2}$ / CA417	0	GJ	2760	313	1076	454	917
429	Total off-load heat	$CA399 + CA428$	0	GJ	34089	4063	12540	5319	12167
430	Heat for excess superheat spray	$\frac{(CA359 - 1) \times CA382 \times 100}{CA417}$	0	GJ	266	266	0	0	0
431	Heat for excess reheat spray	$\frac{(CA360 - 1) \times CA382 \times 100}{CA417}$	0	GJ	−673	−266	0	−155	−252
432	Heat loss from other sources	—	0	GJ	0	0	0	0	0
433	**Unit total accounted heat**	$\Sigma(CA407 + CA427 + CA429 + CA430 + CA431 + CA432)$	0	GJ	8740131	2718120	1853743	1597799	2570469

TABLE 7.59

Station thermal performance factor — losses breakdown summary

STATION..

PERIOD....................M............................

Units Supplied (M19) 903.041 GWh STEP Factor (M28) 97.60% Station Loss/Gain (100% – STEP factor %) 2.40%

Station Thermal Efficiency (M27) 36.63% Station Heat Cost £/GJ 2.02480 Station Cost Loss/Saving (G28(h)) £432451

ITEM	TARGET HEAT BALANCE		ACTUAL HEAT BALANCE		LOSS OR GAIN DUE TO VARIATION FROM TARGET		
	Station Target or Specified Value	Station Target Heat (GJ)	Station Actual Values	Station Actual Heat (GJ)	Heat (GJ)	STEP Loss (%)	Cost (£)
(a)	(b)	(c)	(d)	(e)	(f)	(g)	(h)
TURBINE HEAT CONSUMPTION 1. Basic Turbine Heat Consumption	8316 kJ/kWh	7937277	8376 kJ/kWh	7994926	61452	0.69	124427
2. Deterioration Since Latest Test	–	–	8376 kJ/kWh	0	0	0	0
3. TSV Steam Temperature Deviation	565.6 °C	–	566°C	– 2250	– 2398	0.03	– 4856
4. TSV Steam Pressure Deviation	159.6 bar	–	157.5 bar	0	0	0	0
5. Reheat Steam Temperature Deviation	565.6 °C	–	567°C	– 5842	– 6227	– 0.07	– 12609
6. Reheat Steam Pressure Drop Deviation	7.3%	–	7.3%	0	0	0	0
7. Final Feed Water Temperature Deviation	250°C	–	246°C	8318	8866	0.10	17953
8. Feed Train Abnormalities	–	–	–	531	566	0.01	1146
9. Excess BSAH Duty (Willington 'B' only)	–	–	–	–	–	–	–
10. Exhaust Pressure Deviation	46.22 mbar	– 24094	47.35 mbar	– 12451	12411	0.14	25130
(a) C.W. Temperature		– 24094		– – 24303	– – 223	– 0	– – 451
(b) C.W. Quantity		–		– – 5948	– – 6340	– – 0.07	– – 12838
(c) Fouling, etc.		– –		– 17800	– 18974	– 0.21	– 38419
11. **Total Turbine On-Load Heat Consumption**		(7913183)		(7983232)	–	–	–
SUNDRY HEAT CONSUMPTION 12. Total On-Load Make-Up	82947 tonne	163612	84292 tonne	165968	2511	0.02	5085
(a) Sootblowing	16246 tonne	– 49574	16247 tonne	– 49666	– 98	– 0	– 198
(b) Blowdown	36357 tonne	– 58737	36356 tonne	– 58679	– – 62	– 0	– – 125
(c) Leakage	30345 tonne	– 55301	31689 tonne	– 57623	– 2475	– 0.03	– 5012
(d) Evaporator Inefficiency (Ratcliffe only)		–		– 0	– 0	– 0	– 0
13. Steam and Feed Range Losses		12524		12524	0	0	0
14. Superheater Spray		–		266	266	0	539
15. Reheater Spray		–		– 673	– 673	– 0.01	– 1363
BOILER LOSSES 16. Unburnt Carbon Loss	0.51%	44376	0.39%	34311	– 10065	– 0.11	– 20380
17. Total Dry Flue Gas Loss	4.68%	403531	4.94%	430258	26727	0.30	54117
(a) Excess Air	13.9%	– –	13.3%	– 19748	– 17748	– 0.22	– 39986
(b) Gas Temperature Rise	105.2°C	– –	100.5°C	– 6979	– 6979	– 0.08	– 14131
18. Unburnt Gas (CO) Loss	–		0.006%	2416	2416	0.03	48921
19. Vapour Heat Loss	0.44%	37576	0.34%	29629	– 2947	– 0.09	– 16091
20. Radiation and Other Losses	0.56%	48111	0.56%	48111	0	0	0
21. **Boiler Efficiency/Total Boiler Losses**	93.81%	(533594)	93.74%	(544725)	– 0	– 0	– 0
22. Steam Auxiliaries and Auxy. Boilers		13999		13999	0	0	0
23. Heat Consumption of non-Steam Sets (Drakelow 'C' only)			0		0	0	0
OFF-LOAD HEAT CONSUMPTION 24. Total Off-Load Heat Consumption		34269		34089	– 180	0	– 364
(a) Boiler Cold Starts		– 30000		– 30000	– 0	– 0	– 0
(b) Boiler Hot Starts		– 1329		– 1329	– 0	– 0	– 0
(c) Boiler Standby and Testing		– 0		– 0	– 0	– 0	– 0
(d) Turbine Starts and Testing		– 2940		– 2760	– – 180	– 0	– – 364
25. Loss from other Sources		–		0	0	0	0
26. Total Accounted Heat Consumption		8671210		8754130	87725	0.98	177626
27. Works Electricity Loss	49.430 GWh	–	50.429 GWh	9818	9818	0.11	19880
28. Overall Heat Consumption		8761392		8874969	213577	2.40	432451
29. Unaccounted Heat Consumption				116034	116034	1.31	234945
AS RECEIVED HEAT BASIS MONITORING FOR PREVIOUS MONTH M							
UNACCOUNTED HEAT CONSUMPTION BREAKDOWN							
29. Total Unaccounted Heat Consumption					105262	1.37	211105
(a) Heat Not Received		–	–	–	– – 10112	– – 0.13	– – 20280
(b) Heat Adjustment (for stock control etc.)		–	–	–	– 81747	– 1.06	– 163944
(c) Residual Unaccounted Heat		–	–	–	– 33627	– 0.44	– 67441

Thermal Efficiency %	As Invoiced	As Received
	35.78	35.73

STEP Factor	As Invoiced	As Received
	98.01	97.88

TABLE 7.60

CEGB registered data for order of merit heat rate calculations (form STEP 3E)

FORM STEP 3E

STATION .... CODE .... REVISION DATE.......................................

Net CV Basis

R			PF	PF	PF	PF
	Type of Firing					
	Turbine Generator Groups		1	2	1	3
	Turbine Generator Site Number		1	2	3	4
101	Practical Economic Capacity	MW	487.0	487.0	487.0	487.0
102	Practical Overload Capacity	MW				
103	Part Load Capacity	MW	292.2	292.2	292.2	292.2
104	Standard Boiler Load at Practical Economic Capacity	%	99.3	97.9	99.3	97.9
105	Standard Boiler Load at Practical Overload Capacity	%				
106	Standard Boiler Load at Part Load Capacity	%	60.3	59.5	60.3	59.5
107	Standard Boiler Efficiency at Practical Economic Capacity	%	93.70	93.77	93.70	93.77
108	Standard Boiler Efficiency at Practical Overload Capacity	%				
109	Standard Boiler Efficiency at Part Load Capacity	%	93.18	93.12	93.18	93.12
110	Ancillary Losses at Practical Economic Capacity	GJ/h	94.6	94.6	94.6	94.6
111	Ancillary Losses at Practical Overload Capacity	GJ/h				
112	Ancillary Losses at Part Load Capacity	GJ/h	93.1	93.1	93.1	93.1
113	Fuel Consumption at Practical Economic Capacity	t/h	188.0	188.0	188.0	188.0
114	Fuel Consumption at Practical Overload Capacity	t/h				
115	Fuel Consumption at Part Load Capacity	t/h	120.6	120.6	120.6	120.6
116	Electricity Sent-out at Practical Economic Capacity	MW	462.9	462.9	462.9	462.9
117	Electricity Sent-out at Practical Overload Capacity	MW				
118	Electricity Sent-out at Part Load Capacity	MW	272.0	272.0	272.0	272.0
	Abnormalities					
122/1	Description of Abnormality No.1		BFPTC O/C			
122/2	Description of Abnormality No.2		One bank HP heaters O/C — BFPTI/S			
122/3	Description of Abnormality No.3		One bank HP heaters O/C — BFPTC O/S			
122/4	Description of Abnormality No.4					
122/5	Description of Abnormality No.5		O/C out of commission			
122/6	Description of Abnormality No.6		I/S – in service			
	(a) Economic Load Range					
123/1	Change in Fixed Heat Consumption – Abnormality No.1	GJ/h	– 13.9	– 13.9	– 13.9	– 13.9
123/2	– Abnormality No.2	GJ/h	– 12.1	– 12.1	– 12.1	– 12.1
123/3	– Abnormality No.3	GJ/h	– 25.3	– 25.3	– 25.3	– 25.3
123/4	– Abnormality No.4	GJ/h				
123/5	– Abnormality No.5	GJ/h				
123/6	– Abnormality No.6	GJ/h				
124/1	Change in Incremental Heat Rate – Abnormality No.1	kJ/kWh	– 114	– 114	– 114	– 114
124/2	– Abnormality No.2	kJ/kWh	+ 186	+ 186	+ 186	+ 186
124/3	– Abnormality No.3	kJ/kWh	+ 33	+ 33	+ 33	+ 33
124/4	– Abnormality No.4	kJ/kWh				
124/5	– Abnormality No.5	kJ/kWh				
124/6	– Abnormality No.6	kJ/kWh				
	(b) Overload Range					
125/1	Change in Fixed Heat Consumption – Abnormality No.1	GJ/h				
125/2	– Abnormality No.2	GJ/h				
125/3	– Abnormality No.3	GJ/h				
125/4	– Abnormality No.4	GJ/h				
126/1	Change in Incremental Heat Rate – Abnormality No.1	kJ/kWh				
126/2	– Abnormality No.2	kJ/kWh				
126/3	– Abnormality No.3	kJ/kWh				
126/4	– Abnormality No.4	kJ/kWh				

06/64/853

616

TABLE 7.61

CEGB registered data — latest test Willans line data (form STEP 3F)

FORM STEP 3F

STATION.. CODE... REVISION DATE................................

R			1	2	3	4		
	Turbines Site No.		1	2	3	4		
	(a) Economic Load Range							
128	Fixed Heat Consumption	GJ/h	295.1	290.6	281.6	290.7		
129	Incremental Heat Rate	kJ/kWh	7825	7710	7966	7709		
	(b) Overload Range							
130	Fixed Heat Consumption	GJ/h						
131	Incremental Heat Rate	kJ/kWh						

Revision 2M82

R					3			
	Turbines Site No.				3			
	(a) Economic Load Range							
128	Fixed Heat Consumption	GJ/h			285.1			
129	Incremental Heat Rate	kJ/kWh			7814			
	(b) Overload Range							
130	Fixed Heat Consumption	GJ/h						
131	Incremental Heat Rate	kJ/kWh						

Revision 3M82

R			1	2	3	4		
	Turbines Site No.		1	2	3	4		
	(a) Economic Load Range							
128	Fixed Heat Consumption	GJ/h	292.3	288.0	284.7	288.0		
129	Incremental Heat Rate	kJ/kWh	7778	7663	7802	7663		
	(b) Overload Range							
130	Fixed Heat Consumption	GJ/h						
131	Incremental Heat Rate	kJ/kWh						

Revision 1M83

R						4		
	Turbines Site No.					4		
	(a) Economic Load Range							
128	Fixed Heat Consumption	GJ/h				287.6		
129	Incremental Heat Rate	kJ/kWh				7864		
	(b) Overload Range							
130	Fixed Heat Consumption	GJ/h						
131	Incremental Heat Rate	kJ/kWh						

TABLE 7.62

CEGB registered data for order of merit heat rate calculations (form STEP 3G)

STATION .... CODE .... REVISION DATE ....

FORM STEP 3G

ITEM R127 – Works Power of Electric Feed Pumps
(Bled Steam Pumps Only)

MERIT TURBINE GROUP 1-4

Unit Running Load – % of MCR Capacity		0	10	15	20	25	30	40	50
Works Power	MW	2.5	3.5	4.0	4.5	5.0	5.2	5.6	6.0

Unit Running Load – % of MCR Capacity		55	60	70	75	80	90	100	110
Works Power	MW	7.4	7.8	8.6	9.2	9.8	11.0	12.5	12.5

ITEM R127 – Works Power of Electric Feed Pumps
(Bled Steam Pumps Only)

MERIT TURBINE GROUP

Unit Running Load – % of MCR Capacity		10	15	20	30	40	50
Works Power	MW						

Unit Running Load – % of MCR Capacity		60	70	80	90	100	110
Works Power	MW						

ITEM R132 – Forecast Monthly Circulating Water Inlet Temperature

STATION VALUES

Review Period		1	2	3	4	5	6
CW Inlet Temperature	°C	16.0	14.4	15.1	17.0	18.1	21.8

Review Period		7	8	9	10	11	12
CW Inlet Temperature	°C	22.5	24.6	21.7	18.8	16.2	14.9

Net CV Basis

Item	R133
Forecast CW inlet temperature °C	Forecast percentage power of cooling tower pumps %
0	
2	
4	
6	
8	
10	
12	
14	
16	
18	
20	
22	
24	
26	
28	
30	

Set Number	R119	R120	R121
	Heat Consumption		Standard sent-out Incremental Heat Rate kJ/kWh
	at Practical Economic Capacity GJ/h	at Part Load Capacity GJ/h	
1,3	4455.5	2852.7	8396
2,4	4387.8	2813.8	8245

TABLE 7.63

CEGB STEP scheme — monthly operating data (STEP 4E)

STATION	CODE	MONTH	YEAR
		M	

CARD TYPE	M101 SET NUMBER	M102/1 ELECTRICITY GENERATED GWh	M103/1 DURATION ON-LOAD h	M102/2 ELECTRICITY GENERATED GWh	M103/2 DURATION ON-LOAD h	M102/3 ELECTRICITY GENERATED GW/h	M103/3 DURATION ON-LOAD h
		ABNORMALITY No. 1		ABNORMALITY No. 2		ABNORMALITY No. 3	
10 (15 17)		(24 ...30)	(34 ...37)	(44 ...50)	(54 ...57)	(64 ...70)	(74 ...77)
	1 .	2.405	6.	0.0	0.	0.0	0.
	2 .	155.308	364.	0.0	0.	0.0	0.
	3 .	0.997	3.	0.0	0.	0.0	0.
	4 .	1.622	4.	3.311	7.	2.441	5.

TABLE 7.64

CEGB STEP scheme — monthly operating data (STEP 4F)

STATION	CODE	MONTH	YEAR
		M	

CARD TYPE	M105 SET NUMBER	M106 EXHAUST PRESSURE M.BAR	M107 PERIOD OF RETURN	M108 PERIOD OF RETURN PLUS 2	M109 ELECTRICITY GENERATED GWh	M110 DURATION h	M111
			DETERIORATION FACTOR		WITHOUT STEAM FEED PUMPS		
10 (15 17)		(24 ...29)	(36 ...40)	(46 ...50)	(54 ...60)	(64 ...67)	
Q	1 .	45.440	1.000	1.000	2.405	6.	
	2 .	51.170	1.000	1.000	155.308	364.	
	3 .	51.970	1.000	1.000	0.997	3.	
	4 .	43.730	1.000	1.000	4.063	9.	

CARD TYPE	M112 MONTH	M113 ACTUAL STATION HEAT RATE kJ/kWh	M114 OPERATIONAL STATION HEAT RATE kJ/kWh	M115 ACTUAL STATION HEAT CONSUMPTION GJ	M116 STEP FACTOR	M117 SYMAC STATION HEAT RATE kJ/kWh	M118
10 (15 17)		(22 ...27)	(32 ...37)	(41 ...50)	(54 ...59)	(62 ...67)	
R	.	10048.	9893.	6409469.	97.50	9890.	
	.	10016.	9886.	9427016.	98.00	9898.	
	.	10061.	9953.	7712170.	98.01	9943.	

TABLE 7.65

Merit order heat rate data — sheet 1

STATION _____

DATE _____

UNIT OR SET NUMBER		1-4			
Fixed Heat Consumption (O-CMER or NER)	GJ/h				
1 Normal Fixed Heat Consumption change in (1) due to:	GJ/h				
2 BFPT out of commission	GJ/h	− 8.5			
3 One bank of heaters out of commission and BFPT in service	GJ/h	− 9.3			
4 One bank of heaters out of commission and BFPT out of commission	GJ/h	− 17.9			
5	GJ/h				
6	GJ/h				
7	GJ/h				
Incremental Heat Rate (O-CMER or NER) kJ/kWh					
8 Normal Incremental Heat Rate change in (8) due to:	kJ/kWh				
9 BFPT out of commission BFPT in service	kJ/kWh	+ 84			
10 One bank heaters out of commission BFPT in service	kJ/kWh	+ 203			
11 One bank heaters out of commission BFPT out of commission	kJ/kWh	+ 245			
12	kJ/kWh				
13	kJ/kWh				
14	kJ/kWh				

TABLE 7.66
Merit order heat rate data — sheet 2

STATION_____

DATE_____

Daily Predictions for following day

UNIT OR SET NUMBER		1	2	3	4
20 Last observed generator load	MW	500	o/c	490	510
21 Corresponding exhaust pressure	mbar	52		48	48
22 Corresponding CW inlet temperature	°C	NOT REQUIRED			
23 BFPT out of commission	Yes/No				
24 One bank HP heaters out of commission BFPT in service	Yes/No				
25 One bank HP heaters out of commission BFPT out of commission	Yes/No				
26	Yes/No				
27	Yes/No				
28	Yes/No				

Daily Calculations

29 Normal fixed heat consumption (O-CMER)	GJ/h	5791		573.2	590.4
30 Change due to:- BFPT out of commission	GJ/h				− 8.5
31 One bank HP heaters out of commission BFPT in service	GJ/h				
32 One bank HP heaters out of commission BFPT out of commission	GJ/h				
33	GJ/h				
34	GJ/h				
35	GJ/h				
36 Σ(29) to (35)		5791		573.2	581.9
37 Normal Incremental Heat Rate (0-CMER)	kJ/kWh	8572		8523	8610
38 change due to:-	kJ/kWh				+ 84
39	kJ/kWh				
40	kJ/kWh				
41	kJ/kWh				
42	kJ/kWh				
43	kJ/kWh				
44 Σ(37) to (43)		8572		8523	8694
+ 45 Exhaust pressure correction factor (from table R41)		1.0009		0.9967	0.9967

Items to be reported to Grid Control

* 52 Merit Fixed Heat Consumption (O-CMER) (36) × (45)	GJ/h	579.6		571.3	580.0
* 53 Merit Incremental Heat Rate (O-CMER) (44) × (45)	kJ/kWh	8580		8495	8665
* 54 Merit Incremental Heat Rate (NER-MCR) (51) × (45)	kJ/kWh	NOT REQUIRED			
* 55 Predicted Sent Out Capability	MW	500		475	490

* Report items (52) to (55) only, at time agreed with Grid Control Centre.

Reported by.........................Designation...........................Time..

+ Item (45) to be read from R41 (Table 7.45) at load given in item (20) and exhaust pressure given in item (21)

TABLE 7.67

Merit order heat rate data — sheet 3

STATION ....

MONTH................................

SET No..

NON-STANDARD CONDITION No.				1		2		3		4		5		6	
Description of Non-Standard Condition				BFPT o/c		One Bank of HP heaters o/c		One Bank of HP heaters o/c							
						BFPT i/c		BFPT o/c							
(1)	(2)	(3)	(4)	(5)	(6)	(7)	(8)	(9)	(10)	(11)	(12)	(13)	(14)	(15)	(16)
Day No.	Total Daily Generation	Total Hours Run	Av. Load Gen.	(a) Hours run and				(b) Electricity generated during non-standard condition							
				(a)	(b)	(a)	(b)	(a)	(b)	(a)	(b)	(a)	(b)	(a)	(b)
1	MWh	h	MW	h	MWh	h	MWh	h	MWh	h	MWh	h	MWh	h	MWh
2															
3															
4															
5															
6															
7															
8															
9															
10															
11															
12															
13															
14															
15															
16															
17															
18															
19															
20															
21															
22															
23															
24															
25															
26															
27															
28															
29															
30															
31															
32															
33															
34															
35															
Totals															

Definitions: 1. Column (4)

The Average Load Generated $= \dfrac{\text{col (2) ``total daily generation''}}{\text{col (3) ``total hours run''}}$ (MW)

2. Columns (6), (8), (10), (12), (14) and (16)
If the abnormal condition lasts only part of the daily running hours (col.(3)), then: for each column (b), the electricity generated during non-standard condition = col. (4) × the appropriate column (a)
i.e. column (5), (7), (9), (11), (13) or (15).
If the abnormal conditions persists for all of the daily running hours, then enter the column 2 value.

SUBJECT INDEX